中国科学院科学出版基金资助出版

《国外数学名著系列》(影印版)专家委员会

(按姓氏笔画排序)

丁伟岳　王　元　石钟慈　冯克勤　严加安　李邦河
李大潜　张伟平　张继平　杨　乐　姜伯驹　郭　雷

项目策划

向安全　林　鹏　王春香　吕　虹　范庆奎　王　璐

执行编辑

范庆奎

国外数学名著系列（影印版） 4

Foundations of Modern Probability
Second Edition

现代概率论基础（第二版）

Olav Kallenberg

科学出版社
北京

图字:01-2005-6738

Olav Kallenberg: Foundations of Modern Probability (Second Edition)
© 2002 by the Applied Probability Trust.

This reprint has been authorized by Springer-Verlag(Berlin/Heidelberg/New York) for sale in the People's Republic of China only and not for export therefrom.

本书英文影印版由德国施普林格出版公司授权出版。未经出版者书面许可,不得以任何方式复制或抄袭本书的任何部分。本书仅限在中华人民共和国销售,不得出口。版权所有,翻印必究。

图书在版编目(CIP)数据

现代概率论基础:第2版= Foundations of Modern Probability (Second Edition)/(美)卡伦伯格(Kallenberg,O.)著.影印版.—北京:科学出版社,2006

(国外数学名著系列)

ISBN 978-7-03-016672-2

Ⅰ.现… Ⅱ.卡… Ⅲ.概率论 英文 Ⅳ.O211

中国版本图书馆CIP数据核字(2005)第154381号

责任编辑:李 欣 / 责任校对:彭珍珍
责任印制:赵 博 / 封面设计:陈 敬

科学出版社出版
北京东黄城根北街16号
邮政编码:100717
http://www.sciencep.com

三河市春园印刷有限公司印刷
科学出版社发行 各地新华书店经销

*

2006年1月第 一 版 开本:B5(720×1000)
2025年1月第八次印刷 印张:41 1/2
字数:782 000
定价:198.00元

(如有印装质量问题,我社负责调换)

《国外数学名著系列》(影印版)序

要使我国的数学事业更好地发展起来，需要数学家淡泊名利并付出更艰苦地努力。另一方面，我们也要从客观上为数学家创造更有利的发展数学事业的外部环境，这主要是加强对数学事业的支持与投资力度，使数学家有较好的工作与生活条件，其中也包括改善与加强数学的出版工作。

从出版方面来讲，除了较好较快地出版我们自己的成果外，引进国外的先进出版物无疑也是十分重要与必不可少的。从数学来说，施普林格（Springer）出版社至今仍然是世界上最具权威的出版社。科学出版社影印一批他们出版的好的新书，使我国广大数学家能以较低的价格购买，特别是在边远地区工作的数学家能普遍见到这些书，无疑是对推动我国数学的科研与教学十分有益的事。

这次科学出版社购买了版权，一次影印了23本施普林格出版社出版的数学书，就是一件好事，也是值得继续做下去的事情。大体上分一下，这23本书中，包括基础数学书5本，应用数学书6本与计算数学书12本，其中有些书也具有交叉性质。这些书都是很新的，2000年以后出版的占绝大部分，共计16本，其余的也是1990年以后出版的。这些书可以使读者较快地了解数学某方面的前沿，例如基础数学中的数论、代数与拓扑三本，都是由该领域大数学家编著的"数学百科全书"的分册。对从事这方面研究的数学家了解该领域的前沿与全貌很有帮助。按照学科的特点，基础数学类的书以"经典"为主，应用和计算数学类的书以"前沿"为主。这些书的作者多数是国际知名的大数学家，例如《拓扑学》一书的作者诺维科夫是俄罗斯科学院的院士，曾获"菲尔兹奖"和"沃尔夫数学奖"。这些大数学家的著作无疑将会对我国的科研人员起到非常好的指导作用。

当然，23本书只能涵盖数学的一部分，所以，这项工作还应该继续做下去。更进一步，有些读者面较广的好书还应该翻译成中文出版，使之有更大的读者群。

总之，我对科学出版社影印施普林格出版社的部分数学著作这一举措表示热烈的支持，并盼望这一工作取得更大的成绩。

王 元

2005年12月3日

Praise for the First Edition

"It is truly surprising how much material the author has managed to cover in the book. ... More advanced readers are likely to regard the book as an ideal reference. Indeed, the monograph has the potential to become a (possibly even 'the') major reference book on large parts of probability theory for the next decade or more." —*M. Scheutzow (Berlin)*

"I am often asked by mathematicians ... for literature on 'a broad introduction to modern stochastics.' ... Due to this book, my task for answering is made easier. This is it! A concise, broad overview of the main results and techniques .... From the table of contents it is difficult to believe that behind all these topics a streamlined, readable text is at all possible. It is: Convince yourself. I have no doubt that this text will become a classic. Its main feature of keeping the whole area of probability together and presenting a general overview is a real success. Scores of students ... and indeed researchers will be most grateful!" —*P.A.L. Embrechts (Zürich)*

"The theory of probability has grown exponentially during the second half of the twentieth century, and the idea of writing a single volume that could serve as a general reference ... seems almost foolhardy. Yet this is precisely what Professor Kallenberg has attempted ... and he has accomplished it brilliantly. ... With regard to his primary goal, the author has been more successful than I would have imagined possible. It is astonishing that a single volume of just over five hundred pages could contain so much material presented with complete rigor, and still be at least formally self-contained. ... As a general reference for a good deal of modern probability theory [the book] is outstanding. It should have a place in the library of every probabilist. Professor Kallenberg set himself a very difficult task, and he should be congratulated for carrying it out so well."
—*R.K. Getoor (La Jolla, California)*

"This is a superbly written, high-level introduction to contemporary probability theory. In it, the advanced mathematics student will find basic information, presented in a uniform terminology and notation, essential to gaining access to much present-day research. ... I congratulate Professor Kallenberg on a noteworthy achievement."
—*M.F. Neuts (Tucson, Arizona)*

"This is a very modern, very ambitious, and very well-written book. The scope is greater than I would have thought possible in a book of this length. This is made possible by the extremely efficient treatment, particularly the proofs .... [Kallenberg] has succeeded in his mammoth task beyond all reasonable expectations. I think this book is destined to become a modern classic." —*N.H. Bingham (London)*

"Kallenberg has ably achieved [his] goal and presents all the important results and techniques that every probabilist should know.... We do not doubt that the book ... will be widely used as material for advanced postgraduate courses and seminars on various topics in probability."
—*jste, European Math. Soc. Newsletter*

"This is a very well written book.... Much effort must have been put into simplifying and streamlining proofs, and the results are quite impressive.... I would highly recommend [the book] to anybody who wants a good concise reference text on several very important parts of modern probability theory. For a mathematical sciences library, such a book is a must."
—*K. Borovkov (Melbourne)*

"[This] is an unusual book about a wide range of probability and stochastic processes, written by a single excellent mathematician.... The graduate student will definitely enjoy reading it, and for the researcher it will become a useful reference book and necessary tool for his or her work."
—*T. Mikosch (Groningen)*

"The author has succeeded in writing a text containing—in the spirit of Loève's *Probability Theory*—all the essential results that any probabilist needs to know. Like Loève's classic, this book will become a standard source of study and reference for students and researchers in probability theory."
—*R. Kiesel (London)*

"Kallenberg's present book would have to qualify as the assimilation of probability par excellence. It is a great edifice of material, clearly and ingeniously presented, without any nonmathematical distractions. Readers wishing to venture into it may do so with confidence that they are in very capable hands."
—*F.B. Knight (Urbana, Illinois)*

"The presentation of the material is characterized by a surprising clarity and precision. The author's overview over the various subfields of probability theory and his detailed knowledge are impressive. Through an activity over many years as a researcher, academic teacher, and editor, he has acquired a deep competence in many areas. Wherever one reads, all chapters are carefully worked through and brought in streamlined form. One can imagine what an enormous effort it has cost the author to reach this final state, though no signs of this are visible. His goal, as set forth in the preface, of giving clear and economical proofs of the included theorems has been achieved admirably.... I can't recall that in recent times I have held in my hands a mathematics book so thoroughly worked through."
—*H. Rost (Heidelberg)*

Preface to the Second Edition

For this new edition the entire text has been carefully revised, and some portions are totally rewritten. More importantly, I have inserted more than a hundred pages of new material, in chapters on general measure and ergodic theory, the asymptotics of Markov processes, and large deviations. The expanded size has made it possible to give a self-contained treatment of the underlying measure theory and to include topics like multivariate and ratio ergodic theorems, shift coupling, Palm distributions, entropy and information, Harris recurrence, invariant measures, strong and weak ergodicity, Strassen's law of the iterated logarithm, and the basic large deviation results of Cramér, Sanov, Schilder, and Freidlin and Ventzel.

Unfortunately, the body of knowledge in probability theory keeps growing at an ever increasing rate, and I am painfully aware that I will never catch up in my efforts to survey the entire subject. Many areas are still totally beyond reach, and a comprehensive treatment of the more recent developments would require another volume or two. I am asking for the reader's patience and understanding.

Many colleagues have pointed out errors or provided helpful information. I am especially grateful for some valuable comments from Wlodzimierz Kuperberg, Michael Scheutzow, Josef Teichmann, and Hermann Thorisson. Some of the new material was presented in our probability seminar at Auburn, where I benefited from stimulating discussions with Bill Hudson, Ming Liao, Lisa Peterson, and Hussain Talibi. My greatest thanks are due, as always, to my wife Jinsoo, whose constant love and support have sustained and inspired me throughout many months of hard work.

Olav Kallenberg *March 2001*

Preface to the First Edition

Some thirty years ago it was still possible, as Loève so ably demonstrated, to write a single book in probability theory containing practically everything worth knowing in the subject. The subsequent development has been explosive, and today a corresponding comprehensive coverage would require a whole library. Researchers and graduate students alike seem compelled to a rather extreme degree of specialization. As a result, the subject is threatened by disintegration into dozens or hundreds of subfields.

At the same time the interaction between the areas is livelier than ever, and there is a steadily growing core of key results and techniques that every probabilist needs to know, if only to read the literature in his or her own field. Thus, it seems essential that we all have at least a general overview of the whole area, and we should do what we can to keep the subject together. The present volume is an earnest attempt in that direction.

My original aim was to write a book about "everything." Various space and time constraints forced me to accept more modest and realistic goals for the project. Thus, "foundations" had to be understood in the narrower sense of the early 1970s, and there was no room for some of the more recent developments. I especially regret the omission of topics such as large deviations, Gibbs and Palm measures, interacting particle systems, stochastic differential geometry, Malliavin calculus, SPDEs, measure-valued diffusions, and branching and superprocesses. Clearly plenty of fundamental and intriguing material remains for a possible second volume.

Even with my more limited, revised ambitions, I had to be extremely selective in the choice of material. More importantly, it was necessary to look for the most economical approach to every result I did decide to include. In the latter respect, I was surprised to see how much could actually be done to simplify and streamline proofs, often handed down through generations of textbook writers. My general preference has been for results conveying some new idea or relationship, whereas many propositions of a more technical nature have been omitted. In the same vein, I have avoided technical or computational proofs that give little insight into the proven results. This conforms with my conviction that the logical structure is what matters most in mathematics, even when applications is the ultimate goal.

Though the book is primarily intended as a general reference, it should also be useful for graduate and seminar courses on different levels, ranging from elementary to advanced. Thus, a first-year graduate course in measure-theoretic probability could be based on the first ten or so chapters, while the rest of the book will readily provide material for more advanced courses on various topics. Though the treatment is formally self-contained, as far as measure theory and probability are concerned, the text is intended for a rather sophisticated reader with at least some rudimentary knowledge of subjects like topology, functional analysis, and complex variables.

My exposition is based on experiences from the numerous graduate and seminar courses I have been privileged to teach in Sweden and in the United States, ever since I was a graduate student myself. Over the years I have developed a personal approach to almost every topic, and even experts might find something of interest. Thus, many proofs may be new, and every chapter contains results that are not available in the standard textbook literature. It is my sincere hope that the book will convey some of the excitement I still feel for the subject, which is without a doubt (even apart from its utter usefulness) one of the richest and most beautiful areas of modern mathematics.

Notes and Acknowledgments: My first thanks are due to my numerous Swedish teachers, and especially to Peter Jagers, whose 1971 seminar opened my eyes to modern probability. The idea of this book was raised a few years later when the analysts at Gothenburg asked me to give a short lecture course on "probability for mathematicians." Although I objected to the title, the lectures were promptly delivered, and I became convinced of the project's feasibility. For many years afterward I had a faithful and enthusiastic audience in numerous courses on stochastic calculus, SDEs, and Markov processes. I am grateful for that learning opportunity and for the feedback and encouragement I received from colleagues and graduate students.

Inevitably I have benefited immensely from the heritage of countless authors, many of whom are not even listed in the bibliography. I have further been fortunate to know many prominent probabilists of our time, who have often inspired me through their scholarship and personal example. Two people, Klaus Matthes and Gopi Kallianpur, stand out as particularly important influences in connection with my numerous visits to Berlin and Chapel Hill, respectively.

The great Kai Lai Chung, my mentor and friend from recent years, offered penetrating comments on all aspects of the work: linguistic, historical, and mathematical. My colleague Ming Liao, always a stimulating partner for discussions, was kind enough to check my material on potential theory. Early versions of the manuscript were tested on several groups of graduate students, and Kamesh Casukhela, Davorin Dujmovic, and Hussain Talibi in particular were helpful in spotting misprints. Ulrich Albrecht and Ed Slaminka offered generous help with software problems. I am further grateful to John Kimmel, Karina Mikhli, and the Springer production team for their patience with my last-minute revisions and their truly professional handling of the project.

My greatest thanks go to my family, who is my constant source of happiness and inspiration. Without their love, encouragement, and understanding, this work would not have been possible.

Olav Kallenberg *May 1997*

Contents

Preface to the Second Edition   v

Preface to the First Edition   vii

1. Measure Theory — Basic Notions   1

Measurable sets and functions
measures and integration
monotone and dominated convergence
transformation of integrals
product measures and Fubini's theorem
L^p-*spaces and projection*
approximation
measure spaces and kernels

2. Measure Theory — Key Results   23

Outer measures and extension
Lebesgue and Lebesgue–Stieltjes measures
Jordan–Hahn and Lebesgue decompositions
Radon–Nikodým theorem
Lebesgue's differentiation theorem
functions of finite variation
Riesz' representation theorem
Haar and invariant measures

3. Processes, Distributions, and Independence   45

Random elements and processes
distributions and expectation
independence
zero–one laws
Borel–Cantelli lemma
Bernoulli sequences and existence
moments and continuity of paths

4. Random Sequences, Series, and Averages   62

Convergence in probability and in L^p
uniform integrability and tightness
convergence in distribution
convergence of random series
strong laws of large numbers
Portmanteau theorem
continuous mapping and approximation
coupling and measurability

5. Characteristic Functions and Classical Limit Theorems   83
Uniqueness and continuity theorem
Poisson convergence
positive and symmetric terms
Lindeberg's condition
general Gaussian convergence
weak laws of large numbers
domain of Gaussian attraction
vague and weak compactness

6. Conditioning and Disintegration   103
Conditional expectations and probabilities
regular conditional distributions
disintegration
conditional independence
transfer and coupling
existence of sequences and processes
extension through conditioning

7. Martingales and Optional Times   119
Filtrations and optional times
random time-change
martingale property
optional stopping and sampling
maximum and upcrossing inequalities
martingale convergence, regularity, and closure
limits of conditional expectations
regularization of submartingales

8. Markov Processes and Discrete-Time Chains   140
Markov property and transition kernels
finite-dimensional distributions and existence
space and time homogeneity
strong Markov property and excursions
invariant distributions and stationarity
recurrence and transience
ergodic behavior of irreducible chains
mean recurrence times

9. Random Walks and Renewal Theory   159
Recurrence and transience
dependence on dimension
general recurrence criteria
symmetry and duality
Wiener–Hopf factorization

ladder time and height distribution
stationary renewal process
renewal theorem

10. Stationary Processes and Ergodic Theory 178

Stationarity, invariance, and ergodicity
discrete- and continuous-time ergodic theorems
moment and maximum inequalities
multivariate ergodic theorems
sample intensity of a random measure
subadditivity and products of random matrices
conditioning and ergodic decomposition
shift coupling and the invariant σ-field

11. Special Notions of Symmetry and Invariance 202

Palm distributions and inversion formulas
stationarity and cycle stationarity
local hitting and conditioning
ergodic properties of Palm measures
exchangeable sequences and processes
strong stationarity and predictable sampling
ballot theorems
entropy and information

12. Poisson and Pure Jump-Type Markov Processes 224

Random measures and point processes
Cox processes, randomization, and thinning
mixed Poisson and binomial processes
independence and symmetry criteria
Markov transition and rate kernels
embedded Markov chains and explosion
compound and pseudo-Poisson processes
ergodic behavior of irreducible chains

13. Gaussian Processes and Brownian Motion 249

Symmetries of Gaussian distribution
existence and path properties of Brownian motion
strong Markov and reflection properties
arcsine and uniform laws
law of the iterated logarithm
Wiener integrals and isonormal Gaussian processes
multiple Wiener–Itô integrals
chaos expansion of Brownian functionals

14. Skorohod Embedding and Invariance Principles 270

Embedding of random variables
approximation of random walks
functional central limit theorem
laws of the iterated logarithm
arcsine laws
approximation of renewal processes
empirical distribution functions
embedding and approximation of martingales

15. Independent Increments and Infinite Divisibility 285

Regularity and integral representation
Lévy processes and subordinators
stable processes and first-passage times
infinitely divisible distributions
characteristics and convergence criteria
approximation of Lévy processes and random walks
limit theorems for null arrays
convergence of extremes

16. Convergence of Random Processes, Measures, and Sets 307

Relative compactness and tightness
uniform topology on $C(K,S)$
Skorohod's J_1-topology
equicontinuity and tightness
convergence of random measures
superposition and thinning
exchangeable sequences and processes
simple point processes and random closed sets

17. Stochastic Integrals and Quadratic Variation 329

Continuous local martingales and semimartingales
quadratic variation and covariation
existence and basic properties of the integral
integration by parts and Itô's formula
Fisk–Stratonovich integral
approximation and uniqueness
random time-change
dependence on parameter

18. Continuous Martingales and Brownian Motion 350

Real and complex exponential martingales
martingale characterization of Brownian motion
random time-change of martingales
integral representation of martingales

iterated and multiple integrals
change of measure and Girsanov's theorem
Cameron–Martin theorem
Wald's identity and Novikov's condition

19. Feller Processes and Semigroups   367

Semigroups, resolvents, and generators
closure and core
Hille–Yosida theorem
existence and regularization
strong Markov property
characteristic operator
diffusions and elliptic operators
convergence and approximation

20. Ergodic Properties of Markov Processes   390

transition and contraction operators
ratio ergodic theorem
space-time invariance and tail triviality
mixing and convergence in total variation
Harris recurrence and transience
existence and uniqueness of invariant measure
distributional and pathwise limits

21. Stochastic Differential Equations and Martingale Problems   412

Linear equations and Ornstein–Uhlenbeck processes
strong existence, uniqueness, and nonexplosion criteria
weak solutions and local martingale problems
well-posedness and measurability
pathwise uniqueness and functional solution
weak existence and continuity
transformation of SDEs
strong Markov and Feller properties

22. Local Time, Excursions, and Additive Functionals   428

Tanaka's formula and semimartingale local time
occupation density, continuity and approximation
regenerative sets and processes
excursion local time and Poisson process
Ray–Knight theorem
excessive functions and additive functionals
local time at a regular point
additive functionals of Brownian motion

23. One-dimensional SDEs and Diffusions   450

Weak existence and uniqueness
pathwise uniqueness and comparison
scale function and speed measure
time-change representation
boundary classification
entrance boundaries and Feller properties
ratio ergodic theorem
recurrence and ergodicity

24. Connections with PDEs and Potential Theory   470

Backward equation and Feynman–Kac formula
uniqueness for SDEs from existence for PDEs
harmonic functions and Dirichlet's problem
Green functions as occupation densities
sweeping and equilibrium problems
dependence on conductor and domain
time reversal
capacities and random sets

25. Predictability, Compensation, and Excessive Functions   490

Accessible and predictable times
natural and predictable processes
Doob–Meyer decomposition
quasi-left-continuity
compensation of random measures
excessive and superharmonic functions
additive functionals as compensators
Riesz decomposition

26. Semimartingales and General Stochastic Integration   515

Predictable covariation and L^2-integral
semimartingale integral and covariation
general substitution rule
Doléans' exponential and change of measure
norm and exponential inequalities
martingale integral
decomposition of semimartingales
quasi-martingales and stochastic integrators

27. Large Deviations   537

Legendre–Fenchel transform
Cramér's and Schilder's theorems
large-deviation principle and rate function
functional form of the LDP

continuous mapping and extension
perturbation of dynamical systems
empirical processes and entropy
Strassen's law of the iterated logarithm

Appendices
A1. Advanced Measure Theory   561
Polish and Borel spaces
measurable inverses
projection and sections

A2. Some Special Spaces   562
Function spaces
measure spaces
spaces of closed sets
measure-valued functions
projective limits

Historical and Bibliographical Notes   569

Bibliography   596

Symbol Index   621

Author Index   623

Subject Index   629

Words of Wisdom and Folly

♣ "A mathematician who argues from probabilities in geometry is not worth an ace" — *Socrates* (on the demands of rigor in mathematics)

♣ "[We will travel a road] full of interest of its own. It familiarizes us with the measurement of variability, and with curious laws of chance that apply to a vast diversity of social subjects" — *Francis Galton* (on the wondrous world of probability)

♣ "God doesn't play dice" [i.e., there is no randomness in the universe] — *Albert Einstein* (on quantum mechanics and causality)

♣ "It might be possible to prove certain theorems, but they would not be of any interest, since, in practice, one could not verify whether the assumptions are fulfilled" — *Émile Borel* (on why bothering with probability)

♣ "[The stated result] is a special case of a very general theorem [the strong Markov property]. The measure [theoretic] ideas involved are somewhat glossed over in the proof, in order to avoid complexities out of keeping with the rest of this paper" — *Joseph L. Doob* (on why bothering with generality or mathematical rigor)

♣ "Probability theory [has two hands]: On the right is the rigorous [technical work]; the left hand ... reduces problems to gambling situations, coin-tossing, motions of a physical particle" — *Leo Breiman* (on probabilistic thinking)

♣ "There are good taste and bad taste in mathematics just as in music, literature, or cuisine, and one who dabbles in it must stand judged thereby" — *Kai Lai Chung* (on the art of writing mathematics)

♣ "The traveler often has the choice between climbing a peak or using a cable car" — *William Feller* (on the art of reading mathematics)

♣ "A Catalogue Aria of triumphs is of less benefit [to the student] than an indication of the techniques by which such results are achieved" — *David Williams* (on seduction and the art of discovery)

♣ "One needs [for stochastic integration] a six months course [to cover only] the definitions. What is there to do?" — *Paul-André Meyer* (on the dilemma of modern math education)

♣ "There were very many [bones] in the open valley; and lo, they were very dry. And [God] said unto me, 'Son of man, can these bones live?' And I answered, 'O Lord, thou knowest.'" — *Ezekiel 37:2-3* (on the ultimate reward of hard studies, as quoted by *Chris Rogers* and *David Williams*)

Chapter 1

Measure Theory — Basic Notions

Measurable sets and functions; measures and integration; monotone and dominated convergence; transformation of integrals; product measures and Fubini's theorem; L^p-spaces and projection; approximation; measure spaces and kernels

Modern probability theory is technically a branch of measure theory, and any systematic exposition of the subject must begin with some basic measure-theoretic facts. In this chapter and its sequel we have collected some basic ideas and results from measure theory that will be useful throughout this book. Though most of the quoted propositions may be found in any textbook in real analysis, our emphasis is often somewhat different and has been chosen to suit our special needs. Many readers may prefer to omit these chapters on their first encounter and return for reference when the need arises.

To fix our notation, we begin with some elementary notions from set theory. For subsets $A, A_k, B, \ldots$ of some abstract space Ω, recall the definitions of *union* $A \cup B$ or $\bigcup_k A_k$, *intersection* $A \cap B$ or $\bigcap_k A_k$, *complement* A^c, and *difference* $A \setminus B = A \cap B^c$. The latter is said to be *proper* if $A \supset B$. The *symmetric difference* of A and B is given by $A \triangle B = (A \setminus B) \cup (B \setminus A)$. Among basic set relations, we note in particular the *distributive laws*

$$A \cap \bigcup_k B_k = \bigcup_k (A \cap B_k), \qquad A \cup \bigcap_k B_k = \bigcap_k (A \cup B_k),$$

and *de Morgan's laws*

$$\left\{\bigcup_k A_k\right\}^c = \bigcap_k A_k^c, \qquad \left\{\bigcap_k A_k\right\}^c = \bigcup_k A_k^c,$$

valid for arbitrary (not necessarily countable) unions and intersections. The latter formulas allow us to convert any relation involving unions (intersections) into the dual formula for intersections (unions).

We define a *σ-algebra* or *σ-field* in Ω as a nonempty collection $\mathcal{A}$ of subsets of Ω that is closed under countable unions and intersections as well as under complementation. (For a *field*, closure is required only under finite set operations.) Thus, if $A, A_1, A_2, \cdots \in \mathcal{A}$, then also A^c, $\bigcup_k A_k$, and $\bigcap_k A_k$ lie in $\mathcal{A}$. In particular, the whole space Ω and the empty set $\emptyset$ belong to every σ-field. In any space Ω there is a smallest σ-field $\{\emptyset, \Omega\}$ and a largest one 2^Ω —the class of *all* subsets of Ω. Note that any σ-field $\mathcal{A}$ is closed under monotone limits. Thus, if $A_1, A_2, \cdots \in \mathcal{A}$ with $A_n \uparrow A$ or $A_n \downarrow A$, then also $A \in \mathcal{A}$. A *measurable space* is a pair $(\Omega, \mathcal{A})$, where Ω is a space and $\mathcal{A}$ is a σ-field in Ω.

For any class of σ-fields in Ω, the intersection (but usually not the union) is again a σ-field. If $\mathcal{C}$ is an arbitrary class of subsets of Ω, there is a smallest σ-field in Ω containing $\mathcal{C}$, denoted by $\sigma(\mathcal{C})$ and called the σ-field *generated* or *induced* by $\mathcal{C}$. Note that $\sigma(\mathcal{C})$ can be obtained as the intersection of all σ-fields in Ω that contain $\mathcal{C}$. We endow a metric or topological space S with its *Borel σ-field* $\mathcal{B}(S)$ generated by the *topology* (class of open subsets) in S, unless a σ-field is otherwise specified. The elements of $\mathcal{B}(S)$ are called *Borel sets*. In the case of the real line $\mathbb{R}$, we often write $\mathcal{B}$ instead of $\mathcal{B}(\mathbb{R})$.

More primitive classes than σ-fields often arise in applications. A class $\mathcal{C}$ of subsets of some space Ω is called a *π-system* if it is closed under finite intersections, so that $A, B \in \mathcal{C}$ implies $A \cap B \in \mathcal{C}$. Furthermore, a class $\mathcal{D}$ is a *λ-system* if it contains Ω and is closed under proper differences and increasing limits. Thus, we require that $\Omega \in \mathcal{D}$, that $A, B \in \mathcal{D}$ with $A \supset B$ implies $A \setminus B \in \mathcal{D}$, and that $A_1, A_2, \ldots \in \mathcal{D}$ with $A_n \uparrow A$ implies $A \in \mathcal{D}$.

The following *monotone-class theorem* is often useful to extend an established property or relation from a class $\mathcal{C}$ to the generated σ-field $\sigma(\mathcal{C})$. An application of this result is referred to as a *monotone-class argument*.

Theorem 1.1 *(monotone classes, Sierpiński)* Let $\mathcal{C}$ be a π-system and $\mathcal{D}$ a λ-system in some space Ω such that $\mathcal{C} \subset \mathcal{D}$. Then $\sigma(\mathcal{C}) \subset \mathcal{D}$.

Proof: We may clearly assume that $\mathcal{D} = \lambda(\mathcal{C})$ —the smallest λ-system containing $\mathcal{C}$. It suffices to show that $\mathcal{D}$ is a π-system, since it is then a σ-field containing $\mathcal{C}$ and therefore contains the smallest σ-field $\sigma(\mathcal{C})$ with this property. Thus, we need to show that $A \cap B \in \mathcal{D}$ whenever $A, B \in \mathcal{D}$.

The relation $A \cap B \in \mathcal{D}$ is certainly true when $A, B \in \mathcal{C}$, since $\mathcal{C}$ is a π-system contained in $\mathcal{D}$. We proceed by extension in two steps. First we fix any $B \in \mathcal{C}$ and define $\mathcal{A}_B = \{A \subset \Omega; A \cap B \in \mathcal{D}\}$. Then $\mathcal{A}_B$ is a λ-system containing $\mathcal{C}$, and so it contains the smallest λ-system $\mathcal{D}$ with this property. This shows that $A \cap B \in \mathcal{D}$ for any $A \in \mathcal{D}$ and $B \in \mathcal{C}$. Next we fix any $A \in \mathcal{D}$ and define $\mathcal{B}_A = \{B \subset \Omega; A \cap B \in \mathcal{D}\}$. As before, we note that even $\mathcal{B}_A$ contains $\mathcal{D}$, which yields the desired property. □

For any family of spaces Ω_t, $t \in T$, we define the *Cartesian product* $\times_{t \in T} \Omega_t$ as the class of all collections $(\omega_t; t \in T)$, where $\omega_t \in \Omega_t$ for all t. When $T = \{1, \ldots, n\}$ or $T = \mathbb{N} = \{1, 2, \ldots\}$, we often write the product space as $\Omega_1 \times \cdots \times \Omega_n$ or $\Omega_1 \times \Omega_2 \times \cdots$, respectively; if $\Omega_t = \Omega$ for all t, we use the notation Ω^T, Ω^n, or Ω^∞. In case of topological spaces Ω_t, we endow $\times_t \Omega_t$ with the product topology unless a topology is otherwise specified.

Now assume that each space Ω_t is equipped with a σ-field $\mathcal{A}_t$. In $\times_t \Omega_t$ we may then introduce the *product σ-field* $\bigotimes_t \mathcal{A}_t$, generated by all one-dimensional *cylinder sets* $A_t \times \times_{s \neq t} \Omega_s$, where $t \in T$ and $A_t \in \mathcal{A}_t$. (Note the analogy with the definition of product topologies.) As before, we write $\mathcal{A}_1 \otimes \cdots \otimes \mathcal{A}_n$, $\mathcal{A}_1 \otimes \mathcal{A}_2 \otimes \cdots$, $\mathcal{A}^T$, $\mathcal{A}^n$, or $\mathcal{A}^\infty$ in the appropriate special cases.

Lemma 1.2 *(product and Borel σ-fields)* *If $S_1, S_2, \ldots$ are separable metric spaces, then*
$$\mathcal{B}(S_1 \times S_2 \times \cdots) = \mathcal{B}(S_1) \otimes \mathcal{B}(S_2) \otimes \cdots.$$

Thus, for countable products of separable metric spaces, the product and Borel σ-fields agree. In particular, $\mathcal{B}(\mathbb{R}^d) = (\mathcal{B}(\mathbb{R}))^d = \mathcal{B}^d$, the σ-field generated by all rectangular boxes $I_1 \times \cdots \times I_d$, where $I_1, \ldots, I_d$ are arbitrary real intervals. This special case can also be proved directly.

Proof: The assertion may be written as $\sigma(\mathcal{C}_1) = \sigma(\mathcal{C}_2)$, and it suffices to show that $\mathcal{C}_1 \subset \sigma(\mathcal{C}_2)$ and $\mathcal{C}_2 \subset \sigma(\mathcal{C}_1)$. For $\mathcal{C}_2$ we may choose the class of all cylinder sets $G_k \times \mathsf{X}_{n \neq k} S_n$ with $k \in \mathbb{N}$ and G_k open in S_k. Those sets generate the product topology in $S = \mathsf{X}_n S_n$, and so they belong to $\mathcal{B}(S)$.

Conversely, we note that $S = \mathsf{X}_n S_n$ is again separable. Thus, for any topological base $\mathcal{C}$ in S, the open subsets of S are countable unions of sets in $\mathcal{C}$. In particular, we may choose $\mathcal{C}$ to consist of all finite intersections of cylinder sets $G_k \times \mathsf{X}_{n \neq k} S_n$ as above. It remains to note that the latter sets lie in $\bigotimes_n \mathcal{B}(S_n)$. $\square$

Every point mapping f between two spaces S and T induces a set mapping f^{-1} in the opposite direction, that is, from 2^T to 2^S, given by
$$f^{-1}B = \{s \in S;\ f(s) \in B\}, \quad B \subset T.$$

Note that f^{-1} preserves the basic set operations in the sense that, for any subsets B and B_k of T,
$$f^{-1}B^c = (f^{-1}B)^c, \quad f^{-1}\bigcup_k B_k = \bigcup_k f^{-1}B_k, \quad f^{-1}\bigcap_k B_k = \bigcap_k f^{-1}B_k. \tag{1}$$

The next result shows that f^{-1} also preserves σ-fields, in both directions. For convenience, we write
$$f^{-1}\mathcal{C} = \{f^{-1}B;\ B \in \mathcal{C}\}, \quad \mathcal{C} \subset 2^T.$$

Lemma 1.3 *(induced σ-fields)* *Let f be a mapping between two measurable spaces $(S, \mathcal{S})$ and $(T, \mathcal{T})$. Then*

(i) $\mathcal{S}' = f^{-1}\mathcal{T}$ *is a σ-field in S;*

(ii) $\mathcal{T}' = \{B \subset T;\ f^{-1}B \in \mathcal{S}\}$ *is a σ-field in T.*

Proof: (i) Let $A, A_1, A_2, \ldots \in \mathcal{S}'$. Then there exists some sets $B, B_1, B_2, \ldots \in \mathcal{T}$ with $A = f^{-1}B$ and $A_n = f^{-1}B_n$ for each n. Since $\mathcal{T}$ is a σ-field, the sets $B^c, \bigcup_n B_n$, and $\bigcap_n B_n$ all belong to $\mathcal{T}$, and by (1) we get
$$\begin{aligned} A^c &= (f^{-1}B)^c = f^{-1}B^c \in f^{-1}\mathcal{T} = \mathcal{S}', \\ \bigcup_n A_n &= \bigcup_n f^{-1}B_n = f^{-1}\bigcup_n B_n \in f^{-1}\mathcal{T} = \mathcal{S}', \\ \bigcap_n A_n &= \bigcap_n f^{-1}B_n = f^{-1}\bigcap_n B_n \in f^{-1}\mathcal{T} = \mathcal{S}'. \end{aligned}$$

(ii) Let $B, B_1, B_2, \cdots \in \mathcal{T}'$, so that $f^{-1}B, f^{-1}B_1, f^{-1}B_2, \cdots \in \mathcal{S}$. Using (1) and the fact that $\mathcal{S}$ is a σ-field, we get

$$\begin{aligned} f^{-1}B^c &= (f^{-1}B)^c \in \mathcal{S}, \\ f^{-1}\bigcup_n B_n &= \bigcup_n f^{-1}B_n \in \mathcal{S}, \\ f^{-1}\bigcap_n B_n &= \bigcap_n f^{-1}B_n \in \mathcal{S}, \end{aligned}$$

which shows that B^c, $\bigcup_n B_n$, and $\bigcap_n B_n$ all lie in $\mathcal{T}'$. □

Given two measurable spaces $(S, \mathcal{S})$ and $(T, \mathcal{T})$, a mapping $f: S \to T$ is said to be $\mathcal{S}/\mathcal{T}$-*measurable* or simply *measurable* if $f^{-1}\mathcal{T} \subset \mathcal{S}$, that is, if $f^{-1}B \in \mathcal{S}$ for every $B \in \mathcal{T}$. (Note the analogy with the definition of continuity in terms of topologies on S and T.) By the next result, it is enough to verify the defining condition for a generating subclass.

Lemma 1.4 *(measurable functions)* *Consider a mapping f between two measurable spaces $(S, \mathcal{S})$ and $(T, \mathcal{T})$, and let $\mathcal{C} \subset 2^T$ with $\sigma(\mathcal{C}) = \mathcal{T}$. Then f is $\mathcal{S}/\mathcal{T}$-measurable iff $f^{-1}\mathcal{C} \subset \mathcal{S}$.*

Proof: Let $\mathcal{T}' = \{B \subset T;\ f^{-1}B \in \mathcal{S}\}$. Then $\mathcal{C} \subset \mathcal{T}'$ by hypothesis and $\mathcal{T}'$ is a σ-field by Lemma 1.3 (ii). Hence,

$$\mathcal{T}' = \sigma(\mathcal{T}') \supset \sigma(\mathcal{C}) = \mathcal{T},$$

which shows that $f^{-1}B \in \mathcal{S}$ for all $B \in \mathcal{T}$. □

Lemma 1.5 *(continuity and measurability)* *Let f be a continuous mapping between two topological spaces S and T with Borel σ-fields $\mathcal{S}$ and $\mathcal{T}$. Then f is $\mathcal{S}/\mathcal{T}$-measurable.*

Proof: Let $\mathcal{S}'$ and $\mathcal{T}'$ denote the classes of open sets in S and T. Since f is continuous and $\mathcal{S} = \sigma(\mathcal{S}')$, we have

$$f^{-1}\mathcal{T}' \subset \mathcal{S}' \subset \mathcal{S}.$$

By Lemma 1.4 it follows that f is $\mathcal{S}/\sigma(\mathcal{T}')$-measurable. It remains to note that $\sigma(\mathcal{T}') = \mathcal{T}$. □

We insert a result about subspace topologies and σ-fields that will be needed in Chapter 16. Given a class $\mathcal{C}$ of subsets of S and a set $A \subset S$, we define $A \cap \mathcal{C} = \{A \cap C;\ C \in \mathcal{C}\}$.

Lemma 1.6 *(subspaces)* *Fix a metric space (S, ρ) with topology $\mathcal{T}$ and Borel σ-field $\mathcal{S}$, and let $A \subset S$. Then (A, ρ) has topology $\mathcal{T}_A = A \cap \mathcal{T}$ and Borel σ-field $\mathcal{S}_A = A \cap \mathcal{S}$.*

Proof: The natural embedding $I_A: A \to S$ is continuous and hence measurable, and so $A \cap \mathcal{T} = I_A^{-1}\mathcal{T} \subset \mathcal{T}_A$ and $A \cap \mathcal{S} = I_A^{-1}\mathcal{S} \subset \mathcal{S}_A$. Conversely, given any $B \in \mathcal{T}_A$, we define $G = (B \cup A^c)^\circ$ where the complement and interior are with respect to S and note that $B = A \cap G$. Hence, $\mathcal{T}_A \subset A \cap \mathcal{T}$,

and therefore
$$S_A = \sigma(\mathcal{T}_A) \subset \sigma(A \cap \mathcal{T}) \subset \sigma(A \cap \mathcal{S}) = A \cap \mathcal{S},$$
where the operation $\sigma(\cdot)$ refers to the subspace A. $\square$

As with continuity, we note that even measurability is preserved by composition.

Lemma 1.7 *(composition)* *Fix three measurable spaces $(S, \mathcal{S})$, $(T, \mathcal{T})$, and $(U, \mathcal{U})$, and consider some measurable mappings $f : S \to T$ and $g : T \to U$. Then the composition $h = g \circ f : S \to U$ is again measurable.*

Proof: Let $C \in \mathcal{U}$, and note that $B \equiv g^{-1}C \in \mathcal{T}$ since g is measurable. Noting that $(f \circ g)^{-1} = g^{-1} \circ f^{-1}$ and using the fact that even f is measurable, we get
$$h^{-1}C = (f \circ g)^{-1}C = g^{-1}f^{-1}C = g^{-1}B \in \mathcal{S}. \qquad \square$$

To state the next result, we note that any collection of functions $f_t : \Omega \to S_t$, $t \in T$, defines a mapping $f = (f_t)$ from Ω to $\mathsf{X}_t S_t$ given by
$$f(\omega) = (f_t(\omega); t \in T), \quad \omega \in \Omega. \tag{2}$$
It is often useful to relate the measurability of f to that of the *coordinate mappings* f_t.

Lemma 1.8 *(collections of functions)* *Consider any set of functions $f_t : \Omega \to S_t$, $t \in T$, where $(\Omega, \mathcal{A})$ and $(S_t, \mathcal{S}_t)$, $t \in T$, are measurable spaces, and define $f = (f_t) : \Omega \to \mathsf{X}_t S_t$. Then f is $\mathcal{A}/\bigotimes_t \mathcal{S}_t$-measurable iff f_t is $\mathcal{A}/\mathcal{S}_t$-measurable for every $t \in T$.*

Proof: We may use Lemma 1.4, with $\mathcal{C}$ equal to the class of cylinder sets $A_t \times \mathsf{X}_{s \neq t} S_t$ for arbitrary $t \in T$ and $A_t \in \mathcal{S}_t$. $\square$

Changing our perspective, assume the f_t in (2) to be mappings into some measurable spaces $(S_t, \mathcal{S}_t)$. In Ω we may then introduce the *generated* or *induced* σ-field $\sigma(f) = \sigma\{f_t; t \in T\}$, defined as the smallest σ-field in Ω that makes all the f_t measurable. In other words, $\sigma(f)$ is the intersection of all σ-fields $\mathcal{A}$ in Ω such that f_t is $\mathcal{A}/\mathcal{S}_t$-measurable for every $t \in T$. In this notation, the functions f_t are clearly measurable with respect to a σ-field $\mathcal{A}$ in Ω iff $\sigma(f) \subset \mathcal{A}$. It is further useful to note that $\sigma(f)$ agrees with the σ-field in Ω generated by the collection $\{f_t^{-1}\mathcal{S}_t; t \in T\}$.

For functions on or into a Euclidean space $\mathbb{R}^d$, measurability is understood to be with respect to the Borel σ-field $\mathcal{B}^d$. Thus, a real-valued function f on some measurable space $(\Omega, \mathcal{A})$ is measurable iff $\{\omega; f(\omega) \leq x\} \in \mathcal{A}$ for all $x \in \mathbb{R}$. The same convention applies to functions into the *extended real line* $\overline{\mathbb{R}} = [-\infty, \infty]$ or the *extended half-line* $\overline{\mathbb{R}}_+ = [0, \infty]$, regarded as compactifications of $\mathbb{R}$ and $\mathbb{R}_+ = [0, \infty)$, respectively. Note that $\mathcal{B}(\overline{\mathbb{R}}) = \sigma\{\mathcal{B}, \pm\infty\}$ and $\mathcal{B}(\overline{\mathbb{R}}_+) = \sigma\{\mathcal{B}(\mathbb{R}_+), \infty\}$.

For any set $A \subset \Omega$, we define the associated *indicator function* $1_A : \Omega \to \mathbb{R}$ to be equal to 1 on A and to 0 on A^c. (The term *characteristic function* has a different meaning in probability theory.) For sets $A = \{\omega; f(\omega) \in B\}$, it is often convenient to write $1\{\cdot\}$ instead of $1_{\{\cdot\}}$. Assuming $\mathcal{A}$ to be a σ-field in Ω, we note that 1_A is $\mathcal{A}$-measurable iff $A \in \mathcal{A}$.

Linear combinations of indicator functions are called *simple functions*. Thus, a general simple function $f: \Omega \to \mathbb{R}$ has the form

$$f = c_1 1_{A_1} + \cdots + c_n 1_{A_n},$$

where $n \in \mathbb{Z}_+ = \{0, 1, \ldots\}$, $c_1, \ldots, c_n \in \mathbb{R}$, and $A_1, \ldots, A_n \subset \Omega$. Here we may clearly take $c_1, \ldots, c_n$ to be the distinct nonzero values attained by f and define $A_k = f^{-1}\{c_k\}$, $k = 1, \ldots, n$. With this choice of representation, we note that f is measurable with respect to a given σ-field $\mathcal{A}$ in Ω iff $A_1, \ldots, A_n \in \mathcal{A}$.

We proceed to show that the class of measurable functions is closed under the basic finite or countable operations occurring in analysis.

Lemma 1.9 *(bounds and limits)* Let $f_1, f_2, \ldots$ be measurable functions from some measurable space $(\Omega, \mathcal{A})$ into $\overline{\mathbb{R}}$. Then $\sup_n f_n$, $\inf_n f_n$, $\limsup_n f_n$, and $\liminf_n f_n$ are again measurable.

Proof: To see that $\sup_n f_n$ is measurable, write

$$\{\omega; \sup_n f_n(\omega) \leq t\} = \bigcap_n \{\omega; f_n(\omega) \leq t\} = \bigcap_n f_n^{-1}[-\infty, t] \in \mathcal{A},$$

and use Lemma 1.4. The measurability of the other three functions follows easily if we write $\inf_n f_n = -\sup_n(-f_n)$ and note that

$$\limsup_{n \to \infty} f_n = \inf_n \sup_{k \geq n} f_k, \qquad \liminf_{n \to \infty} f_n = \sup_n \inf_{k \geq n} f_k. \qquad \square$$

Since $f_n \to f$ iff $\limsup_n f_n = \liminf_n f_n = f$, it follows easily that both the set of convergence and the possible limit are measurable. The next result gives an extension to functions with values in more general spaces.

Lemma 1.10 *(convergence and limits)* Let $f_1, f_2, \ldots$ be measurable functions from a measurable space $(\Omega, \mathcal{A})$ into some metric space (S, ρ). Then

(i) $\{\omega; f_n(\omega) \text{ converges}\} \in \mathcal{A}$ when S is complete;

(ii) $f_n \to f$ on Ω implies that f is measurable.

Proof: (i) Since S is complete, the convergence of f_n is equivalent to the Cauchy convergence

$$\lim_{n \to \infty} \sup_{m \geq n} \rho(f_m, f_n) = 0.$$

Here the left-hand side is measurable by Lemmas 1.5 and 1.9.

(ii) If $f_n \to f$, we have $g \circ f_n \to g \circ f$ for any continuous function $g: S \to \mathbb{R}$, and so $g \circ f$ is measurable by Lemmas 1.5 and 1.9. Fixing any open set $G \subset S$, we may choose some continuous functions $g_1, g_2, \ldots: S \to \mathbb{R}_+$ with $g_n \uparrow 1_G$ and conclude from Lemma 1.9 that $1_G \circ f$ is measurable. Thus, $f^{-1}G \in \mathcal{A}$ for all G, and so f is measurable by Lemma 1.4. $\square$

Many results in measure theory are proved by a simple approximation, based on the following observation.

Lemma 1.11 *(approximation) For any measurable function $f: (\Omega, \mathcal{A}) \to \mathbb{R}_+$, there exist some simple measurable functions $f_1, f_2, \ldots : \Omega \to \mathbb{R}_+$ with $0 \leq f_n \uparrow f$.*

Proof: We may define
$$f_n(\omega) = 2^{-n}[2^n f(\omega)] \wedge n, \quad \omega \in \Omega, \, n \in \mathbb{N}. \qquad \Box$$

To illustrate the method, we may use the last lemma to prove the measurability of the basic arithmetic operations.

Lemma 1.12 *(elementary operations) Fix any measurable functions $f, g: (\Omega, \mathcal{A}) \to \mathbb{R}$ and constants $a, b \in \mathbb{R}$. Then $af + bg$ and fg are again measurable, and so is f/g when $g \neq 0$ on Ω.*

Proof: By Lemma 1.11 applied to $f_\pm = (\pm f) \vee 0$ and $g_\pm = (\pm g) \vee 0$, we may approximate by simple measurable functions $f_n \to f$ and $g_n \to g$. Here $af_n + bg_n$ and $f_n g_n$ are again simple measurable functions. Since they converge to $af + bg$ and fg, respectively, even the latter functions are measurable by Lemma 1.9. The same argument applies to the ratio f/g, provided we choose $g_n \neq 0$.

An alternative argument is to write $af + bg$, fg, or f/g as a composition $\psi \circ \varphi$, where $\varphi = (f, g): \Omega \to \mathbb{R}^2$, and $\psi(x, y)$ is defined as $ax + by$, xy, or x/y, respectively. The desired measurability then follows by Lemmas 1.2, 1.5, and 1.8. In the case of ratios, we may use the continuity of the mapping $(x, y) \mapsto x/y$ on $\mathbb{R} \times (\mathbb{R} \setminus \{0\})$. $\qquad \Box$

For many statements in measure theory and probability, it is convenient first to give a proof for the real line and then to extend the result to more general spaces. In this context, it is useful to identify pairs of measurable spaces S and T that are *Borel isomorphic*, in the sense that there exists a bijection $f: S \to T$ such that both f and f^{-1} are measurable. A space S that is Borel isomorphic to a Borel subset of $[0, 1]$ is called a *Borel space*. In particular, any Polish space endowed with its Borel σ-field is known to be Borel (cf. Theorem A1.2). (Recall that a topological space is said to be *Polish* if it admits a separable and complete metrization.)

The next result gives a useful functional representation of measurable functions. Given any two functions f and g on the same space Ω, we say that f is *g-measurable* if the induced σ-fields are related by $\sigma(f) \subset \sigma(g)$.

Lemma 1.13 *(functional representation, Doob) Fix two measurable functions f and g from a space Ω into some measurable spaces $(S, \mathcal{S})$ and $(T, \mathcal{T})$, where the former is Borel. Then f is g-measurable iff there exists some measurable mapping $h: T \to S$ with $f = h \circ g$.*

Proof: Since S is Borel, we may assume that $S \in \mathcal{B}([0, 1])$. By a suitable modification of h, we may further reduce to the case when $S = [0, 1]$. If

$f = 1_A$ with a g-measurable $A \subset \Omega$, then by Lemma 1.3 there exists some set $B \in \mathcal{T}$ with $A = g^{-1}B$. In this case $f = 1_A = 1_B \circ g$, and we may choose $h = 1_B$. The result extends by linearity to any simple g-measurable function f. In the general case, there exist by Lemma 1.11 some simple g-measurable functions $f_1, f_2, \ldots$ with $0 \leq f_n \uparrow f$, and we may choose associated $\mathcal{T}$-measurable functions $h_1, h_2, \ldots : T \to [0, 1]$ with $f_n = h_n \circ g$. Then $h = \sup_n h_n$ is again $\mathcal{T}$-measurable by Lemma 1.9, and we note that

$$h \circ g = (\sup_n h_n) \circ g = \sup_n (h_n \circ g) = \sup_n f_n = f. \qquad \square$$

Given any measurable space $(\Omega, \mathcal{A})$, a function $\mu : \mathcal{A} \to \overline{\mathbb{R}}_+$ is said to be *countably additive* if

$$\mu \bigcup\nolimits_{k \geq 1} A_k = \sum\nolimits_{k \geq 1} \mu A_k, \quad A_1, A_2, \cdots \in \mathcal{A} \text{ disjoint}. \tag{3}$$

A *measure* on $(\Omega, \mathcal{A})$ is defined as a function $\mu : \mathcal{A} \to \overline{\mathbb{R}}_+$ with $\mu \emptyset = 0$ and satisfying (3). A triple $(\Omega, \mathcal{A}, \mu)$ as above, where μ is a measure, is called a *measure space*. From (3) we note that any measure is finitely additive and nondecreasing. This implies in turn the *countable subadditivity*

$$\mu \bigcup\nolimits_{k \geq 1} A_k \leq \sum\nolimits_{k \geq 1} \mu A_k, \quad A_1, A_2, \cdots \in \mathcal{A}.$$

We note the following basic continuity properties.

Lemma 1.14 *(continuity)* *Let μ be a measure on $(\Omega, \mathcal{A})$, and assume that $A_1, A_2, \cdots \in \mathcal{A}$. Then*

(i) *$A_n \uparrow A$ implies $\mu A_n \uparrow \mu A$;*

(ii) *$A_n \downarrow A$ with $\mu A_1 < \infty$ implies $\mu A_n \downarrow \mu A$.*

Proof: For (i) we may apply (3) to the differences $D_n = A_n \setminus A_{n-1}$ with $A_0 = \emptyset$. To get (ii), apply (i) to the sets $B_n = A_1 \setminus A_n$. $\qquad \square$

The simplest measures on a measurable space $(\Omega, \mathcal{A})$ are the unit masses or *Dirac measures* δ_x, $x \in \Omega$, given by $\delta_x A = 1_A(x)$. For any countable set $A = \{x_1, x_2, \ldots\}$, we may form the associated *counting measure* $\mu = \sum_n \delta_{x_n}$. More generally, we may form countable linear combinations of arbitrary measures on Ω, as follows.

Proposition 1.15 *(series of measures)* *For any measures $\mu_1, \mu_2, \ldots$ on $(\Omega, \mathcal{A})$ and constants $c_1, c_2, \cdots \geq 0$, the sum $\mu = \sum_n c_n \mu_n$ is again a measure.*

Proof: We need the fact that, for any array of constants $c_{ij} \geq 0$, $i, j \in \mathbb{N}$,

$$\sum\nolimits_i \sum\nolimits_j c_{ij} = \sum\nolimits_j \sum\nolimits_i c_{ij}. \tag{4}$$

This is trivially true for finite sums. In general, let $m, n \in \mathbb{N}$ and write

$$\sum\nolimits_i \sum\nolimits_j c_{ij} \geq \sum\nolimits_{i \leq m} \sum\nolimits_{j \leq n} c_{ij} = \sum\nolimits_{j \leq n} \sum\nolimits_{i \leq m} c_{ij}.$$

Letting $m \to \infty$ and then $n \to \infty$, we obtain (4) with the inequality $\geq$. The same argument yields the reverse relation, and the equality follows.

Now consider any disjoint sets $A_1, A_2, \cdots \in \mathcal{A}$. Using (4) and the countable additivity of each μ_n, we get

$$\mu \bigcup_k A_k = \sum_n c_n \mu_n \bigcup_k A_k = \sum_n \sum_k c_n \mu_n A_k$$
$$= \sum_k \sum_n c_n \mu_n A_k = \sum_k \mu A_k. \qquad \square$$

The last result may be restated in terms of monotone sequences.

Corollary 1.16 *(monotone limits) Let $\mu_1, \mu_2, \ldots$ be measures on a measurable space $(\Omega, \mathcal{A})$ such that either $\mu_n \uparrow \mu$, or $\mu_n \downarrow \mu$ with μ_1 bounded. Then μ is again a measure on $(\Omega, \mathcal{A})$.*

Proof: In the increasing case, we may apply Proposition 1.15 to the sum $\mu = \sum_n (\mu_n - \mu_{n-1})$, where $\mu_0 = 0$. For decreasing sequences, the previous case applies to the increasing measures $\mu_1 - \mu_n$. $\qquad \square$

For any measure μ on $(\Omega, \mathcal{A})$ and set $B \in \mathcal{A}$, the function $\nu : A \mapsto \mu(A \cap B)$ is again a measure on $(\Omega, \mathcal{A})$, called the *restriction* of μ to B. Given any countable partition of Ω into disjoint sets $A_1, A_2, \cdots \in \mathcal{A}$, we note that $\mu = \sum_n \mu_n$, where μ_n denotes the restriction of μ to A_n. The measure μ is said to be σ-*finite* if the partition can be chosen such that $\mu A_n < \infty$ for all n. In that case the restrictions μ_n are clearly bounded.

A measure μ on some topological space S with Borel σ-field $\mathcal{S}$ is said to be *locally finite* if every point $s \in S$ has a neighborhood where μ is finite. A locally finite measure on a σ-compact space is clearly σ-finite. It is often useful to identify simple *measure-determining* classes $\mathcal{C} \subset \mathcal{S}$ such that a measure on S is uniquely determined by its values on $\mathcal{C}$. For locally finite measures on a Euclidean space $\mathbb{R}^d$, we may take $\mathcal{C} = \mathcal{I}^d$, the class of all bounded rectangles.

Lemma 1.17 *(uniqueness) Let μ and ν be bounded measures on some measurable space $(\Omega, \mathcal{A})$ and let $\mathcal{C}$ be a π-system in Ω such that $\Omega \in \mathcal{C}$ and $\sigma(\mathcal{C}) = \mathcal{A}$. Then $\mu = \nu$ iff $\mu A = \nu A$ for all $A \in \mathcal{C}$.*

Proof: Assuming $\mu = \nu$ on $\mathcal{C}$, let $\mathcal{D}$ denote the class of sets $A \in \mathcal{A}$ with $\mu A = \nu A$. Using the condition $\Omega \in \mathcal{C}$, the finite additivity of μ and ν, and Lemma 1.14, we see that $\mathcal{D}$ is a λ-system. Moreover, $\mathcal{C} \subset \mathcal{D}$ by hypothesis. Hence, Theorem 1.1 yields $\mathcal{D} \supset \sigma(\mathcal{C}) = \mathcal{A}$, which means that $\mu = \nu$. The converse assertion is obvious. $\qquad \square$

For any measure μ on a topological space S, the *support* $\mathrm{supp}\, \mu$ is defined as the smallest closed set $F \subset S$ with $\mu F^c = 0$. If $|\mathrm{supp}\, \mu| \leq 1$, then μ is said to be *degenerate*, and we note that $\mu = c\delta_s$ for some $s \in S$ and $c \geq 0$. More generally, a measure μ is said to have an *atom* at $s \in S$ if $\{s\} \in \mathcal{S}$ and $\mu\{s\} > 0$. For any locally finite measure μ on some σ-compact metric space S, the set $A = \{s \in S;\ \mu\{s\} > 0\}$ is clearly measurable, and we may define the *atomic* and *diffuse* components μ_a and μ_d of μ as the restrictions of μ to A and its complement. We further say that μ is *diffuse* if $\mu_a = 0$ and *purely atomic* if $\mu_d = 0$.

In the important special case when μ is locally finite and integer valued, the set A above is clearly locally finite and hence closed. By Lemma 1.14 we further have $\operatorname{supp}\mu \subset A$, and so μ is purely atomic. Hence, in this case $\mu = \sum_{s \in A} c_s \delta_s$ for some integers c_s. In particular, μ is said to be *simple* if $c_s = 1$ for all $s \in A$. Then clearly μ agrees with the counting measure on its support A.

Any measurable mapping f between two measurable spaces $(S, \mathcal{S})$ and $(T, \mathcal{T})$ induces a mapping of measures on S into measures on T. More precisely, given any measure μ on $(S, \mathcal{S})$, we may define a measure $\mu \circ f^{-1}$ on $(T, \mathcal{T})$ by

$$(\mu \circ f^{-1})B = \mu(f^{-1}B) = \mu\{s \in S;\ f(s) \in B\}, \quad B \in \mathcal{T}.$$

Here the countable additivity of $\mu \circ f^{-1}$ follows from that for μ together with the fact that f^{-1} preserves unions and intersections.

Our next aim is to define the *integral*

$$\mu f = \int f d\mu = \int f(\omega) \mu(d\omega)$$

of a real-valued, measurable function f on some measure space $(\Omega, \mathcal{A}, \mu)$. First assume that f is simple and nonnegative, hence of the form $c_1 1_{A_1} + \cdots + c_n 1_{A_n}$ for some $n \in \mathbb{Z}_+$, $A_1, \ldots, A_n \in \mathcal{A}$, and $c_1, \ldots, c_n \in \mathbb{R}_+$, and define

$$\mu f = c_1 \mu A_1 + \cdots + c_n \mu A_n.$$

(Throughout measure theory we are following the convention $0 \cdot \infty = 0$.) Using the finite additivity of μ, it is easy to verify that μf is independent of the choice of representation of f. It is further clear that the mapping $f \mapsto \mu f$ is *linear* and *nondecreasing*, in the sense that

$$\begin{aligned} \mu(af + bg) &= a\mu f + b\mu g, \quad a, b \geq 0, \\ f \leq g &\Rightarrow \mu f \leq \mu g. \end{aligned}$$

To extend the integral to any nonnegative measurable function f, we may choose as in Lemma 1.11 some simple measurable functions $f_1, f_2, \ldots$ with $0 \leq f_n \uparrow f$, and define $\mu f = \lim_n \mu f_n$. The following result shows that the limit is independent of the choice of approximating sequence (f_n).

Lemma 1.18 *(consistency)* *Fix any measurable function $f \geq 0$ on some measure space $(\Omega, \mathcal{A}, \mu)$, and let $f_1, f_2, \ldots$ and g be simple measurable functions satisfying $0 \leq f_n \uparrow f$ and $0 \leq g \leq f$. Then $\lim_n \mu f_n \geq \mu g$.*

Proof: By the linearity of μ, it is enough to consider the case when $g = 1_A$ for some $A \in \mathcal{A}$. Then fix any $\varepsilon > 0$, and define

$$A_n = \{\omega \in A;\ f_n(\omega) \geq 1 - \varepsilon\}, \quad n \in \mathbb{N}.$$

Here $A_n \uparrow A$, and so
$$\mu f_n \geq (1-\varepsilon)\mu A_n \uparrow (1-\varepsilon)\mu A = (1-\varepsilon)\mu g.$$
It remains to let $\varepsilon \to 0$. □

The linearity and monotonicity extend immediately to arbitrary $f \geq 0$, since if $f_n \uparrow f$ and $g_n \uparrow g$, then $af_n + bg_n \uparrow af + bg$, and if also $f \leq g$, then $f_n \leq (f_n \vee g_n) \uparrow g$. We are now ready to prove the basic continuity property of the integral.

Theorem 1.19 *(monotone convergence, Levi)* *Let $f, f_1, f_2 \ldots$ be measurable functions on $(\Omega, \mathcal{A}, \mu)$ with $0 \leq f_n \uparrow f$. Then $\mu f_n \uparrow \mu f$.*

Proof: For each n we may choose some simple measurable functions g_{nk}, with $0 \leq g_{nk} \uparrow f_n$ as $k \to \infty$. The functions $h_{nk} = g_{1k} \vee \cdots \vee g_{nk}$ have the same properties and are further nondecreasing in both indices. Hence,
$$f \geq \lim_{k \to \infty} h_{kk} \geq \lim_{k \to \infty} h_{nk} = f_n \uparrow f,$$
and so $0 \leq h_{kk} \uparrow f$. Using the definition and monotonicity of the integral, we obtain
$$\mu f = \lim_{k \to \infty} \mu h_{kk} \leq \lim_{k \to \infty} \mu f_k \leq \mu f.$$
□

The last result leads to the following key inequality.

Lemma 1.20 *(Fatou)* *For any measurable functions $f_1, f_2, \cdots \geq 0$ on $(\Omega, \mathcal{A}, \mu)$, we have*
$$\liminf_{n \to \infty} \mu f_n \geq \mu \liminf_{n \to \infty} f_n.$$

Proof: Since $f_m \geq \inf_{k \geq n} f_k$ for all $m \geq n$, we have
$$\inf_{k \geq n} \mu f_k \geq \mu \inf_{k \geq n} f_k, \quad n \in \mathbb{N}.$$
Letting $n \to \infty$, we get by Theorem 1.19
$$\liminf_{k \to \infty} \mu f_k \geq \lim_{n \to \infty} \mu \inf_{k \geq n} f_k = \mu \liminf_{k \to \infty} f_k.$$
□

A measurable function f on $(\Omega, \mathcal{A}, \mu)$ is said to be *integrable* if $\mu|f| < \infty$. In that case f may be written as the difference of two nonnegative, integrable functions g and h (e.g., as $f_+ - f_-$, where $f_\pm = (\pm f) \vee 0$), and we may define $\mu f = \mu g - \mu h$. It is easy to check that the extended integral is independent of the choice of representation $f = g - h$ and that μf satisfies the basic linearity and monotonicity properties (the former with arbitrary real coefficients).

We are now ready to state the basic condition that allows us to take limits under the integral sign. For $g_n \equiv g$ the result reduces to *Lebesgue's dominated convergence theorem*, a key result in analysis.

Theorem 1.21 *(dominated convergence, Lebesgue)* Let $f, f_1, f_2, \ldots$ and $g, g_1, g_2, \ldots$ be measurable functions on $(\Omega, \mathcal{A}, \mu)$ with $|f_n| \leq g_n$ for all n, and such that $f_n \to f$, $g_n \to g$, and $\mu g_n \to \mu g < \infty$. Then $\mu f_n \to \mu f$.

Proof: Applying Fatou's lemma to the functions $g_n \pm f_n \geq 0$, we get

$$\mu g + \liminf_{n \to \infty}(\pm \mu f_n) = \liminf_{n \to \infty} \mu(g_n \pm f_n) \geq \mu(g \pm f) = \mu g \pm \mu f.$$

Subtracting $\mu g < \infty$ from each side gives

$$\mu f \leq \liminf_{n \to \infty} \mu f_n \leq \limsup_{n \to \infty} \mu f_n \leq \mu f. \qquad \Box$$

The next result shows how integrals are transformed by measurable mappings.

Lemma 1.22 *(substitution)* Consider a measure space $(\Omega, \mathcal{A}, \mu)$, a measurable space $(S, \mathcal{S})$, and two measurable mappings $f : \Omega \to S$ and $g : S \to \mathbb{R}$. Then

$$\mu(g \circ f) = (\mu \circ f^{-1})g \qquad (5)$$

whenever either side exists. (In other words, if one side exists, then so does the other and the two are equal.)

Proof: If g is an indicator function, then (5) reduces to the definition of $\mu \circ f^{-1}$. From here on we may extend by linearity and monotone convergence to any measurable function $g \geq 0$. For general g it follows that $\mu|g \circ f| = (\mu \circ f^{-1})|g|$, and so the integrals in (5) exist at the same time. When they do, we get (5) by taking differences on both sides. $\qquad \Box$

Turning to the other basic transformation of measures and integrals, fix any measurable function $f \geq 0$ on some measure space $(\Omega, \mathcal{A}, \mu)$, and define a function $f \cdot \mu$ on $\mathcal{A}$ by

$$(f \cdot \mu)A = \mu(1_A f) = \int_A f d\mu, \qquad A \in \mathcal{A},$$

where the last relation defines the integral over a set A. It is easy to check that $\nu = f \cdot \mu$ is again a measure on $(\Omega, \mathcal{A})$. Here f is referred to as the μ-*density* of ν. The corresponding transformation rule is as follows.

Lemma 1.23 *(chain rule)* For any measure space $(\Omega, \mathcal{A}, \mu)$ and measurable functions $f: \Omega \to \mathbb{R}_+$ and $g: \Omega \to \mathbb{R}$, we have

$$\mu(fg) = (f \cdot \mu)g$$

whenever either side exists.

Proof: As in the last proof, we may begin with the case when g is an indicator function and then extend in steps to the general case. $\qquad \Box$

Given a measure space $(\Omega, \mathcal{A}, \mu)$, a set $A \in \mathcal{A}$ is said to be μ-*null* or simply *null* if $\mu A = 0$. A relation between functions on Ω is said to hold *almost everywhere* with respect to μ (abbreviated as a.e. μ or μ-a.e.) if it holds for all $\omega \in \Omega$ outside some μ-null set. The following frequently used result explains the relevance of null sets.

Lemma 1.24 *(null sets and functions) For any measurable function $f \geq 0$ on some measure space $(\Omega, \mathcal{A}, \mu)$, we have $\mu f = 0$ iff $f = 0$ a.e. μ.*

Proof: The statement is obvious when f is simple. In the general case, we may choose some simple measurable functions f_n with $0 \leq f_n \uparrow f$, and note that $f = 0$ a.e. iff $f_n = 0$ a.e. for every n, that is, iff $\mu f_n = 0$ for all n. Here the latter integrals converge to μf, and so the last condition is equivalent to $\mu f = 0$. □

The last result shows that two integrals agree when the integrands are a.e. equal. We may then allow integrands that are undefined on some μ-null set. It is also clear that the conclusions of Theorems 1.19 and 1.21 remain valid if the hypotheses are only fulfilled outside some null set.

In the other direction, we note that if two σ-finite measures μ and ν are related by $\nu = f \cdot \mu$ for some density f, then the latter is μ-a.e. unique, which justifies the notation $f = d\nu/d\mu$. It is further clear that any μ-null set is also a null set for ν. For measures μ and ν with the latter property, we say that ν is *absolutely continuous* with respect to μ and write $\nu \ll \mu$. The other extreme case is when μ and ν are *mutually singular* or *orthogonal* (written as $\mu \perp \nu$), in the sense that $\mu A = 0$ and $\nu A^c = 0$ for some set $A \in \mathcal{A}$.

Given a measure space $(\Omega, \mathcal{A}, \mu)$ and a σ-field $\mathcal{F} \subset \mathcal{A}$, we define the *$\mu$-completion of $\mathcal{F}$ in $\mathcal{A}$* as the σ-field $\mathcal{F}^\mu = \sigma(\mathcal{F}, \mathcal{N}_\mu)$, where $\mathcal{N}_\mu$ denotes the class of all subsets of arbitrary μ-null sets in $\mathcal{A}$. The description of $\mathcal{F}^\mu$ can be made more explicit, as follows.

Lemma 1.25 *(completion) Consider a measure space $(\Omega, \mathcal{A}, \mu)$, a σ-field $\mathcal{F} \subset \mathcal{A}$, and a Borel space $(S, \mathcal{S})$. Then a function $f : \Omega \to S$ is $\mathcal{F}^\mu$-measurable iff there exists some $\mathcal{F}$-measurable function g satisfying $f = g$ a.e. μ.*

Proof: Beginning with indicator functions, let $\mathcal{G}$ be the class of subsets $A \subset \Omega$ such that $A \triangle B \in \mathcal{N}_\mu$ for some $B \in \mathcal{F}$. Then $A \setminus B$ and $B \setminus A$ are again in $\mathcal{N}_\mu$, which implies $\mathcal{G} \subset \mathcal{F}^\mu$. Conversely, $\mathcal{F}^\mu \subset \mathcal{G}$ since both $\mathcal{F}$ and $\mathcal{N}_\mu$ are trivially contained in $\mathcal{G}$. Combining the two relations gives $\mathcal{G} = \mathcal{F}^\mu$, which shows that $A \in \mathcal{F}^\mu$ iff $1_A = 1_B$ a.e. for some $B \in \mathcal{F}$.

In the general case, we may clearly assume that $S = [0, 1]$. For any $\mathcal{F}^\mu$-measurable function f, we may then choose some simple $\mathcal{F}^\mu$-measurable functions f_n such that $0 \leq f_n \uparrow f$. By the result for indicator functions, we may next choose some simple $\mathcal{F}$-measurable functions g_n such that $f_n = g_n$ a.e. for each n. Since a countable union of null sets is again a null set, the function $g = \limsup_n g_n$ has the desired property. □

Any measure μ on $(\Omega, \mathcal{A})$ has a unique extension to the σ-field $\mathcal{A}^\mu$. Indeed, for any $A \in \mathcal{A}^\mu$ there exist by Lemma 1.25 some sets $A_\pm \in \mathcal{A}$ with

$A_- \subset A \subset A_+$ and $\mu(A_+ \setminus A_-) = 0$, and any extension must satisfy $\mu A = \mu A_\pm$. With this choice, it is easy to check that μ remains a measure on $\mathcal{A}^\mu$.

Our next aims are to construct product measures and to establish the basic condition for changing the order of integration. This requires a preliminary technical lemma.

Lemma 1.26 *(sections) Fix two measurable spaces $(S, \mathcal{S})$ and $(T, \mathcal{T})$, a measurable function $f: S \times T \to \mathbb{R}_+$, and a σ-finite measure μ on S. Then*

(i) $f(s,t)$ *is $\mathcal{S}$-measurable in $s \in S$ for each $t \in T$;*

(ii) $\int f(s,t)\mu(ds)$ *is $\mathcal{T}$-measurable in $t \in T$.*

Proof: We may assume that μ is bounded. Both statements are obvious when $f = 1_A$ with $A = B \times C$ for some $B \in \mathcal{S}$ and $C \in \mathcal{T}$, and they extend by a monotone class argument to any indicator functions of sets in $\mathcal{S} \otimes \mathcal{T}$. The general case follows by linearity and monotone convergence. □

We are now ready to state the main result involving product measures, commonly referred to as *Fubini's theorem*.

Theorem 1.27 *(product measures and iterated integrals, Lebesgue, Fubini, Tonelli) For any σ-finite measure spaces $(S, \mathcal{S}, \mu)$ and $(T, \mathcal{T}, \nu)$, there exists a unique measure $\mu \otimes \nu$ on $(S \times T, \mathcal{S} \otimes \mathcal{T})$ satisfying*

$$(\mu \otimes \nu)(B \times C) = \mu B \cdot \nu C, \quad B \in \mathcal{S}, \, C \in \mathcal{T}. \tag{6}$$

Furthermore, for any measurable function $f: S \times T \to \mathbb{R}_+$,

$$(\mu \otimes \nu)f = \int \mu(ds) \int f(s,t)\nu(dt) = \int \nu(dt) \int f(s,t)\mu(ds). \tag{7}$$

The last relation remains valid for any measurable function $f: S \times T \to \mathbb{R}$ with $(\mu \otimes \nu)|f| < \infty$.

Note that the iterated integrals in (7) are well defined by Lemma 1.26, although the inner integrals $\nu f(s,\cdot)$ and $\mu f(\cdot, t)$ may fail to exist on some null sets in S and T, respectively.

Proof: By Lemma 1.26 we may define

$$(\mu \otimes \nu)A = \int \mu(ds) \int 1_A(s,t)\nu(dt), \quad A \in \mathcal{S} \otimes \mathcal{T}, \tag{8}$$

which is clearly a measure on $S \times T$ satisfying (6). By a monotone class argument there can be at most one such measure. In particular, (8) remains true with the order of integration reversed, which proves (7) for indicator functions f. The formula extends by linearity and monotone convergence to arbitrary measurable functions $f \geq 0$.

In the general case, we note that (7) holds with f replaced by $|f|$. If $(\mu \otimes \nu)|f| < \infty$, it follows that $N_S = \{s \in S; \nu|f(s,\cdot)| = \infty\}$ is a μ-null set in S whereas $N_T = \{t \in T; \mu|f(\cdot, t)| = \infty\}$ is a ν-null set in T. By Lemma 1.24 we may redefine $f(s,t)$ to be zero when $s \in N_S$ or $t \in N_T$. Then (7) follows for f by subtraction of the formulas for f_+ and f_-. □

The measure $\mu \otimes \nu$ in Theorem 1.27 is called the *product measure* of μ and ν. Iterating the construction in finitely many steps, we obtain product measures $\mu_1 \otimes \cdots \otimes \mu_n = \bigotimes_k \mu_k$ satisfying higher-dimensional versions of (7). If $\mu_k = \mu$ for all k, we often write the product as $\mu^{\otimes n}$ or μ^n.

By a *measurable group* we mean a group G endowed with a σ-field $\mathcal{G}$ such that the group operations in G are $\mathcal{G}$-measurable. If $\mu_1, \ldots, \mu_n$ are σ-finite measures on G, we may define the *convolution* $\mu_1 * \cdots * \mu_n$ as the image of the product measure $\mu_1 \otimes \cdots \otimes \mu_n$ on G^n under the iterated group operation $(x_1, \ldots, x_n) \mapsto x_1 \cdots x_n$. The convolution is said to be *associative* if $(\mu_1 * \mu_2) * \mu_3 = \mu_1 * (\mu_2 * \mu_3)$ whenever both $\mu_1 * \mu_2$ and $\mu_2 * \mu_3$ are σ-finite and *commutative* if $\mu_1 * \mu_2 = \mu_2 * \mu_1$.

A measure μ on G is said to be *right* or *left invariant* if $\mu \circ T_g^{-1} = \mu$ for all $g \in G$, where T_g denotes the right or left *shift* $x \mapsto xg$ or $x \mapsto gx$. When G is Abelian, the shift is called a *translation*. We may also consider spaces of the form $G \times S$, in which case translations are defined to be mappings of the form $T_g: (x, s) \mapsto (x + g, s)$.

Lemma 1.28 (*convolution*) *The convolution of σ-finite measures on a measurable group $(G, \mathcal{G})$ is associative, and for Abelian G it is also commutative. In the latter case,*

$$(\mu * \nu)B = \int \mu(B - s)\nu(ds) = \int \nu(B - s)\mu(ds), \quad B \in \mathcal{G}.$$

If $\mu = f \cdot \lambda$ and $\nu = g \cdot \lambda$ for some invariant measure λ, then $\mu * \nu$ has the λ-density

$$(f * g)(s) = \int f(s - t)g(t)\lambda(dt) = \int f(t)g(s - t)\lambda(dt), \quad s \in G.$$

Proof: Use Fubini's theorem. $\square$

Given a measure space $(\Omega, \mathcal{A}, \mu)$ and a $p > 0$, we write $L^p = L^p(\Omega, \mathcal{A}, \mu)$ for the class of all measurable functions $f : \Omega \to \mathbb{R}$ with

$$\|f\|_p \equiv (\mu|f|^p)^{1/p} < \infty.$$

Lemma 1.29 (*Hölder and Minkowski inequalities*) *For any measurable functions f and g on some measure space $(\Omega, \mathcal{A}, \mu)$, we have*

(i) $\|fg\|_r \leq \|f\|_p \|g\|_q$ *for all $p, q, r > 0$ with $p^{-1} + q^{-1} = r^{-1}$,*
(ii) $\|f + g\|_p^{p \wedge 1} \leq \|f\|_p^{p \wedge 1} + \|g\|_p^{p \wedge 1}$ *for all $p > 0$.*

Proof: (i) It is clearly enough to take $r = 1$ and $\|f\|_p = \|g\|_q = 1$. The relation $p^{-1} + q^{-1} = 1$ implies $(p-1)(q-1) = 1$, and so the equations $y = x^{p-1}$ and $x = y^{q-1}$ are equivalent for $x, y \geq 0$. By calculus,

$$|fg| \leq \int_0^{|f|} x^{p-1}dx + \int_0^{|g|} y^{q-1}dy = p^{-1}|f|^p + q^{-1}|g|^q,$$

and so

$$\|fg\|_1 \leq p^{-1}\int |f|^p d\mu + q^{-1}\int |g|^q d\mu = p^{-1} + q^{-1} = 1.$$

(ii) The relation holds for $p \leq 1$ by the concavity of x^p on $\mathbb{R}_+$. For $p > 1$, we get by (i) with $q = p/(1-p)$ and $r = 1$

$$\|f + g\|_p^p \leq \int |f| |f + g|^{p-1} d\mu + \int |g| |f + g|^{p-1} d\mu$$
$$\leq \|f\|_p \|f + g\|_p^{p-1} + \|g\|_p \|f + g\|_p^{p-1}. \qquad \Box$$

The inequality in (ii) is often needed in the following extended form.

Corollary 1.30 (*extended Minkowski inequality*) *Let μ, ν, and f be such as in Theorem 1.27, and assume that $\mu f(t) = \int f(s,t) \mu(ds)$ exists for $t \in T$ a.e. ν. Write $\|f\|_p(s) = (\nu|f(s,\cdot)|^p)^{1/p}$. Then*

$$\|\mu f\|_p \leq \mu \|f\|_p, \quad p \geq 1.$$

Proof: Since $|\mu f| \leq \mu |f|$, we may assume that $f \geq 0$, and we may also assume that $\|\mu f\|_p \in (0, \infty)$. For $p > 1$, we get by Fubini's theorem and Hölder's inequality

$$\|\mu f\|_p^p = \nu(\mu f)^p = \nu(\mu f (\mu f)^{p-1}) = \mu \nu(f(\mu f)^{p-1})$$
$$\leq \mu \|f\|_p \|(\mu f)^{p-1}\|_q = \mu \|f\|_p \|\mu f\|_p^{p-1},$$

and it remains to divide by $\|\mu f\|_p^{p-1}$. The proof for $p = 1$ is similar but simpler. $\quad \Box$

In particular, Lemma 1.29 shows that $\|\cdot\|_p$ becomes a norm for $p \geq 1$ if we identify functions that agree a.e. For any $p > 0$ and $f, f_1, f_2, \cdots \in L^p$, we write $f_n \to f$ in L^p if $\|f_n - f\|_p \to 0$ and say that (f_n) is *Cauchy in L^p* if $\|f_m - f_n\|_p \to 0$ as $m, n \to \infty$.

Lemma 1.31 (*completeness*) *Let (f_n) be a Cauchy sequence in L^p, where $p > 0$. Then $\|f_n - f\|_p \to 0$ for some $f \in L^p$.*

Proof: Choose a subsequence $(n_k) \subset \mathbb{N}$ with $\sum_k \|f_{n_{k+1}} - f_{n_k}\|_p^{p \wedge 1} < \infty$. By Lemma 1.29 and monotone convergence we get $\|\sum_k |f_{n_{k+1}} - f_{n_k}|\|_p^{p \wedge 1} < \infty$, and so $\sum_k |f_{n_{k+1}} - f_{n_k}| < \infty$ a.e. Hence, (f_{n_k}) is a.e. Cauchy in $\mathbb{R}$, and so Lemma 1.10 yields $f_{n_k} \to f$ a.e. for some measurable function f. By Fatou's lemma,

$$\|f - f_n\|_p \leq \liminf_{k \to \infty} \|f_{n_k} - f_n\|_p \leq \sup_{m \geq n} \|f_m - f_n\|_p \to 0, \quad n \to \infty,$$

which shows that $f_n \to f$ in L^p. $\quad \Box$

The next result gives a useful criterion for convergence in L^p.

Lemma 1.32 (L^p-*convergence*) *For any $p > 0$, let $f, f_1, f_2, \cdots \in L^p$ with $f_n \to f$ a.e. Then $f_n \to f$ in L^p iff $\|f_n\|_p \to \|f\|_p$.*

Proof: If $f_n \to f$ in L^p, we get by Lemma 1.29

$$\big| \|f_n\|_p^{p \wedge 1} - \|f\|_p^{p \wedge 1} \big| \leq \|f_n - f\|_p^{p \wedge 1} \to 0,$$

and so $\|f_n\|_p \to \|f\|_p$. Now assume instead the latter condition, and define

$$g_n = 2^p(|f_n|^p + |f|^p), \qquad g = 2^{p+1}|f|^p.$$

Then $g_n \to g$ a.e. and $\mu g_n \to \mu g < \infty$ by hypotheses. Since also $|g_n| \geq |f_n - f|^p \to 0$ a.e., Theorem 1.21 yields $\|f_n - f\|_p^p = \mu|f_n - f|^p \to 0$. □

Taking $p = q = 2$ and $r = 1$ in Lemma 1.29 (i), we get the *Cauchy-Buniakovsky* or *Schwarz inequality*

$$\|fg\|_1 \leq \|f\|_2 \|g\|_2.$$

In particular, we note that, for any $f, g \in L^2$, the *inner product* $\langle f, g \rangle = \mu(fg)$ exists and satisfies $|\langle f, g \rangle| \leq \|f\|_2 \|g\|_2$. From the obvious bilinearity of the inner product, we get the *parallelogram identity*

$$\|f+g\|^2 + \|f-g\|^2 = 2\|f\|^2 + 2\|g\|^2, \quad f, g \in L^2. \tag{9}$$

Two functions $f, g \in L^2$ are said to be *orthogonal* (written as $f \perp g$) if $\langle f, g \rangle = 0$. Orthogonality between two subsets $A, B \subset L^2$ means that $f \perp g$ for all $f \in A$ and $g \in B$. A subspace $M \subset L^2$ is said to be *linear* if $af + bg \in M$ for any $f, g \in M$ and $a, b \in \mathbb{R}$, and *closed* if $f \in M$ whenever f is the L^2-limit of a sequence in M.

Theorem 1.33 *(orthogonal projection)* Let M be a closed linear subspace of L^2. Then any function $f \in L^2$ has an a.e. unique decomposition $f = g + h$ with $g \in M$ and $h \perp M$.

Proof: Fix any $f \in L^2$, and define $d = \inf\{\|f - g\|; g \in M\}$. Choose $g_1, g_2, \cdots \in M$ with $\|f - g_n\| \to d$. Using the linearity of M, the definition of d, and (9), we get as $m, n \to \infty$,

$$\begin{aligned} 4d^2 + \|g_m - g_n\|^2 &\leq \|2f - g_m - g_n\|^2 + \|g_m - g_n\|^2 \\ &= 2\|f - g_m\|^2 + 2\|f - g_n\|^2 \to 4d^2. \end{aligned}$$

Thus, $\|g_m - g_n\| \to 0$, and so the sequence (g_n) is Cauchy in L^2. By Lemma 1.31 it converges toward some $g \in L^2$, and since M is closed we have $g \in M$. Noting that $h = f - g$ has norm d, we get for any $l \in M$,

$$d^2 \leq \|h + tl\|^2 = d^2 + 2t\langle h, l \rangle + t^2\|l\|^2, \quad t \in \mathbb{R},$$

which implies $\langle h, l \rangle = 0$. Hence, $h \perp M$, as required.

To prove the uniqueness, let $g' + h'$ be another decomposition with the stated properties. Then $g - g' \in M$ and also $g - g' = h' - h \perp M$, so $g - g' \perp g - g'$, which implies $\|g - g'\|^2 = \langle g - g', g - g' \rangle = 0$, and hence $g = g'$ a.e. □

We proceed with a basic approximation property of sets.

Lemma 1.34 *(regularity)* *Let μ be a bounded measure on some metric space S with Borel σ-field $\mathcal{S}$. Then*

$$\mu B = \sup_{F \subset B} \mu F = \inf_{G \supset B} \mu G, \quad B \in \mathcal{S},$$

with F and G restricted to the classes of closed and open subsets of S, respectively.

Proof: For any open set G there exist some closed sets $F_n \uparrow G$, and by Lemma 1.14 we get $\mu F_n \uparrow \mu G$. This proves the statement for B belonging to the π-system $\mathcal{G}$ of all open sets. Letting $\mathcal{D}$ denote the class of all sets B with the stated property, we further note that $\mathcal{D}$ is a λ-system. Hence, Theorem 1.1 shows that $\mathcal{D} \supset \sigma(\mathcal{G}) = \mathcal{S}$. □

The last result leads to a basic approximation property for functions.

Lemma 1.35 *(approximation)* *Given a metric space S with Borel σ-field $\mathcal{S}$, a bounded measure μ on $(S, \mathcal{S})$, and a constant $p > 0$, the set of bounded, continuous functions on S is dense in $L^p(S, \mathcal{S}, \mu)$. Thus, for any $f \in L^p$ there exist some bounded, continuous functions $f_1, f_2, \ldots : S \to \mathbb{R}$ with $\|f_n - f\|_p \to 0$.*

Proof: If $f = 1_A$ with $A \subset S$ open, we may choose some continuous functions f_n with $0 \leq f_n \uparrow f$, and then $\|f_n - f\|_p \to 0$ by dominated convergence. By Lemma 1.34 the result remains true for arbitrary $A \in \mathcal{S}$. The further extension to simple measurable functions is immediate. For general $f \in L^p$ we may choose some simple measurable functions $f_n \to f$ with $|f_n| \leq |f|$. Since $|f_n - f|^p \leq 2^{p+1}|f|^p$, we get $\|f_n - f\|_p \to 0$ by dominated convergence. □

The next result shows how the pointwise convergence of a sequence of measurable functions is almost uniform.

Lemma 1.36 *(near uniformity, Egorov)* *Let $f, f_1, f_2, \ldots$ be measurable functions on some finite measure space $(\Omega, \mathcal{A}, \mu)$ such that $f_n \to f$ on Ω. Then for any $\varepsilon > 0$ there exists some $A \in \mathcal{A}$ with $\mu A^c < \varepsilon$ such that $f_n \to f$ uniformly on A.*

Proof: Define

$$A_{m,n} = \bigcap_{k \geq n} \{x \in \Omega;\ |f_k(x) - f(x)| < m^{-1}\}, \quad m, n \in \mathbb{N}.$$

As $n \to \infty$ for fixed m, we have $A_{m,n} \uparrow \Omega$ and hence $\mu A_{m,n}^c \to 0$. Given any $\varepsilon > 0$, we may then choose $n_1, n_2, \cdots \in \mathbb{N}$ so large that $\mu A_{m,n_m}^c < \varepsilon 2^{-m}$ for all m. Letting $A = \bigcap_m A_{m,n_m}$, we get

$$\mu A^c \leq \mu \bigcup_m A_{m,n_m}^c < \varepsilon \sum_m 2^{-m} = \varepsilon,$$

and we note that $f_n \to f$ uniformly on A. □

The last two results may be combined to show that every measurable function is almost continuous.

Lemma 1.37 *(near continuity, Lusin) Let f be a measurable function on some compact metric space S with Borel σ-field $\mathcal{S}$ and a bounded measure μ. Then there exist some continuous functions $f_1, f_2, \ldots$ on S such that $\mu\{x; f_n(x) \neq f(x)\} \to 0$.*

Proof: We may clearly assume that f is bounded. By Lemma 1.35 we may choose some continuous functions $g_1, g_2, \ldots$ on S such that $\mu|g_k - f| \leq 2^{-k}$. By Fubini's theorem, we get

$$\mu \sum_k |g_k - f| = \sum_k \mu|g_k - f| \leq \sum_k 2^{-k} = 1,$$

and so $\sum_k |g_k - f| < \infty$ a.e., which implies $g_k \to f$ a.e. By Lemma 1.36, we may next choose $A_1, A_2, \ldots \in \mathcal{S}$ with $\mu A_n^c \to 0$ such that the convergence is uniform on each A_n. Since each g_k is uniformly continuous on S, we conclude that f is uniformly continuous on each A_n. By Tietze's extension theorem, the restriction $f|_{A_n}$ then admits a continuous extension f_n to S. □

For any measurable space $(S, \mathcal{S})$, we may introduce the class $\mathcal{M}(S)$ of σ-finite measures on S. The set $\mathcal{M}(S)$ becomes a measurable space in its own right when endowed with the σ-field induced by the mappings $\pi_B \colon \mu \mapsto \mu B$, $B \in \mathcal{S}$. Note in particular that the class $\mathcal{P}(S)$ of probability measures on S is a measurable subset of $\mathcal{M}(S)$. In the next two lemmas we state some less obvious measurability properties, which will be needed in subsequent chapters.

Lemma 1.38 *(measurability of products) For any measurable spaces $(S, \mathcal{S})$ and $(T, \mathcal{T})$, the mapping $(\mu, \nu) \mapsto \mu \otimes \nu$ is measurable from $\mathcal{P}(S) \times \mathcal{P}(T)$ to $\mathcal{P}(S \times T)$.*

Proof: Note that $(\mu \otimes \nu)A$ is measurable whenever $A = B \times C$ with $B \in \mathcal{S}$ and $C \in \mathcal{T}$, and extend by a monotone class argument. □

In the context of separable metric spaces S, we assume the measures $\mu \in \mathcal{M}(S)$ to be *locally finite*, in the sense that $\mu B < \infty$ for any bounded Borel set B.

Lemma 1.39 *(diffuse and atomic parts) For any separable metric space S,*
  (i) *the set $D \subset \mathcal{M}(S)$ of degenerate measures on S is measurable;*
  (ii) *the diffuse and purely atomic components μ_d and μ_a are measurable functions of $\mu \in \mathcal{M}(S)$.*

Proof: (i) Choose a countable topological base $B_1, B_2, \ldots$ in S, and define $J = \{(i,j); B_i \cap B_j = \emptyset\}$. Then, clearly,

$$D = \left\{ \mu \in M(S); \sum_{(i,j) \in J} (\mu B_i)(\mu B_j) = 0 \right\}.$$

(ii) Choose a nested sequence of countable partitions $\mathcal{B}_n$ of S into Borel sets of diameter less than n^{-1}. For any $\varepsilon > 0$ and $n \in \mathbb{N}$ we introduce the sets $U_n^\varepsilon = \bigcup \{B \in \mathcal{B}_n; \mu B \geq \varepsilon\}$, $U^\varepsilon = \{s \in S; \mu\{s\} \geq \varepsilon\}$, and $U = \{s \in S; \mu\{s\} > 0\}$. It is easily seen that $U_n^\varepsilon \downarrow U^\varepsilon$ as $n \to \infty$ and $U^\varepsilon \uparrow U$ as $\varepsilon \to 0$. By dominated convergence, the restrictions $\mu_n^\varepsilon = \mu(U_n^\varepsilon \cap \cdot)$ and $\mu^\varepsilon = \mu(U^\varepsilon \cap \cdot)$ satisfy locally $\mu_n^\varepsilon \downarrow \mu^\varepsilon$ and $\mu^\varepsilon \uparrow \mu_a$. Since μ_n^ε is clearly a measurable function of μ, the asserted measurability of μ_a and μ_d now follows by Lemma 1.10. □

Given two measurable spaces $(S, \mathcal{S})$ and $(T, \mathcal{T})$, a mapping $\mu: S \times \mathcal{T} \to \overline{\mathbb{R}}_+$ is called a *(probability) kernel* from S to T if the function $\mu_s B = \mu(s, B)$ is $\mathcal{S}$-measurable in $s \in S$ for fixed $B \in \mathcal{T}$ and a (probability) measure in $B \in \mathcal{T}$ for fixed $s \in S$. Any kernel μ determines an associated operator that maps suitable functions $f: T \to \mathbb{R}$ into their integrals $\mu f(s) = \int \mu(s, dt) f(t)$. Kernels play an important role in probability theory, where they may appear in the guises of random measures, conditional distributions, Markov transition functions, and potentials.

The following characterizations of the kernel property are often useful. For simplicity we restrict our attention to probability kernels.

Lemma 1.40 *(kernels)* *Fix two measurable spaces $(S, \mathcal{S})$ and $(T, \mathcal{T})$, a π-system $\mathcal{C}$ with $\sigma(\mathcal{C}) = \mathcal{T}$, and a family $\mu = \{\mu_s; s \in S\}$ of probability measures on T. Then these conditions are equivalent:*

(i) μ *is a probability kernel from S to T;*

(ii) μ *is a measurable mapping from S to $\mathcal{P}(T)$;*

(iii) $s \mapsto \mu_s B$ *is a measurable mapping from S to $[0, 1]$ for every $B \in \mathcal{C}$.*

Proof: Since $\pi_B : \mu \mapsto \mu B$ is measurable on $\mathcal{P}(T)$ for every $B \in \mathcal{T}$, condition (ii) implies (iii) by Lemma 1.7. Furthermore, (iii) implies (i) by a straightforward application of Theorem 1.1. Finally, under (i) we have $\mu^{-1} \pi_B^{-1}[0, x] \in \mathcal{S}$ for all $B \in \mathcal{T}$ and $x \geq 0$, and (ii) follows by Lemma 1.4. □

Let us now introduce a third measurable space $(U, \mathcal{U})$, and consider two kernels μ and ν, one from S to T and the other from $S \times T$ to U. Imitating the construction of product measures, we may attempt to combine μ and ν into a kernel $\mu \otimes \nu$ from S to $T \times U$ given by

$$(\mu \otimes \nu)(s, B) = \int \mu(s, dt) \int \nu(s, t, du) 1_B(t, u), \quad B \in \mathcal{T} \otimes \mathcal{U}.$$

The following lemma justifies the formula and provides some further useful information.

Lemma 1.41 *(kernels and functions)* *Fix three measurable spaces $(S, \mathcal{S})$, $(T, \mathcal{T})$, and $(U, \mathcal{U})$. Let μ and ν be probability kernels from S to T and from $S \times T$ to U, respectively, and consider two measurable functions $f: S \times T \to \mathbb{R}_+$ and $g: S \times T \to U$. Then*

(i) $\mu_s f(s, \cdot)$ *is a measurable function of $s \in S$;*
(ii) $\mu_s \circ (g(s, \cdot))^{-1}$ *is a kernel from S to U;*
(iii) $\mu \otimes \nu$ *is a kernel from S to $T \times U$.*

Proof: Assertion (i) is obvious when f is the indicator function of a set $A = B \times C$ with $B \in \mathcal{S}$ and $C \in \mathcal{T}$. From here on, we may extend to general $A \in \mathcal{S} \otimes \mathcal{T}$ by a monotone class argument and then to arbitrary f by linearity and monotone convergence. The statements in (ii) and (iii) are easy consequences. □

For any measurable function $f \geq 0$ on $T \times U$, we get as in Theorem 1.27

$$(\mu \otimes \nu)_s f = \int \mu(s, dt) \int \nu(s, t, du) f(t, u), \quad s \in S,$$

or simply $(\mu \otimes \nu)f = \mu(\nu f)$. By iteration we may combine any kernels μ_k from $S_0 \times \cdots \times S_{k-1}$ to S_k, $k = 1, \ldots, n$, into a kernel $\mu_1 \otimes \cdots \otimes \mu_n$ from S_0 to $S_1 \times \cdots \times S_n$, given by

$$(\mu_1 \otimes \cdots \otimes \mu_n) f = \mu_1(\mu_2(\cdots(\mu_n f)\cdots))$$

for any measurable function $f \geq 0$ on $S_1 \times \cdots \times S_n$.

In applications we may often encounter kernels μ_k from S_{k-1} to S_k, $k = 1, \ldots, n$, in which case the *composition* $\mu_1 \cdots \mu_n$ is defined as a kernel from S_0 to S_n given for measurable $B \subset S_n$ by

$$\begin{aligned}(\mu_1 \cdots \mu_n)_s B &= (\mu_1 \otimes \cdots \otimes \mu_n)_s (S_1 \times \cdots \times S_{n-1} \times B) \\ &= \int \mu_1(s, ds_1) \int \mu_2(s_1, ds_2) \cdots \\ &\quad \cdots \int \mu_{n-1}(s_{n-2}, ds_{n-1}) \mu_n(s_{n-1}, B).\end{aligned}$$

Exercises

1. Prove the triangle inequality $\mu(A \triangle C) \leq \mu(A \triangle B) + \mu(B \triangle C)$. (*Hint:* Note that $1_{A \triangle B} = |1_A - 1_B|$.)

2. Show that Lemma 1.9 is false for uncountable index sets. (*Hint:* Show that every measurable set depends on countably many coordinates.)

3. For any space S, let μA denote the cardinality of the set $A \subset S$. Show that μ is a measure on $(S, 2^S)$.

4. Let $\mathcal{K}$ be the class of compact subsets of some metric space S, and let μ be a bounded measure such that $\inf_{K \in \mathcal{K}} \mu K^c = 0$. Show for any $B \in \mathcal{B}(S)$ that $\mu B = \sup_{K \in \mathcal{K} \cap B} \mu K$.

5. Show that any absolutely convergent series can be written as an integral with respect to counting measure on N. State series versions of Fatou's lemma and the dominated convergence theorem, and give direct elementary proofs.

6. Give an example of integrable functions $f, f_1, f_2, \ldots$ on some probability space $(\Omega, \mathcal{A}, \mu)$ such that $f_n \to f$ but $\mu f_n \not\to \mu f$.

7. Fix two σ-finite measures μ and ν on some measurable space $(\Omega, \mathcal{F})$ with sub-σ-field $\mathcal{G}$. Show that if $\mu \ll \nu$ holds on $\mathcal{F}$, it is also true on $\mathcal{G}$. Further show by an example that the converse may fail.

8. Fix two measurable spaces $(S, \mathcal{S})$ and $(T, \mathcal{T})$, a measurable function $f: S \to T$, and a measure μ on S with image $\nu = \mu \circ f^{-1}$. Show that f remains measurable w.r.t. the completions $\mathcal{S}^\mu$ and $\mathcal{T}^\nu$.

9. Fix a measure space $(S, \mathcal{S}, \mu)$ and a σ-field $\mathcal{T} \subset \mathcal{S}$, let $\mathcal{S}^\mu$ denote the μ-completion of $\mathcal{S}$, and let $\mathcal{T}^\mu$ be the σ-field generated by $\mathcal{T}$ and the μ-null sets of $\mathcal{S}^\mu$. Show that $A \in \mathcal{T}^\mu$ iff there exist some $B \in \mathcal{T}$ and $N \in \mathcal{S}^\mu$ with $A \triangle B \subset N$ and $\mu N = 0$. Also, show by an example that $\mathcal{T}^\mu$ may be strictly greater than the μ-completion of $\mathcal{T}$.

10. State Fubini's theorem for the case where μ is any σ-finite measure and ν is the counting measure on N. Give a direct proof of this result.

11. Let $f_1, f_2, \ldots$ be μ-integrable functions on some measurable space S such that $g = \sum_k f_k$ exists a.e., and put $g_n = \sum_{k \leq n} f_k$. Restate the dominated convergence theorem for the integrals μg_n in terms of the functions f_k, and compare with the result of the preceding exercise.

12. Extend Theorem 1.27 to the product of n measures.

13. Let λ denote Lebesgue measure on $\mathbb{R}_+$, and fix any $p > 0$. Show that the class of step functions with bounded support and finitely many jumps is dense in $L^p(\lambda)$. Generalize to $\mathbb{R}_+^d$.

14. Let $M \supset N$ be closed linear subspaces of L^2. Show that if $f \in L^2$ has projections g onto M and h onto N, then g has projection h onto N.

15. Let M be a closed linear subspace of L^2, and let $f, g \in L^2$ with M-projections $\hat{f}$ and $\hat{g}$. Show that $\langle \hat{f}, g \rangle = \langle f, \hat{g} \rangle = \langle \hat{f}, \hat{g} \rangle$.

16. Let $\mu_1, \mu_2, \ldots$ be kernels between two measurable spaces S and T. Show that the function $\mu = \sum_n \mu_n$ is again a kernel.

17. Fix a function f between two measurable spaces S and T, and define $\mu(s, B) = 1_B \circ f(s)$. Show that μ is a kernel iff f is measurable.

18. Show that if $\mu \ll \nu$ and $\nu f = 0$ with $f \geq 0$, then also $\mu f = 0$. (Hint: Use Lemma 1.24.)

19. For any σ-finite measures $\mu_1 \ll \mu_2$ and $\nu_1 \ll \nu_2$, show that $\mu_1 \otimes \nu_1 \ll \mu_2 \otimes \nu_2$. (Hint: Use Fubini's theorem and Lemma 1.24.)

Chapter 2

Measure Theory — Key Results

Outer measures and extension; Lebesgue and Lebesgue–Stieltjes measures; Jordan–Hahn and Lebesgue decompositions; Radon–Nikodým theorem; Lebesgue's differentiation theorem; functions of finite variation; Riesz' representation theorem; Haar and invariant measures

We continue our introduction to measure theory with a detailed discussion of some basic results of the subject, all of special relevance to probability theory. Again the hurried or impatient reader may skip to the next chapter and return for reference when need arises.

Most important, by far, of the quoted results is the existence of Lebesgue measure, which lies at the heart of most probabilistic constructions, often via a use of the Daniell–Kolmogorov theorem of Chapter 6. A similar role is played by the construction of Haar and other invariant measures, which ensures the existence of uniform distributions or homogeneous Poisson processes on spheres and other manifolds. Other key results include Riesz' representation theorem, which will enable us in Chapter 19 to construct Markov processes with a given generator, via the resolvents and the associated semigroup of transition operators. We may also mention the Radon–Nikodým theorem, of relevance to the theory of conditioning in Chapter 6, Lebesgue's differentiation theorem, instrumental for proving the general ballot theorem in Chapter 11, and various results on functions of bounded variation, important for the theory of predictable processes and general semimartingales in Chapters 25 and 26.

We begin with an ingenious technical result that will play a crucial role for our construction of Lebesgue measure in Theorem 2.2 and for the proof of Riesz' representation Theorem 2.22. By an *outer measure* on a space Ω we mean a nondecreasing and countably subadditive set function $\mu\colon 2^\Omega \to \overline{\mathbb{R}}_+$ with $\mu\emptyset = 0$. Given an outer measure μ on Ω, we say that a set $A \subset \Omega$ is μ-*measurable* if

$$\mu E = \mu(E \cap A) + \mu(E \cap A^c), \quad E \subset \Omega. \tag{1}$$

Note that the inequality $\leq$ holds automatically by subadditivity. The following result gives the basic construction of measures from outer measures.

Theorem 2.1 *(restriction of outer measure, Carathéodory) Let μ be an outer measure on Ω, and write $\mathcal{A}$ for the class of μ-measurable sets. Then $\mathcal{A}$ is a σ-field and the restriction of μ to $\mathcal{A}$ is a measure.*

Proof: Since $\mu \emptyset = 0$, we have for any set $E \subset \Omega$

$$\mu(E \cap \emptyset) + \mu(E \cap \Omega) = \mu \emptyset + \mu E = \mu E,$$

which shows that $\emptyset \in \mathcal{A}$. Also note that trivially $A \in \mathcal{A}$ implies $A^c \in \mathcal{A}$.

Next assume that $A, B \in \mathcal{A}$. Using (1) for A and B together with the subadditivity of μ, we get for any $E \subset \Omega$

$$\begin{aligned}\mu E &= \mu(E \cap A) + \mu(E \cap A^c) \\ &= \mu(E \cap A \cap B) + \mu(E \cap A \cap B^c) + \mu(E \cap A^c) \\ &\geq \mu(E \cap (A \cap B)) + \mu(E \cap (A \cap B)^c),\end{aligned}$$

which shows that even $A \cap B \in \mathcal{A}$. It follows easily that $\mathcal{A}$ is a field. If $A, B \in \mathcal{A}$ are disjoint, we also get by (1) for any $E \subset \Omega$

$$\begin{aligned}\mu(E \cap (A \cup B)) &= \mu(E \cap (A \cup B) \cap A) + \mu(E \cap (A \cup B) \cap A^c) \\ &= \mu(E \cap A) + \mu(E \cap B). \quad (2)\end{aligned}$$

Finally, consider any disjoint sets $A_1, A_2, \cdots \in \mathcal{A}$, and put $U_n = \bigcup_{k \leq n} A_k$ and $U = \bigcup_n U_n$. Using (2) recursively along with the monotonicity of μ, we get

$$\mu(E \cap U) \geq \mu(E \cap U_n) = \sum_{k \leq n} \mu(E \cap A_k).$$

Letting $n \to \infty$ and combining with the subadditivity of μ, we obtain

$$\mu(E \cap U) = \sum_k \mu(E \cap A_k). \quad (3)$$

In particular, for $E = \Omega$ we see that μ is countably additive on $\mathcal{A}$. Noting that $U_n \in \mathcal{A}$ and using (3) twice along with the monotonicity of μ, we also get

$$\begin{aligned}\mu E &= \mu(E \cap U_n) + \mu(E \cap U_n^c) \\ &\geq \sum_{k \leq n} \mu(E \cap A_k) + \mu(E \cap U^c) \\ &\to \mu(E \cap U) + \mu(E \cap U^c),\end{aligned}$$

which shows that $U \in \mathcal{A}$. Thus, $\mathcal{A}$ is a σ-field. $\square$

We are now ready to introduce *Lebesgue measure* λ on $\mathbb{R}$. The length of an interval $I \subset \mathbb{R}$ is denoted by $|I|$.

Theorem 2.2 *(Lebesgue measure, Borel) There exists a unique measure λ on $(\mathbb{R}, \mathcal{B})$ such that $\lambda I = |I|$ for every interval $I \subset \mathbb{R}$.*

As a first step in the proof, we show that the length $|I|$ of intervals $I \subset \mathbb{R}$ admits an extension to an outer measure on $\mathbb{R}$. Then define

$$\lambda A = \inf_{\{I_k\}} \sum_k |I_k|, \quad A \subset \mathbb{R}, \quad (4)$$

where the infimum extends over all countable covers of A by open intervals $I_1, I_2, \ldots$. We show that (4) provides the desired extension.

Lemma 2.3 *(outer Lebesgue measure) The function λ in (4) is an outer measure on $\mathbb{R}$. Moreover, $\lambda I = |I|$ for every interval I.*

Proof: The set function λ is clearly nonnegative and nondecreasing with $\lambda \emptyset = 0$. To prove the countable subadditivity, let $A_1, A_2, \cdots \subset \mathbb{R}$ be arbitrary. For any $\varepsilon > 0$ and $n \in \mathbb{N}$, we may choose some open intervals $I_{n1}, I_{n2}, \ldots$ such that

$$A_n \subset \bigcup_k I_{nk}, \quad \lambda A_n \geq \sum_k |I_{nk}| - \varepsilon 2^{-n}, \qquad n \in \mathbb{N}.$$

Then

$$\bigcup_n A_n \subset \bigcup_n \bigcup_k I_{nk},$$
$$\lambda \bigcup_n A_n \leq \sum_n \sum_k |I_{nk}| \leq \sum_n \lambda A_n + \varepsilon,$$

and the desired relation follows as we let $\varepsilon \to 0$.

To prove the second assertion, we may assume that $I = [a, b]$ for some finite numbers $a < b$. Since $I \subset (a - \varepsilon, b + \varepsilon)$ for every $\varepsilon > 0$, we get $\lambda I \leq |I| + 2\varepsilon$, and so $\lambda I \leq |I|$. To obtain the reverse relation, we need to prove that if $I \subset \bigcup_k I_k$ for some open intervals $I_1, I_2, \ldots$, then $|I| \leq \sum_k |I_k|$. By the Heine–Borel theorem, I remains covered by finitely many intervals $I_1, \ldots, I_n$, and it suffices to show that $|I| \leq \sum_{k \leq n} |I_k|$. This reduces the assertion to the case of finitely many covering intervals $I_1, \ldots, I_n$.

The statement is clearly true for a single covering interval. Proceeding by induction, we assume the assertion to be true for $n - 1$ covering intervals and turn to the case of covering by $I_1, \ldots, I_n$. Then b belongs to some $I_k = (a_k, b_k)$, and so the interval $I'_k = I \setminus I_k$ is covered by the remaining intervals I_j, $j \neq k$. By the induction hypothesis, we get

$$\begin{aligned} |I| &= b - a \leq (b - a_k) + (a_k - a) \\ &\leq |I_k| + |I'_k| \leq |I_k| + \sum_{j \neq k} |I_j| = \sum_j |I_j|, \end{aligned}$$

as required. □

The next result ensures that the class of measurable sets in Lemma 2.3 is large enough to contain all Borel sets.

Lemma 2.4 *(measurability of intervals) Let λ denote the outer measure in Lemma 2.3. Then the interval $(-\infty, a]$ is λ-measurable for every $a \in \mathbb{R}$.*

Proof: For any set $E \subset \mathbb{R}$ and constant $\varepsilon > 0$, we may cover E by some open intervals $I_1, I_2, \ldots$ such that $\lambda E \geq \sum_n |I_n| - \varepsilon$. Writing $I = (-\infty, a]$

and using the subadditivity of λ and Lemma 2.3, we get

$$\begin{aligned}
\lambda E + \varepsilon &\geq \sum_n |I_n| = \sum_n |I_n \cap I| + \sum_n |I_n \cap I^c| \\
&= \sum_n \lambda(I_n \cap I) + \sum_n \lambda(I_n \cap I^c) \\
&\geq \lambda(E \cap I) + \lambda(E \cap I^c).
\end{aligned}$$

Since ε was arbitrary, it follows that I is λ-measurable. $\square$

Proof of Theorem 2.2: Define λ as in (4). Then Lemma 2.3 shows that λ is an outer measure such that $\lambda I = |I|$ for every interval I. Furthermore, Theorem 2.1 shows that λ is a measure on the σ-field $\mathcal{A}$ of all λ-measurable sets. Finally, Lemma 2.4 shows that $\mathcal{A}$ contains all intervals $(-\infty, a]$ with $a \in \mathbb{R}$. Since the latter sets generate the Borel σ-field $\mathcal{B}$, we have $\mathcal{B} \subset \mathcal{A}$.

To prove the uniqueness, consider any measure μ with the stated properties, and put $I_n = [-n, n]$ for $n \in \mathbb{N}$. Using Lemma 1.17 with $\mathcal{C}$ equal to the set of intervals, we see that

$$\lambda(B \cap I_n) = \mu(B \cap I_n), \quad B \in \mathcal{B}, \, n \in \mathbb{N}.$$

Letting $n \to \infty$ and using Lemma 1.14, we get $\lambda B = \mu B$ for all $B \in \mathcal{B}$, as required. $\square$

Before proceeding to a more detailed study of Lebesgue measure, we state an abstract extension theorem that can be proved by essentially the same arguments. Here a nonempty class $\mathcal{I}$ of subsets of a space Ω is called a *semiring* if for any $I, J \in \mathcal{I}$ we have $I \cap J \in \mathcal{I}$ and the set $I \cap J^c$ can be written as a union of finitely many disjoint sets $I_1, \ldots, I_n \in \mathcal{I}$.

Theorem 2.5 (extension, Carathéodory) *Let μ be a finitely additive and countably subadditive set function on a semiring $\mathcal{I}$ such that $\mu \emptyset = 0$. Then μ extends to a measure on $\sigma(\mathcal{I})$.*

Proof: Define a set function μ^* on 2^Ω by

$$\mu^* A = \inf_{\{I_k\}} \sum_k \mu I_k, \quad A \subset \Omega,$$

where the infimum extends over all covers of A by sets $I_1, I_2, \cdots \in \mathcal{I}$. Let $\mu^* A = \infty$ when no such cover exists. Proceeding as in the proof of Lemma 2.3, we see that μ^* is an outer measure on Ω. To check that μ^* extends μ, fix any $I \in \mathcal{I}$, and consider an arbitrary cover $I_1, I_2, \cdots \in \mathcal{I}$ of I. Using both the subadditivity and the finite additivity of μ, we get

$$\mu^* I \leq \mu I \leq \sum_k (I \cap I_k) \leq \sum_k \mu I_k,$$

which implies $\mu^* I = \mu I$. By Theorem 2.1, it remains to show that every set $I \in \mathcal{I}$ is μ^*-measurable. Then let $A \subset \Omega$ be covered by some sets $I_1, I_2, \cdots \in \mathcal{I}$ with $\mu^* A \geq \sum_k \mu I_k - \varepsilon$, and proceed as in the proof of Lemma 2.4, noting that $I_n \cap I^c$ is a finite disjoint union of some sets $I_{nj} \in \mathcal{I}$, and therefore $\mu(I_n \cap I^c) = \sum_j \mu I_{nj}$ by the finite additivity of μ. $\square$

Using Theorem 1.27, we may construct the product measure $\lambda^d = \lambda \otimes \cdots \otimes \lambda$ on $\mathbb{R}^d$ for every $d \in \mathbb{N}$. We call λ^d the *d-dimensional Lebesgue measure*. Note that λ^d generalizes the ordinary notion of *area* (when $d = 2$) or *volume* (when $d \geq 3$). The following result shows that λ^d is invariant under arbitrary translations (or *shifts*) and rotations. We shall also see that the shift invariance characterizes λ^d up to a constant factor.

Theorem 2.6 *(invariance of Lebesgue measure)* *Fix any measurable space $(S, \mathcal{S})$ and a measure μ on $\mathbb{R}^d \times S$ with σ-finite projection $\nu = \mu((0, 1]^d \times \cdot)$ onto S. Then μ is invariant under shifts in $\mathbb{R}^d$ iff $\mu = \lambda^d \otimes \nu$, in which case μ remains invariant under arbitrary rigid motions of $\mathbb{R}^d$.*

Proof: First assume that μ is invariant under shifts in $\mathbb{R}^d$. Let $\mathcal{I}$ denote the class of intervals $I = (a, b]$ with rational endpoints, and note that for any $I_1, \ldots, I_d \in \mathcal{I}$ and $C \in \mathcal{S}$ with $\nu C < \infty$,

$$\mu(I_1 \times \cdots \times I_d \times C) = |I_1| \cdots |I_d| \nu C$$
$$= (\lambda^d \otimes \nu)(I_1 \times \cdots \times I_d \times C).$$

For fixed $I_2, \ldots, I_d$ and C, the relation extends by monotonicity to arbitrary intervals I_1 and then, by the uniqueness in Theorem 2.2, to any $B_1 \in \mathcal{B}$. Proceeding recursively in d steps, we get for arbitrary $B_1, \ldots, B_d \in \mathcal{B}$

$$\mu(B_1 \times \cdots \times B_d \times C) = (\lambda^d \otimes \nu)(B_1 \times \cdots \times B_d \times C),$$

and so $\mu = \lambda^d \otimes \nu$ by the uniqueness in Theorem 1.27.

Conversely, let $\mu = \lambda^d \otimes \nu$. For any $h = (h_1, \ldots, h_d) \in \mathbb{R}^d$, we define the *shift operator* $T_h \colon \mathbb{R}^d \to \mathbb{R}^d$ by $T_h x = x + h$ for all $x \in \mathbb{R}^d$. For any intervals $I_1, \ldots, I_d$ and sets $C \in \mathcal{S}$, we have

$$\mu(I_1 \times \cdots \times I_d \times C) = |I_1| \cdots |I_d| \nu C$$
$$= \mu \circ T_h^{-1}(I_1 \times \cdots \times I_d \times C),$$

where $T_h(x, s) = (x + h, s)$. As before, it follows that $\mu = \mu \circ T_h^{-1}$.

It remains to show that μ is invariant under arbitrary orthogonal transformations P on $\mathbb{R}^d$. Then note that, for any $x, h \in \mathbb{R}^d$,

$$T_h P x = P x + h = P(x + P^{-1} h)$$
$$= P(x + h') = P T_{h'} x,$$

where $h' = P^{-1} h$. Since μ is shift-invariant, we obtain

$$\mu \circ P^{-1} \circ T_h^{-1} = \mu \circ T_{h'}^{-1} \circ P^{-1} = \mu \circ P^{-1},$$

where $P(x, s) = (Px, s)$. Thus, even $\mu \circ P^{-1}$ is shift-invariant and hence of the form $\lambda^d \otimes \nu'$. Writing B for the unit ball in $\mathbb{R}^d$, we get for any $C \in \mathcal{S}$

$$\lambda^d B \cdot \nu' C = \mu \circ P^{-1}(B \times C) = \mu(P^{-1} B \times C)$$
$$= \mu(B \times C) = \lambda^d B \cdot \nu C.$$

Dividing by $\lambda^d B$ yields $\nu' C = \nu C$. Hence, $\nu' = \nu$, and so $\mu \circ P^{-1} = \mu$. □

We proceed to show that integrable functions on $\mathbb{R}^d$ are continuous in a specified average sense.

Lemma 2.7 *(mean continuity) Let f be a measurable function on $\mathbb{R}^d$ with $\lambda^d|f| < \infty$. Then*

$$\lim_{h \to 0} \int |f(x+h) - f(x)|\, dx = 0.$$

Proof: By Lemma 1.35 and a simple truncation, we may choose some continuous functions $f_1, f_2, \ldots$ with bounded supports such that $\lambda^d|f_n - f| \to 0$. By the triangle inequality, we get for $n \in \mathbb{N}$ and $h \in \mathbb{R}^d$

$$\int |f(x+h) - f(x)|\, dx \leq \int |f_n(x+h) - f_n(x)|\, dx + 2\lambda^d|f_n - f|.$$

Since the f_n are bounded, the right-hand side tends to 0 by dominated convergence as $h \to 0$ and then $n \to \infty$. □

By a *bounded signed measure* on a measurable space $(\Omega, \mathcal{A})$ we mean a bounded function $\nu \colon \mathcal{A} \to \mathbb{R}$ such that $\nu \bigcup_n A_n = \sum_n \nu A_n$ for any disjoint sets $A_1, A_2, \cdots \in \mathcal{A}$, where the series converges absolutely. We say that two measures μ and ν on $(\Omega, \mathcal{A})$ are *(mutually) singular* or *orthogonal* and write $\mu \perp \nu$ if there exists some set $A \in \mathcal{A}$ with $\mu A = \nu A^c = 0$. Note that this A may not be unique. The following result gives the basic decomposition of a signed measure into positive components.

Theorem 2.8 *(Hahn decomposition) Any bounded signed measure ν can be written uniquely as a difference of two bounded, nonnegative, and mutually singular measures ν_+ and ν_-.*

Proof: Put $c = \sup\{\nu A;\ A \in \mathcal{A}\}$ and note that, if $A, A' \in \mathcal{A}$ with $\nu A \geq c - \varepsilon$ and $\nu A' \geq c - \varepsilon'$, then

$$\begin{aligned}\nu(A \cup A') &= \nu A + \nu A' - \nu(A \cap A') \\ &\geq (c - \varepsilon) + (c - \varepsilon') - c = c - \varepsilon - \varepsilon'.\end{aligned}$$

Choosing $A_1, A_2, \cdots \in \mathcal{A}$ with $\nu A_n \geq c - 2^{-n}$, we get by iteration and countable additivity

$$\nu \bigcup_{k>n} A_k \geq c - \sum_{k>n} 2^{-k} = c - 2^{-n}, \quad n \in \mathbb{N}.$$

Define $A_+ = \bigcap_n \bigcup_{k>n} A_k$ and $A_- = A_+^c$. Using the countable additivity again, we get $\nu A_+ = c$. Hence, for sets $B \in \mathcal{A}$,

$$\begin{aligned}\nu B &= \nu A_+ - \nu(A_+ \setminus B) \geq 0, & B \subset A_+, \\ \nu B &= \nu(A_+ \cup B) - \nu A_+ \leq 0, & B \subset A_-.\end{aligned}$$

We may then define some measures ν_+ and ν_- by

$$\nu_+ B = \nu(B \cap A_+), \quad \nu_- B = -\nu(B \cap A_-), \qquad B \in \mathcal{A}.$$

To prove the uniqueness, assume also that $\nu = \mu_+ - \mu_-$ for some positive measures $\mu_+ \perp \mu_-$. Choose a set $B_+ \in \mathcal{A}$ with $\mu_- B_+ = \mu_+ B_+^c = 0$. Then

ν is both positive and negative on the sets $A_+ \setminus B_+$ and $B_+ \setminus A_+$, and therefore $\nu = 0$ on $A_+ \Delta B_+$. Hence, for any $C \in \mathcal{A}$

$$\mu_+ C = \mu_+(B_+ \cap C) = \nu(B_+ \cap C) = \nu(A_+ \cap C) = \nu_+ C,$$

which shows that $\mu_+ = \nu_+$. Then also

$$\mu_- = \mu_+ - \nu = \nu_+ - \nu = \nu_-. \qquad \square$$

The last result can be used to construct the *maximum* $\mu \vee \nu$ and *minimum* $\mu \wedge \nu$ of two σ-finite measures μ and ν.

Corollary 2.9 *(maximum and minimum)* *For any σ-finite measures μ and ν on a common measurable space, there exists a largest measure $\mu \wedge \nu$ bounded by μ and ν and a smallest measure $\mu \vee \nu$ bounding μ and ν. Furthermore,*

$$\mu - \mu \wedge \nu \perp \nu - \mu \wedge \nu, \qquad \mu \wedge \nu + \mu \vee \nu = \mu + \nu.$$

Proof: We may assume that μ and ν are bounded. Letting $\rho_+ - \rho_-$ be the Hahn decomposition of $\mu - \nu$, we put

$$\mu \wedge \nu = \mu - \rho_+, \qquad \mu \vee \nu = \mu + \rho_-. \qquad \square$$

For any two measures μ and ν on $(\Omega, \mathcal{A})$, we say that ν is *absolutely continuous* with respect to μ and write $\nu \ll \mu$ if $\mu A = 0$ implies $\nu A = 0$ for all $A \in \mathcal{A}$. The following result gives a fundamental decomposition of a measure into an absolutely continuous and a singular component; at the same time it provides a basic representation of the former part.

Theorem 2.10 *(Lebesgue decomposition, Radon–Nikodým theorem)* *For any σ-finite measures μ and ν on Ω, there exist some unique measures $\nu_a \ll \mu$ and $\nu_s \perp \mu$ such that $\nu = \nu_a + \nu_s$. Furthermore, $\nu_a = f \cdot \mu$ for some μ-a.e. unique measurable function $f \geq 0$ on Ω.*

Two lemmas will be needed for the proof.

Lemma 2.11 *(closure)* *Fix two measures μ and ν on Ω and some measurable functions $f_1, f_2, \ldots \geq 0$ on Ω with $f_n \cdot \mu \leq \nu$. Then even $f \cdot \mu \leq \nu$, where $f = \sup_n f_n$.*

Proof: First assume that $f \cdot \mu \leq \nu$ and $g \cdot \mu \leq \nu$, and put $h = f \vee g$. Writing $A = \{f \geq g\}$, we get

$$h \cdot \mu = 1_A h \cdot \mu + 1_{A^c} h \cdot \mu = 1_A f \cdot \mu + 1_{A^c} g \cdot \mu \leq 1_A \cdot \nu + 1_{A^c} \cdot \nu = \nu.$$

Thus, we may assume that $f_n \uparrow f$. But then $\nu \geq f_n \cdot \mu \uparrow f \cdot \mu$ by monotone convergence, and so $f \cdot \mu \leq \nu$. $\qquad \square$

Lemma 2.12 *(partial density)* *Let μ and ν be bounded measures on Ω with $\mu \not\perp \nu$. Then there exists a measurable function $f \geq 0$ on Ω such that $\mu f > 0$ and $f \cdot \mu \leq \nu$.*

Proof: For each $n \in \mathbb{N}$ we introduce the signed measure $\chi_n = \nu - n^{-1}\mu$. By Theorem 2.8 we may choose some $A_n^+ \in \mathcal{A}$ with complement A_n^- such that $\pm\chi_n \geq 0$ on $A_n^\pm$. Since the χ_n are nondecreasing, we may assume that $A_1^+ \subset A_2^+ \subset \cdots$. Writing $A = \bigcup_n A_n^+$ and noting that $A^c = \bigcap_n A_n^- \subset A_n^-$, we obtain

$$\nu A^c \leq \nu A_n^- = \chi_n A_n^- + n^{-1}\mu A_n^- \leq n^{-1}\mu\Omega \to 0,$$

and so $\nu A^c = 0$. Since $\mu \not\perp \nu$, we get $\mu A > 0$. Furthermore, $A_n^+ \uparrow A$ implies $\mu A_n^+ \uparrow \mu A > 0$, and we may choose n so large that $\mu A_n^+ > 0$. Putting $f = n^{-1}1_{A_n^+}$, we obtain $\mu f = n^{-1}\mu A_n^+ > 0$ and

$$f \cdot \mu = n^{-1}1_{A_n^+} \cdot \mu = 1_{A_n^+} \cdot \nu - 1_{A_n^+} \cdot \chi_n \leq \nu. \qquad \Box$$

Proof of Theorem 2.10: We may assume that μ and ν are bounded. Let $\mathcal{C}$ denote the class of measurable functions $f \geq 0$ on Ω with $f \cdot \mu \leq \nu$, and define $c = \sup\{\mu f;\ f \in \mathcal{C}\}$. Choose $f_1, f_2, \cdots \in \mathcal{C}$ with $\mu f_n \to c$. Then $f \equiv \sup_n f_n \in \mathcal{C}$ by Lemma 2.11 and $\mu f = c$ by monotone convergence. Define $\nu_a = f \cdot \mu$ and $\nu_s = \nu - \nu_a$, and note that $\nu_a \ll \mu$. If $\nu_s \not\perp \mu$, then by Lemma 2.12 there exists a measurable function $g \geq 0$ with $\mu g > 0$ and $g \cdot \mu \leq \nu_s$. But then $f + g \in \mathcal{C}$ with $\mu(f+g) > c$, which contradicts the definition of c. Thus, $\nu_s \perp \mu$.

To prove the uniqueness of ν_a and ν_s, assume that also $\nu = \nu_a' + \nu_s'$ for some measures $\nu_a' \ll \mu$ and $\nu_s' \perp \mu$. Choose $A, B \in \mathcal{A}$ with $\nu_s A = \mu A^c = \nu_s' B = \mu B^c = 0$. Then clearly

$$\nu_s(A \cap B) = \nu_s'(A \cap B) = \nu_a(A^c \cup B^c) = \nu_a'(A^c \cup B^c) = 0,$$

and so

$$\begin{aligned} \nu_a &= 1_{A \cap B} \cdot \nu_a = 1_{A \cap B} \cdot \nu = 1_{A \cap B} \cdot \nu_a' = \nu_a', \\ \nu_s &= \nu - \nu_a = \nu - \nu_a' = \nu_s'. \end{aligned}$$

To see that f is a.e. unique, assume that also $\nu_a = g \cdot \mu$ for some measurable function $g \geq 0$. Writing $h = f - g$ and noting that $h \cdot \mu = 0$, we get

$$\mu|h| = \int_{\{h > 0\}} h\, d\mu - \int_{\{h < 0\}} h\, d\mu = 0,$$

and so $h = 0$ a.e. by Lemma 1.24. $\qquad \Box$

We insert a simple corollary that will be useful in Chapter 10.

Corollary 2.13 *(splitting)* *Consider two finite measure spaces $(S, \mathcal{S}, \mu)$ and $(T, \mathcal{T}, \nu)$ and a measurable map $f: S \to T$ such that $\nu \leq \mu \circ f^{-1}$. Then there exists a measure $\mu' \leq \mu$ on S such that $\nu = \mu' \circ f^{-1}$.*

Proof: Put $\mu' = (g \circ f) \cdot \mu$ with $g = d\nu/d(\mu \circ f^{-1})$, and use Lemma 1.22. $\qquad \Box$

A measure μ on $\mathbb{R}$ is said to be *locally finite* if $\mu I < \infty$ for every bounded interval I. The following result gives a basic correspondence between locally finite measures and nondecreasing functions.

Proposition 2.14 *(Lebesgue–Stieltjes measures) The relation*
$$\mu(a,b] = F(b) - F(a), \quad -\infty < a < b < \infty, \tag{5}$$
defines a one-to-one correspondence between the locally finite measures μ on $\mathbb{R}$ and the right-continuous, nondecreasing functions F on $\mathbb{R}$ with $F(0) = 0$.

Proof: Given a locally finite measure μ on $\mathbb{R}$, we define the function F on $\mathbb{R}$ by
$$F(x) = \begin{cases} \mu(0,x], & x \geq 0, \\ -\mu(x,0], & x < 0. \end{cases}$$
Then F is right-continuous and nondecreasing with $F(0) = 0$, and it is clearly the unique such function satisfying (5).

Conversely, given a function F as stated, we define the left-continuous, generalized inverse $g \colon \mathbb{R} \to \overline{\mathbb{R}}$ by
$$g(t) = \inf\{s \in \mathbb{R};\ F(s) \geq t\}, \quad t \in \mathbb{R}.$$
Since g is again nondecreasing, the set $g^{-1}(-\infty, s]$ is an extended interval for each $s \in \mathbb{R}$, and so g is measurable by Lemma 1.4. We may then define a measure μ on $\overline{\mathbb{R}}$ by $\mu = \lambda \circ g^{-1}$, where λ denotes Lebesgue measure on $\mathbb{R}$. Noting that $g(t) \leq x$ iff $t \leq F(x)$, we get for any $a < b$
$$\mu(a,b] = \lambda\{t;\ g(t) \in (a,b]\}$$
$$= \lambda(F(a), F(b)] = F(b) - F(a).$$
Thus, the restriction of μ to $\mathbb{R}$ satisfies (5). The uniqueness of μ may be proved in the same way as for λ in Theorem 2.2. $\square$

We now specialize Theorem 2.10 to the case when μ equals Lebesgue measure and ν is a locally finite measure on $\mathbb{R}$, defined as in Proposition 2.14 in terms of some nondecreasing, right-continuous function F. The Lebesgue decomposition and Radon–Nikodým property may be expressed in terms of F as
$$F = F_a + F_s = \int f + F_s, \tag{6}$$
where F_a and F_s correspond to the absolutely continuous and singular components of ν, respectively, and we assume that $F_a(0) = 0$. Here $\int f$ denotes the function $\int_0^x f(t)\,dt$, where the *Lebesgue density* f is a locally integrable function on $\mathbb{R}$. The following result extends the *fundamental theorem of calculus* for Riemann integrals of continuously differentiable functions—the fact that differentiation and integration are mutually inverse operations.

Theorem 2.15 *(differentiation, Lebesgue) Any nondecreasing and right-continuous function $F = \int f + F_s$ is differentiable a.e. with derivative $F' = f$.*

Thus, the two parts of the fundamental theorem generalize to $(\int f)' = f$ a.e. and $\int F' = F_a$. In other words, the density of an integral can still be recovered a.e. through differentiation, whereas integration of a derivative yields only the absolutely continuous component of the underlying function. In particular, F is absolutely continuous iff $\int F' = F - F(0)$ and singular iff $F' = 0$ a.e.

The last result extends trivially to any difference $F = F_+ - F_-$ between two nondecreasing, right-continuous functions F_+ and F_-. However, it fails for more general functions, already because the derivative may not exist. For example, the paths of Brownian motion introduced in Chapter 13 are a.s. nowhere differentiable.

Two lemmas will be helpful for the proof of the last theorem.

Lemma 2.16 (interval selection) *Let $\mathcal{I}$ be a class of open intervals with union G. If $\lambda G < \infty$, there exist some disjoint sets $I_1, \ldots, I_n \in \mathcal{I}$ with $\sum_k |I_k| \geq \lambda G/4$.*

Proof: Choose a compact set $K \subset G$ with $\lambda K \geq 3\lambda G/4$. By compactness we may cover K by finitely many intervals $J_1, \ldots, J_m \in \mathcal{I}$. We now define $I_1, I_2, \ldots$ recursively, by letting I_k be the longest interval J_r not yet chosen such that $J_r \cap I_j = \emptyset$ for all $j < k$. The selection terminates when no such interval exists.

If an interval J_r is not selected, it must intersect a longer interval I_k. Writing $\hat{I}_k$ for the interval centered at I_k with length $3|I_k|$, we obtain

$$K \subset \bigcup_r J_r \subset \bigcup_k \hat{I}_k,$$

and so

$$(3/4)\lambda G \leq \lambda K \leq \lambda \bigcup_k \hat{I}_k \leq \sum_k |\hat{I}_k| = 3 \sum_k |I_k|. \qquad \square$$

Lemma 2.17 (differentiation on null sets) *Let $F(x) \equiv \mu(0, x]$ for some locally finite measure μ on $\mathbb{R}$, and let $A \in \mathcal{B}$ with $\mu A = 0$. Then $F' = 0$ a.e. λ on A.*

Proof: By Lemma 1.34 there exists for every $\delta > 0$ some open set $G_\delta \supset A$ with $\mu G_\delta < \delta$. Define

$$A_\varepsilon = \left\{ x \in A;\ \limsup_{h \to 0} \frac{\mu(x-h, x+h)}{h} > \varepsilon \right\}, \quad \varepsilon > 0,$$

and note that each A_ε is measurable since the lim sup may be taken along the rationals. For every $x \in A_\varepsilon$ there exists some interval $I = (x-h, x+h) \subset G_\delta$ with $2\mu I > \varepsilon|I|$, and we note that the class $\mathcal{I}_{\varepsilon,\delta}$ of such intervals covers A_ε. Hence, by Lemma 2.16 we may choose some disjoint sets $I_1, \ldots, I_n \in \mathcal{I}_{\varepsilon,\delta}$ with $\sum_k |I_k| \geq \lambda A_\varepsilon/4$. Then

$$\lambda A_\varepsilon \leq 4 \sum_k |I_k| \leq \frac{8}{\varepsilon} \sum_k \mu I_k \leq \frac{8\mu G_\delta}{\varepsilon} < \frac{8\delta}{\varepsilon}.$$

As $\delta \to 0$, we get $\lambda A_\varepsilon = 0$. Thus, $\limsup \mu(x-h, x+h)/h \leq \varepsilon$ a.e. λ on A, and the assertion follows since ε is arbitrary. $\qquad \square$

Proof of Theorem 2.15: Since $F'_s = 0$ a.e. λ by Lemma 2.17, we may assume that $F = \int f$. Define

$$F^\wedge(x) = \limsup_{h \to 0} h^{-1}(F(x+h) - F(x)),$$
$$F^\vee(x) = \liminf_{h \to 0} h^{-1}(F(x+h) - F(x)),$$

and note that $F^\wedge = 0$ a.e. on the set $\{f = 0\} = \{x;\ f(x) = 0\}$ by Lemma 2.17. Applying this to the function $F_r = \int (f - r)_+$ for arbitrary $r \in \mathbb{R}$ and noting that $f \leq (f - r)_+ + r$, we get $F^\wedge \leq r$ a.e. on $\{f \leq r\}$. Thus, for r restricted to the rationals,

$$\lambda\{f < F^\wedge\} = \lambda \bigcup_r \{f \leq r < F^\wedge\}$$
$$\leq \sum_r \lambda\{f \leq r < F^\wedge\} = 0,$$

which shows that $F^\wedge \leq f$ a.e. Applying this result to $-F = \int(-f)$ yields $F^\vee = -(-F)^\wedge \geq f$ a.e. Thus, $F^\wedge = F^\vee = f$ a.e., and so F' exists a.e. and equals f. □

For any function $F \colon \mathbb{R} \to \mathbb{R}$, we define the *total variation* of F on the interval $[a, b]$ as

$$\|F\|_a^b = \sup_{\{t_k\}} \sum_k |F(t_k) - F(t_{k-1})|,$$

where the supremum extends over all finite partitions $a = t_0 < t_1 < \cdots < t_n = b$. Similarly, the *positive and negative variations* of F are defined by the same expression with the absolute value $|\cdot|$ replaced by the positive and negative parts $(\cdot)^\pm$. Here $x^\pm = (\pm x) \vee 0$, so that $x = x^+ - x^-$ and $|x| = x^+ + x^-$. We also write $\Delta_a^b F = F(b) - F(a)$.

The following result gives a basic decomposition of functions of locally finite variation, similar to the Hahn decomposition in Theorem 2.8.

Proposition 2.18 *(Jordan decomposition) A function F on $\mathbb{R}$ has locally finite variation iff it is a difference of two nondecreasing functions F_+ and F_-. In that case,*

$$\|F\|_s^t \leq \Delta_s^t F_+ + \Delta_s^t F_-, \quad s < t, \tag{7}$$

with equality iff the increments $\Delta_s^t F_\pm$ agree with the positive and negative variations of F on $(s, t]$.

Proof: For any $s < t$ we have

$$(\Delta_s^t F)^+ = (\Delta_s^t F)^- + \Delta_s^t F,$$
$$|\Delta_s^t F| = (\Delta_s^t F)^+ + (\Delta_s^t F)^- = 2(\Delta_s^t F)^- + \Delta_s^t F.$$

Summing over the intervals in an arbitrary partition $s = t_0 < t_1 < \cdots < t_n = t$ and taking the supremum of each side, we obtain

$$\Delta_s^t F_+ = \Delta_s^t F_- + \Delta_s^t F,$$
$$\|F\|_s^t = 2\Delta_s^t F_- + \Delta_s^t F = \Delta_s^t F_+ + \Delta_s^t F_-,$$

where $F_\pm(x)$ denote the positive and negative variations of F on $[0,x]$ (or minus the variations on $[x,0]$ when $x<0$). Thus, $F = F(0) + F_+ - F_-$, and (7) holds with equality. If also $F = G_+ - G_-$ for some nondecreasing functions $G_\pm$, then $(\Delta_s^t F)^\pm \leq \Delta_s^t G_\pm$, and so $\Delta_s^t F_\pm \leq \Delta_s^t G_\pm$. Thus, $\|F\|_s^t \leq \Delta_s^t G_+ + \Delta_s^t G_-$, and equality holds iff $\Delta_s^t F_\pm = \Delta_s^t G_\pm$. □

Next we give another useful decomposition of finite-variation functions.

Proposition 2.19 *(left and right continuity)* Any function F of locally finite variation can be written as $F_r + F_l$, where F_r is right-continuous with left-hand limits and F_l is left-continuous with right-hand limits. If F is right-continuous, then so are the minimal components $F_\pm$ in Proposition 2.18.

Proof: By Proposition 2.18 we may assume that F is nondecreasing. The right- and left-hand limits $F^\pm(s)$ then exist at every point s, and we note that $F^-(s) \leq F(s) \leq F^+(s)$. Also note that F has at most countably many jump discontinuities. For $t > 0$, we define

$$F_l(t) = \sum_{s \in [0,t)} (F^+(s) - F(s)), \qquad F_r(t) = F(t) - F_l(t);$$

when $t \leq 0$ we need to take the negative of the corresponding sum on $(t, 0]$. It is easy to check that F_l is left-continuous and F_r is right-continuous, and that both functions are nondecreasing.

To prove the last assertion, assume that F is right-continuous at some point s. If $\|F\|_s^t \to c > 0$ as $t \downarrow s$, we may choose $t - s$ so small that $\|F\|_s^t < 4c/3$. Next we may choose a partition $s = t_0 < t_1 < \cdots < t_n = t$ of $[s,t]$ such that the corresponding F-increments δ_k satisfy $\sum_k |\delta_k| > 2c/3$. By the right continuity of F at s, we may assume that $t_1 - s$ is small enough that $\delta_1 = |F(t_1) - F(s)| < c/3$. Then $\|F\|_{t_1}^t > c/3$, and so

$$4c/3 > \|F\|_s^t = \|F\|_s^{t_1} + \|F\|_{t_1}^t > c + c/3 = 4c/3,$$

a contradiction. Hence $c = 0$. Assuming $F_\pm$ to be minimal, we obtain

$$\Delta_s^t F_\pm \leq \|F\|_s^t \to 0, \quad t \downarrow s. \qquad \square$$

Justified by the last theorem, we may assume our finite-variation functions to be right-continuous. In that case, we have the following basic relation to signed measures. Here we only require the latter to be *locally bounded*.

Proposition 2.20 *(finite-variation functions and signed measures)* For any right-continuous function F of locally finite variation, there exists a unique signed measure ν on $\mathbb{R}$ such that $\nu(s,t] = \Delta_s^t F$ for all $s < t$. Furthermore, the Hahn decomposition $\nu = \nu_+ - \nu_-$ and the Jordan decomposition $F = F_+ - F_-$ into minimal components are related by $\nu_\pm(s,t] \equiv \Delta_s^t F_\pm$.

Proof: The positive and negative variations $F_\pm$ are right-continuous by Proposition 2.19. Hence, by Proposition 2.14 there exist some locally finite measures $\mu_\pm$ on $\mathbb{R}$ such that $\mu_\pm(s,t] \equiv \Delta_s^t F_\pm$, and we may take $\nu = \mu_+ - \mu_-$.

To see that this agrees with the Hahn decomposition $\nu = \nu_+ - \nu_-$, choose $A \in \mathcal{A}$ such that $\nu_+ A^c = \nu_- A = 0$. For any $B \in \mathcal{B}$, we get

$$\mu_+ B \geq \mu_+(B \cap A) \geq \nu(B \cap A) = \nu_+(B \cap A) = \nu_+ B,$$

which shows that $\mu_+ \geq \nu_+$. Then also $\mu_- \geq \nu_-$. If the equality fails on some interval $(s,t]$, then

$$\|F\|_s^t = \mu_+(s,t] + \mu_-(s,t] > \nu_+(s,t] + \nu_-(s,t],$$

which contradicts Proposition 2.18. Hence, $\mu_\pm = \nu_\pm$. □

A function $F: \mathbb{R} \to \mathbb{R}$ is said to be *absolutely continuous* if for any $a < b$ and $\varepsilon > 0$ there exists some $\delta > 0$ such that, for any finite collection of disjoint intervals $(a_k, b_k] \subset (a,b]$ with $\sum_k |b_k - a_k| < \delta$, we have $\sum_k |F(b_k) - F(a_k)| < \varepsilon$. In particular, we note that every absolutely continuous function is continuous and has locally finite variation.

Given a function F of locally finite variation, we say that F is *singular* if for any $a < b$ and $\varepsilon > 0$ there exist finitely many disjoint intervals $(a_k, b_k] \subset (a,b]$ such that $\sum_k |b_k - a_k| < \varepsilon$ and $\|F\|_a^b < \sum_k |F(b_k) - F(a_k)| + \varepsilon$.

We say that a locally finite signed measure ν on $\mathbb{R}$ is absolutely continuous or singular if the components $\nu_\pm$ of the associated Hahn decomposition satisfy $\nu_\pm \ll \lambda$ or $\nu_\pm \perp \lambda$, respectively. The following result relates the notions of absolute continuity and singularity for functions and measures.

Proposition 2.21 *(absolutely continuous and singular functions) Let F be a right-continuous function on $\mathbb{R}$ of locally finite variation, and let ν be the associated signed measure on $\mathbb{R}$ with $\nu(s,t] \equiv \Delta_s^t F$. Then F is absolutely continuous or singular iff the corresponding property holds for ν.*

Proof: If F is absolutely continuous or singular, then the corresponding property holds for the total variation function $\|F\|_a^x$ with arbitrary a and hence also for the minimal components $F_\pm$ in Proposition 2.20. Thus, we may assume that F is nondecreasing, so that ν is a positive and locally finite measure on $\mathbb{R}$.

First assume that F is absolutely continuous. If $\nu \not\ll \lambda$, there exists a bounded interval $I = (a,b)$ with a subset $A \in \mathcal{B}$ such that $\lambda A = 0$ but $\nu A > 0$. Taking $\varepsilon = \nu A/2$, we choose a corresponding $\delta > 0$ as in the definition of absolute continuity. Since A is measurable and has outer Lebesgue measure 0, we may next choose an open set G with $A \subset G \subset I$ such that $\lambda G < \delta$. But then $\nu A \leq \nu G < \varepsilon = \nu A/2$, a contradiction. This shows that $\nu \ll \lambda$.

Next assume that F is singular, and fix any bounded interval $I = (a,b]$. Given any $\varepsilon > 0$, we may choose some Borel sets $A_1, A_2, \cdots \subset I$ such that $\lambda A_n < \varepsilon 2^{-n}$ and $\nu A_n \to \nu I$. Then $B = \bigcup_n A_n$ satisfies $\lambda B < \varepsilon$ and $\nu B = \nu I$. Next we may choose some Borel sets $B_n \subset I$ with $\lambda B_n \to 0$ and

$\nu B_n = \nu I$. Then $C = \bigcap_n B_n$ satisfies $\lambda C = 0$ and $\nu C = \nu I$, which shows that $\nu \perp \lambda$ on I.

Conversely, assume that $\nu \ll \lambda$, so that $\nu = f \cdot \lambda$ for some locally integrable function $f \geq 0$. Fix any bounded interval I and put $A_n = \{x \in I; f(x) > n\}$. Fix any $\varepsilon > 0$. Since $\nu A_n \to 0$ by Lemma 1.14, we may choose n so large that $\nu A_n < \varepsilon/2$. Put $\delta = \varepsilon/2n$. For any Borel set $B \subset I$ with $\lambda B < \delta$ we obtain

$$\nu B = \nu(B \cap A_n) + \nu(B \cap A_n^c) \leq \nu A_n + n\lambda B < \tfrac{1}{2}\varepsilon + n\delta = \varepsilon.$$

In particular, this applies to any finite union B of intervals $(a_k, b_k] \subset I$, and so we may conclude that F is absolutely continuous.

Finally, assume that $\nu \perp \lambda$. Fix any finite interval $I = (a, b]$, and choose a Borel set $A \subset I$ such that $\lambda A = 0$ and $\nu A = \nu I$. For any $\varepsilon > 0$ we may choose some open set $G \supset A$ with $\lambda G < \varepsilon$. Letting (a_n, b_n) denote the connected components of G and writing $I_n = (a_n, b_n]$, we get $\sum_n |I_n| < \varepsilon$ and $\sum_n \nu(I \cap I_n) = \nu I$. This shows that F is singular. □

From now on, we assume the basic space S to be *locally compact, second countable, and Hausdorff* (abbreviated lcscH). Let $\mathcal{G}$, $\mathcal{F}$, and $\mathcal{K}$ denote the classes of open, closed, and compact sets in S, and put $\hat{\mathcal{G}} = \{G \in \mathcal{G}; \overline{G} \in \mathcal{K}\}$. Let $\hat{C}_+ = \hat{C}_+(S)$ denote the class of continuous functions $f: S \to \mathbb{R}_+$ with compact support, where the latter is defined as the closure of the set $\{x \in S; f(x) > 0\}$. Relations such as $U \prec f \prec V$ mean that $f \in \hat{C}_+$ with $0 \leq f \leq 1$ and satisfies $f = 1$ on U and $\operatorname{supp} f \subset V^\circ$.

By a *positive linear functional* on $\hat{C}_+$ we mean a mapping $\mu: \hat{C}_+ \to \mathbb{R}_+$ such that $\mu(f + g) = \mu f + \mu g$ for all $f, g \in \hat{C}_+$. This clearly implies the homogeneity $\mu(cf) = c\mu f$ for any $f \in \hat{C}_+$ and $c \in \mathbb{R}_+$. A *Radon measure* on S is defined as a measure μ on the Borel σ-field $\mathcal{S} = \mathcal{B}(S)$ such that $\mu K < \infty$ for every $K \in \mathcal{K}$. The following result gives the basic extension of positive linear functionals to measures.

Theorem 2.22 *(Riesz representation) If S is lcscH, then every positive linear functional μ on $\hat{C}_+(S)$ extends uniquely to a Radon measure on S.*

Several lemmas will be needed for the proof, and we begin with a simple topological fact.

Lemma 2.23 *(partition of unity) For any open cover $G_1, \ldots, G_n$ of a compact set $K \subset S$, there exist some functions $f_1, \ldots, f_n \in \hat{C}_+(S)$ with $f_k \prec G_k$ such that $\sum_k f_k = 1$ on K.*

Proof: For any $x \in K$ we may choose some $k \leq n$ and $V \in \hat{\mathcal{G}}$ with $x \in V$ and $\overline{V} \subset G_k$. By compactness, K is covered by finitely many such sets $V_1, \ldots, V_m$. For each $k \leq n$, let U_k be the union of all sets V_j with $\overline{V}_j \subset G_k$. Then $\overline{U}_k \subset G_k$, and so we may choose $g_1, \ldots, g_n \in \hat{C}_+$ with $U_k \prec g_k \prec G_k$. Define

$$f_k = g_k(1 - g_1) \cdots (1 - g_{k-1}), \quad k = 1, \ldots, n.$$

Then $f_k \prec G_k$ for all k, and by induction
$$f_1 + \cdots + f_n = 1 - (1 - g_1) \cdots (1 - g_n).$$
It remains to note that $\prod_k (1 - g_k) = 0$ on K since $K \subset \bigcup_k U_k$. □

By an *inner content* on an lcscH space S we mean a nondecreasing function $\mu : \mathcal{G} \to \overline{\mathbb{R}}_+$, finite on $\hat{\mathcal{G}}$, such that μ is both finitely additive and countably subadditive, and also satisfies the *inner continuity*
$$\mu G = \sup\{\mu U;\ U \in \hat{\mathcal{G}}, \overline{U} \subset G\}, \quad G \in \mathcal{G}. \tag{8}$$

Lemma 2.24 (*inner approximation*) *For any positive linear functional μ on $\hat{C}_+(S)$, we may define an inner content ν on S by*
$$\nu G = \sup\{\mu f;\ f \prec G\}, \quad G \in \mathcal{G}.$$

Proof: Note that ν is nondecreasing with $\nu \emptyset = 0$ and that $\nu G < \infty$ for bounded G. It is also clear that ν is inner continuous in the sense of (8).

To show that ν is countably subadditive, fix any $G_1, G_2, \cdots \in \mathcal{G}$ and let $f \prec \bigcup_k G_k$. By compactness, $f \prec \bigcup_{k \le n} G_k$ for some finite n, and by Lemma 2.23 we may choose some functions $g_k \prec G_k$ such that $\sum_k g_k = 1$ on supp f. Then the products $f_k = g_k f$ satisfy $f_k \prec G_k$ and $\sum_k f_k = f$, and so
$$\mu f = \sum_{k \le n} \mu f_k \le \sum_{k \le n} \nu G_k \le \sum_k \nu G_k.$$
Since $f \prec \bigcup_k G_k$ was arbitrary, we obtain $\nu \bigcup_k G_k \le \sum_k \nu G_k$, as required.

To show that ν is finitely additive, fix any disjoint sets $G, G' \in \mathcal{G}$. If $f \prec G$ and $f' \prec G'$, then $f + f' \prec G \cup G'$, and so
$$\mu f + \mu f' = \mu(f + f') \le \nu(G \cup G') \le \nu G + \nu G'.$$
Taking the supremum over all f and f' gives $\nu G + \nu G' = \nu(G \cup G')$, as required. □

An outer measure μ on S is said to be *regular* if it is finitely additive on $\mathcal{G}$ and enjoys the *outer* and *inner regularity*
$$\mu A = \inf\{\mu G;\ G \in \mathcal{G}, G \supset A\}, \quad A \subset S, \tag{9}$$
$$\mu G = \sup\{\mu K;\ K \in \mathcal{K}, K \subset G\}, \quad G \in \mathcal{G}. \tag{10}$$

Lemma 2.25 (*outer approximation*) *Every inner content μ on S admits an extension to a regular outer measure.*

Proof: We may define the extension by (9), since the right-hand side equals μA when $A \in \mathcal{G}$. By the finite additivity on $\mathcal{G}$ we have $2\mu \emptyset = \mu \emptyset < \infty$, which implies $\mu \emptyset = 0$. To prove the countable subadditivity, fix any $A_1, A_2, \cdots \subset S$. For any $\varepsilon > 0$ we may choose some $G_1, G_2, \cdots \in \mathcal{G}$ with $G_n \supset A_n$ and $\mu G_n \le \mu A_n + \varepsilon 2^{-n}$. Since μ is subadditive on $\mathcal{G}$, we get
$$\mu \bigcup_n A_n \le \mu \bigcup_n G_n \le \sum_n \mu G_n \le \sum_n \mu A_n + \varepsilon.$$

The desired relation follows since ε was arbitrary. Thus, the extension is an outer measure on S. Finally, the inner regularity in (10) follows from (8) and the monotonicity of μ. □

Lemma 2.26 *(measurability)* *If μ is a regular outer measure on S, then every Borel set in S is μ-measurable.*

Proof: Fix any $F \in \mathcal{F}$ and $A \subset G \in \mathcal{G}$. By the inner regularity in (10), we may choose $G_1, G_2, \cdots \in \mathcal{G}$ with $\overline{G}_n \subset G \setminus F$ and $\mu G_n \to \mu(G \setminus F)$. Since μ is nondecreasing and finitely additive on $\mathcal{G}$, we get

$$\begin{aligned}\mu G &\geq \mu(G \setminus \partial G_n) = \mu G_n + \mu(G \setminus \overline{G}_n) \\ &\geq \mu G_n + \mu(G \cap F) \\ &\to \mu(G \setminus F) + \mu(G \cap F) \\ &\geq \mu(A \setminus F) + \mu(A \cap F).\end{aligned}$$

Using the outer regularity in (9) gives

$$\mu A \geq \mu(A \setminus F) + \mu(A \cap F), \quad F \in \mathcal{F}, \ A \subset S.$$

Hence, every closed set is measurable, and by Theorem 2.1 the measurability extends to $\sigma(\mathcal{F}) = \mathcal{B}(S) = \mathcal{S}$. □

Proof of Theorem 2.22: Construct an inner content ν as in Lemma 2.24, and conclude from Lemma 2.25 that ν admits an extension to a regular outer measure on S. By Theorem 2.1 and Lemma 2.26, the restriction of the latter to $\mathcal{S} = \mathcal{B}(S)$ is a Radon measure on S, here still denoted by ν.

To see that $\mu = \nu$ on $\hat{C}_+$, fix any $f \in \hat{C}_+$. For $n \in \mathbb{N}$ and $k \in \mathbb{Z}_+$, let

$$\begin{aligned}f_k^n(x) &= (nf(x) - k)_+ \wedge 1, \\ G_k^n &= \{nf > k\} = \{f_k^n > 0\}.\end{aligned}$$

Noting that $\overline{G}_{k+1}^n \subset \{f_k^n = 1\}$ and using the definition of ν and the outer regularity in (9), we get for appropriate k

$$\nu f_{k+1}^n \leq \nu G_{k+1}^n \leq \mu f_k^n \leq \nu \overline{G}_k^n \leq \nu f_{k-1}^n.$$

Writing $G_0 = G_0^n = \{f > 0\}$ and noting that $nf = \sum_k f_k^n$, we obtain

$$n\nu f - \nu G_0 \leq n\mu f \leq n\nu f + \nu \overline{G}_0.$$

Here $\nu \overline{G}_0 < \infty$ since G_0 is bounded. Dividing by n and letting $n \to \infty$ gives $\mu f = \nu f$.

To prove the asserted uniqueness, let μ and ν be Radon measues on S with $\mu f = \nu f$ for all $f \in \hat{C}_+$. By an inner approximation, we have $\mu G = \nu G$ for every $G \in \mathcal{G}$, and a monotone-class argument yields $\mu = \nu$. □

By a *topological group* we mean a group endowed with a topology that renders the group operations continuous. Thus, the mapping $(f, g) \mapsto fg$ is continuous from G^2 to G, whereas the mapping $g \mapsto g^{-1}$ is continuous from G to G. In the former case, G^2 is equipped with the product topology.

Introducing the Borel σ-field $\mathcal{G} = \mathcal{B}(G)$, we obtain a measurable group $(G, \mathcal{G})$, and we note that the group operations are measurable when G is lcscH. A measure μ on G is said to be *left-invariant* if $\mu(gB) = \mu B$ for all $g \in G$ and $B \in \mathcal{G}$, where $gB = \{gb;\, b \in B\}$, the left translate of B by g. This is clearly equivalent to $\int f(gk)\mu(dk) = \mu f$ for any measurable function $f: G \to \mathbb{R}_+$ and element $g \in G$. The definition of *right-invariant* measures is similar.

We may now state the basic existence and uniqueness theorem for invariant measures on groups.

Theorem 2.27 *(Haar measure) On every lcscH group G there exists, uniquely up to a normalization, a left-invariant Radon measure $\lambda \neq 0$. If G is compact, then λ is also right-invariant.*

Proof (Weil): For any $f, g \in \hat{C}_+$ we define $|f|_g = \inf \sum_k c_k$, where the infimum extends over all finite sets of constants $c_1, \ldots, c_n \geq 0$ such that
$$f(x) \leq \sum_{k \leq n} c_k g(s_k x), \quad x \in G,$$
for some $s_1, \ldots, s_n \in G$. By compactness, $|f|_g < \infty$ when $g \neq 0$. We also note that $|f|_g$ is nondecreasing and translation invariant in f, and that it satisfies the subadditivity and homogeneity properties
$$|f + f'|_g \leq |f|_g + |f'|_g, \qquad |cf|_g = c|f|_g, \tag{11}$$
as well as the inequalities
$$\frac{\|f\|}{\|g\|} \leq |f|_g \leq |f|_h |h|_g. \tag{12}$$

We may normalize $|f|_g$ by fixing an $f_0 \in \hat{C}_+ \setminus \{0\}$ and putting
$$\lambda_g f = |f|_g / |f_0|_g, \quad f, g \in \hat{C}_+,\ g \neq 0.$$
From (11) and (12) we note that
$$\lambda_g(f + f') \leq \lambda_g f + \lambda_g f', \qquad \lambda_g(cf) = c\lambda_g f, \tag{13}$$
$$|f_0|_f^{-1} \leq \lambda_g f \leq |f|_{f_0}. \tag{14}$$

Conversely, λ_g is nearly superadditive in the following sense.

Lemma 2.28 *(near superadditivity) For any $f, f' \in \hat{C}_+$ and $\varepsilon > 0$, there exists an open set $U \neq \emptyset$ such that*
$$\lambda_g f + \lambda_g f' \leq \lambda_g(f + f') + \varepsilon, \quad 0 \neq g \prec U.$$

Proof: Fix any $h \in \hat{C}_+$ with $h = 1$ on $\mathrm{supp}(f + f')$, and define for $\delta > 0$
$$f_\delta = f + f' + \delta h, \qquad h_\delta = f/f_\delta, \qquad h'_\delta = f'/f_\delta,$$
so that $h_\delta, h'_\delta \in \hat{C}_+$. By compactness we may choose a neighborhood U of the identity element $e \in G$ such that
$$|h_\delta(x) - h_\delta(y)| < \delta, \quad |h'_\delta(x) - h'_\delta(y)| < \delta, \qquad x^{-1}y \in U. \tag{15}$$

Now assume $0 \neq g \prec U$, and let $f_\delta(x) \leq \sum_k c_k g(s_k x)$ for some $s_1, \ldots, s_n \in G$ and $c_1, \ldots, c_n \geq 0$. Since $g(s_k x) \neq 0$ implies $s_k x \in U$, we have by (15)

$$\begin{aligned} f(x) = f_\delta(x) h_\delta(x) &\leq \sum_k c_k g(s_k x) h_\delta(x) \\ &\leq \sum_k c_k g(s_k x) \left\{ h_\delta(s_k^{-1}) + \delta \right\}, \end{aligned}$$

and similarly for f'. Noting that $h_\delta + h'_\delta \leq 1$, we get

$$|f|_g + |f'|_g \leq \sum_k c_k (1 + 2\delta).$$

Taking the infimum over all dominating sums for f_δ and using (11), we conclude that

$$|f|_g + |f'|_g \leq |f_\delta|_g (1 + 2\delta) \leq \{|f + f'|_g + \delta |h|_g\}(1 + 2\delta).$$

Now divide by $|f_0|_g$, and use (14) to obtain

$$\begin{aligned} \lambda_g f + \lambda_g f' &\leq \{\lambda_g(f + f') + \delta \lambda_g h\}(1 + 2\delta) \\ &\leq \lambda_g(f + f') + 2\delta |f + f'|_{f_0} + \delta(1 + 2\delta)|h|_{f_0}, \end{aligned}$$

which tends to $\lambda_g(f + f')$ as $\delta \to 0$. □

Returning to the proof of Theorem 2.27, we may consider the functionals λ_g as elements of the product space $\Lambda = \mathbb{R}_+^{\hat{C}_+}$. For any neighborhood U of e, let Λ_U denote the closure in Λ of the set $\{\lambda_g; 0 \neq g \prec U\}$. Since $\lambda_g f \leq |f|_{f_0} < \infty$ for all $f \in \hat{C}_+$ by (14), the Λ_U are compact by Tychonov's theorem. Furthermore, the family $\{\Lambda_U; e \in U\}$ has the finite intersection property since $U \subset V$ implies $\Lambda_U \subset \Lambda_V$. We may then choose an element $\lambda \in \bigcap_U \Lambda_U$, here regarded as a functional on $\hat{C}_+$. From (14) we note that $\lambda \neq 0$.

To see that λ is linear, fix any $f, f' \in \hat{C}_+$ and $a, b \geq 0$, and choose some $g_1, g_2, \cdots \in \hat{C}_+$ with $\operatorname{supp} g_n \downarrow \{e\}$ such that

$$\lambda_{g_n} f \to \lambda f, \quad \lambda_{g_n} f' \to \lambda f', \quad \lambda_{g_n}(af + bf') \to \lambda(af + bf').$$

By (13) and Lemma 2.28 we obtain $\lambda(af + bf') = a\lambda f + b\lambda f'$. Thus, λ is a nontrivial, positive linear functional on $\hat{C}_+$, and so by Theorem 2.22 it extends uniquely to a Radon measure on S. The invariance of the functionals λ_g clearly carries over to λ.

Now consider any left-invariant Radon measure $\lambda \neq 0$ on G. Fixing a right-invariant Radon measure $\mu \neq 0$ and a function $h \in \hat{C}_+ \setminus \{0\}$, we define

$$p(x) = \int h(y^{-1} x) \mu(dy), \quad x \in G,$$

and we note that $p > 0$ on G. Using the invariance of λ and μ together with Fubini's theorem, we get for any $f \in \hat{C}_+$

$$\begin{aligned}
(\lambda h)(\mu f) &= \int h(x)\,\lambda(dx) \int f(y)\,\mu(dy) \\
&= \int h(x)\,\lambda(dx) \int f(yx)\,\mu(dy) \\
&= \int \mu(dy) \int h(x)f(yx)\,\lambda(dx) \\
&= \int \mu(dy) \int h(y^{-1}x)f(x)\,\lambda(dx) \\
&= \int f(x)\,\lambda(dx) \int h(y^{-1}x)\,\mu(dy) = \lambda(fp).
\end{aligned}$$

Since f was arbitrary, we conclude that $(\lambda h)\mu = p \cdot \lambda$ or, equivalently, $\lambda/\lambda h = p^{-1} \cdot \mu$. Here the right-hand side is independent of λ, and the asserted uniqueness follows. If S is compact, we may choose $h \equiv 1$ to obtain $\lambda/\lambda S = \mu/\mu S$. □

Given a group G and an abstract space S, we define a *left action* of G on S as a mapping $(g,s) \mapsto gs$ from $G \times S$ to S such that $es = s$ and $(gh)s = g(hs)$ for any $g, h \in G$ and $s \in S$, where e denotes the identity element in G. Similarly, a *right action* is a mapping $(s,g) \mapsto sg$ such that $se = s$ and $s(gh) = (sg)h$ for all s, g, h as above. The action is said to be *transitive* if for any $s, t \in S$ there exists some $g \in G$ such that $gs = t$ or $sg = t$, respectively. All actions are henceforth assumed to be from the left.

If G is a topological group and S is a topological space, we assume the action $(x,s) \mapsto xs$ to be continuous from $G \times S$ to S. A function $h: G \to S$ is said to be *proper* if $h^{-1}K$ is compact in G for any compact set $K \subset S$; if this holds for every mapping $\pi_s(x) = xs$, $s \in S$, we say that the group action is proper. Finally, a measure μ on S is G-*invariant* if $\mu(xB) = \mu B$ for any $x \in G$ and $B \in \mathcal{S}$. This is clearly equivalent to the relation $\int f(xs)\mu(ds) = \mu f$ for any measurable function $f: S \to \mathbb{R}_+$ and element $x \in G$.

We may now state the basic existence and uniqueness result for invariant measures on a general lcscH space. The existence of Haar measures in Theorem 2.27 is a special case.

Theorem 2.29 *(invariant measure)* *Consider an lcscH group G that acts transitively and properly on an lcscH space S. Then there exists, uniquely up to a normalization, a G-invariant Radon measure $\mu \neq 0$ on S.*

Proof: Fix any $p \in S$, and let π denote the mapping $x \mapsto xp$ from G to S. Letting λ be a left Haar measure on G, we define $\mu = \lambda \circ \pi^{-1}$. Since π is proper, we note that μ is a Radon measure on S. To see that μ is

G-invariant, let $f \in \hat{C}_+$ be arbitrary, and note that for any $x \in G$

$$\int_S f(xs)\mu(ds) = \int_G f(xyp)\lambda(dy) = \int_G f(yp)\lambda(dy) = \mu f,$$

by the invariance of λ.

To prove the uniqueness, let μ be an arbitrary G-invariant Radon measure on S. Introduce the subgroup

$$K = \{x \in G;\ xp = p\} = \pi^{-1}\{p\},$$

and note that K is compact since π is proper. Let ν be the normalized Haar measure on K, and define

$$\bar{f}(x) = \int_K f(xk)\nu(dk), \quad x \in G,\ f \in \hat{C}_+(G).$$

If $xp = yp$, we have $y^{-1}xp = p$, and so $y^{-1}x \equiv h \in K$, which implies $x = yh$. Hence, the left invariance of ν yields

$$\bar{f}(x) = \bar{f}(yh) = \int_K f(yhk)\nu(dk) = \int_K f(yk)\nu(dk) = \bar{f}(y).$$

We may then define a mapping $f \mapsto f^*$ by

$$f^*(s) = \bar{f}(x), \quad s = xp \in S,\ x \in G,\ f \in \hat{C}_+(G).$$

For any subset $B \subset (0,\infty)$, we note that

$$(f^*)^{-1}B = \pi(\bar{f}^{-1}B) \subset \pi[(\operatorname{supp} f) \cdot K].$$

Here the right-hand side is compact since the sets $\operatorname{supp} f$ and K are compact, and since π and the group operation in G are both continuous. Thus, f^* has bounded support. Furthermore, $\bar{f}$ is continuous by dominated convergence, and so $\bar{f}^{-1}[t,\infty)$ is closed and hence compact for every $t > 0$. By the continuity of π it follows that even $(f^*)^{-1}[t,\infty)$ is compact. In particular, f^* is measurable.

We may now define a functional λ on $\hat{C}_+(G)$ by

$$\lambda f = \mu f^*, \quad f \in \hat{C}_+(G).$$

The linearity and positivity of λ are clear from the corresponding properties of the mapping $f \mapsto f^*$ and the measure μ. We also note that λ is finite on $\hat{C}_+(G)$ since μ is locally finite. By Theorem 2.22, we may then extend λ to a Radon measure on G.

To see that λ is left-invariant, let $f \in \hat{C}_+(G)$ be arbitrary and define $f_y(x) = f(yx)$. For any $s = xp \in S$ and $y \in G$ we get

$$f_y^*(s) = \bar{f}_y(x) = \int_K f(yxk)\nu(dk) = \bar{f}(yx) = f^*(ys).$$

Hence, by the invariance of μ,

$$\int_G f(yx)\lambda(dx) = \lambda f_y = \mu f_y^* = \int_S f^*(ys)\mu(ds) = \mu f^* = \lambda f.$$

Now fix any $g \in \hat{C}_+(S)$, and put
$$f(x) = g(xp) = g \circ \pi(x), \quad x \in G.$$
Then $f \in \hat{C}_+(G)$ because $\{f > 0\} \subset \pi^{-1}\text{supp}\, g$, which is compact since π is proper. By the definition of K, we have for any $s = xp \in S$
$$\begin{aligned}f^*(s) &= \bar{f}(x) = \int_K f(xk)\nu(dk) = \int_K g(xkp)\nu(dk) \\ &= \int_K g(xp)\nu(dk) = g(s),\end{aligned}$$
and so
$$\mu g = \mu f^* = \lambda f = \lambda(g \circ \pi) = (\lambda \circ \pi^{-1})g,$$
which shows that $\mu = \lambda \circ \pi^{-1}$. Since λ is unique up to a normalization, the same thing is true for μ. □

Exercises

1. Show that if $\mu_1 = f_1 \cdot \mu$ and $\mu_2 = f_2 \cdot \mu$, then $\mu_1 \vee \mu_2 = (f_1 \vee f_2) \cdot \mu$ and $\mu_1 \wedge \mu_2 = (f_1 \wedge f_2) \cdot \mu$. In particular, we may take $\mu = \mu_1 + \mu_2$. Extend the result to sequences $\mu_1, \mu_2, \ldots$.

2. Consider an arbitrary family μ_i, $i \in I$, of σ-finite measures on some measurable space S. Show that there exists a largest measure $\mu = \bigwedge_n \mu_n$ such that $\mu \leq \mu_i$ for all $i \in I$. Show also that if the μ_i are bounded by some σ-finite measure ν, there exists a smallest measure $\hat{\mu} = \bigvee_n \mu_i$ such that $\mu_i \leq \hat{\mu}$ for all i. (*Hint:* Use Zorn's lemma.)

3. Show that any countably additive set function $\mu \geq 0$ on a field $\mathcal{A}$ with $\mu\emptyset = 0$ extends to a measure on $\sigma(\mathcal{A})$. Show also that the extension is unique whenever μ is bounded.

4. Extend the first assertion of Theorem 2.6 to the context of general invariant measures, as in Theorem 2.29.

5. Construct d-dimensional Lebesgue measure λ_d directly, by the method of Theorem 2.2. Then show that $\lambda_d = \lambda^d$.

6. Derive the existence of d-dimensional Lebesgue measure from Riesz' representation theorem and the basic properties of the Riemann integral.

7. Extend the mean continuity in Lemma 2.7 to general invariant measures.

8. For any bounded, signed measure ν on $(\Omega, \mathcal{A})$, show that there exists a smallest measure $|\nu|$ such that $|\nu A| \leq |\nu| A$ for all $A \in \mathcal{A}$. Show also that $|\nu| = \nu_+ + \nu_-$, where $\nu_\pm$ are the components in the Hahn decomposition of ν. Finally, for any bounded, measurable function f on Ω, show that $|\nu f| \leq |\nu| |f|$.

9. Extend the last result to complex-valued measures $\chi = \mu + i\nu$, where μ and ν are bounded, signed measures on $(\Omega, \mathcal{A})$. Introducing the complex-valued Radon–Nikodým density $f = d\chi/d(|\mu| + |\nu|)$, show that $|\chi| = |f| \cdot (|\mu| + |\nu|)$.

10. Show by an example that the uniqueness in Theorem 2.29 may fail if the group action is not transitive.

Chapter 3

Processes, Distributions, and Independence

Random elements and processes; distributions and expectation; independence; zero–one laws; Borel–Cantelli lemma; Bernoulli sequences and existence; moments and continuity of paths

Armed with the basic notions and results of measure theory from the previous chapter, we may now embark on our study of probability theory itself. The dual purpose of this chapter is to introduce the basic terminology and notation and to prove some fundamental results, many of which are used throughout the remainder of this book.

In modern probability theory it is customary to relate all objects of study to a basic probability space $(\Omega, \mathcal{A}, P)$, which is nothing more than a normalized measure space. Random variables may then be defined as measurable functions ξ on Ω, and their expected values as the integrals $E\xi = \int \xi dP$. Furthermore, independence between random quantities reduces to a kind of orthogonality between the induced sub-σ-fields. It should be noted, however, that the reference space Ω is introduced only for technical convenience, to provide a consistent mathematical framework. Indeed, the actual choice of Ω plays no role, and the interest focuses instead on the various induced distributions $\mathcal{L}(\xi) = P \circ \xi^{-1}$.

The notion of independence is fundamental for all areas of probability theory. Despite its simplicity, it has some truly remarkable consequences. A particularly striking result is Kolmogorov's 0–1 law, which states that every tail event associated with a sequence of independent random elements has probability zero or one. As a consequence, any random variable that depends only on the "tail" of the sequence must be a.s. constant. This result and the related Hewitt–Savage 0–1 law convey much of the flavor of modern probability: Although the individual elements of a random sequence are erratic and unpredictable, the long-term behavior may often conform to deterministic laws and patterns. Our main objective is to uncover the latter. Here the classical Borel–Cantelli lemma is a useful tool, among others.

To justify our study, we need to ensure the existence of the random objects under discussion. For most purposes, it suffices to use the Lebesgue unit interval $([0,1], \mathcal{B}, \lambda)$ as the basic probability space. In this chapter the existence will be proved only for independent random variables with prescribed distributions; we postpone the more general discussion until

Chapter 6. As a key step, we use the binary expansion of real numbers to construct a so-called Bernoulli sequence, consisting of independent random digits 0 or 1 with probabilities $1-p$ and p, respectively. Such sequences may be regarded as discrete-time counterparts of the fundamental Poisson process, to be introduced and studied in Chapter 12.

The distribution of a random process X is determined by the finite-dimensional distributions, and those are not affected if we change each value X_t on a null set. It is then natural to look for versions of X with suitable regularity properties. As another striking result, we shall provide a moment condition that ensures the existence of a continuous modification of the process. Regularizations of various kinds are important throughout modern probability theory, as they may enable us to deal with events depending on the values of a process at uncountably many times.

To begin our systematic exposition of the theory, we may fix an arbitrary *probability space* $(\Omega, \mathcal{A}, P)$, where P, the *probability measure*, has total mass 1. In the probabilistic context the sets $A \in \mathcal{A}$ are called *events*, and $PA = P(A)$ is called the *probability* of A. In addition to results valid for all measures, there are properties that depend on the boundedness or normalization of P, such as the relation $PA^c = 1 - PA$ and the fact that $A_n \downarrow A$ implies $PA_n \to PA$.

Some infinite set operations have special probabilistic significance. Thus, given any sequence of events $A_1, A_2, \cdots \in \mathcal{A}$, we may be interested in the sets $\{A_n \text{ i.o.}\}$, where A_n happens *infinitely often*, and $\{A_n \text{ ult.}\}$, where A_n happens *ultimately* (i.e., for all but finitely many n). Those occurrences are events in their own right, expressible in terms of the A_n as

$$\{A_n \text{ i.o.}\} = \left\{\sum_n 1_{A_n} = \infty\right\} = \bigcap_n \bigcup_{k \geq n} A_k, \tag{1}$$

$$\{A_n \text{ ult.}\} = \left\{\sum_n 1_{A_n^c} < \infty\right\} = \bigcup_n \bigcap_{k \geq n} A_k. \tag{2}$$

From here on, we omit the argument ω from our notation when there is no risk for confusion. For example, the expression $\{\sum_n 1_{A_n} = \infty\}$ is used as a convenient shorthand form of the unwieldy $\{\omega \in \Omega;\ \sum_n 1_{A_n}(\omega) = \infty\}$.

The indicator functions of the events in (1) and (2) may be expressed as

$$1\{A_n \text{ i.o.}\} = \limsup_{n \to \infty} 1_{A_n}, \qquad 1\{A_n \text{ ult.}\} = \liminf_{n \to \infty} 1_{A_n},$$

where, for typographical convenience, we write $1\{\cdot\}$ instead of $1_{\{\cdot\}}$. Applying Fatou's lemma to the functions 1_{A_n} and $1_{A_n^c}$, we get

$$P\{A_n \text{ i.o.}\} \geq \limsup_{n \to \infty} PA_n, \qquad P\{A_n \text{ ult.}\} \leq \liminf_{n \to \infty} PA_n.$$

Using the continuity and subadditivity of P, we further see from (1) that

$$P\{A_n \text{ i.o.}\} = \lim_{n \to \infty} P\bigcup_{k \geq n} A_k \leq \lim_{n \to \infty} \sum_{k \geq n} PA_k.$$

If $\sum_n PA_n < \infty$, we get zero on the right, and it follows that $P\{A_n \text{ i.o.}\} = 0$. The resulting implication constitutes the easy part of the *Borel–Cantelli lemma*, to be reconsidered in Theorem 3.18.

Any measurable mapping ξ of Ω into some measurable space $(S, \mathcal{S})$ is called a *random element* in S. If $B \in \mathcal{S}$, then $\{\xi \in B\} = \xi^{-1} B \in \mathcal{A}$, and we may consider the associated probabilities

$$P\{\xi \in B\} = P(\xi^{-1} B) = (P \circ \xi^{-1}) B, \quad B \in \mathcal{S}.$$

The set function $\mathcal{L}(\xi) = P \circ \xi^{-1}$ is a probability measure on the range space S of ξ, called the *distribution* or *law* of ξ. We shall also use the term *distribution* as synonomous to probability measure, even when no generating random element has been introduced.

Random elements are of interest in a wide variety of spaces. A random element in S is called a *random variable* when $S = \mathbb{R}$, a *random vector* when $S = \mathbb{R}^d$, a *random sequence* when $S = \mathbb{R}^\infty$, a *random* or *stochastic process* when S is a function space, and a *random measure* or *set* when S is a class of measures or sets, respectively. A metric or topological space S will be endowed with its Borel σ-field $\mathcal{B}(S)$ unless a σ-field is otherwise specified. For any separable metric space S, it is clear from Lemma 1.2 that $\xi = (\xi_1, \xi_2, \ldots)$ is a random element in S^∞ iff $\xi_1, \xi_2, \ldots$ are random elements in S.

If $(S, \mathcal{S})$ is a measurable space, then any subset $A \subset S$ becomes a measurable space in its own right when endowed with the σ-field $A \cap \mathcal{S} = \{A \cap B;\ B \in \hat{\mathcal{S}}\}$. By Lemma 1.6 we note in particular that if S is a metric space with Borel σ-field $\mathcal{S}$, then $A \cap \mathcal{S}$ is the Borel σ-field in A. Any random element in $(A, A \cap \mathcal{S})$ may clearly be regarded, alternatively, as a random element in S. Conversely, if ξ is a random element in S such that $\xi \in A$ a.s. (almost surely or with probability 1) for some $A \in \mathcal{S}$, then $\xi = \eta$ a.s. for some random element η in A.

Fixing a measurable space $(S, \mathcal{S})$ and an abstract index set T, we shall write S^T for the class of functions $f : T \to S$, and let $\mathcal{S}^T$ denote the σ-field in S^T generated by all *evaluation maps* $\pi_t : S^T \to S$, $t \in T$, given by $\pi_t f = f(t)$. If $X : \Omega \to U \subset S^T$, then clearly $X_t = \pi_t \circ X$ maps Ω into S. Thus, X may also be regarded as a function $X(t, \omega) = X_t(\omega)$ from $T \times \Omega$ to S.

Lemma 3.1 *(measurability)* *Fix a measurable space $(S, \mathcal{S})$, an index set T, and a subset $U \subset S^T$. Then a function $X : \Omega \to U$ is $U \cap \mathcal{S}^T$-measurable iff $X_t : \Omega \to S$ is $\mathcal{S}$-measurable for every $t \in T$.*

Proof: Since X is U-valued, the $U \cap \mathcal{S}^T$-measurability is equivalent to measurability with respect to $\mathcal{S}^T$. The result now follows by Lemma 1.4 from the fact that $\mathcal{S}^T$ is generated by the mappings π_t. □

A mapping X with the properties in Lemma 3.1 is called an *S-valued (random) process* on T with *paths* in U. By the lemma it is equivalent to regard X as a collection of random elements X_t in the *state space* S.

For any random elements ξ and η in a common measurable space, the equality $\xi \stackrel{d}{=} \eta$ means that ξ and η have the same distribution, or $\mathcal{L}(\xi) = \mathcal{L}(\eta)$. If X is a random process on some index set T, the associated *finite-dimensional distributions* are given by

$$\mu_{t_1,\ldots,t_n} = \mathcal{L}(X_{t_1},\ldots,X_{t_n}), \quad t_1,\ldots,t_n \in T,\ n \in \mathbb{N}.$$

The following result shows that the distribution of a process is determined by the set of finite-dimensional distributions.

Proposition 3.2 *(finite-dimensional distributions) Fix any S, T, and U as in Lemma 3.1, and let X and Y be processes on T with paths in U. Then $X \stackrel{d}{=} Y$ iff*

$$(X_{t_1},\ldots,X_{t_n}) \stackrel{d}{=} (Y_{t_1},\ldots,Y_{t_n}), \quad t_1,\ldots,t_n \in T,\ n \in \mathbb{N}. \qquad (3)$$

Proof: Assume (3). Let $\mathcal{D}$ denote the class of sets $A \in \mathcal{S}^T$ with $P\{X \in A\} = P\{Y \in A\}$, and let $\mathcal{C}$ consist of all sets

$$A = \{f \in S^T;\ (f_{t_1},\ldots,f_{t_n}) \in B\}, \quad t_1,\ldots,t_n \in T,\ B \in \mathcal{S}^n,\ n \in \mathbb{N}.$$

Then $\mathcal{C}$ is a π-system and $\mathcal{D}$ a λ-system, and furthermore $\mathcal{C} \subset \mathcal{D}$ by hypothesis. Hence, $\mathcal{S}^T = \sigma(\mathcal{C}) \subset \mathcal{D}$ by Theorem 1.1, which means that $X \stackrel{d}{=} Y$. □

For any random vector $\xi = (\xi_1,\ldots,\xi_d)$ in $\mathbb{R}^d$, we define the associated *distribution function* F by

$$F(x_1,\ldots,x_d) = P\bigcap_{k\leq d}\{\xi_k \leq x_k\}, \quad x_1,\ldots,x_d \in \mathbb{R}.$$

The next result shows that F determines the distribution of ξ.

Lemma 3.3 *(distribution functions) Let ξ and η be random vectors in $\mathbb{R}^d$ with distribution functions F and G. Then $\xi \stackrel{d}{=} \eta$ iff $F = G$.*

Proof: Use Theorem 1.1. □

The *expected value*, *expectation*, or *mean* of a random variable ξ is defined as

$$E\xi = \int_\Omega \xi\,dP = \int_\mathbb{R} x(P \circ \xi^{-1})(dx) \qquad (4)$$

whenever either integral exists. The last equality then holds by Lemma 1.22. By the same result we note that, for any random element ξ in some measurable space S and for an arbitrary measurable function $f: S \to \mathbb{R}$,

$$\begin{aligned} Ef(\xi) &= \int_\Omega f(\xi)\,dP = \int_S f(s)(P \circ \xi^{-1})(ds) \\ &= \int_\mathbb{R} x(P \circ (f \circ \xi)^{-1})(dx), \end{aligned} \qquad (5)$$

provided that at least one of the three integrals exists. Integrals over a measurable subset $A \subset \Omega$ are often denoted by

$$E[\xi; A] = E(\xi 1_A) = \int_A \xi \, dP, \quad A \in \mathcal{A}.$$

For any random variable ξ and constant $p > 0$, the integral $E|\xi|^p = \|\xi\|_p^p$ is called the *p*th *absolute moment* of ξ. By Hölder's inequality (or by Jensen's inequality in Lemma 3.5) we have $\|\xi\|_p \le \|\xi\|_q$ for $p \le q$, so the corresponding L^p-spaces are nonincreasing in p. If $\xi \in L^p$ and either $p \in \mathbb{N}$ or $\xi \ge 0$, we may further define the *p*th *moment* of ξ as $E\xi^p$.

The following result gives a useful relationship between moments and tail probabilities.

Lemma 3.4 *(moments and tails)* *For any random variable $\xi \ge 0$,*

$$E\xi^p = p \int_0^\infty P\{\xi > t\} \, t^{p-1} dt = p \int_0^\infty P\{\xi \ge t\} \, t^{p-1} dt, \quad p > 0.$$

Proof: By calculus and Fubini's theorem,

$$\begin{aligned} E\xi^p &= pE \int_0^\xi t^{p-1} dt = pE \int_0^\infty 1\{\xi > t\} \, t^{p-1} dt \\ &= p \int_0^\infty P\{\xi > t\} \, t^{p-1} dt. \end{aligned}$$

The proof of the second expression is similar. □

A random vector $\xi = (\xi_1, \ldots, \xi_d)$ or process $X = (X_t)$ is said to be *integrable* if integrability holds for every component ξ_k or value X_t, in which case we may write $E\xi = (E\xi_1, \ldots, E\xi_d)$ or $EX = (EX_t)$. Recall that a function $f: \mathbb{R}^d \to \mathbb{R}$ is said to be *convex* if

$$f(px + (1-p)y) \le pf(x) + (1-p)f(y), \quad x, y \in \mathbb{R}^d, \; p \in [0,1]. \quad (6)$$

The relation may be written as $f(E\xi) \le Ef(\xi)$, where ξ is a random vector in $\mathbb{R}^d$ with $P\{\xi = x\} = 1 - P\{\xi = y\} = p$. The following extension to arbitrary integrable random vectors is known as *Jensen's inequality*.

Lemma 3.5 *(convex maps, Hölder, Jensen)* *For any integrable random vector ξ in $\mathbb{R}^d$ and convex function $f: \mathbb{R}^d \to \mathbb{R}$, we have*

$$Ef(\xi) \ge f(E\xi).$$

Proof: By a version of the Hahn–Banach theorem, the convexity condition (6) is equivalent to the existence for every $s \in \mathbb{R}^d$ of a *supporting* affine function $h_s(x) = ax + b$ with $f \ge h_s$ and $f(s) = h_s(s)$. Taking $s = E\xi$ gives

$$Ef(\xi) \ge Eh_s(\xi) = h_s(E\xi) = f(E\xi). \quad □$$

The *covariance* of two random variables $\xi, \eta \in L^2$ is given by

$$\mathrm{cov}(\xi, \eta) = E(\xi - E\xi)(\eta - E\eta) = E\xi\eta - E\xi \cdot E\eta.$$

The resulting functional is *bilinear*, in the sense that

$$\mathrm{cov}\left(\sum\nolimits_{j\le m} a_j\xi_j,\ \sum\nolimits_{k\le n} b_k\eta_k\right) = \sum\nolimits_{j\le m}\sum\nolimits_{k\le n} a_j b_k \mathrm{cov}(\xi_j,\eta_k).$$

Taking $\xi = \eta \in L^2$ yields the *variance*

$$\mathrm{var}(\xi) = \mathrm{cov}(\xi,\xi) = E(\xi - E\xi)^2 = E\xi^2 - (E\xi)^2,$$

and we note that, by the Cauchy–Buniakovsky inequality,

$$|\mathrm{cov}(\xi,\eta)| \le \{\mathrm{var}(\xi)\,\mathrm{var}(\eta)\}^{1/2}.$$

Two random variables ξ and η are said to be *uncorrelated* if $\mathrm{cov}(\xi,\eta) = 0$.

For any collection of random variables $\xi_t \in L^2$, $t \in T$, the associated *covariance function* $\rho_{s,t} = \mathrm{cov}(\xi_s,\xi_t)$, $s,t \in T$, is *nonnegative definite*, in the sense that $\sum_{ij} a_i a_j \rho_{t_i,t_j} \ge 0$ for any $n \in \mathbb{N}$, $t_1,\ldots t_n \in T$, and $a_1,\ldots,a_n \in \mathbb{R}$. This is clear if we write

$$\sum\nolimits_{i,j} a_i a_j \rho_{t_i,t_j} = \sum\nolimits_{i,j} a_i a_j \mathrm{cov}(\xi_{t_i},\xi_{t_j}) = \mathrm{var}\left\{\sum\nolimits_i a_i \xi_{t_i}\right\} \ge 0.$$

The events $A_t \in \mathcal{A}$, $t \in T$, are said to be *(mutually) independent* if, for any distinct indices $t_1,\ldots,t_n \in T$,

$$P\bigcap\nolimits_{k\le n} A_{t_k} = \prod\nolimits_{k\le n} PA_{t_k}. \tag{7}$$

More generally, we say that the families $\mathcal{C}_t \subset \mathcal{A}$, $t \in T$, are independent if independence holds between the events A_t for arbitrary $A_t \in \mathcal{C}_t$, $t \in T$. Finally, the random elements ξ_t, $t \in T$, are independent if independence holds between the generated σ-fields $\sigma(\xi_t)$, $t \in T$. Pairwise independence between two objects $\mathcal{A}$ and $\mathcal{B}$, ξ and η, or $\mathcal{B}$ and $\mathcal{C}$ is often denoted by $\mathcal{A} \perp\!\!\!\perp \mathcal{B}$, $\xi \perp\!\!\!\perp \eta$, or $\mathcal{B} \perp\!\!\!\perp \mathcal{C}$, respectively.

The following result is often useful to prove extensions of the independence property.

Lemma 3.6 *(extension)* *If the π-systems $\mathcal{C}_t$, $t \in T$, are independent, then so are the generated σ-fields $\mathcal{F}_t = \sigma(\mathcal{C}_t)$, $t \in T$.*

Proof: We may clearly assume that $\mathcal{C}_t \ne \emptyset$ for all t. Fix any distinct indices $t_1,\ldots,t_n \in T$, and note that (7) holds for arbitrary $A_{t_k} \in \mathcal{C}_{t_k}$, $k = 1,\ldots,n$. For fixed $A_{t_2},\ldots,A_{t_n}$, we introduce the class $\mathcal{D}$ of sets $A_{t_1} \in \mathcal{A}$ satisfying (7). Then $\mathcal{D}$ is a λ-system containing $\mathcal{C}_{t_1}$, and so $\mathcal{D} \supset \sigma(\mathcal{C}_{t_1}) = \mathcal{F}_{t_1}$ by Theorem 1.1. Thus, (7) holds for arbitrary $A_{t_1} \in \mathcal{F}_{t_1}$ and $A_{t_k} \in \mathcal{C}_{t_k}$, $k = 2,\ldots,n$. Proceeding recursively in n steps, we obtain the desired extension to arbitrary $A_{t_k} \in \mathcal{F}_{t_k}$, $k = 1,\ldots,n$. □

As an immediate consequence, we obtain the following basic grouping property. Here and in the sequel we shall often write $\mathcal{F} \vee \mathcal{G} = \sigma\{\mathcal{F},\mathcal{G}\}$ and $\mathcal{F}_S = \bigvee_{t\in S} \mathcal{F}_t = \sigma\{\mathcal{F}_t;\ t \in S\}$.

Corollary 3.7 *(grouping) Let $\mathcal{F}_t$, $t \in T$, be independent σ-fields, and let $\mathcal{T}$ be a disjoint partition of T. Then the σ-fields $\mathcal{F}_S = \bigvee_{t \in S} \mathcal{F}_t$, $S \in \mathcal{T}$, are again independent.*

Proof: For any $S \in \mathcal{T}$, let $\mathcal{C}_S$ denote the class of all finite intersections of sets in $\bigcup_{t \in S} \mathcal{F}_t$. Then the classes $\mathcal{C}_S$ are independent π-systems, and by Lemma 3.6 the independence extends to the generated σ-fields $\mathcal{F}_S$. □

Though independence between more than two σ-fields is clearly stronger than pairwise independence, we shall see how the full independence may be reduced to the pairwise notion in various ways. Given any set T, we say that a class $\mathcal{T} \subset 2^T$ is *separating*, if for any $s \neq t$ in T there exists some $S \in \mathcal{T}$ such that exactly one of the elements s and t lies in S.

Lemma 3.8 *(pairwise independence)*

(i) *The σ-fields $\mathcal{F}_1, \mathcal{F}_2, \ldots$ are independent iff $\bigvee_{k \leq n} \mathcal{F}_k \perp\!\!\!\perp \mathcal{F}_{n+1}$ for all n.*

(ii) *The σ-fields $\mathcal{F}_t$, $t \in T$, are independent iff $\mathcal{F}_S \perp\!\!\!\perp \mathcal{F}_{S^c}$ for all sets S in some separating class $\mathcal{T} \subset 2^T$.*

Proof: The necessity of the two conditions follows from Corollary 3.7. As for the sufficiency, we consider only part (ii), the proof for (i) being similar. Under the stated condition, we need to show that, for any finite subset $S \subset T$, the σ-fields $\mathcal{F}_s$, $s \in S$, are independent. Let $|S|$ denote the cardinality of S, and assume the statement to be true for $|S| \leq n$. Proceeding to the case when $|S| = n+1$, we may choose $U \in \mathcal{T}$ such that $S' = S \cap U$ and $S'' = S \setminus U$ are nonempty. Since $\mathcal{F}_{S'} \perp\!\!\!\perp \mathcal{F}_{S''}$, we get for any sets $A_s \in \mathcal{F}_s$, $s \in S$,

$$P\bigcap_{s \in S} A_s = \left(P\bigcap_{s \in S'} A_s\right)\left(P\bigcap_{s \in S''} A_s\right) = \prod_{s \in S} PA_s,$$

where the last relation follows from the induction hypothesis. □

A σ-field $\mathcal{F}$ is said to be *P-trivial* if $PA = 0$ or 1 for every $A \in \mathcal{F}$. We further say that a random element is *a.s. degenerate* if its distribution is a degenerate probability measure.

Lemma 3.9 *(triviality and degeneracy) A σ-field $\mathcal{F}$ is P-trivial iff $\mathcal{F} \perp\!\!\!\perp \mathcal{F}$. In that case, any $\mathcal{F}$-measurable random element ξ taking values in a separable metric space is a.s. degenerate.*

Proof: If $\mathcal{F} \perp\!\!\!\perp \mathcal{F}$, then for any $A \in \mathcal{F}$ we have $PA = P(A \cap A) = (PA)^2$, and so $PA = 0$ or 1. Conversely, assume that $\mathcal{F}$ is P-trivial. Then for any two sets $A, B \in \mathcal{F}$ we have $P(A \cap B) = PA \wedge PB = PA \cdot PB$, which means that $\mathcal{F} \perp\!\!\!\perp \mathcal{F}$.

Now assume that $\mathcal{F}$ is P-trivial, and let ξ be as stated. For each n we may partition S into countably many disjoint Borel sets B_{nj} of diameter $< n^{-1}$. Since $P\{\xi \in B_{nj}\} = 0$ or 1, we have $\xi \in B_{nj}$ a.s. for exactly one j, say for $j = j_n$. Hence, $\xi \in \bigcap_n B_{n,j_n}$ a.s. The latter set has diameter 0, so it consists of exactly one point s, and we get $\xi = s$ a.s. □

The next result gives the basic relation between independence and product measures.

Lemma 3.10 *(product measures) Let $\xi_1, \ldots, \xi_n$ be random elements in some measurable spaces $S_1, \ldots, S_n$ with distributions $\mu_1, \ldots, \mu_n$. Then the ξ_k are independent iff $\xi = (\xi_1, \ldots, \xi_n)$ has distribution $\mu_1 \otimes \cdots \otimes \mu_n$.*

Proof: Assuming the independence, we get for any measurable product set $B = B_1 \times \cdots \times B_n$

$$P\{\xi \in B\} = \prod_{k \leq n} P\{\xi_k \in B_k\} = \prod_{k \leq n} \mu_k B_k = \bigotimes_{k \leq n} \mu_k B.$$

This extends by Theorem 1.1 to arbitrary sets in the product σ-field. □

In conjunction with Fubini's theorem, the last result leads to a useful method of computing expected values.

Lemma 3.11 *(conditioning) Let ξ and η be independent random elements in some measurable spaces S and T, and let the function $f: S \times T \to \mathbb{R}$ be measurable with $E(E|f(s,\eta)|)_{s=\xi} < \infty$. Then $Ef(\xi,\eta) = E(Ef(s,\eta))_{s=\xi}$.*

Proof: Let μ and ν denote the distributions of ξ and η, respectively. Assuming that $f \geq 0$ and writing $g(s) = Ef(s, \eta)$, we get, by Lemma 1.22 and Fubini's theorem,

$$Ef(\xi, \eta) = \int f(s,t)(\mu \otimes \nu)(dsdt)$$

$$= \int \mu(ds) \int f(s,t)\nu(dt) = \int g(s)\mu(ds) = Eg(\xi).$$

For general f, this applies to the function $|f|$, and so $E|f(\xi, \eta)| < \infty$. The desired relation then follows as before. □

In particular, for any independent random variables $\xi_1, \ldots, \xi_n$, we have

$$E\prod_k \xi_k = \prod_k E\xi_k, \quad \text{var} \sum_k \xi_k = \sum_k \text{var}\, \xi_k,$$

whenever the expressions on the right exist.

If ξ and η are random elements in a measurable group G, then the product $\xi\eta$ is again a random element in G. The following result gives the connection between independence and the convolutions of Lemma 1.28.

Corollary 3.12 *(convolution) Let ξ and η be independent random elements with distributions μ and ν, respectively, in some measurable group G. Then the product $\xi\eta$ has distribution $\mu * \nu$.*

Proof: For any measurable set $B \subset G$, we get by Lemma 3.10 and the definition of convolution

$$P\{\xi\eta \in B\} = (\mu \otimes \nu)\{(x,y) \in G^2;\ xy \in B\} = (\mu * \nu)B. \quad \square$$

3. Processes, Distributions, and Independence

Given any sequence of σ-fields $\mathcal{F}_1, \mathcal{F}_2, \ldots$, we introduce the associated *tail σ-field*

$$\mathcal{T} = \bigcap_n \bigvee_{k>n} \mathcal{F}_k = \bigcap_n \sigma\{\mathcal{F}_k;\; k > n\}.$$

The following remarkable result shows that $\mathcal{T}$ is trivial whenever the $\mathcal{F}_n$ are independent. An extension appears in Corollary 7.25.

Theorem 3.13 *(Kolmogorov's 0-1 law)* Let $\mathcal{F}_1, \mathcal{F}_2, \ldots$ be independent σ-fields. Then the tail σ-field $\mathcal{T} = \bigcap_n \bigvee_{k>n} \mathcal{F}_k$ is P-trivial.

Proof: For each $n \in \mathbb{N}$, define $\mathcal{T}_n = \bigvee_{k>n} \mathcal{F}_k$, and note that $\mathcal{F}_1, \ldots, \mathcal{F}_n, \mathcal{T}_n$ are independent by Corollary 3.7. Hence, so are the σ-fields $\mathcal{F}_1, \ldots, \mathcal{F}_n, \mathcal{T}$, and then also $\mathcal{F}_1, \mathcal{F}_2, \ldots, \mathcal{T}$. By the same theorem we obtain $\mathcal{T}_0 \perp\!\!\!\perp \mathcal{T}$, and so $\mathcal{T} \perp\!\!\!\perp \mathcal{T}$. Thus, $\mathcal{T}$ is P-trivial by Lemma 3.9. □

We shall consider some simple illustrations of the last theorem.

Corollary 3.14 *(sums and averages)* Let $\xi_1, \xi_2, \ldots$ be independent random variables, and put $S_n = \xi_1 + \cdots + \xi_n$. Then each of the sequences (S_n) and (S_n/n) is either a.s. convergent or a.s. divergent. For the latter sequence, the possible limit is a.s. degenerate.

Proof: Define $\mathcal{F}_n = \sigma\{\xi_n\}$, $n \in \mathbb{N}$, and note that the associated tail σ-field $\mathcal{T}$ is P-trivial by Theorem 3.13. Since the sets of convergence of (S_n) and (S_n/n) are $\mathcal{T}$-measurable by Lemma 1.9, the first assertion follows. The second assertion is obtained from Lemma 3.9. □

By a *finite permutation* of $\mathbb{N}$ we mean a bijective map $p: \mathbb{N} \to \mathbb{N}$ such that $p_n = n$ for all but finitely many n. For any space S, a finite permutation p of $\mathbb{N}$ induces a permutation T_p on S^∞ given by

$$T_p(s) = s \circ p = (s_{p_1}, s_{p_2}, \ldots), \quad s = (s_1, s_2, \ldots) \in S^\infty.$$

A set $I \subset S^\infty$ is said to be *symmetric* (under finite permutations) if

$$T_p^{-1} I \equiv \{s \in S^\infty;\; s \circ p \in I\} = I$$

for every finite permutation p of $\mathbb{N}$. If $(S, \mathcal{S})$ is a measurable space, the symmetric sets $I \in \mathcal{S}^\infty$ form a sub-σ-field $\mathcal{I} \subset \mathcal{S}^\infty$, called the *permutation invariant σ-field* in S^∞.

We may now state the other basic 0-1 law, which refers to sequences of random elements that are independent and identically distributed (often abbreviated as *i.i.d.*).

Theorem 3.15 *(Hewitt–Savage 0-1 law)* Let ξ be an infinite sequence of i.i.d. random elements in some measurable space $(S, \mathcal{S})$, and let $\mathcal{I}$ denote the permutation invariant σ-field in S^∞. Then the σ-field $\xi^{-1} \mathcal{I}$ is P-trivial.

Our proof is based on a simple approximation. Write

$$A \triangle B = (A \setminus B) \cup (B \setminus A),$$

and note that
$$P(A\triangle B) = P(A^c \triangle B^c) = E|1_A - 1_B|, \quad A, B \in \mathcal{A}. \tag{8}$$

Lemma 3.16 *(approximation)* Given any σ-fields $\mathcal{F}_1 \subset \mathcal{F}_2 \subset \cdots$ and a set $A \in \bigvee_n \mathcal{F}_n$, there exist some $A_1, A_2, \cdots \in \bigcup_n \mathcal{F}_n$ with $P(A\triangle A_n) \to 0$.

Proof: Define $\mathcal{C} = \bigcup_n \mathcal{F}_n$, and let $\mathcal{D}$ denote the class of sets $A \in \bigvee_n \mathcal{F}_n$ with the stated property. Then $\mathcal{C}$ is a π-system and $\mathcal{D}$ a λ-system containing $\mathcal{C}$. By Theorem 1.1 we get $\bigvee_n \mathcal{F}_n = \sigma(\mathcal{C}) \subset \mathcal{D}$. □

Proof of Theorem 3.15: Define $\mu = \mathcal{L}(\xi)$, put $\mathcal{F}_n = \mathcal{S}^n \times \mathcal{S}^\infty$, and note that $\mathcal{I} \subset \mathcal{S}^\infty = \bigvee_n \mathcal{F}_n$. For any $I \in \mathcal{I}$ there exist by Lemma 3.16 some $B_n \in \mathcal{S}^n$ such that the corresponding cylinder sets $I_n = B_n \times \mathcal{S}^\infty$ satisfy $\mu(I \triangle I_n) \to 0$. Writing $\tilde{I}_n = \mathcal{S}^n \times B_n \times \mathcal{S}^\infty$, it is clear from the symmetry of μ and I that $\mu \tilde{I}_n = \mu I_n \to \mu I$ and $\mu(I \triangle \tilde{I}_n) = \mu(I \triangle I_n) \to 0$. Hence, by (8),
$$\mu(I \triangle (I_n \cap \tilde{I}_n)) \leq \mu(I \triangle I_n) + \mu(I \triangle \tilde{I}_n) \to 0.$$
Since moreover $I_n \perp\!\!\!\perp \tilde{I}_n$ under μ, we get
$$\mu I \leftarrow \mu(I_n \cap \tilde{I}_n) = (\mu I_n)(\mu \tilde{I}_n) \to (\mu I)^2.$$
Thus, $\mu I = (\mu I)^2$, and so $P \circ \xi^{-1} I = \mu I = 0$ or 1. □

The next result lists some typical applications. Say that a random variable ξ is *symmetric* if $\xi \stackrel{d}{=} -\xi$.

Corollary 3.17 *(random walk)* Let $\xi_1, \xi_2, \ldots$ be i.i.d., nondegenerate random variables, and put $S_n = \xi_1 + \cdots + \xi_n$. Then
- (i) $P\{S_n \in B \text{ i.o.}\} = 0$ or 1 for any $B \in \mathcal{B}$;
- (ii) $\limsup_n S_n = \infty$ a.s. or $-\infty$ a.s.;
- (iii) $\limsup_n(\pm S_n) = \infty$ a.s. if the ξ_n are symmetric.

Proof: Statement (i) is immediate from Theorem 3.15, since for any finite permutation p of $\mathbb{N}$ we have $x_{p_1} + \cdots + x_{p_n} = x_1 + \cdots + x_n$ for all but finitely many n. To prove (ii), conclude from Theorem 3.15 and Lemma 3.9 that $\limsup_n S_n = c$ a.s. for some constant $c \in \overline{\mathbb{R}} = [-\infty, \infty]$. Hence, a.s.,
$$c = \limsup_n S_{n+1} = \limsup_n(S_{n+1} - \xi_1) + \xi_1 = c + \xi_1.$$
If $|c| < \infty$, we get $\xi_1 = 0$ a.s., which contradicts the nondegeneracy of ξ_1. Thus, $|c| = \infty$. In case (iii), we have
$$c = \limsup_n S_n \geq \liminf_n S_n = -\limsup_n(-S_n) = -c,$$
and so $-c \leq c \in \{\pm\infty\}$, which implies $c = \infty$. □

Using a suitable zero–one law, one can often rather easily see that a given event has probability zero or one. Determining which alternative actually occurs is often harder. The following classical result, known as the

Borel–Cantelli lemma, may then be helpful, especially when the events are independent. An extension to the general case appears in Corollary 7.20.

Theorem 3.18 (Borel, Cantelli) *Let $A_1, A_2, \ldots \in \mathcal{A}$. Then $\sum_n PA_n < \infty$ implies $P\{A_n \text{ i.o.}\} = 0$, and the two conditions are equivalent when the A_n are independent.*

Here the first assertion was proved earlier as an application of Fatou's lemma. The use of expected values allows a more transparent argument.

Proof: If $\sum_n PA_n < \infty$, we get by monotone convergence
$$E\sum_n 1_{A_n} = \sum_n E 1_{A_n} = \sum_n PA_n < \infty.$$

Thus, $\sum_n 1_{A_n} < \infty$ a.s., which means that $P\{A_n \text{ i.o.}\} = 0$.

Next assume that the A_n are independent and satisfy $\sum_n PA_n = \infty$. Noting that $1 - x \leq e^{-x}$ for all x, we get
$$\begin{aligned}
P\bigcup_{k \geq n} A_k &= 1 - P\bigcap_{k \geq n} A_k^c = 1 - \prod_{k \geq n} PA_k^c \\
&= 1 - \prod_{k \geq n}(1 - PA_k) \geq 1 - \prod_{k \geq n} \exp(-PA_k) \\
&= 1 - \exp\left\{-\sum_{k \geq n} PA_k\right\} = 1.
\end{aligned}$$

Hence, as $n \to \infty$,
$$1 = P\bigcup_{k \geq n} A_k \downarrow P\bigcap_n \bigcup_{k \geq n} A_k = P\{A_n \text{ i.o.}\},$$
and so the probability on the right equals 1. □

For many purposes it is sufficient to use the *Lebesgue unit interval* $([0,1], \mathcal{B}[0,1], \lambda)$ as the basic probability space. In particular, the following result ensures the existence on $[0,1]$ of some independent random variables $\xi_1, \xi_2, \ldots$ with arbitrarily prescribed distributions. The present statement is only preliminary. Thus, we shall remove the independence assumption in Theorem 6.14, prove an extension to arbitrary index sets in Theorem 6.16, and eliminate the restriction on the spaces in Theorem 6.17.

Theorem 3.19 (existence, Borel) *For any probability measures $\mu_1, \mu_2, \ldots$ on some Borel spaces $S_1, S_2, \ldots$, there exist some independent random elements $\xi_1, \xi_2, \ldots$ on $([0,1], \lambda)$ with distributions $\mu_1, \mu_2, \ldots$.*

As a consequence, there exists a probability measure μ on $S_1 \times S_2 \times \cdots$ satisfying
$$\mu \circ (\pi_1, \ldots, \pi_n)^{-1} = \mu_1 \otimes \cdots \otimes \mu_n, \quad n \in \mathbb{N}.$$

For the proof, we first consider two special cases of independent interest.

By a *Bernoulli sequence* with rate *rate p* we mean a sequence of i.i.d. random variables $\xi_1, \xi_2, \ldots$ such that $P\{\xi_n = 1\} = 1 - P\{\xi_n = 0\} = p$. Furthermore, we say that a random variable ϑ is *uniformly distributed* on $[0,1]$ (written as $U(0,1)$) if its distribution $\mathcal{L}(\vartheta)$ equals Lebesgue measure λ on $[0,1]$. Every number $x \in [0,1]$ has a *binary expansion* $r_1, r_2, \ldots \in \{0,1\}$ satisfying $x = \sum_n r_n 2^{-n}$, and to ensure uniqueness we assume that $\sum_n r_n = \infty$ when $x > 0$. The following result provides a simple construction of a Bernoulli sequence on the Lebesgue unit interval.

Lemma 3.20 *(Bernoulli sequence) Let ϑ be a random variable in $[0,1]$ with binary expansion $\xi_1, \xi_2, \ldots$. Then ϑ is $U(0,1)$ iff the ξ_n form a Bernoulli sequence with rate $\tfrac{1}{2}$.*

Proof: If ϑ is $U(0,1)$, then $P \bigcap_{j \leq n} \{\xi_j = k_j\} = 2^{-n}$ for all $k_1, \ldots, k_n \in \{0,1\}$. Summing over $k_1, \ldots, k_{n-1}$ gives $P\{\xi_n = k\} = \tfrac{1}{2}$ for $k = 0$ and 1. A similar calculation yields the asserted independence.

Now assume instead that the ξ_n form a Bernoulli sequence with rate $\tfrac{1}{2}$. Letting $\tilde{\vartheta}$ be $U(0,1)$ with binary expansion $\tilde{\xi}_1, \tilde{\xi}_2, \ldots$, we get $(\xi_n) \stackrel{d}{=} (\tilde{\xi}_n)$. Thus,

$$\vartheta = \sum_n \xi_n 2^{-n} \stackrel{d}{=} \sum_n \tilde{\xi}_n 2^{-n} = \tilde{\vartheta}. \qquad \square$$

The next result shows how a single $U(0,1)$ random variable can be used to generate a whole sequence.

Lemma 3.21 *(reproduction) There exist some measurable functions $f_1, f_2, \ldots$ on $[0,1]$ such that whenever ϑ is $U(0,1)$, the random variables $\vartheta_n = f_n(\vartheta)$ are i.i.d. $U(0,1)$.*

Proof: For any $x \in [0,1]$ we introduce the associated binary expansion $g_1(x), g_2(x), \ldots$ and note that the g_k are measurable. Rearranging the g_k into a two-dimensional array h_{nj}, $n, j \in \mathbb{N}$, we define

$$f_n(x) = \sum_j 2^{-j} h_{nj}(x), \quad x \in [0,1], \ n \in \mathbb{N}.$$

By Lemma 3.20 the random variables $g_k(\vartheta)$ form a Bernoulli sequence with rate $\tfrac{1}{2}$, and the same result shows that the variables $\vartheta_n = f_n(\vartheta)$ are $U(0,1)$. The latter are further independent by Corollary 3.7. $\square$

Finally, we need to construct a random element with given distribution from an arbitrary randomization variable. The required lemma is stated in a version for kernels, to meet the needs of Chapters 6, 8, and 14.

Lemma 3.22 *(kernels and randomization) Let μ be a probability kernel from a measurable space S to a Borel space T. Then there exists a measurable function $f: S \times [0,1] \to T$ such that if ϑ is $U(0,1)$, then $f(s, \vartheta)$ has distribution $\mu(s, \cdot)$ for every $s \in S$.*

Proof: We may assume that T is a Borel subset of $[0,1]$, in which case we may easily reduce to the case when $T = [0,1]$. Define

$$f(s,t) = \sup\{x \in [0,1]; \ \mu(s, [0,x]) < t\}, \quad s \in S, \ t \in [0,1], \qquad (9)$$

and note that f is product measurable on $S \times [0,1]$, since the set $\{(s,t); \mu(s,[0,x]) < t\}$ is measurable for each x by Lemma 1.12, and the supremum in (9) can be restricted to rational x. If ϑ is $U(0,1)$, we get

$$P\{f(s,\vartheta) \le x\} = P\{\vartheta \le \mu(s,[0,x])\} = \mu(s,[0,x]), \quad x \in [0,1],$$

and so $f(s,\vartheta)$ has distribution $\mu(s,\cdot)$ by Lemma 3.3. □

Proof of Theorem 3.19: By Lemma 3.22 there exist some measurable functions $f_n : [0,1] \to S_n$ such that $\lambda \circ f_n^{-1} = \mu_n$. Letting ϑ be the identity mapping on $[0,1]$ and choosing $\vartheta_1, \vartheta_2, \ldots$ as in Lemma 3.21, we note that the functions $\xi_n = f_n(\vartheta_n)$, $n \in \mathbb{N}$, have the desired joint distribution. □

Next we consider the regularization and sample path properties of random processes. Say that two processes X and Y on the same index set T are *versions* of each other if $X_t = Y_t$ a.s. for each $t \in T$. In the special case when $T = \mathbb{R}^d$ or $\mathbb{R}_+$, we note that two continuous or right-continuous versions X and Y of the same process are *indistinguishable*, in the sense that $X \equiv Y$ a.s. In general, the latter notion is clearly stronger.

For any function f between two metric spaces (S, ρ) and (S', ρ'), the associated *modulus of continuity* $w_f = w(f,\cdot)$ is given by

$$w_f(r) = \sup\{\rho'(f_s, f_t);\ s,t \in S,\ \rho(s,t) \le r\}, \quad r > 0.$$

Note that f is uniformly continuous iff $w_f(r) \to 0$ as $r \to 0$. Say that f is *Hölder continuous with exponent* c if $w_f(r) \lesssim r^c$ as $r \to 0$. The property is said to hold *locally* if it is true on every bounded set. (Here and in the sequel, the relation $f \lesssim g$ between positive functions means that $f \le cg$ for some constant $c < \infty$.)

A simple moment condition ensures the existence of a Hölder-continuous version of a given process on $\mathbb{R}^d$. Important applications are given in Theorems 13.5, 21.3, and 22.4, and a related tightness criterion appears in Corollary 16.9.

Theorem 3.23 *(moments and continuity, Kolmogorov, Loève, Chentsov) Let X be a process on $\mathbb{R}^d$ with values in a complete metric space (S, ρ), and assume for some $a, b > 0$ that*

$$E\{\rho(X_s, X_t)\}^a \lesssim |s-t|^{d+b}, \quad s,t \in \mathbb{R}^d. \tag{10}$$

Then X has a continuous version, and the latter is a.s. locally Hölder continuous with exponent c for any $c \in (0, b/a)$.

Proof: It is clearly enough to consider the restriction of X to $[0,1]^d$. Define

$$D_n = \{(k_1, \ldots, k_d) 2^{-n};\ k_1, \ldots, k_n \in \{1, \ldots, 2^n\}\}, \quad n \in \mathbb{N},$$

and let

$$\xi_n = \max\{\rho(X_s, X_t);\ s, t \in D_n,\ |s-t| = 2^{-n}\}, \quad n \in \mathbb{N}.$$

Since

$$|\{(s,t) \in D_n^2;\ |s-t| = 2^{-n}\}| \le d 2^{dn}, \quad n \in \mathbb{N},$$

we get by (10), for any $c \in (0, b/a)$,

$$E\sum_n (2^{cn}\xi_n)^a = \sum_n 2^{acn} E\xi_n^a \lesssim \sum_n 2^{acn} 2^{dn}(2^{-n})^{d+b} = \sum_n 2^{(ac-b)n} < \infty.$$

The sum on the left is then a.s. convergent, and therefore $\xi_n \lesssim 2^{-cn}$ a.s. Now any two points $s, t \in \bigcup_n D_n$ with $|s-t| \leq 2^{-m}$ can be connected by a piecewise linear path involving, for each $n \geq m$, at most $2d$ steps between nearest neighbors in D_n. Thus, for $r \in [2^{-m-1}, 2^{-m}]$,

$$\sup\left\{\rho(X_s, X_t);\ s, t \in \bigcup_n D_n,\ |s-t| \leq r\right\}$$
$$\lesssim \sum_{n \geq m} \xi_n \lesssim \sum_{n \geq m} 2^{-cn} \leq 2^{-cm} \leq r^c,$$

which shows that X is a.s. Hölder continuous on $\bigcup_n D_n$ with exponent c.

In particular, there exists a continuous process Y on $[0,1]^d$ that agrees with X a.s. on $\bigcup_n D_n$, and it is easily seen that the Hölder continuity of Y on $\bigcup_n D_n$ extends with the same exponent c to the entire cube $[0,1]^d$. To show that Y is a version of X, fix any $t \in [0,1]^d$ and choose $t_1, t_2, \cdots \in \bigcup_n D_n$ with $t_n \to t$. Then $X_{t_n} = Y_{t_n}$ a.s. for each n. Furthermore, $X_{t_n} \xrightarrow{P} X_t$ by (10) and $Y_{t_n} \to Y_t$ a.s. by continuity, so $X_t = Y_t$ a.s. □

The next result shows how regularity of the paths may sometimes be established by comparison with a regular process.

Lemma 3.24 (*transfer of regularity*) *Let $X \stackrel{d}{=} Y$ be random processes on some index set T, taking values in a separable metric space S, and assume that the paths of Y lie in a set $U \subset S^T$ that is Borel for the σ-field $\mathcal{U} = (\mathcal{B}(S))^T \cap U$. Then X has a version with paths in U.*

Proof: For clarity we may write $\tilde{Y}$ for the path of Y, regarded as a random element in U. Then $\tilde{Y}$ is Y-measurable, and by Lemma 1.13 there exists a measurable mapping $f: S^T \to U$ such that $\tilde{Y} = f(Y)$ a.s. Define $\tilde{X} = f(X)$, and note that $(\tilde{X}, X) \stackrel{d}{=} (\tilde{Y}, Y)$. Since the diagonal in S^2 is measurable, we get in particular

$$P\{\tilde{X}_t = X_t\} = P\{\tilde{Y}_t = Y_t\} = 1, \quad t \in T. \qquad \Box$$

We conclude this chapter with a characterization of distribution functions in $\mathbb{R}^d$, required in Chapter 5. For any vectors $x = (x_1, \ldots, x_d)$ and $y = (y_1, \ldots, y_d)$, write $x \leq y$ for the componentwise inequality $x_k \leq y_k$, $k = 1, \ldots, d$, and similarly for $x < y$. In particular, the distribution function F of a probability measure μ on $\mathbb{R}^d$ is given by $F(x) = \mu\{y;\ y \leq x\}$. Similarly, let $x \vee y$ denote the componentwise maximum. Put $\mathbf{1} = (1, \ldots, 1)$ and $\infty = (\infty, \ldots, \infty)$.

For any rectangular box $(x, y] = \{u;\ x < u \leq y\} = (x_1, y_1] \times \cdots \times (x_d, y_d]$, we note that $\mu(x, y] = \sum_u s(u) F(u)$ where $s(u) = (-1)^p$ with $p = \sum_k 1\{u_k = y_k\}$ and the summation extends over all corners u of $(x, y]$. Let $F(x, y]$ denote the stated sum and say that F has *nonnegative increments* if

$F(x, y] \geq 0$ for all pairs $x < y$. Let us further say that F is *right-continuous* if $F(x_n) \to F(x)$ as $x_n \downarrow x$ and *proper* if $F(x) \to 1$ or 0 as $\min_k x_k \to \pm\infty$, respectively.

The following result characterizes distribution functions in terms of the mentioned properties.

Theorem 3.25 *(distribution functions)* *A function $F: \mathbb{R}^d \to [0, 1]$ is the distribution function of some probability measure μ on $\mathbb{R}^d$ iff it is right-continuous and proper with nonnegative increments.*

Proof: Assume that F has the stated properties, and note that the associated set function $F(x, y]$ is finitely additive. Since F is proper, we further have $F(x, y] \to 1$ as $x \to -\infty$ and $y \to \infty$, that is, as $(x, y] \uparrow (-\infty, \infty) = \mathbb{R}^d$. Hence, for every $n \in \mathbb{N}$ there exists a probability measure μ_n on $(2^{-n}\mathbb{Z})^d$ with $\mathbb{Z} = \{\ldots, -1, 0, 1, \ldots\}$ such that
$$\mu_n\{2^{-n}k\} = F(2^{-n}(k-1), 2^{-n}k], \quad k \in \mathbb{Z}^d, \ n \in \mathbb{N},$$
and from the finite additivity of $F(x, y]$ we obtain
$$\mu_m(2^{-m}(k-1, k]) = \mu_n(2^{-m}(k-1, k]), \quad k \in \mathbb{Z}^d, \ m < n \text{ in } \mathbb{N}. \tag{11}$$

In view of (11), we may split the Lebesgue unit interval $([0, 1], \mathcal{B}[0, 1], \lambda)$ recursively to construct some random vectors $\xi_1, \xi_2, \ldots$ with distributions $\mu_1, \mu_2, \ldots$ such that $\xi_m - 2^{-m} < \xi_n \leq \xi_m$ for all $m < n$. In particular, $\xi_1 \geq \xi_2 \geq \cdots \geq \xi_1 - 1$, and so ξ_n converges pointwise to some random vector ξ. Define $\mu = \lambda \circ \xi^{-1}$.

To see that μ has distribution function F, we note that since F is proper,
$$\lambda\{\xi_n \leq 2^{-n}k\} = \mu_n(-\infty, 2^{-n}k] = F(2^{-n}k), \quad k \in \mathbb{Z}^d, \ n \in \mathbb{N}.$$
Since also $\xi_n \downarrow \xi$ a.s., Fatou's lemma yields for dyadic $x \in \mathbb{R}^d$
$$\begin{aligned}\lambda\{\xi < x\} &= \lambda\{\xi_n < x \text{ ult.}\} \leq \liminf_n \lambda\{\xi_n < x\} \\ &\leq F(x) = \limsup_n \lambda\{\xi_n \leq x\} \\ &\leq \lambda\{\xi_n \leq x \text{ i.o.}\} \leq \lambda\{\xi \leq x\},\end{aligned}$$
and so
$$F(x) \leq \lambda\{\xi \leq x\} \leq F(x + 2^{-n}1), \quad n \in \mathbb{N}.$$
Letting $n \to \infty$ and using the right-continuity of F, we get $\lambda\{\xi \leq x\} = F(x)$, which extends to any $x \in \mathbb{R}^d$ by the right-continuity of both sides. □

The last result has the following version for unbounded measures.

Corollary 3.26 *(unbounded measures)* *Let the function F on $\mathbb{R}^d$ be right-continuous with nonnegative increments. Then there exists a measure μ on $\mathbb{R}^d$ such that $\mu(x, y] = F(x, y]$ for all $x \leq y$ in $\mathbb{R}^d$.*

Proof: For any $a \in \mathbb{R}^d$, we may apply Theorem 3.25 to suitably normalized versions of the function $F_a(x) = F(a, a \vee x]$ to obtain a measure μ_a

on $[a, \infty)$ with $\mu_a(a, x] = F(a, x]$ for all $x > a$. Then clearly $\mu_a = \mu_b$ on $(a \vee b, \infty)$ for any a and b, and so the set function $\mu = \sup_a \mu_a$ is a measure with the required property. □

Exercises

1. Give an example of two processes X and Y with different distributions such that $X_t \stackrel{d}{=} Y_t$ for all t.

2. Let X and Y be $\{0, 1\}$-valued processes on some index set T. Show that $X \stackrel{d}{=} Y$ iff $P\{X_{t_1} + \cdots + X_{t_n} > 0\} = P\{Y_{t_1} + \cdots + Y_{t_n} > 0\}$ for all $n \in \mathbb{N}$ and $t_1, \ldots, t_n \in T$.

3. Let F be a right-continuous function of bounded variation and with $F(-\infty) = 0$. Show for any random variable ξ that $EF(\xi) = \int P\{\xi \geq t\} F(dt)$. (Hint: First take F to be the distribution function of some random variable $\eta \perp\!\!\!\perp \xi$, and use Lemma 3.11.)

4. Consider a random variable $\xi \in L^1$ and a strictly convex function f on $\mathbb{R}$. Show that $Ef(\xi) = f(E\xi)$ iff $\xi = E\xi$ a.s.

5. Assume that $\xi = \sum_j a_j \xi_j$ and $\eta = \sum_j b_j \eta_j$, where the sums converge in L^2. Show that $\mathrm{cov}(\xi, \eta) = \sum_{i,j} a_i b_j \mathrm{cov}(\xi_i, \eta_j)$, where the double series on the right is absolutely convergent.

6. Let the σ-fields $\mathcal{F}_{t,n}$, $t \in T$, $n \in \mathbb{N}$, be nondecreasing in n for each t and independent in t for each n. Show that the independence extends to the σ-fields $\mathcal{F}_t = \bigvee_n \mathcal{F}_{t,n}$.

7. For each $t \in T$, let $\xi^t, \xi_1^t, \xi_2^t, \ldots$ be random elements in some metric space S_t with $\xi_n^t \to \xi^t$ a.s., and assume for each $n \in \mathbb{N}$ that the random elements ξ_n^t are independent. Show that the independence extends to the limits ξ^t. (Hint: First show that $E \prod_{t \in S} f_t(\xi^t) = \prod_{t \in S} Ef_t(\xi^t)$ for any bounded, continuous functions f_t on S_t and for finite subsets $S \subset T$.)

8. Give an example of three events that are pairwise independent but not independent.

9. Give an example of two random variables that are uncorrelated but not independent.

10. Let $\xi_1, \xi_2, \ldots$ be i.i.d. random elements with distribution μ in some measurable space $(S, \mathcal{S})$. Fix a set $A \in \mathcal{S}$ with $\mu A > 0$, and put $\tau = \inf\{k; \xi_k \in A\}$. Show that ξ_τ has distribution $\mu[\cdot | A] = \mu(\cdot \cap A)/\mu A$.

11. Let $\xi_1, \xi_2, \ldots$ be independent random variables taking values in $[0, 1]$. Show that $E \prod_n \xi_n = \prod_n E\xi_n$. In particular, show that $P \bigcap_n A_n = \prod_n PA_n$ for any independent events $A_1, A_2, \ldots$.

12. Let $\xi_1, \xi_2, \ldots$ be arbitrary random variables. Show that there exist some constants $c_1, c_2, \cdots > 0$ such that the series $\sum_n c_n \xi_n$ converges a.s.

13. Let $\xi_1, \xi_2, \ldots$ be random variables with $\xi_n \to 0$ a.s. Show that there exists some measurable function $f > 0$ with $\sum_n f(\xi_n) < \infty$ a.s. Also show that the conclusion fails if we only assume L^1-convergence.

14. Give an example of events $A_1, A_2, \ldots$ such that $P\{A_n \text{ i.o.}\} = 0$ but $\sum_n P A_n = \infty$.

15. Extend Lemma 3.20 to a correspondence between $U(0,1)$ random variables ϑ and Bernoulli sequences $\xi_1, \xi_2, \ldots$ with rate $p \in (0,1)$.

16. Give an elementary proof of Theorem 3.25 for $d = 1$. (*Hint:* Define $\xi = F^{-1}(\vartheta)$, where ϑ is $U(0,1)$, and note that ξ has distribution function F.)

17. Let $\xi_1, \xi_2, \ldots$ be random variables such that $P\{\xi_n \neq 0 \text{ i.o.}\} = 1$. Show that there exist some constants $c_n \in \mathbb{R}$ such that $P\{|c_n \xi_n| > 1 \text{ i.o.}\} = 1$. (*Hint:* Note that $P\{\sum_{k \leq n} |\xi_k| > 0\} \to 1$.)

Chapter 4

Random Sequences, Series, and Averages

Convergence in probability and in L^p; uniform integrability and tightness; convergence in distribution; convergence of random series; strong laws of large numbers; Portmanteau theorem; continuous mapping and approximation; coupling and measurability

The first goal of this chapter is to introduce and compare the basic modes of convergence of random quantities. For random elements ξ and $\xi_1, \xi_2, \ldots$ in a metric or topological space S, the most commonly used notions are those of almost sure convergence, $\xi_n \to \xi$ a.s., and convergence in probability, $\xi_n \xrightarrow{P} \xi$, corresponding to the general notions of convergence a.e. and in measure, respectively. When $S = \mathbb{R}$, we have the additional concept of L^p-convergence, familiar from Chapter 1. Those three notions are used throughout this book. For a special purpose in Chapter 25, we shall also need the notion of weak L^1-convergence.

For our second main topic, we shall study the very different concept of convergence in distribution, $\xi_n \xrightarrow{d} \xi$, defined by the condition $Ef(\xi_n) \to Ef(\xi)$ for all bounded, continuous functions f on S. This is clearly equivalent to weak convergence of the associated distributions $\mu_n = \mathcal{L}(\xi_n)$ and $\mu = \mathcal{L}(\xi)$, written as $\mu_n \xrightarrow{w} \mu$ and defined by the condition $\mu_n f \to \mu f$ for every f as above. In this chapter we shall only establish the most basic results of weak convergence theory, such as the "Portmanteau" theorem, the continuous mapping and approximation theorems, and the Skorohod coupling. Our development of the general theory continues in Chapters 5 and 16, and further distributional limit theorems appear in Chapters 8, 9, 12, 14, 15, 19, and 23.

Our third main theme is to characterize the convergence of series $\sum_k \xi_k$ and averages $n^{-c} \sum_{k \leq n} \xi_k$, where $\xi_1, \xi_2, \ldots$ are independent random variables and c is a positive constant. The two problems are related by the elementary Kronecker lemma, and the main results are the basic three-series criterion and the strong law of large numbers. The former result is extended in Chapter 7 to the powerful martingale convergence theorem, whereas extensions and refinements of the latter result are proved in Chapters 10 and 14. The mentioned theorems are further related to certain weak convergence results presented in Chapters 5 and 15.

Before beginning our systematic study of the various notions of convergence, we consider a couple of elementary but useful inequalities.

Lemma 4.1 *(moments and tails, Bienaymé, Chebyshev, Paley and Zygmund)* Let ξ be an $\mathbb{R}_+$-valued random variable with $0 < E\xi < \infty$. Then

$$(1-r)_+^2 \frac{(E\xi)^2}{E\xi^2} \leq P\{\xi > rE\xi\} \leq \frac{1}{r}, \quad r > 0. \tag{1}$$

The second relation in (1) is often referred to as *Chebyshev's* or *Markov's inequality*. Assuming that $E\xi^2 < \infty$, we get in particular the well-known estimate

$$P\{|\xi - E\xi| > \varepsilon\} \leq \varepsilon^{-2}\mathrm{var}(\xi), \quad \varepsilon > 0.$$

Proof of Lemma 4.1: We may clearly assume that $E\xi = 1$. The upper bound then follows as we take expectations in the inequality $r1\{\xi > r\} \leq \xi$. To get the lower bound, we note that for any $r, t > 0$

$$t^2 1\{\xi > r\} \geq (\xi - r)(2t + r - \xi) = 2\xi(r+t) - r(2t+r) - \xi^2.$$

Taking expected values, we get for $r \in (0, 1)$

$$t^2 P\{\xi > r\} \geq 2(r+t) - r(2t+r) - E\xi^2 \geq 2t(1-r) - E\xi^2.$$

Now choose $t = E\xi^2/(1-r)$. □

For random elements ξ and $\xi_1, \xi_2, \ldots$ in a metric space (S, ρ), we say that ξ_n *converges in probability* to ξ (written as $\xi_n \xrightarrow{P} \xi$) if

$$\lim_{n\to\infty} P\{\rho(\xi_n, \xi) > \varepsilon\} = 0, \quad \varepsilon > 0.$$

By Chebyshev's inequality it is equivalent that $E[\rho(\xi_n, \xi) \wedge 1] \to 0$. This notion of convergence is related to the a.s. version as follows.

Lemma 4.2 *(subsequence criterion)* Let $\xi, \xi_1, \xi_2, \ldots$ be random elements in a metric space (S, ρ). Then $\xi_n \xrightarrow{P} \xi$ iff every subsequence $N' \subset \mathbb{N}$ has a further subsequence $N'' \subset N'$ such that $\xi_n \to \xi$ a.s. along N''. In particular, $\xi_n \to \xi$ a.s. implies $\xi_n \xrightarrow{P} \xi$.

This shows in particular that the notion of convergence in probability depends only on the topology and is independent of the metrization ρ.

Proof: Assume that $\xi_n \xrightarrow{P} \xi$, and fix an arbitrary subsequence $N' \subset \mathbb{N}$. We may then choose a further subsequence $N'' \subset N'$ such that

$$E \sum_{n \in N''} \{\rho(\xi_n, \xi) \wedge 1\} = \sum_{n \in N''} E[\rho(\xi_n, \xi) \wedge 1] < \infty,$$

where the equality holds by monotone convergence. The series on the left then converges a.s., which implies $\xi_n \to \xi$ a.s. along N''.

Now assume instead the stated condition. If $\xi_n \not\xrightarrow{P} \xi$, there exists some $\varepsilon > 0$ such that $E[\rho(\xi_n,\xi) \wedge 1] > \varepsilon$ along a subsequence $N' \subset \mathbb{N}$. By hypothesis, $\xi_n \to \xi$ a.s. along a further subsequence $N'' \subset N'$, and by dominated convergence we get $E[\rho(\xi_n,\xi) \wedge 1] \to 0$ along N'', a contradiction. □

For a first application, we shall see how convergence in probability is preserved by continuous mappings.

Lemma 4.3 *(continuous mapping) For any metric spaces S and T, let $\xi, \xi_1, \xi_2, \ldots$ be random elements in S with $\xi_n \xrightarrow{P} \xi$, and let the mapping $f: S \to T$ be measurable and a.s. continuous at ξ. Then $f(\xi_n) \xrightarrow{P} f(\xi)$.*

Proof: Fix any subsequence $N' \subset \mathbb{N}$. By Lemma 4.2 we have $\xi_n \to \xi$ a.s. along some further subsequence $N'' \subset N'$, and by continuity we get $f(\xi_n) \to f(\xi)$ a.s. along N''. Hence, $f(\xi_n) \xrightarrow{P} f(\xi)$ by Lemma 4.2. □

Now consider a sequence of metric spaces (S_k, ρ_k), and introduce the product space $S = \mathsf{X}_k S_k = S_1 \times S_2 \times \cdots$ endowed with the product topology, a convenient metrization of which is given by

$$\rho(x,y) = \sum_k 2^{-k}\{\rho_k(x_k, y_k) \wedge 1\}, \quad x, y \in \mathsf{X}_k S_k. \tag{2}$$

If each S_k is separable, then $\mathcal{B}(S) = \bigotimes_k \mathcal{B}(S_k)$ by Lemma 1.2, and so a random element in S is simply a sequence of random elements in S_k, $k \in \mathbb{N}$.

Lemma 4.4 *(random sequences) For any separable metric spaces $S_1, S_2, \ldots$, let $\xi = (\xi_1, \xi_2, \ldots)$ and $\xi^n = (\xi_1^n, \xi_2^n, \ldots)$, $n \in \mathbb{N}$, be random elements in $\mathsf{X}_k S_k$. Then $\xi^n \xrightarrow{P} \xi$ iff $\xi_k^n \xrightarrow{P} \xi_k$ in S_k for each k.*

Proof: With ρ as in (2), we get for each $n \in \mathbb{N}$

$$E[\rho(\xi^n, \xi) \wedge 1] = E\rho(\xi^n, \xi) = \sum_k 2^{-k} E[\rho_k(\xi_k^n, \xi_k) \wedge 1].$$

Thus, by dominated convergence $E[\rho(\xi^n, \xi) \wedge 1] \to 0$ iff $E[\rho_k(\xi_k^n, \xi_k) \wedge 1] \to 0$ for all k. □

Combining the last two lemmas, it is easy to see how convergence in probability is preserved by the basic arithmetic operations.

Corollary 4.5 *(elementary operations) Let $\xi, \xi_1, \xi_2, \ldots$ and $\eta, \eta_1, \eta_2, \ldots$ be random variables with $\xi_n \xrightarrow{P} \xi$ and $\eta_n \xrightarrow{P} \eta$. Then $a\xi_n + b\eta_n \xrightarrow{P} a\xi + b\eta$ for all $a, b \in \mathbb{R}$, and $\xi_n \eta_n \xrightarrow{P} \xi\eta$. Furthermore, $\xi_n/\eta_n \xrightarrow{P} \xi/\eta$ whenever a.s. $\eta \neq 0$ and $\eta_n \neq 0$ for all n.*

Proof: By Lemma 4.4 we have $(\xi_n, \eta_n) \xrightarrow{P} (\xi, \eta)$ in $\mathbb{R}^2$, so the results for linear combinations and products follow by Lemma 4.3. To prove the last assertion, we may apply Lemma 4.3 to the function $f: (x,y) \mapsto (x/y)1\{y \neq 0\}$, which is clearly a.s. continuous at (ξ, η). □

Let us next examine the associated completeness properties. For any random elements $\xi_1, \xi_2, \ldots$ in a metric space (S, ρ), we say that (ξ_n) is Cauchy (convergent) in probability if $\rho(\xi_m, \xi_n) \xrightarrow{P} 0$ as $m, n \to \infty$, in the sense that $E[\rho(\xi_m, \xi_n) \wedge 1] \to 0$.

Lemma 4.6 (completeness) *Let $\xi_1, \xi_2, \ldots$ be random elements in a complete metric space (S, ρ). Then (ξ_n) is Cauchy in probability or a.s. iff $\xi_n \xrightarrow{P} \xi$ or $\xi_n \to \xi$ a.s., respectively, for some random element ξ in S.*

Proof: The a.s. case is immediate from Lemma 1.10. Assuming $\xi_n \xrightarrow{P} \xi$, we get
$$E[\rho(\xi_m, \xi_n) \wedge 1] \leq E[\rho(\xi_m, \xi) \wedge 1] + E[\rho(\xi_n, \xi) \wedge 1] \to 0,$$
which means that (ξ_n) is Cauchy in probability.

Now assume instead the latter condition. Define
$$n_k = \inf\left\{n \geq k;\; \sup_{m \geq n} E[\rho(\xi_m, \xi_n) \wedge 1] \leq 2^{-k}\right\}, \quad k \in \mathbb{N}.$$

The n_k are finite and satisfy
$$E \sum_k \{\rho(\xi_{n_k}, \xi_{n_{k+1}}) \wedge 1\} \leq \sum_k 2^{-k} < \infty,$$
and so $\sum_k \rho(\xi_{n_k}, \xi_{n_{k+1}}) < \infty$ a.s. The sequence (ξ_{n_k}) is then a.s. Cauchy and converges a.s. toward some measurable limit ξ. To see that $\xi_n \xrightarrow{P} \xi$, write
$$E[\rho(\xi_m, \xi) \wedge 1] \leq E[\rho(\xi_m, \xi_{n_k}) \wedge 1] + E[\rho(\xi_{n_k}, \xi) \wedge 1],$$
and note that the right-hand side tends to zero as $m, k \to \infty$, by the Cauchy convergence of (ξ_n) and dominated convergence. □

Next consider any probability measures μ and $\mu_1, \mu_2, \ldots$ on some metric space (S, ρ) with Borel σ-field $\mathcal{S}$, and say that μ_n converges weakly to μ (written as $\mu_n \xrightarrow{w} \mu$) if $\mu_n f \to \mu f$ for every $f \in C_b(S)$, the class of bounded, continuous functions $f: S \to \mathbb{R}$. If ξ and $\xi_1, \xi_2, \ldots$ are random elements in S, we further say that ξ_n converges in distribution to ξ (written as $\xi_n \xrightarrow{d} \xi$) if $\mathcal{L}(\xi_n) \xrightarrow{w} \mathcal{L}(\xi)$, that is, if $Ef(\xi_n) \to Ef(\xi)$ for all $f \in C_b(S)$. Note that the latter mode of convergence depends only on the distributions and that ξ and the ξ_n need not even be defined on the same probability space. To motivate the definition, note that $x_n \to x$ in a metric space S iff $f(x_n) \to f(x)$ for all continuous functions $f: S \to \mathbb{R}$, and also that $\mathcal{L}(\xi)$ is determined by the integrals $Ef(\xi)$ for all $f \in C_b(S)$.

The following result gives a connection between convergence in probability and in distribution.

Lemma 4.7 *(convergence in probability and in distribution)* Let $\xi, \xi_1, \xi_2, \ldots$ be random elements in a metric space (S, ρ). Then $\xi_n \xrightarrow{P} \xi$ implies $\xi_n \xrightarrow{d} \xi$, and the two conditions are equivalent when ξ is a.s. constant.

Proof: Assume $\xi_n \xrightarrow{P} \xi$. For any $f \in C_b(S)$ we need to show that $Ef(\xi_n) \to Ef(\xi)$. If the convergence fails, we may choose some subsequence $N' \subset \mathbb{N}$ such that $\inf_{n \in N'} |Ef(\xi_n) - Ef(\xi)| > 0$. By Lemma 4.2 there exists a further subsequence $N'' \subset N'$ such that $\xi_n \to \xi$ a.s. along N''. By continuity and dominated convergence we get $Ef(\xi_n) \to Ef(\xi)$ along N'', a contradiction.

Conversely, assume that $\xi_n \xrightarrow{d} s \in S$. Since $\rho(x, s) \wedge 1$ is a bounded and continuous function of x, we get $E[\rho(\xi_n, s) \wedge 1] \to E[\rho(s, s) \wedge 1] = 0$, and so $\xi_n \xrightarrow{P} s$. □

A family of random vectors ξ_t, $t \in T$, in $\mathbb{R}^d$ is said to be *tight* if
$$\lim_{r \to \infty} \sup_{t \in T} P\{|\xi_t| > r\} = 0.$$

For sequences (ξ_n) the condition is clearly equivalent to
$$\lim_{r \to \infty} \limsup_{n \to \infty} P\{|\xi_n| > r\} = 0, \tag{3}$$
which is often easier to verify. Tightness plays an important role for the compactness methods developed in Chapters 5 and 16. For the moment we note only the following simple connection with weak convergence.

Lemma 4.8 *(weak convergence and tightness)* Let $\xi, \xi_1, \xi_2, \ldots$ be random vectors in $\mathbb{R}^d$ satisfying $\xi_n \xrightarrow{d} \xi$. Then (ξ_n) is tight.

Proof: Fix any $r > 0$, and define $f(x) = (1 - (r - |x|)_+)_+$. Then
$$\limsup_{n \to \infty} P\{|\xi_n| > r\} \le \lim_{n \to \infty} Ef(\xi_n) = Ef(\xi) \le P\{|\xi| > r - 1\}.$$

Here the right-hand side tends to 0 as $r \to \infty$, and (3) follows. □

We may further note the following simple relationship between tightness and convergence in probability.

Lemma 4.9 *(tightness and convergence in probability)* Let $\xi_1, \xi_2, \ldots$ be random vectors in $\mathbb{R}^d$. Then (ξ_n) is tight iff $c_n \xi_n \xrightarrow{P} 0$ for any constants $c_1, c_2, \ldots \ge 0$ with $c_n \to 0$.

Proof: Assume (ξ_n) to be tight, and let $c_n \to 0$. Fixing any $r, \varepsilon > 0$, and noting that $c_n r \le \varepsilon$ for all but finitely many $n \in \mathbb{N}$, we get
$$\limsup_{n \to \infty} P\{|c_n \xi_n| > \varepsilon\} \le \limsup_{n \to \infty} P\{|\xi_n| > r\}.$$

Here the right-hand side tends to 0 as $r \to \infty$, and so $P\{|c_n \xi_n| > \varepsilon\} \to 0$. Since ε was arbitrary, we get $c_n \xi_n \xrightarrow{P} 0$. If instead (ξ_n) is not tight, we may choose a subsequence $(n_k) \subset \mathbb{N}$ such that $\inf_k P\{|\xi_{n_k}| > k\} > 0$. Letting $c_n = \sup\{k^{-1}; n_k \geq n\}$, we note that $c_n \to 0$ and yet $P\{|c_{n_k}\xi_{n_k}| > 1\} \not\to 0$. Thus, the stated condition fails. □

We turn to a related notion for expected values. A family of random variables ξ_t, $t \in T$, is said to be *uniformly integrable* if

$$\lim_{r \to \infty} \sup_{t \in T} E[|\xi_t|; |\xi_t| > r] = 0. \tag{4}$$

For sequences (ξ_n) in L^1, this is clearly equivalent to

$$\lim_{r \to \infty} \limsup_{n \to \infty} E[|\xi_n|; |\xi_n| > r] = 0. \tag{5}$$

Condition (4) holds in particular if the ξ_t are L^p-*bounded* for some $p > 1$, in the sense that $\sup_t E|\xi_t|^p < \infty$. To see this, it suffices to write

$$E[|\xi_t|; |\xi_t| > r] \leq r^{-p+1} E|\xi_t|^p, \quad r, p > 0.$$

The next result gives a useful characterization of uniform integrability. For motivation we note that if ξ is an integrable random variable, then $E[|\xi|; A] \to 0$ as $PA \to 0$, by Lemma 4.2 and dominated convergence. The latter condition means that $\sup_{A \in \mathcal{A}, PA < \varepsilon} E[|\xi|; A] \to 0$ as $\varepsilon \to 0$.

Lemma 4.10 *(uniform integrability)* *The random variables ξ_t, $t \in T$, are uniformly integrable iff $\sup_t E|\xi_t| < \infty$ and*

$$\lim_{PA \to 0} \sup_{t \in T} E[|\xi_t|; A] = 0. \tag{6}$$

Proof: Assume the ξ_t to be uniformly integrable, and write

$$E[|\xi_t|; A] \leq rPA + E[|\xi_t|; |\xi_t| > r], \quad r > 0.$$

Here (6) follows as we let $PA \to 0$ and then $r \to \infty$. To get the boundedness in L^1, it suffices to take $A = \Omega$ and choose $r > 0$ large enough.

Conversely, let the ξ_t be L^1-bounded and satisfy (6). By Chebyshev's inequality we get as $r \to \infty$

$$\sup_t P\{|\xi_t| > r\} \leq r^{-1} \sup_t E|\xi_t| \to 0,$$

and so (4) follows from (6) with $A = \{|\xi_t| > r\}$. □

The relevance of uniform integrability for the convergence of moments is clear from the following result, which also contains a weak convergence version of Fatou's lemma.

Lemma 4.11 *(convergence of means)* *Let $\xi, \xi_1, \xi_2, \ldots$ be $\mathbb{R}_+$-valued random variables with $\xi_n \xrightarrow{d} \xi$. Then $E\xi \leq \liminf_n E\xi_n$, and we have $E\xi_n \to E\xi < \infty$ iff (5) holds.*

Proof: For any $r > 0$ the function $x \mapsto x \wedge r$ is bounded and continuous on $\mathbb{R}_+$. Thus,

$$\liminf_{n \to \infty} E\xi_n \geq \lim_{n \to \infty} E(\xi_n \wedge r) = E(\xi \wedge r),$$

and the first assertion follows as we let $r \to \infty$. Next assume (5), and note in particular that $E\xi \leq \liminf_n E\xi_n < \infty$. For any $r > 0$ we get

$$|E\xi_n - E\xi| \leq |E\xi_n - E(\xi_n \wedge r)| + |E(\xi_n \wedge r) - E(\xi \wedge r)|$$
$$+ |E(\xi \wedge r) - E\xi|.$$

Letting $n \to \infty$ and then $r \to \infty$, we obtain $E\xi_n \to E\xi$. Now assume instead that $E\xi_n \to E\xi < \infty$. Keeping $r > 0$ fixed, we get as $n \to \infty$

$$E[\xi_n; \xi_n > r] \leq E[\xi_n - \xi_n \wedge (r - \xi_n)_+] \to E[\xi - \xi \wedge (r - \xi)_+].$$

Since $x \wedge (r - x)_+ \uparrow x$ as $r \to \infty$, the right-hand side tends to zero by dominated convergence, and (5) follows. □

We may now examine the relationship between convergence in L^p and in probability.

Proposition 4.12 (L^p-convergence) *Fix any $p > 0$, and let $\xi, \xi_1, \xi_2, \cdots \in L^p$ with $\xi_n \xrightarrow{P} \xi$. Then these conditions are equivalent:*

(i) $\xi_n \to \xi$ in L^p;

(ii) $\|\xi_n\|_p \to \|\xi\|_p$;

(iii) *the variables $|\xi_n|^p$, $n \in \mathbb{N}$, are uniformly integrable.*

Conversely, (i) implies $\xi_n \xrightarrow{P} \xi$.

Proof: First assume that $\xi_n \to \xi$ in L^p. Then $\|\xi_n\|_p \to \|\xi\|_p$ by Lemma 1.29, and by Lemma 4.1 we have, for any $\varepsilon > 0$,

$$P\{|\xi_n - \xi| > \varepsilon\} = P\{|\xi_n - \xi|^p > \varepsilon^p\} \leq \varepsilon^{-p} \|\xi_n - \xi\|_p^p \to 0.$$

Thus, $\xi_n \xrightarrow{P} \xi$. For the remainder of the proof we may assume that $\xi_n \xrightarrow{P} \xi$. In particular, $|\xi_n|^p \xrightarrow{d} |\xi|^p$ by Lemmas 4.3 and 4.7, and so (ii) and (iii) are equivalent by Lemma 4.11. Next assume (ii). If (i) fails, there exists some subsequence $N' \subset \mathbb{N}$ with $\inf_{n \in N'} \|\xi_n - \xi\|_p > 0$. By Lemma 4.2 we may choose a further subsequence $N'' \subset N'$ such that $\xi_n \to \xi$ a.s. along N''. But then Lemma 1.32 yields $\|\xi_n - \xi\|_p \to 0$ along N'', a contradiction. Thus, (ii) implies (i), and so all three conditions are equivalent. □

We shall briefly consider yet another notion of convergence of random variables. Assuming $\xi, \xi_1, \cdots \in L^p$ for some $p \in [1, \infty)$, we say that $\xi_n \to \xi$ *weakly in L^p* if $E\xi_n \eta \to E\xi\eta$ for every $\eta \in L^q$, where $p^{-1} + q^{-1} = 1$. Taking $\eta = |\xi|^{p-1} \text{sgn}\,\xi$ gives $\|\eta\|_q = \|\xi\|_p^{p-1}$, and so by Hölder's inequality

$$\|\xi\|_p^p = E\xi\eta = \lim_{n\to\infty} E\xi_n\eta \leq \|\xi\|_p^{p-1} \liminf_{n\to\infty} \|\xi_n\|_p,$$

which shows that $\|\xi\|_p \leq \liminf_n \|\xi_n\|_p$.

Now recall the well-known fact that any L^2-bounded sequence has a subsequence that converges weakly in L^2. The following related criterion for weak compactness in L^1 will be needed in Chapter 25.

Lemma 4.13 *(weak L^1-compactness, Dunford)* *Every uniformly integrable sequence of random variables has a subsequence that converges weakly in L^1.*

Proof: Let (ξ_n) be uniformly integrable. Define $\xi_n^k = \xi_n 1\{|\xi_n| \le k\}$, and note that (ξ_n^k) is L^2-bounded in n for each k. By the compactness in L^2 and a diagonal argument, there exist a subsequence $N' \subset \mathbb{N}$ and some random variables $\eta_1, \eta_2, \ldots$ such that $\xi_n^k \to \eta_k$ holds weakly in L^2 and then also in L^1, as $n \to \infty$ along N' for fixed k.

Now $\|\eta_k - \eta_l\|_1 \le \liminf_n \|\xi_n^k - \xi_n^l\|_1$, and by uniform integrability the right-hand side tends to zero as $k, l \to \infty$. Thus, the sequence (η_k) is Cauchy in L^1, and so it converges in L^1 toward some ξ. By approximation it follows easily that $\xi_n \to \xi$ weakly in L^1 along N'. □

We now derive criteria for the convergence of random series, beginning with an important special case.

Proposition 4.14 *(series with positive terms)* *Let $\xi_1, \xi_2, \ldots$ be independent $\mathbb{R}_+$-valued random variables. Then $\sum_n \xi_n < \infty$ a.s. iff $\sum_n E[\xi_n \wedge 1] < \infty$.*

Proof: Assuming the stated condition, we get $E \sum_n (\xi_n \wedge 1) < \infty$ by Fubini's theorem, so $\sum_n (\xi_n \wedge 1) < \infty$ a.s. In particular, $\sum_n 1\{\xi_n > 1\} < \infty$ a.s., so the series $\sum_n (\xi_n \wedge 1)$ and $\sum_n \xi_n$ differ by at most finitely many terms, and we get $\sum_n \xi_n < \infty$ a.s.

Conversely, assume that $\sum_n \xi_n < \infty$ a.s. Then also $\sum_n (\xi_n \wedge 1) < \infty$ a.s., so we may assume that $\xi_n \le 1$ for all n. Noting that $1 - x \le e^{-x} \le 1 - ax$ for $x \in [0, 1]$ where $a = 1 - e^{-1}$, we get

$$0 < E \exp\left\{-\sum_n \xi_n\right\} = \prod_n E e^{-\xi_n}$$
$$\le \prod_n (1 - aE\xi_n) \le \prod_n e^{-aE\xi_n} = \exp\left\{-a\sum_n E\xi_n\right\},$$

and so $\sum_n E\xi_n < \infty$. □

To handle more general series, we need the following strengthened version of the Bienaymé–Chebyshev inequality. A further extension appears as Proposition 7.15.

Lemma 4.15 *(maximum inequality, Kolmogorov)* *Let $\xi_1, \xi_2, \ldots$ be independent random variables with mean zero, and put $S_n = \xi_1 + \cdots + \xi_n$. Then*

$$P\{\sup_n |S_n| > r\} \le r^{-2} \sum_n E\xi_n^2, \quad r > 0.$$

Proof: We may assume that $\sum_n E\xi_n^2 < \infty$. Writing $\tau = \inf\{n;\ |S_n| > r\}$ and noting that $S_k 1\{\tau = k\} \perp\!\!\!\perp (S_n - S_k)$ for $k \leq n$, we get

$$\begin{aligned}
\sum_{k \leq n} E\xi_k^2 &= ES_n^2 \geq \sum_{k \leq n} E[S_n^2;\ \tau = k] \\
&\geq \sum_{k \leq n} \{E[S_k^2;\ \tau = k] + 2E[S_k(S_n - S_k);\ \tau = k]\} \\
&= \sum_{k \leq n} E[S_k^2;\ \tau = k] \geq r^2 P\{\tau \leq n\}.
\end{aligned}$$

As $n \to \infty$, we obtain

$$\sum_k E\xi_k^2 \geq r^2 P\{\tau < \infty\} = r^2 P\{\sup_k |S_k| > r\}. \qquad \square$$

The last result leads easily to the following sufficient condition for the a.s. convergence of random series with independent terms. Conditions that are both necessary and sufficient are given in Theorem 4.18.

Lemma 4.16 *(variance criterion for series, Khinchin and Kolmogorov)* Let $\xi_1, \xi_2, \ldots$ be independent random variables with mean 0 and $\sum_n E\xi_n^2 < \infty$. Then $\sum_n \xi_n$ converges a.s.

Proof: Write $S_n = \xi_1 + \cdots + \xi_n$. By Lemma 4.15 we get for any $\varepsilon > 0$

$$P\{\sup_{k \geq n} |S_n - S_k| > \varepsilon\} \leq \varepsilon^{-2} \sum_{k \geq n} E\xi_k^2.$$

Hence, $\sup_{k \geq n} |S_n - S_k| \xrightarrow{P} 0$ as $n \to \infty$, and Lemma 4.2 yields $\sup_{k \geq n} |S_n - S_k| \to 0$ a.s. along a subsequence. Since the last supremum is nonincreasing in n, the a.s. convergence extends to the entire sequence, which means that (S_n) is a.s. Cauchy convergent. Thus, S_n converges a.s. by Lemma 4.6. $\square$

The next result gives the basic connection between series with positive and symmetric terms. By $\xi_n \xrightarrow{P} \infty$ we mean that $P\{\xi_n > r\} \to 1$ for every $r > 0$.

Theorem 4.17 *(positive and symmetric terms)* Let $\xi_1, \xi_2, \ldots$ be independent, symmetric random variables. Then these conditions are equivalent:

(i) $\sum_n \xi_n$ converges a.s.;

(ii) $\sum_n \xi_n^2 < \infty$ a.s.;

(iii) $\sum_n E(\xi_n^2 \wedge 1) < \infty$.

If the conditions fail, then $|\sum_{k \leq n} \xi_k| \xrightarrow{P} \infty$.

Proof: Conditions (ii) and (iii) are equivalent by Proposition 4.14. Next assume (iii), and conclude from Lemma 4.16 that $\sum_n \xi_n 1\{|\xi_n| \leq 1\}$ converges a.s. From (iii) and Fubini's theorem we note that also $\sum_n 1\{|\xi_n| > 1\} < \infty$ a.s. Hence, the series $\sum_n \xi_n 1\{|\xi_n| \leq 1\}$ and $\sum_n \xi_n$ differ by at most finitely many terms, and so even the latter series converges a.s. Thus, (iii) implies (i). To see that (i) implies (ii), assume instead that (ii) fails. Then $\sum_n \xi_n^2 = \infty$ a.s. by Kolmogorov's 0–1 law, and so $|S_n| \xrightarrow{P} \infty$ where

$S_n = \sum_{k\le n} \xi_k$. Since the latter condition implies $|S_n| \to \infty$ a.s. along some subsequence, we conclude that even (i) fails. This shows that (i)–(iii) are are equivalent.

To prove the final assertion, we introduce an independent sequence of i.i.d. random variables ϑ_n with $P\{\vartheta_n = \pm 1\} = \frac{1}{2}$, and note that the sequences (ξ_n) and $(\vartheta_n|\xi_n|)$ have the same distribution. Letting μ denote the distribution of the sequence $(|\xi_n|)$, we get by Lemma 3.11

$$P\{|S_n| > r\} = \int P\left\{\left|\sum_{k\le n} \vartheta_k x_k\right| > r\right\} \mu(dx), \quad r > 0,$$

and by dominated convergence it is enough to show that the integrand on the right tends to 0 for μ-almost every $x = (x_1, x_2, \dots)$. Since $\sum_n x_n^2 = \infty$ a.e., this reduces the argument to the case of nonrandom $|\xi_n| = c_n$, $n \in \mathbb{N}$.

First assume that the c_n are unbounded. For any $r > 0$ we may recursively construct a subsequence $(n_k) \subset \mathbb{N}$ such that $c_{n_1} > r$ and $c_{n_k} > 4\sum_{j<k} c_{n_j}$ for each k. Then clearly $P\{\sum_{j\le k} \xi_{n_j} \in I\} \le 2^{-k}$ for every interval I of length $2r$. By convolution we get $P\{|S_n| \le r\} \le 2^{-k}$ for all $n \ge n_k$, which implies $P\{|S_n| \le r\} \to 0$.

Next assume that $c_n \le c < \infty$ for all n. Choosing $a > 0$ so small that $\cos x \le e^{-ax^2}$ for $|x| \le 1$, we get for $0 < |t| \le c^{-1}$

$$0 \le Ee^{itS_n} = \prod_{k\le n} \cos(tc_k) \le \prod_{k\le n} \exp(-at^2 c_k^2) = \exp\left\{-at^2 \sum_{k\le n} c_k^2\right\} \to 0.$$

Anticipating the elementary Lemma 5.1 of the next chapter, we again get $P\{|S_n| \le r\} \to 0$ for each $r > 0$. □

The problem of characterizing the convergence, a.s. or in distribution, of a series of independent random variables is solved completely by the following result. Here we write $\text{var}[\xi; A] = \text{var}(\xi 1_A)$.

Theorem 4.18 *(three-series criterion, Kolmogorov, Lévy)* Let $\xi_1, \xi_2, \dots$ *be independent random variables. Then $\sum_n \xi_n$ converges a.s. iff it converges in distribution, and also iff these conditions are fulfilled:*

(i) $\sum_n P\{|\xi_n| > 1\} < \infty$;
(ii) $\sum_n E[\xi_n; |\xi_n| \le 1]$ *converges;*
(iii) $\sum_n \text{var}[\xi_n; |\xi_n| \le 1] < \infty$.

For the proof we need the following simple *symmetrization inequalities*. Say that m is a *median* of the random variable ξ if $P\{\xi > m\} \vee P\{\xi < m\} \le \frac{1}{2}$. A *symmetrization* of ξ is defined as a random variable of the form $\tilde{\xi} = \xi - \xi'$ with $\xi' \perp\!\!\!\perp \xi$ and $\xi' \stackrel{d}{=} \xi$. For symmetrized versions of the random variables $\xi_1, \xi_2, \dots$, we require the same properties for the whole sequences (ξ_n) and (ξ'_n).

Lemma 4.19 *(symmetrization) Let $\tilde{\xi}$ be a symmetrization of a random variable ξ with median m. Then*

$$\tfrac{1}{2}P\{|\xi - m| > r\} \leq P\{|\tilde{\xi}| > r\} \leq 2P\{|\xi| > r/2\}, \quad r \geq 0.$$

Proof: Assume $\tilde{\xi} = \xi - \xi'$ as above, and write

$$\{\xi - m > r,\ \xi' \leq m\} \cup \{\xi - m < -r,\ \xi' \geq m\}$$
$$\subset \{|\tilde{\xi}| > r\} \subset \{|\xi| > r/2\} \cup \{|\xi'| > r/2\}. \quad \Box$$

We also need a simple centering lemma.

Lemma 4.20 *(centering) Let the random variables $\xi_1, \xi_2, \ldots$ and constants $c_1, c_2, \ldots$ be such that both ξ_n and $\xi_n + c_n$ converge in distribution. Then even c_n converges.*

Proof: Assume that $\xi_n \xrightarrow{d} \xi$. If $c_n \to \pm\infty$ along some subsequence $N' \subset \mathbb{N}$, then clearly $\xi_n + c_n \xrightarrow{P} \pm\infty$ along N', which contradicts the tightness of $\xi_n + c_n$. Thus, the c_n are bounded. Now assume that $c_n \to a$ and $c_n \to b$ along two subsequences $N_1, N_2 \subset \mathbb{N}$. Then $\xi_n + c_n \xrightarrow{d} \xi + a$ along N_1 and $\xi_n + c_n \xrightarrow{d} \xi + b$ along N_2, so $\xi + a \stackrel{d}{=} \xi + b$. Iterating this relation, we get $\xi + n(b-a) \stackrel{d}{=} \xi$ for arbitrary $n \in \mathbb{Z}$, which is impossible unless $a = b$. Thus, all limit points of (c_n) agree, and c_n converges. $\quad \Box$

Proof of Theorem 4.18: Assume conditions (i) through (iii), and define $\xi'_n = \xi_n 1\{|\xi_n| \leq 1\}$. By (iii) and Lemma 4.16 the series $\sum_n (\xi'_n - E\xi'_n)$ converges a.s., so by (ii) the same thing is true for $\sum_n \xi'_n$. Finally, $P\{\xi_n \neq \xi'_n \text{ i.o.}\} = 0$ by (i) and the Borel–Cantelli lemma, so $\sum_n (\xi_n - \xi'_n)$ has a.s. finitely many nonzero terms. Hence, even $\sum_n \xi_n$ converges a.s.

Conversely, assume that $\sum_n \xi_n$ converges in distribution. Then Lemma 4.19 shows that the sequence of symmetrized partial sums $\sum_{k \leq n} \tilde{\xi}_k$ is tight, and so $\sum_n \tilde{\xi}_n$ converges a.s. by Theorem 4.17. In particular, $\tilde{\xi}_n \to 0$ a.s. For any $\varepsilon > 0$ we obtain $\sum_n P\{|\tilde{\xi}_n| > \varepsilon\} < \infty$ by the Borel–Cantelli lemma. Hence, $\sum_n P\{|\xi_n - m_n| > \varepsilon\} < \infty$ by Lemma 4.19, where $m_1, m_2, \ldots$ are medians of $\xi_1, \xi_2, \ldots$. Using the Borel–Cantelli lemma again, we get $\xi_n - m_n \to 0$ a.s.

Now let $c_1, c_2, \ldots$ be arbitrary with $m_n - c_n \to 0$. Then even $\xi_n - c_n \to 0$ a.s. Putting $\eta_n = \xi_n 1\{|\xi_n - c_n| \leq 1\}$, we get a.s. $\xi_n = \eta_n$ for all but finitely many n, and similarly for the symmetrized variables $\tilde{\xi}_n$ and $\tilde{\eta}_n$. Thus, even $\sum_n \tilde{\eta}_n$ converges a.s. Since the $\tilde{\eta}_n$ are bounded and symmetric, Theorem 4.17 yields $\sum_n \text{var}(\eta_n) = \tfrac{1}{2}\sum_n \text{var}(\tilde{\eta}_n) < \infty$. Thus, $\sum_n (\eta_n - E\eta_n)$ converges a.s. by Lemma 4.16, as does the series $\sum_n (\xi_n - E\eta_n)$. Comparing with the distributional convergence of $\sum_n \xi_n$, we conclude from Lemma 4.20 that $\sum_n E\eta_n$ converges. In particular, $E\eta_n \to 0$ and $\eta_n - E\eta_n \to 0$ a.s., so $\eta_n \to 0$ a.s., and then also $\xi_n \to 0$ a.s. Hence, $m_n \to 0$, so we may take $c_n = 0$ in the previous argument, and conditions (i)–(iii) follow. $\quad \Box$

4. Random Sequences, Series, and Averages

A sequence of random variables $\xi_1, \xi_2, \ldots$ with partial sums S_n is said to obey the *strong law of large numbers* if S_n/n converges a.s. to a constant. The *weak law* is defined by the corresponding condition with convergence in probability. The following elementary proposition enables us to convert convergence results for random series into laws of large numbers.

Lemma 4.21 *(series and averages, Kronecker)* If $\sum_n n^{-c} a_n$ converges for some $a_1, a_2, \ldots \in \mathbb{R}$ and $c > 0$, then $n^{-c} \sum_{k \le n} a_k \to 0$.

Proof: Put $b_n = n^{-c} a_n$, and assume that $\sum_n b_n = b$. By dominated convergence as $n \to \infty$,

$$\sum_{k \le n} b_k - n^{-c} \sum_{k \le n} a_k = \sum_{k \le n}(1 - (k/n)^c) b_k = c \sum_{k \le n} b_k \int_{k/n}^1 x^{c-1} dx$$

$$= c \int_0^1 x^{c-1} dx \sum_{k \le nx} b_k \to bc \int_0^1 x^{c-1} dx = b,$$

and the assertion follows since the first term on the left tends to b. □

The following simple result illustrates the method.

Corollary 4.22 *(variance criterion for averages, Kolmogorov)* Let $\xi_1, \xi_2, \ldots$ be independent random variables with zero mean such that $\sum_n n^{-2c} E\xi_n^2 < \infty$ for some $c > 0$. Then $n^{-c} \sum_{k \le n} \xi_k \to 0$ a.s.

Proof: The series $\sum_n n^{-c} \xi_n$ converges a.s. by Lemma 4.16, and the assertion follows by Lemma 4.21. □

In particular, we note that if $\xi, \xi_1, \xi_2, \ldots$ are i.i.d. with $E\xi = 0$ and $E\xi^2 < \infty$, then $n^{-c} \sum_{k \le n} \xi_k \to 0$ a.s. for any $c > \frac{1}{2}$. The statement fails for $c = \frac{1}{2}$, as may be seen by taking ξ to be $N(0,1)$. The best possible normalization is given in Corollary 14.8. The next result characterizes the stated convergence for arbitrary $c > \frac{1}{2}$. For $c = 1$ we recognize the strong law of large numbers. Corresponding criteria for the weak law are given in Theorem 5.16.

Theorem 4.23 *(strong laws of large numbers, Kolmogorov, Marcinkiewicz and Zygmund)* Let $\xi, \xi_1, \xi_2, \ldots$ be i.i.d. random variables, and fix any $p \in (0,2)$. Then $n^{-1/p} \sum_{k \le n} \xi_k$ converges a.s. iff $E|\xi|^p < \infty$ and either $p \le 1$ or $E\xi = 0$. In that case the limit equals $E\xi$ for $p = 1$ and is otherwise 0.

Proof: Assume that $E|\xi|^p < \infty$ and also, for $p \ge 1$, that $E\xi = 0$. Define $\xi'_n = \xi_n 1\{|\xi_n| \le n^{1/p}\}$, and note that by Lemma 3.4

$$\sum_n P\{\xi'_n \ne \xi_n\} = \sum_n P\{|\xi|^p > n\} \le \int_0^\infty P\{|\xi|^p > t\} dt = E|\xi|^p < \infty.$$

By the Borel–Cantelli lemma we get $P\{\xi'_n \ne \xi_n \text{ i.o.}\} = 0$, and so $\xi'_n = \xi_n$ for all but finitely many $n \in \mathbb{N}$ a.s. It is then equivalent to show that

$n^{-1/p}\sum_{k\leq n}\xi'_k \to 0$ a.s. By Lemma 4.21 it suffices to prove instead that $\sum_n n^{-1/p}\xi'_n$ converges a.s.

For $p < 1$, this is clear if we write

$$E\sum_n n^{-1/p}|\xi'_n| = \sum_n n^{-1/p} E[|\xi|;\ |\xi| \leq n^{1/p}]$$
$$\lesssim \int_0^\infty t^{-1/p} E[|\xi|;\ |\xi| \leq t^{1/p}]dt$$
$$= E\Big[|\xi|\int_{|\xi|^p}^\infty t^{-1/p}dt\Big] \leq E|\xi|^p < \infty.$$

If instead $p > 1$, it suffices by Theorem 4.18 to prove that $\sum_n n^{-1/p}E\xi'_n$ converges and $\sum_n n^{-2/p}\mathrm{var}(\xi'_n) < \infty$. Since $E\xi'_n = -E[\xi;\ |\xi| > n^{1/p}]$, we have for the former series

$$\sum_n n^{-1/p}|E\xi'_n| \leq \sum_n n^{-1/p} E[|\xi|;\ |\xi| > n^{1/p}]$$
$$\lesssim \int_0^\infty t^{-1/p} E[|\xi|;\ |\xi| > t^{1/p}]dt$$
$$= E\Big[|\xi|\int_0^{|\xi|^p} t^{-1/p}dt\Big] \leq E|\xi|^p < \infty.$$

As for the latter series, we get

$$\sum_n n^{-2/p}\mathrm{var}(\xi'_n) \leq \sum_n n^{-2/p} E(\xi'_n)^2$$
$$= \sum_n n^{-2/p} E[\xi^2;\ |\xi| \leq n^{1/p}]$$
$$\lesssim \int_0^\infty t^{-2/p} E[\xi^2;\ |\xi| \leq t^{1/p}]dt$$
$$= E\Big[\xi^2\int_{|\xi|^p}^\infty t^{-2/p}dt\Big] \leq E|\xi|^p < \infty.$$

If $p = 1$, then $E\xi'_n = E[\xi;\ |\xi| \leq n] \to 0$ by dominated convergence. Thus, $n^{-1}\sum_{k\leq n} E\xi'_k \to 0$, and we may prove instead that $n^{-1}\sum_{k\leq n}\xi''_k \to 0$ a.s. where $\xi''_n = \xi'_n - E\xi'_n$. By Lemma 4.21 and Theorem 4.18 it is then enough to show that $\sum_n n^{-2}\mathrm{var}(\xi'_n) < \infty$, which may be seen as before.

Conversely, assume that $n^{-1/p}S_n = n^{-1/p}\sum_{k\leq n}\xi_k$ converges a.s. Then

$$\frac{\xi_n}{n^{1/p}} = \frac{S_n}{n^{1/p}} - \Big(\frac{n-1}{n}\Big)^{1/p}\frac{S_{n-1}}{(n-1)^{1/p}} \to 0 \text{ a.s.,}$$

and in particular $P\{|\xi_n|^p > n \text{ i.o.}\} = 0$. Hence, by Lemma 3.4 and the Borel–Cantelli lemma,

$$E|\xi|^p = \int_0^\infty P\{|\xi|^p > t\}dt \leq 1 + \sum_{n\geq 1} P\{|\xi|^p > n\} < \infty.$$

For $p > 1$, the direct assertion yields $n^{-1/p}(S_n - nE\xi) \to 0$ a.s., and so $n^{1-1/p}E\xi$ converges, which implies $E\xi = 0$. □

For a simple application of the law of large numbers, consider an arbitrary sequence of random variables $\xi_1, \xi_2, \ldots$, and define the associated *empirical distributions* as the random probability measures $\hat{\mu}_n = n^{-1} \sum_{k \leq n} \delta_{\xi_k}$. The corresponding *empirical distribution functions* $\hat{F}_n$ are given by

$$\hat{F}_n(x) = \hat{\mu}_n(-\infty, x] = n^{-1} \sum_{k \leq n} 1\{\xi_k \leq x\}, \quad x \in \mathbb{R}, \ n \in \mathbb{N}.$$

Proposition 4.24 *(empirical distribution functions, Glivenko, Cantelli)* Let $\xi_1, \xi_2, \ldots$ be i.i.d. random variables with distribution function F and empirical distribution functions $\hat{F}_1, \hat{F}_2, \ldots$. Then

$$\lim_{n \to \infty} \sup_x |\hat{F}_n(x) - F(x)| = 0 \quad \text{a.s.} \tag{7}$$

Proof: By the law of large numbers we have $\hat{F}_n(x) \to F(x)$ a.s. for every $x \in \mathbb{R}$. Now fix a finite partition $-\infty = x_1 < x_2 < \cdots < x_m = \infty$. By the monotonicity of F and $\hat{F}_n$

$$\sup_x |\hat{F}_n(x) - F(x)| \leq \max_k |\hat{F}_n(x_k) - F(x_k)| + \max_k |F(x_{k+1}) - F(x_k)|.$$

Letting $n \to \infty$ and refining the partition indefinitely, we get in the limit

$$\limsup_{n \to \infty} \sup_x |\hat{F}_n(x) - F(x)| \leq \sup_x \Delta F(x) \quad \text{a.s.,}$$

which proves (7) when F is continuous.

For general F, let $\vartheta_1, \vartheta_2, \ldots$ be i.i.d. $U(0,1)$, and define $\eta_n = g(\vartheta_n)$ for each n, where $g(t) = \sup\{x; F(x) < t\}$. Then $\eta_n \leq x$ iff $\vartheta_n \leq F(x)$, and so $(\eta_n) \stackrel{d}{=} (\xi_n)$. We may then assume that $\xi_n \equiv \eta_n$. Writing $\hat{G}_1, \hat{G}_2, \ldots$ for the empirical distribution functions of $\vartheta_1, \vartheta_2, \ldots$, we see that also $\hat{F}_n = \hat{G}_n \circ F$. Writing $A = F(\mathbb{R})$ and using the result for continuous F, we get a.s.

$$\sup_x |\hat{F}_n(x) - F(x)| = \sup_{t \in A} |\hat{G}_n(t) - t| \leq \sup_{t \in [0,1]} |\hat{G}_n(t) - t| \to 0. \quad \square$$

We turn to a systematic study of convergence in distribution. Although we are currently mostly interested in distributions on Euclidean spaces, it is crucial for future applications that we consider the more general setting of an abstract metric space. In particular, the theory is applied in Chapter 16 to random elements in various function spaces.

Theorem 4.25 *(Portmanteau theorem, Alexandrov)* For any random elements $\xi, \xi_1, \xi_2, \ldots$ in a metric space S, these conditions are equivalent:

(i) $\xi_n \stackrel{d}{\to} \xi$;
(ii) $\liminf_n P\{\xi_n \in G\} \geq P\{\xi \in G\}$ for any open set $G \subset S$;
(iii) $\limsup_n P\{\xi_n \in F\} \leq P\{\xi \in F\}$ for any closed set $F \subset S$;
(iv) $P\{\xi_n \in B\} \to P\{\xi \in B\}$ for any $B \in \mathcal{B}(S)$ with $\xi \notin \partial B$ a.s.

A set $B \in \mathcal{B}(S)$ with $\xi \notin \partial B$ a.s. is often called a ξ-*continuity set*.

Proof: Assume (i), and fix any open set $G \subset S$. Letting f be continuous with $0 \leq f \leq 1_G$, we get $Ef(\xi_n) \leq P\{\xi_n \in G\}$, and (ii) follows as we let

$n \to \infty$ and then $f \uparrow 1_G$. The equivalence between (ii) and (iii) is clear from taking complements. Now assume (ii) and (iii). For any $B \in \mathcal{B}(S)$,

$$P\{\xi \in B^\circ\} \leq \liminf_{n \to \infty} P\{\xi_n \in B\} \leq \limsup_{n \to \infty} P\{\xi_n \in B\} \leq P\{\xi \in \overline{B}\}.$$

Here the extreme members agree when $\xi \notin \partial B$ a.s., and (iv) follows.

Conversely, assume (iv) and fix any closed set $F \subset S$. Write $F^\varepsilon = \{s \in S;\ \rho(s, F) \leq \varepsilon\}$. Then the sets $\partial F^\varepsilon \subset \{s;\ \rho(s, F) = \varepsilon\}$ are disjoint, and so $\xi \notin \partial F^\varepsilon$ for almost every $\varepsilon > 0$. For such an ε we may write $P\{\xi_n \in F\} \leq P\{\xi \in F^\varepsilon\}$, and (iii) follows as we let $n \to \infty$ and then $\varepsilon \to 0$. Finally, assume (ii) and let $f \geq 0$ be continuous. By Lemma 3.4 and Fatou's lemma,

$$Ef(\xi) = \int_0^\infty P\{f(\xi) > t\}dt \leq \int_0^\infty \liminf_{n \to \infty} P\{f(\xi_n) > t\}dt$$

$$\leq \liminf_{n \to \infty} \int_0^\infty P\{f(\xi_n) > t\}dt = \liminf_{n \to \infty} Ef(\xi_n). \tag{8}$$

Now let f be continuous with $|f| \leq c < \infty$. Applying (8) to $c \pm f$ yields $Ef(\xi_n) \to Ef(\xi)$, which proves (i). □

For an easy application, we insert a simple lemma that is needed in Chapter 16.

Lemma 4.26 *(subspaces) For any metric space (S, ρ) with subspace $A \subset S$, let $\xi, \xi_1, \xi_2, \ldots$ be random elements in (A, ρ). Then $\xi_n \xrightarrow{d} \xi$ in (A, ρ) iff the same convergence holds in (S, ρ).*

Proof: Since $\xi, \xi_1, \xi_2, \cdots \in A$, condition (ii) of Theorem 4.25 is equivalent to

$$\liminf_{n \to \infty} P\{\xi_n \in A \cap G\} \geq P\{\xi \in A \cap G\}, \quad G \subset S \text{ open}.$$

By Lemma 1.6, this is precisely condition (ii) of Theorem 4.25 for the subspace A. □

It is clear directly from the definitions that convergence in distribution is preserved by continuous mappings. The following more general statement is a key result of weak convergence theory.

Theorem 4.27 *(continuous mapping, Mann and Wald, Prohorov, Rubin) For any metric spaces S and T, let $\xi, \xi_1, \xi_2, \ldots$ be random elements in S with $\xi_n \xrightarrow{d} \xi$, and consider some measurable mappings $f, f_1, f_2, \ldots : S \to T$ and a measurable set $C \subset S$ with $\xi \in C$ a.s. such that $f_n(s_n) \to f(s)$ as $s_n \to s \in C$. Then $f_n(\xi_n) \xrightarrow{d} f(\xi)$.*

In particular, we note that if $\xi_n \xrightarrow{d} \xi$ in S and $f: S \to T$ is a.s. continuous at ξ, then $f(\xi_n) \xrightarrow{d} f(\xi)$. This frequently used statement is commonly referred to as the *continuous mapping theorem*.

Proof: Fix any open set $G \subset T$, and let $s \in f^{-1}G \cap C$. By hypothesis there exist an integer $m \in \mathbb{N}$ and some neighborhood N of s such that $f_k(s') \in G$ for all $k \geq m$ and $s' \in N$. Thus, $N \subset \bigcap_{k \geq m} f_k^{-1} G$, and so

$$f^{-1}G \cap C \subset \bigcup_m \{\bigcap_{k \geq m} f_k^{-1} G\}^\circ.$$

Now let $\mu, \mu_1, \mu_2, \ldots$ denote the distributions of $\xi, \xi_2, \xi_2, \ldots$. By Theorem 4.25 we get

$$\mu(f^{-1}G) \leq \mu \bigcup_m \{\bigcap_{k \geq m} f_k^{-1} G\}^\circ = \sup_m \mu \{\bigcap_{k \geq m} f_k^{-1} G\}^\circ$$
$$\leq \sup_m \liminf_{n \to \infty} \mu_n \bigcap_{k \geq m} f_k^{-1} G \leq \liminf_{n \to \infty} \mu_n(f_n^{-1} G).$$

Using the same theorem again gives $\mu_n \circ f_n^{-1} \xrightarrow{w} \mu \circ f^{-1}$, which means that $f_n(\xi_n) \xrightarrow{d} f(\xi)$. □

We will now prove an equally useful approximation theorem. Here the idea is to prove $\xi_n \xrightarrow{d} \xi$ by choosing approximations η_n of ξ_n and η of ξ such that $\eta_n \xrightarrow{d} \eta$. The desired convergence will follow if we can ensure that the approximation errors are uniformly small.

Theorem 4.28 *(approximation) Let $\xi, \xi_n, \eta^k,$ and η_n^k be random elements in a metric space (S, ρ) such that $\eta_n^k \xrightarrow{d} \eta^k$ as $n \to \infty$ for fixed k and also $\eta^k \xrightarrow{d} \xi$. Then $\xi_n \xrightarrow{d} \xi$ holds under the further condition*

$$\lim_{k \to \infty} \limsup_{n \to \infty} E[\rho(\eta_n^k, \xi_n) \wedge 1] = 0. \tag{9}$$

Proof: For any closed set $F \subset S$ and constant $\varepsilon > 0$ we have

$$P\{\xi_n \in F\} \leq P\{\eta_n^k \in F^\varepsilon\} + P\{\rho(\eta_n^k, \xi_n) > \varepsilon\},$$

where $F^\varepsilon = \{s \in S; \rho(s, F) \leq \varepsilon\}$. By Theorem 4.25 we get as $n \to \infty$

$$\limsup_{n \to \infty} P\{\xi_n \in F\} \leq P\{\eta^k \in F^\varepsilon\} + \limsup_{n \to \infty} P\{\rho(\eta_n^k, \xi_n) > \varepsilon\}.$$

Now let $k \to \infty$, and conclude from Theorem 4.25 together with (9) that

$$\limsup_{n \to \infty} P\{\xi_n \in F\} \leq P\{\xi \in F^\varepsilon\}.$$

As $\varepsilon \to 0$, the right-hand side tends to $P\{\xi \in F\}$. Since F was arbitrary, we get $\xi_n \xrightarrow{d} \xi$ by Theorem 4.25. □

Next we consider convergence in distribution on product spaces.

Theorem 4.29 *(random sequences)* *For any separable metric spaces $S_1, S_2, \ldots$, let $\xi = (\xi^1, \xi^2, \ldots)$ and $\xi_n = (\xi_n^1, \xi_n^2, \ldots)$, $n \in \mathbb{N}$, be random elements in $\mathsf{X}_k S_k$. Then $\xi_n \xrightarrow{d} \xi$ iff for any functions $f_k \in C_b(S_k)$,*

$$E[f_1(\xi_n^1) \cdots f_m(\xi_n^m)] \to E[f_1(\xi^1) \cdots f_m(\xi^m)], \quad m \in \mathbb{N}. \qquad (10)$$

In particular, we note that $\xi_n \xrightarrow{d} \xi$ follows from the finite-dimensional convergence

$$(\xi_n^1, \ldots, \xi_n^m) \xrightarrow{d} (\xi^1, \ldots, \xi^m), \quad m \in \mathbb{N}. \qquad (11)$$

If ξ and the ξ_n have independent components, it is even sufficient that $\xi_n^k \xrightarrow{d} \xi^k$ for every k.

Proof: The necessity of the condition is clear from the continuity of the projections $s \mapsto s_k$. To prove the sufficiency, we first assume that (10) holds for a fixed m. Writing $S_k' = \{B \in \mathcal{B}(S_k); \xi^k \notin \partial B \text{ a.s.}\}$ and applying Theorem 4.25 m times, we obtain

$$P\{(\xi_n^1, \ldots, \xi_n^m) \in B\} \to P\{(\xi^1, \ldots, \xi^m) \in B\}, \qquad (12)$$

for any set $B = B^1 \times \cdots \times B^m$ such that $B^k \in S_k'$ for all k. Since the S_k are separable, we may choose some countable bases $\mathcal{C}_k \subset S_k'$, and we note that $\mathcal{C}_1 \times \cdots \times \mathcal{C}_m$ is then a countable base in $S_1 \times \cdots \times S_m$. Hence, any open set $G \subset S_1 \times \cdots \times S_m$ can be written as a countable union of measurable rectangles $B_j = B_j^1 \times \cdots \times B_j^m$ with $B_j^k \in S_k'$ for all k. Since the S_k' are fields, we may easily reduce to the case when the sets B_j are disjoint. By Fatou's lemma and (12) we obtain

$$\begin{aligned}
\liminf_{n \to \infty} P\{(\xi_n^1, \ldots, \xi_n^m) \in G\} &= \liminf_{n \to \infty} \sum_j P\{(\xi_n^1, \ldots, \xi_n^m) \in B_j\} \\
&\geq \sum_j P\{(\xi^1, \ldots, \xi^m) \in B_j\} \\
&= P\{(\xi^1, \ldots, \xi^m) \in G\},
\end{aligned}$$

and so (11) holds by Theorem 4.25.

To see that (11) implies $\xi_n \xrightarrow{d} \xi$, fix any $a_k \in S_k$, $k \in \mathbb{N}$, and note that the mapping $(s_1, \ldots, s_m) \mapsto (s_1, \ldots, s_m, a_{m+1}, a_{m+2}, \ldots)$ is continuous on $S_1 \times \cdots \times S_m$ for each $m \in \mathbb{N}$. By (11) it follows that

$$(\xi_n^1, \ldots, \xi_n^m, a_{m+1}, \ldots) \xrightarrow{d} (\xi^1, \ldots, \xi^m, a_{m+1}, \ldots), \quad m \in \mathbb{N}. \qquad (13)$$

Writing η_n^m and η^m for the sequences in (13) and letting ρ be the metric in (2), we also note that $\rho(\xi, \eta^m) \leq 2^{-m}$ and $\rho(\xi_n, \eta_n^m) \leq 2^{-m}$ for all m and n. The convergence $\xi_n \xrightarrow{d} \xi$ now follows by Theorem 4.28. □

In discussions involving distributional convergence of a random sequence $\xi_1, \xi_2, \ldots$, the relationship between the elements ξ_n is often irrelevant. It is then natural to look for a more convenient representation, which may lead to simpler and more transparent proofs.

Theorem 4.30 *(coupling, Skorohod, Dudley) Let $\xi, \xi_1, \xi_2, \ldots$ be random elements in a separable metric space (S, ρ) such that $\xi_n \xrightarrow{d} \xi$. Then there exists a probability space with some random elements $\eta \stackrel{d}{=} \xi$ and $\eta_n \stackrel{d}{=} \xi_n$, $n \in \mathbb{N}$, such that $\eta_n \to \eta$ a.s.*

In the course of the proof, we need to introduce families of independent random elements with given distributions. The existence of such families is ensured, in general, by Corollary 6.18. When S is complete, we may instead rely on the more elementary Theorem 3.19.

Proof: First assume that $S = \{1, \ldots, m\}$, and put $p_k = P\{\xi = k\}$ and $p_k^n = P\{\xi_n = k\}$. Assuming ϑ to be $U(0,1)$ and independent of ξ, we may easily construct some random elements $\tilde{\xi}_n \stackrel{d}{=} \xi_n$ such that $\tilde{\xi}_n = k$ whenever $\xi = k$ and $\vartheta \leq p_k^n/p_k$. Since $p_k^n \to p_k$ for each k, we get $\tilde{\xi}_n \to \xi$ a.s.

For general S, fix any $p \in \mathbb{N}$, and choose a partition of S into ξ-continuity sets $B_1, B_2, \cdots \in \mathcal{B}(S)$ of diameter $< 2^{-p}$. Next choose m so large that $P\{\xi \notin \bigcup_{k \leq m} B_k\} < 2^{-p}$, and put $B_0 = \bigcap_{k \leq m} B_k^c$. For $k = 0, \ldots, m$, define $\kappa = k$ when $\xi \in B_k$ and $\kappa_n = k$ when $\xi_n \in B_k$, $n \in \mathbb{N}$. Then $\kappa_n \xrightarrow{d} \kappa$, and by the result for finite S we may choose some $\tilde{\kappa}_n \stackrel{d}{=} \kappa_n$ with $\tilde{\kappa}_n \to \kappa$ a.s. Let us further introduce some independent random elements ζ_n^k in S with distributions $P[\xi_n \in \cdot | \xi_n \in B_k]$ and define $\tilde{\xi}_n^p = \sum_k \zeta_n^k 1\{\tilde{\kappa}_n = k\}$, so that $\tilde{\xi}_n^p \stackrel{d}{=} \xi_n$ for each n.

From the construction it is clear that
$$\left\{\rho(\tilde{\xi}_n^p, \xi) > 2^{-p}\right\} \subset \{\tilde{\kappa}_n \neq \kappa\} \cup \{\xi \in B_0\}, \quad n, p \in \mathbb{N}.$$

Since $\tilde{\kappa}_n \to \kappa$ a.s. and $P\{\xi \in B_0\} < 2^{-p}$, there exists for every p some $n_p \in \mathbb{N}$ with
$$P\bigcup_{n \geq n_p}\left\{\rho(\tilde{\xi}_n^p, \xi) > 2^{-p}\right\} < 2^{-p}, \quad p \in \mathbb{N},$$
and we may further assume that $n_1 < n_2 < \cdots$. By the Borel–Cantelli lemma we get a.s. $\sup_{n \geq n_p} \rho(\tilde{\xi}_n^p, \xi) \leq 2^{-p}$ for all but finitely many p. Now define $\eta_n = \tilde{\xi}_n^p$ for $n_p \leq n < n_{p+1}$, and note that $\xi_n \stackrel{d}{=} \eta_n \to \xi$ a.s. □

We conclude this chapter with a result on functional representations of limits, needed in Chapters 17 and 21. To motivate the problem, recall from Lemma 4.6 that if $\xi_n \xrightarrow{P} \eta$ for some random elements in a complete metric space S, then $\eta = f(\xi)$ a.s. for some measurable function $f : S^\infty \to S$, where $\xi = (\xi_n)$. Here f depends on the distribution μ of ξ, so a universal representation must be of the form $\eta = f(\xi, \mu)$. For certain purposes, it is crucial to choose a measurable version even of the latter function. To allow constructions by repeated approximation in probability, we need to consider the more general case when $\eta_n \xrightarrow{P} \eta$ for some random elements $\eta_n = f_n(\xi, \mu)$.

For a precise statement of the result, let $\mathcal{P}(S)$ denote the space of probability measures μ on S, endowed with the σ-field induced by all evaluation maps $\mu \mapsto \mu B$, $B \in \mathcal{B}(S)$.

Proposition 4.31 *(representation of limits)* *Fix a complete metric space (S, ρ), a measurable space U, and some measurable functions $f_1, f_2, \ldots : U \times \mathcal{P}(U) \to S$. Then there exist a measurable set $A \subset \mathcal{P}(U)$ and a measurable function $f : U \times A \to S$ such that, whenever ξ is a random element in U with distribution μ, the sequence $\eta_n = f_n(\xi, \mu)$ converges in probability iff $\mu \in A$, in which case the limit equals $f(\xi, \mu)$.*

Proof: For sequences $s = (s_1, s_2, \ldots)$ in S, define $l(s) = \lim_k s_k$ when the limit exists and put $l(s) = s_\infty$ otherwise, where $s_\infty \in S$ is arbitrary. By Lemma 1.10 we note that l is a measurable mapping from S^∞ to S. Next consider a sequence $\eta = (\eta_1, \eta_2, \ldots)$ of random elements in S, and put $\nu = \mathcal{L}(\eta)$. Define $n_1, n_2, \ldots$ as in the proof of Lemma 4.6, and note that each $n_k = n_k(\nu)$ is a measurable function of ν. Let C be the set of measures ν such that $n_k(\nu) < \infty$ for all k, and note that η_n converges in probability iff $\nu \in C$. Introduce the measurable function

$$g(s, \nu) = l(s_{n_1(\nu)}, s_{n_2(\nu)}, \ldots), \quad s = (s_1, s_2, \ldots) \in S^\infty, \nu \in \mathcal{P}(S^\infty).$$

If $\nu \in C$, we see from the proof of Lemma 4.6 that $\eta_{n_k(\nu)}$ converges a.s., and so $\eta_n \xrightarrow{P} g(\eta, \nu)$.

Now assume that $\eta_n = f_n(\xi, \mu)$ for some random element ξ in U with distribution μ and some measurable functions f_n. It remains to show that ν is a measurable function of μ. But this is clear from Lemma 1.41 (ii) applied to the kernel $K(\mu, \cdot) = \mu$ from $\mathcal{P}(U)$ to U and the function $F = (f_1, f_2, \ldots) : U \times \mathcal{P}(U) \to S^\infty$. □

As a simple consequence, we may consider limits in probability of measurable processes. The resulting statement will be useful in Chapter 17.

Corollary 4.32 *(measurability of limits, Stricker and Yor)* *For any measurable space T and complete metric space S, let $X^1, X^2, \ldots$ be S-valued measurable processes on T. Then there exist a measurable set $A \subset T$ and a measurable process X on A such that X_t^n converges in probability iff $t \in A$, in which case $X_t^n \xrightarrow{P} X_t$.*

Proof: Define $\xi_t = (X_t^1, X_t^2, \ldots)$ and $\mu_t = \mathcal{L}(\xi_t)$. By Proposition 4.31 there exist a measurable set $C \subset \mathcal{P}(S^\infty)$ and a measurable function $f : S^\infty \times C \to S$ such that X_t^n converges in probability iff $\mu_t \in C$, in which case $X_t^n \xrightarrow{P} f(\xi_t, \mu_t)$. It remains to note that the mapping $t \mapsto \mu_t$ is measurable, which is clear from Lemmas 1.4 and 1.26. □

Exercises

1. Let $\xi_1, \ldots, \xi_n$ be independent symmetric random variables. Show that $P\{(\sum_k \xi_k)^2 \geq r \sum_k \xi_k^2\} \geq (1-r)^2/3$ for any $r \in (0,1)$. (*Hint:* Reduce by means of Lemma 3.11 to the case of nonrandom $|\xi_k|$, and use Lemma 4.1.)

2. Let $\xi_1, \ldots, \xi_n$ be independent symmetric random variables. Show that $P\{\max_k |\xi_k| > r\} \leq 2P\{|S| > r\}$ for all $r > 0$, where $S = \sum_k \xi_k$. (*Hint:* Let η be the first term ξ_k where $\max_k |\xi_k|$ is attained, and check that $(\eta, S - \eta) \stackrel{d}{=} (\eta, \eta - S)$.)

3. Let $\xi_1, \xi_2, \ldots$ be i.i.d. random variables with $P\{|\xi_n| > t\} > 0$ for all $t > 0$. Show that there exist some constants $c_1, c_2, \ldots$ such that $c_n \xi_n \to 0$ in probability but not a.s.

4. Show that a family of random variables ξ_t is tight iff $\sup_t Ef(|\xi_t|) < \infty$ for some increasing function $f: \mathbb{R}_+ \to \mathbb{R}_+$ with $f(\infty) = \infty$.

5. Consider some random variables ξ_n and η_n such that (ξ_n) is tight and $\eta_n \xrightarrow{P} 0$. Show that even $\xi_n \eta_n \xrightarrow{P} 0$.

6. Show that the random variables ξ_t are uniformly integrable iff $\sup_t Ef(|\xi_t|) < \infty$ for some increasing function $f: \mathbb{R}_+ \to \mathbb{R}_+$ with $f(x)/x \to \infty$ as $x \to \infty$.

7. Show that the condition $\sup_t E|\xi_t| < \infty$ in Lemma 4.10 can be omitted if $\mathcal{A}$ is nonatomic.

8. Let $\xi_1, \xi_2, \cdots \in L^1$. Show that the ξ_n are uniformly integrable iff the condition in Lemma 4.10 holds with $\sup_n$ replaced by $\limsup_n$.

9. Deduce the dominated convergence theorem from Lemma 4.11.

10. Show that if $\{|\xi_t|^p\}$ and $\{|\eta_t|^p\}$ are uniformly integrable for some $p > 0$, then so is $\{|a\xi_t + b\eta_t|^p\}$ for any $a, b \in \mathbb{R}$. (*Hint:* Use Lemma 4.10.) Use this fact to deduce Proposition 4.12 from Lemma 4.11.

11. Give examples of random variables $\xi, \xi_1, \xi_2, \cdots \in L^2$ such that $\xi_n \to \xi$ holds a.s. but not in L^2, in L^2 but not a.s., or in L^1 but not in L^2.

12. Let $\xi_1, \xi_2, \ldots$ be independent random variables in L^2. Show that $\sum_n \xi_n$ converges in L^2 iff $\sum_n E\xi_n$ and $\sum_n \mathrm{var}(\xi_n)$ both converge.

13. Give an example of independent symmetric random variables $\xi_1, \xi_2, \ldots$ such that $\sum_n \xi_n$ is a.s. conditionally (nonabsolutely) convergent.

14. Let ξ_n and η_n be symmetric random variables with $|\xi_n| \leq |\eta_n|$ such that the pairs (ξ_n, η_n) are independent. Show that $\sum_n \xi_n$ converges whenever $\sum_n \eta_n$ does.

15. Let $\xi_1, \xi_2, \ldots$ be independent symmetric random variables. Show that $E[(\sum_n \xi_n)^2 \wedge 1] \leq \sum_n E[\xi_n^2 \wedge 1]$ whenever the latter series converges. (*Hint:* Integrate over the sets where $\sup_n |\xi_n| \leq 1$ or > 1, respectively.)

16. Consider some independent sequences of symmetric random variables $\xi_k, \eta_k^1, \eta_k^2, \ldots$ with $|\eta_k^n| \leq |\xi_k|$ such that $\sum_k \xi_k$ converges, and assume $\eta_k^n \xrightarrow{P}$

η_k for each k. Show that $\sum_k \eta_k^n \xrightarrow{P} \sum_k \eta_k$. (*Hint:* Use a truncation based on the preceding exercise.)

17. Let $\sum_n \xi_n$ be a convergent series of independent random variables. Show that the sum is a.s. independent of the order of terms iff $\sum_n |E[\xi_n; |\xi_n| \leq 1]| < \infty$.

18. Let the random variables ξ_{nj} be symmetric and independent for each n. Show that $\sum_j \xi_{nj} \xrightarrow{P} 0$ iff $\sum_j E[\xi_{nj}^2 \wedge 1] \to 0$.

19. Let $\xi_n \xrightarrow{d} \xi$ and $a_n \xi_n \xrightarrow{d} \xi$ for some nondegenerate random variable ξ and some constants $a_n > 0$. Show that $a_n \to 1$. (*Hint:* Turning to subsequences, we may assume that $a_n \to a$.)

20. Let $\xi_n \xrightarrow{d} \xi$ and $a_n \xi_n + b_n \xrightarrow{d} \xi$ for some nondegenerate random variable ξ, where $a_n > 0$. Show that $a_n \to 1$ and $b_n \to 0$. (*Hint:* Symmetrize.)

21. Let $\xi_1, \xi_2, \ldots$ be independent random variables such that $a_n \sum_{k \leq n} \xi_k$ converges in probability for some constants $a_n \to 0$. Show that the limit is degenerate.

22. Show that Theorem 4.23 is false for $p = 2$ by taking the ξ_k to be independent and $N(0, 1)$.

23. Let $\xi_1, \xi_2, \ldots$ be i.i.d. and such that $n^{-1/p} \sum_{k \leq n} \xi_k$ is a.s. bounded for some $p \in (0, 2)$. Show that $E|\xi_1|^p < \infty$. (*Hint:* Argue as in the proof of Theorem 4.23.)

24. Show for $p \leq 1$ that the a.s. convergence in Theorem 4.23 remains valid in L^p. (*Hint:* Truncate the ξ_k.)

25. Give an elementary proof of the strong law of large numbers when $E|\xi|^4 < \infty$. (*Hint:* Assuming $E\xi = 0$, show that $E \sum_n (S_n/n)^4 < \infty$.)

26. Show by examples that Theorem 4.25 is false without the stated restrictions on the sets G, F, and B.

27. Use Theorem 4.30 to give a simple proof of Theorem 4.27 when S is separable. Generalize to random elements ξ and ξ_n in Borel sets C and C_n, respectively, assuming only $f_n(x_n) \to f(x)$ for $x_n \in C_n$ and $x \in C$ with $x_n \to x$. Extend the original proof to that case.

28. Give a short proof of Theorem 4.30 when $S = \mathbb{R}$. (*Hint:* Note that the distribution functions F_n and F satisfy $F_n^{-1} \to F^{-1}$ a.e. on $[0, 1]$.)

Chapter 5

Characteristic Functions and Classical Limit Theorems

Uniqueness and continuity theorem; Poisson convergence; positive and symmetric terms; Lindeberg's condition; general Gaussian convergence; weak laws of large numbers; domain of Gaussian attraction; vague and weak compactness

In this chapter we continue the treatment of weak convergence from Chapter 4 with a detailed discussion of probability measures on Euclidean spaces. Our first aim is to develop the theory of characteristic functions and Laplace transforms. In particular, the basic uniqueness and continuity theorem will be established by simple equicontinuity and approximation arguments. The traditional compactness approach—in higher dimensions a highly nontrivial route—is required only for the case when the limiting function is not known in advance to be a characteristic function. The compactness theory also serves as a crucial bridge to the general theory of weak convergence presented in Chapter 16.

Our second aim is to establish the basic distributional limit theorems in the case of Poisson or Gaussian limits. We shall then consider triangular arrays of random variables ξ_{nj}, assumed to be independent for each n and such that $\xi_{nj} \xrightarrow{P} 0$ as $n \to \infty$ uniformly in j. In this setting, general criteria will be obtained for the convergence of $\sum_j \xi_{nj}$ toward a Poisson or Gaussian distribution. Specializing to the case of suitably centered and normalized partial sums from a single i.i.d. sequence $\xi_1, \xi_2, \ldots$, we may deduce the ultimate versions of the weak law of large numbers and the central limit theorem, including a complete description of the domain of attraction of the Gaussian law.

The mentioned limit theorems lead in Chapters 12 and 13 to some basic characterizations of Poisson and Gaussian processes, which in turn are needed to describe the general independent increment processes in Chapter 15. Even the limit theorems themselves are generalized in various ways in subsequent chapters. Thus, the Gaussian convergence is extended in Chapter 14 to suitable martingales, and the result is strengthened to uniform approximation of the summation process by the path of a Brownian motion. Similarly, the Poisson convergence is extended in Chapter 16 to a general limit theorem for point processes. A complete solution to the general limit problem for triangular arrays is given in Chapter 15, in connection with our treatment of Lévy processes.

In view of the crucial role of the independence assumption for the methods in this chapter, it may come as a surprise that the scope of the method of characteristic functions and Laplace transforms extends far beyond the present context. Thus, exponential martingales based on characteristic functions play a crucial role in Chapters 15 and 18, whereas Laplace functionals of random measures are used extensively in Chapters 12 and 16. Even more importantly, Laplace transforms play a key role in Chapters 19 and 22, in the guises of resolvents and potentials for Markov processes and their additive functionals, and also in connection with the large deviation theory of Chapter 27.

To begin with the basic definitions, consider a random vector ξ in $\mathbb{R}^d$ with distribution μ. The associated *characteristic function* $\hat{\mu}$ is given by

$$\hat{\mu}(t) = \int e^{itx} \mu(dx) = E e^{it\xi}, \quad t \in \mathbb{R}^d,$$

where tx denotes the inner product $t_1 x_1 + \cdots + t_d x_d$. For distributions μ on $\mathbb{R}_+^d$, it is often more convenient to consider the *Laplace transform* $\tilde{\mu}$, given by

$$\tilde{\mu}(u) = \int e^{-ux} \mu(dx) = E e^{-u\xi}, \quad u \in \mathbb{R}_+^d.$$

Finally, for distributions μ on $\mathbb{Z}_+$, it is often preferable to use the *(probability) generating function* ψ, given by

$$\psi(s) = \sum_{n \geq 0} s^n P\{\xi = n\} = E s^\xi, \quad s \in [0, 1].$$

Formally, $\tilde{\mu}(u) = \hat{\mu}(iu)$ and $\hat{\mu}(t) = \tilde{\mu}(-it)$, and so the functions $\hat{\mu}$ and $\tilde{\mu}$ are essentially the same, apart from domain. Furthermore, the generating function ψ is related to the Laplace transform $\tilde{\mu}$ by $\tilde{\mu}(u) = \psi(e^{-u})$ or $\psi(s) = \tilde{\mu}(-\log s)$. Though the characteristic function always exists, it may not be extendable to an analytic function in the complex plane.

For any distribution μ on $\mathbb{R}^d$, we note that the characteristic function $\varphi = \hat{\mu}$ is uniformly continuous with $|\varphi(t)| \leq \varphi(0) = 1$. It is also seen to be *Hermitian* in the sense that $\varphi(-t) = \bar{\varphi}(t)$, where the bar denotes complex conjugation. If ξ has characteristic function φ, then the linear combination $a\xi = a_1\xi_1 + \cdots + a_d\xi_d$ has characteristic function $t \mapsto \varphi(ta)$. Also note that if ξ and η are independent random vectors with characteristic functions φ and ψ, then the characteristic function of the pair (ξ, η) is given by the tensor product $\varphi \otimes \psi : (s, t) \mapsto \varphi(s)\psi(t)$. In particular, $\xi + \eta$ has characteristic function $\varphi\psi$, and the characteristic function of the symmetrized variable $\xi - \xi'$ equals $|\varphi|^2$.

Whenever applicable, the quoted statements carry over to Laplace transforms and generating functions. The latter functions have the further

5. Characteristic Functions and Classical Limit Theorems

advantage of being positive, monotone, convex, and analytic—properties that simplify many arguments.

The following result contains some elementary but useful estimates involving characteristic functions. The second inequality was used in the proof of Theorem 4.17, and the remaining relations will be useful in the sequel to establish tightness.

Lemma 5.1 *(tail estimates)* For any probability measure μ on $\mathbb{R}$, we have

$$\mu\{x;\ |x| \geq r\} \leq \frac{r}{2} \int_{-2/r}^{2/r} (1 - \hat{\mu}_t)\,dt, \quad r > 0, \tag{1}$$

$$\mu[-r, r] \leq 2r \int_{-1/r}^{1/r} |\hat{\mu}_t|\,dt, \quad r > 0. \tag{2}$$

If μ is supported by $\mathbb{R}_+$, then also

$$\mu[r, \infty) \leq 2(1 - \tilde{\mu}(1/r)), \quad r > 0. \tag{3}$$

Proof: Using Fubini's theorem and noting that $\sin x \leq x/2$ for $x \geq 2$, we get for any $c > 0$

$$\int_{-c}^{c} (1 - \hat{\mu}_t)\,dt = \int \mu(dx) \int_{-c}^{c} (1 - e^{itx})\,dt$$

$$= 2c \int \left\{ 1 - \frac{\sin cx}{cx} \right\} \mu(dx) \geq c\mu\{x;\ |cx| \geq 2\},$$

and (1) follows as we take $c = 2/r$. To prove (2), we may write

$$\tfrac{1}{2}\mu[-r, r] \leq 2 \int \frac{1 - \cos(x/r)}{(x/r)^2} \mu(dx)$$

$$= r \int \mu(dx) \int (1 - r|t|)_+ e^{ixt}\,dt$$

$$= r \int (1 - r|t|)_+ \hat{\mu}_t\,dt \leq r \int_{-1/r}^{1/r} |\hat{\mu}_t|\,dt.$$

To obtain (3), we note that $e^{-x} < \tfrac{1}{2}$ for $x \geq 1$. Thus, for $t > 0$,

$$1 - \tilde{\mu}_t = \int (1 - e^{-tx})\mu(dx) \geq \tfrac{1}{2}\mu\{x;\ tx \geq 1\}. \qquad \square$$

Recall that a family of probability measures μ_α on $\mathbb{R}^d$ is said to be *tight* if

$$\lim_{r \to \infty} \sup_\alpha \mu_\alpha\{x;\ |x| > r\} = 0.$$

The following lemma describes tightness in terms of characteristic functions.

Lemma 5.2 *(equicontinuity and tightness) A family $\{\mu_\alpha\}$ of probability measures on $\mathbb{R}^d$ is tight iff $\{\hat{\mu}_\alpha\}$ is equicontinuous at 0, and then $\{\hat{\mu}_\alpha\}$ is uniformly equicontinuous on $\mathbb{R}^d$. A similar statement holds for the Laplace transforms of distributions on $\mathbb{R}_+^d$.*

Proof: The sufficiency is immediate from Lemma 5.1, applied separately in each coordinate. To prove the necessity, let ξ_α denote a random vector with distribution μ_α, and write for any $s, t \in \mathbb{R}^d$

$$|\hat{\mu}_\alpha(s) - \hat{\mu}_\alpha(t)| \leq E|e^{is\xi_\alpha} - e^{it\xi_\alpha}| = E|1 - e^{i(t-s)\xi_\alpha}|$$
$$\leq 2E[|(t-s)\xi_\alpha| \wedge 1].$$

If $\{\xi_\alpha\}$ is tight, then by Lemma 4.9 the right-hand side tends to 0 as $t - s \to 0$, uniformly in α, and the asserted uniform equicontinuity follows. The proof for Laplace transforms is similar. $\square$

For any probability measures $\mu, \mu_1, \mu_2, \ldots$ on $\mathbb{R}^d$, we recall that the weak convergence $\mu_n \xrightarrow{w} \mu$ holds by definition iff $\mu_n f \to \mu f$ for any bounded, continuous function f on $\mathbb{R}^d$, where μf denotes the integral $\int f d\mu$. The usefulness of characteristic functions is mainly due to the following basic result.

Theorem 5.3 *(uniqueness and continuity, Lévy) For any probability measures $\mu, \mu_1, \mu_2, \ldots$ on $\mathbb{R}^d$, we have $\mu_n \xrightarrow{w} \mu$ iff $\hat{\mu}_n(t) \to \hat{\mu}(t)$ for every $t \in \mathbb{R}^d$, and then $\hat{\mu}_n \to \hat{\mu}$ uniformly on every bounded set. A corresponding statement holds for the Laplace transforms of distributions on $\mathbb{R}_+^d$.*

In particular, we may take $\mu_n \equiv \nu$ and conclude that a probability measure μ on $\mathbb{R}^d$ is uniquely determined by its characteristic function $\hat{\mu}$. Similarly, a probability measure μ on $\mathbb{R}_+^d$ is seen to be determined by its Laplace transform $\tilde{\mu}$.

For the proof of Theorem 5.3, we need the following simple cases or consequences of the Stone–Weierstrass approximation theorem. Here $[0, \infty]$ denotes the compactification of $\mathbb{R}_+$.

Lemma 5.4 *(approximation) Every continuous function $f: \mathbb{R}^d \to \mathbb{R}$ with period 2π in each coordinate admits a uniform approximation by linear combinations of $\cos kx$ and $\sin kx$, $k \in \mathbb{Z}_+^d$. Similarly, every continuous function $g: [0, \infty]^d \to \mathbb{R}_+$ can be approximated uniformly by linear combinations of the functions e^{-kx}, $k \in \mathbb{Z}_+^d$.*

Proof of Theorem 5.3: We consider only the case of characteristic functions, the proof for Laplace transforms being similar. If $\mu_n \xrightarrow{w} \mu$, then $\hat{\mu}_n(t) \to \hat{\mu}(t)$ for every t, by the definition of weak convergence. By Lemmas 4.8 and 5.2, the latter convergence is uniform on every bounded set.

Conversely, assume that $\hat{\mu}_n(t) \to \hat{\mu}(t)$ for every t. By Lemma 5.1 and dominated convergence we get, for any $a \in \mathbb{R}^d$ and $r > 0$,

$$\limsup_{n\to\infty} \mu_n\{x; |ax| > r\} \leq \lim_{n\to\infty} \frac{r}{2} \int_{-2/r}^{2/r} (1 - \hat{\mu}_n(ta)) dt$$

$$= \frac{r}{2} \int_{-2/r}^{2/r} (1 - \hat{\mu}(ta)) dt.$$

Since $\hat{\mu}$ is continuous at 0, the right-hand side tends to 0 as $r \to \infty$, which shows that the sequence (μ_n) is tight. Given any $\varepsilon > 0$, we may then choose $r > 0$ so large that $\mu_n\{|x| > r\} \leq \varepsilon$ for all n and $\mu\{|x| > r\} \leq \varepsilon$.

Now fix any bounded, continuous function $f: \mathbb{R}^d \to \mathbb{R}$, say with $|f| \leq m < \infty$. Let f_r denote the restriction of f to the ball $\{|x| \leq r\}$, and extend f_r to a continuous function $\tilde{f}$ on $\mathbb{R}^d$ with $|\tilde{f}| \leq m$ and period $2\pi r$ in each coordinate. By Lemma 5.4 there exists some linear combination g of the functions $\cos(kx/r)$ and $\sin(kx/r)$, $k \in \mathbb{Z}_+^d$, such that $|\tilde{f} - g| \leq \varepsilon$. Writing $\|\cdot\|$ for the supremum norm, we get for any $n \in \mathbb{N}$

$$|\mu_n f - \mu_n g| \leq \mu_n\{|x| > r\}\|f - \tilde{f}\| + \|\tilde{f} - g\| \leq (2m+1)\varepsilon,$$

and similarly for μ. Thus,

$$|\mu_n f - \mu f| \leq |\mu_n g - \mu g| + 2(2m+1)\varepsilon, \quad n \in \mathbb{N}.$$

Letting $n \to \infty$ and then $\varepsilon \to 0$, we obtain $\mu_n f \to \mu f$. Since f was arbitrary, this proves that $\mu_n \xrightarrow{w} \mu$. □

The next result provides a way of reducing the d-dimensional case to that of one dimension.

Corollary 5.5 *(one-dimensional projections, Cramér and Wold)* Let ξ and $\xi_1, \xi_2, \ldots$ be random vectors in $\mathbb{R}^d$. Then $\xi_n \xrightarrow{d} \xi$ iff $t\xi_n \xrightarrow{d} t\xi$ for all $t \in \mathbb{R}^d$. For random vectors in $\mathbb{R}_+^d$, it suffices that $u\xi_n \xrightarrow{d} u\xi$ for all $u \in \mathbb{R}_+^d$.

Proof: If $t\xi_n \xrightarrow{d} t\xi$, then $Ee^{it\xi_n} \to Ee^{it\xi}$ by the definition of weak convergence, and so $\xi_n \xrightarrow{d} \xi$ by Theorem 5.3. The proof for random vectors in $\mathbb{R}_+^d$ is similar. □

The last result contains in particular a basic uniqueness result, the fact that $\xi \stackrel{d}{=} \eta$ iff $t\xi \stackrel{d}{=} t\eta$ for all $t \in \mathbb{R}^d$ or $\mathbb{R}_+^d$, respectively. In other words, a probability measure on $\mathbb{R}^d$ is uniquely determined by its one-dimensional projections.

We now apply the continuity theorem to prove some classical limit theorems, and we begin with the case of Poisson convergence. For an introduction, consider for each $n \in \mathbb{N}$ some i.i.d. random variables $\xi_{n1}, \ldots, \xi_{nn}$ with distribution

$$P\{\xi_{nj} = 1\} = 1 - P\{\xi_{nj} = 0\} = c_n, \quad n \in \mathbb{N},$$

and assume that $nc_n \to c < \infty$. Then the sums $S_n = \xi_{n1} + \cdots + \xi_{nn}$ have generating functions

$$\psi_n(s) = (1 - (1-s)c_n)^n \to e^{-c(1-s)} = e^{-c} \sum_{n \geq 0} \frac{c^n s^n}{n!}, \quad s \in [0,1].$$

The limit $\psi(s) = e^{-c(1-s)}$ is the generating function of the *Poisson distribution* with parameter c, the distribution of a random variable η with probabilities $P\{\eta = n\} = e^{-c}c^n/n!$ for $n \in \mathbb{Z}_+$. Note that the corresponding expected value equals $E\eta = \psi'(1) = c$. Since $\psi_n \to \psi$, it is clear from Theorem 5.3 that $S_n \xrightarrow{d} \eta$.

Before turning to more general cases of Poisson convergence, we need to introduce the notion of a *null array*. By this we mean a triangular array of random variables or vectors ξ_{nj}, $1 \leq j \leq m_n$, $n \in \mathbb{N}$, such that the ξ_{nj} are independent for each n and satisfy

$$\sup_j E[|\xi_{nj}| \wedge 1] \to 0. \tag{4}$$

The latter condition may be thought of as the convergence $\xi_{nj} \xrightarrow{P} 0$ as $n \to \infty$, uniformly in j. When $\xi_{nj} \geq 0$ for all n and j, we may allow the m_n to be infinite.

The following lemma characterizes null arrays in terms of the associated characteristic functions or Laplace transforms.

Lemma 5.6 (null arrays) *Consider a triangular array of random vectors ξ_{nj} with characteristic functions φ_{nj} or Laplace transforms ψ_{nj}. Then (4) holds iff, respectively,*

$$\sup_j |1 - \varphi_{nj}(t)| \to 0, \quad t \in \mathbb{R}^d, \tag{5}$$

$$\inf_j \psi_{nj}(u) \to 1, \quad u \in \mathbb{R}_+^d.$$

Proof: Relation (4) holds iff $\xi_{n,j_n} \xrightarrow{P} 0$ for all sequences (j_n). By Theorem 5.3 this is equivalent to $\varphi_{n,j_n}(t) \to 1$ for all t and (j_n), which in turn is equivalent to (5). The proof for Laplace transforms is similar. □

We now give a general criterion for Poisson convergence of the row sums in a null array of integer-valued random variables. The result will be extended in Lemmas 15.15 and 15.24 to more general limiting distributions and in Theorem 16.18 to the context of point processes.

Theorem 5.7 (Poisson convergence) *Let (ξ_{nj}) be a null array of $\mathbb{Z}_+$-valued random variables, and let ξ be Poisson distributed with mean c. Then $\sum_j \xi_{nj} \xrightarrow{d} \xi$ iff these conditions hold:*

 (i) $\sum_j P\{\xi_{nj} > 1\} \to 0$;
 (ii) $\sum_j P\{\xi_{nj} = 1\} \to c$.

Moreover, (i) is equivalent to $\sup_j \xi_{nj} \vee 1 \xrightarrow{P} 1$. If $\sum_j \xi_{nj}$ converges in distribution, then (i) holds iff the limit is Poisson.

We need the following frequently used lemma.

Lemma 5.8 *(sums and products) Consider a null array of constants $c_{nj} \geq 0$, and fix any $c \in [0, \infty]$. Then $\prod_j (1 - c_{nj}) \to e^{-c}$ iff $\sum_j c_{nj} \to c$.*

Proof: Since $\sup_j c_{nj} < 1$ for large n, the first relation is equivalent to $\sum_j \log(1 - c_{nj}) \to -c$, and the assertion follows from the fact that $\log(1-x) = -x + o(x)$ as $x \to 0$. □

Proof of Theorem 5.7: Let ψ_{nj} denote the generating function of ξ_{nj}. By Theorem 5.3 the convergence $\sum_j \xi_{nj} \xrightarrow{d} \xi$ is equivalent to $\prod_j \psi_{nj}(s) \to e^{-c(1-s)}$ for arbitrary $s \in [0,1]$, which holds by Lemmas 5.6 and 5.8 iff

$$\sum_j (1 - \psi_{nj}(s)) \to c(1-s), \quad s \in [0,1]. \tag{6}$$

By an easy computation, the sum on the left equals

$$(1-s)\sum_j P\{\xi_{nj} > 0\} + \sum_{k>1}(s - s^k)\sum_j P\{\xi_{nj} = k\} = T_1 + T_2, \tag{7}$$

and we also note that

$$s(1-s)\sum_j P\{\xi_{nj} > 1\} \leq T_2 \leq s\sum_j P\{\xi_{nj} > 1\}. \tag{8}$$

Assuming (i) and (ii), it is clear that (6) follows from (7) and (8). Now assume instead that (6) holds. For $s = 0$ we get $\sum_j P\{\xi_{nj} > 0\} \to c$, and so in general $T_1 \to c(1-s)$. But then (6) implies $T_2 \to 0$, and (i) follows by (8). Finally, (ii) is obtained by subtraction.

To prove that (i) is equivalent to $\sup_j \xi_{nj} \vee 1 \xrightarrow{P} 1$, we note that

$$P\{\sup_j \xi_{nj} \leq 1\} = \prod_j P\{\xi_{nj} \leq 1\} = \prod_j (1 - P\{\xi_{nj} > 1\}).$$

By Lemma 5.8 the right-hand side tends to 1 iff $\sum_j P\{\xi_{nj} > 1\} \to 0$, which is the stated equivalence.

To prove the last assertion, put $c_{nj} = P\{\xi_{nj} > 0\}$ and write

$$E \exp\left\{-\sum_j \xi_{nj}\right\} - P\{\sup_j \xi_{nj} > 1\} \leq E \exp\left\{-\sum_j (\xi_{nj} \wedge 1)\right\}$$
$$= \prod_j E \exp\{-(\xi_{nj} \wedge 1)\} = \prod_j \{1 - (1 - e^{-1})c_{nj}\}$$
$$\leq \prod_j \exp\{-(1-e^{-1})c_{nj}\} = \exp\left\{-(1-e^{-1})\sum_j c_{nj}\right\}.$$

If (i) holds and $\sum_j \xi_{nj} \xrightarrow{d} \eta$, then the left-hand side tends to $Ee^{-\eta} > 0$, and so the sums $c_n = \sum_j c_{nj}$ are bounded. Hence, c_n converges along a subsequence $N' \subset \mathbb{N}$ toward some constant c. But then (i) and (ii) hold along N', and the first assertion shows that η is Poisson with mean c. □

Next consider some i.i.d. random variables $\xi_1, \xi_2, \ldots$ with $P\{\xi_k = \pm 1\} = \frac{1}{2}$, and write $S_n = \xi_1 + \cdots + \xi_n$. Then $n^{-1/2}S_n$ has characteristic function

$$\varphi_n(t) = \cos^n(n^{-1/2}t) = (1 - \tfrac{1}{2}t^2 n^{-1} + O(n^{-2}))^n \to e^{-t^2/2} = \varphi(t).$$

By a classical computation, the function $e^{-x^2/2}$ has Fourier transform

$$\int_{-\infty}^{\infty} e^{itx} e^{-x^2/2} dx = (2\pi)^{1/2} e^{-t^2/2}, \quad t \in \mathbb{R}.$$

Hence, φ is the characteristic function of a probability measure on $\mathbb{R}$ with density $(2\pi)^{-1/2} e^{-x^2/2}$. This is the standard *normal* or *Gaussian* distribution $N(0,1)$, and Theorem 5.3 shows that $n^{-1/2} S_n \xrightarrow{d} \zeta$, where ζ is $N(0,1)$. The general Gaussian law $N(m, \sigma^2)$ is defined as the distribution of the random variable $\eta = m + \sigma\zeta$, and we note that η has mean m and variance σ^2. From the form of the characteristic functions together with the uniqueness property, it is clear that any linear combination of independent Gaussian random variables is again Gaussian.

The convergence to a Gaussian limit generalizes easily to a more general setting, as in the following classical result. The present statement is only preliminary, and a more general version is obtained by different methods in Theorem 5.17.

Proposition 5.9 *(central limit theorem, Lindeberg, Lévy)* *Let $\xi, \xi_1, \xi_2, \ldots$ be i.i.d. random variables with $E\xi = 0$ and $E\xi^2 = 1$, and let ζ be $N(0,1)$. Then $n^{-1/2} \sum_{k \leq n} \xi_k \xrightarrow{d} \zeta$.*

The proof may be based on a simple Taylor expansion.

Lemma 5.10 *(Taylor expansion)* *Let φ be the characteristic function of a random variable ξ with $E|\xi|^n < \infty$. Then*

$$\varphi(t) = \sum_{k=0}^{n} \frac{(it)^k E\xi^k}{k!} + o(t^n), \quad t \to 0.$$

Proof: Noting that $|e^{it} - 1| \leq t$ for all $t \in \mathbb{R}$, we get recursively by dominated convergence

$$\varphi^{(k)}(t) = E(i\xi)^k e^{it\xi}, \quad t \in \mathbb{R}, \ 0 \leq k \leq n.$$

In particular, $\varphi^{(k)}(0) = E(i\xi)^k$ for $k \leq n$, and the result follows from Taylor's formula. □

Proof of Proposition 5.9: Let the ξ_k have characteristic function φ. By Lemma 5.10, the characteristic function of $n^{-1/2} S_n$ equals

$$\varphi_n(t) = \left(\varphi(n^{-1/2}t)\right)^n = (1 - \tfrac{1}{2}t^2 n^{-1} + o(n^{-1}))^n \to e^{-t^2/2},$$

where the convergence holds as $n \to \infty$ for fixed t. □

Our next aim is to examine the relationship between null arrays of symmetric and positive random variables. In this context, we may also derive criteria for convergence toward Gaussian and degenerate limits, respectively.

5. Characteristic Functions and Classical Limit Theorems

Theorem 5.11 *(positive and symmetric terms)* *Let (ξ_{nj}) be a null array of symmetric random variables, and let ξ be $N(0,c)$ for some $c \geq 0$. Then $\sum_j \xi_{nj} \xrightarrow{d} \xi$ iff $\sum_j \xi_{nj}^2 \xrightarrow{P} c$, and also iff these conditions hold:*

(i) $\sum_j P\{|\xi_{nj}| > \varepsilon\} \to 0$ for all $\varepsilon > 0$;

(ii) $\sum_j E(\xi_{nj}^2 \wedge 1) \to c$.

Moreover, (i) is equivalent to $\sup_j |\xi_{nj}| \xrightarrow{P} 0$. If $\sum_j \xi_{nj}$ or $\sum_j \xi_{nj}^2$ converges in distribution, then (i) holds iff the limit is Gaussian or degenerate, respectively.

Here the necessity of condition (i) is a remarkable fact that plays a crucial role in our proof of the more general Theorem 5.15. It is instructive to compare the present statement with the corresponding result for random series in Theorem 4.17. Note also the extended version appearing in Proposition 15.23.

Proof: First assume that $\sum_j \xi_{nj} \xrightarrow{d} \xi$. By Theorem 5.3 and Lemmas 5.6 and 5.8 it is equivalent that

$$\sum_j E(1 - \cos t\xi_{nj}) \to \tfrac{1}{2}ct^2, \quad t \in \mathbb{R}, \tag{9}$$

where the convergence is uniform on every bounded interval. Comparing the integrals of (9) over $[0,1]$ and $[0,2]$, we get $\sum_j Ef(\xi_{nj}) \to 0$, where $f(0) = 0$ and

$$f(x) = 3 - \frac{4\sin x}{x} + \frac{\sin 2x}{2x}, \quad x \in \mathbb{R} \setminus \{0\}.$$

Now f is continuous with $f(x) \to 3$ as $|x| \to \infty$, and furthermore $f(x) > 0$ for $x \neq 0$. Indeed, the last relation is equivalent to $8\sin x - \sin 2x < 6x$ for $x > 0$, which is obvious when $x \geq \pi/2$ and follows by differentiation twice when $x \in (0, \pi/2)$. Writing $g(x) = \inf_{y > x} f(y)$ and letting $\varepsilon > 0$ be arbitrary, we get

$$\sum_j P\{|\xi_{nj}| > \varepsilon\} \leq \sum_j P\{f(\xi_{nj}) > g(\varepsilon)\} \leq \sum_j Ef(\xi_{nj})/g(\varepsilon) \to 0,$$

which proves (i).

If instead $\sum_j \xi_{nj}^2 \xrightarrow{P} c$, the corresponding symmetrized variables η_{nj} satisfy $\sum_j \eta_{nj} \xrightarrow{P} 0$, and we get $\sum_j P\{|\eta_{nj}| > \varepsilon\} \to 0$ as before. By Lemma 4.19 it follows that $\sum_j P\{|\xi_{nj}^2 - m_{nj}| > \varepsilon\} \to 0$, where the m_{nj} are medians of ξ_{nj}^2, and since $\sup_j m_{nj} \to 0$, condition (i) follows again. Using Lemma 5.8, we further note that (i) is equivalent to $\sup_j |\xi_{nj}| \xrightarrow{P} 0$. Thus, we may henceforth assume that (i) is fulfilled.

Next we note that, for any $t \in \mathbb{R}$ and $\varepsilon > 0$,

$$\sum_j E[1 - \cos t\xi_{nj}; |\xi_{nj}| \leq \varepsilon] = \tfrac{1}{2}t^2(1 - O(t^2\varepsilon^2)) \sum_j E[\xi_{nj}^2; |\xi_{nj}| \leq \varepsilon].$$

Assuming (i), the equivalence between (9) and (ii) now follows as we let $n \to \infty$ and then $\varepsilon \to 0$. To get the corresponding result for the variables

ξ_{nj}^2, we may instead write

$$\sum_j E[1 - e^{-t\xi_{nj}^2}; \xi_{nj}^2 \leq \varepsilon] = t(1 - O(t\varepsilon))\sum_j E[\xi_{nj}^2; \xi_{nj}^2 \leq \varepsilon], \quad t, \varepsilon > 0,$$

and proceed as before. This completes the proof of the first assertion.

Finally, assume that (i) holds and $\sum_j \xi_{nj} \xrightarrow{d} \eta$. Then the same relation holds for the truncated variables $\xi_{nj}1\{|\xi_{nj}| \leq 1\}$, and so we may assume that $|\xi_{nj}| \leq 1$ for all j and k. Define $c_n = \sum_j E\xi_{nj}^2$. If $c_n \to \infty$ along some subsequence, then the distribution of $c_n^{-1/2} \sum_j \xi_{nj}$ tends to $N(0,1)$ by the first assertion, which is impossible by Lemmas 4.8 and 4.9. Thus, (c_n) is bounded and converges along some subsequence. By the first assertion, $\sum_j \xi_{nj}$ then tends to some Gaussian limit, so even η is Gaussian. □

The following result gives the basic criterion for Gaussian convergence, under a normalization by second moments.

Theorem 5.12 *(Gaussian convergence under classical normalization, Lindeberg, Feller)* *Let (ξ_{nj}) be a triangular array of rowwise independent random variables with mean 0 and $\sum_j E\xi_{nj}^2 \to 1$, and let ξ be $N(0,1)$. Then these conditions are equivalent:*

(i) $\sum_j \xi_{nj} \xrightarrow{d} \xi$ and $\sup_j E\xi_{nj}^2 \to 0$;

(ii) $\sum_j E[\xi_{nj}^2; |\xi_{nj}| > c] \to 0$ for all $\varepsilon > 0$.

Here (ii) is the celebrated *Lindeberg condition*. Our proof is based on two elementary lemmas.

Lemma 5.13 *(comparison of products)* *For any complex numbers $z_1, \ldots, z_n$ and $z'_1, \ldots, z'_n$ of modulus ≤ 1, we have*

$$\left|\prod_k z_k - \prod_k z'_k\right| \leq \sum_k |z_k - z'_k|.$$

Proof: For $n = 2$ we get

$$|z_1 z_2 - z'_1 z'_2| \leq |z_1 z_2 - z'_1 z_2| + |z'_1 z_2 - z'_1 z'_2| \leq |z_1 - z'_1| + |z_2 - z'_2|,$$

and the general result follows by induction. □

Lemma 5.14 *(Taylor expansion)* *For any $t \in \mathbb{R}$ and $n \in \mathbb{Z}_+$, we have*

$$\left|e^{it} - \sum_{k=0}^n \frac{(it)^k}{k!}\right| \leq \frac{2|t|^n}{n!} \wedge \frac{|t|^{n+1}}{(n+1)!}.$$

Proof: Letting $h_n(t)$ denote the difference on the left, we get

$$h_n(t) = i\int_0^t h_{n-1}(s)ds, \quad t > 0, \, n \in \mathbb{Z}_+.$$

Starting from the obvious relations $|h_{-1}| \equiv 1$ and $|h_0| \leq 2$, it follows by induction that $|h_{n-1}(t)| \leq |t|^n/n!$ and $|h_n(t)| \leq 2|t|^n/n!$. □

5. Characteristic Functions and Classical Limit Theorems

We return to the proof of Theorem 5.12. At this point we shall prove only the sufficiency of the Lindeberg condition (ii), which is needed for the proof of the main Theorem 5.15. To avoid repetition, we postpone the proof of the necessity part until after the proof of that theorem.

Proof of Theorem 5.12, (ii) $\Rightarrow$ (i): Write $c_{nj} = E\xi_{nj}^2$ and $c_n = \sum_j c_{nj}$. First we note that for any $\varepsilon > 0$

$$\sup_j c_{nj} \leq \varepsilon^2 + \sup_j E[\xi_{nj}^2; |\xi_{nj}| > \varepsilon] \leq \varepsilon^2 + \sum_j E[\xi_{nj}^2; |\xi_{nj}| > \varepsilon],$$

which tends to 0 under (ii), as $n \to \infty$ and then $\varepsilon \to 0$.

Now introduce some independent random variables ζ_{nj} with distributions $N(0, c_{nj})$, and note that $\zeta_n = \sum_j \zeta_{nj}$ is $N(0, c_n)$. Hence, $\zeta_n \xrightarrow{d} \xi$. Letting φ_{nj} and ψ_{nj} denote the characteristic functions of ξ_{nj} and ζ_{nj}, respectively, it remains by Theorem 5.3 to show that $\prod_j \varphi_{nj} - \prod_j \psi_{nj} \to 0$. Then conclude from Lemmas 5.13 and 5.14 that, for fixed $t \in \mathbb{R}$,

$$\left|\prod_j \varphi_{nj}(t) - \prod_j \psi_{nj}(t)\right| \leq \sum_j |\varphi_{nj}(t) - \psi_{nj}(t)|$$

$$\leq \sum_j |\varphi_{nj}(t) - 1 + \tfrac{1}{2}t^2 c_{nj}| + \sum_j |\psi_{nj}(t) - 1 + \tfrac{1}{2}t^2 c_{nj}|$$

$$\lesssim \sum_j E\xi_{nj}^2 (1 \wedge |\xi_{nj}|) + \sum_j E\zeta_{nj}^2 (1 \wedge |\zeta_{nj}|).$$

For any $\varepsilon > 0$, we have

$$\sum_j E\xi_{nj}^2 (1 \wedge |\xi_{nj}|) \leq \varepsilon \sum_j c_{nj} + \sum_j E[\xi_{nj}^2; |\xi_{nj}| > \varepsilon],$$

which tends to 0 by (ii), as $n \to \infty$ and then $\varepsilon \to 0$. Further note that

$$\sum_j E\zeta_{nj}^2 (1 \wedge |\zeta_{nj}|) \leq \sum_j E|\zeta_{nj}|^3 = \sum_j c_{nj}^{3/2} E|\xi|^3 \leq c_n \sup_j c_{nj}^{1/2} \to 0$$

by the first part of the proof. $\square$

The problem of characterizing the convergence to a Gaussian limit is solved completely by the following result. The reader should notice the striking resemblance between the present conditions and those of the three-series criterion in Theorem 4.18. A far-reaching extension of the present result is obtained by different methods in Chapter 15. As before $\mathrm{var}[\xi; A] = \mathrm{var}(\xi 1_A)$.

Theorem 5.15 *(Gaussian convergence, Feller, Lévy) Let (ξ_{nj}) be a null array of random variables, and let ξ be $N(b,c)$ for some constants b and c. Then $\sum_j \xi_{nj} \xrightarrow{d} \xi$ iff these conditions hold:*

(i) $\sum_j P\{|\xi_{nj}| > \varepsilon\} \to 0$ *for all $\varepsilon > 0$;*

(ii) $\sum_j E[\xi_{nj}; |\xi_{nj}| \leq 1] \to b$;

(iii) $\sum_j \mathrm{var}[\xi_{nj}; |\xi_{nj}| \leq 1] \to c$.

Moreover, (i) is equivalent to $\sup_j |\xi_{nj}| \xrightarrow{P} 0$. If $\sum_j \xi_{nj}$ converges in distribution, then (i) holds iff the limit is Gaussian.

Proof: To see that (i) is equivalent to $\sup_j |\xi_{nj}| \xrightarrow{P} 0$, we note that

$$P\{\sup_j |\xi_{nj}| > \varepsilon\} = 1 - \prod_j (1 - P\{|\xi_{nj}| > \varepsilon\}), \quad \varepsilon > 0.$$

Since $\sup_j P\{|\xi_{nj}| > \varepsilon\} \to 0$ under both conditions, the assertion follows by Lemma 5.8.

Now assume $\sum_{nj} \xi_{nj} \xrightarrow{d} \xi$. Introduce medians m_{nj} and symmetrizations $\tilde{\xi}_{nj}$ of the variables ξ_{nj}, and note that $m_n \equiv \sup_j |m_{nj}| \to 0$ and $\sum_j \tilde{\xi}_{nj} \xrightarrow{d} \tilde{\xi}$, where $\tilde{\xi}$ is $N(0, 2c)$. By Lemma 4.19 and Theorem 5.11, we get for any $\varepsilon > 0$

$$\sum_j P\{|\xi_{nj}| > \varepsilon\} \leq \sum_j P\{|\xi_{nj} - m_{nj}| > \varepsilon - m_n\}$$
$$\leq 2 \sum_j P\{|\tilde{\xi}_{nj}| > \varepsilon - m_n\} \to 0.$$

Thus, we may henceforth assume condition (i) and hence that $\sup_j |\xi_{nj}| \xrightarrow{P} 0$. But then $\sum_j \xi_{nj} \xrightarrow{d} \eta$ is equivalent to $\sum_j \xi'_{nj} \xrightarrow{d} \eta$, where $\xi'_{nj} = \xi_{nj} 1\{|\xi_{nj}| \leq 1\}$, and so we may further assume that $|\xi_{nj}| \leq 1$ a.s. for all n and j. In this case (ii) and (iii) reduce to $b_n \equiv \sum_j E\xi_{nj} \to b$ and $c_n \equiv \sum_j \mathrm{var}(\xi_{nj}) \to c$, respectively.

Write $b_{nj} = E\xi_{nj}$, and note that $\sup_j |b_{nj}| \to 0$ because of (i). Assuming (ii) and (iii), we get $\sum_j \xi_{nj} - b_n \xrightarrow{d} \xi - b$ by Theorem 5.12, and so $\sum_j \xi_{nj} \xrightarrow{d} \xi$. Conversely, $\sum_j \xi_{nj} \xrightarrow{d} \xi$ implies $\sum_j \tilde{\xi}_{nj} \xrightarrow{d} \tilde{\xi}$, and (iii) follows by Theorem 5.11. But then $\sum_j \xi_{nj} - b_n \xrightarrow{d} \xi - b$, so Lemma 4.20 shows that b_n converges toward some b'. Hence, $\sum_j \xi_{nj} \xrightarrow{d} \xi + b' - b$, and so $b' = b$, which means that even (ii) is fulfilled.

It remains to prove that, under condition (i), any limiting distribution is Gaussian. Then assume $\sum_j \xi_{nj} \xrightarrow{d} \eta$, and note that $\sum_j \tilde{\xi}_{nj} \xrightarrow{d} \tilde{\eta}$, where $\tilde{\eta}$ denotes a symmetrization of η. If $c_n \to \infty$ along some subsequence, then $c_n^{-1/2} \sum_j \tilde{\xi}_{nj}$ tends to $N(0, 2)$ by the first assertion, which is impossible by Lemma 4.9. Thus, (c_n) is bounded, and we have convergence $c_n \to c$ along some subsequence. But then $\sum_{nj} \xi_{nj} - b_n$ tends to $N(0, c)$, again by the first assertion, and Lemma 4.20 shows that even b_n converges toward some limit b. Hence, $\sum_{nj} \xi_{nj}$ tends to $N(b, c)$, which is then the distribution of η. □

Proof of Theorem 5.12, (i) $\Rightarrow$ (ii): The second condition in (i) implies that (ξ_{nj}) is a null array. Furthermore, we have for any $\varepsilon > 0$

$$\sum_j \mathrm{var}[\xi_{nj}; |\xi_{nj}| \leq \varepsilon] \leq \sum_j E[\xi_{nj}^2; |\xi_{nj}| \leq \varepsilon] \leq \sum_j E\xi_{nj}^2 \to 1.$$

By Theorem 5.15 even the left-hand side tends to 1, and (ii) follows. □

As a first application of Theorem 5.15, we shall prove the following ultimate version of the weak law of large numbers. The result should be compared with the corresponding strong law established in Theorem 4.23.

Theorem 5.16 *(weak laws of large numbers)* *Let $\xi, \xi_1, \xi_2, \ldots$ be i.i.d. random variables, and fix any $p \in (0,2)$ and $c \in \mathbb{R}$. Then $n^{-1/p}\sum_{k \leq n} \xi_k \xrightarrow{P} c$ iff the following condition holds as $r \to \infty$, depending on the value of p:*

$p < 1:$ $r^p P\{|\xi| > r\} \to 0$ and $c = 0$;
$p = 1:$ $rP\{|\xi| > r\} \to 0$ and $E[\xi; |\xi| \leq r] \to c$;
$p > 1:$ $r^p P\{|\xi| > r\} \to 0$ and $E\xi = c = 0$.

Proof: Applying Theorem 5.15 to the null array of random variables $\xi_{nj} = n^{-1/p}\xi_j$, $j \leq n$, we note that the stated convergence is equivalent to the three conditions

(i) $nP\{|\xi| > n^{1/p}\varepsilon\} \to 0$ for all $\varepsilon > 0$,

(ii) $n^{1-1/p}E[\xi; |\xi| \leq n^{1/p}] \to c$,

(iii) $n^{1-2/p}\text{var}[\xi; |\xi| \leq n^{1/p}] \to 0$.

By the monotonicity of $P\{|\xi| > r^{1/p}\}$, condition (i) is equivalent to $r^p P\{|\xi| > r\} \to 0$. Furthermore, Lemma 3.4 yields for any $r > 0$

$$r^{p-2}\text{var}[\xi; |\xi| \leq r] \leq r^p E[(\xi/r)^2 \wedge 1] = r^p \int_0^1 P\{|\xi| \geq r\sqrt{t}\}dt,$$

$$r^{p-1}|E[\xi; |\xi| \leq r]| \leq r^p E(|\xi/r| \wedge 1) = r^p \int_0^1 P\{|\xi| \geq rt\}dt.$$

Since t^{-a} is integrable on $[0,1]$ for any $a < 1$, it follows by dominated convergence that (i) implies (iii) and also that (i) implies (ii) with $c = 0$ when $p < 1$.

If instead $p > 1$, we see from (i) and Lemma 3.4 that

$$E|\xi| = \int_0^\infty P\{|\xi| > r\}dr \lesssim \int_0^\infty (1 \wedge r^{-p})dr < \infty.$$

Thus, $E[\xi; |\xi| \leq r] \to E\xi$, and (ii) implies $E\xi = 0$. Moreover, we get from (i)

$$r^{p-1}E[|\xi|; |\xi| > r] = r^p P\{|\xi| > r\} + r^{p-1}\int_r^\infty P\{|\xi| > t\}dt \to 0.$$

Under the further assumption that $E\xi = 0$, we obtain (ii) with $c = 0$.
Finally, let $p = 1$, and conclude from (i) that

$$E[|\xi|; n < |\xi| \leq n+1] \leq nP\{|\xi| > n\} \to 0.$$

Hence, under (i), condition (ii) is equivalent to $E[\xi; |\xi| \leq r] \to c$. □

We next extend the central limit theorem in Proposition 5.9 by characterizing convergence of suitably normalized partial sums from a single i.i.d. sequence toward a Gaussian limit. Here a nondecreasing function $L \geq 0$ is said to *vary slowly at* ∞ if $\sup_x L(x) > 0$ and moreover $L(cx) \sim L(x)$ as $x \to \infty$ for each $c > 0$. This holds in particular when L is bounded, but it is also true for many unbounded functions, such as $\log(x \vee 1)$.

Theorem 5.17 *(domain of Gaussian attraction, Lévy, Feller, Khinchin)* Let $\xi, \xi_1, \xi_2, \ldots$ be i.i.d. nondegenerate random variables, and let ζ be $N(0,1)$. Then $a_n \sum_{k \leq n} (\xi_k - m_n) \xrightarrow{d} \zeta$ for some constants a_n and m_n iff the function $L(x) = E[\xi^2; |\xi| \leq x]$ varies slowly at ∞, in which case we may take $m_n \equiv E\xi$. In particular, the stated convergence holds with $a_n \equiv n^{-1/2}$ and $m_n \equiv 0$ iff $E\xi = 0$ and $E\xi^2 = 1$.

Even other so-called stable distributions may occur as limits, but the conditions for convergence are too restrictive to be of much interest for applications. Our proof of Theorem 5.17 is based on the following result.

Lemma 5.18 *(slow variation, Karamata)* Let ξ be a nondegenerate random variable such that $L(x) = E[\xi^2; |\xi| \leq x]$ varies slowly at ∞. Then so does the function $L_m(x) = E[(\xi - m)^2; |\xi - m| \leq x]$ for every $m \in \mathbb{R}$, and moreover

$$\lim_{x \to \infty} x^{2-p} E[|\xi|^p; |\xi| > x]/L(x) = 0, \quad p \in [0, 2). \tag{10}$$

Proof: Fix any constant $r \in (1, 2^{2-p})$, and choose $x_0 > 0$ so large that $L(2x) \leq rL(x)$ for all $x \geq x_0$. For such an x, we get

$$\begin{aligned}
x^{2-p} E[|\xi|^p; |\xi| > x] &= x^{2-p} \sum_{n \geq 0} E[|\xi|^p; |\xi|/x \in (2^n, 2^{n+1}]] \\
&\leq \sum_{n \geq 0} 2^{(p-2)n} E[\xi^2; |\xi|/x \in (2^n, 2^{n+1}]] \\
&\leq \sum_{n \geq 0} 2^{(p-2)n} (r-1) r^n L(x) \\
&= (r-1) L(x)/(1 - 2^{p-2} r).
\end{aligned}$$

Now (10) follows, as we divide by $L(x)$ and let $x \to \infty$ and then $r \to 1$.

In particular, we note that $E|\xi|^p < \infty$ for all $p < 2$. If even $E\xi^2 < \infty$, then $E(\xi - m)^2 < \infty$, and the first assertion is obvious. If instead $E\xi^2 = \infty$, we may write

$$L_m(x) = E[\xi^2; |\xi - m| \leq x] + mE[m - 2\xi; |\xi - m| \leq x].$$

Here the last term is bounded, and the first term lies between the bounds $L(x \pm m) \sim L(x)$. Thus, $L_m(x) \sim L(x)$, and the slow variation of L_m follows from that of L. □

Proof of Theorem 5.17: Assume that L varies slowly at ∞. By Lemma 5.18 this is also true for the function $L_m(x) = E[(\xi - m)^2; |\xi - m| > x]$, where $m = E\xi$, and so we may assume that $E\xi = 0$. Now define

$$c_n = 1 \vee \sup\{x > 0; nL(x) \geq x^2\}, \quad n \in \mathbb{N},$$

5. Characteristic Functions and Classical Limit Theorems

and note that $c_n \uparrow \infty$. From the slow variation of L it is further clear that $c_n < \infty$ for all n and that, moreover, $nL(c_n) \sim c_n^2$. In particular, $c_n \sim n^{1/2}$ iff $L(c_n) \sim 1$, that is, iff $\mathrm{var}(\xi) = 1$.

We shall verify the conditions of Theorem 5.15 with $b = 0$, $c = 1$, and $\xi_{nj} = \xi_j/c_n$, $j \le n$. Beginning with (i), let $\varepsilon > 0$ be arbitrary, and conclude from Lemma 5.18 that

$$nP\{|\xi/c_n| > \varepsilon\} \sim \frac{c_n^2 P\{|\xi| > c_n\varepsilon\}}{L(c_n)} \sim \frac{c_n^2 P\{|\xi| > c_n\varepsilon\}}{L(c_n\varepsilon)} \to 0.$$

Recalling that $E\xi = 0$, we get by the same lemma

$$n|E[\xi/c_n; |\xi/c_n| \le 1]| \le \frac{n}{c_n} E[|\xi|; |\xi| > c_n] \sim \frac{c_n E[|\xi|; |\xi| > c_n]}{L(c_n)} \to 0, \tag{11}$$

which proves (ii). To obtain (iii), we note that in view of (11)

$$n \mathrm{var}[\xi/c_n; |\xi/c_n| \le 1] = \frac{n}{c_n^2} L(c_n) - n(E[\xi/c_n; |\xi| \le c_n])^2 \to 1.$$

By Theorem 5.15 the required convergence follows with $a_n = c_n^{-1}$ and $m_n \equiv 0$.

Now assume instead that the stated convergence holds for suitable constants a_n and m_n. Then a corresponding result holds for the symmetrized variables $\tilde{\xi}, \tilde{\xi}_1, \tilde{\xi}_2, \ldots$ with constants $a_n/\sqrt{2}$ and 0, and so we may assume that $c_n^{-1} \sum_{k \le n} \tilde{\xi}_k \xrightarrow{d} \zeta$. Here, clearly, $c_n \to \infty$ and, moreover, $c_{n+1} \sim c_n$, since even $c_{n+1}^{-1} \sum_{k \le n} \tilde{\xi}_k \xrightarrow{d} \zeta$ by Theorem 4.28. Now define for $x > 0$

$$\tilde{T}(x) = P\{|\tilde{\xi}| > x\}, \quad \tilde{L}(x) = E[\tilde{\xi}^2; |\tilde{\xi}| \le x], \quad \tilde{U}(x) = E(\tilde{\xi}^2 \wedge x^2).$$

By Theorem 5.15 we have $n\tilde{T}(c_n\varepsilon) \to 0$ for all $\varepsilon > 0$, and also $nc_n^{-2}\tilde{L}(c_n) \to 1$. Thus, $c_n^2 \tilde{T}(c_n\varepsilon)/\tilde{L}(c_n) \to 0$, which extends by monotonicity to

$$\frac{x^2 \tilde{T}(x)}{\tilde{U}(x)} \le \frac{x^2 \tilde{T}(x)}{\tilde{L}(x)} \to 0, \quad x \to \infty.$$

Next define for any $x > 0$

$$T(x) = P\{|\xi| > x\}, \quad U(x) = E(\xi^2 \wedge x^2).$$

By Lemma 4.19 we have $T(x + |m|) \le 2\tilde{T}(x)$ for any median m of ξ. Furthermore, by Lemmas 3.4 and 4.19, we get

$$\tilde{U}(x) = \int_0^{x^2} P\{\tilde{\xi}^2 > t\}dt \le 2\int_0^{x^2} P\{4\xi^2 > t\}dt = 8U(x/2).$$

Hence, as $x \to \infty$,

$$\frac{L(2x) - L(x)}{L(x)} \le \frac{4x^2 T(x)}{U(x) - x^2 T(x)} \le \frac{8x^2 \tilde{T}(x - |m|)}{8^{-1}\tilde{U}(2x) - 2x^2 \tilde{T}(x - |m|)} \to 0,$$

which shows that L is slowly varying.

Finally, assume that $n^{-1/2}\sum_{k\le n}\xi_k \xrightarrow{d} \zeta$. By the previous argument with $c_n = n^{1/2}$, we get $\tilde{L}(n^{1/2}) \to 2$, which implies $E\tilde{\xi}^2 = 2$ and hence $\operatorname{var}(\xi) = 1$. But then $n^{-1/2}\sum_{k\le n}(\xi_k - E\xi) \xrightarrow{d} \zeta$, and so by comparison $E\xi = 0$. $\square$

We return to the general problem of characterizing the weak convergence of a sequence of probability measures μ_n on $\mathbb{R}^d$ in terms of the associated characteristic functions $\hat{\mu}_n$ or Laplace transforms $\tilde{\mu}_n$. Suppose that $\hat{\mu}_n$ or $\tilde{\mu}_n$ converges toward some continuous limit φ, which is not recognized as a characteristic function or Laplace transform. To conclude that μ_n converges weakly toward some measure μ, we need an extended version of Theorem 5.3, which in turn requires a compactness argument for its proof.

As a preparation, consider the space $\mathcal{M} = \mathcal{M}(\mathbb{R}^d)$ of locally finite measures on $\mathbb{R}^d$. On $\mathcal{M}$ we may introduce the *vague topology*, generated by the mappings $\mu \mapsto \mu f = \int f d\mu$ for all $f \in C_K^+$, the class of continuous functions $f: \mathbb{R}^d \to \mathbb{R}_+$ with compact support. In particular, μ_n converges vaguely to μ (written as $\mu_n \xrightarrow{v} \mu$) iff $\mu_n f \to \mu f$ for all $f \in C_K^+$. If the μ_n are probability measures, then clearly $\mu \mathbb{R}^d \le 1$. The following version of *Helly's selection theorem* shows that the set of probability measures on $\mathbb{R}^d$ is vaguely relatively sequentially compact.

Theorem 5.19 (*vague sequential compactness, Helly*) *Any sequence of probability measures on $\mathbb{R}^d$ has a vaguely convergent subsequence.*

Proof: Fix any probability measures $\mu_1, \mu_2, \ldots$ on $\mathbb{R}^d$, and let $F_1, F_2, \ldots$ denote the corresponding distribution functions. Write $\mathbb{Q}$ for the set of rational numbers. By a diagonal argument, the functions F_n converge on $\mathbb{Q}^d$ toward some limit G, along a suitable subsequence $N' \subset \mathbb{N}$, and we may define

$$F(x) = \inf\{G(r); r \in \mathbb{Q}^d, r > x\}, \quad x \in \mathbb{R}^d. \tag{12}$$

Since each F_n has nonnegative increments, the same thing is true for G and hence also for F. From (12) and the monotonicity of G, it is further clear that F is right-continuous. Hence, by Corollary 3.26 there exists some measure μ on $\mathbb{R}^d$ with $\mu(x, y] = F(x, y]$ for any bounded rectangular box $(x, y] \subset \mathbb{R}^d$, and it remains to show that $\mu_n \xrightarrow{v} \mu$ along N'.

Then note that $F_n(x) \to F(x)$ at every continuity point x of F. By the monotonicity of F there exist some countable sets $D_1, \ldots, D_d \subset \mathbb{R}$ such that F is continuous on $C = D_1^c \times \cdots \times D_d^c$. Then $\mu_n U \to \mu U$ for every finite union U of rectangular boxes with corners in C, and by a simple approximation we get for any bounded Borel set $B \subset \mathbb{R}^d$

$$\mu B^\circ \le \liminf_{n\to\infty} \mu_n B \le \limsup_{n\to\infty} \mu_n B \le \mu \overline{B}. \tag{13}$$

For any bounded μ-continuity set B, we may consider functions $f \in C_K^+$ supported by B, and proceed as in the proof of Theorem 4.25 to show that $\mu_n f \to \mu f$. Thus, $\mu_n \xrightarrow{v} \mu$. $\square$

5. Characteristic Functions and Classical Limit Theorems

If $\mu_n \xrightarrow{v} \mu$ for some probability measures μ_n on $\mathbb{R}^d$, we may still have $\mu \mathbb{R}^d < 1$, due to an escape of mass to infinity. To exclude this possibility, we need to assume that (μ_n) be tight.

Lemma 5.20 *(vague and weak convergence)* *For any probability measures $\mu_1, \mu_2, \ldots$ on $\mathbb{R}^d$ with $\mu_n \xrightarrow{v} \mu$ for some measure μ, we have $\mu \mathbb{R}^d = 1$ iff (μ_n) is tight, and then $\mu_n \xrightarrow{w} \mu$.*

Proof: By a simple approximation, the vague convergence implies (13) for every bounded Borel set B, and in particular for the balls $B_r = \{x \in \mathbb{R}^d;\ |x| \leq r\}$, $r > 0$. If $\mu \mathbb{R}^d = 1$, then $\mu B_r^\circ \to 1$ as $r \to \infty$, and the first inequality shows that (μ_n) is tight. Conversely, if (μ_n) is tight, then $\limsup_n \mu_n B_r \to 1$, and the last inequality yields $\mu \mathbb{R}^d = 1$.

Now assume that (μ_n) is tight, and fix any bounded continuous function $f: \mathbb{R}^d \to \mathbb{R}$. For any $r > 0$, we may choose some $g_r \in C_K^+$ with $1_{B_r} \leq g_r \leq 1$ and note that

$$|\mu_n f - \mu f| \leq |\mu_n f - \mu_n f g_r| + |\mu_n f g_r - \mu f g_r| + |\mu f g_r - \mu f|$$
$$\leq |\mu_n f g_r - \mu f g_r| + \|f\|(\mu_n + \mu)B_r^c.$$

Here the right-hand side tends to zero as $n \to \infty$ and then $r \to \infty$, so $\mu_n f \to \mu f$. Hence, in this case $\mu_n \xrightarrow{w} \mu$. □

Combining the last two results, we may easily show that the notions of tightness and weak sequential compactness are equivalent. The result is extended in Theorem 16.3, which forms a starting point for the theory of weak convergence on function spaces.

Proposition 5.21 *(tightness and weak sequential compactness)* *A sequence of probability measures on $\mathbb{R}^d$ is tight iff every subsequence has a weakly convergent further subsequence.*

Proof: Fix any probability measures $\mu_1, \mu_2, \ldots$ on $\mathbb{R}^d$. By Theorem 5.19 every subsequence has a vaguely convergent further subsequence. If (μ_n) is tight, then by Lemma 5.20 the convergence holds even in the weak sense.

Now assume instead that (μ_n) has the stated property. If it fails to be tight, we may choose a sequence $n_k \to \infty$ and some constant $\varepsilon > 0$ such that $\mu_{n_k} B_k^c > \varepsilon$ for all $k \in \mathbb{N}$. By hypothesis there exists some probability measure μ on $\mathbb{R}^d$ such that $\mu_{n_k} \xrightarrow{w} \mu$ along a subsequence $N' \subset \mathbb{N}$. The sequence $(\mu_{n_k}; k \in N')$ is then tight by Lemma 4.8, and in particular there exists some $r > 0$ with $\mu_{n_k} B_r^c \leq \varepsilon$ for all $k \in N'$. For $k > r$ this is a contradiction, and the asserted tightness follows. □

We may now prove the desired extension of Theorem 5.3.

Theorem 5.22 *(extended continuity theorem, Lévy, Bochner)* *Let $\mu_1, \mu_2, \ldots$ be probability measures on $\mathbb{R}^d$ with $\hat{\mu}_n(t) \to \varphi(t)$ for every $t \in \mathbb{R}^d$, where the limit φ is continuous at 0. Then $\mu_n \xrightarrow{w} \mu$ for some probability measure μ on $\mathbb{R}^d$ with $\hat{\mu} = \varphi$. A corresponding statement holds for the Laplace transforms of measures on $\mathbb{R}_+^d$.*

Proof: Assume that $\hat{\mu}_n \to \varphi$, where the limit is continuous at 0. As in the proof of Theorem 5.3, we may conclude that (μ_n) is tight. Hence, by Proposition 5.21 there exists some probability measure μ on $\mathbb{R}^d$ such that $\mu_n \xrightarrow{w} \mu$ along a subsequence $N' \subset \mathbb{N}$. By continuity we get $\hat{\mu}_n \to \hat{\mu}$ along N', and so $\varphi = \hat{\mu}$. Finally, the convergence $\mu_n \xrightarrow{w} \mu$ extends to $\mathbb{N}$ by Theorem 5.3. The proof for Laplace transforms is similar. □

Exercises

1. Show that if ξ and η are independent Poisson random variables, then $\xi + \eta$ is again Poisson. Also show that the Poisson property is preserved under convergence in distribution.

2. Show that any linear combination of independent Gaussian random variables is again Gaussian. Also show that the class of Gaussian distributions is preserved under weak convergence.

3. Show that $\varphi_r(t) = (1 - t/r)_+$ is a characteristic functions for every $r > 0$. (*Hint:* Compute the Fourier transform $\hat{\psi}_r$ of the function $\psi_r(t) = 1\{|t| \leq r\}$, and note that the Fourier transform $\hat{\psi}_r^2$ of ψ_r^{*2} is integrable. Now use Fourier inversion.)

4. Let φ be a real, even function that is convex on $\mathbb{R}_+$ and satisfies $\varphi(0) = 1$ and $\varphi(\infty) \in [0, 1]$. Show that φ is the characteristic function of some symmetric distribution on $\mathbb{R}$. In particular, $\varphi(t) = e^{-|t|^c}$ is a characteristic function for every $c \in [0, 1]$. (*Hint:* Approximate by convex combinations of functions φ_r as above, and use Theorem 5.22.)

5. Show that if $\hat{\mu}$ is integrable, then μ has a bounded and continuous density. (*Hint:* Let φ_r be the triangular density above. Then $(\hat{\varphi}_r)\check{} = 2\pi\varphi_r$, and so $\int e^{-itu} \hat{\mu}_t \hat{\varphi}_r(t) dt = 2\pi \int \varphi_r(x-u)\mu(dx)$. Now let $r \to 0$.)

6. Show that a distribution μ is supported by some set $a\mathbb{Z} + b$ iff $|\hat{\mu}_t| = 1$ for some $t \neq 0$.

7. Give an elementary proof of the continuity theorem for generating functions of distributions on $\mathbb{Z}_+$. (*Hint:* Note that if $\mu_n \xrightarrow{v} \mu$ for some distributions on $\mathbb{R}_+$, then $\tilde{\mu}_n \to \tilde{\mu}$ on $(0, \infty)$.)

8. The *moment-generating function* of a distribution μ on $\mathbb{R}$ is given by $\tilde{\mu}_t = \int e^{tx} \mu(dx)$. Assuming $\tilde{\mu}_t < \infty$ for all t in some nondegenerate interval I, show that $\tilde{\mu}$ is analytic in the strip $\{z \in \mathbb{C}; \Re z \in I^\circ\}$. (*Hint:* Approximate by measures with bounded support.)

9. Let $\mu, \mu_1, \mu_2, \ldots$ be distributions on $\mathbb{R}$ with moment-generating functions $\tilde{\mu}, \tilde{\mu}_1, \tilde{\mu}_2, \ldots$ such that $\tilde{\mu}_n \to \tilde{\mu} < \infty$ on some nondegenerate interval I. Show that $\mu_n \xrightarrow{w} \mu$. (*Hint:* If $\mu_n \xrightarrow{v} \nu$ along some subsequence N', then $\tilde{\mu}_n \to \tilde{\nu}$ on I° along N', and so $\tilde{\nu} = \tilde{\mu}$ on I. By the preceding exercise we get $\nu\mathbb{R} = 1$ and $\hat{\nu} = \hat{\mu}$. Thus, $\nu = \mu$.)

10. Let μ and ν be distributions on $\mathbb{R}$ with finite moments $\int x^n \mu(dx) = \int x^n \nu(dx) = m_n$, where $\sum_n t^n |m_n|/n! < \infty$ for some $t > 0$. Show that $\mu = \nu$. (*Hint:* The absolute moments satisfy the same relation for any smaller value of t, so the moment-generating functions exist and agree on $(-t, t)$.)

11. For each $n \in \mathbb{N}$, let μ_n be a distribution on $\mathbb{R}$ with finite moments m_n^k, $k \in \mathbb{N}$, such that $\lim_n m_n^k = a_k$ for some constants a_k with $\sum_k t^k |a_k|/k! < \infty$ for some $t > 0$. Show that $\mu_n \xrightarrow{w} \mu$ for some distribution μ with moments a_k. (*Hint:* Each function x^k is uniformly integrable with respect to the measures μ_n. In particular, (μ_n) is tight. If $\mu_n \xrightarrow{w} \nu$ along some subsequence, then ν has moments a_k.)

12. Given a distribution μ on $\mathbb{R} \times \mathbb{R}_+$, introduce the mixed transform $\varphi(s, t) = \int e^{isx - ty} \mu(dx\, dy)$, where $s \in \mathbb{R}$ and $t \geq 0$. Prove versions for φ of the continuity Theorems 5.3 and 5.22.

13. Consider a null array of random vectors $\xi_{nj} = (\xi_{nj}^1, \ldots, \xi_{nj}^d)$ in $\mathbb{Z}_+^d$, let $\xi^1, \ldots, \xi^d$ be independent Poisson variables with means $c_1, \ldots, c_d$, and put $\xi = (\xi^1, \ldots, \xi^d)$. Show that $\sum_j \xi_{nj} \xrightarrow{d} \xi$ iff $\sum_j P\{\xi_{nj}^k = 1\} \to c_k$ for all k and $\sum_j P\{\sum_k \xi_{nj}^k > 1\} \to 0$. (*Hint:* Introduce independent random variables $\eta_{nj}^k \stackrel{d}{=} \xi_{nj}^k$, and note that $\sum_j \xi_{nj} \xrightarrow{d} \xi$ iff $\sum_j \eta_{nj} \xrightarrow{d} \xi$.)

14. Consider some random variables $\xi \perp\!\!\!\perp \eta$ with finite variance such that the distribution of (ξ, η) is rotationally invariant. Show that ξ is centered Gaussian. (*Hint:* Let $\xi_1, \xi_2, \ldots$ be i.i.d. and distributed as ξ, and note that $n^{-1/2} \sum_{k \leq n} \xi_k$ has the same distribution for all n. Now use Proposition 5.9.)

15. Prove a multivariate version of the Taylor expansion in Lemma 5.10.

16. Let μ have a finite nth moment m_n. Show that $\hat{\mu}$ is n times continuously differentiable and satisfies $\hat{\mu}_0^{(n)} = i^n m_n$. (*Hint:* Differentiate n times under the integral sign.)

17. For μ and m_n as above, show that $\hat{\mu}_0^{(2n)}$ exists iff $m_{2n} < \infty$. Also, characterize the distributions such that $\hat{\mu}_0^{(2n-1)}$ exists. (*Hint:* For $\hat{\mu}_0''$ proceed as in the proof of Proposition 5.9, and use Theorem 5.17. For $\hat{\mu}_0'$ use Theorem 5.16. Extend by induction to $n > 1$.)

18. Let μ be a distribution on $\mathbb{R}_+$ with moments m_n. Show that $\tilde{\mu}_0^{(n)} = (-1)^n m_n$ whenever either side exists and is finite. (*Hint:* Prove the statement for $n = 1$, and extend by induction.)

19. Deduce Proposition 5.9 from Theorem 5.12.

20. Let the random variables ξ and ξ_{nj} be such as in Theorem 5.12, and assume that $\sum_j E|\xi_{nj}|^c \to 0$ for some $c > 2$. Show that $\sum_j \xi_{nj} \xrightarrow{d} \xi$.

21. Extend Theorem 5.12 to random vectors in $\mathbb{R}^d$, with the condition $\sum_j E\xi_{nj}^2 \to 1$ replaced by $\sum_j \text{cov}(\xi_{nj}) \to a$, with ξ as $N(0, a)$, and with ξ_{nj}^2 replaced by $|\xi_{nj}|^2$. (*Hint:* Use Corollary 5.5 to reduce to one dimension.)

22. Show that Theorem 5.15 remains true for random vectors in $\mathbb{R}^d$, with $\text{var}[\xi_{nj}; |\xi_{nj}| \leq 1]$ replaced by the corresponding covariance matrix. (*Hint:* If $a, a_1, a_2, \ldots$ are symmetric, nonnegative definite matrices, then $a_n \to a$ iff $u'a_n u \to u'au$ for all $u \in \mathbb{R}^d$. To see this, use a compactness argument.)

23. Show that Theorems 5.7 and 5.15 remain valid for possibly infinite row-sums $\sum_j \xi_{nj}$. (*Hint:* Use Theorem 4.17 or 4.18 together with Theorem 4.28.)

24. Let $\xi, \xi_1, \xi_2, \ldots$ be i.i.d. random variables. Show that $n^{-1/2} \sum_{k \leq n} \xi_k$ converges in probability iff $\xi = 0$ a.s. (*Hint:* Use condition (iii) in Theorem 5.15.)

25. Let $\xi_1, \xi_2, \ldots$ be i.i.d. μ, and fix any $p \in (0, 2)$. Find a μ such that $n^{-1/p} \sum_{k \leq n} \xi_k \to 0$ in probability but not a.s.

26. Let $\xi_1, \xi_2, \ldots$ be i.i.d., and let $p > 0$ be such that $n^{-1/p} \sum_{k \leq n} \xi_k \to 0$ in probability but not a.s. Show that $\limsup_n n^{-1/p} |\sum_{k \leq n} \xi_k| = \infty$ a.s. (*Hint:* Note that $E|\xi_1|^p = \infty$.)

27. Give an example of a distribution with infinite second moment in the domain of attraction of the Gaussian law, and find the corresponding normalization.

Chapter 6

Conditioning and Disintegration

Conditional expectations and probabilities; regular conditional distributions; disintegration; conditional independence; transfer and coupling; existence of sequences and processes; extension through conditioning

Modern probability theory can be said to begin with the notions of conditioning and disintegration. In particular, conditional expectations and distributions are needed already for the *definitions* of martingales and Markov processes, the two basic dependence structures beyond independence and stationarity. Even in other areas and throughout probability theory, conditioning is constantly used as a basic tool to describe and analyze systems involving randomness. The notion may be thought of in terms of averaging, projection, and disintegration—viewpoints that are all essential for a proper understanding.

In all but the most elementary contexts, one defines conditioning with respect to a σ-field rather than a single event. In general, the result of the operation is not a constant but a random variable, measurable with respect to the given σ-field. The idea is familiar from elementary constructions of the conditional expectation $E[\xi|\eta]$, in cases where (ξ, η) is a random vector with a nice density, and the result is obtained as a suitable function of η. This corresponds to conditioning on the σ-field $\mathcal{F} = \sigma(\eta)$.

The simplest and most intuitive *general* approach to conditioning is via projection. Here $E[\xi|\mathcal{F}]$ is defined for any $\xi \in L^2$ as the orthogonal Hilbert space projection of ξ onto the linear subspace of $\mathcal{F}$-measurable random variables. The L^2-version extends immediately, by continuity, to arbitrary $\xi \in L^1$. From the orthogonality of the projection one gets the relation $E(\xi - E[\xi|\mathcal{F}])\zeta = 0$ for any bounded, $\mathcal{F}$-measurable random variable ζ. This leads in particular to the familiar averaging characterization of $E[\xi|\mathcal{F}]$ as a version of the density $d(\xi \cdot P)/dP$ on the σ-field $\mathcal{F}$, the existence of which can also be inferred from the Radon–Nikodým theorem.

The conditional expectation is defined only up to a null set, in the sense that any two versions agree a.s. It is then natural to look for versions of the conditional probabilities $P[A|\mathcal{F}] = E[1_A|\mathcal{F}]$ that combine into a random probability measure on Ω. In general, such regular versions exist only for A restricted to suitable sub-σ-fields. The basic case is when ξ is a random element in some Borel space S, and the conditional distribution $P[\xi \in \cdot|\mathcal{F}]$ may be constructed as an $\mathcal{F}$-measurable random measure on

S. If we further assume that $\mathcal{F} = \sigma(\eta)$ for a random element η in some space T, we may write $P[\xi \in B|\eta] = \mu(\eta, B)$ for some probability kernel μ from T to S. This leads to a decomposition of the distribution of (ξ, η) according to the values of η. The result is formalized in the disintegration theorem—a powerful extension of Fubini's theorem that is often used in subsequent chapters, especially in combination with the (strong) Markov property.

Using conditional distributions, we shall further establish the basic transfer theorem, which may be used to convert any distributional equivalence $\xi \stackrel{d}{=} f(\eta)$ into a corresponding a.s. representation $\xi = f(\tilde{\eta})$ with a suitable $\tilde{\eta} \stackrel{d}{=} \eta$. From the latter result, one easily obtains the fundamental Daniell–Kolmogorov theorem, which ensures the existence of random sequences and processes with specified finite-dimensional distributions. A different approach is required for the more general Ionescu Tulcea extension, where the measure is specified by a sequence of conditional distributions.

Further topics treated in this chapter include the notion of conditional independence, which is fundamental for both Markov processes and exchangeability and also plays an important role in Chapter 21, in connection with SDEs. Especially useful in those contexts is the elementary but powerful chain rule. Let us finally call attention to the local property of conditional expectations, which in particular leads to simple and transparent proofs of the strong Markov and optional sampling theorems.

Returning to our construction of conditional expectations, let us fix a probability space $(\Omega, \mathcal{A}, P)$ and consider an arbitrary sub-σ-field $\mathcal{F} \subset \mathcal{A}$. In $L^2 = L^2(\mathcal{A})$ we may introduce the closed linear subspace M, consisting of all random variables $\eta \in L^2$ that agree a.s. with some element of $L^2(\mathcal{F})$. By the Hilbert space projection Theorem 1.33, there exists for every $\xi \in L^2$ an a.s. unique random variable $\eta \in M$ with $\xi - \eta \perp M$, and we define $E^{\mathcal{F}}\xi = E[\xi|\mathcal{F}]$ as an arbitrary $\mathcal{F}$-measurable version of η.

The L^2-projection $E^{\mathcal{F}}$ is easily extended to L^1, as follows.

Theorem 6.1 (*conditional expectation, Kolmogorov*) *For any σ-field $\mathcal{F} \subset \mathcal{A}$ there exists an a.s. unique linear operator $E^{\mathcal{F}}: L^1 \to L^1(\mathcal{F})$ such that*

(i) $E[E^{\mathcal{F}}\xi; A] = E[\xi; A]$, $\xi \in L^1$, $A \in \mathcal{F}$.

The following additional properties hold whenever the corresponding expressions exist for the absolute values:

(ii) $\xi \geq 0$ *implies* $E^{\mathcal{F}}\xi \geq 0$ *a.s.;*

(iii) $E|E^{\mathcal{F}}\xi| \leq E|\xi|$;

(iv) $0 \leq \xi_n \uparrow \xi$ *implies* $E^{\mathcal{F}}\xi_n \uparrow E^{\mathcal{F}}\xi$ *a.s.;*

(v) $E^{\mathcal{F}}\xi\eta = \xi E^{\mathcal{F}}\eta$ *a.s. when ξ is $\mathcal{F}$-measurable;*

(vi) $E(\xi E^{\mathcal{F}}\eta) = E(\eta E^{\mathcal{F}}\xi) = E(E^{\mathcal{F}}\xi \cdot E^{\mathcal{F}}\eta)$;

(vii) $E^{\mathcal{F}}E^{\mathcal{G}}\xi = E^{\mathcal{F}}\xi$ *a.s. for all $\mathcal{F} \subset \mathcal{G}$.*

In particular, we note that $E^{\mathcal{F}}\xi = \xi$ a.s. iff ξ has an $\mathcal{F}$-measurable version and that $E^{\mathcal{F}}\xi = E\xi$ a.s. when $\xi \perp\!\!\!\perp \mathcal{F}$. We shall often refer to (i) as the *averaging property*, to (ii) as the *positivity*, to (iii) as the L^1-*contractivity*, to (iv) as the *monotone convergence property*, to (v) as the *pull-out property*, to (vi) as the *self-adjointness*, and to (vii) as the *chain rule*. Since the operator $E^{\mathcal{F}}$ is both self-adjoint by (vi) and idempotent by (vii), it may be thought of as a generalized projection on L^1.

The existence of $E^{\mathcal{F}}$ is an immediate consequence of the Radon–Nikodým Theorem 2.10. However, we prefer the following elementary construction from the L^2-version.

Proof of Theorem 6.1: First assume that $\xi \in L^2$, and define $E^{\mathcal{F}}\xi$ by projection as above. For any $A \in \mathcal{F}$ we get $\xi - E^{\mathcal{F}}\xi \perp 1_A$, and (i) follows. Taking $A = \{E^{\mathcal{F}}\xi \geq 0\}$, we get in particular

$$E|E^{\mathcal{F}}\xi| = E[E^{\mathcal{F}}\xi; A] - E[E^{\mathcal{F}}\xi; A^c] = E[\xi; A] - E[\xi; A^c] \leq E|\xi|,$$

which proves (iii). Thus, the mapping $E^{\mathcal{F}}$ is uniformly L^1-continuous on L^2. Also note that L^2 is dense in L^1 by Lemma 1.11 and that L^1 is complete by Lemma 1.31. Hence, $E^{\mathcal{F}}$ extends a.s. uniquely to a linear and continuous mapping on L^1.

Properties (i) and (iii) extend by continuity to L^1, and from Lemma 1.24 we note that $E^{\mathcal{F}}\xi$ is a.s. determined by (i). If $\xi \geq 0$, we see from (i) with $A = \{E^{\mathcal{F}}\xi \leq 0\}$ together with Lemma 1.24 that $E^{\mathcal{F}}\xi \geq 0$ a.s., which proves (ii). If $0 \leq \xi_n \uparrow \xi$, then $\xi_n \to \xi$ in L^1 by dominated convergence, so by (iii) we get $E^{\mathcal{F}}\xi_n \to E^{\mathcal{F}}\xi$ in L^1. Now the sequence $(E^{\mathcal{F}}\xi_n)$ is a.s. nondecreasing by (ii), and so by Lemma 4.2 the convergence remains true in the a.s. sense. This proves (iv).

Property (vi) is obvious when $\xi, \eta \in L^2$, and it extends to the general case by means of (iv). To prove (v), we note from the characterization in (i) that $E^{\mathcal{F}}\xi = \xi$ a.s. when ξ is $\mathcal{F}$-measurable. In the general case we need to show that

$$E[\xi\eta; A] = E[\xi E^{\mathcal{F}}\eta; A], \quad A \in \mathcal{F},$$

which follows immediately from (vi). Finally, property (vii) is obvious for $\xi \in L^2$ since $L^2(\mathcal{F}) \subset L^2(\mathcal{G})$, and it extends to the general case by means of (iv). □

The next result shows that the conditional expectation $E^{\mathcal{F}}\xi$ is *local* in both ξ and $\mathcal{F}$, an observation that simplifies many proofs. Given two σ-fields $\mathcal{F}$ and $\mathcal{G}$, we say that $\mathcal{F} = \mathcal{G}$ on A if $A \in \mathcal{F} \cap \mathcal{G}$ and $A \cap \mathcal{F} = A \cap \mathcal{G}$.

Lemma 6.2 *(local property) Let the σ-fields $\mathcal{F}, \mathcal{G} \subset \mathcal{A}$ and functions $\xi, \eta \in L^1$ be such that $\mathcal{F} = \mathcal{G}$ and $\xi = \eta$ a.s. on some set $A \in \mathcal{F} \cap \mathcal{G}$. Then $E^{\mathcal{F}}\xi = E^{\mathcal{G}}\eta$ a.s. on A.*

Proof: Since $1_A E^{\mathcal{F}}\xi$ and $1_A E^{\mathcal{G}}\eta$ are $\mathcal{F} \cap \mathcal{G}$-measurable, we get $B \equiv A \cap \{E^{\mathcal{F}}\xi > E^{\mathcal{G}}\eta\} \in \mathcal{F} \cap \mathcal{G}$, and the averaging property yields

$$E[E^{\mathcal{F}}\xi; B] = E[\xi; B] = E[\eta; B] = E[E^{\mathcal{G}}\eta; B].$$

Hence, $E^{\mathcal{F}}\xi \le E^{\mathcal{G}}\eta$ a.s. on A by Lemma 1.24. The opposite inequality is obtained by interchanging the roles of $(\xi, \mathcal{F})$ and $(\eta, \mathcal{G})$. □

The *conditional probability* of an event $A \in \mathcal{A}$, given a σ-field $\mathcal{F}$, is defined as

$$P^{\mathcal{F}}A = E^{\mathcal{F}}1_A \quad \text{or} \quad P[A|\mathcal{F}] = E[1_A|\mathcal{F}], \quad A \in \mathcal{A}.$$

Thus, $P^{\mathcal{F}}A$ is the a.s. unique random variable in $L^1(\mathcal{F})$ satisfying

$$E[P^{\mathcal{F}}A; B] = P(A \cap B), \quad B \in \mathcal{F}.$$

Note that $P^{\mathcal{F}}A = PA$ a.s. iff $A \perp\!\!\!\perp \mathcal{F}$ and that $P^{\mathcal{F}}A = 1_A$ a.s. iff A agrees a.s. with a set in $\mathcal{F}$. The positivity of $E^{\mathcal{F}}$ implies $0 \le P^{\mathcal{F}}A \le 1$ a.s., and the monotone convergence property gives

$$P^{\mathcal{F}} \bigcup_n A_n = \sum_n P^{\mathcal{F}}A_n \text{ a.s.}, \quad A_1, A_2, \cdots \in \mathcal{A} \text{ disjoint.} \tag{1}$$

However, the random set function $P^{\mathcal{F}}$ is not a measure in general since the exceptional null set in (1) may depend on the sequence (A_n).

If η is a random element in some measurable space $(S, \mathcal{S})$, we define conditioning on η as conditioning with respect to the induced σ-field $\sigma(\eta)$. Thus,

$$E^{\eta}\xi = E^{\sigma(\eta)}\xi, \quad P^{\eta}A = P^{\sigma(\eta)}A,$$

or

$$E[\xi|\eta] = E[\xi|\sigma(\eta)], \quad P[A|\eta] = P[A|\sigma(\eta)].$$

By Lemma 1.13, the η-measurable function $E^{\eta}\xi$ may be represented in the form $f(\eta)$, where f is a measurable function on S, determined a.e. $\mathcal{L}(\eta)$ by the averaging property

$$E[f(\eta); \eta \in B] = E[\xi; \eta \in B], \quad B \in \mathcal{S}.$$

In particular, the function f depends only on the distribution of (ξ, η). The situation for $P^{\eta}A$ is similar. Conditioning with respect to a σ-field $\mathcal{F}$ is the special case when η is the identity map from $(\Omega, \mathcal{A})$ to $(\Omega, \mathcal{F})$.

Motivated by (1), we proceed to examine the existence of measure-valued versions of the functions $P^{\mathcal{F}}$ and P^{η}. Then recall from Chapter 1 that a *kernel* between two measurable spaces $(T, \mathcal{T})$ and $(S, \mathcal{S})$ is a function $\mu: T \times \mathcal{S} \to \overline{\mathbb{R}}_+$ such that $\mu(t, B)$ is $\mathcal{T}$-measurable in $t \in T$ for fixed $B \in \mathcal{S}$ and a measure in $B \in \mathcal{S}$ for fixed $t \in T$. Say that μ is a *probability kernel* if $\mu(t, S) = 1$ for all t. Kernels on the basic probability space Ω are called *random measures*.

Now fix a σ-field $\mathcal{F} \subset \mathcal{A}$ and a random element ξ in some measurable space $(S, \mathcal{S})$. By a *regular conditional distribution of ξ, given $\mathcal{F}$*, we mean a version of the function $P[\xi \in \cdot | \mathcal{F}]$ on $\Omega \times \mathcal{S}$ which is a probability kernel from $(\Omega, \mathcal{F})$ to $(S, \mathcal{S})$, hence an $\mathcal{F}$-measurable random probability measure on S. More generally, if η is another random element in some measurable space $(T, \mathcal{T})$, a regular conditional distribution of ξ, given η, is defined as a random measure of the form

$$\mu(\eta, B) = P[\xi \in B|\eta] \quad \text{a.s.}, \quad B \in \mathcal{S}, \tag{2}$$

where μ is a probability kernel from T to S. In the extreme cases when ξ is $\mathcal{F}$-measurable or independent of $\mathcal{F}$, we note that $P[\xi \in B|\mathcal{F}]$ has the regular version $1\{\xi \in B\}$ or $P\{\xi \in B\}$, respectively. The general case requires some regularity conditions on the space S.

Theorem 6.3 *(conditional distribution)* *For any Borel space S and measurable space T, let ξ and η be random elements in S and T, respectively. Then there exists a probability kernel μ from T to S satisfying $P[\xi \in \cdot |\eta] = \mu(\eta, \cdot)$ a.s., and μ is unique a.e. $\mathcal{L}(\eta)$.*

Proof: We may assume that $S \in \mathcal{B}(\mathbb{R})$. For every $r \in \mathbb{Q}$ we may choose some measurable function $f_r = f(\cdot, r) : T \to [0, 1]$ such that

$$f(\eta, r) = P[\xi \le r|\eta] \quad \text{a.s.}, \quad r \in \mathbb{Q}. \tag{3}$$

Let A be the set of all $t \in T$ such that $f(t, r)$ is nondecreasing in $r \in \mathbb{Q}$ with limits 1 and 0 at $\pm\infty$. Since A is specified by countably many measurable conditions, each of which holds a.s. at η, we have $A \in \mathcal{T}$ and $\eta \in A$ a.s. Now define

$$F(t, x) = 1_A(t) \inf_{r > x} f(t, r) + 1_{A^c}(t) 1\{x \ge 0\}, \quad x \in \mathbb{R}, \ t \in T,$$

and note that $F(t, \cdot)$ is a distribution function on $\mathbb{R}$ for every $t \in T$. Hence, by Proposition 2.14 there exist some probability measures $m(t, \cdot)$ on $\mathbb{R}$ with

$$m(t, (-\infty, x]) = F(t, x), \quad x \in \mathbb{R}, \ t \in T.$$

The function $F(t, x)$ is clearly measurable in t for each x, and by a monotone class argument it follows that m is a kernel from T to $\mathbb{R}$.

By (3) and the monotone convergence property of E^η, we have

$$m(\eta, (-\infty, x]) = F(\eta, x) = P[\xi \le x|\eta] \quad \text{a.s.}, \quad x \in \mathbb{R}.$$

Using a monotone class argument based on the a.s. monotone convergence property, we may extend the last relation to

$$m(\eta, B) = P[\xi \in B|\eta] \quad \text{a.s.}, \quad B \in \mathcal{B}(\mathbb{R}). \tag{4}$$

In particular, we get $m(\eta, S^c) = 0$ a.s., and so (4) remains true on $\mathcal{S} = \mathcal{B} \cap S$ with m replaced by the kernel

$$\mu(t, \cdot) = m(t, \cdot) 1\{m(t, S) = 1\} + \delta_s 1\{m(t, S) < 1\}, \quad t \in T,$$

where $s \in S$ is arbitrary. If μ' is another kernel with the stated property, then

$$\mu(\eta, (-\infty, r]) = P[\xi \le r|\eta] = \mu'(\eta, (-\infty, r]) \quad \text{a.s.}, \quad r \in \mathbb{Q},$$

and a monotone class argument yields $\mu(\eta, \cdot) = \mu'(\eta, \cdot)$ a.s. $\square$

Our next aim is to extend Fubini's theorem, by showing how ordinary and conditional expectations can be computed by integration with respect to suitable conditional distributions. The result may be regarded as a *disintegration* of measures on a product space into their one-dimensional components.

Theorem 6.4 *(disintegration)* *Fix two measurable spaces S and T, a σ-field $\mathcal{F} \subset \mathcal{A}$, and a random element ξ in S such that $P[\xi \in \cdot | \mathcal{F}]$ has a regular version ν. Further consider an $\mathcal{F}$-measurable random element η in T and a measurable function f on $S \times T$ with $E|f(\xi, \eta)| < \infty$. Then*

$$E[f(\xi, \eta)|\mathcal{F}] = \int \nu(ds) f(s, \eta) \quad a.s. \tag{5}$$

The a.s. existence and $\mathcal{F}$-measurability of the integral on the right should be regarded as part of the assertion. In the special case when $\mathcal{F} = \sigma(\eta)$ and $P[\xi \in \cdot | \eta] = \mu(\eta, \cdot)$ for some probability kernel μ from T to S, (5) becomes

$$E[f(\xi, \eta)|\eta] = \int \mu(\eta, ds) f(s, \eta) \quad \text{a.s.} \tag{6}$$

Integrating (5) and (6), we get the commonly used formulas

$$Ef(\xi, \eta) = E \int \nu(ds) f(s, \eta) = E \int \mu(\eta, ds) f(s, \eta). \tag{7}$$

If $\xi \perp\!\!\!\perp \eta$, we may take $\mu(\eta, \cdot) \equiv \mathcal{L}(\xi)$, and (7) reduces to the relation in Lemma 3.11.

Proof of Theorem 6.4: If $B \in \mathcal{S}$ and $C \in \mathcal{T}$, we may use the averaging property of conditional expectations to get

$$\begin{aligned} P\{\xi \in B, \eta \in C\} &= E[P[\xi \in B|\mathcal{F}]; \eta \in C] = E[\nu B; \eta \in C] \\ &= E \int \nu(ds) 1\{s \in B, \eta \in C\}, \end{aligned}$$

which proves the first relation in (7) for $f = 1_{B \times C}$. The formula extends, along with the measurability of the inner integral on the right, first by a monotone class argument to all measurable indicator functions, and then by linearity and monotone convergence to any measurable function $f \geq 0$.

Now fix a measurable function $f : S \times T \to \mathbb{R}_+$ with $Ef(\xi, \eta) < \infty$, and let $A \in \mathcal{F}$ be arbitrary. Regarding $(\eta, 1_A)$ as an $\mathcal{F}$-measurable random element in $T \times \{0, 1\}$, we may conclude from (7) that

$$E[f(\xi, \eta); A] = E \int \nu(ds) f(s, \eta) 1_A, \quad A \in \mathcal{F}.$$

This proves (5) for $f \geq 0$, and the general result follows by taking differences. $\square$

Applying (7) to functions of the form $f(\xi)$, we may extend many properties of ordinary expectations to a conditional setting. In particular, such extensions hold for the Jensen, Hölder, and Minkowski inequalities. The first of those implies the L^p-*contractivity*

$$\|E^{\mathcal{F}}\xi\|_p \leq \|\xi\|_p, \quad \xi \in L^p, \ p \geq 1.$$

Considering conditional distributions of entire sequences $(\xi, \xi_1, \xi_2, \ldots)$, we may further derive conditional versions of the basic continuity properties of ordinary integrals.

The following result plays an important role in Chapter 7.

Lemma 6.5 *(uniform integrability, Doob) For any $\xi \in L^1$, the conditional expectations $E[\xi|\mathcal{F}]$, $\mathcal{F} \subset \mathcal{A}$, are uniformly integrable.*

Proof: By Jensen's inequality and the self-adjointness property,

$$E[|E^{\mathcal{F}}\xi|; A] \leq E[E^{\mathcal{F}}|\xi|; A] = E[|\xi|P^{\mathcal{F}}A], \quad A \in \mathcal{A},$$

and by Lemma 4.10 we need to show that this tends to zero as $PA \to 0$, uniformly in $\mathcal{F}$. By dominated convergence along subsequences, it is then enough to show that $P^{\mathcal{F}_n} A_n \xrightarrow{P} 0$ for any σ-fields $\mathcal{F}_n \subset \mathcal{A}$ and sets $A_n \in \mathcal{A}$ with $PA_n \to 0$. But this is clear, since $EP^{\mathcal{F}_n} A_n = PA_n \to 0$. $\square$

Turning to the topic of conditional independence, consider any sub-σ-fields $\mathcal{F}_1, \ldots, \mathcal{F}_n, \mathcal{G} \subset \mathcal{A}$. Imitating the definition of ordinary independence, we say that $\mathcal{F}_1, \ldots, \mathcal{F}_n$ are *conditionally independent, given $\mathcal{G}$*, if

$$P^{\mathcal{G}} \bigcap\nolimits_{k \leq n} B_k = \prod\nolimits_{k \leq n} P^{\mathcal{G}} B_k \text{ a.s.}, \quad B_k \in \mathcal{F}_k, \ k = 1, \ldots, n.$$

For infinite collections of σ-fields $\mathcal{F}_t$, $t \in T$, the same property is required for every finite subcollection $\mathcal{F}_{t_1}, \ldots, \mathcal{F}_{t_n}$ with distinct indices $t_1, \ldots, t_n \in T$. We use the symbol $\perp\!\!\!\perp_{\mathcal{G}}$ to denote pairwise conditional independence, given some σ-field $\mathcal{G}$. Conditional independence involving events A_t or random elements ξ_t, $t \in T$, is defined as before in terms of the induced σ-fields $\sigma(A_t)$ or $\sigma(\xi_t)$, respectively, and the notation involving $\perp\!\!\!\perp$ carries over to this case.

In particular, we note that any $\mathcal{F}$-measurable random elements ξ_t are conditionally independent, given $\mathcal{F}$. If the ξ_t are instead independent of $\mathcal{F}$, then their conditional independence, given $\mathcal{F}$, is equivalent to ordinary independence between the ξ_t. By Theorem 6.3, any general statement or formula involving independencies between countably many random elements in some Borel spaces has a conditional counterpart. For example, we see from Lemma 3.8 that the σ-fields $\mathcal{F}_1, \mathcal{F}_2, \ldots$ are conditionally independent, given some $\mathcal{G}$, iff

$$(\mathcal{F}_1, \ldots, \mathcal{F}_n) \underset{\mathcal{G}}{\perp\!\!\!\perp} \mathcal{F}_{n+1}, \quad n \in \mathbb{N}.$$

Much more can be said in the conditional case, and we begin with a fundamental characterization. Here and below, $\mathcal{F}, \mathcal{G}, \ldots$ with or without subscripts denote sub-σ-fields of $\mathcal{A}$.

Proposition 6.6 *(conditional independence, Doob)* For any σ-fields $\mathcal{F}$, $\mathcal{G}$, and $\mathcal{H}$, we have $\mathcal{F} \perp\!\!\!\perp_{\mathcal{G}} \mathcal{H}$ iff

$$P[H|\mathcal{F},\mathcal{G}] = P[H|\mathcal{G}] \text{ a.s.}, \quad H \in \mathcal{H}. \tag{8}$$

Proof: Assuming (8) and using the chain and pull-out properties of conditional expectations, we get for any $F \in \mathcal{F}$ and $H \in \mathcal{H}$

$$\begin{aligned} P^{\mathcal{G}}(F \cap H) &= E^{\mathcal{G}} P^{\mathcal{F} \vee \mathcal{G}}(F \cap H) = E^{\mathcal{G}}[P^{\mathcal{F} \vee \mathcal{G}} H; F] \\ &= E^{\mathcal{G}}[P^{\mathcal{G}} H; F] = (P^{\mathcal{G}} F)(P^{\mathcal{G}} H), \end{aligned}$$

which shows that $\mathcal{F} \perp\!\!\!\perp_{\mathcal{G}} \mathcal{H}$. Conversely, assuming $\mathcal{F} \perp\!\!\!\perp_{\mathcal{G}} \mathcal{H}$ and using the chain and pull-out properties, we get for any $F \in \mathcal{F}$, $G \in \mathcal{G}$, and $H \in \mathcal{H}$

$$\begin{aligned} E[P^{\mathcal{G}} H; F \cap G] &= E[(P^{\mathcal{G}} F)(P^{\mathcal{G}} H); G] \\ &= E[P^{\mathcal{G}}(F \cap H); G] = P(F \cap G \cap H). \end{aligned}$$

By a monotone class argument, this extends to

$$E[P^{\mathcal{G}} H; A] = P(H \cap A), \quad A \in \mathcal{F} \vee \mathcal{G},$$

and (8) follows by the averaging characterization of $P^{\mathcal{F} \vee \mathcal{G}} H$. □

From the last result we may easily deduce some further useful properties. Let $\overline{\mathcal{G}}$ denote the *completion* of $\mathcal{G}$ with respect to the basic σ-field $\mathcal{A}$, generated by $\mathcal{G}$ and the family $\mathcal{N} = \{N \subset A; A \in \mathcal{A}, PA = 0\}$.

Corollary 6.7 For any σ-fields $\mathcal{F}$, $\mathcal{G}$, and $\mathcal{H}$, we have
 (i) $\mathcal{F} \perp\!\!\!\perp_{\mathcal{G}} \mathcal{H}$ iff $\mathcal{F} \perp\!\!\!\perp_{\mathcal{G}} (\mathcal{G}, \mathcal{H})$;
 (ii) $\mathcal{F} \perp\!\!\!\perp_{\mathcal{G}} \mathcal{F}$ iff $\mathcal{F} \subset \overline{\mathcal{G}}$.

Proof: (i) By Proposition 6.6, both relations are equivalent to

$$P[F|\mathcal{G}, \mathcal{H}] = P[F|\mathcal{G}] \text{ a.s.}, \quad F \in \mathcal{F}.$$

(ii) If $\mathcal{F} \perp\!\!\!\perp_{\mathcal{G}} \mathcal{F}$, then by Proposition 6.6

$$1_F = P[F|\mathcal{F}, \mathcal{G}] = P[F|\mathcal{G}] \text{ a.s.}, \quad F \in \mathcal{F},$$

which implies $\mathcal{F} \subset \overline{\mathcal{G}}$. Conversely, the latter relation yields

$$P[F|\mathcal{G}] = P[F|\overline{\mathcal{G}}] = 1_F = P[F|\mathcal{F}, \mathcal{G}] \text{ a.s.}, \quad F \in \mathcal{F},$$

and so $\mathcal{F} \perp\!\!\!\perp_{\mathcal{G}} \mathcal{F}$ by Proposition 6.6. □

The following result is often applied in both directions.

Proposition 6.8 *(chain rule) For any σ-fields $\mathcal{G}$, $\mathcal{H}$, and $\mathcal{F}_1, \mathcal{F}_2, \ldots$, these conditions are equivalent:*

(i) $\mathcal{H} \underset{\mathcal{G}}{\perp\!\!\!\perp} (\mathcal{F}_1, \mathcal{F}_2, \ldots)$;

(ii) $\mathcal{H} \underset{\mathcal{G}, \mathcal{F}_1, \ldots, \mathcal{F}_n}{\perp\!\!\!\perp} \mathcal{F}_{n+1}, \quad n \geq 0.$

In particular, we have the commonly used equivalence

$$\mathcal{H} \perp\!\!\!\perp_{\mathcal{G}} (\mathcal{F}, \mathcal{F}') \iff \mathcal{H} \perp\!\!\!\perp_{\mathcal{G}} \mathcal{F}, \ \mathcal{H} \perp\!\!\!\perp_{\mathcal{G}, \mathcal{F}} \mathcal{F}'$$

Proof: Assuming (i), we get by Proposition 6.6 for any $H \in \mathcal{H}$ and $n \geq 0$

$$P[H|\mathcal{G}, \mathcal{F}_1, \ldots, \mathcal{F}_n] = P[H|\mathcal{G}] = P[H|\mathcal{G}, \mathcal{F}_1, \ldots, \mathcal{F}_{n+1}],$$

and (ii) follows by another application of Proposition 6.6.

Now assume (ii) instead, and conclude by Proposition 6.6 that for any $H \in \mathcal{H}$

$$P[H|\mathcal{G}, \mathcal{F}_1, \ldots, \mathcal{F}_n] = P[H|\mathcal{G}, \mathcal{F}_1, \ldots, \mathcal{F}_{n+1}], \quad n \geq 0.$$

Summing over $n < m$ gives

$$P[H|\mathcal{G}] = P[H|\mathcal{G}, \mathcal{F}_1, \ldots, \mathcal{F}_m], \quad m \geq 1,$$

and so by Proposition 6.6 we have $\mathcal{H} \perp\!\!\!\perp_{\mathcal{G}}(\mathcal{F}_1, \ldots, \mathcal{F}_m)$ for all $m \geq 1$, which extends to (i) by a monotone class argument. $\square$

The last result is even useful for establishing ordinary independence. In fact, taking $\mathcal{G} = \{\emptyset, \Omega\}$ in Proposition 6.8, we see that $\mathcal{H} \perp\!\!\!\perp (\mathcal{F}_1, \mathcal{F}_2, \ldots)$ iff

$$\mathcal{H} \underset{\mathcal{F}_1, \ldots, \mathcal{F}_n}{\perp\!\!\!\perp} \mathcal{F}_{n+1}, \quad n \geq 0.$$

Our next aim is to show how regular conditional distributions can be used to construct random elements with desired properties. This may require an extension of the basic probability space. By an *extension* of $(\Omega, \mathcal{A}, P)$ we mean a product space $(\hat{\Omega}, \hat{\mathcal{A}}) = (\Omega \times S, \mathcal{A} \otimes \mathcal{S})$, equipped with a probability measure $\hat{P}$ satisfying $\hat{P}(\cdot \times S) = P$. Any random element ξ on Ω may be regarded as a function on $\hat{\Omega}$. Thus, we may formally replace ξ by the random element $\hat{\xi}(\omega, s) = \xi(\omega)$, which clearly has the same distribution. For extensions of this type, we may retain our original notation and write P and ξ instead of $\hat{P}$ and $\hat{\xi}$.

We begin with an elementary extension suggested by Theorem 6.4. The result is needed for various constructions in Chapter 12.

Lemma 6.9 *(extension)* Fix a probability kernel μ between two measurable spaces S and T, and let ξ be a random element in S. Then there exists a random element η in T, defined on some extension of the original probability space Ω, such that $P[\eta \in \cdot|\xi] = \mu(\xi, \cdot)$ a.s. and also $\eta \perp\!\!\!\perp_\xi \zeta$ for every random element ζ on Ω.

Proof: Put $(\hat{\Omega}, \hat{\mathcal{A}}) = (\Omega \times T, \mathcal{A} \otimes \mathcal{T})$, where $\mathcal{T}$ denotes the σ-field in T, and define a probability measure $\hat{P}$ on $\hat{\Omega}$ by

$$\hat{P}A = E \int 1_A(\cdot, t)\mu(\xi, dt), \quad A \in \hat{\mathcal{A}}.$$

Then clearly $\hat{P}(\cdot \times T) = P$, and the random element $\eta(\omega, t) \equiv t$ on $\hat{\Omega}$ satisfies $\hat{P}[\eta \in \cdot|\mathcal{A}] = \mu(\xi, \cdot)$ a.s. In particular, we get $\eta \perp\!\!\!\perp_\xi \mathcal{A}$ by Proposition 6.6, and so $\eta \perp\!\!\!\perp_\xi \zeta$. □

For most constructions we need only a single *randomization variable*. By this we mean a $U(0, 1)$ random variable ϑ that is independent of all previously introduced random elements and σ-fields. The basic probability space is henceforth assumed to be rich enough to support any randomization variables we may need. This involves no essential loss of generality, since we can always get the condition fulfilled by a simple extension of the original space. In fact, it suffices to take

$$\hat{\Omega} = \Omega \times [0, 1], \quad \hat{\mathcal{A}} = \mathcal{A} \otimes \mathcal{B}[0, 1], \quad \hat{P} = P \otimes \lambda,$$

where λ denotes Lebesgue measure on $[0, 1]$. Then $\vartheta(\omega, t) \equiv t$ is $U(0, 1)$ on $\hat{\Omega}$ and $\vartheta \perp\!\!\!\perp \mathcal{A}$. By Lemma 3.21 we may use ϑ to produce a whole sequence of independent randomization variables $\vartheta_1, \vartheta_2, \ldots$ if required.

The following basic result shows how a probabilistic structure can be carried over from one context to another by means of a suitable randomization. Constructions of this type are frequently employed in the sequel.

Theorem 6.10 *(transfer)* For any measurable space S and Borel space T, let $\xi \stackrel{d}{=} \tilde{\xi}$ and η be random elements in S and T, respectively. Then there exists a random element $\tilde{\eta}$ in T with $(\tilde{\xi}, \tilde{\eta}) \stackrel{d}{=} (\xi, \eta)$. More precisely, there exists a measurable function $f: S \times [0, 1] \to T$ such that we may take $\tilde{\eta} = f(\tilde{\xi}, \vartheta)$ whenever $\vartheta \perp\!\!\!\perp \tilde{\xi}$ is $U(0, 1)$.

Proof: By Theorem 6.3 there exists a probability kernel μ from S to T satisfying

$$\mu(\xi, B) = P[\eta \in B|\xi], \quad B \in \mathcal{B}[0, 1],$$

and by Lemma 3.22 we may choose a measurable function $f: S \times [0, 1] \to T$ such that $f(s, \vartheta)$ has distribution $\mu(s, \cdot)$ for every $s \in S$. Define $\tilde{\eta} = f(\tilde{\xi}, \vartheta)$. Using Lemmas 1.22 and 3.11 together with Theorem 6.4, we get for any measurable function $g: S \times [0,1] \to \mathbb{R}_+$

$$Eg(\tilde{\xi}, \tilde{\eta}) = Eg(\tilde{\xi}, f(\tilde{\xi}, \vartheta)) = E \int g(\xi, f(\xi, u)) du$$
$$= E \int g(\xi, t) \mu(\xi, dt) = Eg(\xi, \eta),$$

which shows that $(\tilde{\xi}, \tilde{\eta}) \stackrel{d}{=} (\xi, \eta)$. □

The following version of the last result is often useful to transfer representations of random objects.

Corollary 6.11 *(stochastic equations) Fix two Borel spaces S and T, a measurable mapping $f: T \to S$, and some random elements ξ in S and η in T with $\xi \stackrel{d}{=} f(\eta)$. Then there exists a random element $\tilde{\eta} \stackrel{d}{=} \eta$ in T with $\xi = f(\tilde{\eta})$ a.s.*

Proof: By Theorem 6.10 there exists some random element $\tilde{\eta}$ in T with $(\xi, \tilde{\eta}) \stackrel{d}{=} (f(\eta), \eta)$. In particular, $\tilde{\eta} \stackrel{d}{=} \eta$ and $(\xi, f(\tilde{\eta})) \stackrel{d}{=} (f(\eta), f(\eta))$. Since the diagonal in S^2 is measurable, we get $P\{\xi = f(\tilde{\eta})\} = P\{f(\eta) = f(\eta)\} = 1$, and so $\xi = f(\tilde{\eta})$ a.s. □

The last result leads in particular to a useful extension of Theorem 4.30.

Corollary 6.12 *(extended Skorohod coupling) Let $f, f_1, f_2, \ldots$ be measurable functions from a Borel space S to a Polish space T, and let $\xi, \xi_1, \xi_2, \ldots$ be random elements in S with $f_n(\xi_n) \stackrel{d}{\to} f(\xi)$. Then there exist some random elements $\tilde{\xi} \stackrel{d}{=} \xi$ and $\tilde{\xi}_n \stackrel{d}{=} \xi_n$ such that $f_n(\tilde{\xi}_n) \to f(\tilde{\xi})$ a.s.*

Proof: By Theorem 4.30 there exist some $\eta \stackrel{d}{=} f(\xi)$ and $\eta_n \stackrel{d}{=} f_n(\xi_n)$ with $\eta_n \to \eta$ a.s. By Corollary 6.11 we may further choose some $\tilde{\xi} \stackrel{d}{=} \xi$ and $\tilde{\xi}_n \stackrel{d}{=} \xi_n$ such that a.s. $f(\tilde{\xi}) = \eta$ and $f_n(\tilde{\xi}_n) = \eta_n$ for all n. But then $f_n(\tilde{\xi}_n) \to f(\tilde{\xi})$ a.s. □

The next result clarifies the relationship between randomizations and conditional independence. Important applications appear in Chapters 8, 12, and 21.

Proposition 6.13 *(conditional independence and randomization) Let ξ, η, and ζ be random elements in some measurable spaces S, T, and U, respectively, where S is Borel. Then $\xi \perp\!\!\!\perp_\eta \zeta$ iff $\xi = f(\eta, \vartheta)$ a.s. for some measurable function $f: T \times [0,1] \to S$ and some $U(0,1)$ random variable $\vartheta \perp\!\!\!\perp (\eta, \zeta)$.*

Proof: First assume that $\xi = f(\eta, \vartheta)$ a.s., where f is measurable and $\vartheta \perp\!\!\!\perp (\eta, \zeta)$. Then Proposition 6.8 yields $\vartheta \perp\!\!\!\perp_\eta \zeta$, and so $(\eta, \vartheta) \perp\!\!\!\perp_\eta \zeta$ by Corollary 6.7, which implies $\xi \perp\!\!\!\perp_\eta \zeta$.

Conversely, assume that $\xi \perp\!\!\!\perp_\eta \zeta$, and let $\vartheta \perp\!\!\!\perp (\eta, \zeta)$ be $U(0,1)$. By Theorem 6.10 there exists some measurable function $f: T \times [0,1] \to S$ such that the random element $\tilde{\xi} = f(\eta, \vartheta)$ satisfies $\tilde{\xi} \stackrel{d}{=} \xi$ and $(\tilde{\xi}, \eta) \stackrel{d}{=} (\xi, \eta)$. By the sufficiency part, we further note that $\tilde{\xi} \perp\!\!\!\perp_\eta \zeta$. Hence, by Proposition 6.6,

$$P[\tilde{\xi} \in \cdot \,|\eta, \zeta] = P[\tilde{\xi} \in \cdot \,|\eta] = P[\xi \in \cdot \,|\eta] = P[\xi \in \cdot \,|\eta, \zeta],$$

and so $(\tilde{\xi}, \eta, \zeta) \stackrel{d}{=} (\xi, \eta, \zeta)$. By Theorem 6.10 we may choose some $\tilde{\vartheta} \stackrel{d}{=} \vartheta$ with $(\xi, \eta, \zeta, \tilde{\vartheta}) \stackrel{d}{=} (\tilde{\xi}, \eta, \zeta, \vartheta)$. In particular, $\tilde{\vartheta} \perp\!\!\!\perp (\eta, \zeta)$ and $(\xi, f(\eta, \tilde{\vartheta})) \stackrel{d}{=} (\tilde{\xi}, f(\eta, \vartheta))$. Since $\tilde{\xi} = f(\eta, \vartheta)$ and the diagonal in S^2 is measurable, we get $\xi = f(\eta, \tilde{\vartheta})$ a.s., and so the stated condition holds with $\tilde{\vartheta}$ in place of ϑ. □

We may use the transfer theorem to construct random sequences or processes with given finite-dimensional distributions. Given any measurable spaces $S_1, S_2, \ldots$, we say that a sequence of probability measures μ_n on $S_1 \times \cdots \times S_n$, $n \in \mathbb{N}$, is *projective* if

$$\mu_{n+1}(\cdot \times S_{n+1}) = \mu_n, \quad n \in \mathbb{N}. \tag{9}$$

Theorem 6.14 *(existence of random sequences, Daniell)* Given a projective sequence of probability measures μ_n on $S_1 \times \cdots \times S_n$, $n \in \mathbb{N}$, where $S_2, S_3, \ldots$ are Borel, there exist some random elements ξ_n in S_n, $n \in \mathbb{N}$, such that $\mathcal{L}(\xi_1, \ldots, \xi_n) = \mu_n$ for all n.

Proof: By Lemmas 3.10 and 3.21 there exist some independent random variables $\xi_1, \vartheta_2, \vartheta_3, \ldots$ such that $\mathcal{L}(\xi_1) = \mu_1$ and the ϑ_n are i.i.d. $U(0,1)$. We proceed to construct recursively $\xi_2, \xi_2, \ldots$ with the desired properties such that each ξ_n is a measurable function of $\xi_1, \vartheta_2, \ldots, \vartheta_n$. Assuming that $\xi_1, \ldots, \xi_n$ have already been constructed, let $\eta_1, \ldots, \eta_{n+1}$ be arbitrary with joint distribution μ_{n+1}. The projective property yields $(\xi_1, \ldots, \xi_n) \stackrel{d}{=} (\eta_1, \ldots, \eta_n)$, and so by Theorem 6.10 we may form ξ_{n+1} as a measurable function of $\xi_1, \ldots, \xi_n, \vartheta_{n+1}$ such that $(\xi_1, \ldots, \xi_{n+1}) \stackrel{d}{=} (\eta_1, \ldots, \eta_{n+1})$. This completes the recursion. □

The last theorem may be used to extend a process from bounded to unbounded domains. We state the result in an abstract form, designed to fulfill the needs of Chapters 18 and 24. Let I denote the identity mapping on any space.

Corollary 6.15 *(projective limit)* For any Borel spaces $S, S_1, S_2, \ldots$, consider some measurable mappings $\pi_n : S \to S_n$ and $\pi_k^n : S_n \to S_k$, $k \leq n$, such that

$$\pi_k^n = \pi_k^m \circ \pi_m^n, \quad k \leq m \leq n. \tag{10}$$

Let $\overline{S}$ denote the set of sequences $(s_1, s_2, \ldots) \in S_1 \times S_2 \times \cdots$ with $\pi_k^n s_n = s_k$ for all $k \leq n$, and suppose there exists a measurable mapping $h : \overline{S} \to S$ satisfying $(\pi_1, \pi_2, \ldots) \circ h = I$ on $\overline{S}$. Then for any probability measures μ_n on S_n with $\mu_n \circ (\pi_k^n)^{-1} = \mu_k$ for all $k \leq n$, there exists a probability measure μ on S such that $\mu \circ \pi_n^{-1} = \mu_n$ for all n.

Proof: Introduce the measures

$$\bar{\mu}_n = \mu_n \circ (\pi_1^n, \ldots, \pi_n^n)^{-1}, \quad n \in \mathbb{N}, \qquad (11)$$

and conclude from (10) and the relation between the μ_n that

$$\begin{aligned}
\bar{\mu}_{n+1}(\cdot \times S_{n+1}) &= \mu_{n+1} \circ (\pi_1^{n+1}, \ldots, \pi_n^{n+1})^{-1} \\
&= \mu_{n+1} \circ (\pi_n^{n+1})^{-1} \circ (\pi_1^n, \ldots, \pi_n^n)^{-1} \\
&= \mu_n \circ (\pi_1^n, \ldots, \pi_n^n)^{-1} = \bar{\mu}_n.
\end{aligned}$$

By Theorem 6.14 there exists some measure $\bar{\mu}$ on $S_1 \times S_2 \times \cdots$ with

$$\bar{\mu} \circ (\bar{\pi}_1, \ldots, \bar{\pi}_n)^{-1} = \bar{\mu}_n, \quad n \in \mathbb{N}, \qquad (12)$$

where $\bar{\pi}_1, \bar{\pi}_2, \ldots$ denote the coordinate projections in $S_1 \times S_2 \times \cdots$. From (10)–(12) we see that $\bar{\mu}$ is restricted to $\overline{S}$, which allows us to define $\mu = \bar{\mu} \circ h^{-1}$. It remains to note that

$$\mu \circ \pi_n^{-1} = \bar{\mu} \circ (\pi_n h)^{-1} = \bar{\mu} \circ \bar{\pi}_n^{-1} = \bar{\mu}_n \circ \bar{\pi}_n^{-1} = \mu_n \circ (\pi_n^n)^{-1} = \mu_n. \qquad \square$$

We often need a version of Theorem 6.14 for processes on an arbitrary index set T. For any collection of spaces S_t, $t \in T$, define $S_I = \times_{t \in I} S_t$, $I \subset T$. Similarly, if each S_t is endowed with a σ-field $\mathcal{S}_t$, let $\mathcal{S}_I$ denote the product σ-field $\otimes_{t \in I} \mathcal{S}_t$. Finally, if each ξ_t is a random element in S_t, write ξ_I for the restriction of the process (ξ_t) to the index set I.

Now let $\hat{T}$ and $\overline{T}$ denote the classes of finite and countable subsets of T, respectively. A family of probability measures μ_I, $I \in \hat{T}$ or $\overline{T}$, is said to be *projective* if

$$\mu_J(\cdot \times S_{J \setminus I}) = \mu_I, \quad I \subset J \text{ in } \hat{T} \text{ or } \overline{T}. \qquad (13)$$

Theorem 6.16 *(existence of processes, Kolmogorov)* *For any set of Borel spaces S_t, $t \in T$, consider a projective family of probability measures μ_I on S_I, $I \in \hat{T}$. Then there exist some random elements X_t in S_t, $t \in T$, such that $\mathcal{L}(X_I) = \mu_I$ for all $I \in \hat{T}$.*

Proof: Recall that the product σ-field $\mathcal{S}_T$ in S_T is generated by all coordinate projections π_t, $t \in T$, and hence consists of all countable *cylinder* sets $B \times S_{T \setminus U}$, $B \in \mathcal{S}_U$, $U \in \overline{T}$. For each $U \in \overline{T}$, there exists by Theorem 6.14 some probability measure μ_U on S_U satisfying

$$\mu_U(\cdot \times S_{U \setminus I}) = \mu_I, \quad I \in \hat{U},$$

and by Proposition 3.2 the family μ_U, $U \in \overline{T}$, is again projective. We may then define a function $\mu \colon \mathcal{S}_T \to [0, 1]$ by

$$\mu(\cdot \times S_{T \setminus U}) = \mu_U, \quad U \in \overline{T}.$$

To check the countable additivity of μ, consider any disjoint sets $A_1, A_2, \cdots \in \mathcal{S}_T$. For each n we have $A_n = B_n \times S_{T \setminus U_n}$ for some $U_n \in \overline{T}$

and $B_n \in \mathcal{S}_{U_n}$. Writing $U = \bigcup_n U_n$ and $C_n = B_n \times \mathcal{S}_{U \setminus U_n}$, we get

$$\mu \bigcup_n A_n = \mu_U \bigcup_n C_n = \sum_n \mu_U C_n = \sum_n \mu A_n.$$

We may now define the process $X = (X_t)$ as the identity mapping on the probability space $(\mathcal{S}_T, \mathcal{S}_T, \mu)$. □

If the projective sequence in Theorem 6.14 is defined recursively in terms of a sequence of conditional distributions, then no regularity condition is needed on the state spaces. For a precise statement, define the product $\mu \otimes \nu$ of two kernels μ and ν as in Chapter 1.

Theorem 6.17 *(extension by conditioning, Ionescu Tulcea)* *For any measurable spaces $(S_n, \mathcal{S}_n)$ and probability kernels μ_n from $S_1 \times \cdots \times S_{n-1}$ to S_n, $n \in \mathbb{N}$, there exist some random elements ξ_n in S_n, $n \in \mathbb{N}$, such that $\mathcal{L}(\xi_1, \ldots, \xi_n) = \mu_1 \otimes \cdots \otimes \mu_n$ for all n.*

Proof: Put $\mathcal{F}_n = \mathcal{S}_1 \otimes \cdots \otimes \mathcal{S}_n$ and $T_n = S_{n+1} \times S_{n+2} \times \cdots$, and note that the class $\mathcal{C} = \bigcup_n (\mathcal{F}_n \times T_n)$ is a field in T_0 generating the σ-field $\mathcal{F}_\infty$. Define an additive function μ on $\mathcal{C}$ by

$$\mu(A \times T_n) = (\mu_1 \otimes \cdots \otimes \mu_n) A, \quad A \in \mathcal{F}_n, \ n \in \mathbb{N}, \tag{14}$$

which is clearly independent of the representation $C = A \times T_n$. We need to extend μ to a probability measure on $\mathcal{F}_\infty$. By Theorem 2.5, it is then enough to show that μ is continuous at $\emptyset$.

For any sequence $C_1, C_2, \cdots \in \mathcal{C}$ with $C_n \downarrow \emptyset$, we need to show that $\mu C_n \to 0$. Renumbering if necessary, we may assume for each n that $C_n = A_n \times T_n$ with $A_n \in \mathcal{F}_n$. Now define

$$f_k^n = (\mu_{k+1} \otimes \cdots \otimes \mu_n) 1_{A_n}, \quad k \leq n, \tag{15}$$

with the understanding that $f_n^n = 1_{A_n}$ for $k = n$. By Lemma 1.41 (i) and (iii), each f_k^n is an $\mathcal{F}_k$-measurable function on $S_1 \times \cdots \times S_k$, and from (15) we note that

$$f_k^n = \mu_{k+1} f_{k+1}^n, \quad 0 \leq k < n. \tag{16}$$

Since $C_n \downarrow \emptyset$, the functions f_k^n are nonincreasing in n for fixed k, say with limits g_k. By (16) and dominated convergence,

$$g_k = \mu_{k+1} g_{k+1}, \quad k \geq 0. \tag{17}$$

Combining (14) and (15), we get $\mu C_n = f_0^n \downarrow g_0$. If $g_0 > 0$, then by (17) there exists some $s_1 \in S_1$ with $g_1(s_1) > 0$. Continuing recursively, we may construct a sequence $\bar{s} = (s_1, s_2, \ldots) \in T_0$ such that $g_n(s_1, \ldots, s_n) > 0$ for all n. Then

$$1_{C_n}(\bar{s}) = 1_{A_n}(s_1, \ldots, s_n) = f_n^n(s_1, \ldots, s_n) \geq g_n(s_1, \ldots, s_n) > 0,$$

and so $\bar{s} \in \bigcap_n C_n$, which contradicts the hypothesis $C_n \downarrow \emptyset$. Thus, $g_0 = 0$, which means that $\mu C_n \to 0$. □

As a simple application, we may deduce the existence of independent random elements with arbitrary distributions. The result extends the elementary Theorem 3.19.

Corollary 6.18 *(infinite product measures, Lomnicki and Ulam)* *For any collection of probability spaces $(S_t, \mathcal{S}_t, \mu_t)$, $t \in T$, there exist some independent random elements ξ_t in S_t with distributions μ_t, $t \in T$.*

Proof: For any countable subset $I \subset T$, the associated product measure $\mu_I = \bigotimes_{t \in I} \mu_t$ exists by Theorem 6.17. Now proceed as in the proof of Theorem 6.16. □

Exercises

1. Show that $(\xi, \eta) \stackrel{d}{=} (\xi', \eta)$ iff $P[\xi \in B|\eta] = P[\xi' \in B|\eta]$ a.s. for any measurable set B.

2. Show that $E^{\mathcal{F}}\xi = E^{\mathcal{G}}\xi$ a.s. for all $\xi \in L^1$ iff $\overline{\mathcal{F}} = \overline{\mathcal{G}}$.

3. Show that the averaging property implies the other properties of conditional expectations listed in Theorem 6.1.

4. Let $0 \leq \xi_n \uparrow \xi$ and $0 \leq \eta \leq \xi$, where $\xi_1, \xi_2, \ldots, \eta \in L^1$, and fix a σ-field $\mathcal{F}$. Show that $E^{\mathcal{F}}\eta \leq \sup_n E^{\mathcal{F}}\xi_n$. (*Hint:* Apply the monotone convergence property to $E^{\mathcal{F}}(\xi_n \wedge \eta)$.)

5. For any $[0, \infty]$-valued random variable ξ, define $E^{\mathcal{F}}\xi = \sup_n E^{\mathcal{F}}(\xi \wedge n)$. Show that this extension of $E^{\mathcal{F}}$ satisfies the monotone convergence property. (*Hint:* Use the preceding result.)

6. Show that the above extension of $E^{\mathcal{F}}$ remains characterized by the averaging property and that $E^{\mathcal{F}}\xi < \infty$ a.s. iff the measure $\xi \cdot P = E[\xi; \cdot]$ is σ-finite on $\mathcal{F}$. Extend $E^{\mathcal{F}}\xi$ to any random variable ξ such that the measure $|\xi| \cdot P$ is σ-finite on $\mathcal{F}$.

7. Let $\xi_1, \xi_2, \ldots$ be $[0, \infty]$-valued random variables, and fix any σ-field $\mathcal{F}$. Show that $\liminf_n E^{\mathcal{F}}\xi_n \geq E^{\mathcal{F}} \liminf_n \xi_n$ a.s.

8. Fix any σ-field $\mathcal{F}$, and let $\xi, \xi_1, \xi_2, \ldots$ be random variables with $\xi_n \to \xi$ and $E^{\mathcal{F}} \sup_n |\xi_n| < \infty$ a.s. Show that $E^{\mathcal{F}}\xi_n \to E^{\mathcal{F}}\xi$ a.s.

9. Let $\mathcal{F}$ be the σ-field generated by some partition $A_1, A_2, \cdots \in \mathcal{A}$ of Ω. Show for any $\xi \in L^1$ that $E[\xi|\mathcal{F}] = E[\xi|A_k] = E[\xi; A_k]/PA_k$ on A_k whenever $PA_k > 0$.

10. For any σ-field $\mathcal{F}$, event A, and random variable $\xi \in L^1$, show that $E[\xi|\mathcal{F}, 1_A] = E[\xi; A|\mathcal{F}]/P[A|\mathcal{F}]$ a.s. on A.

11. Let the random variables $\xi_1, \xi_2, \cdots \geq 0$ and σ-fields $\mathcal{F}_1, \mathcal{F}_2, \ldots$ be such that $E[\xi_n|\mathcal{F}_n] \stackrel{P}{\to} 0$. Show that $\xi_n \stackrel{P}{\to} 0$. (*Hint:* Consider the random variables $\xi_n \wedge 1$.)

12. Let $(\xi, \eta) \stackrel{d}{=} (\tilde{\xi}, \tilde{\eta})$, where $\xi \in L^1$. Show that $E[\xi|\eta] \stackrel{d}{=} E[\tilde{\xi}|\tilde{\eta}]$. (*Hint:* If $E[\xi|\eta] = f(\eta)$, then $E[\tilde{\xi}|\tilde{\eta}] = f(\tilde{\eta})$ a.s.)

13. Let (ξ, η) be a random vector in $\mathbb{R}^2$ with probability density f, put $F(y) = \int f(x, y)dx$, and let $g(x, y) = f(x, y)/F(y)$. Show that $P[\xi \in B|\eta] = \int_B g(x, \eta)dx$ a.s.

14. Use conditional distributions to deduce the monotone and dominated convergence theorems for conditional expectations from the corresponding unconditional results.

15. Assume that $E^{\mathcal{F}}\xi \stackrel{d}{=} \xi$ for some $\xi \in L^1$. Show that ξ is a.s. $\mathcal{F}$-measurable. (*Hint:* Choose a strictly convex function f with $Ef(\xi) < \infty$, and apply the strict Jensen inequality to the conditional distributions.)

16. Assume that $(\xi, \eta) \stackrel{d}{=} (\xi, \zeta)$, where η is ζ-measurable. Show that $\xi \perp\!\!\!\perp_\eta \zeta$. (*Hint:* Show as above that $P[\xi \in B|\eta] \stackrel{d}{=} P[\xi \in B|\zeta]$, and deduce the corresponding a.s. equality.)

17. Let ξ be a random element in some separable metric space S. Show that $P[\xi \in \cdot|\mathcal{F}]$ is a.s. degenerate iff ξ is a.s. $\mathcal{F}$-measurable. (*Hint:* Reduce to the case when $P[\xi \in \cdot|\mathcal{F}]$ is degenerate everywhere and hence equal to δ_η for some $\mathcal{F}$-measurable random element η in S. Then show that $\xi = \eta$ a.s.)

18. Assuming $\xi \perp\!\!\!\perp_\eta \zeta$ and $\gamma \perp\!\!\!\perp (\xi, \eta, \zeta)$, show that $\xi \perp\!\!\!\perp_{\eta, \gamma} \zeta$ and $\xi \perp\!\!\!\perp_\eta (\zeta, \gamma)$.

19. Extend Lemma 3.6 to the context of conditional independence. Also show that Corollary 3.7 and Lemma 3.8 remain valid for the conditional independence, given some σ-field $\mathcal{H}$.

20. Fix any σ-field $\mathcal{F}$ and random element ξ in some Borel space, and define $\eta = P[\xi \in \cdot|\mathcal{F}]$. Show that $\xi \perp\!\!\!\perp_\eta \mathcal{F}$.

21. Let ξ and η be random elements in some Borel space S. Prove the existence of a measurable function $f : S \times [0, 1] \to S$ and some $U(0, 1)$ random variable $\gamma \perp\!\!\!\perp \eta$ such that $\xi = f(\eta, \gamma)$ a.s. (*Hint:* Choose f with $(f(\eta, \vartheta), \eta) \stackrel{d}{=} (\xi, \eta)$ for any $U(0, 1)$ random variable $\vartheta \perp\!\!\!\perp (\xi, \eta)$, and then let $(\gamma, \tilde{\eta}) \stackrel{d}{=} (\vartheta, \eta)$ with $(\xi, \eta) = (f(\gamma, \tilde{\eta}), \tilde{\eta})$ a.s.)

22. Let ξ and η be random elements in some Borel space S. Show that we may choose a random element $\tilde{\eta}$ in S with $(\xi, \eta) \stackrel{d}{=} (\xi, \tilde{\eta})$ and $\eta \perp\!\!\!\perp_\xi \tilde{\eta}$.

23. Let the probability measures P and Q on $(\Omega, \mathcal{A})$ be related by $Q = \xi \cdot P$ for some random variable $\xi \geq 0$, and consider any σ-field $\mathcal{F} \subset \mathcal{A}$. Show that $Q = E_P[\xi|\mathcal{F}] \cdot P$ on $\mathcal{F}$.

24. Assume as before that $Q = \xi \cdot P$ on $\mathcal{A}$, and let $\mathcal{F} \subset \mathcal{A}$. Show that $E_Q[\eta|\mathcal{F}] = E_P[\xi\eta|\mathcal{F}]/E_P[\xi|\mathcal{F}]$ a.s. Q for any random variable $\eta \geq 0$.

Chapter 7

Martingales and Optional Times

Filtrations and optional times; random time-change; martingale property; optional stopping and sampling; maximum and upcrossing inequalities; martingale convergence, regularity, and closure; limits of conditional expectations; regularization of submartingales

The importance of martingale methods and ideas can hardly be exaggerated. Indeed, martingales and the associated notions of filtrations and optional times are constantly used in all areas of modern probability; they appear frequently throughout the remainder of this book.

In discrete time a martingale is simply a sequence of integrable random variables centered at the successive conditional means, a centering that can always be achieved by the elementary Doob decomposition. More precisely, given any discrete filtration $\mathcal{F} = (\mathcal{F}_n)$, that is, an increasing sequence of σ-fields in Ω, we say that a sequence $M = (M_n)$ forms a martingale with respect to $\mathcal{F}$ if $E[M_n|\mathcal{F}_{n-1}] = M_{n-1}$ a.s. for all n. A special role is played by the class of uniformly integrable martingales, which can be represented in the form $M_n = E[\xi|\mathcal{F}_n]$ for some integrable random variables ξ.

Martingale theory owes its usefulness to a number of powerful general results, such as the optional sampling theorem, the submartingale convergence theorem, and a wide range of maximum inequalities. The applications discussed in this chapter include extensions of the Borel–Cantelli lemma and Kolmogorov's 0–1 law. Martingales can also be used to establish the existence of measurable densities and to give a short proof of the law of large numbers.

Much of the discrete-time theory extends immediately to continuous time, thanks to the fundamental regularization theorem, which ensures that every continuous-time martingale with respect to a right-continuous filtration has a right-continuous version with left-hand limits. The implications of this result extend far beyond martingale theory. In particular, it will enable us in Chapters 15 and 19 to obtain right-continuous versions of independent-increment and Feller processes.

The theory of continuous-time martingales is continued in Chapters 17, 18, 25, and 26 with studies of quadratic variation, random time-change, integral representations, removal of drift, additional maximum inequalities, and various decomposition theorems. Martingales also play a basic role for especially the Skorohod embedding in Chapter 14, the stochastic integration in Chapters 17 and 26, and the theories of Feller processes, SDEs, and diffusions in Chapters 19, 21, and 23.

As for the closely related notion of optional times, our present treatment is continued with a more detailed study in Chapter 25. Optional times are fundamental not only for martingale theory but also for various models involving Markov processes. In the latter context they appear frequently in the sequel, especially in Chapters 8, 9, 12, 13, 14, 19, and 22–25.

To begin our systematic exposition of the theory, we may fix an arbitrary index set $T \subset \overline{\mathbb{R}}$. A *filtration* on T is defined as a nondecreasing family of σ-fields $\mathcal{F}_t \subset \mathcal{A}$, $t \in T$. We say that a process X on T is *adapted* to $\mathcal{F} = (\mathcal{F}_t)$ if X_t is $\mathcal{F}_t$-measurable for every $t \in T$. The smallest filtration with this property, namely $\mathcal{F}_t = \sigma\{X_s;\, s \leq t\}$, $t \in T$, is called the *induced* or *generated* filtration. Here "smallest" is understood in the sense of set inclusion for every fixed t.

By a *random time* we mean a random element τ in $\overline{T} = T \cup \{\sup T\}$. We say that τ is $\mathcal{F}$-*optional* or an $\mathcal{F}$-*stopping time* if $\{\tau \leq t\} \in \mathcal{F}_t$ for every $t \in T$, that is, if the process $X_t = 1\{\tau \leq t\}$ is adapted. (Here and in similar cases, we often omit the prefix $\mathcal{F}$ when there is no risk for confusion.) If T is countable, it is clearly equivalent that $\{\tau = t\} \in \mathcal{F}_t$ for every $t \in T$. For any optional times σ and τ we note that even $\sigma \vee \tau$ and $\sigma \wedge \tau$ are optional.

With every optional time τ we may associate a σ-field

$$\mathcal{F}_\tau = \{A \in \mathcal{A};\, A \cap \{\tau \leq t\} \in \mathcal{F}_t,\, t \in T\}.$$

Some basic properties of optional times and the associated σ-fields are listed below.

Lemma 7.1 *(optional times)* For any optional times σ and τ, we have

(i) τ is $\mathcal{F}_\tau$-measurable;
(ii) $\mathcal{F}_\tau = \mathcal{F}_t$ on $\{\tau = t\}$ for all $t \in T$;
(iii) $\mathcal{F}_\sigma \cap \{\sigma \leq \tau\} \subset \mathcal{F}_{\sigma \wedge \tau} = \mathcal{F}_\sigma \cap \mathcal{F}_\tau$.

In particular, we see from (iii) that $\{\sigma \leq \tau\} \in \mathcal{F}_\sigma \cap \mathcal{F}_\tau$, that $\mathcal{F}_\sigma = \mathcal{F}_\tau$ on $\{\sigma = \tau\}$, and that $\mathcal{F}_\sigma \subset \mathcal{F}_\tau$ whenever $\sigma \leq \tau$.

Proof: (iii) For any $A \in \mathcal{F}_\sigma$ and $t \in T$, we have

$$A \cap \{\sigma \leq \tau\} \cap \{\tau \leq t\} = (A \cap \{\sigma \leq t\}) \cap \{\tau \leq t\} \cap \{\sigma \wedge t \leq \tau \wedge t\},$$

which belongs to $\mathcal{F}_t$ since $\sigma \wedge t$ and $\tau \wedge t$ are both $\mathcal{F}_t$-measurable. Hence,

$$\mathcal{F}_\sigma \cap \{\sigma \leq \tau\} \subset \mathcal{F}_\tau.$$

The first relation now follows as we replace τ by $\sigma \wedge \tau$. Replacing σ and τ by the pairs $(\sigma \wedge \tau, \sigma)$ and $(\sigma \wedge \tau, \tau)$, we obtain $\mathcal{F}_{\sigma \wedge \tau} \subset \mathcal{F}_\sigma \cap \mathcal{F}_\tau$. To prove the reverse relation, we note that for any $A \in \mathcal{F}_\sigma \cap \mathcal{F}_\tau$ and $t \in T$

$$A \cap \{\sigma \wedge \tau \leq t\} = (A \cap \{\sigma \leq t\}) \cup (A \cap \{\tau \leq t\}) \in \mathcal{F}_t,$$

whence $A \in \mathcal{F}_{\sigma \wedge \tau}$.

(i) Applying (iii) to the pair (τ, t) gives $\{\tau \leq t\} \in \mathcal{F}_\tau$ for all $t \in T$, which extends immediately to any $t \in \mathbb{R}$. Now use Lemma 1.4.

(ii) First assume that $\tau \equiv t$. Then $\mathcal{F}_\tau = \mathcal{F}_t \cap \{\tau \leq t\} \subset \mathcal{F}_t$. Conversely, assume that $A \in \mathcal{F}_t$ and $s \in T$. If $s \geq t$ we get $A \cap \{\tau \leq s\} = A \in \mathcal{F}_t \subset \mathcal{F}_s$, and for $s < t$ we have $A \cap \{\tau \leq s\} = \emptyset \in \mathcal{F}_s$. Thus, $A \in \mathcal{F}_\tau$. This shows that $\mathcal{F}_\tau = \mathcal{F}_t$ when $\tau \equiv t$. The general case now follows by part (iii). □

Given an arbitrary filtration $\mathcal{F}$ on $\mathbb{R}_+$, we may define a new filtration $\mathcal{F}^+$ by $\mathcal{F}_t^+ = \bigcap_{u>t} \mathcal{F}_u$, $t \geq 0$, and we say that $\mathcal{F}$ is *right-continuous* if $\mathcal{F}^+ = \mathcal{F}$. In particular, $\mathcal{F}^+$ is right-continuous for any filtration $\mathcal{F}$. We say that a random time τ is *weakly $\mathcal{F}$-optional* if $\{\tau < t\} \in \mathcal{F}_t$ for every $t > 0$. In that case $\tau + h$ is clearly $\mathcal{F}$-optional for every $h > 0$, and we may define $\mathcal{F}_{\tau+} = \bigcap_{h>0} \mathcal{F}_{\tau+h}$. When the index set is $\mathbb{Z}_+$, we take $\mathcal{F}^+ = \mathcal{F}$ and make no difference between strictly and weakly optional times.

The following result shows that the notions of optional and weakly optional times agree when $\mathcal{F}$ is right-continuous.

Lemma 7.2 *(weakly optional times)* A random time τ is weakly $\mathcal{F}$-optional iff it is $\mathcal{F}^+$-optional, in which case
$$\mathcal{F}_{\tau+} = \mathcal{F}_\tau^+ = \{A \in \mathcal{A}; A \cap \{\tau < t\} \in \mathcal{F}_t, t > 0\}. \tag{1}$$

Proof: For any $t \geq 0$, we note that
$$\{\tau \leq t\} = \bigcap_{r>t} \{\tau < r\}, \qquad \{\tau < t\} = \bigcup_{r<t} \{\tau \leq r\}, \tag{2}$$
where r may be restricted to the rationals. If $A \cap \{\tau \leq t\} \in \mathcal{F}_{t+}$ for all t, we get by (2) for any $t > 0$
$$A \cap \{\tau < t\} = \bigcup_{r<t} (A \cap \{\tau \leq r\}) \in \mathcal{F}_t.$$

Conversely, if $A \cap \{\tau < t\} \in \mathcal{F}_t$ for all t, then (2) yields for any $t \geq 0$ and $h > 0$
$$A \cap \{\tau \leq t\} = \bigcap_{r \in (t, t+h)} (A \cap \{\tau < r\}) \in \mathcal{F}_{t+h},$$

and so $A \cap \{\tau \leq t\} \in \mathcal{F}_{t+}$. For $A = \Omega$ this proves the first assertion, and for general $A \in \mathcal{A}$ it proves the second relation in (1).

To prove the first relation, we note that $A \in \mathcal{F}_{\tau+}$ iff $A \in \mathcal{F}_{\tau+h}$ for each $h > 0$, that is, iff $A \cap \{\tau + h \leq t\} \in \mathcal{F}_t$ for all $t \geq 0$ and $h > 0$. But this is equivalent to $A \cap \{\tau \leq t\} \in \mathcal{F}_{t+h}$ for all $t \geq 0$ and $h > 0$, hence to $A \cap \{\tau \leq t\} \in \mathcal{F}_{t+}$ for every $t \geq 0$, which means that $A \in \mathcal{F}_\tau^+$. □

We have already seen that the maximum and minimum of two optional times are again optional. The result extends to countable collections as follows.

Lemma 7.3 *(closure properties)* *For any random times $\tau_1, \tau_2, \ldots$ and filtration $\mathcal{F}$ on $\mathbb{R}_+$ or $\mathbb{Z}_+$, we have:*

(i) *If the τ_n are $\mathcal{F}$-optional, then so is $\sigma = \sup_n \tau_n$.*

(ii) *If the τ_n are weakly $\mathcal{F}$-optional, then so is $\tau = \inf_n \tau_n$, and we have $\mathcal{F}_\tau^+ = \bigcap_n \mathcal{F}_{\tau_n}^+$.*

Proof: To prove (i) and the first assertion in (ii), we note that

$$\{\sigma \le t\} = \bigcap_n \{\tau_n \le t\}, \qquad \{\tau < t\} = \bigcup_n \{\tau_n < t\}, \qquad (3)$$

where the strict inequalities may be replaced by $\le$ for the index set $T = \mathbb{Z}_+$. To prove the second assertion in (ii), we note that $\mathcal{F}_\tau^+ \subset \bigcap_n \mathcal{F}_{\tau_n}^+$ by Lemma 7.1. Conversely, assuming $A \in \bigcap_n \mathcal{F}_{\tau_n}^+$, we get by (3) for any $t \ge 0$

$$A \cap \{\tau < t\} = A \cap \bigcup_n \{\tau_n < t\} = \bigcup_n (A \cap \{\tau_n < t\}) \in \mathcal{F}_t,$$

with the indicated modification for $T = \mathbb{Z}_+$. Thus, $A \in \mathcal{F}_\tau^+$. $\square$

Part (ii) of the last result is often useful in connection with the following approximation of optional times from the right.

Lemma 7.4 *(discrete approximation)* *For any weakly optional time τ in $\overline{\mathbb{R}}_+$, there exist some countably valued optional times $\tau_n \downarrow \tau$.*

Proof: We may define

$$\tau_n = 2^{-n}[2^n \tau + 1], \quad n \in \mathbb{N}.$$

Then $\tau_n \in 2^{-n}\overline{\mathbb{N}}$ for all n, and $\tau_n \downarrow \tau$. Also note that the τ_n are optional since $\{\tau_n \le k2^{-n}\} = \{\tau < k2^{-n}\} \in \mathcal{F}_{k2^{-n}}$. $\square$

It is now time to relate the optional times to random processes. We say that a process X on $\mathbb{R}_+$ is *progressively measurable* or simply *progressive* if its restriction to $\Omega \times [0, t]$ is $\mathcal{F}_t \otimes \mathcal{B}[0, t]$-measurable for every $t \ge 0$. Note that any progressive process is adapted by Lemma 1.26. Conversely, a simple approximation from the left or right shows that any adapted and left- or right-continuous process is progressive. A set $A \subset \Omega \times \mathbb{R}_+$ is said to be progressive if the corresponding indicator function 1_A has this property, and we note that the progressive sets form a σ-field.

Lemma 7.5 *(optional evaluation)* *Fix a filtration $\mathcal{F}$ on an index set T, let X be a process on T with values in a measurable space $(S, \mathcal{S})$, and let τ be an optional time in T. Then X_τ is $\mathcal{F}_\tau$-measurable under each of these conditions:*

(i) *T is countable and X is adapted;*

(ii) *$T = \mathbb{R}_+$ and X is progressive.*

Proof: In both cases, we need to show that

$$\{X_\tau \in B, \tau \le t\} \in \mathcal{F}_t, \quad t \ge 0, \ B \in \mathcal{S}.$$

This is clear in case (i) if we write

$$\{X_\tau \in B\} = \bigcup_{s \leq t}\{X_s \in B, \tau = s\} \in \mathcal{F}_t, \quad B \in \mathcal{S}.$$

In case (ii) it is enough to show that $X_{\tau \wedge t}$ is $\mathcal{F}_t$-measurable for every $t \geq 0$. We may then assume $\tau \leq t$ and prove instead that X_τ is $\mathcal{F}_t$-measurable. Writing $X_\tau = X \circ \psi$ where $\psi(\omega) = (\omega, \tau(\omega))$, we note that ψ is measurable from $\mathcal{F}_t$ to $\mathcal{F}_t \otimes \mathcal{B}[0,t]$ whereas X is measurable on $\Omega \times [0,t]$ from $\mathcal{F}_t \otimes \mathcal{B}[0,t]$ to $\mathcal{S}$. The required measurability of X_τ now follows by Lemma 1.7. □

Given a process X on $\mathbb{R}_+$ or $\mathbb{Z}_+$ and a set B in the range space of X, we introduce the *hitting time*

$$\tau_B = \inf\{t > 0;\ X_t \in B\}.$$

It is often important to decide whether τ_B is optional. The following elementary result covers the most commonly occurring cases.

Lemma 7.6 *(hitting times) Fix a filtration $\mathcal{F}$ on $T = \mathbb{R}_+$ or $\mathbb{Z}_+$, let X be an $\mathcal{F}$-adapted process on T with values in a measurable space $(S, \mathcal{S})$, and let $B \in \mathcal{S}$. Then τ_B is weakly optional under each of these conditions:*

(i) $T = \mathbb{Z}_+$;

(ii) $T = \mathbb{R}_+$, S *is a metric space, B is closed, and X is continuous;*

(iii) $T = \mathbb{R}_+$, S *is a topological space, B is open, and X is right-continuous.*

Proof: In case (i) it is enough to write

$$\{\tau_B \leq n\} = \bigcup_{k \in [1,n]} \{X_k \in B\} \in \mathcal{F}_n, \quad n \in \mathbb{N}.$$

In case (ii) we get for any $t > 0$

$$\{\tau_B \leq t\} = \bigcup_{h > 0} \bigcap_{n \in \mathbb{N}} \bigcup_{r \in \mathbb{Q} \cap [h,t]} \{\rho(X_r, B) \leq n^{-1}\} \in \mathcal{F}_t,$$

where ρ denotes the metric in S. Finally, in case (iii) we get

$$\{\tau_B < t\} = \bigcup_{r \in \mathbb{Q} \cap (0,t)} \{X_r \in B\} \in \mathcal{F}_t, \quad t > 0,$$

which suffices by Lemma 7.2. □

For special purposes we need the following more general but much deeper result, known as the *debut theorem*. Here and below, a filtration $\mathcal{F}$ is said to be *complete* if the basic σ-field $\mathcal{A}$ is complete and each $\mathcal{F}_t$ contains all P-null sets in $\mathcal{A}$.

Theorem 7.7 *(first entry, Doob, Hunt) Let the set $A \subset \mathbb{R}_+ \times \Omega$ be progressive with respect to some right-continuous and complete filtration $\mathcal{F}$. Then the time $\tau(\omega) = \inf\{t \geq 0;\ (t,\omega) \in A\}$ is $\mathcal{F}$-optional.*

Proof: Since A is progressive, we have $A \cap [0,t) \in \mathcal{F}_t \otimes \mathcal{B}([0,t])$ for every $t > 0$. Noting that $\{\tau < t\}$ is the projection of $A \cap [0,t)$ onto Ω, we get $\{\tau < t\} \in \mathcal{F}_t$ by Theorem A1.4, and so τ is optional by Lemma 7.2. □

In applications of the last result and for other purposes, we may need to extend a given filtration $\mathcal{F}$ on $\mathbb{R}_+$ to make it both right-continuous and complete. Writing $\overline{\mathcal{A}}$ for the completion of $\mathcal{A}$, we put $\mathcal{N} = \{A \in \overline{\mathcal{A}};\ PA = 0\}$ and define $\overline{\mathcal{F}}_t = \sigma\{\mathcal{F}_t, \mathcal{N}\}$. Then $\overline{\mathcal{F}} = (\overline{\mathcal{F}}_t)$ is the smallest complete extension of $\mathcal{F}$. Similarly, $\mathcal{F}^+ = (\mathcal{F}_{t+})$ is the smallest right-continuous extension of $\mathcal{F}$. We show that the two operations commute and can be combined into a smallest right-continuous and complete extension, known as the *(usual) augmentation* of $\mathcal{F}$.

Lemma 7.8 *(augmented filtration) Every filtration $\mathcal{F}$ on $\mathbb{R}_+$ has a smallest right-continuous and complete extension $\mathcal{G}$, given by*

$$\mathcal{G}_t = \overline{\mathcal{F}_{t+}} = \overline{\mathcal{F}}_{t+}, \quad t \geq 0. \tag{4}$$

Proof: First we note that

$$\overline{\mathcal{F}_{t+}} \subset \overline{\mathcal{F}}_{t+} \subset \overline{\mathcal{F}_{t+}}, \quad t \geq 0.$$

Conversely, assume that $A \in \overline{\mathcal{F}}_{t+}$. Then $A \in \overline{\mathcal{F}}_{t+h}$ for every $h > 0$, and so, as in Lemma 1.25, there exist some sets $A_h \in \mathcal{F}_{t+h}$ with $P(A \Delta A_h) = 0$. Now choose $h_n \to 0$, and define $A' = \{A_{h_n}\ \text{i.o.}\}$. Then $A' \in \mathcal{F}_{t+}$ and $P(A \Delta A') = 0$, so $A \in \overline{\mathcal{F}_{t+}}$. Thus, $\overline{\mathcal{F}}_{t+} \subset \overline{\mathcal{F}_{t+}}$, which proves the second relation in (4).

In particular, the filtration $\mathcal{G}$ in (4) contains $\mathcal{F}$ and is both right-continuous and complete. For any filtration $\mathcal{H}$ with those properties, we have

$$\mathcal{G}_t = \overline{\mathcal{F}}_{t+} \subset \overline{\mathcal{H}}_{t+} = \mathcal{H}_{t+} = \mathcal{H}_t, \quad t \geq 0,$$

which proves the required minimality of $\mathcal{G}$. □

The next result shows how the σ-fields $\mathcal{F}_\tau$ arise naturally in connection with a random time-change.

Proposition 7.9 *(random time-change) Let $X \geq 0$ be a nondecreasing, right-continuous process adapted to some right-continuous filtration $\mathcal{F}$. Then*

$$\tau_s = \inf\{t > 0;\ X_t > s\}, \quad s \geq 0,$$

is a right-continuous process of optional times, generating a right-continuous filtration $\mathcal{G}_s = \mathcal{F}_{\tau_s}$, $s \geq 0$. If X is continuous and the time τ is $\mathcal{F}$-optional, then X_τ is $\mathcal{G}$-optional and $\mathcal{F}_\tau \subset \mathcal{G}_{X_\tau}$. If X is further strictly increasing, then $\mathcal{F}_\tau = \mathcal{G}_{X_\tau}$.

In the latter case, we have in particular $\mathcal{F}_t = \mathcal{G}_{X_t}$ for all t, so the processes (τ_s) and (X_t) play symmetric roles.

Proof: The times τ_s are optional by Lemmas 7.2 and 7.6, and since (τ_s) is right-continuous, so is $(\mathcal{G}_s)$ by Lemma 7.3. If X is continuous, then by Lemma 7.1 we get for any $\mathcal{F}$-optional time $\tau > 0$ and set $A \in \mathcal{F}_\tau$

$$A \cap \{X_\tau \leq s\} = A \cap \{\tau \leq \tau_s\} \in \mathcal{F}_{\tau_s} = \mathcal{G}_s, \quad s \geq 0.$$

For $A = \Omega$ it follows that X_τ is $\mathcal{G}$-optional, and for general A we get $A \in \mathcal{G}_{X_\tau}$. Thus, $\mathcal{F}_\tau \subset \mathcal{G}_{X_\tau}$. Both statements extend by Lemma 7.3 to arbitrary τ.

Now assume that X is also strictly increasing. For any $A \in \mathcal{G}_{X_t}$ with $t > 0$ we have

$$A \cap \{t \leq \tau_s\} = A \cap \{X_t \leq s\} \in \mathcal{G}_s = \mathcal{F}_{\tau_s}, \quad s \geq 0,$$

and so

$$A \cap \{t \leq \tau_s \leq u\} \in \mathcal{F}_u, \quad s \geq 0, \, u > t.$$

Taking the union over all $s \in \mathbb{Q}_+$—the set of nonnegative rationals—gives $A \in \mathcal{F}_u$, and as $u \downarrow t$ we get $A \in \mathcal{F}_{t+} = \mathcal{F}_t$. Hence, $\mathcal{F}_t = \mathcal{G}_{X_t}$, which extends as before to $t = 0$. By Lemma 7.1 we now obtain for any $A \in \mathcal{G}_{X_\tau}$

$$A \cap \{\tau \leq t\} = A \cap \{X_\tau \leq X_t\} \in \mathcal{G}_{X_t} = \mathcal{F}_t, \quad t \geq 0,$$

and so $A \in \mathcal{F}_\tau$. Thus, $\mathcal{G}_{X_\tau} \subset \mathcal{F}_\tau$, so the two σ-fields agree. □

To motivate the introduction of martingales, we may fix a random variable $\xi \in L^1$ and a filtration $\mathcal{F}$ on some index set T, and put

$$M_t = E[\xi | \mathcal{F}_t], \quad t \in T.$$

The process M is clearly integrable (for each t) and adapted, and by the chain rule for conditional expectations we note that

$$M_s = E[M_t | \mathcal{F}_s] \text{ a.s.}, \quad s \leq t. \tag{5}$$

Any integrable and adapted process M satisfying (5) is called a *martingale with respect to* $\mathcal{F}$, or an $\mathcal{F}$-*martingale*. When $T = \mathbb{Z}_+$, it suffices to require (5) for $t = s + 1$, so in that case the condition becomes

$$E[\Delta M_n | \mathcal{F}_{n-1}] = 0 \text{ a.s.}, \quad n \in \mathbb{N}, \tag{6}$$

where $\Delta M_n = M_n - M_{n-1}$. A process $M = (M^1, \ldots, M^d)$ in $\mathbb{R}^d$ is said to be a martingale if $M^1, \ldots, M^d$ are one-dimensional martingales.

Replacing the equality in (5) or (6) by an inequality, we arrive at the notions of sub- and supermartingales. Thus, a *submartingale* is defined as an integrable and adapted process X with

$$X_s \leq E[X_t | \mathcal{F}_s] \text{ a.s.}, \quad s \leq t; \tag{7}$$

reversing the inequality sign yields the notion of a *supermartingale*. In particular, the mean is nondecreasing for submartingales and nonincreasing for supermartingales. (The sign convention is suggested by analogy with sub- and superharmonic functions.)

Given a filtration $\mathcal{F}$ on $\mathbb{Z}_+$, we say that a random sequence $A = (A_n)$ with $A_0 = 0$ is *predictable with respect to $\mathcal{F}$*, or *$\mathcal{F}$-predictable*, if A_n is $\mathcal{F}_{n-1}$-measurable for every $n \in \mathbb{N}$, that is, if the shifted sequence $\theta A = (A_{n+1})$ is adapted. The following elementary result, known as the *Doob decomposition*, is useful to deduce results for submartingales from the corresponding martingale versions. An extension to continuous time is proved in Chapter 25.

Lemma 7.10 *(centering)* Any integrable and $\mathcal{F}$-adapted process X on $\mathbb{Z}_+$ has an a.s. unique decomposition $M + A$, where M is an $\mathcal{F}$-martingale and A is an $\mathcal{F}$-predictable process with $A_0 = 0$. In particular, X is a submartingale iff A is a.s. nondecreasing.

Proof: If $X = M + A$ for some processes M and A as stated, then clearly $\Delta A_n = E[\Delta X_n | \mathcal{F}_{n-1}]$ a.s. for all $n \in \mathbb{N}$, and so

$$A_n = \sum_{k \leq n} E[\Delta X_k | \mathcal{F}_{k-1}] \quad \text{a.s.}, \quad n \in \mathbb{Z}_+, \tag{8}$$

which proves the required uniqueness. In general, we may define a predictable process A by (8). Then $M = X - A$ is a martingale, since

$$E[\Delta M_n | \mathcal{F}_{n-1}] = E[\Delta X_n | \mathcal{F}_{n-1}] - \Delta A_n = 0 \quad \text{a.s.}, \quad n \in \mathbb{N}. \qquad \square$$

We proceed to show how the martingale and submartingale properties are preserved under various transformations.

Lemma 7.11 *(convex maps)* Let M be a martingale in $\mathbb{R}^d$, and consider a convex function $f: \mathbb{R}^d \to \mathbb{R}$ such that $X = f(M)$ is integrable. Then X is a submartingale. The statement remains true for any real submartingale M, provided that f is also nondecreasing.

Proof: In the martingale case, the conditional version of Jensen's inequality yields

$$f(M_s) = f(E[M_t | \mathcal{F}_s]) \leq E[f(M_t) | \mathcal{F}_s] \quad \text{a.s.}, \quad s \leq t, \tag{9}$$

which shows that $f(M)$ is a submartingale. If instead M is a submartingale and f is nondecreasing, the first relation in (9) becomes $f(M_s) \leq f(E[M_t | \mathcal{F}_s])$, and the conclusion remains valid. $\qquad \square$

The last result is often applied with $f(x) = |x|^p$ for some $p \geq 1$ or, for $d = 1$, with $f(x) = x_+ = x \vee 0$.

We say that an optional time τ is *bounded* if $\tau \leq u$ a.s. for some $u \in T$. This is always true when T has a last element. The following result is an elementary version of the basic *optional sampling theorem*. An extension to continuous-time submartingales appears as Theorem 7.29.

Theorem 7.12 *(optional sampling, Doob)* *Let M be a martingale on some countable index set T with filtration $\mathcal{F}$, and consider two optional times σ and τ, where τ is bounded. Then M_τ is integrable, and*
$$M_{\sigma \wedge \tau} = E[M_\tau | \mathcal{F}_\sigma] \quad a.s.$$

Proof: By Lemmas 6.2 and 7.1 we get for any $t \leq u$ in T
$$E[M_u | \mathcal{F}_\tau] = E[M_u | \mathcal{F}_t] = M_t = M_\tau \quad a.s. \text{ on } \{\tau = t\},$$
and so $E[M_u | \mathcal{F}_\tau] = M_\tau$ a.s. whenever $\tau \leq u$ a.s. If $\sigma \leq \tau \leq u$, then $\mathcal{F}_\sigma \subset \mathcal{F}_\tau$ by Lemma 7.1, and we get
$$E[M_\tau | \mathcal{F}_\sigma] = E[E[M_u | \mathcal{F}_\tau] | \mathcal{F}_\sigma] = E[M_u | \mathcal{F}_\sigma] = M_\sigma \quad a.s.$$
On the other hand, clearly $E[M_\tau | \mathcal{F}_\sigma] = M_\tau$ a.s. when $\tau \leq \sigma \wedge u$. In the general case, the previous results combine by means of Lemmas 6.2 and 7.1 into
$$E[M_\tau | \mathcal{F}_\sigma] = E[M_\tau | \mathcal{F}_{\sigma \wedge \tau}] = M_{\sigma \wedge \tau} \quad a.s. \text{ on } \{\sigma \leq \tau\},$$
$$E[M_\tau | \mathcal{F}_\sigma] = E[M_{\sigma \wedge \tau} | \mathcal{F}_\sigma] = M_{\sigma \wedge \tau} \quad a.s. \text{ on } \{\sigma > \tau\}. \qquad \square$$

In particular, we note that if M is a martingale on an arbitrary time scale T with filtration $\mathcal{F}$ and (τ_s) is a nondecreasing family of bounded, optional times that take countably many values, then the process (M_{τ_s}) is a martingale with respect to the filtration $(\mathcal{F}_{\tau_s})$. In this sense, the martingale property is preserved by a random time-change.

From the last theorem we note that every martingale M satisfies $EM_\sigma = EM_\tau$, for any bounded optional times σ and τ that take only countably many values. An even weaker property characterizes the class of martingales.

Lemma 7.13 *(martingale criterion)* *Let M be an integrable, adapted process on some index set T. Then M is a martingale iff $EM_\sigma = EM_\tau$ for any T-valued optional times σ and τ that take at most two values.*

Proof: If $s < t$ in T and $A \in \mathcal{F}_s$, then $\tau = s 1_A + t 1_{A^c}$ is optional, and so
$$0 = EM_t - EM_\tau = EM_t - E[M_s; A] - E[M_t; A^c] = E[M_t - M_s; A].$$
Since A is arbitrary, it follows that $E[M_t - M_s | \mathcal{F}_s] = 0$ a.s. $\qquad \square$

The following predictable transformation of martingales is basic for the theory of stochastic integration.

Corollary 7.14 *(martingale transform)* *Let M be a martingale on some index set T with filtration $\mathcal{F}$, fix an optional time τ that takes countably many values, and let η be a bounded, $\mathcal{F}_\tau$-measurable random variable. Then the process $N_t = \eta(M_t - M_{t \wedge \tau})$ is again a martingale.*

Proof: The integrability follows from Theorem 7.12, and the adaptedness is clear if we replace η by $\eta 1\{\tau \leq t\}$ in the expression for N_t. Now fix any bounded, optional time σ taking countably many values. By Theorem 7.12 and the pull-out property of conditional expectations, we get a.s.

$$E[N_\sigma|\mathcal{F}_\tau] = \eta E[M_\sigma - M_{\sigma\wedge\tau}|\mathcal{F}_\tau] = \eta(M_{\sigma\wedge\tau} - M_{\sigma\wedge\tau}) = 0,$$

and so $EN_\sigma = 0$. Thus, N is a martingale by Lemma 7.13. $\square$

In particular, we note that *optional stopping* preserves the martingale property, in the sense that the *stopped* process $M_t^\tau = M_{\tau\wedge t}$ is a martingale whenever M is a martingale and τ is an optional time that takes countably many values.

More generally, we may consider *predictable step processes* of the form

$$V_t = \sum_{k\leq n}\eta_k 1\{t > \tau_k\}, \quad t \in T,$$

where $\tau_1 \leq \cdots \leq \tau_n$ are optional times, and each η_k is a bounded, $\mathcal{F}_{\tau_k}$-measurable random variable. For any process X, we may introduce the associated *elementary stochastic integral*

$$(V \cdot X)_t \equiv \int_0^t V_s dX_s = \sum_{k\leq n}\eta_k(X_t - X_{t\wedge\tau_k}), \quad t \in T.$$

From Corollary 7.14 we note that $V \cdot X$ is a martingale whenever X is a martingale and each τ_k takes countably many values. In discrete time we may clearly allow V to be any bounded, predictable sequence, in which case

$$(V \cdot X)_n = \sum_{k\leq n} V_k \Delta X_k, \quad n \in \mathbb{Z}_+.$$

The result for martingales extends in an obvious way to submartingales X, provided that the predictable sequence V is nonnegative.

Our next aim is to derive some basic martingale inequalities. We begin with an extension of Kolmogorov's maximum inequality in Lemma 4.15.

Proposition 7.15 (*maximum inequalities, Bernstein, Lévy*) *Let X be a submartingale on a countable index set T. Then for any $r \geq 0$ and $u \in T$,*

$$rP\{\sup\nolimits_{t\leq u} X_t \geq r\} \leq E[X_u; \sup\nolimits_{t\leq u} X_t \geq r] \leq EX_u^+, \qquad (10)$$
$$rP\{\sup\nolimits_t |X_t| \geq r\} \leq 3\sup\nolimits_t E|X_t|. \qquad (11)$$

Proof: By dominated convergence it is enough to consider finite index sets, so we may assume that $T = \mathbb{Z}_+$. Define $\tau = u \wedge \inf\{t; X_t \geq r\}$ and $B = \{\max_{t\leq u} X_t \geq r\}$. Then τ is an optional time bounded by u, and we note that $B \in \mathcal{F}_\tau$ and $X_\tau \geq r$ on B. Hence, by Lemma 7.10 and Theorem 7.12,

$$rPB \leq E[X_\tau; B] \leq E[X_u; B] \leq EX_u^+,$$

which proves (10). Letting $M + A$ be the Doob decomposition of X and applying (10) to $-M$, we further get

$$\begin{aligned} rP\{\min_{t\leq u} X_t \leq -r\} &\leq rP\{\min_{t\leq u} M_t \leq -r\} \leq EM_u^- \\ &= EM_u^+ - EM_u \leq EX_u^+ - EX_0 \\ &\leq 2\max_{t\leq u} E|X_t|. \end{aligned}$$

Combining this with (10) yields (11). □

We proceed to derive a basic norm inequality. For processes X on some index set T, we define

$$X_t^* = \sup_{s\leq t} |X_s|, \qquad X^* = \sup_{t\in T} |X_t|.$$

Proposition 7.16 *(norm inequality, Doob) Let M be a martingale on a countable index set T, and fix any $p, q > 1$ with $p^{-1} + q^{-1} = 1$. Then*

$$\|M_t^*\|_p \leq q\|M_t\|_p, \quad t \in T.$$

Proof: By monotone convergence we may assume that $T = \mathbb{Z}_+$. If $\|M_t\|_p < \infty$, then $\|M_s\|_p < \infty$ for all $s \leq t$ by Jensen's inequality, and so we may assume that $0 < \|M_t^*\|_p < \infty$. Applying Proposition 7.15 to the submartingale $|M|$, we get

$$rP\{M_t^* > r\} \leq E[|M_t|; M_t^* > r], \quad r > 0.$$

Hence, by Lemma 3.4, Fubini's theorem, and Hölder's inequality,

$$\begin{aligned} \|M_t^*\|_p^p &= p\int_0^\infty P\{M_t^* > r\} r^{p-1} dr \\ &\leq p\int_0^\infty E[|M_t|; M_t^* > r] r^{p-2} dr \\ &= pE|M_t| \int_0^{M_t^*} r^{p-2} dr = q\, E|M_t|\, M_t^{*(p-1)} \\ &\leq q\|M_t\|_p \left\|M_t^{*(p-1)}\right\|_q = q\|M_t\|_p \|M_t^*\|_p^{p-1}. \end{aligned}$$

It remains to divide by the last factor on the right. □

The next inequality is needed to prove the basic Theorem 7.18. For any function f on T and constants $a < b$, the *number of $[a,b]$-crossings* of f up to time t is defined as the supremum of all $n \in \mathbb{Z}_+$ such that there exist times $s_1 < t_1 < s_2 < t_2 < \cdots < s_n < t_n \leq t$ in T with $f(s_k) \leq a$ and $f(t_k) \geq b$ for all k. The supremum may clearly be infinite.

Lemma 7.17 *(upcrossing inequality, Doob, Snell) Let X be a submartingale on a countable index set T, and let $N_a^b(t)$ denote the number of $[a,b]$-crossings of X up to time t. Then*

$$EN_a^b(t) \leq \frac{E(X_t - a)^+}{b - a}, \quad t \in T, \ a < b \text{ in } \mathbb{R}.$$

Proof: As before, we may assume that $T = \mathbb{Z}_+$. Since $Y = (X - a)^+$ is again a submartingale by Lemma 7.11 and the $[a,b]$-crossings of X correspond to $[0, b-a]$-crossings of Y, we may assume that $X \geq 0$ and $a = 0$. Now define recursively the optional times $0 = \tau_0 \leq \sigma_1 < \tau_1 < \sigma_2 < \cdots$ by

$$\sigma_k = \inf\{n \geq \tau_{k-1}; X_n = 0\}, \quad \tau_k = \inf\{n \geq \sigma_k; X_n \geq b\}, \quad k \in \mathbb{N},$$

and introduce the predictable process

$$V_n = \sum_{k \geq 1} 1\{\sigma_k < n \leq \tau_k\}, \quad n \in \mathbb{N}.$$

Then $(1 - V) \cdot X$ is again a submartingale by Corollary 7.14, and so

$$E((1 - V) \cdot X)_t \geq E((1 - V) \cdot X)_0 = 0, \quad t \geq 0.$$

Since also $(V \cdot X)_t \geq b N_0^b(t)$, we get

$$b E N_0^b(t) \leq E(V \cdot X)_t \leq E(1 \cdot X)_t = EX_t - EX_0 \leq EX_t. \qquad \Box$$

We may now state the fundamental regularity and convergence theorem for submartingales.

Theorem 7.18 *(regularity and convergence, Doob) Let X be an L^1-bounded submartingale on a countable index set T. Then X_t converges along every increasing or decreasing sequence in T, outside some fixed P-null set A.*

Proof: By Proposition 7.15 we have $X^* < \infty$ a.s., and Lemma 7.17 shows that X has a.s. finitely many upcrossings of every interval $[a, b]$ with rational $a < b$. Outside the null set A where any of these conditions fails, it is clear that X has the asserted property. $\qquad \Box$

The following is an interesting and useful application.

Proposition 7.19 *(one-sided bounds) Let M be a martingale on $\mathbb{Z}_+$ with $\Delta M \leq c$ a.s. for some constant $c < \infty$. Then a.s.*

$$\{M_n \text{ converges}\} = \{\sup_n M_n < \infty\}.$$

Proof: Since $M - M_0$ is again a martingale, we may assume that $M_0 = 0$. Introduce the optional times

$$\tau_m = \inf\{n; M_n \geq m\}, \quad m \in \mathbb{N}.$$

The processes M^{τ_m} are again martingales by Corollary 7.14. Since $M^{\tau_m} \leq m + c$ a.s., we have $E|M^{\tau_m}| \leq 2(m + c) < \infty$, and so M^{τ_m} converges a.s.

by Theorem 7.18. Hence, M converges a.s. on
$$\{\sup_n M_n < \infty\} = \bigcup_m \{M \equiv M^{\tau_m}\}.$$
The reverse implication is obvious, since every convergent sequence in $\mathbb{R}$ is bounded. □

From the last result we may easily derive the following useful extension of the Borel–Cantelli lemma in Theorem 3.18.

Corollary 7.20 *(extended Borel–Cantelli lemma, Lévy)* *For any filtration $\mathcal{F}$ on $\mathbb{Z}_+$, let $A_n \in \mathcal{F}_n$, $n \in \mathbb{N}$. Then a.s.*
$$\{A_n \text{ i.o.}\} = \left\{\sum_n P[A_n|\mathcal{F}_{n-1}] = \infty\right\}.$$

Proof: The sequence
$$M_n = \sum_{k \leq n} (1_{A_k} - P[A_k|\mathcal{F}_{k-1}]), \quad n \in \mathbb{Z}_+,$$
is a martingale with $|\Delta M_n| \leq 1$, and so by Proposition 7.19
$$P\{M_n \to \infty\} = P\{M_n \to -\infty\} = 0.$$
Hence, a.s.
$$\{A_n \text{ i.o.}\} = \left\{\sum_n 1_{A_n} = \infty\right\} = \left\{\sum_n P[A_n|\mathcal{F}_{n-1}] = \infty\right\}. \quad \square$$

A martingale M or submartingale X is said to be *closed* if $u = \sup T$ belongs to T. In the former case, clearly $M_t = E[M_u|\mathcal{F}_t]$ a.s. for all $t \in T$. If instead $u \notin T$, we say that M is *closable* if it can be extended to a martingale on $\overline{T} = T \cup \{u\}$. If $M_t = E[\xi|\mathcal{F}_t]$ for some $\xi \in L^1$, we may clearly choose $M_u = \xi$. The next result gives general criteria for closability. An extension to continuous-time submartingales appears as part of Theorem 7.29.

Theorem 7.21 *(uniform integrability and closure, Doob)* *For any martingale M on an unbounded index set T, these conditions are equivalent:*

(i) *M is uniformly integrable;*

(ii) *M is closable at $\sup T$;*

(iii) *M is L^1-convergent at $\sup T$.*

Under those conditions, M is closable by the limit in (iii).

Proof: First note that (ii) implies (i) by Lemma 6.5. Next (i) implies (iii) by Theorem 7.18 and Proposition 4.12. Finally, assume that $M_t \to \xi$ in L^1 as $t \to u \equiv \sup T$. Using the L^1-contractivity of conditional expectations, we get as $t \to u$ for fixed s,
$$M_s = E[M_t|\mathcal{F}_s] \to E[\xi|\mathcal{F}_s] \text{ in } L^1.$$
Thus, $M_s = E[\xi|\mathcal{F}_s]$ a.s., and we may take $M_u = \xi$. This shows that (iii) implies (ii). □

For comparison, we may examine the case of L^p-convergence for $p > 1$.

Corollary 7.22 (*L^p-convergence*) *Let M be a martingale on an unbounded index set T, and fix any $p > 1$. Then M converges in L^p iff it is L^p-bounded.*

Proof: We may clearly assume that T is countable. If M is L^p-bounded, it converges in L^1 by Theorem 7.18. Since $|M|^p$ is also uniformly integrable by Proposition 7.16, the convergence extends to L^p by Proposition 4.12. Conversely, if M converges in L^p, it is L^p-bounded by Lemma 7.11. □

We now consider the convergence of martingales of the special form $M_t = E[\xi|\mathcal{F}_t]$, as t increases or decreases along some sequence. Without loss of generality, we may assume that the index set T is unbounded above or below, and define respectively

$$\mathcal{F}_\infty = \bigvee_{t \in T} \mathcal{F}_t, \qquad \mathcal{F}_{-\infty} = \bigcap_{t \in T} \mathcal{F}_t.$$

Theorem 7.23 (*conditioning limits, Jessen, Lévy*) *Let $\mathcal{F}$ be a filtration on a countable index set $T \subset \mathbb{R}$ that is unbounded above or below. Then for any $\xi \in L^1$, we have as $t \to \pm\infty$*

$$E[\xi|\mathcal{F}_t] \to E[\xi|\mathcal{F}_{\pm\infty}] \quad \text{a.s. and in } L^1.$$

Proof: By Theorems 7.18 and 7.21, the martingale $M_t = E[\xi|\mathcal{F}_t]$ converges a.s. and in L^1 as $t \to \pm\infty$, and the limit $M_{\pm\infty}$ may clearly be taken to be $\mathcal{F}_{\pm\infty}$-measurable. To see that $M_{\pm\infty} = E[\xi|\mathcal{F}_{\pm\infty}]$ a.s., we need to verify the relations

$$E[M_{\pm\infty}; A] = E[\xi; A], \qquad A \in \mathcal{F}_{\pm\infty}. \tag{12}$$

Then note that, by the definition of M,

$$E[M_t; A] = E[\xi; A], \qquad A \in \mathcal{F}_s, \; s \leq t. \tag{13}$$

This clearly remains true for $s = -\infty$, and as $t \to -\infty$ we get the "minus" version of (12). To get the "plus" version, let $t \to \infty$ in (13) for fixed s, and extend by a monotone class argument to arbitrary $A \in \mathcal{F}_\infty$. □

In particular, we note the following useful special case.

Corollary 7.24 (*Lévy*) *For any filtration $\mathcal{F}$ on $\mathbb{Z}_+$, we have*

$$P[A|\mathcal{F}_n] \to 1_A \quad \text{a.s.}, \qquad A \in \mathcal{F}_\infty.$$

For a simple application, we consider an extension of Kolmogorov's 0–1 law in Theorem 3.13. Say that two σ-fields *agree a.s.* if they have the same completion with respect to the basic σ-field.

Corollary 7.25 *(tail σ-field)* If $\mathcal{F}_1, \mathcal{F}_2, \ldots$ and $\mathcal{G}$ are independent σ-fields, then
$$\bigcap_n \sigma\{\mathcal{F}_n, \mathcal{F}_{n+1}, \ldots; \mathcal{G}\} = \mathcal{G} \quad a.s.$$

Proof: Let $\mathcal{T}$ denote the σ-field on the left, and note that $\mathcal{T} \perp\!\!\!\perp_\mathcal{G} (\mathcal{F}_1 \vee \cdots \vee \mathcal{F}_n)$ by Proposition 6.8. Using Proposition 6.6 and Corollary 7.24, we get for any $A \in \mathcal{T}$
$$P[A|\mathcal{G}] = P[A|\mathcal{G}, \mathcal{F}_1, \ldots, \mathcal{F}_n] \to 1_A \quad a.s.,$$
which shows that $\mathcal{T} \subset \mathcal{G}$ a.s. The converse relation is obvious. $\square$

The last theorem can be used to give a short proof of the law of large numbers. Then let $\xi_1, \xi_2, \ldots$ be i.i.d. random variables in L^1, put $S_n = \xi_1 + \cdots + \xi_n$, and define $\mathcal{F}_{-n} = \sigma\{S_n, S_{n+1}, \ldots\}$. Here $\mathcal{F}_{-\infty}$ is trivial by Theorem 3.15, and for any $k \leq n$ we have $E[\xi_k|\mathcal{F}_{-n}] = E[\xi_1|\mathcal{F}_{-n}]$ a.s., since $(\xi_k, S_n, S_{n+1}, \ldots) \stackrel{d}{=} (\xi_1, S_n, S_{n+1}, \ldots)$. Hence, by Theorem 7.23,
$$\begin{aligned} n^{-1} S_n &= E[n^{-1} S_n | \mathcal{F}_{-n}] = n^{-1} \sum_{k \leq n} E[\xi_k | \mathcal{F}_{-n}] \\ &= E[\xi_1 | \mathcal{F}_{-n}] \to E[\xi_1 | \mathcal{F}_{-\infty}] = E\xi_1. \end{aligned}$$

As a further application of Theorem 7.23, we consider a kernel version of the regularization Theorem 6.3. The result is needed in Chapter 21.

Proposition 7.26 *(regular densities)* For any measurable space $(S, \mathcal{S})$ and Borel spaces $(T, \mathcal{T})$ and $(U, \mathcal{U})$, let μ be a probability kernel from S to $T \times U$. Then the densities
$$\nu(s, t, B) = \frac{\mu(s, dt \times B)}{\mu(s, dt \times U)}, \quad s \in S, t \in T, B \in \mathcal{U}, \tag{14}$$
have versions that form a probability kernel from $S \times T$ to U.

Proof: We may assume T and U to be Borel subsets of $\mathbb{R}$, in which case μ can be regarded as a probability kernel from S to $\mathbb{R}^2$. Letting $\mathcal{D}_n$ denote the σ-field in $\mathbb{R}$ generated by the intervals $I_{nk} = [(k-1)2^{-n}, k2^{-n})$, $k \in \mathbb{Z}$, we define
$$M_n(s, t, B) = \sum_k \frac{\mu(s, I_{nk} \times B)}{\mu(s, I_{nk} \times U)} 1\{t \in I_{nk}\}, \quad s \in S, t \in T, B \in \mathcal{B},$$
under the convention $0/0 = 0$. Then $M_n(s, \cdot, B)$ is a version of the density in (14) with respect to $\mathcal{D}_n$, and for fixed s and B it is also a martingale with respect to $\mu(s, \cdot \times U)$. By Theorem 7.23 we get $M_n(s, \cdot, B) \to \nu(s, \cdot, B)$ a.e. $\mu(s, \cdot \times U)$. Thus, a product-measurable version of ν is given by
$$\nu(s, t, B) = \limsup_{n \to \infty} M_n(s, t, B), \quad s \in S, t \in T, B \in \mathcal{U}.$$

It remains to find a version of ν that is a probability measure on U for fixed s and t. Then proceed as in the proof of Theorem 6.3, noting that in each step the exceptional (s,t)-set A lies in $\mathcal{S} \otimes \mathcal{T}$ and is such that the sections $A_s = \{t \in T;\ (s,t) \in A\}$ satisfy $\mu(s, A_s \times U) = 0$ for all $s \in S$. $\square$

In order to extend the previous theory to martingales on $\mathbb{R}_+$, we need to choose suitably regular versions of the studied processes. The next result provides two closely related regularizations of a given submartingale. Say that a process X on $\mathbb{R}_+$ is *right-continuous with left-hand limits* (abbreviated as *rcll*) if $X_t = X_{t+}$ for all $t \geq 0$ and the left-hand limits X_{t-} exist and are finite for all $t > 0$. For any process Y on $\mathbb{Q}_+$, we write Y^+ for the process of right-hand limits Y_{t+}, $t \geq 0$, provided that the latter exist.

Theorem 7.27 *(regularization, Doob)* For any $\mathcal{F}$-submartingale X on $\mathbb{R}_+$ with restriction Y to $\mathbb{Q}_+$, we have:

(i) Y^+ exists and is rcll outside some fixed P-null set A, and $Z = 1_{A^c} Y^+$ is a submartingale with respect to the augmented filtration $\overline{\mathcal{F}^+}$.

(ii) If $\mathcal{F}$ is right-continuous, then X has an rcll version iff EX is right-continuous; this holds in particular when X is a martingale.

The proof requires an extension of Theorem 7.21 to suitable submartingales.

Lemma 7.28 *(uniform integrability)* A submartingale X on $\mathbb{Z}_-$ is uniformly integrable iff EX is bounded.

Proof: Let EX be bounded. Introduce the predictable sequence
$$\alpha_n = E[\Delta X_n | \mathcal{F}_{n-1}] \geq 0, \quad n \leq 0,$$
and note that
$$E \sum\nolimits_{n \leq 0} \alpha_n = EX_0 - \inf\nolimits_{n \leq 0} EX_n < \infty.$$
Hence, $\sum_n \alpha_n < \infty$ a.s., and so we may define
$$A_n = \sum\nolimits_{k \leq n} \alpha_k, \quad M_n = X_n - A_n, \quad n \leq 0.$$
Since $EA^* < \infty$ and M is a martingale closed at 0, both A and M are uniformly integrable. $\square$

Proof of Theorem 7.27: (i) By Lemma 7.11 the process $Y \vee 0$ is L^1-bounded on bounded intervals, and so the same thing is true for Y. Thus, by Theorem 7.18, the right- and left-hand limits $Y_{t\pm}$ exist outside some fixed P-null set A, and so $Z = 1_{A^c} Y^+$ is rcll. Also note that Z is adapted to $\overline{\mathcal{F}^+}$.

To prove that Z is an $\overline{\mathcal{F}^+}$-submartingale, fix any times $s < t$, and choose $s_n \downarrow s$ and $t_n \downarrow t$ in $\mathbb{Q}_+$ with $s_n < t$. Then $Y_{s_m} \leq E[Y_{t_n} | \mathcal{F}_{s_m}]$ a.s. for all m and n, and as $m \to \infty$ we get $Z_s \leq E[Y_{t_n} | \mathcal{F}_{s+}]$ a.s. by Theorem 7.23. Since $Y_{t_n} \to Z_t$ in L^1 by Lemma 7.28, it follows that $Z_s \leq E[Z_t | \mathcal{F}_{s+}] = E[Z_t | \overline{\mathcal{F}_{s+}}]$ a.s.

(ii) For any $t < t_n \in \mathbb{Q}_+$,
$$(EX)_{t_n} = E(Y_{t_n}), \qquad X_t \leq E[Y_{t_n}|\mathcal{F}_t] \text{ a.s.},$$
and as $t_n \downarrow t$ we get, by Lemma 7.28 and the right-continuity of $\mathcal{F}$,
$$(EX)_{t+} = EZ_t, \qquad X_t \leq E[Z_t|\mathcal{F}_t] = Z_t \text{ a.s.} \qquad (15)$$
If X has a right-continuous version, then clearly $Z_t = X_t$ a.s. Hence, (15) yields $(EX)_{t+} = EX_t$, which shows that EX is right-continuous. If instead EX is right-continuous, then (15) gives $E|Z_t - X_t| = EZ_t - EX_t = 0$, and so $Z_t = X_t$ a.s., which means that Z is a version of X. □

Justified by the last theorem, we henceforth assume all submartingales to be rcll, unless otherwise specified, and also that the underlying filtration is right-continuous and complete. Most of the previously quoted results for submartingales on a countable index set extend immediately to such a context. In particular, this is true for the convergence Theorem 7.18 and the inequalities in Proposition 7.15 and Lemma 7.17. We proceed to show how Theorems 7.12 and 7.21 extend to submartingales in continuous time.

Theorem 7.29 (*optional sampling and closure, Doob*) *Let X be an $\mathcal{F}$-submartingale on $\mathbb{R}_+$, where X and $\mathcal{F}$ are right-continuous, and consider two optional times σ and τ, where τ is bounded. Then X_τ is integrable, and*
$$X_{\sigma \wedge \tau} \leq E[X_\tau|\mathcal{F}_\sigma] \text{ a.s.} \qquad (16)$$
The statement extends to unbounded times τ iff X^+ is uniformly integrable.

Proof: Introduce the optional times $\sigma_n = 2^{-n}[2^n\sigma + 1]$ and $\tau_n = 2^{-n}[2^n\tau + 1]$, and conclude from Lemma 7.10 and Theorem 7.12 that
$$X_{\sigma_m \wedge \tau_n} \leq E[X_{\tau_n}|\mathcal{F}_{\sigma_m}] \text{ a.s.}, \qquad m, n \in \mathbb{N}.$$
As $m \to \infty$, we get by Lemma 7.3 and Theorem 7.23
$$X_{\sigma \wedge \tau_n} \leq E[X_{\tau_n}|\mathcal{F}_\sigma] \text{ a.s.}, \qquad n \in \mathbb{N}. \qquad (17)$$
By the result for the index sets $2^{-n}\mathbb{Z}_+$, the random variables $X_0; \ldots, X_{\tau_2}, X_{\tau_1}$ form a submartingale with bounded mean and are therefore uniformly integrable by Lemma 7.28. Thus, (16) follows as we let $n \to \infty$ in (17).

If X^+ is uniformly integrable, then X is L^1-bounded and hence converges a.s. toward some $X_\infty \in L^1$. By Proposition 4.12 we get $X_t^+ \to X_\infty^+$ in L^1, and so $E[X_t^+|\mathcal{F}_s] \to E[X_\infty^+|\mathcal{F}_s]$ in L^1 for each s. Letting $t \to \infty$ along a sequence, we get by Fatou's lemma
$$\begin{aligned} X_s &\leq \lim_t E[X_t^+|\mathcal{F}_s] - \liminf_t E[X_t^-|\mathcal{F}_s] \\ &\leq E[X_\infty^+|\mathcal{F}_s] - E[X_\infty^-|\mathcal{F}_s] = E[X_\infty|\mathcal{F}_s]. \end{aligned}$$
We may now approximate as before to obtain (16) for arbitrary σ and τ.

Conversely, the stated condition implies that there exists some $X_\infty \in L^1$ with $X_s \leq E[X_\infty|\mathcal{F}_s]$ a.s. for all $s > 0$, and so $X_s^+ \leq E[X_\infty^+|\mathcal{F}_s]$ a.s. by Lemma 7.11. Hence, X^+ is uniformly integrable by Lemma 6.5. □

For a simple application, we consider the hitting probabilities of a continuous martingale. The result will be useful in Chapters 14, 17, and 23.

Corollary 7.30 *(first hit)* Let M be a continuous martingale with $M_0 = 0$ and $P\{M^* > 0\} > 0$, and define $\tau_x = \inf\{t > 0;\ M_t = x\}$. Then

$$P[\tau_a < \tau_b| M^* > 0] \le \frac{b}{b-a} \le P[\tau_a \le \tau_b| M^* > 0], \quad a < 0 < b.$$

Proof: Since $\tau = \tau_a \wedge \tau_b$ is optional by Lemma 7.6, Theorem 7.29 yields $EM_{\tau \wedge t} = 0$ for all $t > 0$, and so by dominated convergence $EM_\tau = 0$. Hence,

$$\begin{aligned}
0 &= aP\{\tau_a < \tau_b\} + bP\{\tau_b < \tau_a\} + E[M_\infty;\ \tau = \infty] \\
&\le aP\{\tau_a < \tau_b\} + bP\{\tau_b \le \tau_a,\ M^* > 0\} \\
&= bP\{M^* > 0\} - (b - a)P\{\tau_a < \tau_b\},
\end{aligned}$$

which implies the first inequality. The second one follows by taking complements. □

The next result plays a crucial role in Chapter 19.

Lemma 7.31 *(absorption)* Let $X \ge 0$ be a right-continuous supermartingale, and put $\tau = \inf\{t \ge 0;\ X_t \wedge X_{t-} = 0\}$. Then $X = 0$ a.s. on $[\tau, \infty)$.

Proof: By Theorem 7.27 the process X remains a supermartingale with respect to the right-continuous filtration $\mathcal{F}^+$. The times $\tau_n = \inf\{t \ge 0;\ X_t < n^{-1}\}$ are $\mathcal{F}^+$-optional by Lemma 7.6, and by the right-continuity of X we have $X_{\tau_n} \le n^{-1}$ on $\{\tau_n < \infty\}$. Hence, by Theorem 7.29,

$$E[X_t;\ \tau_n \le t] \le E[X_{\tau_n};\ \tau_n \le t] \le n^{-1}, \quad t \ge 0,\ n \in \mathbb{N}.$$

Noting that $\tau_n \uparrow \tau$, we get by dominated convergence $E[X_t;\ \tau \le t] = 0$, and so $X_t = 0$ a.s. on $\{\tau \le t\}$. The assertion now follows, as we apply this result to all $t \in \mathbb{Q}_+$ and use the right-continuity of X. □

We proceed to show how the right-continuity of an increasing sequence of supermartingales extends to the limit. The result is needed in Chapter 25.

Theorem 7.32 *(increasing limits of supermartingales, Meyer)* Let $X^1 \le X^2 \le \cdots$ be right-continuous supermartingales with $\sup_n EX_0^n < \infty$. Then $X_t = \sup_n X_t^n$, $t \ge 0$, is again an a.s. right-continuous supermartingale.

Proof (Doob): By Theorem 7.27 we may assume the filtration to be right-continuous. The supermartingale property carries over to X by monotone convergence. To prove the asserted right-continuity, we may assume that X^1 is bounded below by an integrable random variable; otherwise consider the processes obtained by optional stopping at the times $m \wedge \inf\{t;\ X_t^1 < -m\}$ for arbitrary $m > 0$.

Now fix any $\varepsilon > 0$, let $\mathcal{T}$ denote the class of optional times τ with

$$\limsup\nolimits_{u \downarrow t}|X_u - X_t| \leq 2\varepsilon, \quad t < \tau,$$

and put $p = \inf_{\tau \in \mathcal{T}} Ee^{-\tau}$. Choose $\sigma_1, \sigma_2, \cdots \in \mathcal{T}$ with $Ee^{-\sigma_n} \to p$, and note that $\sigma \equiv \sup_n \sigma_n \in \mathcal{T}$ with $Ee^{-\sigma} = p$. We need to show that $\sigma = \infty$ a.s. Then introduce the optional times

$$\tau_n = \inf\{t > \sigma; \ |X_t^n - X_\sigma| > \varepsilon\}, \quad n \in \mathbb{N},$$

and put $\tau = \limsup_n \tau_n$. Noting that

$$|X_t - X_\sigma| = \liminf_{n \to \infty} |X_t^n - X_\sigma| \leq \varepsilon, \quad t \in [\sigma, \tau),$$

we obtain $\tau \in \mathcal{T}$.

By the right-continuity of X^n, we note that $|X_{\tau_n}^n - X_\sigma| \geq \varepsilon$ on $\{\tau_n < \infty\}$ for every n. Furthermore, on the set $A = \{\sigma = \tau < \infty\}$ we have

$$\liminf_{n \to \infty} X_{\tau_n}^n \geq \sup\nolimits_k \lim\nolimits_n X_{\tau_n}^k = \sup\nolimits_k X_\sigma^k = X_\sigma,$$

and so $\liminf_n X_{\tau_n}^n \geq X_\sigma + \varepsilon$ on A. Since $A \in \mathcal{F}_\sigma$ by Lemma 7.1, we get by Fatou's lemma, optional sampling, and monotone convergence,

$$\begin{aligned} E[X_\sigma + \varepsilon; A] &\leq E[\liminf\nolimits_n X_{\tau_n}^n; A] \leq \liminf\nolimits_n E[X_{\tau_n}^n; A] \\ &\leq \lim\nolimits_n E[X_\sigma^n; A] = E[X_\sigma; A]. \end{aligned}$$

Thus, $PA = 0$, and so $\tau > \sigma$ a.s. on $\{\sigma < \infty\}$. If $p > 0$, we get the contradiction $Ee^{-\tau} < p$, so $p = 0$. Hence, $\sigma = \infty$ a.s. □

Exercises

1. Show for any optional times σ and τ that $\{\sigma = \tau\} \in \mathcal{F}_\sigma \cap \mathcal{F}_\tau$ and $\mathcal{F}_\sigma = \mathcal{F}_\tau$ on $\{\sigma = \tau\}$. However, $\mathcal{F}_\tau$ and $\mathcal{F}_\infty$ may differ on $\{\tau = \infty\}$.

2. Show that if σ and τ are optional times on the time scale $\mathbb{R}_+$ or $\mathbb{Z}_+$, then so is $\sigma + \tau$.

3. Give an example of a random time that is weakly optional but not optional. (Hint: Let $\mathcal{F}$ be the filtration induced by the process $X_t = \vartheta t$ with $P\{\vartheta = \pm 1\} = \frac{1}{2}$, and take $\tau = \inf\{t; \ X_t > 0\}$.)

4. Fix a random time τ and a random variable ξ in $\mathbb{R} \setminus \{0\}$. Show that the process $X_t = \xi 1\{\tau \leq t\}$ is adapted to a given filtration $\mathcal{F}$ iff τ is $\mathcal{F}$-optional and ξ is $\mathcal{F}_\tau$-measurable. Give corresponding conditions for the process $Y_t = \xi 1\{\tau < t\}$.

5. Let $\mathcal{P}$ denote the class of sets $A \in \mathbb{R}_+ \times \Omega$ such that the process 1_A is progressive. Show that $\mathcal{P}$ is a σ-field and that a process X is progressive iff it is $\mathcal{P}$-measurable.

6. Let X be a progressive process with induced filtration $\mathcal{F}$, and fix any optional time $\tau < \infty$. Show that $\sigma\{\tau, X^\tau\} \subset \mathcal{F}_\tau \subset \mathcal{F}_\tau^+ \subset \sigma\{\tau, X^{\tau+h}\}$ for every $h > 0$. (*Hint:* The first relation becomes an equality when τ takes only countably many values.) Note that the result may fail when $P\{\tau = \infty\} > 0$.

7. Let M be an $\mathcal{F}$-martingale on some countable index set, and fix an optional time τ. Show that $M - M^\tau$ remains a martingale conditionally on $\mathcal{F}_\tau$. (*Hint:* Use Theorem 7.12 and Lemma 7.13.) Extend the result to continuous time.

8. Show that any submartingale remains a submartingale with respect to the induced filtration.

9. Let $X^1, X^2, \ldots$ be submartingales such that the process $X = \sup_n X^n$ is integrable. Show that X is again a submartingale. Also show that $\limsup_n X^n$ is a submartingale when even $\sup_n |X^n|$ is integrable.

10. Show that the Doob decomposition of an integrable random sequence $X = (X_n)$ depends on the filtration unless X is a.s. X_0-measurable. (*Hint:* Compare the filtrations induced by X and by the sequence $Y_n = (X_0, X_{n+1})$.)

11. Fix a random time τ and a random variable $\xi \in L^1$, and define $M_t = \xi 1\{\tau \le t\}$. Show that M is a martingale with respect to the induced filtration $\mathcal{F}$ iff $E[\xi; \tau \le t | \tau > s] = 0$ for any $s < t$. (*Hint:* The set $\{\tau > s\}$ is an atom of $\mathcal{F}_s$.)

12. Let $\mathcal{F}$ and $\mathcal{G}$ be filtrations on a common probability space. Show that every $\mathcal{F}$-martingale is a $\mathcal{G}$-martingale iff $\mathcal{F}_t \subset \mathcal{G}_t \perp\!\!\!\perp_{\mathcal{F}_t} \mathcal{F}_\infty$ for every $t \ge 0$. (*Hint:* For the necessity, consider $\mathcal{F}$-martingales of the form $M_s = E[\xi | \mathcal{F}_s]$ with $\xi \in L^1(\mathcal{F}_t)$.)

13. Show for any rcll supermartingale $X \ge 0$ and constant $r \ge 0$ that $rP\{\sup_t X_t \ge r\} \le EX_0$.

14. Let M be an L^2-bounded martingale on $\mathbb{Z}_+$. Imitate the proof of Lemma 4.16 to show that M_n converges a.s. and in L^2.

15. Give an example of a martingale that is L^1-bounded but not uniformly integrable. (*Hint:* Every positive martingale is L^1-bounded.)

16. Show that if $\mathcal{G} \perp\!\!\!\perp_{\mathcal{F}_n} \mathcal{H}$ for some increasing σ-fields $\mathcal{F}_n$, then $\mathcal{G} \perp\!\!\!\perp_{\mathcal{F}_\infty} \mathcal{H}$.

17. Let $\xi_n \to \xi$ in L^1. Show for any increasing σ-fields $\mathcal{F}_n$ that $E[\xi_n | \mathcal{F}_n] \to E[\xi | \mathcal{F}_\infty]$ in L^1.

18. Let $\xi, \xi_1, \xi_2, \cdots \in L^1$ with $\xi_n \uparrow \xi$ a.s. Show for any increasing σ-fields $\mathcal{F}_n$ that $E[\xi_n | \mathcal{F}_n] \to E[\xi | \mathcal{F}_\infty]$ a.s. (*Hint:* By Proposition 7.15 we have $\sup_m E[\xi - \xi_n | \mathcal{F}_m] \xrightarrow{P} 0$. Now use the monotonicity.)

19. Show that any right-continuous submartingale is a.s. rcll.

20. Let σ and τ be optional times with respect to some right-continuous filtration $\mathcal{F}$. Show that the operators $E^{\mathcal{F}_\sigma}$ and $E^{\mathcal{F}_\tau}$ commute on L^1 with product $E^{\mathcal{F}_{\sigma \wedge \tau}}$. (*Hint:* For any $\xi \in L^1$, apply the optional sampling theorem to a right-continuous version of the martingale $M_t = E[\xi | \mathcal{F}_t]$.)

21. Let $X \geq 0$ be a supermartingale on $\mathbb{Z}_+$, and let $\tau_0 \leq \tau_1 \leq \cdots$ be optional times. Show that the sequence (X_{τ_n}) is again a supermartingale. (*Hint:* Truncate the times τ_n, and use the conditional Fatou lemma.) Show by an example that the result fails for submartingales.

22. For any random time $\tau \geq 0$ and right-continuous filtration $\mathcal{F} = (\mathcal{F}_t)$, show that the process $X_t = P[\tau \leq t | \mathcal{F}_t]$ has a right-continuous version. (*Hint:* Use Theorem 7.27 (ii).)

Chapter 8

Markov Processes and Discrete-Time Chains

Markov property and transition kernels; finite-dimensional distributions and existence; space and time homogeneity; strong Markov property and excursions; invariant distributions and stationarity; recurrence and transience; ergodic behavior of irreducible chains; mean recurrence times

A Markov process may be described informally as a randomized dynamical system, a description that explains the fundamental role that Markov processes play both in theory and in a wide range of applications. Processes of this type appear more or less explicitly throughout the remainder of this book.

To make the above description precise, let us fix any Borel space S and filtration $\mathcal{F}$. An adapted process X in S is said to be Markov if for any times $s < t$ we have $X_t = f_{s,t}(X_s, \vartheta_{s,t})$ a.s. for some measurable function $f_{s,t}$ and some $U(0,1)$ random variable $\vartheta_{s,t} \perp\!\!\!\perp \mathcal{F}_s$. The stated condition is equivalent to the less transparent conditional independence $X_t \perp\!\!\!\perp_{X_s} \mathcal{F}_s$. The process is said to be time-homogeneous if we can take $f_{s,t} \equiv f_{0,t-s}$ and space-homogeneous (when $S = \mathbb{R}^d$) if $f_{s,t}(x, \cdot) \equiv f_{s,t}(0, \cdot) + x$. A more convenient description of the evolution is in terms of the transition kernels $\mu_{s,t}(x, \cdot) = P\{f_{s,t}(x, \vartheta) \in \cdot\}$, which are easily seen to satisfy an a.s. version of the Chapman–Kolmogorov relation $\mu_{s,t}\mu_{t,u} = \mu_{s,u}$. In the usual axiomatic treatment, the latter equation is assumed to hold identically.

This chapter is devoted to some of the most basic and elementary portions of Markov process theory. Thus, the space homogeneity will be shown to be equivalent to the independence of the increments, which motivates our discussion of random walks and Lévy processes in Chapters 9 and 15. In the time-homogeneous case we shall establish a primitive form of the strong Markov property and see how the result simplifies when the process is also space-homogeneous. Next we shall see how invariance of the initial distribution implies stationarity of the process, which motivates our treatment of stationary processes in Chapter 10. Finally, we shall discuss the classification of states and examine the ergodic behavior of discrete-time Markov chains on a countable state space. The analogous but less elementary theory for continuous-time chains is postponed until Chapter 12.

The general theory of Markov processes is more advanced and is not continued until Chapter 19, which develops the basic theory of Feller processes. In the meantime we shall consider several important subclasses, such as the pure jump-type processes in Chapter 12, Brownian motion and related processes in Chapters 13 and 18, and the above-mentioned random walks and Lévy processes in Chapters 9 and 15. A detailed discussion of diffusion processes appears in Chapters 21 and 23, and additional aspects of Brownian motion are considered in Chapters 22, 24, and 25.

To begin our systematic study of Markov processes, consider an arbitrary time scale $T \subset \mathbb{R}$, equipped with a filtration $\mathcal{F} = (\mathcal{F}_t)$, and fix a measurable space $(S, \mathcal{S})$. An S-valued process X on T is said to be a *Markov process* if it is adapted to $\mathcal{F}$ and such that

$$\mathcal{F}_t \perp\!\!\!\perp_{X_t} X_u, \quad t \leq u \text{ in } T. \tag{1}$$

Just as for the martingale property, we note that even the Markov property depends on the choice of filtration, with the weakest version obtained for the filtration induced by X. The simple property in (1) may be strengthened as follows.

Lemma 8.1 *(extended Markov property)* If X satisfies (1), then

$$\mathcal{F}_t \perp\!\!\!\perp_{X_t} \{X_u; u \geq t\}, \quad t \in T. \tag{2}$$

Proof: Fix any $t = t_0 \leq t_1 \leq \cdots$ in T. By (1) we have $\mathcal{F}_{t_n} \perp\!\!\!\perp_{X_{t_n}} X_{t_{n+1}}$ for every $n \geq 0$, and so by Proposition 6.8

$$\mathcal{F}_t \perp\!\!\!\perp_{X_{t_0}, \ldots, X_{t_n}} X_{t_{n+1}}, \quad n \geq 0.$$

By the same proposition, this is equivalent to

$$\mathcal{F}_t \perp\!\!\!\perp_{X_t} (X_{t_1}, X_{t_2}, \ldots),$$

and (2) follows by a monotone class argument. □

For any times $s \leq t$ in T, we assume the existence of some regular conditional distributions

$$\mu_{s,t}(X_s, B) = P[X_t \in B | X_s] = P[X_t \in B | \mathcal{F}_s] \text{ a.s.}, \quad B \in \mathcal{S}. \tag{3}$$

In particular, we note that the *transition kernels* $\mu_{s,t}$ exist by Theorem 6.3 when S is Borel. We may further introduce the one-dimensional distributions $\nu_t = \mathcal{L}(X_t)$, $t \in T$. When T begins at 0, we shall prove that the distribution of X is uniquely determined by the kernels $\mu_{s,t}$ together with the *initial distribution* ν_0.

For a precise statement, it is convenient to use the kernel operations introduced in Chapter 1. Note in particular that if μ and ν are kernels on

S, then $\mu \otimes \nu$ and $\mu\nu$ are kernels from S to S^2 and S, respectively, given for $s \in S$ by

$$(\mu \otimes \nu)(s, B) = \int \mu(s, dt) \int \nu(t, du) 1_B(t, u), \qquad B \in S^2,$$
$$(\mu\nu)(s, B) = (\mu \otimes \nu)(s, S \times B) = \int \mu(s, dt)\nu(t, B), \qquad B \in S.$$

Proposition 8.2 *(finite-dimensional distributions)* Let X be a Markov process on T with one-dimensional distributions ν_t and transition kernels $\mu_{s,t}$. Then for any $t_0 \leq \cdots \leq t_n$ in T,

$$\mathcal{L}(X_{t_0}, \ldots, X_{t_n}) = \nu_{t_0} \otimes \mu_{t_0, t_1} \otimes \cdots \otimes \mu_{t_{n-1}, t_n}, \qquad (4)$$
$$P[(X_{t_1}, \ldots, X_{t_n}) \in \cdot | \mathcal{F}_{t_0}] = (\mu_{t_0, t_1} \otimes \cdots \otimes \mu_{t_{n-1}, t_n})(X_{t_0}, \cdot). \qquad (5)$$

Proof: Formula (4) is clearly true for $n = 0$. Proceeding by induction, assume (4) to be true with n replaced by $n-1$, and fix any bounded measurable function f on S^{n+1}. Noting that $X_{t_0}, \ldots, X_{t_{n-1}}$ are $\mathcal{F}_{t_{n-1}}$-measurable, we get by Theorem 6.4 and the induction hypothesis

$$\begin{aligned}
Ef(X_{t_0}, \ldots, X_{t_n}) &= E\,E[f(X_{t_0}, \ldots, X_{t_n})|\mathcal{F}_{t_{n-1}}] \\
&= E\int f(X_{t_0}, \ldots, X_{t_{n-1}}, x_n) \mu_{t_{n-1}, t_n}(X_{t_{n-1}}, dx_n) \\
&= (\nu_{t_0} \otimes \mu_{t_0, t_1} \otimes \cdots \otimes \mu_{t_{n-1}, t_n}) f,
\end{aligned}$$

as desired. This completes the proof of (4).

In particular, for any $B \in S$ and $C \in S^n$ we get

$$P\{(X_{t_0}, \ldots, X_{t_n}) \in B \times C\}$$
$$= \int_B \nu_{t_0}(dx)(\mu_{t_0, t_1} \otimes \cdots \otimes \mu_{t_{n-1}, t_n})(x, C)$$
$$= E[(\mu_{t_0, t_1} \otimes \cdots \otimes \mu_{t_{n-1}, t_n})(X_{t_0}, C);\ X_{t_0} \in B],$$

and (5) follows by Theorem 6.1 and Lemma 8.1. $\square$

An obvious consistency requirement leads to the following basic so-called *Chapman–Kolmogorov relation* between the transition kernels. Here we say that two kernels μ and μ' agree a.s. if $\mu(x, \cdot) = \mu'(x, \cdot)$ for almost every x.

Corollary 8.3 *(Chapman, Smoluchovsky)* For any Markov process in a Borel space S, we have

$$\mu_{s,u} = \mu_{s,t}\mu_{t,u} \quad \text{a.s. } \nu_s, \qquad s \leq t \leq u.$$

Proof: By Proposition 8.2 we have a.s. for any $B \in S$

$$\begin{aligned}
\mu_{s,u}(X_s, B) &= P[X_u \in B | \mathcal{F}_s] = P[(X_t, X_u) \in S \times B | \mathcal{F}_s] \\
&= (\mu_{s,t} \otimes \mu_{t,u})(X_s, S \times B) = (\mu_{s,t}\mu_{t,u})(X_s, B).
\end{aligned}$$

Since S is Borel, we may choose a common null set for all B. $\square$

8. Markov Processes and Discrete-Time Chains

We henceforth assume that the Chapman–Kolmogorov relation holds *identically*, so that

$$\mu_{s,u} = \mu_{s,t}\mu_{t,u}, \quad s \leq t \leq u. \tag{6}$$

Thus, we define a Markov process by condition (3), in terms of some transition kernels $\mu_{s,t}$ satisfying (6). In *discrete time*, when $T = \mathbb{Z}_+$, the latter relation is no restriction, since we may then start from any versions of the kernels $\mu_n = \mu_{n-1,n}$, and define $\mu_{m,n} = \mu_{m+1} \cdots \mu_n$ for arbitrary $m < n$.

Given such a family of transition kernels $\mu_{s,t}$ and an arbitrary initial distribution ν, we need to show that an associated Markov process exists. This is ensured, under weak restrictions, by the following result.

Theorem 8.4 (existence, Kolmogorov) *Fix a time scale T starting at 0, a Borel space $(S, \mathcal{S})$, a probability measure ν on S, and a family of probability kernels $\mu_{s,t}$ on S, $s \leq t$ in T, satisfying (6). Then there exists an S-valued Markov process X on T with initial distribution ν and transition kernels $\mu_{s,t}$.*

Proof: Introduce the probability measures

$$\nu_{t_1,\ldots,t_n} = \nu\mu_{t_0,t_1} \otimes \cdots \otimes \mu_{t_{n-1},t_n}, \quad 0 = t_0 \leq t_1 \leq \cdots \leq t_n, \ n \in \mathbb{N}.$$

To see that the family $(\nu_{t_0,\ldots,t_n})$ is projective, let $B \in \mathcal{S}^{n-1}$ be arbitrary, and define for any $k \in \{1,\ldots,n\}$ the set

$$B_k = \{(x_1,\ldots,x_n) \in S^n; \ (x_1,\ldots,x_{k-1}, x_{k+1},\ldots,x_n) \in B\}.$$

Then by (6)

$$\begin{aligned}\nu_{t_1,\ldots,t_n} B_k &= (\nu\mu_{t_0,t_1} \otimes \cdots \otimes \mu_{t_{k-1},t_{k+1}} \otimes \cdots \otimes \mu_{t_{n-1},t_n}) B \\ &= \nu_{t_1,\ldots,t_{k-1},t_{k+1},\ldots,t_n} B,\end{aligned}$$

as desired. By Theorem 6.16 there exists an S-valued process X on T with

$$\mathcal{L}(X_{t_1},\ldots,X_{t_n}) = \nu_{t_1,\ldots,t_n}, \quad t_1 \leq \cdots \leq t_n, \ n \in \mathbb{N}, \tag{7}$$

and, in particular, $\mathcal{L}(X_0) = \nu_0 = \nu$.

To see that X is Markov with transition kernels $\mu_{s,t}$, fix any times $s_1 \leq \cdots \leq s_n = s \leq t$ and sets $B \in \mathcal{S}^n$ and $C \in \mathcal{S}$, and conclude from (7) that

$$\begin{aligned}P\{(X_{s_1},\ldots,X_{s_n},X_t) \in B \times C\} &= \nu_{s_1,\ldots,s_n,t}(B \times C) \\ &= E[\mu_{s,t}(X_s,C); (X_{s_1},\ldots,X_{s_n}) \in B].\end{aligned}$$

Writing $\mathcal{F}$ for the filtration induced by X, we get by a monotone class argument

$$P[X_t \in C; A] = E[\mu_{s,t}(X_s,C); A], \quad A \in \mathcal{F}_s,$$

and so $P[X_t \in C | \mathcal{F}_s] = \mu_{s,t}(X_s, C)$ a.s. □

Now assume that S is a measurable Abelian group. A kernel μ on S is then said to be *homogeneous* if
$$\mu(x, B) = \mu(0, B - x), \quad x \in S, \ B \in \mathcal{S}.$$
An S-valued Markov process with homogeneous transition kernels $\mu_{s,t}$ is said to be *space-homogeneous*. Furthermore, we say that a process X in S has *independent increments* if, for any times $t_0 \leq \cdots \leq t_n$, the increments $X_{t_k} - X_{t_{k-1}}$ are mutually independent and independent of X_0. More generally, given any filtration $\mathcal{F}$ on T, we say that X has $\mathcal{F}$-*independent increments* if X is adapted to $\mathcal{F}$ and such that $X_t - X_s \perp\!\!\!\perp \mathcal{F}_s$ for all $s \leq t$ in T. Note that the elementary notion of independence corresponds to the case when $\mathcal{F}$ is induced by X.

Proposition 8.5 (*independent increments and homogeneity*) *Consider a measurable Abelian group S, a filtration $\mathcal{F}$ on some time scale T, and an S-valued and $\mathcal{F}$-adapted process X on T. Then X is space-homogeneous $\mathcal{F}$-Markov iff it has $\mathcal{F}$-independent increments, in which case the transition kernels are given by*
$$\mu_{s,t}(x, B) = P\{X_t - X_s \in B - x\}, \quad x \in S, \ B \in \mathcal{S}, \ s \leq t \text{ in } T. \tag{8}$$

Proof: First assume that X is Markov with transition kernels
$$\mu_{s,t}(x, B) = \mu_{s,t}(B - x), \quad x \in S, \ B \in \mathcal{S}, \ s \leq t \text{ in } T. \tag{9}$$
By Theorem 6.4, for any $s \leq t$ in T and $B \in \mathcal{S}$ we get
$$\begin{aligned} P[X_t - X_s \in B | \mathcal{F}_s] &= P[X_t \in B + X_s | \mathcal{F}_s] \\ &= \mu_{s,t}(X_s, B + X_s) = \mu_{s,t} B. \end{aligned}$$
Thus, $X_t - X_s$ is independent of $\mathcal{F}_s$ with distribution $\mu_{s,t}$, and (8) follows by means of (9).

Conversely, assume that $X_t - X_s$ is independent of $\mathcal{F}_s$ with distribution $\mu_{s,t}$. Defining the associated kernel $\mu_{s,t}$ by (9), we get by Theorem 6.4, for any s, t, and B as before,
$$\begin{aligned} P[X_t \in B | \mathcal{F}_s] &= P[X_t - X_s \in B - X_s | \mathcal{F}_s] \\ &= \mu_{s,t}(B - X_s) = \mu_{s,t}(X_s, B). \end{aligned}$$
Thus, X is Markov with the homogeneous transition kernels in (9). $\square$

We may now specialize to the *time-homogeneous* case—when $T = \mathbb{R}_+$ or $\mathbb{Z}_+$ and the transition kernels are of the form $\mu_{s,t} = \mu_{t-s}$, so that
$$P[X_t \in B | \mathcal{F}_s] = \mu_{t-s}(X_s, B) \text{ a.s.}, \quad B \in \mathcal{S}, \ s \leq t \text{ in } T.$$
Introducing the initial distribution $\nu = \mathcal{L}(X_0)$, we may write the formulas of Proposition 8.2 as
$$\begin{aligned} \mathcal{L}(X_{t_0}, \ldots, X_{t_n}) &= \nu \mu_{t_0} \otimes \mu_{t_1 - t_0} \otimes \cdots \otimes \mu_{t_n - t_{n-1}}, \\ P[(X_{t_1}, \ldots, X_{t_n}) \in \cdot | \mathcal{F}_{t_0}] &= (\mu_{t_1 - t_0} \otimes \cdots \otimes \mu_{t_n - t_{n-1}})(X_{t_0}, \cdot). \end{aligned}$$

The Chapman–Kolmogorov relation now becomes

$$\mu_{s+t} = \mu_s \mu_t, \quad s, t \in T,$$

which is again assumed to hold identically. We often refer to the family (μ_t) as a *semigroup* of transition kernels.

The following result justifies the interpretation of a discrete-time Markov process as a randomized dynamical system.

Proposition 8.6 *(recursion)* Let X be a process on $\mathbb{Z}_+$ with values in a Borel space S. Then X is Markov iff there exist some measurable functions $f_1, f_2, \ldots : S \times [0,1] \to S$ and i.i.d. $U(0,1)$ random variables $\vartheta_1, \vartheta_2, \ldots \perp\!\!\!\perp X_0$ such that $X_n = f_n(X_{n-1}, \vartheta_n)$ a.s. for all $n \in \mathbb{N}$. Here we may choose $f_1 = f_2 = \cdots = f$ iff X is time-homogeneous.

Proof: Let X have the stated representation and introduce the kernels $\mu_n(x, \cdot) = P\{f_n(x, \vartheta) \in \cdot\}$, where ϑ is $U(0,1)$. Writing $\mathcal{F}$ for the filtration induced by X, we get by Theorem 6.4 for any $B \in \mathcal{S}$

$$\begin{aligned} P[X_n \in B | \mathcal{F}_{n-1}] &= P[f_n(X_{n-1}, \vartheta_n) \in B | \mathcal{F}_{n-1}] \\ &= \lambda\{t;\, f_n(X_{n-1}, t) \in B\} = \mu_n(X_{n-1}, B), \end{aligned}$$

which shows that X is Markov with transition kernels μ_n.

Now assume instead the latter condition. By Lemma 3.22 we may choose some associated functions f_n as above. Let $\tilde{\vartheta}_1, \tilde{\vartheta}_2, \ldots$ be i.i.d. $U(0,1)$ and independent of $\tilde{X}_0 \stackrel{d}{=} X_0$, and define recursively $\tilde{X}_n = f_n(\tilde{X}_{n-1}, \tilde{\vartheta}_n)$ for $n \in \mathbb{N}$. As before, $\tilde{X}$ is Markov with transition kernels μ_n. Hence, $\tilde{X} \stackrel{d}{=} X$ by Proposition 8.2, and so by Theorem 6.10 there exist some random variables ϑ_n with $(X, (\vartheta_n)) \stackrel{d}{=} (\tilde{X}, (\tilde{\vartheta}_n))$. Since the diagonal in S^2 is measurable, the desired representation follows. The last assertion is obvious from the construction. □

Now fix a transition semigroup (μ_t) on some Borel space S. For any probability measure ν on S, there exists by Theorem 8.4 an associated Markov process X_ν, and by Proposition 3.2 the corresponding distribution P_ν is uniquely determined by ν. Note that P_ν is a probability measure on the path space $(S^T, \mathcal{S}^T)$. For degenerate initial distributions δ_x, we may write P_x instead of P_{δ_x}. Integration with respect to P_ν or P_x is denoted by E_ν or E_x, respectively.

Lemma 8.7 *(mixtures)* The measures P_x form a probability kernel from S to S^T, and for any initial distribution ν we have

$$P_\nu A = \int_S P_x(A)\, \nu(dx), \quad A \in \mathcal{S}^T. \tag{10}$$

Proof: Both the measurability of $P_x A$ and formula (10) are obvious for cylinder sets of the form $A = (\pi_{t_1}, \ldots, \pi_{t_n})^{-1} B$. The general case follows easily by a monotone class argument. □

Rather than considering one Markov process X_ν for each initial distribution ν, it is more convenient to introduce the *canonical* process X, defined as the identity mapping on the path space $(S^T, \mathcal{S}^T)$, and equip the latter space with the different probability measures P_ν. Then X_t agrees with the evaluation map $\pi_t : \omega \mapsto \omega_t$ on S^T, which is measurable by the definition of $\mathcal{S}^T$. For our present purposes, it is sufficient to endow the path space S^T with the *canonical filtration* $\mathcal{F}$ induced by X.

On S^T we may also introduce the *shift operators* $\theta_t : S^T \to S^T$, $t \in T$, given by

$$(\theta_t \omega)_s = \omega_{s+t}, \quad s, t \in T, \; \omega \in S^T,$$

and we note that the θ_t are measurable with respect to $\mathcal{S}^T$. In the canonical case it is further clear that $\theta_t X = \theta_t = X \circ \theta_t$.

Optional times with respect to a Markov process are often constructed recursively in terms of shifts on the underlying path space. Thus, for any pair of optional times σ and τ on the canonical space, we may consider the random time $\gamma = \sigma + \tau \circ \theta_\sigma$, with the understanding that $\gamma = \infty$ when $\sigma = \infty$. Under weak restrictions on space and filtration, we show that γ is again optional. Let $C(S)$ and $D(S)$ denote the spaces of continuous or rcll functions, respectively, from $\mathbb{R}_+$ to S.

Proposition 8.8 *(compound optional times)* *For any metric space S, let σ and τ be optional times on the canonical space S^∞, $C(S)$, or $D(S)$, endowed with the right-continuous, induced filtration. Then even $\gamma = \sigma + \tau \circ \theta_\sigma$ is optional.*

Proof: Since $\sigma \wedge n + \tau \circ \theta_{\sigma \wedge n} \uparrow \gamma$, we may assume by Lemma 7.3 that σ is bounded. Let X denote the canonical process with induced filtration $\mathcal{F}$. Since X is $\mathcal{F}^+$-progressive, $X_{\sigma+s} = X_s \circ \theta_\sigma$ is $\mathcal{F}^+_{\sigma+s}$-measurable for every $s \geq 0$ by Lemma 7.5. Fixing any $t \geq 0$, it follows that all sets $A = \{X_s \in B\}$ with $s \leq t$ and $B \in \mathcal{S}$ satisfy $\theta_\sigma^{-1} A \in \mathcal{F}^+_{\sigma+t}$. The sets A with the latter property form a σ-field, and therefore

$$\theta_\sigma^{-1} \mathcal{F}_t \subset \mathcal{F}^+_{\sigma+t}, \quad t \geq 0. \tag{11}$$

Now fix any $t \geq 0$, and note that

$$\{\gamma < t\} = \bigcup_{r \in \mathbb{Q} \cap (0,t)} \{\sigma < r, \tau \circ \theta_\sigma < t - r\}. \tag{12}$$

For every $r \in (0, t)$ we have $\{\tau < t - r\} \in \mathcal{F}_{t-r}$, so $\theta_\sigma^{-1}\{\tau < t-r\} \in \mathcal{F}^+_{\sigma+t-r}$ by (11), and Lemma 7.2 yields

$$\{\sigma < r, \tau \circ \theta_\sigma < t - r\} = \{\sigma + t - r < t\} \cap \theta_\sigma^{-1}\{\tau < t - r\} \in \mathcal{F}_t.$$

Thus, $\{\gamma < t\} \in \mathcal{F}_t$ by (12), and so γ is $\mathcal{F}^+$-optional by Lemma 7.2. □

We proceed to show how the elementary Markov property may be extended to suitable optional times. The present statement is only preliminary, and stronger versions are obtained under further conditions in Theorems 12.14, 13.11, and 19.17.

Proposition 8.9 *(strong Markov property) Fix a time-homogeneous Markov process X on $T = \mathbb{R}_+$ or $\mathbb{Z}_+$, and let τ be an optional time taking countably many values. Then*

$$P[\theta_\tau X \in A | \mathcal{F}_\tau] = P_{X_\tau} A \quad a.s. \text{ on } \{\tau < \infty\}, \quad A \in \mathcal{S}^T. \tag{13}$$

If X is canonical, it is equivalent that

$$E_\nu[\xi \circ \theta_\tau | \mathcal{F}_\tau] = E_{X_\tau} \xi, \quad P_\nu\text{-}a.s. \text{ on } \{\tau < \infty\}, \tag{14}$$

for any distribution ν on S and bounded or nonnegative random variable ξ.

Since $\{\tau < \infty\} \in \mathcal{F}_\tau$, we note that (13) and (14) make sense by Lemma 6.2, although $\theta_\tau X$ and P_{X_τ} are defined only for $\tau < \infty$.

Proof: By Lemmas 6.2 and 7.1 we may assume that $\tau = t$ is finite and nonrandom. For sets A of the form

$$A = (\pi_{t_1}, \ldots, \pi_{t_n})^{-1} B, \quad t_1 \leq \cdots \leq t_n, \ B \in \mathcal{S}^n, \ n \in \mathbb{N}, \tag{15}$$

Proposition 8.2 yields

$$\begin{aligned} P[\theta_t X \in A | \mathcal{F}_t] &= P[(X_{t+t_1}, \ldots, X_{t+t_n}) \in B | \mathcal{F}_t] \\ &= (\mu_{t_1} \otimes \mu_{t_2 - t_1} \otimes \cdots \otimes \mu_{t_n - t_{n-1}})(X_t, B) = P_{X_t} A, \end{aligned}$$

which extends by a monotone class argument to arbitrary $A \in \mathcal{S}^T$.

In the canonical case we note that (13) is equivalent to (14) with $\xi = 1_A$, since in that case $\xi \circ \theta_\tau = 1\{\theta_\tau X \in A\}$. The result extends by linearity and monotone convergence to general ξ. $\square$

When X is both space- and time-homogeneous, the strong Markov property can be stated without reference to the family (P_x).

Theorem 8.10 *(space and time homogeneity) Let X be a space- and time-homogeneous Markov process in a measurable Abelian group S. Then*

$$P_x A = P_0(A - x), \quad x \in S, \ A \in \mathcal{S}^T. \tag{16}$$

Furthermore, (13) holds for a given optional time $\tau < \infty$ iff X_τ is a.s. $\mathcal{F}_\tau$-measurable and

$$X - X_0 \stackrel{d}{=} \theta_\tau X - X_\tau \perp\!\!\!\perp \mathcal{F}_\tau. \tag{17}$$

Proof: By Proposition 8.2 we get for any set A as in (15)

$$\begin{aligned}
P_x A &= P_x \circ (\pi_{t_1}, \ldots, \pi_{t_n})^{-1} B \\
&= (\mu_{t_1} \otimes \mu_{t_2 - t_1} \otimes \cdots \otimes \mu_{t_n - t_{n-1}})(x, B) \\
&= (\mu_{t_1} \otimes \mu_{t_2 - t_1} \otimes \cdots \otimes \mu_{t_n - t_{n-1}})(0, B - x) \\
&= P_0 \circ (\pi_{t_1}, \ldots, \pi_{t_n})^{-1}(B - x) = P_0(A - x),
\end{aligned}$$

which extends to (16) by a monotone class argument.

Next assume (13). Letting $A = \pi_0^{-1} B$ with $B \in \mathcal{S}$, we get

$$1_B(X_\tau) = P_{X_\tau}\{\pi_0 \in B\} = P[X_\tau \in B | \mathcal{F}_\tau] \quad \text{a.s.},$$

and so X_τ is a.s. $\mathcal{F}_\tau$-measurable. By (16) and Theorem 6.4 we have

$$P[\theta_\tau X - X_\tau \in A | \mathcal{F}_\tau] = P_{X_\tau}(A + X_\tau) = P_0 A, \quad A \in \mathcal{S}^T, \qquad (18)$$

which shows that $\theta_\tau X - X_\tau$ is independent of $\mathcal{F}_\tau$ with distribution P_0. For $\tau = 0$ we get in particular $\mathcal{L}(X - X_0) = P_0$, and (17) follows.

Next assume (17). To deduce (13), let $A \in \mathcal{S}^T$ be arbitrary, and conclude from (16) and Theorem 6.4 that

$$\begin{aligned}
P[\theta_\tau X \in A | \mathcal{F}_\tau] &= P[\theta_\tau X - X_\tau \in A - X_\tau | \mathcal{F}_\tau] \\
&= P_0(A - X_\tau) = P_{X_\tau} A. \qquad \square
\end{aligned}$$

If a time-homogeneous Markov process X has initial distribution ν, then the distribution at time $t \in T$ equals $\nu_t = \nu \mu_t$, or

$$\nu_t B = \int \nu(dx) \mu_t(x, B), \quad B \in \mathcal{S}, \, t \in T.$$

A distribution ν is said to be *invariant* for the semigroup (μ_t) if ν_t is independent of t, so that $\nu \mu_t = \nu$ for all $t \in T$. We also say that a process X on T is *stationary* if $\theta_t X \stackrel{d}{=} X$ for all $t \in T$. The two notions are related as follows.

Lemma 8.11 *(stationarity and invariance) Let X be a time-homogeneous Markov process on T with transition kernels μ_t and initial distribution ν. Then X is stationary iff ν is invariant for (μ_t).*

Proof: Assuming ν to be invariant, we get by Proposition 8.2

$$(X_{t+t_1}, \ldots, X_{t+t_n}) \stackrel{d}{=} (X_{t_1}, \ldots, X_{t_n}), \quad t, \, t_1 \leq \cdots \leq t_n \text{ in } T,$$

and the stationarity of X follows by Proposition 3.2. $\square$

For processes X in discrete time, we may consider the sequence of successive visits to a fixed state $y \in S$. Assuming the process to be canonical, we may introduce the hitting time $\tau_y = \inf\{n \in \mathbb{N}; X_n = y\}$ and then define recursively

$$\tau_y^{k+1} = \tau_y^k + \tau_y \circ \theta_{\tau_y^k}, \quad k \in \mathbb{Z}_+,$$

starting from $\tau_y^0 = 0$. Let us further introduce the *occupation times*

$$\kappa_y = \sup\{k;\ \tau_y^k < \infty\} = \sum_{n\geq 1} 1\{X_n = y\}, \quad y \in S.$$

The next result expresses the distribution of κ_y in terms of the hitting probabilities

$$r_{xy} = P_x\{\tau_y < \infty\} = P_x\{\kappa_y > 0\}, \quad x, y \in S.$$

Proposition 8.12 *(occupation times)* For any $x, y \in S$ and $k \in \mathbb{N}$,

$$P_x\{\kappa_y \geq k\} = P_x\{\tau_y^k < \infty\} = r_{xy} r_{yy}^{k-1}, \tag{19}$$

$$E_x \kappa_y = \frac{r_{xy}}{1 - r_{yy}}. \tag{20}$$

Proof: By the strong Markov property, we get for any $k \in \mathbb{N}$

$$P_x\{\tau_y^{k+1} < \infty\} = P_x\left\{\tau_y^k < \infty,\ \tau_y \circ \theta_{\tau_y^k} < \infty\right\}$$
$$= P_x\{\tau_y^k < \infty\} P_y\{\tau_y < \infty\} = r_{yy} P_x\{\tau_y^k < \infty\},$$

and the second relation in (19) follows by induction on k. The first relation is clear from the fact that $\kappa_y \geq k$ iff $\tau_y^k < \infty$. To deduce (20), conclude from (19) and Lemma 3.4 that

$$E_x \kappa_y = \sum_{k\geq 1} P_x\{\kappa_y \geq k\} = \sum_{k\geq 1} r_{xy} r_{yy}^{k-1} = \frac{r_{xy}}{1 - r_{yy}}. \quad \square$$

For $x = y$ the last result yields

$$P_x\{\kappa_x \geq k\} = P_x\{\tau_x^k < \infty\} = r_{xx}^k, \quad k \in \mathbb{N}.$$

Thus, under P_x, the number of visits to x is either a.s. infinite or geometrically distributed with mean $E_x \kappa_x + 1 = (1 - r_{xx})^{-1} < \infty$. This leads to a corresponding classification of the states into *recurrent* and *transient* ones.

Recurrence can often be deduced from the existence of an invariant distribution. Here and below we write $p_{xy}^n = \mu_n(x, \{y\})$.

Proposition 8.13 *(invariant distributions and recurrence)* If an invariant distribution ν exists, then any state x with $\nu\{x\} > 0$ is recurrent.

Proof: By the invariance of ν,

$$0 < \nu\{x\} = \int \nu(dy) p_{yx}^n, \quad n \in \mathbb{N}. \tag{21}$$

Thus, by Proposition 8.12 and Fubini's theorem,

$$\infty = \sum_{n\geq 1} \int \nu(dy) p_{yx}^n = \int \nu(dy) \sum_{n\geq 1} p_{yx}^n = \int \nu(dy) \frac{r_{yx}}{1 - r_{xx}} \leq \frac{1}{1 - r_{xx}}.$$

Hence, $r_{xx} = 1$, and so x is recurrent. $\quad\square$

The *period* d_x of a state x is defined as the greatest common divisor of the set $\{n \in \mathbb{N}; p_{xx}^n > 0\}$, and we say that x is *aperiodic* if $d_x = 1$.

Proposition 8.14 *(positivity)* *If $x \in S$ has period $d < \infty$, then $p_{xx}^{nd} > 0$ for all but finitely many n.*

Proof: Define $S = \{n \in \mathbb{N}; p_{xx}^{nd} > 0\}$, and conclude from the Chapman–Kolmogorov relation that S is closed under addition. Since S has greatest common divisor 1, the generated additive group equals $\mathbb{Z}$. In particular, there exist some $n_1, \ldots, n_k \in S$ and $z_1, \ldots, z_k \in \mathbb{Z}$ with $\sum_j z_j n_j = 1$. Writing $m = n_1 \sum_j |z_j| n_j$, we note that any number $n \geq m$ can be represented, for suitable $h \in \mathbb{Z}_+$ and $r \in \{0, \ldots, n_1 - 1\}$, as

$$n = m + hn_1 + r = hn_1 + \sum_{j \leq k} (n_1|z_j| + rz_j) n_j \in S. \qquad \Box$$

For each $x \in S$, the successive *excursions* of X from x are given by

$$Y_n = X^{\tau_x} \circ \theta_{\tau_x^n}, \quad n \in \mathbb{Z}_+,$$

as long as $\tau_x^n < \infty$. To allow for infinite excursions, we may introduce an extraneous element $\delta \notin S$, and define $Y_n = \bar{\delta} \equiv (\delta, \delta, \ldots)$ whenever $\tau_x^n = \infty$. Conversely, X may be recovered from the Y_n through the formulas

$$\tau_n = \sum_{k<n} \inf\{t > 0; Y_k(t) = x\}, \tag{22}$$

$$X_t = Y_n(t - \tau_n), \quad \tau_n \leq t < \tau_{n+1}, \; n \in \mathbb{Z}_+. \tag{23}$$

The distribution $\nu_x = P_x \circ Y_0^{-1}$ is called the *excursion law* at x. When x is recurrent and $r_{yx} = 1$, Proposition 8.9 shows that $Y_1, Y_2, \ldots$ are i.i.d. ν_x under P_y. The result extends to the general case, as follows.

Proposition 8.15 *(excursions)* *Consider a discrete-time Markov process X in a Borel space S, and fix any $x \in S$. Then there exist some independent processes $Y_0, Y_1, \ldots$ in S, all but Y_0 with distribution ν_x, such that X is a.s. given by (22) and (23).*

Proof: Put $\tilde{Y}_0 \stackrel{d}{=} Y_0$, and let $\tilde{Y}_1, \tilde{Y}_2, \ldots$ be independent of $\tilde{Y}_0$ and i.i.d. ν_x. Construct associated random times $\tilde{\tau}_0, \tilde{\tau}_1, \ldots$ as in (22), and define a process $\tilde{X}$ as in (23). By Corollary 6.11, it is enough to show that $X \stackrel{d}{=} \tilde{X}$. Writing

$$\kappa = \sup\{n \geq 0; \tau_n < \infty\}, \qquad \tilde{\kappa} = \sup\{n \geq 0; \tilde{\tau}_n < \infty\},$$

it is equivalent to show that

$$(Y_0, \ldots, Y_\kappa, \bar{\delta}, \bar{\delta}, \ldots) \stackrel{d}{=} (\tilde{Y}_0, \ldots, \tilde{Y}_{\tilde{\kappa}}, \bar{\delta}, \bar{\delta}, \ldots). \tag{24}$$

Using the strong Markov property on the left and the independence of the $\tilde{Y}_n$ on the right, it is easy to check that both sides are Markov processes in $S^{\mathbb{Z}_+} \cup \{\bar{\delta}\}$ with the same initial distribution and transition kernel. Hence, (24) holds by Proposition 8.2. $\qquad\Box$

By a *discrete-time Markov chain* we mean a Markov process on the time scale $\mathbb{Z}_+$, taking values in a countable state space S. In this case the transition kernels of X are determined by the n-step transition probabilities $p_{ij}^n = \mu_n(i, \{j\})$, $i, j \in S$, and the Chapman–Kolmogorov relation becomes

$$p_{ik}^{m+n} = \sum_j p_{ij}^m p_{jk}^n, \quad i, k \in S, \ m, n \in \mathbb{N}, \tag{25}$$

or in matrix notation, $p^{m+n} = p^m p^n$. Thus, p^n is the nth power of the matrix $p = p^1$, which justifies our notation. Regarding the initial distribution ν as a row vector (ν_i), we may write the distribution at time n as νp^n.

As before, we define $r_{ij} = P_i\{\tau_j < \infty\}$, where $\tau_j = \inf\{n > 0; X_n = j\}$. A Markov chain in S is said to be *irreducible* if $r_{ij} > 0$ for all $i, j \in S$, so that every state can be reached from any other state. For irreducible chains, all states have the same recurrence and periodicity properties.

Proposition 8.16 *(irreducible chains) For any irreducible Markov chain,*

(i) *the states are either all recurrent or all transient;*

(ii) *all states have the same period;*

(iii) *if ν is invariant, then $\nu_i > 0$ for all i.*

For the proof of (i) we need the following lemma.

Lemma 8.17 *(recurrence classes) Let $i \in S$ be recurrent, and define $S_i = \{j \in S; r_{ij} > 0\}$. Then $r_{jk} = 1$ for any $j, k \in S_i$, and all states in S_i are recurrent.*

Proof: By the recurrence of i and the strong Markov property, we get for any $j \in S_i$

$$\begin{aligned} 0 &= P_i\{\tau_j < \infty, \tau_i \circ \theta_{\tau_j} = \infty\} \\ &= P_i\{\tau_j < \infty\} P_j\{\tau_i = \infty\} = r_{ij}(1 - r_{ji}). \end{aligned}$$

Since $r_{ij} > 0$ by hypothesis, we obtain $r_{ji} = 1$. Fixing any $m, n \in \mathbb{N}$ with $p_{ij}^m, p_{ji}^n > 0$, we get by (25)

$$E_j \kappa_j \geq \sum_{s>0} p_{jj}^{m+n+s} \geq \sum_{s>0} p_{ji}^n p_{ii}^s p_{ij}^m = p_{ji}^n p_{ij}^m E_i \kappa_i = \infty,$$

and so j is recurrent by Proposition 8.12. Reversing the roles of i and j gives $r_{ij} = 1$. Finally, we get for any $j, k \in S_i$

$$r_{jk} \geq P_j\{\tau_i < \infty, \tau_k \circ \theta_{\tau_i} < \infty\} = r_{ji} r_{ik} = 1. \qquad \square$$

Proof of Proposition 8.16: (i) This is clear from Lemma 8.17.

(ii) Fix any $i, j \in S$, and choose $m, n \in \mathbb{N}$ with $p_{ij}^m, p_{ji}^n > 0$. By (25),

$$p_{jj}^{m+h+n} \geq p_{ji}^n p_{ii}^h p_{ij}^m, \quad h \geq 0.$$

For $h = 0$ we get $p_{jj}^{m+n} > 0$, and so $d_j|(m+n)$ (d_j divides $m+n$). Hence, in general, $p_{ii}^h > 0$ implies $d_j|h$, and we get $d_j \leq d_i$. Reversing the roles of i and j yields the opposite inequality.

(iii) Fix any $i \in S$. Choosing $j \in S$ with $\nu_j > 0$ and then $n \in \mathbb{N}$ with $p_{ji}^n > 0$, we see from (21) that even $\nu_i > 0$. □

We may now state the basic ergodic theorem for irreducible Markov chains. Related results will appear in Chapters 12, 19, and 23. For any signed measure μ we define $\|\mu\| = \sup_A |\mu A|$.

Theorem 8.18 *(ergodic behavior, Markov, Kolmogorov, Orey)* *For any irreducible, aperiodic Markov chain in S, exactly one of these cases occurs:*

(i) *There exists a unique invariant distribution ν, the latter satisfies $\nu_i > 0$ for all $i \in S$, and for any distribution μ on S we have*

$$\lim_{n \to \infty} \|P_\mu \circ \theta_n^{-1} - P_\nu\| = 0. \tag{26}$$

(ii) *No invariant distribution exists, and we have*

$$\lim_{n \to \infty} p_{ij}^n = 0, \quad i, j \in S. \tag{27}$$

A Markov chain satisfying (i) is clearly recurrent, whereas one that satisfies (ii) may be either recurrent or transient. This leads to the further classification of the irreducible, aperiodic, and recurrent Markov chains into *positive recurrent* and *null-recurrent* ones, depending on whether (i) or (ii) applies.

We shall prove Theorem 8.18 by the powerful method of *coupling*. Here the general idea is to compare the distributions of two processes X and Y, by constructing copies $\tilde{X} \stackrel{d}{=} X$ and $\tilde{Y} \stackrel{d}{=} Y$ on a common probability space. By a suitable choice of joint distribution, one may sometimes reduce the original problem to a pathwise comparison. The coupling approach often leads to simple and transparent proofs; we shall see further applications of the method in Chapters 9, 14, 15, 16, 20, and 23. For our present needs, an elementary coupling by independence is sufficient.

Lemma 8.19 *(coupling)* *Let X and Y be independent Markov chains in S and T with transition matrices $(p_{ii'})$ and $(q_{jj'})$, respectively. Then (X,Y) is a Markov chain in $S \times T$ with transition matrix $r_{ij,i'j'} = p_{ii'}q_{jj'}$. If X and Y are irreducible and aperiodic, then so is (X,Y); in that case (X,Y) is recurrent whenever invariant distributions exist for both X and Y.*

Proof: The first assertion is easily proved by computation of the finite-dimensional distributions of (X,Y) for an arbitrary initial distribution $\mu \otimes \nu$ on $S \times T$, using Proposition 8.2. Now assume that X and Y are irreducible and aperiodic. Fixing any $i, i' \in S$ and $j, j' \in T$, we see from Proposition 8.14 that $r_{ij,i'j'}^n = p_{ii'}^n q_{jj'}^n > 0$ for all but finitely many $n \in \mathbb{N}$, and so even (X,Y) has the stated properties. Finally, if μ and ν are invariant

8. Markov Processes and Discrete-Time Chains

distributions for X and Y, respectively, then $\mu \otimes \nu$ is invariant for (X, Y), and the last assertion follows by Proposition 8.13. □

The point of the construction is that, if the coupled processes eventually meet, their distributions will agree asymptotically.

Lemma 8.20 *(strong ergodicity)* If the Markov chain in S^2 with transition matrix $p_{ii'}p_{jj'}$ is irreducible and recurrent, then for any distributions μ and ν on S,

$$\lim_{n \to \infty} \|P_\mu \circ \theta_n^{-1} - P_\nu \circ \theta_n^{-1}\| = 0. \tag{28}$$

Proof (Doeblin): Let X and Y be independent with distributions P_μ and P_ν. By Lemma 8.19 the pair (X, Y) is again Markov with respect to the induced filtration $\mathcal{F}$, and by Proposition 8.9 it satisfies the strong Markov property at every finite optional time τ. Taking $\tau = \inf\{n \geq 0;\ X_n = Y_n\}$, we get for any measurable set $A \subset S^\infty$

$$P[\theta_\tau X \in A | \mathcal{F}_\tau] = P_{X_\tau} A = P_{Y_\tau} A = P[\theta_\tau Y \in A | \mathcal{F}_\tau].$$

In particular, $(\tau, X^\tau, \theta_\tau X) \stackrel{d}{=} (\tau, X^\tau, \theta_\tau Y)$. Defining $\tilde{X}_n = X_n$ for $n \leq \tau$ and $\tilde{X}_n = Y_n$ otherwise, we obtain $\tilde{X} \stackrel{d}{=} X$, and so for any A as above

$$\begin{aligned}
|P\{\theta_n X \in A\} - P\{\theta_n Y \in A\}| \\
= |P\{\theta_n \tilde{X} \in A\} - P\{\theta_n Y \in A\}| \\
= |P\{\theta_n \tilde{X} \in A, \tau > n\} - P\{\theta_n Y \in A, \tau > n\}| \\
\leq P\{\tau > n\} \to 0.
\end{aligned}$$
□

The next result ensures the existence of an invariant distribution. Here a coupling argument is again useful.

Lemma 8.21 *(existence)* If (27) fails, there exists an invariant distribution.

Proof: Assume that (27) fails, so that $\limsup_n p_{i_0,j_0}^n > 0$ for some $i_0, j_0 \in S$. By a diagonal argument we may choose a subsequence $N' \subset \mathbb{N}$ and some constants c_j with $c_{j_0} > 0$ such that $p_{i_0,j}^n \to c_j$ along N' for every $j \in S$. Note that $0 < \sum_j c_j \leq 1$ by Fatou's lemma.

To extend the convergence to arbitrary i, let X and Y be independent processes with the given transition matrix (p_{ij}), and conclude from Lemma 8.19 that (X, Y) is an irreducible Markov chain on S^2 with transition probabilities $q_{ij,i'j'} = p_{ii'}p_{jj'}$. If (X, Y) is transient, then by Proposition 8.12

$$\sum_n (p_{ij}^n)^2 = \sum_n q_{ii,jj}^n < \infty, \quad i, j \in S,$$

and (27) follows. The pair (X, Y) is then recurrent and Lemma 8.20 yields $p_{ij}^n - p_{i_0,j}^n \to 0$ for all $i, j \in I$. Hence, $p_{ij}^n \to c_j$ along N' for all i and j.

Next conclude from the Chapman–Kolmogorov relation that

$$p_{ik}^{n+1} = \sum_j p_{ij}^n p_{jk} = \sum_j p_{ij} p_{jk}^n, \quad i, k \in S.$$

Using Fatou's lemma on the left and dominated convergence on the right, we get as $n \to \infty$ along N'

$$\sum_j c_j p_{jk} \leq \sum_j p_{ij} c_k = c_k, \quad k \in S. \tag{29}$$

Summing over k gives $\sum_j c_j \leq 1$ on both sides, and so (29) holds with equality. Thus, (c_i) is invariant, and we get an invariant distribution ν by taking $\nu_i = c_i / \sum_j c_j$. □

Proof of Theorem 8.18: If no invariant distribution exists, then (27) holds by Lemma 8.21. Now let ν be an invariant distribution, and note that $\nu_i > 0$ for all i by Proposition 8.16. By Lemma 8.19 the coupled chain in Lemma 8.20 is irreducible and recurrent, so (28) holds for any initial distribution μ, and (26) follows since $P_\nu \circ \theta_n^{-1} = P_\nu$ by Lemma 8.11. If even ν' is invariant, then (26) yields $P_{\nu'} = P_\nu$, and so $\nu' = \nu$. □

The limits in Theorem 8.18 may be expressed in terms of the *mean recurrence times* $E_j \tau_j$, as follows.

Theorem 8.22 *(mean recurrence times, Kolmogorov) For any Markov chain in S and states $i, j \in S$ with j aperiodic, we have*

$$\lim_{n \to \infty} p_{ij}^n = \frac{P_i\{\tau_j < \infty\}}{E_j \tau_j}. \tag{30}$$

Proof: First take $i = j$. If j is transient, then $p_{jj}^n \to 0$ and $E_j \tau_j = \infty$, and so (30) is trivially true. If instead j is recurrent, then the restriction of X to the set $S_j = \{i; r_{ji} > 0\}$ is irreducible recurrent by Lemma 8.17 and aperiodic by Proposition 8.16. Hence, p_{jj}^n converges by Theorem 8.18.

To identify the limit, define

$$L_n = \sup\{k \in \mathbb{Z}_+; \tau_j^k \leq n\} = \sum_{k=1}^n 1\{X_k = j\}, \quad n \in \mathbb{N}.$$

The τ_j^n form a random walk under P_j, and so, by the law of large numbers,

$$\frac{L(\tau_j^n)}{\tau_j^n} = \frac{n}{\tau_j^n} \to \frac{1}{E_j \tau_j} \quad \text{a.s. } P_j.$$

By the monotonicity of L_k and τ_j^n it follows that $L_n / n \to (E_j \tau_j)^{-1}$ a.s. P_j. Noting that $L_n \leq n$, we get by dominated convergence

$$\frac{1}{n} \sum_{k=1}^n p_{jj}^k = \frac{E_j L_n}{n} \to \frac{1}{E_j \tau_j},$$

and (30) follows.

Now let $i \neq j$. Using the strong Markov property, the disintegration theorem, and dominated convergence, we get

$$\begin{aligned} p_{ij}^n &= P_i\{X_n = j\} = P_i\{\tau_j \leq n,\ (\theta_{\tau_j} X)_{n-\tau_j} = j\} \\ &= E_i[p_{jj}^{n-\tau_j};\ \tau_j \leq n] \to P_i\{\tau_j < \infty\}/E_j\tau_j. \end{aligned}$$ □

We return to continuous time and a general state space, to clarify the nature of the strong Markov property of a process X at finite optional times τ. The condition is clearly a combination of the conditional independence $\theta_\tau X \perp\!\!\!\perp_{X_\tau} \mathcal{F}_\tau$ and the *strong homogeneity*

$$P[\theta_\tau X \in \cdot | X_\tau] = P_{X_\tau} \quad \text{a.s.} \tag{31}$$

Though (31) appears to be weaker than (13), the two properties are in fact equivalent, under suitable regularity conditions on X and $\mathcal{F}$.

Theorem 8.23 *(strong homogeneity)* *Fix a separable metric space (S, ρ), a probability kernel (P_x) from S to $D(S)$, and a right-continuous filtration $\mathcal{F}$ on $\mathbb{R}_+$. Let X be an $\mathcal{F}$-adapted rcll process in S such that (31) holds for all bounded optional times τ. Then X satisfies the strong Markov property.*

Our proof is based on a 0–1 law for absorption probabilities, involving the sets

$$I = \{w \in D;\ w_t \equiv w_0\}, \qquad A = \{x \in S;\ P_x I = 1\}. \tag{32}$$

Lemma 8.24 *(absorption)* *For X as in Theorem 8.23 and for any optional time $\tau < \infty$, we have*

$$P_{X_\tau} I = 1_I(\theta_\tau X) = 1_A(X_\tau) \quad \text{a.s.} \tag{33}$$

Proof: We may clearly assume that τ is bounded, say by $n \in \mathbb{N}$. Fix any $h > 0$, and divide S into disjoint Borel sets $B_1, B_2, \ldots$ of diameter $< h$. For each $k \in \mathbb{N}$, define

$$\tau_k = n \wedge \inf\{t > \tau;\ \rho(X_\tau, X_t) > h\} \text{ on } \{X_\tau \in B_k\}, \tag{34}$$

and put $\tau_k = \tau$ otherwise. The times τ_k are again bounded and optional, and we note that

$$\{X_{\tau_k} \in B_k\} \subset \{X_\tau \in B_k,\ \sup_{t \in [\tau, n]} \rho(X_\tau, X_t) \leq h\}. \tag{35}$$

Using (31) and (35), we get as $n \to \infty$ and $h \to 0$

$$E[P_{X_\tau} I^c; \theta_\tau X \in I] = \sum_k E[P_{X_\tau} I^c; \theta_\tau X \in I, X_\tau \in B_k]$$
$$\leq \sum_k E[P_{X_{\tau_k}} I^c; X_{\tau_k} \in B_k]$$
$$= \sum_k P\{\theta_{\tau_k} X \notin I, X_{\tau_k} \in B_k\}$$
$$\leq \sum_k P\{\theta_\tau X \notin I, X_\tau \in B_k, \sup_{t \in [\tau,n]} \rho(X_\tau, X_t) \leq h\}$$
$$\to P\{\theta_\tau X \notin I, \sup_{t \geq \tau} \rho(X_\tau, X_t) = 0\} = 0,$$

and so $P_{X_\tau} I = 1$ a.s. on $\{\theta_\tau X \in I\}$. Since also $EP_{X_\tau} I = P\{\theta_\tau X \in I\}$ by (31), we obtain the first relation in (33). The second relation follows by the definition of A. $\square$

Proof of Theorem 8.23: Define I and A as in (32). To prove (13) on $\{X_\tau \in A\}$, fix any times $t_1 < \cdots < t_n$ and Borel sets $B_1, \ldots, B_n$, write $B = \bigcap_k B_k$, and conclude from (31) and Lemma 8.24 that

$$P\left[\bigcap_k \{X_{\tau+t_k} \in B_k\} \Big| \mathcal{F}_\tau\right] = P[X_\tau \in B | \mathcal{F}_\tau] = 1\{X_\tau \in B\}$$
$$= P[X_\tau \in B | X_\tau] = P_{X_\tau}\{w_0 \in B\}$$
$$= P_{X_\tau} \bigcap_k \{w_{t_k} \in B_k\}.$$

This extends to (13) by a monotone class argument.

To prove (13) on $\{X_\tau \notin A\}$, we may assume that $\tau \leq n$ a.s., and divide A^c into disjoint Borel sets B_k of diameter $< h$. Fix any $F \in \mathcal{F}_\tau$ with $F \subset \{X_\tau \notin A\}$. For each $k \in \mathbb{N}$, define τ_k as in (34) on the set $F^c \cap \{X_\tau \in B_k\}$, and let $\tau_k = \tau$ otherwise. Note that (35) remains true on F^c. Using (31), (35), and Lemma 8.24, we get as $n \to \infty$ and $h \to 0$

$$|P[\theta_\tau X \in \cdot ; F] - E[P_{X_\tau}; F]|$$
$$= \left|\sum_k E[1\{\theta_\tau X \in \cdot\} - P_{X_\tau}; X_\tau \in B_k, F]\right|$$
$$= \left|\sum_k E[1\{\theta_{\tau_k} X \in \cdot\} - P_{X_{\tau_k}}; X_{\tau_k} \in B_k, F]\right|$$
$$= \left|\sum_k E[1\{\theta_{\tau_k} X \in \cdot\} - P_{X_{\tau_k}}; X_{\tau_k} \in B_k, F^c]\right|$$
$$\leq \sum_k P[X_{\tau_k} \in B_k; F^c]$$
$$\leq \sum_k P\{X_\tau \in B_k, \sup_{t \in [\tau,n]} \rho(X_\tau, X_t) \leq h\}$$
$$\to P\{X_\tau \notin A, \sup_{t \geq \tau} \rho(X_\tau, X_t) = 0\} = 0.$$

Hence, the left-hand side is zero. $\square$

Exercises

1. Let X be a process with $X_s \perp\!\!\!\perp_{X_t} \{X_u, u \geq t\}$ for all $s < t$. Show that X is Markov with respect to the induced filtration.

2. Let X be a Markov process in some space S, and fix a measurable function f on S. Show by an example that the process $Y_t = f(X_t)$ need not be Markov. (*Hint:* Let X be a simple symmetric random walk on $\mathbb{Z}$, and take $f(x) = [x/2]$.)

3. Let X be a Markov process in $\mathbb{R}$ with transition functions μ_t satisfying $\mu_t(x, B) = \mu_t(-x, -B)$. Show that the process $Y_t = |X_t|$ is again Markov.

4. Fix any process X on $\mathbb{R}_+$, and define $Y_t = X^t = \{X_{s \wedge t}; s \geq 0\}$. Show that Y is Markov with respect to the induced filtration.

5. Consider a random element ξ in some Borel space and a filtration $\mathcal{F}$ with $\mathcal{F}_\infty \subset \sigma\{\xi\}$. Show that the measure-valued process $X_t = P[\xi \in \cdot | \mathcal{F}_t]$ is Markov. (*Hint:* Note that $\xi \perp\!\!\!\perp_{X_t} \mathcal{F}_t$ for all t.)

6. For any Markov process X on $\mathbb{R}_+$ and time $u > 0$, show that the reversed process $Y_t = X_{u-t}$, $t \in [0, u]$, is Markov with respect to the induced filtration. Also show by an example that a possible time homogeneity of X need not carry over to Y.

7. Let X be a time-homogeneous Markov process in some Borel space S. Show that there exist some measurable functions $f_h : S \times [0, 1] \to S$, $h \geq 0$, and $U(0, 1)$ random variables $\vartheta_{t,h} \perp\!\!\!\perp X^t$, $t, h \geq 0$, such that $X_{t+h} = f_h(X_t, \vartheta_{t,h})$ a.s. for all $t, h \geq 0$.

8. Let X be a time-homogeneous and rcll Markov process in some Polish space S. Show that there exist a measurable function $f : S \times [0, 1] \to D(\mathbb{R}_+, S)$ and some $U(0, 1)$ random variables $\vartheta_t \perp\!\!\!\perp X^t$ such that $\theta_t X = f(X_t, \vartheta_t)$ a.s. Extend the result to optional times taking countably many values.

9. Let X be a process on $\mathbb{R}_+$ with state space S, and define $Y_t = (X_t, t)$, $t \geq 0$. Show that X and Y are simultanously Markov, and that Y is then time-homogeneous. Give a relation between the transition kernels for X and Y. Express the strong Markov property of Y at a random time τ in terms of the process X.

10. Let X be a discrete-time Markov process in S with invariant distribution ν. Show for any measurable set $B \subset S$ that $P_\nu\{X_n \in B \text{ i.o.}\} \geq \nu B$. Use the result to give an alternative proof of Proposition 8.13. (*Hint:* Use Fatou's lemma.)

11. Fix an irreducible Markov chain in S with period d. Show that S has a unique partition into subsets $S_1, \ldots, S_d$ such that $p_{ij} = 0$ unless $i \in S_k$ and $j \in S_{k+1}$ for some $k \in \{1, \ldots, d\}$, where the addition is defined modulo d.

12. Let X be an irreducible Markov chain with period d, and define $S_1, \ldots, S_d$ as above. Show that the restrictions of (X_{nd}) to $S_1, \ldots, S_d$ are

irreducible, aperiodic and either all positive recurrent or all null recurrent. In the former case, show that the original chain has a unique invariant distribution ν. Further show that (26) holds iff $\mu S_k = 1/d$ for all k. (*Hint:* If (X_{nd}) has an invariant distribution ν^k in S_k, then $\nu_j^{k+1} = \sum_i \nu_i^k p_{ij}$ form an invariant distribution in S_{k+1}.)

13. Given a Markov chain X on S, define the classes C_i as in Lemma 8.17. Show that if $j \in C_i$ but $i \notin C_j$ for some $i, j \in S$, then i is transient. If instead $i \in C_j$ for every $j \in C_i$, show that C_i is irreducible (i.e., the restriction of X to C_i is an irreducible Markov chain). Further show that the irreducible sets are disjoint and that every state outside all irreducible sets is transient.

14. For an arbitrary Markov chain, show that (26) holds iff $\sum_j |p_{ij}^n - \nu_j| \to 0$ for all i.

15. Let X be an irreducible, aperiodic Markov chain in $\mathbb{N}$. Show that X is transient iff $X_n \to \infty$ a.s. under any initial distribution and is null recurrent iff the same divergence holds in probability but not a.s.

16. For every irreducible, positive recurrent subset $S_k \subset S$, there exists a unique invariant distribution ν_k restricted to S_k, and every invariant distribution is a convex combination $\sum_k c_k \nu_k$.

17. Show that a Markov chain on a finite state space S has at least one irreducible set and one invariant distribution. (*Hint:* Starting from any $i_0 \in S$, choose $i_1 \in C_{i_0}$, $i_2 \in C_{i_1}$, etc. Then $\bigcap_n C_{i_n}$ is irreducible.)

18. Let X and Y be independent Markov processes with transition kernels $\mu_{s,t}$ and $\nu_{s,t}$. Show that (X, Y) is again Markov with transition kernels $\mu_{s,t}(x, \cdot) \otimes \nu_{s,t}(y, \cdot)$. (*Hint:* Compute the finite-dimensional distributions from Proposition 8.2, or use Proposition 6.8 with no computations.)

19. Let X and Y be independent, irreducible Markov chains with periods d_1 and d_2. Show that $Z = (X, Y)$ is irreducible iff d_1 and d_2 have greatest common divisor 1 and that Z then has period $d_1 d_2$.

20. State and prove a discrete-time version of Theorem 8.23. Further simplify the continuous-time proof when S is countable.

Chapter 9

Random Walks and Renewal Theory

Recurrence and transience; dependence on dimension; general recurrence criteria; symmetry and duality; Wiener–Hopf factorization; ladder time and height distribution; stationary renewal process; renewal theorem

A *random walk* in $\mathbb{R}^d$ is defined as a discrete-time random process (S_n) evolving by i.i.d. steps $\xi_n = \Delta S_n = S_n - S_{n-1}$. For most purposes we may take $S_0 = 0$, so that $S_n = \xi_1 + \cdots + \xi_n$ for all n. Random walks may be regarded as the simplest of all Markov processes. Indeed, we recall from Chapter 8 that random walks are precisely the discrete-time Markov processes in $\mathbb{R}^d$ that are both space- and time-homogeneous. (In continuous time, a similar role is played by the so-called Lévy processes, to be studied in Chapter 15.) Despite their simplicity, random walks exhibit many basic features of Markov processes in discrete time and hence may serve as a good introduction to the general subject. We shall further see how random walks enter naturally into the discussion of certain continuous-time phenomena.

Some basic facts about random walks were obtained in previous chapters. Thus, we established some simple 0–1 laws in Chapter 3, and in Chapters 4 and 5 we proved the ultimate versions of the laws of large numbers and the central limit theorem, both of which deal with the asymptotic behavior of $n^{-c} S_n$ for suitable constants $c > 0$. More sophisticated limit theorems of this type will be derived in Chapters 14–16 and 27, often through approximation by a Brownian motion or some other Lévy process.

Random walks in $\mathbb{R}^d$ are either recurrent or transient, and our first major task is to derive a recurrence criterion in terms of the transition distribution μ. We proceed with some striking connections between maximum and return times, anticipating the arcsine laws of Chapters 13, 14, and 15. This is followed by a detailed study of ladder times and heights for one-dimensional random walks, culminating with the Wiener–Hopf factorization and Baxter's formula. Finally, we prove a two-sided version of the renewal theorem, which describes the asymptotic behavior of the occupation measure and associated intensity for a transient random walk.

In addition to the already mentioned connections to other chapters, we note the relevance of renewal theory for the study of continuous-time Markov chains, as considered in Chapter 12. Renewal processes may further be regarded as constituting an elementary subclass of the regenerative sets, to be studied in full generality in Chapter 22 in connection with local time and excursion theory.

To begin our systematic discussion of random walks, assume as before that $S_n = \xi_1 + \cdots + \xi_n$ for all $n \in \mathbb{Z}_+$, where the ξ_n are i.i.d. random vectors in $\mathbb{R}^d$. The distribution of (S_n) is then determined by the common distribution $\mu = \mathcal{L}(\xi_n)$ of the increments. By the *effective dimension* of (S_n) we mean the dimension of the linear subspace spanned by the support of μ. For most purposes, we may assume that the effective dimension agrees with the dimension of the underlying space, since we may otherwise restrict our attention to the generated subspace.

The *occupation measure* of (S_n) is defined as the random measure

$$\eta B = \sum_{n \geq 0} 1\{S_n \in B\}, \quad B \in \mathcal{B}^d.$$

We also need to consider the corresponding intensity measure

$$(E\eta)B = E(\eta B) = \sum_{n \geq 0} P\{S_n \in B\}, \quad B \in \mathcal{B}^d.$$

Writing $B_x^\varepsilon = \{y; |x - y| < \varepsilon\}$, we may introduce the *accessible set* A, the *mean recurrence set* M, and the *recurrence set* R, given by

$$A = \bigcap_{\varepsilon > 0} \{x \in \mathbb{R}^d;\ E\eta B_x^\varepsilon > 0\},$$
$$M = \bigcap_{\varepsilon > 0} \{x \in \mathbb{R}^d;\ E\eta B_x^\varepsilon = \infty\},$$
$$R = \bigcap_{\varepsilon > 0} \{x \in \mathbb{R}^d;\ \eta B_x^\varepsilon = \infty\ \text{a.s.}\}.$$

The following result gives the basic dichotomy for random walks in $\mathbb{R}^d$.

Theorem 9.1 *(recurrence dichotomy)* Let (S_n) be a random walk in $\mathbb{R}^d$, and define A, M, and R as above. Then exactly one of these conditions holds:

(i) $R = M = A$, which is then a closed additive subgroup of $\mathbb{R}^d$;

(ii) $R = M = \emptyset$, and $|S_n| \to \infty$ a.s.

A random walk is said to be *recurrent* if (i) holds and to be *transient* otherwise.

Proof: Since trivially $R \subset M \subset A$, the relations in (i) and (ii) are equivalent to $A \subset R$ and $M = \emptyset$, respectively. Further note that A is a closed additive semigroup.

First assume $P\{|S_n| \to \infty\} < 1$, so that $P\{|S_n| < r\ \text{i.o.}\} > 0$ for some $r > 0$. Fix any $\varepsilon > 0$, cover the r-ball around 0 by finitely many open balls $B_1, \ldots, B_n$ of radius $\varepsilon/2$, and note that $P\{S_n \in B_k\ \text{i.o.}\} > 0$ for at least one k. By the Hewitt–Savage 0–1 law, the latter probability equals 1. Thus, the optional time $\tau = \inf\{n \geq 0;\ S_n \in B_k\}$ is a.s. finite, and the strong Markov property at τ yields

$$1 = P\{S_n \in B_k\ \text{i.o.}\} \leq P\{|S_{\tau+n} - S_\tau| < \varepsilon\ \text{i.o.}\} = P\{|S_n| < \varepsilon\ \text{i.o.}\}.$$

Hence, $0 \in R$ in this case.

To extend the latter relation to $A \subset R$, fix any $x \in A$ and $\varepsilon > 0$. By the strong Markov property at $\sigma = \inf\{n \geq 0; |S_n - x| < \varepsilon/2\}$,

$$P\{|S_n - x| < \varepsilon \text{ i.o.}\} \geq P\{\sigma < \infty, |S_{\sigma+n} - S_\sigma| < \varepsilon/2 \text{ i.o.}\}$$
$$= P\{\sigma < \infty\}P\{|S_n| < \varepsilon/2 \text{ i.o.}\} > 0,$$

and by the Hewitt–Savage 0–1 law the probability on the left equals 1. Thus, $x \in R$. The asserted group property will follow if we can prove that even $-x \in A$. This is clear if we write

$$P\{|S_n + x| < \varepsilon \text{ i.o.}\} = P\{|S_{\sigma+n} - S_\sigma + x| < \varepsilon \text{ i.o.}\}$$
$$\geq P\{|S_n| < \varepsilon/2 \text{ i.o.}\} = 1.$$

Next assume that $|S_n| \to \infty$ a.s. Fix any $m, k \in \mathbb{N}$, and conclude from the Markov property at m that

$$P\{|S_m| < r, \inf_{n \geq k}|S_{m+n}| \geq r\}$$
$$\geq P\{|S_m| < r, \inf_{n \geq k}|S_{m+n} - S_m| \geq 2r\}$$
$$= P\{|S_m| < r\}P\{\inf_{n \geq k}|S_n| \geq 2r\}.$$

Here the event on the left can occur for at most k different values of m, and therefore

$$P\{\inf_{n \geq k}|S_n| \geq 2r\}\sum_m P\{|S_m| < r\} < \infty, \quad k \in \mathbb{N}.$$

As $k \to \infty$, the probability on the left tends to 1. Hence, the sum converges, and we get $E\eta B < \infty$ for any bounded set B. This shows that $M = \emptyset$. □

The next result gives some easily verified recurrence criteria.

Theorem 9.2 (*recurrence for $d = 1, 2$*) *A random walk (S_n) in $\mathbb{R}^d$ is recurrent under each of these conditions:*

(i) $d = 1$ and $n^{-1}S_n \xrightarrow{P} 0$;

(ii) $d = 2$, $E\xi_1 = 0$, and $E|\xi_1|^2 < \infty$.

In (i) we recognize the weak law of large numbers, which is characterized in Theorem 5.16. In particular, the condition is fulfilled when $E\xi_1 = 0$. By contrast, $E\xi_1 \in (0, \infty]$ implies $S_n \to \infty$ a.s. by the strong law of large numbers, so in that case (S_n) is transient.

Our proof of Theorem 9.2 is based on the following scaling relation. As before, $a \lesssim b$ means that $a \leq cb$ for some constant $c > 0$.

Lemma 9.3 (*scaling*) *For any random walk (S_n) in $\mathbb{R}^d$,*

$$\sum_{n \geq 0} P\{|S_n| \leq r\varepsilon\} \lesssim r^d \sum_{n \geq 0} P\{|S_n| \leq \varepsilon\}, \quad r \geq 1, \varepsilon > 0.$$

Proof: Cover the ball $\{x; |x| \leq r\varepsilon\}$ by balls $B_1, \ldots, B_m$ of radius $\varepsilon/2$, and note that we can make $m \lesssim r^d$. Introduce the optional times $\tau_k =$

$\inf\{n; S_n \in B_k\}$, $k = 1, \ldots, m$, and conclude from the strong Markov property that

$$\begin{aligned}\sum_n P\{|S_n| \le r\varepsilon\} &\le \sum_k \sum_n P\{S_n \in B_k\} \\ &\le \sum_k \sum_n P\{|S_{\tau_k+n} - S_{\tau_k}| \le \varepsilon; \tau_k < \infty\} \\ &= \sum_k P\{\tau_k < \infty\} \sum_n P\{|S_n| \le \varepsilon\} \\ &\le r^d \sum_n P\{|S_n| \le \varepsilon\}.\end{aligned}$$
□

Proof of Theorem 9.2 (Chung and Ornstein): (i) Fix any $\varepsilon > 0$ and $r \ge 1$, and conclude from Lemma 9.3 that

$$\sum_n P\{|S_n| \le \varepsilon\} \ge r^{-1} \sum_n P\{|S_n| \le r\varepsilon\} = \int_0^\infty P\{|S_{[rt]}| \le r\varepsilon\} dt.$$

Here the integrand on the right tends to 1 as $r \to \infty$, so the integral tends to ∞ by Fatou's lemma, and the recurrence of (S_n) follows by Theorem 9.1.

(ii) We may assume that (S_n) is two-dimensional, since the one-dimensional case is already covered by part (i). By the central limit theorem we have $n^{-1/2} S_n \xrightarrow{d} \zeta$, where the random vector ζ has a nondegenerate normal distribution. In particular, $P\{|\zeta| \le c\} \ge c^2$ for bounded $c > 0$. Now fix any $\varepsilon > 0$ and $r \ge 1$, and conclude from Lemma 9.3 that

$$\sum_n P\{|S_n| \le \varepsilon\} \ge r^{-2} \sum_n P\{|S_n| \le r\varepsilon\} = \int_0^\infty P\{|S_{[r^2 t]}| \le r\varepsilon\} dt.$$

As $r \to \infty$, we get by Fatou's lemma

$$\sum_n P\{|S_n| \le \varepsilon\} \ge \int_0^\infty P\{|\zeta| \le \varepsilon t^{-1/2}\} dt \ge \varepsilon^2 \int_1^\infty t^{-1} dt = \infty,$$

and the recurrence follows again by Theorem 9.1. □

Our next aim is to derive a general recurrence criterion, stated in terms of the characteristic function $\hat{\mu}$ of μ. Write $B_\varepsilon = \{x \in \mathbb{R}^d; |x| < \varepsilon\}$.

Theorem 9.4 (*recurrence criterion, Chung and Fuchs*) *Let (S_n) be a random walk in $\mathbb{R}^d$ based on some distribution μ, and fix any $\varepsilon > 0$. Then (S_n) is recurrent iff*

$$\sup_{0 < r < 1} \int_{B_\varepsilon} \Re \frac{1}{1 - r\hat{\mu}_t} dt = \infty. \tag{1}$$

The proof is based on an elementary identity.

Lemma 9.5 (*Parseval*) *Let μ and ν be probability measures on $\mathbb{R}^d$ with characteristic functions $\hat{\mu}$ and $\hat{\nu}$. Then $\int \hat{\mu} d\nu = \int \hat{\nu} d\mu$.*

Proof: Use Fubini's theorem. □

Proof of Theorem 9.4: The function $f(s) = (1 - |s|)_+$ has Fourier transform $\hat{f}(t) = 2t^{-2}(1 - \cos t)$, so the tensor product $f^{\otimes d}(s) = \prod_{k \leq d} f(s_k)$ on $\mathbb{R}^d$ has Fourier transform $\hat{f}^{\otimes d}(t) = \prod_{k \leq d} \hat{f}(t_k)$. Writing $\mu^{*n} = \mathcal{L}(S_n)$, we get by Lemma 9.5 for any $a > 0$ and $n \in \mathbb{Z}_+$

$$\int f^{\otimes d}(x/a) \mu^{*n}(dx) = a^d \int \hat{f}^{\otimes d}(at) \hat{\mu}_t^n dt.$$

By Fubini's theorem it follows that, for any $r \in (0, 1)$,

$$\int f^{\otimes d}(x/a) \sum_{n \geq 0} r^n \mu^{*n}(dx) = a^d \int \frac{\hat{f}^{\otimes d}(at)}{1 - r\hat{\mu}_t} dt. \qquad (2)$$

Now assume that (1) is false. Taking $\delta = \varepsilon^{-1} d^{1/2}$, we get by (2)

$$\sum_n P\{|S_n| < \delta\} = \sum_n \mu^{*n}(B_\delta) \lesssim \int f^{\otimes d}(x/\delta) \sum_n \mu^{*n}(dx)$$

$$= \delta^d \sup_{r < 1} \int \frac{\hat{f}^{\otimes d}(\delta t)}{1 - r\hat{\mu}_t} dt \lesssim \varepsilon^{-d} \sup_{r < 1} \int_{B_\varepsilon} \frac{dt}{1 - r\hat{\mu}_t} < \infty,$$

and so (S_n) is transient by Theorem 9.1.

To prove the converse, we note that $\hat{f}^{\otimes d}$ has Fourier transform $(2\pi)^d f^{\otimes d}$. Hence, (2) remains true with f and $\hat{f}$ interchanged, apart from a factor $(2\pi)^d$ on the left. If (S_n) is transient, then for any $\varepsilon > 0$ with $\delta = \varepsilon^{-1} d^{1/2}$ we get

$$\sup_{r < 1} \int_{B_\varepsilon} \frac{dt}{1 - r\hat{\mu}_t} \lesssim \sup_{r < 1} \int \frac{\hat{f}^{\otimes d}(t/\varepsilon)}{1 - r\hat{\mu}_t} dt$$

$$\lesssim \varepsilon^d \int f^{\otimes d}(\varepsilon x) \sum_n \mu^{*n}(dx)$$

$$\leq \varepsilon^d \sum_n \mu^{*n}(B_\delta) < \infty. \qquad \Box$$

In particular, we note that if μ is *symmetric* in the sense that $\xi_1 \stackrel{d}{=} -\xi_1$, then $\hat{\mu}$ is real valued, and the last criterion reduces to

$$\int_{B_\varepsilon} \frac{dt}{1 - \hat{\mu}_t} = \infty.$$

By a *symmetrization* of (S_n) we mean a random walk $\tilde{S}_n = S_n - S_n'$, $n \geq 0$, where (S_n') is an independent copy of (S_n). The following result relates the recurrence behavior of (S_n) and $(\tilde{S}_n)$.

Corollary 9.6 *(symmetrization)* *If a random walk (S_n) is recurrent, then so is the symmetrized version $(\tilde{S}_n)$.*

Proof: Noting that $(\Re z)(\Re z^{-1}) \leq 1$ for any complex number $z \neq 0$, we get

$$\Re \frac{1}{1 - r\hat{\mu}^2} \leq \frac{1}{1 - r\Re\hat{\mu}^2} \leq \frac{1}{1 - r|\hat{\mu}|^2}.$$

Thus, if $(\tilde{S}_n)$ is transient, then so is the random walk (S_{2n}) by Theorem 9.4. But then $|S_{2n}| \to \infty$ a.s. by Theorem 9.1, and so $|S_{2n+1}| \to \infty$ a.s. By combination, $|S_n| \to \infty$ a.s., which means that (S_n) is transient. □

The following sufficient conditions for recurrence or transience are often more convenient for applications.

Corollary 9.7 *(sufficient conditions)* *Fix any $\varepsilon > 0$. Then (S_n) is recurrent if*

$$\int_{B_\varepsilon} \Re \frac{1}{1 - \hat{\mu}_t} dt = \infty \qquad (3)$$

and transient if

$$\int_{B_\varepsilon} \frac{dt}{1 - \Re\hat{\mu}_t} < \infty. \qquad (4)$$

Proof: First assume (3). By Fatou's lemma, we get for any sequence $r_n \uparrow 1$

$$\liminf_{n \to \infty} \int_{B_\varepsilon} \Re \frac{1}{1 - r_n \hat{\mu}} \geq \int_{B_\varepsilon} \lim_{n \to \infty} \Re \frac{1}{1 - r_n \hat{\mu}} = \int_{B_\varepsilon} \Re \frac{1}{1 - \hat{\mu}} = \infty.$$

Thus, (1) holds, and (S_n) is recurrent.

Now assume (4) instead. Decreasing ε if necessary, we may further assume that $\Re\hat{\mu} \geq 0$ on B_ε. As before, we get

$$\int_{B_\varepsilon} \Re \frac{1}{1 - r\hat{\mu}} \leq \int_{B_\varepsilon} \frac{1}{1 - r\Re\hat{\mu}} \leq \int_{B_\varepsilon} \frac{1}{1 - \Re\hat{\mu}} < \infty,$$

and so (1) fails. Thus, (S_n) is transient. □

The last result enables us to supplement Theorem 9.2 with some conclusive information for $d \geq 3$.

Theorem 9.8 *(transience for $d \geq 3$)* *Any random walk of effective dimension $d \geq 3$ is transient.*

Proof: We may assume that the symmetrized distribution is again d-dimensional, since μ is otherwise supported by some hyperplane outside the origin, and the transience follows by the strong law of large numbers. By Corollary 9.6, it is enough to prove that the symmetrized random walk $(\tilde{S}_n)$ is transient, and so we may assume that μ is symmetric. Considering the conditional distributions on B_r and B_r^c for large enough $r > 0$, we may write μ as a convex combination $c\mu_1 + (1-c)\mu_2$, where μ_1 is symmetric and d-dimensional with bounded support. Letting (r_{ij}) denote the covariance matrix of μ_1, we get as in Lemma 5.10

$$\hat{\mu}_1(t) = 1 - \tfrac{1}{2} \sum_{i,j} r_{ij} t_i t_j + o(|t|^2), \quad t \to 0.$$

Since the matrix (r_{ij}) is positive definite, it follows that $1 - \hat{\mu}_1(t) \gtrsim |t|^2$ for small enough $|t|$, say for $t \in B_\varepsilon$. A similar relation then holds for $\hat{\mu}$, and so

$$\int_{B_\varepsilon} \frac{dt}{1 - \hat{\mu}_t} \lesssim \int_{B_\varepsilon} \frac{dt}{|t|^2} \lesssim \int_0^\varepsilon r^{d-3} dr < \infty.$$

Thus, (S_n) is transient by Theorem 9.4. □

We turn to a more detailed study of the one-dimensional random walk $S_n = \xi_1 + \cdots + \xi_n$, $n \in \mathbb{Z}_+$. Say that (S_n) is *simple* if $|\xi_1| = 1$ a.s. For a simple, symmetric random walk (S_n) we note that

$$u_n \equiv P\{S_{2n} = 0\} = 2^{-2n}\binom{2n}{n}, \quad n \in \mathbb{Z}_+. \tag{5}$$

The following result gives a surprising connection between the probabilities u_n and the distribution of last return to the origin.

Proposition 9.9 *(last return, Feller)* Let (S_n) be a simple, symmetric random walk in $\mathbb{Z}$, put $\sigma_n = \max\{k \leq n;\ S_{2k} = 0\}$, and define u_n by (5). Then

$$P\{\sigma_n = k\} = u_k u_{n-k}, \quad 0 \leq k \leq n.$$

Our proof will be based on a simple symmetry property, which will also appear in a continuous-time version as Lemma 13.14.

Lemma 9.10 *(reflection principle, André)* For any symmetric random walk (S_n) and optional time τ, we have $(\tilde{S}_n) \stackrel{d}{=} (S_n)$, where

$$\tilde{S}_n = S_{n \wedge \tau} - (S_n - S_{n \wedge \tau}), \quad n \geq 0.$$

Proof: We may clearly assume that $\tau < \infty$ a.s. Writing $S'_n = S_{\tau+n} - S_\tau$, $n \in \mathbb{Z}_+$, we get by the strong Markov property $S \stackrel{d}{=} S' \perp\!\!\!\perp (S^\tau, \tau)$, and by symmetry $-S' \stackrel{d}{=} S'$. Hence, by combination $(-S', S^\tau, \tau) \stackrel{d}{=} (S', S^\tau, \tau)$, and the assertion follows by suitable assembly. □

Proof of Proposition 9.9: By the Markov property at time $2k$, we get

$$P\{\sigma_n = k\} = P\{S_{2k} = 0\} P\{\sigma_{n-k} = 0\}, \quad 0 \leq k \leq n,$$

which reduces the proof to the case when $k = 0$. Thus, it remains to show that

$$P\{S_2 \neq 0, \ldots, S_{2n} \neq 0\} = P\{S_{2n} = 0\}, \quad n \in \mathbb{N}.$$

By the Markov property at time 1, the left-hand side equals

$$\tfrac{1}{2} P\{\min_{k < 2n} S_k = 0\} + \tfrac{1}{2} P\{\max_{k < 2n} S_k = 0\} = P\{M_{2n-1} = 0\},$$

where $M_n = \max_{k \leq n} S_k$. Using Lemma 9.10 with $\tau = \inf\{k;\ S_k = 1\}$, we get

$$\begin{aligned}
1 - P\{M_{2n-1} = 0\} &= P\{M_{2n-1} \geq 1\} \\
&= P\{M_{2n-1} \geq 1,\ S_{2n-1} \geq 1\} + P\{M_{2n-1} \geq 1,\ S_{2n-1} \leq 0\} \\
&= P\{S_{2n-1} \geq 1\} + P\{S_{2n-1} \geq 2\} \\
&= 1 - P\{S_{2n-1} = 1\} = 1 - P\{S_{2n} = 0\}.
\end{aligned}$$
□

We continue with an even more striking connection between the maximum of a symmetric random walk and the last return probabilities in

Proposition 9.9. Related results for Brownian motion and more general random walks will appear in Theorems 13.16 and 14.11.

Theorem 9.11 *(first maximum, Sparre-Andersen) Let (S_n) be a random walk based on a symmetric, diffuse distribution, put $M_n = \max_{k \leq n} S_k$, and write $\tau_n = \min\{k \geq 0; S_k = M_n\}$. Define σ_n as in Proposition 9.9 in terms of a simple, symmetric random walk. Then $\tau_n \stackrel{d}{=} \sigma_n$ for every $n \geq 0$.*

Here and below, we shall use the relation

$$(S_1, \ldots, S_n) \stackrel{d}{=} (S_n - S_{n-1}, \ldots, S_n - S_0), \quad n \in \mathbb{N}, \tag{6}$$

valid for any random walk (S_n). The formula is obvious from the fact that $(\xi_1, \ldots, \xi_n) \stackrel{d}{=} (\xi_n, \ldots, \xi_1)$.

Proof of Theorem 9.11: By the symmetry of (S_n) together with (6), we have

$$v_k \equiv P\{\tau_k = 0\} = P\{\tau_k = k\}, \quad k \geq 0. \tag{7}$$

Using the Markov property at time k, we hence obtain

$$P\{\tau_n = k\} = P\{\tau_k = k\}P\{\tau_{n-k} = 0\} = v_k v_{n-k}, \quad 0 \leq k \leq n. \tag{8}$$

Clearly $\sigma_0 = \tau_0 = 0$. Proceeding by induction, assume that $\sigma_k \stackrel{d}{=} \tau_k$ and hence $u_k = v_k$ for all $k < n$. Comparing (8) with Proposition 9.9, we obtain $P\{\sigma_n = k\} = P\{\tau_n = k\}$ for $0 < k < n$, and by (7) the equality extends to $k = 0$ and n. Thus, $\sigma_n \stackrel{d}{=} \tau_n$. □

For a general one-dimensional random walk (S_n), we may introduce the *ascending ladder times* $\tau_1, \tau_2, \ldots$, given recursively by

$$\tau_n = \inf\{k > \tau_{n-1}; S_k > S_{\tau_{n-1}}\}, \quad n \in \mathbb{N}, \tag{9}$$

starting with $\tau_0 = 0$. The associated *ascending ladder heights* are defined as the random variables S_{τ_n}, $n \in \mathbb{N}$, where S_∞ may be interpreted as ∞. In a similar way, we may define the *descending* ladder times τ_n^- and heights $S_{\tau_n^-}$, $n \in \mathbb{N}$. The times τ_n and τ_n^- are clearly optional. By the strong Markov property, we conclude that the pairs (τ_n, S_{τ_n}) and $(\tau_n^-, S_{\tau_n^-})$ form possibly terminating random walks in $\overline{\mathbb{R}}^2$.

Replacing the relation $S_k > S_{\tau_{n-1}}$ in (9) by $S_k \geq S_{\tau_{n-1}}$, we obtain the *weak ascending ladder times* σ_n and *heights* S_{σ_n}. Similarly, we may introduce the *weak descending* ladder times σ_n^- and heights $S_{\sigma_n^-}$. The mentioned sequences are connected by a pair of simple but powerful duality relations.

Lemma 9.12 *(duality)* Let η, η', ζ, and ζ' denote the occupation measures of the sequences (S_{τ_n}), (S_{σ_n}), $(S_n; n < \tau_1^-)$, and $(S_n; n < \sigma_1^-)$, respectively. Then $E\eta = E\zeta'$ and $E\eta' = E\zeta$.

Proof: By (6) we have for any $B \in \mathcal{B}(0, \infty)$ and $n \in \mathbb{N}$

$$P\{S_1 \wedge \cdots \wedge S_{n-1} > 0, S_n \in B\} = P\{S_1 \vee \cdots \vee S_{n-1} < S_n \in B\}$$
$$= \sum_k P\{\tau_k = n, S_{\tau_k} \in B\}. \quad (10)$$

Summing over $n \geq 1$ gives $E\zeta'B = E\eta B$, and the first assertion follows. The proof of the second assertion is similar. $\square$

The last lemma yields some interesting information. For example, in a simple symmetric random walk, the expected number of visits to an arbitrary state $k \neq 0$ before the first return to 0 is constant and equal to 1. In particular, the mean recurrence time is infinite, and so (S_n) is a null-recurrent Markov chain.

The following result shows how the asymptotic behavior of a random walk is related to the expected values of the ladder times.

Proposition 9.13 *(fluctuations and mean ladder times)* For any nondegenerate random walk (S_n) in $\mathbb{R}$, exactly one of these cases occurs:

(i) $S_n \to \infty$ a.s. and $E\tau_1 < \infty$;
(ii) $S_n \to -\infty$ a.s. and $E\tau_1^- < \infty$;
(iii) $\limsup_n (\pm S_n) = \infty$ a.s. and $E\sigma_1 = E\sigma_1^- = \infty$.

Proof: By Corollary 3.17 there are only three possibilities: $S_n \to \infty$ a.s., $S_n \to -\infty$ a.s., and $\limsup_n (\pm S_n) = \infty$ a.s. In the first case $\sigma_n^- < \infty$ for finitely many n, say for $n < \kappa < \infty$. Here κ is geometrically distributed, and so $E\tau_1 = E\kappa < \infty$ by Lemma 9.12. The proof in case (ii) is similar. In case (iii) the variables τ_n and τ_n^- are all finite, and Lemma 9.12 yields $E\sigma_1 = E\sigma_1^- = \infty$. $\square$

Next we shall see how the asymptotic behavior of a random walk is related to the expected values of ξ_1 and S_{τ_1}. Here we define $E\xi = E\xi^+ - E\xi^-$ whenever $E\xi^+ \wedge E\xi^- < \infty$.

Proposition 9.14 *(fluctuations and mean ladder heights)* If (S_n) is a nondegenerate random walk in $\mathbb{R}$, then

(i) $E\xi_1 = 0$ implies $\limsup_n (\pm S_n) = \infty$ a.s.;
(ii) $E\xi_1 \in (0, \infty]$ implies $S_n \to \infty$ a.s. and $ES_{\tau_1} = E\tau_1 E\xi_1$;
(iii) $E\xi_1^+ = E\xi_1^- = \infty$ implies $ES_{\tau_1} = -ES_{\tau_1^-} = \infty$.

The first assertion is an immediate consequence of Theorem 9.2 (i). It can also be obtained more directly, as follows.

Proof: (i) By symmetry, we may assume that $\limsup_n S_n = \infty$ a.s. If $E\tau_1 < \infty$, then the law of large numbers applies to each of the three ratios in the equation

$$\frac{S_{\tau_n}}{\tau_n}\frac{\tau_n}{n} = \frac{S_{\tau_n}}{n}, \quad n \in \mathbb{N},$$

and we get $0 = E\xi_1 E\tau_1 = ES_{\tau_1} > 0$. The contradiction shows that $E\tau_1 = \infty$, and so $\liminf_n S_n = -\infty$ by Proposition 9.13.

(ii) In this case $S_n \to \infty$ a.s. by the law of large numbers, and the formula $ES_{\tau_1} = E\tau_1 E\xi_1$ follows as before.

(iii) This is clear from the relations $S_{\tau_1} \geq \xi_1^+$ and $S_{\tau_1^-} \leq -\xi_1^-$. $\square$

We proceed with a celebrated factorization, which provides some more detailed information about the distributions of ladder times and heights. Here we write $\chi^\pm$ for the possibly defective distributions of the pairs (τ_1, S_{τ_1}) and $(\tau_1^-, S_{\tau_1^-})$, respectively, and let $\psi^\pm$ denote the corresponding distributions of (σ_1, S_{σ_1}) and $(\sigma_1^-, S_{\sigma_1^-})$. Put $\chi_n^\pm = \chi^\pm(\{n\} \times \cdot)$ and $\psi_n^\pm = \psi^\pm(\{n\} \times \cdot)$. Let us finally introduce the measure χ^0 on $\mathbb{N}$, given by

$$\begin{aligned}\chi_n^0 &= P\{S_1 \wedge \cdots \wedge S_{n-1} > 0 = S_n\} \\ &= P\{S_1 \vee \cdots \vee S_{n-1} < 0 = S_n\}, \quad n \in \mathbb{N},\end{aligned}$$

where the second equality holds by (6).

Theorem 9.15 *(Wiener–Hopf factorization)* *For any random walk in $\mathbb{R}$ based on some distribution μ, we have*

$$\delta_0 - \delta_1 \otimes \mu = (\delta_0 - \chi^+) * (\delta_0 - \psi^-) = (\delta_0 - \psi^+) * (\delta_0 - \chi^-), \quad (11)$$
$$\delta_0 - \psi^\pm = (\delta_0 - \chi^\pm) * (\delta_0 - \chi^0). \quad (12)$$

Note that the convolutions in (11) are defined on the space $\mathbb{Z}_+ \times \mathbb{R}$, whereas those in (12) can be regarded as defined on $\mathbb{Z}_+$. Alternatively, we may consider χ^0 as a measure on $\mathbb{N} \times \{0\}$, and interpret all convolutions as defined on $\mathbb{Z}_+ \times \mathbb{R}$.

Proof: Define the measures $\rho_1, \rho_2, \ldots$ on $(0, \infty)$ by

$$\begin{aligned}\rho_n B &= P\{S_1 \wedge \cdots \wedge S_{n-1} > 0, S_n \in B\} \\ &= E\sum_k 1\{\tau_k = n, S_{\tau_k} \in B\}, \quad n \in \mathbb{N}, B \in \mathcal{B}(0, \infty), \quad (13)\end{aligned}$$

where the second equality holds by (10). Put $\rho_0 = \delta_0$, and regard the sequence $\rho = (\rho_n)$ as a measure on $\mathbb{Z}_+ \times (0, \infty)$. Noting that the corresponding measures on $\mathbb{R}$ equal $\rho_n + \psi_n^-$ and using the Markov property at time $n-1$, we get

$$\rho_n + \psi_n^- = \rho_{n-1} * \mu = (\rho * (\delta_1 \otimes \mu))_n, \quad n \in \mathbb{N}. \quad (14)$$

Applying the strong Markov property at τ_1 to the second expression in (13), we see that also

$$\rho_n = \sum_{k=1}^n \chi_k^+ * \rho_{n-k} = (\chi^+ * \rho)_n, \quad n \in \mathbb{N}. \quad (15)$$

Recalling the values at zero, we get from (14) and (15)

$$\rho + \psi^- = \delta_0 + \rho * (\delta_1 \otimes \mu), \qquad \rho = \delta_0 + \chi^+ * \rho.$$

Eliminating ρ between the two equations yields the first relation in (11), and the second relation follows by symmetry.

To prove (12), we note that the restriction of ψ^+ to $(0, \infty)$ equals $\psi_n^+ - \chi_n^0$. Thus, for any $B \in \mathcal{B}(0, \infty)$,

$$(\chi_n^+ - \psi_n^+ + \chi_n^0)B = P\{\max_{k<n} S_k = 0,\ S_n \in B\}.$$

Decomposing the event on the right according to the time of first return to 0, we get

$$\chi_n^+ - \psi_n^+ + \chi_n^0 = \sum_{k=1}^{n-1} \chi_k^0 \chi_{n-k}^+ = (\chi^0 * \chi^+)_n, \quad n \in \mathbb{N},$$

and so $\chi^+ - \psi^+ + \chi^0 = \chi^0 * \chi^+$, which is equivalent to the plus-sign version of (12). The minus-sign version follows by symmetry. □

The preceding factorization yields in particular an explicit formula for the joint distribution of the first ladder time and height.

Theorem 9.16 (*ladder distributions, Sparre-Andersen, Baxter*) *If (S_n) is a random walk in $\mathbb{R}$, then for $|s| < 1$ and $u \geq 0$,*

$$E\, s^{\tau_1} \exp(-u S_{\tau_1}) = 1 - \exp\left\{ -\sum_{n=1}^{\infty} \frac{s^n}{n} E[e^{-uS_n};\ S_n > 0] \right\}. \tag{16}$$

A similar relation holds for (σ_1, S_{σ_1}) with $S_n > 0$ replaced by $S_n \geq 0$.

Proof: Introduce the mixed generating and characteristic functions

$$\hat{\chi}_{s,t}^+ = E\, s^{\tau_1} \exp(it S_{\tau_1}), \qquad \hat{\psi}_{s,t}^- = E\, s^{\sigma_1} \exp(it S_{\sigma_1^-}),$$

and note that the first relation in (11) is equivalent to

$$1 - s\hat{\mu}_t = (1 - \hat{\chi}_{s,t}^+)(1 - \hat{\psi}_{s,t}^-), \qquad |s| < 1,\ t \in \mathbb{R}.$$

Taking logarithms and expanding in Taylor series, we obtain

$$\sum_n n^{-1}(s\hat{\mu}_t)^n = \sum_n n^{-1}(\hat{\chi}_{s,t}^+)^n + \sum_n n^{-1}(\hat{\psi}_{s,t}^-)^n.$$

For fixed $s \in (-1, 1)$, this equation is of the form $\hat{\nu} = \hat{\nu}^+ + \hat{\nu}^-$, where ν and $\nu^\pm$ are bounded signed measures on $\mathbb{R}$, $(0, \infty)$, and $(-\infty, 0]$, respectively. By the uniqueness theorem for characteristic functions we get $\nu = \nu^+ + \nu^-$. In particular, ν^+ equals the restriction of ν to $(0, \infty)$. Thus, the corresponding Laplace transforms agree, and (16) follows by summation of a Taylor series for the logarithm. A similar argument yields the formula for (σ_1, S_{σ_1}). □

From the last result we may easily obtain expressions for the probability that a random walk stays negative or nonpositive, and also deduce criteria for its divergence to $-\infty$.

Corollary 9.17 *(negativity and divergence to $-\infty$)* *For any random walk (S_n) in $\mathbb{R}$, we have*

$$P\{\tau_1 = \infty\} = (E\sigma_1^-)^{-1} = \exp\left\{-\sum_{n\geq 1} n^{-1}P\{S_n > 0\}\right\}, \quad (17)$$

$$P\{\sigma_1 = \infty\} = (E\tau_1^-)^{-1} = \exp\left\{-\sum_{n\geq 1} n^{-1}P\{S_n \geq 0\}\right\}. \quad (18)$$

Furthermore, each of these two conditions is equivalent to $S_n \to -\infty$ a.s.:

$$\sum_{n\geq 1} n^{-1}P\{S_n > 0\} < \infty, \qquad \sum_{n\geq 1} n^{-1}P\{S_n \geq 0\} < \infty.$$

Proof: The last expression for $P\{\tau_1 = \infty\}$ follows from (16) with $u = 0$ as we let $s \to 1$. Similarly, the formula for $P\{\sigma_1 = \infty\}$ is obtained from the version of (16) for the pair (σ_1, S_{σ_1}). In particular, $P\{\tau_1 = \infty\} > 0$ iff the series in (17) converges, and similarly for the condition $P\{\sigma_1 = \infty\} > 0$ in terms of the series in (18). Since both conditions are equivalent to $S_n \to -\infty$ a.s., the last assertion follows. Finally, the first equalities in (17) and (18) are obtained most easily from Lemma 9.12, if we note that the number of strict or weak ladder times $\tau_n < \infty$ or $\sigma_n < \infty$ is geometrically distributed. $\square$

We turn to a detailed study of the occupation measure $\eta = \sum_{n\geq 0} \delta_{S_n}$ of a transient random walk on $\mathbb{R}$, based on transition and initial distributions μ and ν. Recall from Theorem 9.1 that the associated intensity measure $E\eta = \nu * \sum_n \mu^{*n}$ is locally finite. By the strong Markov property, the sequence $(S_{\tau+n} - S_\tau)$ has the same distribution for every finite optional time τ. Thus, a similar invariance holds for the occupation measure, and the associated intensities must agree. A *renewal* is then said to occur at time τ, and the whole subject is known as *renewal theory*. In the special case when μ and ν are supported by $\mathbb{R}_+$, we refer to η as a *renewal process* based on μ and ν, and to $E\eta$ as the associated *renewal measure*. For most purposes, we may assume that $\nu = \delta_0$; if this is not the case, we say that η is *delayed*.

The occupation measure η is clearly a *random measure* on $\mathbb{R}$, in the sense that ηB is a random variable for every bounded Borel set B. From Lemma 12.1 we anticipate the simple fact that the distribution of a random measure on $\mathbb{R}_+$ is determined by the distributions of the integrals $\eta f = \int f d\eta$ for all $f \in C_K^+(\mathbb{R}_+)$, the space of continuous functions $f: \mathbb{R}_+ \to \mathbb{R}_+$ with bounded support. For any measure μ on $\mathbb{R}$ and constant $t \geq 0$, we may introduce the *shifted* measure $\theta_t \mu$ on $\mathbb{R}_+$, given by $(\theta_t \mu) B = \mu(B + t)$ for arbitrary $B \in \mathcal{B}(\mathbb{R}_+)$. A random measure η on $\mathbb{R}$ is said to be *stationary on $\mathbb{R}_+$* if $\theta_t \eta \stackrel{d}{=} \theta_0 \eta$.

Given a renewal process η based on some distribution μ, we say that the delayed process $\tilde{\eta} = \delta_\alpha * \eta$ is a *stationary version* of η, if the delay distribution $\nu = \mathcal{L}(\alpha)$ is such that the random measure $\tilde{\eta}$ becomes stationary on $\mathbb{R}_+$. We proceed to show that such a version exists iff μ has finite mean, in which case ν is uniquely determined by μ. Write λ for Lebesgue measure on $\mathbb{R}_+$.

Proposition 9.18 *(stationary renewal process)* *Let η be a renewal process based on some distribution μ on $\mathbb{R}_+$ with mean c. Then η has a stationary version $\tilde{\eta}$ iff $c \in (0, \infty)$. In that case $E\tilde{\eta} = c^{-1}\lambda$, and the delay distribution of $\tilde{\eta}$ is uniquely given by $\nu = c^{-1}(\delta_0 - \mu) * \lambda$, or*

$$\nu[0, t] = c^{-1} \int_0^t \mu(s, \infty) ds, \quad t \geq 0. \tag{19}$$

Proof: By Fubini's theorem,

$$E\eta = E\sum_n \delta_{S_n} = \sum_n \mathcal{L}(S_n) = \sum_n \nu * \mu^{*n}$$
$$= \nu + \mu * \sum_n \nu * \mu^{*n} = \nu + \mu * E\eta,$$

and so $\nu = (\delta_0 - \mu) * E\eta$. If η is stationary, then $E\eta$ is shift invariant, and Theorem 2.6 yields $E\eta = a\lambda$ for some constant $a > 0$. Thus, $\nu = a(\delta_0 - \mu) * \lambda$, and (19) holds with c^{-1} replaced by a. As $t \to \infty$, we get $1 = ac$ by Lemma 3.4, which implies $c \in (0, \infty)$ and $a = c^{-1}$.

Conversely, assume that $c \in (0, \infty)$, and let ν be given by (19). Then

$$E\eta = \nu * \sum_n \mu^{*n} = c^{-1}(\delta_0 - \mu) * \lambda * \sum_n \mu^{*n}$$
$$= c^{-1}\lambda * \left\{\sum_{n\geq 0} \mu^{*n} - \sum_{n\geq 1} \mu^{*n}\right\} = c^{-1}\lambda.$$

By the strong Markov property, the shifted random measure $\theta_t \eta$ is again a renewal process based on μ, say with delay distribution ν_t. As before,

$$\nu_t = (\delta_0 - \mu) * (\theta_t E\eta) = (\delta_0 - \mu) * E\eta = \nu,$$

which implies the asserted stationarity of η. $\square$

From the last result we may deduce a corresponding statement for the occupation measure of a general random walk.

Proposition 9.19 *(stationary occupation measure)* *Let η be the occupation measure of a random walk in $\mathbb{R}$ based on some distributions μ and ν, where μ has mean $c \in (0, \infty)$ and ν is defined as in (19) in terms of the ladder height distribution $\tilde{\mu}$ and its mean $\tilde{c}$. Then η is stationary on $\mathbb{R}_+$ with intensity $\tilde{c}^{-1}$.*

Proof: Since $S_n \to \infty$ a.s., Propositions 9.13 and 9.14 show that the ladder times τ_n and heights $H_n = S_{\tau_n}$ have finite mean, and by Proposition 9.18 the renewal process $\zeta = \sum_n \delta_{H_n}$ is stationary for the prescribed choice of ν. Fixing $t \geq 0$ and putting $\sigma_t = \inf\{n \in \mathbb{Z}_+; S_n \geq t\}$, we note in particular that $S_{\sigma_t} - t$ has distribution ν. By the strong Markov property at σ_t, the sequence $S_{\sigma_t + n} - t$, $n \in \mathbb{Z}_+$, has then the same distribution as (S_n). Since $S_k < t$ for $k < \sigma_t$, we get $\theta_t \eta \stackrel{d}{=} \eta$ on $\mathbb{R}_+$, which proves the asserted stationarity.

To identify the intensity, let η_n denote the occupation measure of the sequence $S_k - H_n$, $\tau_n \leq k < \tau_{n+1}$, and note that $H_n \perp\!\!\!\perp \eta_n \stackrel{d}{=} \eta_0$ for each n, by the strong Markov property. Hence, by Fubini's theorem,

$$E\eta = E\sum_n \eta_n * \delta_{H_n} = \sum_n E(\delta_{H_n} * E\eta_n)$$
$$= E\eta_0 * E\sum_n \delta_{H_n} = E\eta_0 * E\zeta.$$

Noting that $E\zeta = \tilde{c}^{-1}\lambda$ by Proposition 9.18, that $E\eta_0(0,\infty) = 0$, and that $\tilde{c} = cE\tau_1$ by Proposition 9.14, we get on $\mathbb{R}_+$

$$E\eta = \frac{E\eta_0 \mathbb{R}_-}{\tilde{c}}\lambda = \frac{E\tau_1}{\tilde{c}}\lambda = c^{-1}\lambda. \qquad \square$$

The next result describes the asymptotic behavior of the occupation measure η and its intensity $E\eta$. Under weak restrictions on μ, we shall see how $\theta_t \eta$ approaches the corresponding stationary version $\tilde{\eta}$, whereas $E\eta$ is asymptotically proportional to Lebesgue measure. For simplicity, we assume that the mean of μ exists in $\overline{\mathbb{R}}$. Thus, if ξ is a random variable with distribution μ, we assume that $E(\xi^+ \wedge \xi^-) < \infty$ and define $E\xi = E\xi^+ - E\xi^-$.

It is natural to state the result in terms of vague convergence for measures on $\mathbb{R}_+$, and the corresponding notion of distributional convergence for random measures. Recall that, for locally finite measures $\nu, \nu_1, \nu_2, \ldots$ on $\mathbb{R}_+$, the vague convergence $\nu_n \xrightarrow{v} \nu$ means that $\nu_n f \to \nu f$ for all $f \in C_K^+(\mathbb{R}_+)$. Similarly, if $\eta, \eta_1, \eta_2, \ldots$ are random measures on $\mathbb{R}_+$, we define the distributional convergence $\eta_n \xrightarrow{d} \eta$ by the condition $\eta_n f \xrightarrow{d} \eta f$ for every $f \in C_K^+(\mathbb{R}_+)$. (The latter notion of convergence will be studied in detail in Chapter 16.) A measure μ on $\mathbb{R}$ is said to be *nonarithmetic* if the additive subgroup generated by $\operatorname{supp} \mu$ is dense in $\mathbb{R}$.

Theorem 9.20 (*two-sided renewal theorem, Blackwell, Feller and Orey*) *Let η be the occupation measure of a random walk in $\mathbb{R}$ based on some distributions μ and ν, where μ is nonarithmetic with mean $c \in \overline{\mathbb{R}} \setminus \{0\}$. If $c \in (0, \infty)$, let $\tilde{\eta}$ be the stationary version in Proposition 9.19; otherwise, put $\tilde{\eta} = 0$. Then as $t \to \infty$,*

(i) $\theta_t \eta \xrightarrow{d} \tilde{\eta}$,

(ii) $\theta_t E\eta \xrightarrow{v} E\tilde{\eta} = (c^{-1} \vee 0)\lambda$.

Our proof is based on two lemmas. First we consider the distribution ν_t of the first nonnegative ladder height for the shifted process $(S_n - t)$. For $c \in (0, \infty)$, the key step is to show that ν_t converges weakly toward the corresponding distribution $\tilde{\nu}$ for the stationary version. This will be accomplished by a coupling argument.

Lemma 9.21 (*asymptotic delay*) *If $c \in (0, \infty)$, then $\nu_t \xrightarrow{w} \tilde{\nu}$ as $t \to \infty$.*

Proof: Let α and α' be independent random variables with distributions ν and $\tilde{\nu}$. Choose some i.i.d. sequences $(\xi_k) \perp\!\!\!\perp (\vartheta_k)$ independent of α and α'

such that $\mathcal{L}(\xi_k) = \mu$ and $P\{\vartheta_k = \pm 1\} = \frac{1}{2}$. Then

$$\tilde{S}_n = \alpha' - \alpha - \sum_{k \leq n} \vartheta_k \xi_k, \quad n \in \mathbb{Z}_+,$$

is a random walk based on a nonarithmetic distribution with mean 0, and so by Theorems 9.1 and 9.2 the set $\{\tilde{S}_n\}$ is a.s. dense in $\mathbb{R}$. For any $\varepsilon > 0$, the optional time $\sigma = \inf\{n \geq 0; \tilde{S}_n \in [0, \varepsilon]\}$ is then a.s. finite.

Now define $\vartheta'_k = (-1)^{1\{k \leq \sigma\}} \vartheta_k$, $k \in \mathbb{N}$, and note as in Lemma 9.10 that $\{\alpha', (\xi_k, \vartheta'_k)\} \stackrel{d}{=} \{\alpha', (\xi_k, \vartheta_k)\}$. Let $\kappa_1 < \kappa_2 < \cdots$ be the values of k with $\vartheta_k = 1$, and define $\kappa'_1 < \kappa'_2 < \cdots$ similarly in terms of (ϑ'_k). By a simple conditioning argument, the sequences

$$S_n = \alpha + \sum_{j \leq n} \xi_{\kappa_j}, \quad S'_n = \alpha' + \sum_{j \leq n} \xi_{\kappa'_j}, \quad n \in \mathbb{Z}_+,$$

are random walks based on μ and the initial distributions ν and $\tilde{\nu}$, respectively. Writing $\sigma_\pm = \sum_{k \leq \sigma} 1\{\vartheta_k = \pm 1\}$, we note that

$$S'_{\sigma_- + n} - S_{\sigma_+ + n} = \tilde{S}_\sigma \in [0, \varepsilon], \quad n \in \mathbb{Z}_+.$$

Putting $\gamma = S^*_{\sigma_+} \vee S'^*_{\sigma_-}$, and considering the first entry of (S_n) and (S'_n) into the interval $[t, \infty)$, we obtain

$$\tilde{\nu}[\varepsilon, x] - P\{\gamma \geq t\} \leq \nu_t[0, x] \leq \tilde{\nu}[0, x + \varepsilon] + P\{\gamma \geq t\}.$$

Letting $t \to \infty$ and then $\varepsilon \to 0$, and noting that $\tilde{\nu}\{0\} = 0$ by stationarity, we get $\nu_t[0, x] \to \tilde{\nu}[0, x]$. □

The following simple statement will be needed to deduce (ii) from (i) in the main theorem.

Lemma 9.22 *(uniform integrability)* *Let η be the occupation measure of a transient random walk (S_n) in $\mathbb{R}^d$ with arbitrary initial distribution, and fix any bounded set $B \in \mathcal{B}^d$. Then the random variables $\eta(B + x)$, $x \in \mathbb{R}^d$, are uniformly integrable.*

Proof: Fix any $x \in \mathbb{R}^d$, and put $\tau = \inf\{t \geq 0; S_n \in B + x\}$. Letting η_0 denote the occupation measure of an independent random walk starting at 0, we get by the strong Markov property

$$\eta(B + x) \stackrel{d}{=} \eta_0(B + x - S_\tau) 1\{\tau < \infty\} \leq \eta_0(B - B).$$

In remains to note that $E\eta_0(B - B) < \infty$ by Theorem 9.1, since (S_n) is transient. □

Proof of Theorem 9.20 ($c < \infty$): By Lemma 9.22 it is enough to prove (i). If $c < 0$, then $S_n \to -\infty$ a.s. by the law of large numbers, so $\theta_t \eta = 0$ for sufficiently large t, and (i) follows. If instead $c \in (0, \infty)$, then $\nu_t \stackrel{w}{\to} \tilde{\nu}$ by Lemma 9.21, and we may choose some random variables α_t and α with distributions ν_t and ν, respectively, such that $\alpha_t \to \alpha$ a.s. We may also introduce the occupation measure η_0 of an independent random walk starting at 0.

Now fix any $f \in C_K^+(\mathbb{R}_+)$, and extend f to $\mathbb{R}$ by putting $f(x) = 0$ for $x < 0$. Since $\tilde{\nu} \ll \lambda$, we have $\eta_0\{-\alpha\} = 0$ a.s. Hence, by the strong Markov property and dominated convergence,

$$(\theta_t \eta) f \stackrel{d}{=} \int f(\alpha_t + x) \eta_0(dx) \to \int f(\alpha + x) \eta_0(dx) \stackrel{d}{=} \tilde{\eta} f.$$

$(c = \infty)$: In this case it is clearly enough to prove (ii). Then note that $E\eta = \nu * E\chi * E\zeta$, where χ is the occupation measure of the ladder height sequence of $(S_n - S_0)$, and ζ is the occupation measure of the same process prior to the first ladder time. Here $E\zeta \mathbb{R}_- < \infty$ by Proposition 9.13, and so by dominated convergence it suffices to show that $\theta_t E\chi \stackrel{v}{\to} 0$. Since the mean of the ladder height distribution is again infinite by Proposition 9.14, we may henceforth take $\nu = \delta_0$ and let μ be an arbitrary distribution on $\mathbb{R}_+$ with infinite mean.

Put $I = [0, 1]$, and note that $E\eta(I + t)$ is bounded by Lemma 9.22. Define $b = \limsup_t E\eta(I+t)$, and choose some $t_k \to \infty$ with $E\eta(I+t_k) \to b$. Subtracting the finite measures μ^{*j} for $j < m$, we get $(\mu^{*m} * E\eta)(I+t_k) \to b$ for all $m \in \mathbb{Z}_+$. Using the reverse Fatou lemma, we obtain for any $B \in \mathcal{B}(\mathbb{R}_+)$

$$\liminf_{k \to \infty} E\eta(I - B + t_k) \mu^{*m} B$$

$$\geq \liminf_{k \to \infty} \int_B E\eta(I - x + t_k) \mu^{*m}(dx)$$

$$= b - \limsup_{k \to \infty} \int_{B^c} E\eta(I - x + t_k) \mu^{*m}(dx)$$

$$\geq b - \int_{B^c} \limsup_{k \to \infty} E\eta(I - x + t_k) \mu^{*m}(dx) \geq b \mu^{*m} B. \quad (20)$$

Now fix any $h > 0$ with $\mu(0, h] > 0$. Noting that $E\eta[r, r+h] > 0$ for all $r \geq 0$ and writing $J = [0, a]$ with $a = h + 1$, we get by (20)

$$\liminf_{k \to \infty} E\eta(J + t_k - r) \geq b, \quad r \geq a. \quad (21)$$

Next conclude from the identity $\delta_0 = (\delta_0 - \mu) * E\eta$ that

$$1 = \int_0^{t_k} \mu(t_k - x, \infty) E\eta(dx) \geq \sum_{n \geq 1} \mu(na, \infty) E\eta(J + t_k - na).$$

As $k \to \infty$, we get by (21) and Fatou's lemma $1 \geq b \sum_{k \geq 1} \mu(na, \infty)$. Since the sum diverges by Lemma 3.4, it follows that $b = 0$. □

We may use the preceding theory to study the *renewal equation* $F = f + F * \mu$, which often arises in applications. Here the convolution $F * \mu$ is defined by

$$(F * \mu)_t = \int_0^t F(t-s) \mu(ds), \quad t \geq 0,$$

whenever the integrals on the right exist. Under suitable regularity conditions, the renewal equation has the unique solution $F = f * \bar{\mu}$, where $\bar{\mu}$ denotes the renewal measure $\sum_{n \geq 0} \mu^{*n}$. Additional conditions ensure the solution F to converge at ∞.

A precise statement requires some further terminology. By a *regular step function* we mean a function on $\mathbb{R}_+$ of the form

$$f_t = \sum_{j \geq 1} a_j 1_{[j-1,j)}(t/h), \quad t \geq 0, \tag{22}$$

where $h > 0$ and $a_1, a_2, \ldots \in \mathbb{R}$. A measurable function f on $\mathbb{R}_+$ is said to be *directly Riemann integrable* if $\lambda|f| < \infty$ and there exist some regular step functions $f_n^\pm$ with $f_n^- \leq f \leq f_n^+$ and $\lambda(f_n^+ - f_n^-) \to 0$.

Corollary 9.23 *(renewal equation) Fix a distribution $\mu \neq \delta_0$ on $\mathbb{R}_+$ with associated renewal measure $\bar{\mu}$, and let f be a locally bounded and measurable function on $\mathbb{R}_+$. Then the equation $F = f + F * \mu$ has the unique, locally bounded solution $F = f * \bar{\mu}$. If f is also directly Riemann integrable and if μ is nonarithmetic with mean c, then $F_t \to c^{-1} \lambda f$ as $t \to \infty$.*

Proof: Iterating the renewal equation gives

$$F = \sum_{k < n} f * \mu^{*k} + F * \mu^{*n}, \quad n \in \mathbb{N}. \tag{23}$$

Now $\mu^{*n}[0,t] \to 0$ as $n \to \infty$ for fixed $t \geq 0$ by the weak law of large numbers, and so for a locally bounded F we have $F * \mu^{*n} \to 0$. If even f is locally bounded, then by (23) and Fubini's theorem,

$$F = \sum_{k \geq 0} f * \mu^{*k} = f * \sum_{k \geq 0} \mu^{*k} = f * \bar{\mu}.$$

Conversely, $f + f * \bar{\mu} * \mu = f * \bar{\mu}$, which shows that $F = f * \bar{\mu}$ solves the given equation.

Now let μ be nonarithmetic. If f is a regular step function as in (22), then by Theorem 9.20 and dominated convergence we get as $t \to \infty$

$$F_t = \int_0^t f(t-s) \bar{\mu}(ds) = \sum_{j \geq 1} a_j \bar{\mu}((0, h] + t - jh)$$
$$\to c^{-1} h \sum_{j \geq 1} a_j = c^{-1} \lambda f.$$

In the general case, we may introduce some regular step functions $f_n^\pm$ with $f_n^- \leq f \leq f_n^+$ and $\lambda(f_n^+ - f_n^-) \to 0$, and note that

$$(f_n^- * \bar{\mu})_t \leq F_t \leq (f_n^+ * \bar{\mu})_t, \quad t \geq 0, \; n \in \mathbb{N}.$$

Letting $t \to \infty$ and then $n \to \infty$, we obtain $F_t \to c^{-1} \lambda f$. □

Exercises

1. Show that if (S_n) is recurrent, then so is the random walk (S_{nk}) for each $k \in \mathbb{N}$. (*Hint*: If (S_{nk}) is transient, then so is (S_{nk+j}) for any $j > 0$.)

2. For any nondegenerate random walk (S_n) in $\mathbb{R}^d$, show that $|S_n| \xrightarrow{P} \infty$. (*Hint:* Use Lemma 5.1.)

3. Let (S_n) be a random walk in $\mathbb{R}$ based on a symmetric, nondegenerate distribution with bounded support. Show that (S_n) is recurrent, using the fact that $\limsup_n(\pm S_n) = \infty$ a.s.

4. Show that the accessible set A equals the closed semigroup generated by $\mathrm{supp}\,\mu$. Also show by examples that A may or may not be a group.

5. Let ν be an invariant measure on the accessible set of a recurrent random walk in $\mathbb{R}^d$. Show by examples that $E\eta$ may or may not be of the form $\infty\cdot\nu$.

6. Show that a nondegenerate random walk in $\mathbb{R}^d$ has no invariant distribution. (*Hint:* If ν is invariant, then $\mu * \nu = \nu$.)

7. Show by examples that the conditions in Theorem 9.2 are not necessary. (*Hint:* For $d=2$, consider mixtures of $N(0,\sigma^2)$ and use Lemma 5.18.)

8. Consider a random walk (S_n) based on the symmetric p-stable distribution on $\mathbb{R}$ with characteristic function $e^{-|t|^p}$. Show that (S_n) is recurrent for $p \geq 1$ and transient for $p < 1$.

9. Let (S_n) be a random walk in $\mathbb{R}^2$ based on the distribution μ^2, where μ is symmetric p-stable. Show that (S_n) is recurrent for $p=2$ and transient for $p<2$.

10. Let $\mu = c\mu_1 + (1-c)\mu_2$, where μ_1 and μ_2 are symmetric distributions on $\mathbb{R}^d$ and c is a constant in $(0,1)$. Show that a random walk based on μ is recurrent iff recurrence holds for the random walks based on μ_1 and μ_2.

11. Let $\mu = \mu_1 * \mu_2$, where μ_1 and μ_2 are symmetric distributions on $\mathbb{R}^d$. Show that if a random walk based on μ is recurrent, then so are the random walks based on μ_1 and μ_2. Also show by an example that the converse is false. (*Hint:* For the latter part, let μ_1 and μ_2 be supported by orthogonal subspaces.)

12. For any symmetric, recurrent random walk on $\mathbb{Z}^d$, show that the expected number of visits to an accessible state $k \neq 0$ before return to the origin equals 1. (*Hint:* Compute the distribution, assuming probability p for return before visit to k.)

13. Use Proposition 9.13 to show that any nondegenerate random walk in $\mathbb{Z}^d$ has infinite mean recurrence time. Compare with the preceding problem.

14. Show how part (i) of Proposition 9.14 can be strengthened by means of Theorems 5.16 and 9.2.

15. For a nondegenerate random walk in $\mathbb{R}$, show that $\limsup_n S_n = \infty$ a.s. iff $\sigma_1 < \infty$ a.s. and that $S_n \to \infty$ a.s. iff $E\sigma_1 < \infty$. In both conditions, note that σ_1 can be replaced by τ_1.

16. Let η be a renewal process based on some nonarithmetic distribution on $\mathbb{R}_+$. Show for any $\varepsilon > 0$ that $\sup\{t > 0;\ E\eta[t, t+\varepsilon] = 0\} < \infty$. (*Hint:* Imitate the proof of Proposition 8.14.)

17. Let μ be a distribution on $\mathbb{Z}_+$ such that the group generated by $\operatorname{supp}\mu$ equals $\mathbb{Z}$. Show that Proposition 9.18 remains true with $\nu\{n\} = c^{-1}\mu(n,\infty)$, $n \geq 0$, and prove a corresponding version of Proposition 9.19.

18. Let η be the occupation measure of a random walk on $\mathbb{Z}$ based on some distribution μ with mean $c \in \overline{\mathbb{R}} \setminus \{0\}$ such that the group generated by $\operatorname{supp}\mu$ equals $\mathbb{Z}$. Show as in Theorem 9.20 that $E\eta\{n\} \to c^{-1} \vee 0$.

19. Derive the renewal theorem for random walks on $\mathbb{Z}_+$ from the ergodic theorem for discrete-time Markov chains, and conversely. (*Hint:* Given a distribution μ on $\mathbb{N}$, construct a Markov chain X on $\mathbb{Z}_+$ with $X_{n+1} = X_n + 1$ or 0, and such that the recurrence times at 0 are i.i.d. μ. Note that X is aperiodic iff $\mathbb{Z}$ is the smallest group containing $\operatorname{supp}\mu$.)

20. Fix a distribution μ on $\mathbb{R}$ with symmetrization $\tilde{\mu}$. Note that if $\tilde{\mu}$ is nonarithmetic, then so is μ. Show by an example that the converse is false.

21. Simplify the proof of Lemma 9.21, in the case when even the symmetrization $\tilde{\mu}$ is nonarithmetic. (*Hint:* Let $\xi_1, \xi_2, \ldots$ and $\xi'_1, \xi'_2, \ldots$ be i.i.d. μ, and define $\tilde{S}_n = \alpha' - \alpha + \sum_{k \leq n}(\xi'_k - \xi_k)$.)

22. Show that any monotone and Lebesgue integrable function on $\mathbb{R}_+$ is directly Riemann integrable.

23. State and prove the counterpart of Corollary 9.23 for arithmetic distributions.

24. Let (ξ_n) and (η_n) be independent i.i.d. sequences with distributions μ and ν, put $S_n = \sum_{k \leq n}(\xi_k + \eta_k)$, and define $U = \bigcup_{n \geq 0}[S_n, S_n + \xi_{n+1})$. Show that $F_t = P\{t \in U\}$ satisfies the renewal equation $F = f + F * \mu * \nu$ with $f_t = \mu(t, \infty)$. Assuming μ and ν to have finite means, show also that F_t converges as $t \to \infty$, and identify the limit.

25. Consider a renewal process η based on some nonarithmetic distribution μ with mean $c < \infty$, fix an $h > 0$, and define $F_t = P\{\eta[t, t+h] = 0\}$. Show that $F = f + F * \mu$, where $f_t = \mu(t+h, \infty)$. Also show that F_t converges as $t \to \infty$, and identify the limit. (*Hint:* Consider the first point of η in $(0, t)$, if any.)

26. For η as above, let $\tau = \inf\{t \geq 0;\, \eta[t, t+h] = 0\}$, and put $F_t = P\{\tau \leq t\}$. Show that $F_t = \mu(h, \infty) + \int_0^{h \wedge t} \mu(ds) F_{t-s}$, or $F = f + F * \mu_h$, where $\mu_h = 1_{[0,h]} \cdot \mu$ and $f \equiv \mu(h, \infty)$.

Chapter 10

Stationary Processes and Ergodic Theory

Stationarity, invariance, and ergodicity; discrete- and continuous-time ergodic theorems; moment and maximum inequalities; multivariate ergodic theorems; sample intensity of a random measure; subadditivity and products of random matrices; conditioning and ergodic decomposition; shift coupling and the invariant σ-field

In this chapter we come to the third important dependence structure of probability theory, beside those of martingales and Markov processes, namely stationarity. A stationary process is simply a process whose distribution is invariant under shifts. Stationary processes are important in their own right, and they also arise under broad conditions as steady-state limits of various Markov and renewal-type processes, as we have seen in Chapters 8 and 9 and will see again in Chapters 12, 20, and 23. Our present aim is to present some of the most useful general results for stationary and related processes.

The key result of stationarity theory is Birkhoff's ergodic theorem, which may be regarded as a strong law of large numbers for stationary sequences and processes. After proving the classical ergodic theorems in discrete and continuous time, we turn to the multivariate versions of Zygmund and Wiener, the former in a setting for noncommutative mappings and rectangular regions, the latter in the commutative case but with averages over increasing families of convex sets. Wiener's theorem will also be considered in a version for random measures that will be useful in Chapter 11 for the theory of Palm distributions. We finally present a version of Kingman's subadditive ergodic theorem, along with an important application to random matrices.

In all the mentioned results, the limit is a random variable, measurable with respect to the appropriate invariant σ-field $\mathcal{I}$. Of special interest then is the *ergodic* case, when $\mathcal{I}$ is trivial and the limit reduces to a constant. For general stationary processes, we consider a decomposition of the distribution into ergodic components. The chapter concludes with some basic criteria for coupling and shift coupling of two processes, expressed in terms of the tail and invariant σ-fields $\mathcal{T}$ and $\mathcal{I}$, respectively. Those results will be helpful to prove some ergodic theorems in Chapters 11 and 20.

Our treatment of stationary sequences and processes is continued in Chapter 11 with some important applications and extensions of the present theory. In particular, we will then derive ergodic theorems for Palm distributions, as well as for entropy and information. In Chapter 20 we show how the basic ergodic theorems admit extensions to suitable contraction operators, which leads to a profound unification of the present theory with the ergodic theory for Markov transition operators. Our treatment of the ratio ergodic theorem is also postponed until then.

Let us now return to the basic notions of stationarity and invariance. Then fix an arbitrary measurable space $(S, \mathcal{S})$. Given a measure μ and a measurable transformation T on S, we say that T is μ-*preserving* or *measure-preserving* if $\mu \circ T^{-1} = \mu$. Thus, if ξ is a random element of S with distribution μ, then T is measure-preserving iff $T\xi \stackrel{d}{=} \xi$. In particular, consider a random sequence $\xi = (\xi_0, \xi_1, \ldots)$ in some measurable space $(S', \mathcal{S}')$, and let θ denote the *shift* on $S = (S')^\infty$ given by $\theta(x_0, x_1, \ldots) = (x_1, x_2, \ldots)$. Then ξ is said to be *stationary* if $\theta\xi \stackrel{d}{=} \xi$. We show that the general situation is equivalent to this special case.

Lemma 10.1 *(stationarity and invariance) For any random element ξ in S and measurable transformation T on S, we have $T\xi \stackrel{d}{=} \xi$ iff the sequence $(T^n\xi)$ is stationary, in which case even $(f \circ T^n\xi)$ is stationary for every measurable function f. Conversely, any stationary random sequence admits such a representation.*

Proof: Assuming $T\xi \stackrel{d}{=} \xi$, we get

$$\theta(f \circ T^n \xi) = (f \circ T^{n+1}\xi) = (f \circ T^n T\xi) \stackrel{d}{=} (f \circ T^n \xi),$$

and so $(f \circ T^n \xi)$ is stationary. Conversely, if $\eta = (\eta_0, \eta_1, \ldots)$ is stationary, we may write $\eta_n = \pi_0(\theta^n \eta)$ with $\pi_0(x_0, x_1, \ldots) = x_0$, and we note that $\theta\eta \stackrel{d}{=} \eta$ by the stationarity of η. □

In particular, we note that if $\xi_0, \xi_1, \ldots$ is a stationary sequence of random elements in some measurable space S, and if f is a measurable mapping of S^∞ into some measurable space S', then the random sequence

$$\eta_n = f(\xi_n, \xi_{n+1}, \ldots), \quad n \in \mathbb{Z}_+,$$

is again stationary.

The definition of stationarity extends in the obvious way to random sequences indexed by $\mathbb{Z}$. The two-sided versions have the technical advantage that the associated shift operators form a group, rather than just a semigroup as in the one-sided context. The following result shows that the two cases are essentially equivalent. Here we assume the existence of appropriate randomization variables, as explained in Chapter 6.

Lemma 10.2 *(two-sided extension)* *Any stationary random sequence $\xi_0, \xi_1, \ldots$ in a Borel space admits a stationary extension $\ldots, \xi_{-1}, \xi_0, \xi_1, \ldots$ to the index set $\mathbb{Z}$.*

Proof: Assuming $\vartheta_1, \vartheta_2, \ldots$ to be i.i.d. $U(0,1)$ and independent of $\xi = (\xi_0, \xi_1, \ldots)$, we may construct the ξ_{-n} recursively as functions of ξ and $\vartheta_1, \ldots, \vartheta_n$ such that $(\xi_{-n}, \xi_{-n+1}, \ldots) \stackrel{d}{=} \xi$ for all n. In fact, once $\xi_{-1}, \ldots, \xi_{-n}$ have been chosen, the existence of ξ_{-n-1} is clear from Theorem 6.10 if we note that $(\xi_{-n}, \xi_{-n+1}, \ldots) \stackrel{d}{=} \theta\xi$. Finally, the extended sequence is stationary by Proposition 3.2. □

Now fix a measurable transformation T on some measure space $(S, \mathcal{S}, \mu)$, and let $\mathcal{S}^\mu$ denote the μ-completion of $\mathcal{S}$. We say that a set $I \subset S$ is *invariant* if $T^{-1}I = I$ and *almost invariant* if $T^{-1}I = I$ a.e. μ, in the sense that $\mu(T^{-1}I \triangle I) = 0$. Since inverse mappings preserve the basic set operations, the classes $\mathcal{I}$ and $\mathcal{I}'$ of invariant sets in $\mathcal{S}$ and almost invariant sets in $\mathcal{S}^\mu$ form σ-fields in S, called the *invariant* and *almost invariant σ-fields*, respectively.

A measurable function f on S is said to be *invariant* if $f \circ T \equiv f$ and *almost invariant* if $f \circ T = f$ a.e. μ. The following result gives the basic relationship between invariant or almost invariant sets and functions.

Lemma 10.3 *(invariant sets and functions)* *Fix a measure μ and a measurable transformation T on S, and let f be a measurable mapping of S into a Borel space S'. Then f is invariant or almost invariant iff it is $\mathcal{I}$-measurable or $\mathcal{I}'$-measurable, respectively.*

Proof: We may first apply a Borel isomorphism to reduce to the case when $S' = \mathbb{R}$. If f is invariant or almost invariant, then so is the set $I_x = f^{-1}(-\infty, x)$ for any $x \in \mathbb{R}$, and so $I_x \in \mathcal{I}$ or $\mathcal{I}'$, respectively. Conversely, if f is measurable with respect to $\mathcal{I}$ or $\mathcal{I}'$, then $I_x \in \mathcal{I}$ or $\mathcal{I}'$, respectively, for every $x \in \mathbb{R}$. Hence, the function $f_n(s) = 2^{-n}[2^n f(s)]$, $s \in S$, is invariant or almost invariant for every $n \in \mathbb{N}$, and the invariance or almost invariance carries over to the limit f. □

The next result clarifies the relationship between the invariant and almost invariant σ-fields. Here we write $\mathcal{I}^\mu$ for the μ-completion of $\mathcal{I}$ in $\mathcal{S}^\mu$, the σ-field generated by $\mathcal{I}$ and the μ-null sets in $\mathcal{S}^\mu$.

Lemma 10.4 *(almost invariance)* *For any distribution μ and μ-preserving transformation T on S, the associated invariant and almost invariant σ-fields $\mathcal{I}$ and $\mathcal{I}'$ are related by $\mathcal{I}' = \mathcal{I}^\mu$.*

Proof: If $J \in \mathcal{I}^\mu$, there exists some $I \in \mathcal{I}$ with $\mu(I \triangle J) = 0$. Since T is μ-preserving, we get

$$\mu(T^{-1}J \triangle J) \leq \mu(T^{-1}J \triangle T^{-1}I) + \mu(T^{-1}I \triangle I) + \mu(I \triangle J)$$
$$= \mu \circ T^{-1}(J \triangle I) = \mu(J \triangle I) = 0,$$

which shows that $J \in \mathcal{I}'$. Conversely, given any $J \in \mathcal{I}'$, we may choose some $J' \in \mathcal{S}$ with $\mu(J\Delta J') = 0$ and put $I = \bigcap_n \bigcup_{k \geq n} T^{-n} J'$. Then, clearly, $I \in \mathcal{I}$ and $\mu(I \Delta J) = 0$, and so $J \in \mathcal{I}^\mu$. □

A measure-preserving mapping T on some probability space $(S, \mathcal{S}, \mu)$ is said to be *ergodic for μ* or simply *μ-ergodic* if the invariant σ-field $\mathcal{I}$ is *μ-trivial*, in the sense that $\mu I = 0$ or 1 for every $I \in \mathcal{I}$. Depending on viewpoint, we may prefer to say that μ is ergodic for T, or T-ergodic. The terminology carries over to any random element ξ with distribution μ, which is said to be ergodic whenever this is true for T or μ. Thus, ξ is ergodic iff $P\{\xi \in I\} = 0$ or 1 for any $I \in \mathcal{I}$, that is, if the σ-field $\mathcal{I}_\xi = \xi^{-1}\mathcal{I}$ in Ω is P-trivial. In particular, a stationary sequence $\xi = (\xi_n)$ is ergodic if the shift-invariant σ-field is trivial for the distribution of ξ.

The next result shows how the ergodicity of a random element ξ is related to the ergodicity of the generated stationary sequence.

Lemma 10.5 *(ergodicity)* *Let ξ be a random element in S with distribution μ, and let T be a μ-preserving mapping on S. Then ξ is T-ergodic iff the sequence $(T^n \xi)$ is θ-ergodic, in which case even $\eta = (f \circ T^n \xi)$ is θ-ergodic for every measurable mapping f on S.*

Proof: Fix any measurable mapping $f: S \to S'$, and define $F = (f \circ T^n; n \geq 0)$, so that $F \circ T = \theta \circ F$. If $I \subset (S')^\infty$ is θ-invariant, then $T^{-1} F^{-1} I = F^{-1} \theta^{-1} I = F^{-1} I$, and so $F^{-1} I$ is T-invariant in S. Assuming ξ to be ergodic, we obtain $P\{\eta \in I\} = P\{\xi \in F^{-1} I\} = 0$ or 1, which shows that even η is ergodic.

Conversely, let the sequence $(T^n \xi)$ be ergodic, and fix any T-invariant set I in S. Put $F = (T^n; n \geq 0)$, and define $A = \{s \in S^\infty; s_n \in I \text{ i.o.}\}$. Then $I = F^{-1} A$ and A is θ-invariant. Hence, $P\{\xi \in I\} = P\{(T^n \xi) \in A\} = 0$ or 1, which means that even ξ is ergodic. □

We may now state the fundamental a.s. and mean ergodic theorem for stationary sequences of random variables. Recall that $(S, \mathcal{S})$ denotes an arbitrary measurable space, and write $\mathcal{I}_\xi = \xi^{-1}\mathcal{I}$ for convenience.

Theorem 10.6 *(ergodic theorem, Birkhoff)* *Let ξ be a random element in S with distribution μ, and let T be a μ-preserving map on S with invariant σ-field $\mathcal{I}$. Then for any measurable function $f \geq 0$ on S,*

$$n^{-1} \sum_{k<n} f(T^k \xi) \to E[f(\xi) | \mathcal{I}_\xi] \quad a.s. \tag{1}$$

The same convergence holds in L^p for some $p \geq 1$ when $f \in L^p(\mu)$.

The proof is based on a simple, but ingenious, inequality.

Lemma 10.7 *(maximzal ergodic lemma)* *Let $\xi = (\xi_k)$ be a stationary sequence of integrable random variables, and put $S_n = \xi_1 + \cdots + \xi_n$. Then*

$$E[\xi_1; \sup_n S_n > 0] \geq 0.$$

Proof (Garsia): Put $M_n = S_1 \vee \cdots \vee S_n$. Assuming ξ to be defined on the canonical space $\mathbb{R}^\infty$, we note that
$$S_k = \xi_1 + S_{k-1} \circ \theta \leq \xi_1 + (M_n \circ \theta)_+, \quad k = 1, \ldots, n.$$
Taking maxima yields $M_n \leq \xi_1 + (M_n \circ \theta)_+$ for all $n \in \mathbb{N}$, and so by stationarity
$$\begin{aligned} E[\xi_1; M_n > 0] &\geq E[M_n - (M_n \circ \theta)_+; M_n > 0] \\ &\geq E[(M_n)_+ - (M_n \circ \theta)_+] = 0. \end{aligned}$$
Since $M_n \uparrow \sup_n S_n$, the assertion follows by dominated convergence. □

Proof of Theorem 10.6 (Yosida and Kakutani): First assume that $f \in L^1$, and put $\eta_k = f(T^{k-1}\xi)$ for convenience. Since $E[\eta_1|\mathcal{I}_\xi]$ is an invariant function of ξ by Lemma 10.3, the sequence $\zeta_k = \eta_k - E[\eta_1|\mathcal{I}_\xi]$ is again stationary. Writing $S_n = \zeta_1 + \cdots + \zeta_n$, we define for any $\varepsilon > 0$
$$A_\varepsilon = \{\limsup_n (S_n/n) > \varepsilon\}, \qquad \zeta_n^\varepsilon = (\zeta_n - \varepsilon)1_{A_\varepsilon},$$
and note that the sums $S_n^\varepsilon = \zeta_1^\varepsilon + \cdots + \zeta_n^\varepsilon$ satisfy
$$\begin{aligned} \{\sup_n S_n^\varepsilon > 0\} &= \{\sup_n (S_n^\varepsilon/n) > 0\} \\ &= \{\sup_n (S_n/n) > \varepsilon\} \cap A_\varepsilon = A_\varepsilon. \end{aligned}$$
Since $A_\varepsilon \in \mathcal{I}_\xi$, the sequence (ζ_n^ε) is stationary, and Lemma 10.7 yields
$$\begin{aligned} 0 &\leq E[\zeta_1^\varepsilon; \sup_n S_n^\varepsilon > 0] = E[\zeta - \varepsilon; A_\varepsilon] \\ &= E[E[\zeta|\mathcal{I}_\xi]; A_\varepsilon] - \varepsilon P A_\varepsilon = -\varepsilon P A_\varepsilon, \end{aligned}$$
which implies $PA_\varepsilon = 0$. Thus, $\limsup_n (S_n/n) \leq \varepsilon$ a.s., and ε being arbitrary, we obtain $\limsup_n (S_n/n) \leq 0$ a.s. Applying the same result to $-S_n$ yields $\liminf_n (S_n/n) \geq 0$ a.s., and so by combination $S_n/n \to 0$ a.s.

Next assume that $f \in L^p$ for some $p \geq 1$. Using Jensen's inequality and the stationarity of $T^k\xi$, we get for any $A \in \mathcal{A}$ and $r > 0$
$$\begin{aligned} E 1_A \left| n^{-1} \sum_{k<n} f(T^k\xi) \right|^p &\leq n^{-1} \sum_{k<n} E[|f(T^k\xi)|^p; A] \\ &\leq r^p PA + E[|f(\xi)|^p; |f(\xi)| > r], \end{aligned}$$
which tends to 0 as $PA \to 0$ and then $r \to \infty$. Hence, by Lemma 4.10 the pth powers on the left are uniformly integrable, and the asserted L^p-convergence follows by Proposition 4.12.

Finally, let $f \geq 0$ be arbitrary and put $E[f(\xi)|\mathcal{I}_\xi] = \bar{\eta}$. Conditioning on the event $\{\bar{\eta} \leq r\}$ for arbitrary $r > 0$, we see that (1) holds a.s. on $\{\bar{\eta} < \infty\}$. Next we have a.s. for any $r > 0$
$$\begin{aligned} \liminf_{n\to\infty} n^{-1} \sum_{k\leq n} f(T^k\xi) &\geq \lim_{n\to\infty} n^{-1} \sum_{k\leq n} (f(T^k\xi) \wedge r) \\ &= E[f(\xi) \wedge r|\mathcal{I}_\xi]. \end{aligned}$$
As $r \to \infty$, the right-hand side tends a.s. to $\bar{\eta}$ by the monotone convergence property of conditional expectations. In particular, the left-hand side is a.s. infinite on $\{\bar{\eta} = \infty\}$, as required. □

Write $\mathcal{I}$ and $\mathcal{T}$ for the shift-invariant and tail σ-fields, respectively, in $\mathbb{R}^\infty$ and note that $\mathcal{I} \subset \mathcal{T}$. Thus, for any sequence of random variables $\xi = (\xi_1, \xi_2, \ldots)$, we have $\mathcal{I}_\xi = \xi^{-1}\mathcal{I} \subset \xi^{-1}\mathcal{T}$. By Kolmogorov's 0–1 law, the latter σ-field is trivial when the ξ_n are independent. If they are even i.i.d. and integrable, then Theorem 10.6 yields $n^{-1}(\xi_1 + \cdots + \xi_n) \to E\xi_1$ a.s. and in L^1, in agreement with Theorem 4.23. Hence, the last theorem contains the strong law of large numbers.

It is often useful to allow the function $f = f_{n,k}$ in Theorem 10.6 to depend on n or k. For later needs, we consider a slighty more general situation.

Corollary 10.8 *(approximation, Maker)* *Let ξ be a random element in S with distribution μ, let T be a μ-preserving map on S with invariant σ-field $\mathcal{I}$, and consider some measurable functions f and $f_{m,k}$ on S.*

(i) *If $f_{m,k} \to f$ a.s. and $\sup_{m,k} |f_{m,k}| \in L^1$, then as $m, n \to \infty$,*
$$n^{-1} \sum\nolimits_{k<n} f_{m,k}(T^k \xi) \to E[f(\xi)|\mathcal{I}_\xi] \quad a.s.$$

(ii) *If $f_{m,k} \to f$ in L^p for some $p \geq 1$, the same convergence holds in L^p.*

Proof: (i) By Theorem 10.6 we may assume that $f = 0$. Then put $g_r = \sup_{m,k>r} |f_{m,k}|$, and conclude from the same result that a.s.

$$\limsup_{m,n\to\infty} \left| n^{-1} \sum\nolimits_{k<n} f_{m,k}(T^k \xi) \right| \leq \lim_{n\to\infty} n^{-1} \sum\nolimits_{k<n} g_r(T^k \xi)$$
$$= E[g_r(\xi)|\mathcal{I}_\xi].$$

Here $g_r(\xi) \to 0$ a.s., and so by dominated convergence $E[g_r(\xi)|\mathcal{I}_\xi] \to 0$ a.s.

(ii) Assuming $f = 0$, we get by Minkowski's inequality and the invariance of μ

$$\left\| n^{-1} \sum\nolimits_{k<n} f_{m,k} \circ T^k \right\|_p \leq n^{-1} \sum\nolimits_{k<n} \|f_{m,k}\|_p \to 0. \qquad \Box$$

Our next aim is to extend the ergodic theorem to continuous time. We may then consider a family of transformations T_t on S, $t \geq 0$, satisfying the *semigroup* property $T_{s+t} = T_s T_t$. The semigroup (T_t) is called a *flow* if it is also *measurable*, in the sense that the mapping $(x, t) \mapsto T_t x$ is product measurable from $S \times \mathbb{R}_+$ to S. The *invariant σ-field* $\mathcal{I}$ now consists of all sets $I \in \mathcal{S}$ such that $T_t^{-1} I = I$ for all t. A random element ξ in S is said to be (T_t)-*stationary* if $T_t \xi \stackrel{d}{=} \xi$ for all $t \geq 0$.

Corollary 10.9 *(continuous-time ergodic theorem)* *Let ξ be a random element in S with distribution μ, and let (T_s) be a μ-preserving flow on S with invariant σ-field $\mathcal{I}$. Then for any measurable function $f \geq 0$ on S,*

$$\lim_{t\to\infty} t^{-1} \int_0^t f(T_s \xi)\, ds = E[f(\xi)|\mathcal{I}_\xi] \quad a.s. \qquad (2)$$

The same convergence holds in L^p for some $p \geq 1$ when $f \in L^p(\mu)$.

Proof: In both cases we may assume that $f \geq 0$. Writing $X_s = f(T_s \xi)$, we get by Jensen's inequality and Fubini's theorem

$$E \left| t^{-1} \int_0^t X_s ds \right|^p \leq E t^{-1} \int_0^t X_s^p ds = t^{-1} \int_0^t E X_s^p ds = E X_0^p < \infty.$$

The required convergence now follows as we apply Theorem 10.6 to the function $g(x) = \int_0^1 f(T_s x) ds$ and the discrete shift $T = T_1$.

To identify the limit, we first assume that $f \in L^1$ and introduce the invariant version

$$\bar{f}(\xi) = \lim_{r \to \infty} \limsup_{n \to \infty} n^{-1} \int_r^{r+n} f(T_s \xi) ds,$$

which is also $\mathcal{I}_\xi$-measurable. By the stationarity of $T_s \xi$ we have $E^{\mathcal{I}_\xi} f(T_s \xi) = E^{\mathcal{I}_\xi} f(\xi)$ a.s. for all $s \geq 0$. Using Fubini's theorem, the L^1-convergence in (2), and the contraction property of conditional expectations, we get as $t \to \infty$

$$E^{\mathcal{I}_\xi} f(\xi) = E^{\mathcal{I}_\xi} t^{-1} \int_0^t f(T_s \xi) ds \xrightarrow{P} E^{\mathcal{I}_\xi} \bar{f}(\xi) = \bar{f}(\xi),$$

as required. The result extends as before to arbitrary $f \geq 0$. □

We return to the case when $\xi_1, \xi_2, \ldots$ is a stationary sequence of integrable random variables, and put $S_n = \sum_{k \leq n} \xi_k$. Since S_n/n converges a.s. by Theorem 10.6, we note that the maximum $M = \sup_n (S_n/n)$ is a.s. finite. The following result, relating the moments of ξ and M, is known as the *dominated ergodic theorem*. Here we write $\log_+ x = \log(x \vee 1)$ for convenience.

Proposition 10.10 *(moment inequalities, Hardy and Littlewood, Wiener)* Let $\xi = (\xi_k)$ be a stationary sequence of random variables, and put $S_n = \sum_{k \leq n} \xi_k$ and $M = \sup_n (S_n/n)$. Then
 (i) $E|M|^p \leq E|\xi_1|^p$ for fixed $p > 1$;
 (ii) $E|M| \log_+^m |M| \leq 1 + E|\xi_1| \log_+^{m+1} |\xi_1|$ for fixed $m \geq 0$.

The proof requires a simple estimate related to Lemma 10.7.

Lemma 10.11 *(maximum inequality)* If $\xi = (\xi_k)$ is stationary in L^1, then
$$r P\{\sup_n (S_n/n) > 2r\} \leq E[\xi_1; \xi_1 > r], \quad r > 0.$$

Proof: For any $r > 0$, we put $\xi_k^r = \xi_k 1\{\xi_k > r\}$ and note that $\xi_k \leq \xi_k^r + r$. Assuming ξ to be defined on the canonical space $\mathbb{R}^\infty$ and writing $A_n = S_n/n$, we get

$$A_n - 2r = A_n \circ (\xi - 2r) \leq A_n \circ (\xi^r - r),$$

which implies $M - 2r \leq M \circ (\xi^r - r)$. Applying Lemma 10.7 to the sequence $\xi^r - r$, we obtain

$$\begin{aligned} rP\{M > 2r\} &\leq rP\{M \circ (\xi^r - r) > 0\} \\ &\leq E[\xi_1^r; M \circ (\xi^r - r) > 0] \\ &\leq E\xi_1^r = E[\xi_1; \xi_1 > r]. \end{aligned}$$

□

Proof of Proposition 10.10: We may clearly assume that $\xi_1 \geq 0$ a.s.
(i) By Lemma 10.11, Fubini's theorem, and some calculus,

$$\begin{aligned} EM^p &= pE\int_0^M r^{p-1}dr = p\int_0^\infty P\{M > r\}r^{p-1}dr \\ &\leq 2p\int_0^\infty E[\xi_1; 2\xi_1 > r]r^{p-2}dr \\ &= 2p\, E\, \xi_1 \int_0^{2\xi_1} r^{p-2}dr \\ &= 2p(p-1)^{-1}E\,\xi_1(2\xi_1)^{p-1} \leq E\xi_1^p. \end{aligned}$$

(ii) For $m = 0$, we may write

$$\begin{aligned} EM - 1 &\leq E(M-1)_+ = \int_1^\infty P\{M > r\}dr \\ &\leq 2\int_1^\infty E[\xi_1; 2\xi_1 > r]r^{-1}dr \\ &= 2E\,\xi_1 \int_1^{2\xi_1 \vee 1} r^{-1}dr = 2E\,\xi_1 \log_+ 2\xi_1 \\ &\leq e + 2E[\xi_1 \log 2\xi_1; 2\xi_1 > e] \\ &\leq 1 + E\,\xi_1 \log_+ \xi_1. \end{aligned}$$

For $m > 0$, we instead write

$$\begin{aligned} EM\log_+^m M &= \int_0^\infty P\{M\log_+^m M > r\}dr \\ &= \int_1^\infty P\{M > t\}\left(m\log^{m-1}t + \log^m t\right)dt \\ &\leq 2\int_1^\infty E[\xi_1; 2\xi_1 > t]\left(m\log^{m-1}t + \log^m t\right)t^{-1}dt \\ &= 2E\,\xi_1\int_0^{\log_+ 2\xi_1}\left(mx^{m-1} + x^m\right)dx \\ &= 2E\,\xi_1\left\{\log_+^m 2\xi_1 + \frac{\log_+^{m+1} 2\xi_1}{m+1}\right\} \\ &\leq 2e + 4E\left[\xi_1\log^{m+1}2\xi_1; 2\xi_1 > e\right] \\ &\leq 1 + E\,\xi_1\log_+^{m+1}\xi_1. \end{aligned}$$

□

Given a measure space $(S, \mathcal{S}, \mu)$, we introduce for any $m \geq 0$ the class $L\log^m L(\mu)$ of measurable functions f on S satisfying $\int |f| \log_+^m |f| d\mu < \infty$. Note in particular that $L\log^0 L = L^1$. Using the maximum inequalities of Proposition 10.10, we may prove the following multivariate version of Theorem 10.6 for possibly noncommuting, measure-preserving transformations $T_1, \ldots, T_d$.

Theorem 10.12 *(multivariate ergodic theorem, Zygmund)* Let ξ be a random element in S with distribution μ, let $T_1, \ldots, T_d$ be μ-preserving maps on S with invariant σ-fields $\mathcal{I}_1, \ldots, \mathcal{I}_d$, and put $\mathcal{J}_k = \xi^{-1}\mathcal{I}_k$. Then for any $f \in L\log^{d-1} L(\mu)$, we have as $n_1, \ldots, n_d \to \infty$

$$(n_1 \cdots n_d)^{-1} \sum_{k_1 < n_1} \cdots \sum_{k_d < n_d} f(T_1^{k_1} \cdots T_d^{k_d} \xi) \to E^{\mathcal{J}_d} \cdots E^{\mathcal{J}_1} f(\xi) \quad a.s. \quad (3)$$

The same convergence holds in L^p for some $p \geq 1$ when $f \in L^p(\mu)$.

Proof: Since $E[f(\xi)|\mathcal{J}_k] = \mu[f|\mathcal{I}_k] \circ \xi$ a.s., e.g. by Theorem 10.6, we may take ξ to be the identity mapping on S. For $d = 1$ the result reduces to Theorem 10.6. Now assume the statement to be true up to dimension d. Proceeding by induction, consider any μ-preserving maps $T_1, \ldots, T_{d+1}$ on S and let $f \in L\log^d L$. By the induction hypothesis, the d-dimensional version of (3) holds as stated, and we may write the result in the form $f_m \to \bar{f}$ a.s., where $m = (n_1, \ldots, n_d)$. Iterating Proposition 10.10, we also note that $\mu \sup_m |f_m| < \infty$. Hence, by Corollary 10.8 (i) we have as $m, n \to \infty$

$$n^{-1} \sum_{k<n} f_m \circ T_{d+1}^k \to \mu[\bar{f}|\mathcal{I}_{d+1}] \quad a.s.,$$

as required. The proof of the L^p-version is similar. □

In the commutative case, the last result leads immediately to an interesting relationship between the associated conditional expectations. Let $L^1(\xi)$ denote the set of all integrable, ξ-measurable random variables.

Corollary 10.13 *(commuting maps and expectations)* Assume in Theorem 10.12 that $T_1, \ldots, T_d$ commute, and put $\mathcal{J} = \bigcap_k \mathcal{J}_k$. Then
$$E^{\mathcal{J}_1} \cdots E^{\mathcal{J}_d} = E^{\mathcal{J}} \quad on \ L^1(\xi).$$

Proof: Since even $T_1^{k_1}, \ldots, T_d^{k_d}$ commute for arbitrary $k_1, \ldots, k_d \in \mathbb{Z}_+$, Theorem 10.12 yields

$$E^{\mathcal{J}_1} \cdots E^{\mathcal{J}_d} f(\xi) = E^{\mathcal{J}_{p_1}} \cdots E^{\mathcal{J}_{p_d}} f(\xi) \quad a.s. \quad (4)$$

for any measurable function $f \geq 0$ on S and permutation $p_1, \ldots, p_d$ of $1, \ldots, d$. In particular, the expression in (4) is a.s. $\mathcal{J}_k$-measurable for every k and therefore $\mathcal{J}$-measurable. It remains to note that

$$E[E^{\mathcal{J}_1} \cdots E^{\mathcal{J}_d} f(\xi); A] = E[f(\xi); A], \quad A \in \mathcal{J}. \quad □$$

For commuting mappings $T_1, \ldots, T_d$ on S, we note that the compositions $T^k = T_1^{k_1} \cdots T_d^{k_d}$ form a d-dimensional semigroup indexed by $\mathbb{Z}_+^d$. Similarly, when $(T_1^s), \ldots, (T_d^s)$ are commuting flows on S, the compositions $T^s =$

$T_1^{s_1} \cdots T_d^{s_d}$ form a d-dimensional measurable semigroup or flow indexed by $\mathbb{R}_+^d$. In the continuous parameter case, it may be more natural to consider flows indexed by $\mathbb{R}^d$, corresponding to the case of stationary processes on $\mathbb{R}^d$. In this context, one may also want to average over more general sets than rectangles. Here we consider a basic ergodic theorem for increasing sequences of convex sets. Given such a set B, we define the *inner radius* $r(B)$ as the radius of the largest open ball contained in B. Put $\lambda^d B = |B|$.

Theorem 10.14 *(monotone, multivariate ergodic theorem, Wiener)* Let ξ be a random element in S with distribution μ. Consider a flow of μ-preserving maps T_s, $s \in \mathbb{R}^d$, on S with invariant σ-field $\mathcal{I}$ and fix some bounded, convex sets $B_1 \subset B_2 \subset \cdots$ in $\mathcal{B}^d$ with $r(B_n) \to \infty$. Then for any measurable function $f \geq 0$ on S,
$$|B_n|^{-1} \int_{B_n} f(T_s\xi)\,ds \to E[f(\xi)|\mathcal{I}_\xi] \quad a.s.$$
The same convergence holds in L^p for some $p \geq 1$ when $f \in L^p(\mu)$.

Several lemmas are needed for the proof. We begin with some estimates for convex sets, stated here without proof. Let $\partial_\varepsilon B$ denote the ε-neighborhood of the boundary ∂B, and write binomial coefficients as $(n /\!\!/ k)$.

Lemma 10.15 *(convex sets)* *If $B \subset \mathbb{R}^d$ is convex and $\varepsilon > 0$, then*
(i) $|B - B| \leq (2d /\!\!/ d)|B|$;
(ii) $|\partial_\varepsilon B| \leq 2((1 + \varepsilon/r(B))^d - 1)|B|$.

We continue with a simple geometric estimate.

Lemma 10.16 *(space filling)* *Fix any bounded, convex sets $B_1 \subset \cdots \subset B_m$ in $\mathcal{B}^d$ with $|B_1| > 0$, a bounded set $K \in \mathcal{B}^d$, and a function $p \colon K \to \{1, \ldots, m\}$. Then there exists a finite subset $H \subset K$ such that the sets $B_{p(x)} + x$, $x \in H$, are disjoint and satisfy $|K| \leq (2d /\!\!/ d) \sum_{x \in H} |B_{p(x)}|$.*

Proof: Put $C_x = B_{p(x)} + x$ and choose $x_1, x_2, \cdots \in K$ recursively, as follows. Once $x_1, \ldots, x_{j-1}$ have been selected, we choose $x_j \in K$ with the largest possible $p(x)$ such that $C_{x_i} \cap C_{x_j} = \emptyset$ for all $i < j$. The construction terminates when no such x_j exists. Put $H = \{x_i\}$, and note that the sets C_x with $x \in H$ are disjoint. Now fix any $y \in K$. By the construction of H we have $C_x \cap C_y \neq \emptyset$ for some $x \in H$ with $p(x) \geq p(y)$, and so
$$y \in B_{p(x)} - B_{p(y)} + x \subset B_{p(x)} - B_{p(x)} + x.$$
Hence, $K \subset \bigcup_{x \in H}(B_{p(x)} - B_{p(x)} + x)$, and so by Lemma 10.15 (i)
$$|K| \leq \sum_{x \in H}|B_{p(x)} - B_{p(x)}| \leq (2d /\!\!/ d) \sum_{x \in H}|B_{p(x)}|. \qquad \square$$

We may now establish a multivariate version of Lemma 10.11, stated for convenience in terms of random measures (see the detailed discussion below). For motivation, we note that the set function $\eta B = \int_B f(T_s\xi)ds$ in

Theorem 10.14 is a stationary random measure on $\mathbb{R}^d$ and that the *intensity* m of η, defined by the relation $E\eta = m\lambda^d$, is equal to $Ef(\xi)$.

Lemma 10.17 *(maximum inequality)* *Let ξ be a stationary random measure on $\mathbb{R}^d$ with intensity m, and let $B_1 \subset B_2 \subset \cdots$ be bounded, convex sets in $\mathcal{B}^d$ with $|B_1| > 0$. Then*
$$r\, P\{\sup_k(\xi B_k/|B_k|) > r\} \leq m\,(2d/\!/d), \quad r > 0.$$

Proof: Fix any $r, a > 0$ and $n \in \mathbb{N}$, and define a process ν on $\mathbb{R}^d$ and a random set K in $S_a = \{x \in \mathbb{R}^d;\; |x| \leq a\}$ by
$$\begin{aligned}
\nu(x) &= \inf\{k \in \mathbb{N};\; \xi(B_k + x) > r|B_k|\},\quad x \in \mathbb{R}^d,\\
K &= \{x \in S_a;\; \nu(x) \leq n\}.
\end{aligned}$$

By Lemma 10.16 there exists a finite, random subset $H \subset K$ such that the sets $B_{\nu(x)} + x$, $x \in H$, are disjoint and $|K| \leq (2d/\!/d)\sum_{x \in H}|B_{\nu(x)}|$. Writing $b = \sup\{|x|;\; x \in B_n\}$, we get
$$\xi S_{a+b} \geq \sum_{x\in H}\xi(B_{\nu(x)} + x) \geq r\sum_{x\in H}|B_{\nu(x)}| \geq r|K|/(2d/\!/d).$$

Taking expectations and using Fubini's theorem and the stationarity and measurability of ν, we obtain
$$\begin{aligned}
m\,(2d/\!/d)\,|S_{a+b}| &\geq rE|K| = r\int_{S_a} P\{\nu(x) \leq n\}\,dx\\
&= r|S_a|\,P\{\max_{k\leq n}(\xi B_k/|B_k|) > r\}.
\end{aligned}$$

Now divide by $|S_a|$, and then let $a \to \infty$ and $n \to \infty$ in this order. $\square$

We finally need an elementary Hilbert-space result. Recall that a *contraction* on a Hilbert space H is defined as a linear operator T such that $\|T\xi\| \leq \|\xi\|$ for all $\xi \in H$. For any linear subspace $M \subset H$, we write $M^\perp$ for the orthogonal complement and $\overline{M}$ for the closure of M. The *adjoint* T^* of an operator T is characterized by the identity $\langle \xi, T\eta\rangle = \langle T^*\xi, \eta\rangle$, where $\langle \cdot,\cdot\rangle$ denotes the inner product in H.

Lemma 10.18 *(invariant subspace)* *For any family $\mathcal{T}$ of contractions on a Hilbert space H, let N denote the $\mathcal{T}$-invariant subspace of H, and let R be the linear subspace of H spanned by the set $\{\xi - T\xi;\; \xi \in H,\, T \in \mathcal{T}\}$. Then $N^\perp \subset \overline{R}$.*

Proof: If $\xi \perp R$, then
$$\langle \xi - T^*\xi, \eta\rangle = \langle \xi, \eta - T\eta\rangle = 0,\quad T \in \mathcal{T},\; \eta \in H,$$
which implies $T^*\xi = \xi$ for every $T \in \mathcal{T}$. Hence, for any $T \in \mathcal{T}$ we have $\langle T\xi, \xi\rangle = \langle \xi, T^*\xi\rangle = \|\xi\|^2$, and so by the contraction property,
$$\begin{aligned}
0 &\leq \|T\xi - \xi\|^2 = \|T\xi\|^2 + \|\xi\|^2 - 2\langle T\xi,\xi\rangle\\
&\leq 2\|\xi\|^2 - 2\|\xi\|^2 = 0,
\end{aligned}$$
which implies $T\xi = \xi$. This gives $R^\perp \subset N$, and so $N^\perp \subset (R^\perp)^\perp = \overline{R}$. $\square$

Proof of Theorem 10.14: First assume that $f \in L^1$, and define

$$\tilde{T}_s f = f \circ T_s, \qquad A_n = |B_n|^{-1} \int_{B_n} \tilde{T}_s ds.$$

For any $\varepsilon > 0$, Lemma 10.18 yields a measurable decomposition

$$f = f^\varepsilon + \sum_{k \leq m} (g_k^\varepsilon - \tilde{T}_{s_k} g_k^\varepsilon) + h^\varepsilon,$$

where $f^\varepsilon \in L^2$ is $\tilde{T}_s$-invariant for all $s \in \mathbb{R}^d$, the functions $g_1^\varepsilon, \ldots, g_m^\varepsilon$ are bounded, and $E|h^\varepsilon(\xi)| < \varepsilon$. Here clearly $A_n f^\varepsilon \equiv f^\varepsilon$. Next, we see from Lemma 10.15 (ii) that, as $n \to \infty$ for fixed $k \leq m$ and $\varepsilon > 0$,

$$\begin{aligned}
\|A_n(g_k^\varepsilon - \tilde{T}_{s_k} g_k^\varepsilon)\| &\leq (|(B_n + s_k) \Delta B_n|/|B_n|) \|g_k^\varepsilon\| \\
&\leq 2\left((1 + |s_k|/r(B_n))^d - 1\right) \|g_k^\varepsilon\| \to 0.
\end{aligned}$$

Finally, Lemma 10.17 yields

$$r\, P\{\sup_n A_n |h^\varepsilon(\xi)| \geq r\} \leq (2d/\!/d) E|h^\varepsilon(\xi)| \leq (2d/\!/d)\varepsilon, \qquad r, \varepsilon > 0,$$

which implies $\sup_n A_n |h^\varepsilon(\xi)| \xrightarrow{P} 0$ as $\varepsilon \to 0$. In particular, it follows that $\liminf_n A_n f(\xi) < \infty$ a.s., which justifies the estimate

$$\begin{aligned}
(\limsup_n - \liminf_n) A_n f(\xi) &= (\limsup_n - \liminf_n) A_n h^\varepsilon(\xi) \\
&\leq 2 \sup_n A_n |h^\varepsilon(\xi)| \xrightarrow{P} 0.
\end{aligned}$$

This shows that the left-hand side vanishes a.s., and the required a.s. convergence follows.

When $f \in L^p$ for some $p \geq 1$, the asserted L^p-convergence follows as before from the uniform integrability of the powers $|A_n f(\xi)|^p$. We may now identify the limit, as in the proof of Corollary 10.9, and the a.s. convergence extends to arbitrary $f \geq 0$, as in case of Theorem 10.6. □

We turn to a version of Theorem 10.14 for random measures on $\mathbb{R}^d$. Recall that a *random measure* ξ on $\mathbb{R}^d$ is defined as a locally finite kernel from the basic probability space $(\Omega, \mathcal{A}, P)$ into $\mathbb{R}^d$. In other words, $\xi(\omega, B)$ is required to be a locally finite measure in $B \in \mathcal{B}^d$ for fixed $\omega \in \Omega$ and a random variable in $\omega \in \Omega$ for every bounded set $B \in \mathcal{B}^d$. Alternatively, we may regard ξ as a random element in the space $\mathcal{M}(\mathbb{R}^d)$ of locally finite measures μ on $\mathbb{R}^d$, endowed with the σ-field generated by all evaluation maps $\mu \mapsto \mu B$ with $B \in \mathcal{B}^d$.

We say that ξ is *stationary* if $\theta_s \xi \stackrel{d}{=} \xi$ for every $s \in \mathbb{R}^d$, where the shift operators θ_s on $\mathcal{M}(\mathbb{R}^d)$ are defined by $(\theta_s \mu)B = \mu(B + s)$ for all $B \in \mathcal{B}^d$. The *invariant σ-field of ξ* is given by $\mathcal{I}_\xi = \xi^{-1}\mathcal{I}$, where $\mathcal{I}$ denotes the σ-field of all shift-invariant, measurable sets in $\mathcal{M}(\mathbb{R}^d)$. We may now define the *sample intensity* of ξ as the extended-valued random variable $\bar{\xi} = E[\xi B | \mathcal{I}_\xi]/|B|$, where $B \in \mathcal{B}^d$ is arbitrary with $|B| \in (0, \infty)$. Note that this expression is independent of B, by the stationarity of ξ and Theorem 2.6.

Corollary 10.19 *(sample intensity, Nguyen and Zessin)* Let ξ be a stationary random measure on $\mathbb{R}^d$, and fix some bounded, convex sets $B_1 \subset B_2 \subset \cdots$ in $\mathcal{B}^d$ with $r(B_n) \to \infty$. Then $\xi B_n/|B_n| \to \bar\xi$ a.s., where $\bar\xi\lambda^d = E[\xi|\mathcal{I}_\xi]$. The same convergence holds in L^p for some $p \geq 1$ when $\xi[0,1]^d \in L^p$.

Proof: By Fubini's theorem, we have for any $A, B \in \mathcal{B}^d$

$$\int_B (\theta_s \xi) A\, ds = \int_B ds \int 1_A(t-s)\, \xi(dt)$$

$$= \int \xi(dt) \int_B 1_A(t-s)\, ds = \xi(1_A * 1_B).$$

Assuming $|A| = 1$ and $A \subset S_a = \{s;\ |s| < a\}$, and putting $B^+ = B + S_a$ and $B^- = (B^c + S_a)^c$, we note that also $1_A * 1_{B^-} \leq 1_B \leq 1_A * 1_{B^+}$. Applying this to the sets $B = B_n$ gives

$$\frac{|B_n^-|}{|B_n|} \frac{\xi(1_A * 1_{B_n^-})}{|B_n^-|} \leq \frac{\xi B_n}{|B_n|} \leq \frac{|B_n^+|}{|B_n|} \frac{\xi(1_A * 1_{B_n^+})}{|B_n^+|}.$$

Since $r(B_n) \to \infty$, Lemma 10.15 (ii) yields $|B_n^\pm|/|B_n| \to 1$. Next we may apply Theorem 10.14 to the function $f(\mu) = \mu A$ and the convex sets $B_n^\pm$ to obtain $\xi(1_A * 1_{B_n^\pm})/|B_n^\pm| \to E^{\mathcal{I}_\xi}\xi A = \bar\xi$ in the appropriate sense. $\square$

The L^p-versions of Theorem 10.14 and Corollary 10.19 remain valid under weaker conditions than previously indicated. The following results are adequate for most purposes.

Here we say that the distributions (probability measures) μ_n on $\mathbb{R}^d$ are *asymptotically invariant* if $\|\mu_n - \mu_n * \delta_s\| \to 0$ for every $s \in \mathbb{R}^d$, where $\|\cdot\|$ denotes the total variation norm. Similarly, the *weight functions* (probability densities) f_n on $\mathbb{R}^d$ are said to be asymptotically invariant if $\lambda^d|f_n - \theta_s f_n| \to 0$ for every s. Note that the conclusion of Theorem 10.14 can be written as $\mu_n X \to \bar X$, where $\mu_n = (1_{B_n} \cdot \lambda^d)/|B_n|$, $X_s = f(T_s\xi)$, and $\bar X = E[f(\xi)|\mathcal{I}_\xi]$.

Corollary 10.20 *(mean ergodic theorem)*

(i) *For any $p \geq 1$, consider on $\mathbb{R}^d$ a stationary, measurable, and L^p-valued process X and some asymptotically invariant distributions μ_n. Then $\mu_n X \to \bar X \equiv E[X|\mathcal{I}_X]$ in L^p.*

(ii) *Consider on $\mathbb{R}^d$ a stationary random measure ξ with finite intensity and some asymptotically invariant weight functions f_n. Then $\xi f_n \to \bar\xi$ in L^1, where $\bar\xi\lambda^d = E[\xi|\mathcal{I}_\xi]$.*

Proof: (i) By Theorem 10.14 we may choose some distributions ν_m on $\mathbb{R}^d$ such that $\nu_m X \to \bar X$ in L^p. Using Minkowski's inequality and its extension in Corollary 1.30, along with the stationarity of X, the invariance of $\bar X$, and dominated convergence, we get as $n \to \infty$ and then $m \to \infty$

$$\|\mu_n X - \bar{X}\|_p \leq \|\mu_n X - (\mu_n * \nu_m)X\|_p + \|(\mu_n * \nu_m)X - \bar{X}\|_p$$
$$\leq \|\mu_n - \mu_n * \nu_m\| \|X\|_p + \int \|(\delta_s * \nu_m)X - \bar{X}\|_p \mu_n(ds)$$
$$\leq \|X\|_p \int \|\mu_n - \mu_n * \delta_t\| \nu_m(dt) + \|\nu_m X - \bar{X}\|_p \to 0.$$

(ii) By Corollary 10.19 we may choose some weight functions g_m such that $\xi g_m \to \bar{\xi}$ in L^1. Using Minkowski's inequality, the stationarity of ξ, the invariance of $\bar{\xi}$, and dominated convergence, we get as $n \to \infty$ and then $m \to \infty$

$$\|\xi f_n - \bar{\xi}\|_1 \leq \|\xi f_n - \xi(f_n * g_m)\|_1 + \|\xi(f_n * g_m) - \bar{\xi}\|_1$$
$$\leq E\xi |f_n - f_n * g_m| + \int \|\xi(\delta_s * g_m) - \bar{\xi}\|_1 f_n(s) \, ds$$
$$\leq E\bar{\xi} \int \lambda^d |f_n - \theta_t f_n| g_m(t) \, dt + \|\xi g_m - \bar{\xi}\|_1 \to 0. \quad \square$$

Additional conditions may be needed to ensure L^p-convergence in case (ii) when $\xi B \in L^p$ for bounded sets B. It is certainly enough to require $f_n \leq c g_n$ for some weight functions g_n with $\xi g_n \to \bar{\xi}$ in L^p and some constant $c > 0$.

Our next aim is to prove a subadditive version of Theorem 10.6. For motivation and subsequent needs, we begin with a simple result for nonrandom sequences. Recall that a sequence $c_1, c_2, \cdots \in \mathbb{R}$ is said to be *subadditive* if $c_{m+n} \leq c_m + c_n$ for all $m, n \in \mathbb{N}$.

Lemma 10.21 *(subadditivity)* *For any subadditive sequence $c_1, c_2, \ldots \in \mathbb{R}$, we have*

$$\lim_{n \to \infty} \frac{c_n}{n} = \inf_n \frac{c_n}{n} \in [-\infty, \infty).$$

Proof: Iterating the subadditivity relation, we get for any $k, n \in \mathbb{N}$

$$c_n \leq [n/k]c_k + c_{n-k[n/k]} \leq [n/k]c_k + c_0 \vee \cdots \vee c_{k-1},$$

where $c_0 = 0$. Noting that $[n/k] \sim n/k$ as $n \to \infty$, we get $\limsup_n (c_n/n) \leq c_k/k$ for all k, and so

$$\inf_n \frac{c_n}{n} \leq \liminf_{n \to \infty} \frac{c_n}{n} \leq \limsup_{n \to \infty} \frac{c_n}{n} \leq \inf_n \frac{c_n}{n}. \quad \square$$

We turn to the more general case of a two-dimensional array c_{jk}, $0 \leq j < k$, which is said to be subadditive if $c_{0,n} \leq c_{0,m} + c_{m,n}$ for all $m < n$. The present notion reduces to the previous one when $c_{jk} = c_{k-j}$ for some sequence c_k. We also note that subadditivity holds automatically for arrays of the form $c_{jk} = a_{j+1} + \cdots + a_k$.

We shall now extend the ergodic theorem to subadditive arrays of random variables ξ_{jk}, $0 \leq j < k$. For motivation, we recall from Theorem 10.6 that if $\xi_{jk} = \eta_{j+1} + \cdots + \eta_k$ for some stationary and integrable sequence of random variables η_k, then $\xi_{0,n}/n$ converges a.s. and in L^1. A similar result holds for general subadditive arrays (ξ_{jk}) that are stationary under simultaneous shifts in the two indices, so that $(\xi_{j+1,k+1}) \stackrel{d}{=} (\xi_{j,k})$. To allow for a wider range of applications, we introduce the slightly weaker assumptions

$$(\xi_{k,2k}, \xi_{2k,3k}, \ldots) \stackrel{d}{=} (\xi_{0,k}, \xi_{k,2k}, \ldots), \qquad k \in \mathbb{N}, \qquad (5)$$

$$(\xi_{k,k+1}, \xi_{k,k+2}, \ldots) \stackrel{d}{=} (\xi_{0,1}, \xi_{0,2}, \ldots), \qquad k \in \mathbb{N}. \qquad (6)$$

For convenience of reference, we also restate the subadditivity requirement:

$$\xi_{0,n} \leq \xi_{0,m} + \xi_{m,n}, \qquad 0 < m < n. \qquad (7)$$

Theorem 10.22 (*subadditive ergodic theorem, Kingman*) *Let (ξ_{jk}) be a subadditive array of random variables satisfying (5) and (6), and assume that $E\xi_{0,1}^+ < \infty$. Then $\xi_{0,n}/n$ converges a.s. toward a random variable $\bar{\xi}$ in $[-\infty, \infty)$ with $E\bar{\xi} = \inf_n(E\xi_{0,n}/n) \equiv c$. The same convergence holds in L^1 when $c > -\infty$. If the sequences in (5) are ergodic, then $\bar{\xi}$ is a.s. a constant.*

Proof (Liggett): Put $\xi_{0,n} = \xi_n$ for convenience. By (6) and (7) we have $E\xi_n^+ \leq nE\xi_1^+ < \infty$. We first assume $c > -\infty$, so that the variables $\xi_{m,n}$ are integrable. Iterating (7) gives

$$\frac{\xi_n}{n} \leq \sum_{j=1}^{[n/k]} \xi_{(j-1)k,jk}/n + \sum_{j=k[n/k]+1}^{n} \xi_{j-1,j}/n, \qquad n,k \in \mathbb{N}. \qquad (8)$$

By (5) the sequence $\xi_{(j-1)k,jk}$, $j \in \mathbb{N}$, is stationary for fixed k, and so by Theorem 10.6 we have $n^{-1}\sum_{j\leq n}\xi_{(j-1)k,jk} \to \bar{\xi}_k$ a.s. and in L^1, where $E\bar{\xi}_k = E\xi_k$. Hence, the first term in (8) tends a.s. and in L^1 toward $\bar{\xi}_k/k$. Similarly, $n^{-1}\sum_{j\leq n}\xi_{j-1,j} \to \bar{\xi}_1$ a.s. and in L^1, and so the second term in (8) tends in the same sense to 0. Thus, the right-hand side converges a.s. and in L^1 toward $\bar{\xi}_k/k$, and since k is arbitrary, we get

$$\limsup_n (\xi_n/n) \leq \inf_n(\bar{\xi}_n/n) \equiv \bar{\xi} < \infty \quad \text{a.s.} \qquad (9)$$

The variables ξ_n^+/n are uniformly integrable by Proposition 4.12, and moreover

$$E\limsup_n(\xi_n/n) \leq E\bar{\xi} \leq \inf_n(E\bar{\xi}_n/n) = \inf_n(E\xi_n/n) = c. \qquad (10)$$

To derive a lower bound, let $\kappa_n \perp\!\!\!\perp (\xi_{jk})$ be uniformly distributed over $\{1, \ldots, n\}$ for each n, and define

$$\zeta_k^n = \xi_{\kappa_n, \kappa_n+k}, \quad \eta_k^n = \xi_{\kappa_n+k} - \xi_{\kappa_n+k-1}, \qquad k \in \mathbb{N}.$$

By (6) we have

$$(\zeta_1^n, \zeta_2^n, \ldots) \stackrel{d}{=} (\xi_1, \xi_2, \ldots), \qquad n \in \mathbb{N}. \qquad (11)$$

Moreover, $\eta_k^n \leq \xi_{\kappa_n+k-1,\kappa_n+k} \stackrel{d}{=} \xi_1$ by (6) and (7), and so the variables $(\eta_k^n)^+$ are uniformly integrable. On the other hand, the sequence

$E\xi_1, E\xi_2, \ldots$ is subadditive, and so by Lemma 10.21 we have as $n \to \infty$

$$E\eta_k^n = n^{-1}(E\xi_{n+k} - E\xi_k) \to \inf_n(E\xi_n/n) = c, \quad k \in \mathbb{N}. \quad (12)$$

In particular, $\sup_n E|\eta_k^n| < \infty$, which shows that the sequence $\eta_k^1, \eta_k^2, \ldots$ is tight for each k. Hence, by Theorems 4.29, 5.19, and 6.14, there exist some random variables ζ_k and η_k such that

$$(\zeta_1^n, \zeta_2^n, \ldots; \eta_1^n, \eta_2^n, \ldots) \overset{d}{\to} (\zeta_1, \zeta_2, \ldots; \eta_1, \eta_2, \ldots) \quad (13)$$

along a subsequence. Here $(\zeta_k) \overset{d}{=} (\xi_k)$ by (11), and so by Theorem 6.10 we may assume that $\zeta_k = \xi_k$ for each k.

The sequence $\eta_1, \eta_2, \ldots$ is clearly stationary, and by Lemma 4.11 it is also integrable. From (7) we get

$$\eta_1^n + \cdots + \eta_k^n = \xi_{\kappa_n+k} - \xi_{\kappa_n} \leq \xi_{\kappa_n, \kappa_n+k} = \zeta_k^n,$$

and so in the limit $\eta_1 + \cdots + \eta_k \leq \xi_k$ a.s. Hence, Theorem 10.6 yields

$$\xi_n/n \geq n^{-1} \sum_{k \leq n} \eta_k \to \bar{\eta} \text{ a.s. and in } L^1$$

for some $\bar{\eta} \in L^1$. In particular, the variables ξ_n^-/n are uniformly integrable, and so the same thing is true for ξ_n/n. Using Lemma 4.11 and the uniform integrability of the variables $(\eta_k^n)^+$ together with (10) and (12), we get

$$\begin{aligned} c &= \limsup_n E\eta_1^n \leq E\eta_1 = E\bar{\eta} \\ &\leq E\liminf_n(\xi_n/n) \\ &\leq E\limsup_n(\xi_n/n) \leq E\bar{\xi} \leq c. \end{aligned}$$

Thus, ξ_n/n converges a.s., and by (9) the limit equals $\bar{\xi}$. Furthermore, by Lemma 4.11 the convergence holds even in L^1, and $E\bar{\xi} = c$. If the sequences in (5) are ergodic, then $\bar{\xi}_n = E\xi_n$ a.s. for each n, and we get $\bar{\xi} = c$ a.s.

Now assume instead that $c = -\infty$. Then for each $r \in \mathbb{Z}$, the truncated array $\xi_{m,n} \vee r(n-m)$, $0 \leq m < n$, satisfies the hypotheses of the theorem with c replaced by $c^r = \inf_n(E\xi_n^r/n) \geq r$, where $\xi_n^r = \xi_n \vee rn$. Thus, $\xi_n^r/n = (\xi_n/n) \vee r$ converges a.s. toward some random variable $\bar{\xi}^r$ with mean c^r, and so $\xi_n/n \to \inf_r \bar{\xi}^r \equiv \bar{\xi}$. Finally, $E\bar{\xi} \leq \inf_r c^r = c = -\infty$ by monotone convergence. □

As an application of the last theorem, we may derive a celebrated ergodic theorem for products of random matrices.

Theorem 10.23 *(random matrices, Furstenberg and Kesten) Consider a stationary sequence of random $d \times d$ matrices $X^k = (X_{ij}^k)$ such that $X_{ij}^k > 0$ a.s. and $E|\log X_{ij}^k| < \infty$ for all i and j. Then $n^{-1}\log(X^1 \cdots X^n)_{ij}$ converges a.s. and in L^1 as $n \to \infty$, and the limit is independent of (i,j).*

Proof: First let $i = j = 1$, and define

$$\xi_{m,n} = \log(X^{m+1} \cdots X^n)_{11}, \quad 0 \leq m < n.$$

The array $(-\xi_{m,n})$ is clearly subadditive and jointly stationary, and we have $E|\xi_{0,1}| < \infty$ by hypothesis. Further note that

$$(X^1 \cdots X^n)_{11} \leq d^{n-1} \prod_{k \leq n} \max_{i,j} X_{ij}^k.$$

Hence,

$$\xi_{0,n} - (n-1)\log d \leq \sum_{k \leq n} \log \max_{i,j} X_{ij}^k \leq \sum_{k \leq n} \sum_{i,j} |\log X_{ij}^k|,$$

and so

$$n^{-1} E\xi_{0,n} \leq \log d + \sum_{i,j} E \left|\log X_{i,j}^1\right| < \infty.$$

Thus, by Theorem 10.22 and its proof, there exists an invariant random variable ξ such that $\xi_{0,n}/n \to \xi$ a.s. and in L^1.

To extend the convergence to arbitrary $i, j \in \{1, \ldots, d\}$, we write for any $n \in \mathbb{N}$

$$X_{i1}^2 (X^3 \cdots X^n)_{11} X_{1j}^{n+1} \leq (X^2 \cdots X^{n+1})_{ij}$$
$$\leq (X_{1i}^1 X_{j1}^{n+2})^{-1} (X^1 \cdots X^{n+2})_{11}.$$

Noting that $n^{-1} \log X_{ij}^n \to 0$ a.s. and in L^1 by Theorem 10.6, and using the stationarity of (X^n) and the invariance of ξ, we obtain $n^{-1} \log(X^2 \cdots X^{n+1})_{ij} \to \xi$ a.s. and in L^1. The desired convergence now follows by stationarity. □

We turn to the decomposition of an invariant distribution into ergodic components. For motivation, consider the setting of Theorem 10.6 or 10.14, and assume that S is Borel, to ensure the existence of regular conditional distributions. Writing $\eta = P[\xi \in \cdot | \mathcal{I}_\xi]$, we get

$$\mathcal{L}(\xi) = EP[\xi \in \cdot | \mathcal{I}_\xi] = E\eta = \int m \, P\{\eta \in dm\}. \tag{14}$$

Furthermore,

$$\eta I = P[\xi \in I | \mathcal{I}_\xi] = 1\{\xi \in I\} \text{ a.s.}, \quad I \in \mathcal{I},$$

and so $\eta I = 0$ or 1 a.s. for all $I \in \mathcal{I}$. If we can choose the exceptional null set to be independent of I, it follows that η is a.s. ergodic, and (14) gives the desired ergodic decomposition of $\mu = \mathcal{L}(\xi)$. Though the suggested result is indeed true, the proof requires a different approach.

Proposition 10.24 *(ergodicity by conditioning, Farrell, Varadarajan)*
Let ξ be a random element with distribution μ in a Borel space S, and let $\mathcal{T} = (T_s;\ s \in \mathbb{R}^d)$ be a measurable group of μ-preserving maps on S with invariant σ-field $\mathcal{I}$. Then $\eta = P[\xi \in \cdot | \mathcal{I}_\xi]$ is a.s. invariant and ergodic under $\mathcal{T}$.

For the proof, we fix an increasing sequence of convex sets $B_n \in \mathcal{B}^d$ with $r(R_n) \to \infty$ and introduce on S the probability kernels

$$\mu_n(x, A) = |B_n|^{-1} \int_{B_n} 1_A(T_s x)\, ds, \quad x \in S,\ A \in \mathcal{S},$$

and the associated *empirical distributions* $\eta_n = \mu_n(\xi, \cdot)$. By Theorem 10.14 we note that $\eta_n f \to \eta f$ a.s. for every bounded, measurable function f on S, where $\eta = P[\xi \in \cdot | \mathcal{I}_\xi]$. We say that a class $\mathcal{C} \subset \mathcal{S}$ is *measure-determining* if every probability measure on S is uniquely determined by its values on $\mathcal{C}$.

Lemma 10.25 *(degenerate limit) Let $A_1, A_2, \ldots \in \mathcal{S}$ be measure-determining and such that $\eta_n A_k \to P\{\xi \in A_k\}$ a.s. for each k. Then ξ is ergodic.*

Proof: By Theorem 10.14 we have $\eta_n A \to \eta A \equiv P[\xi \in A | \mathcal{I}_\xi]$ a.s. for every $A \in \mathcal{S}$, and so by comparison $\eta A_k = P\{\xi \in A_k\}$ a.s. for all k. Since the A_k are measure-determining, it follows that $\eta = \mathcal{L}(\xi)$ a.s. Hence, for any $I \in \mathcal{I}$ we have a.s.

$$P\{\xi \in I\} = \eta I = P[\xi \in I | \mathcal{I}_\xi] = 1_I(\xi) \in \{0, 1\},$$

which implies $P\{\xi \in I\} = 0$ or 1. □

Proof of Proposition 10.24: By the stationarity of ξ, we have for any $A \in \mathcal{S}$ and $s \in \mathbb{R}^d$

$$\eta \circ T_s^{-1} A = P[T_s \xi \in A | \mathcal{I}_\xi] = P[\xi \in A | \mathcal{I}_\xi] = \eta A \text{ a.s.}$$

Since S is Borel, we obtain $\eta \circ T_s^{-1} = \eta$ a.s. for every s. Now put $C = [0, 1]^d$, and define $\bar{\eta} = \int_C (\eta \circ T_s^{-1}) ds$. Since η is a.s. invariant under shifts in $\mathbb{Z}^d$, the variable $\bar{\eta}$ is a.s. invariant under arbitrary shifts. Furthermore, by Fubini's theorem,

$$\lambda^d \{s \in [0,1]^d;\ \eta \circ T_s^{-1} = \eta\} = 1 \text{ a.s.,}$$

and therefore $\bar{\eta} = \eta$ a.s. This shows that η is a.s. $\mathcal{T}$-invariant.

Let us now choose a measure-determining sequence $A_1, A_2, \ldots \in \mathcal{S}$, which is possible since S is Borel. Noting that $\eta_n A_k \to \eta A_k$ a.s. for every k by Theorem 10.14, we get by Theorem 6.4

$$\eta \bigcap_k \{x \in S;\ \mu_n(x, A_k) \to \eta A_k\} = P^{\mathcal{I}_\xi} \bigcap_k \{\eta_n A_k \to \eta A_k\} = 1 \text{ a.s.}$$

Since η is a.s. a $\mathcal{T}$-invariant probability measure on S, Lemma 10.25 applies for every $\omega \in \Omega$ outside a P-null set, and we conclude that η is a.s. ergodic. □

We have seen that (14) gives a representation of the distribution $\mu = \mathcal{L}(\xi)$ as a mixture of invariant and ergodic probability measures. The next result shows that this decomposition is unique and characterizes the ergodic measures as extreme points in the convex set of invariant measures.

To explain the terminology, recall that a subset M of a linear space is said to be *convex* if $cm_1 + (1-c)m_2 \in M$ for all $m_1, m_2 \in M$ and $c \in (0,1)$. In that case, we say that $m \in M$ is *extreme* if for any m_1, m_2, and c as above, the relation $m = cm_1 + (1-c)m_2$ implies $m_1 = m_2 = m$. With any set of measures μ on a measurable space $(S, \mathcal{S})$, we associate the σ-field generated by all evaluation maps $\pi_B : \mu \mapsto \mu B$, $B \in \mathcal{S}$.

Theorem 10.26 (*ergodic decomposition, Krylov and Bogolioubov*) *Let $\mathcal{T} = (T_s;\ s \in \mathbb{R}^d)$ be a measurable group of transformations on some Borel space S. Then the $\mathcal{T}$-invariant distributions on S form a convex set M, whose extreme points agree with the ergodic measures in M. Moreover, any measure $\mu \in M$ has a unique representation $\mu = \int m\,\nu(dm)$ with ν restricted to the set of ergodic measures in M.*

Proof: The set M is clearly convex, and by Proposition 10.24 we have for every $\mu \in M$ a representation $\mu = \int m\,\nu(dm)$, where ν is a probability measure on the set of ergodic measures in M. To see that ν is unique, we introduce a regular conditional distribution $\eta = \mu[\cdot|\mathcal{I}]$ a.s. μ on S, and note that $\mu_n A \to \eta A$ a.s. μ for all $A \in \mathcal{S}$ by Theorem 10.14. Thus, for any $A_1, A_2, \cdots \in \mathcal{S}$, we have

$$m\bigcap_k \{x \in S;\ \mu_n(x, A_k) \to \eta(x, A_k)\} = 1 \quad \text{a.e. } \nu.$$

The same relation holds with $\eta(x, A_k)$ replaced by mA_k, since ν is restricted to the class of ergodic measures in M. Assuming the sets A_k to be measure-determining, we conclude that $m\{x;\ \eta(x,\cdot) = m\} = 1$ a.e. ν. Hence, for any measurable set $A \subset M$,

$$\mu\{\eta \in A\} = \int m\{\eta \in A\}\nu(dm) = \int 1_A(m)\nu(dm) = \nu A,$$

which shows that $\nu = \mu \circ \eta^{-1}$.

To prove the equivalence of ergodicity and extremality, fix any measure $\mu \in M$ with ergodic decomposition $\int m\,\nu(dm)$. Let us first assume that μ is extreme. If it is not ergodic, then ν is nondegenerate, and we have $\nu = c\nu_1 + (1-c)\nu_2$ for some $\nu_1 \perp \nu_2$ and $c \in (0,1)$. Since μ is extreme, we obtain $\int m\,\nu_1(dm) = \int m\,\nu_2(dm)$, and so $\nu_1 = \nu_2$ by the uniqueness of the decomposition. The contradiction shows that μ is ergodic.

Next assume μ to be ergodic, so that $\nu = \delta_\mu$, and let $\mu = c\mu_1 + (1-c)\mu_2$ with $\mu_1, \mu_2 \in M$ and $c \in (0,1)$. If $\mu_i = \int m\,\nu_i(dm)$ for $i = 1, 2$, then $\delta_\mu = c\nu_1 + (1-c)\nu_2$ by the uniqueness of the decomposition. Hence, $\nu_1 = \nu_2 = \delta_\mu$, and so $\mu_1 = \mu_2$, which shows that μ is extreme. □

We conclude the chapter with some powerful coupling results that will be needed for our discussion of Palm distributions in Chapter 11 and also for the ergodic theory of Markov processes developed in Chapter 20. First we consider pairs of measurable processes on $\mathbb{R}_+$ with values in an arbitrary measurable space S. In the associated path space, we introduce the invariant σ-field $\mathcal{I}$ and the tail σ-field $\mathcal{T} = \bigcap_t \mathcal{T}_t$, where $\mathcal{T}_t = \sigma(\theta_t)$, and we note that $\mathcal{I} \subset \mathcal{T}$. For any signed measure ν, let $\|\nu\|_{\mathcal{A}}$ denote the total variation of ν on the σ-field $\mathcal{A}$.

Theorem 10.27 *(coupling on $\mathbb{R}_+$, Goldstein, Berbee, Aldous and Thorisson) For any S-valued, measurable processes X and Y on $\mathbb{R}_+$, we have*

(i) $X \stackrel{d}{=} Y$ on $\mathcal{T}$ iff $(\sigma, \theta_\sigma X) \stackrel{d}{=} (\tau, \theta_\tau Y)$ for some random times $\sigma, \tau \geq 0$, and also iff $\|\mathcal{L}(\theta_t X) - \mathcal{L}(\theta_t Y)\| \to 0$ as $t \to \infty$;

(ii) $X \stackrel{d}{=} Y$ on $\mathcal{I}$ iff $\theta_\sigma X \stackrel{d}{=} \theta_\tau Y$ for some random times $\sigma, \tau \geq 0$, and also iff $\|\int_0^1 (\mathcal{L}(\theta_{st} X) - \mathcal{L}(\theta_{st} Y)) ds\| \to 0$ as $t \to \infty$.

If the path space is Borel, we can strengthen the distributional couplings in (i) and (ii) to the a.s. versions $\theta_\tau X = \theta_\tau \tilde{Y}$ and $\theta_\sigma X = \theta_\tau \tilde{Y}$, respectively, for some $\tilde{Y} \stackrel{d}{=} Y$.

Proof of (i): Let μ_1 and μ_2 be the distributions of X and Y, and assume that $\mu_1 = \mu_2$ on $\mathcal{T}$. Write $U = S^{\mathbb{R}_+}$, and define a mapping p on $\mathbb{R}_+ \times U$ by $p(s, x) = (s, \theta_s x)$. Let $\mathcal{C}$ denote the class of all pairs (ν_1, ν_2) of measures on $\mathbb{R}_+ \times U$ such that

$$\nu_1 \circ p^{-1} = \nu_2 \circ p^{-1}, \quad \bar{\nu}_1 \leq \mu_1, \quad \bar{\nu}_2 \leq \mu_2, \tag{15}$$

where $\bar{\nu}_i = \nu_i(\mathbb{R}_+ \times \cdot)$, and regard $\mathcal{C}$ as partially ordered under componentwise inequality. By Corollary 1.16 we note that every linearly ordered subset has an upper bound in $\mathcal{C}$. Hence, Zorn's lemma ensures the existence of a maximal element (ν_1, ν_2).

To see that $\bar{\nu}_1 = \mu_1$ and $\bar{\nu}_2 = \mu_2$, we define $\mu_i' = \mu_i - \bar{\nu}_i$, and conclude from the equality in (15) that

$$\begin{aligned}\|\mu_1' - \mu_2'\|_{\mathcal{T}} &= \|\bar{\nu}_1 - \bar{\nu}_2\|_{\mathcal{T}} \leq \|\bar{\nu}_1 - \bar{\nu}_2\|_{\mathcal{T}_n} \\ &\leq 2\nu_1((n, \infty) \times U) \to 0,\end{aligned} \tag{16}$$

which implies $\mu_1' = \mu_2'$ on $\mathcal{T}$. Next, by Corollary 2.13, there exist some measures $\mu_i^n \leq \mu_i'$ satisfying

$$\mu_1^n = \mu_2^n = \mu_1' \wedge \mu_2' \text{ on } \mathcal{T}_n, \quad n \in \mathbb{N}.$$

Writing $\nu_i^n = \delta_n \otimes \mu_i^n$, we get $\bar{\nu}_i^n \leq \mu_i'$ and $\nu_1^n \circ p^{-1} = \nu_2^n \circ p^{-1}$, and so $(\nu_1 + \nu_1^n, \nu_2 + \nu_2^n) \in \mathcal{C}$. Since (ν_1, ν_2) is maximal, we obtain $\nu_1^n = \nu_2^n = 0$, and so by Corollary 2.9 we have $\mu_1' \perp \mu_2'$ on $\mathcal{T}_n$ for all n. In other words, $\mu_1' A_n = \mu_2' A_n^c = 0$ for some sets $A_n \in \mathcal{T}_n$. But then also $\mu_1' A = \mu_2' A^c = 0$, where $A = \limsup_n A_n \in \mathcal{T}$. Since the μ_i' agree on $\mathcal{T}$, we obtain $\mu_1' = \mu_2' = 0$, which means that $\bar{\nu}_i = \mu_i$. Hence, by Theorem 6.10 there exist some random variables $\sigma, \tau \geq 0$ such that the pairs (σ, X) and (τ, Y) have distributions ν_1 and ν_2, and the desired coupling follows from the equality in (15).

The remaining claims are easy. Thus, the relation $(\sigma, \theta_\sigma X) \stackrel{d}{=} (\tau, \theta_\tau Y)$ implies $\|\mu_1 - \mu_2\|_{\mathcal{T}_n} \to 0$ as in (16), and the latter condition yields $\mu_1 = \mu_2$ on $\mathcal{T}$. When the path space is Borel, the asserted a.s. coupling follows from the distributional version by Theorem 6.10. $\square$

To avoid repetitions, we postpone the proof of part (ii) until after the proof of the next theorem, where we consider a closely related result involving groups G of transformations on an arbitrary measurable space $(S, \mathcal{S})$.

Theorem 10.28 *(group coupling, Thorisson)* *Let the lcscH group G act measurably on a space S, and let ξ and η be random elements in S such that $\xi \stackrel{d}{=} \eta$ on the G-invariant σ-field $\mathcal{I}$. Then $\gamma \xi \stackrel{d}{=} \eta$ for some random element γ in G.*

Proof: Let μ_1 and μ_2 be the distributions of ξ and η. Define $p: G \times S \to S$ by $p(g, s) = gs$ and let $\mathcal{C}$ denote the class of pairs (ν_1, ν_2) of measures on $G \times S$ satisfying (15) with $\bar{\nu}_i = \nu_i(G \times \cdot)$. Using Zorn's lemma as before, we see that $\mathcal{C}$ has a maximal element (ν_1, ν_2), and we claim that $\bar{\nu}_i = \mu_i$ for $i = 1, 2$.

To see this, let λ be a right-invariant Haar measure on G, which exists by Theorem 2.27. Since λ is σ-finite, we may choose a probability measure $\tilde{\lambda} \sim \lambda$ and define

$$\mu'_i = \mu_i - \bar{\nu}_i, \quad \chi_i = \tilde{\lambda} \otimes \mu'_i, \quad i = 1, 2.$$

By Corollary 2.13 there exist some measures $\nu'_i \leq \chi_i$ satisfying

$$\nu'_1 \circ p^{-1} = \nu'_2 \circ p^{-1} = \chi_1 \circ p^{-1} \wedge \chi_2 \circ p^{-1}.$$

Then $\bar{\nu}'_i \leq \mu'_i$ for $i = 1, 2$, and so $(\nu_1 + \nu'_1, \nu_2 + \nu'_2) \in \mathcal{C}$. Since (ν_1, ν_2) is maximal, we have $\nu'_1 = \nu'_2 = 0$, and so $\chi_1 \circ p^{-1} \perp \chi_2 \circ p^{-1}$ by Corollary 2.9. In other words, there exists a set $A_1 = A_2^c \in \mathcal{S}$ such that $\chi_i \circ p^{-1} A_i = 0$ for $i = 1, 2$. Since $\lambda \ll \tilde{\lambda}$, Fubini's theorem gives

$$\int_S \mu'_i(ds) \int_G 1_{A_i}(gs) \lambda(dg) = (\lambda \otimes \mu'_i) \circ p^{-1} A_i = 0. \tag{17}$$

By the right invariance of λ, the inner integral on the left is G-invariant and therefore $\mathcal{I}$-measurable in $s \in S$. Since also $\mu'_1 = \mu'_2$ on $\mathcal{I}$ by (15), equation (17) remains true with A_i replaced by A_i^c. Adding the two formulas gives $\lambda \otimes \mu'_i = 0$, and so $\mu'_i = 0$. Thus, $\bar{\nu}_i = \mu_i$ for $i = 1, 2$.

Since G is Borel, there exist by Theorem 6.10 some random elements σ and τ in G such that (σ, ξ) and (τ, η) have distributions ν_1 and ν_2. By (15) we get $\sigma \xi \stackrel{d}{=} \tau \eta$, and so the same theorem yields a random element $\tilde{\tau}$ in G such that $(\tilde{\tau}, \sigma \xi) \stackrel{d}{=} (\tau, \tau \eta)$. But then $\tilde{\tau}^{-1} \sigma \xi \stackrel{d}{=} \tau^{-1} \tau \eta = \eta$, which proves the desired relation with $\gamma = \tilde{\tau}^{-1} \sigma$. $\square$

Proof of Theorem 10.27 (ii): In the last proof, we replace S by the path space $U = S^{\mathbb{R}_+}$, G by the semigroup of shifts θ_t, $t \geq 0$, and λ by Lebesgue measure on $\mathbb{R}_+$. Assuming $X \stackrel{d}{=} Y$ on $\mathcal{I}$, we may proceed as before up to equation (17), which now takes the form

$$\int_U \mu_i'(dx) \int_0^\infty 1_{A_i}(\theta_t x)\, dt = (\lambda \otimes \mu_i') \circ p^{-1} A_i = 0. \tag{18}$$

Writing $f_i(x)$ for the inner integral on the left, we note that for any $h > 0$

$$f_i(\theta_h x) = \int_h^\infty 1_{A_i}(\theta_t x)\, dt = \int_0^\infty 1_{\theta_h^{-1} A_i}(\theta_t x)\, dt. \tag{19}$$

Hence, (18) remains true with A_i replaced by $\theta_h^{-1} A_i$, and then also for the θ_1-invariant sets

$$\tilde{A}_1 = \limsup_{n\to\infty} \theta_n^{-1} A_1, \qquad \tilde{A}_2 = \liminf_{n\to\infty} \theta_n^{-1} A_2,$$

where $n \to \infty$ along $\mathbb{N}$. Since $\tilde{A}_1 = \tilde{A}_2^c$, we may henceforth assume the A_i in (18) to be θ_1-invariant. Then so are the functions f_i in view of (19). By the monotonicity of $f_i \circ \theta_h$, the f_i are then θ_h-invariant for all $h > 0$ and therefore $\mathcal{I}$-measurable. From this point on, we may argue as before to show that $\theta_\sigma X \stackrel{d}{=} \theta_\tau Y$ for some random variables $\sigma, \tau \geq 0$. The remaining assertions are again routine. □

Exercises

1. State and prove continuous-time, two-sided, and higher-dimensional versions of Lemma 10.1.

2. Consider a stationary random sequence $\xi = (\xi_1, \xi_2, \dots)$. Show that the ξ_n are i.i.d. iff $\xi_1 \perp\!\!\!\perp (\xi_2, \xi_2, \dots)$.

3. Fix a Borel space S, and let X be a stationary array of S-valued random elements in S, indexed by $\mathbb{N}^d$. Show that there exists a stationary array Y indexed by $\mathbb{Z}^d$ such that $X = Y$ a.s. on $\mathbb{N}^d$.

4. Let X be a stationary process on $\mathbb{R}_+$ with values in some Borel space S. Show that there exists a stationary process Y on $\mathbb{R}$ with $X \stackrel{d}{=} Y$ on $\mathbb{R}_+$. Strengthen this to a.s. equality when S is a complete metric space and X is right-continuous.

5. Consider a two-sided, stationary random sequence ξ with restriction η to $\mathbb{N}$. Show that ξ and η are simultaneously ergodic. (*Hint:* For any measurable, invariant set $I \in S^\mathbb{Z}$, there exists some measurable, invariant set $I' \in S^\mathbb{N}$ with $I = S^{\mathbb{Z}_-} \times I'$ a.s. $\mathcal{L}(\xi)$.)

6. Establish two-sided and higher-dimensional versions of Lemmas 10.4 and 10.5 as well as of Theorem 10.9.

7. A measure-preserving transformation T on some probability space $(S, \mathcal{S}, \mu)$ is said to be *mixing* if $\mu(A \cap T^{-n}B) \to \mu A \cdot \mu B$ for all $A, B \in \mathcal{S}$. Prove the counterpart of Lemma 10.5 for mixing. Also, show that any mixing transformation is ergodic. (*Hint:* For the latter assertion, take $A = B$ to be invariant.)

8. Show that it is enough to verify the mixing property for sets in a generating π-system. Use this fact to prove that any i.i.d. sequence is mixing under shifts.

9. Fix any $a \in \mathbb{R}$, and define $Ts = s + a \pmod 1$ on $[0, 1]$. Show that T fails to be mixing but is ergodic iff $a \notin \mathbb{Q}$. (*Hint:* To prove the ergodicity, let $I \subset [0, 1]$ be T-invariant. Then so is the measure $1_I \cdot \lambda$, and since the points ka are dense in $[0, 1]$, it follows that $1_I \cdot \lambda$ is invariant. Now use Theorem 2.6.)

10. (Bohl, Sierpiński, Weyl) For any $a \notin \mathbb{Q}$, let $\mu_n = n^{-1} \sum_{k \leq n} \delta_{ka}$, where ka is defined modulo 1 as a number in $[0, 1]$. Show that $\mu_n \xrightarrow{w} \lambda$. (*Hint:* Apply Theorem 10.6 to the mapping of the previous exercise.)

11. Prove that the transformation $Ts = 2s \pmod 1$ on $[0, 1]$ is mixing. Also show how the mapping of Lemma 3.20 can be generated as in Lemma 10.1 by means of T.

12. Note that Theorem 10.6 remains true for invertible shifts T, with averages taken over increasing index sets $[u_n, b_n]$ with $b_n - a_n \to \infty$. Show by an example that the a.s. convergence may fail without the assumption of monotonicity. (*Hint:* Consider an i.i.d. sequence (ξ_n) and disjoint intervals $[a_n, b_n]$, and use the Borel–Cantelli lemma.)

13. Consider a one- or two-sided stationary random sequence (ξ_n) in some measurable space $(S, \mathcal{S})$, and fix any $B \in \mathcal{S}$. Show that a.s. either $\xi_n \in B^c$ for all n or $\xi_n \in B$ i.o. (*Hint:* Use Theorem 10.6.)

14. (von Neumann) Give a direct proof of the L^2-version of Theorem 10.6. (*Hint:* Define a unitary operator U on $L^2(S)$ by $Uf = f \circ T$. Let M denote the U-invariant subspace of L^2 and put $A = I - U$. Check that $M^\perp = \overline{R}_A$, the closed range of A. By Theorem 1.33 it is enough to take $f \in M$ or $f \in R_A$.) Deduce the general L^p-version, and extend the argument to higher dimensions.

15. In the context of Theorem 10.26, show that the ergodic measures form a measurable subset of M. (*Hint:* Use Lemma 1.41, Proposition 4.31, and Theorem 10.14.)

16. Prove a continuous-time version of Theorem 10.26.

17. Deduce Theorem 4.23 for $p \leq 1$ from Theorem 10.22. (*Hint:* Take $X_{m,n} = |S_n - S_m|^p$, and note that $E|S_n|^p = o(n)$ when $p < 1$.)

18. Let $\xi = (\xi_1, \xi_2, \ldots)$ be a stationary sequence of random variables, fix any $B \in \mathcal{B}(\mathbb{R}^d)$, and let κ_n be the number of indices $k \in \{1, \ldots, n-d\}$ with $(\xi_k, \ldots, \xi_d) \in B$. Prove from Theorem 10.22 that κ_n/n converges a.s. Deduce the same result from Theorem 10.6, by considering suitable subsequences.

19. Show that the inequality in Lemma 10.7 can be strengthened to $E[\xi_1; \sup_n(S_n/n) \geq 0] \geq 0$. (*Hint:* Apply the original result to the variables $\xi_k + \varepsilon$, and let $\varepsilon \to 0$.)

20. Extend Proposition 10.10 to stationary processes on $\mathbb{Z}^d$.

21. Extend Theorem 10.14 to averages over arbitrary rectangles $A_n = [0, a_{n1}] \times \cdots \times [0, a_{nd}]$ such that $a_{nj} \to \infty$ and $\sup_n(a_{ni}/a_{nj}) < \infty$ for all $i \neq j$. (*Hint:* Note that Lemma 10.17 extends to this case.)

22. Derive a version of Theorem 10.14 for stationary processes X on $\mathbb{Z}^d$. (*Hint:* By a suitable randomization, construct an associated stationary process $\tilde{X}$ on $\mathbb{R}^d$, apply Theorem 10.14 to $\tilde{X}$, and estimate the error term as in Corollary 10.19.)

23. Give an example of a stationary, simple point process ξ on $\mathbb{R}^d$ with a.s. infinite sample intensity $\bar{\xi}$.

24. Give an example of two processes X and Y on $\mathbb{R}_+$ such that $X \stackrel{d}{=} Y$ on $\mathcal{I}$ but not on $\mathcal{T}$.

25. Derive a version of Theorem 10.27 for processes on $\mathbb{Z}_+$. Also prove versions for processes on $\mathbb{R}_+^d$ and $\mathbb{Z}_+^d$.

26. Show that Theorem 10.27 (ii) implies a corresponding result for processes on $\mathbb{R}$. (*Hint:* Apply Theorem 10.27 to the processes $\tilde{X}_t = \theta_t X$ and $\tilde{Y} = \theta_t Y$.) Also show how the two-sided statement follows from Theorem 10.28.

27. For processes X on $\mathbb{R}_+$, define $\tilde{X}_t = (X_t, t)$ and let $\tilde{\mathcal{I}}$ be the associated invariant σ-field. Assuming X and Y to be measurable, show that $X \stackrel{d}{=} Y$ on $\mathcal{T}$ iff $\tilde{X} \stackrel{d}{=} \tilde{Y}$ on $\tilde{\mathcal{I}}$. (*Hint:* Use Theorem 10.27.)

28. Prove Lemma 10.15 (ii). (*Hint* (Day): First show that if $S_r \subset B$, then $B + S_\varepsilon \subset (1 + \varepsilon/r)B$, where S_r denotes an r-ball around 0.)

Chapter 11
Special Notions of Symmetry and Invariance

Palm distributions and inversion formulas; stationarity and cycle stationarity; local hitting and conditioning; ergodic properties of Palm measures; exchangeable sequences and processes; strong stationarity and predictable sampling; ballot theorems; entropy and information

This chapter is devoted to some loosely connected topics that are all related to our previous treatment of stationary processes and ergodic theory. We begin with a discussion of Palm distributions of stationary random measures and point processes. In the simplest setting, when ξ is a stationary, simple point process on $\mathbb{R}^d$, we may think of the associated Palm distribution Q_ξ as the conditional distribution, given that ξ has a point at 0. A formal definition is possible when ξ has finite and positive intensity, in which case the mentioned interpretation may be justified by a limit theorem. In the ergodic case, the distributions of the original process and its Palm version agree up to a random shift, which leads to some useful ergodic and averaging relations. Finally, the theory of Palm distributions provides a striking relationship between the notions of stationarity under discrete and continuous shifts.

Asymptotically invariant sampling from a stationary sequence or process leads in the limit to an exchangeable sequence. This is the key observation behind de Finetti's theorem, the fact that exchangeable sequences are mixed i.i.d. It also implies the further equivalence with the notion of spreadability or subsequence invariance, which in turn is equivalent to strong stationarity or invariance in distribution under optional shifts. In the other direction, we consider the striking and useful predictable sampling theorem, the fact that an exchangeable distribution remains invariant under predictable permutations. The latter result will be used in Chapters 13–15 to give simple proofs of the various versions of the arcsine laws.

The chapter concludes with a general so-called ballot theorem for stationary, singular random measures and with a version of the fundamental ergodic theorem of information theory. The former result leads, whenever it applies, to some very precise maximum inequalities, related to those of the preceding chapter and with important applications to queuing theory

11. Special Notions of Symmetry and Invariance

and other areas. The latter result relates ergodic theory to the notion of entropy, of such basic importance in statistical mechanics.

The material in this chapter is related in many ways to other parts of the book. In particular, we may point out some links to various applications and extensions, in Chapters 12, 13, and 16, of results for exchangeable sequences and processes. Furthermore, the predictable sampling theorem is related to some results on random time change appearing in Chapters 18 and 25.

A *random measure* ξ on $\mathbb{R}^d$ is defined as a locally finite kernel from the basic probability space to $\mathbb{R}^d$. It is called a *point process* if ξB is integer-valued for every bounded Borel set B. In the latter case, ξ is said to be *simple* if $\xi\{s\} \leq 1$ for all $s \in \mathbb{R}^d$ outside a fixed P-null set. A more detailed discussion of random measures is given in Chapter 12. We begin the present treatment with a basic general property.

Lemma 11.1 *(zero–infinity law)* *If ξ is a stationary random measure on $\mathbb{R}$ or $\mathbb{Z}$, then $\xi[0,\infty) = \infty$ a.s. on $\{\xi \neq 0\}$.*

Proof: We first consider the case of random measures on $\mathbb{R}$. By the stationarity of ξ and Fatou's lemma, we have for any $t \in \mathbb{R}$ and $h, \varepsilon > 0$

$$\begin{aligned} P\{\xi[t, t+h) > \varepsilon\} &= \limsup_n P\{\xi[(n-1)h, nh) > \varepsilon\} \\ &\leq P\{\xi[(n-1)h, nh) > \varepsilon \text{ i.o.}\} \\ &\leq P\{\xi[0,\infty) = \infty\}. \end{aligned}$$

Letting $\varepsilon \to 0$, $h \to \infty$, and $t \to -\infty$ in this order, we get $P\{\xi \neq 0\} \leq P\{\xi[0,\infty) = \infty\}$. Since trivially $\xi[0,\infty) = \infty$ implies $\xi \neq 0$, we obtain

$$P\{\xi[0,\infty) < \infty, \xi \neq 0\} = P\{\xi \neq 0\} - P\{\xi[0,\infty) = \infty\} \leq 0,$$

and the assertion follows. The result for random measures on $\mathbb{Z}$ may be proved by the same argument with t and h restricted to $\mathbb{Z}$. □

Now consider on $\mathbb{R}^d$ a random measure ξ and a measurable random process X, taking values in an arbitrary measurable space S. We say that ξ and X are *jointly stationary* if $\theta_t(X, \xi) \equiv (\theta_t X, \theta_t \xi) \stackrel{d}{=} (X, \xi)$ for every $t \in \mathbb{R}^d$. By Theorem 2.6 and the stationarity of ξ, we have $E\xi = c\lambda^d$ for some constant $c \in [0, \infty]$, called the *intensity* of ξ, and we note that $c = E\bar{\xi}$, where $\bar{\xi}$ is the sample intensity in Corollary 10.19.

If X and ξ are jointly stationary and ξ has finite and positive intensity, we define the *Palm distribution* $Q_{X,\xi}$ of (X, ξ) with respect to ξ by the formula

$$Q_{X,\xi} f = E \int_B f(\theta_s(X, \xi))\, \xi(ds) / E\xi B, \tag{1}$$

for any set $B \in \mathcal{B}^d$ with $\lambda^d B \in (0, \infty)$ and for measurable functions $f \geq 0$ on $S^{\mathbb{R}^d} \times \mathcal{M}(\mathbb{R}^d)$. The following result shows that the definition is independent of the choice of B.

Lemma 11.2 *(coding) Consider a stationary pair (X,ξ) on $\mathbb{R}^d$, where X is a measurable process in S and ξ is a random measure. Then for any measurable function $f \geq 0$, the stationarity carries over to the random measure*

$$\xi_f B = \int_B f(\theta_s(X,\xi))\,\xi(ds), \quad B \in \mathcal{B}^d.$$

Proof: For any $t \in \mathbb{R}^d$ and $B \in \mathcal{B}^d$, a simple computation gives

$$\begin{aligned}(\theta_t \xi_f) B &= \xi_f(B+t) = \int_{B+t} f(\theta_s(X,\xi))\,\xi(ds) \\ &= \int 1_B(s-t)\,f(\theta_s(X,\xi))\,\xi(ds) \\ &= \int 1_B(u)\,f(\theta_{u+t}(X,\xi))\,\xi(du+t) \\ &= \int_B f(\theta_u \theta_t(X,\xi))\,(\theta_t \xi)(du).\end{aligned}$$

Writing $\xi_f = F(X,\xi)$ and using the stationarity of (X,ξ), we obtain

$$\theta_t \xi_f = F(\theta_t(X,\xi)) \stackrel{d}{=} F(X,\xi) = \xi_f, \quad t \in \mathbb{R}^d. \qquad \Box$$

The mapping in (1) is essentially a one-to-one correspondence, and we proceed to derive some useful inversion formulas. To state the latter, it is suggestive to introduce a random pair (Y,η) with distribution $Q_{X,\xi}$, where in view of (1) the process Y can again be chosen to be measurable. When ξ is a simple point process, then so is η, and we note that $\eta\{0\} = 1$ a.s. The result may then be stated in terms of the associated *Voronoi cells*

$$V_\mu = \{s \in \mathbb{R}^d;\ \mu(S_{|s|} + s) = 0\}, \quad \mu \in \mathcal{N}(\mathbb{R}^d),$$

where $\mathcal{N}(\mathbb{R}^d)$ is the class of locally finite measures on $\mathbb{R}_+$ and S_r denotes the open ball of radius r around the origin. If also $d = 1$, we may enumerate the supporting points of μ in increasing order as $t_n(\mu)$, subject to the convention $t_0(\mu) \leq 0 < t_1(\mu)$. To simplify our statements, we often omit the obvious requirement that the space S and the functions f and g be measurable.

Proposition 11.3 *(uniqueness and inversion) Consider a stationary pair (X,ξ) on $\mathbb{R}^d$, where X is a measurable process in S and ξ is a random measure with $E\bar{\xi} \in (0,\infty)$. Then $P[(X,\xi) \in \cdot | \xi \neq 0]$ is uniquely determined by $\mathcal{L}(Y,\eta) = Q_{X,\xi}$, and the following inversion formulas hold:*

(i) *For any $f \geq 0$ and $g > 0$ with $\lambda^d g < \infty$,*

$$E[f(X,\xi);\ \xi \neq 0] = E\bar{\xi} \cdot E \int \frac{f(\theta_s(Y,\eta))}{(\theta_s \eta)g}\,g(-s)\,ds.$$

(ii) *If ξ is a simple point process, we have for any $f \geq 0$*

$$E[f(X,\xi);\ \xi \neq 0] = E\bar{\xi} \cdot E \int_{V_\eta} f(\theta_s(Y,\eta))\,ds.$$

(iii) If ξ is a simple point process and $d = 1$, we have for any $f \geq 0$
$$E[f(X,\xi); \xi \neq 0] = E\bar{\xi} \cdot E \int_0^{t_1(\eta)} f(\theta_s(Y,\eta))\, ds.$$

To express the conditional distribution $P[f(X,\xi) \in \cdot | \xi \neq 0]$ in terms of $\mathcal{L}(Y,\eta)$, it suffices in each case to divide by the corresponding formula for $f \equiv 1$. The latter equation also expresses $P\{\xi \neq 0\}/E\bar{\xi}$ in terms of $\mathcal{L}(\eta)$. In particular, this ratio equals $E|V_\eta|$ in case (ii) and $Et_1(\eta)$ in case (iii).

Proof: (i) Write (1) in the form
$$E\bar{\xi} \cdot \lambda^d B \cdot E f(Y,\eta) = E \int_B f(\theta_s(X,\xi))\, \xi(ds), \quad B \in \mathcal{B}^d,$$
and extend by a monotone class argument to
$$E\bar{\xi} \cdot E \int h(Y,\eta,s)\, ds = E \int h(\theta_s(X,\xi), s)\, \xi(ds),$$
for any measurable function $h \geq 0$ on the appropriate product space. Applying the latter formula to the function $h(x,\mu,s) = f(\theta_{-s}(x,\mu), s)$ for measurable $f \geq 0$ and substituting $-s$ for s, we get
$$E\bar{\xi} \cdot E \int f(\theta_s(Y,\eta), -s)\, ds = E \int f(X,\xi,s)\, \xi(ds). \quad (2)$$
In particular, we have for measurable $g, h \geq 0$
$$E\bar{\xi} \cdot E \int h(\theta_s(Y,\eta))\, g(-s)\, ds = E\, h(X,\xi)\, \xi g.$$
If $g > 0$ with $\lambda^d g < \infty$, then $\xi g < \infty$ a.s., and the desired relation follows by the further substitution
$$h(x,\mu) = \frac{f(x,\mu)}{\mu g} 1\{\mu g > 0\}.$$

(ii) Here we may apply (2) to the function
$$h(x,\mu,s) = f(x,\mu)\, 1\{\mu\{s\} = 1,\ \mu S_{|s|} = 0\},$$
and note that $(\theta_s\eta)S_{|-s|} = 0$ iff $s \in V_\eta$.

(iii) In this case, we apply (2) to the function
$$h(x,\mu,s) = f(x,\mu)\, 1\{t_0(\mu) = s\},$$
and note that $t_0(\theta_s\eta) = -s$ iff $s \in [0, t_1(\eta))$. □

Now consider a simple point process η on $\mathbb{R}$ and a measurable process Y on $\mathbb{R}$ with values in an arbitrary measurable space $(S, \mathcal{S})$. We say that the pair (Y, η) is *cycle-stationary* if $\eta\{0\} = 1$ and $t_1(\eta) < \infty$ a.s., and if in addition $\theta_{t_1(\eta)}(Y,\eta) \stackrel{d}{=} (Y,\eta)$. The variables $t_n(\eta)$ are then a.s. finite, and the successive differences $\Delta t_n(\eta) = t_{n+1}(\eta) - t_n(\eta)$ along with the shifted processes $Y^n = \theta_{t_n(\eta)} Y$ form a stationary sequence in the space

$(0, \infty) \times S^\mathbb{R}$. The following result gives a striking relationship between the notions of stationarity and cycle stationarity for pairs (X, ξ) and (Y, η).

When $d = 1$ and $\xi \neq 0$ a.s., the definition (1) of the Palm distribution and the inversion formula in Proposition 11.3 (iii) reduce to the nearly symmetric equations

$$Ef(Y, \eta) = E \int_0^1 f(\theta_s(X, \xi)) \, \xi(ds) \big/ E\bar{\xi}, \qquad (3)$$

$$Ef(X, \xi) = E \int_0^{t_1(\eta)} f(\theta_s(Y, \eta)) \, ds \big/ E t_1(\eta). \qquad (4)$$

Theorem 11.4 *(cycle stationarity, Kaplan)* *Equations (3) and (4) provide a one-to-one correspondence between the distributions of all stationary pairs (X, ξ) on $\mathbb{R}$ and all cycle-stationary ones (Y, η), where X and Y are measurable processes in S, and ξ and η are simple point processes with $\xi \neq 0$ a.s., $E\bar{\xi} < \infty$, and $E t_1(\eta) < \infty$.*

Proof: First assume that (X, ξ) is stationary with $\xi \neq 0$ and $E\bar{\xi} < \infty$, put $\sigma_k = t_k(\xi)$, and define $\mathcal{L}(Y, \eta)$ by (3). Then for any $n \in \mathbb{N}$ and for bounded, measurable $f \geq 0$, we have

$$n \, E\bar{\xi} \cdot Ef(Y, \eta) = E \int_0^n f(\theta_s(X, \xi)) \, \xi(ds) = E \sum_{\sigma_k \in (0, n)} f(\theta_{\sigma_k}(X, \xi)).$$

Writing $\tau_k = t_k(\eta)$, we get by a suitable substitution

$$n \, E\bar{\xi} \cdot Ef(\theta_{\tau_1}(Y, \eta)) = E \sum_{\sigma_k \in (0, n)} f(\theta_{\sigma_{k+1}}(X, \xi)),$$

and so by subtraction,

$$|Ef(\theta_{\tau_1}(Y, \eta)) - Ef(Y, \eta)| \leq \frac{2\|f\|}{n \, E\bar{\xi}}.$$

As $n \to \infty$, we obtain $Ef(\theta_{\tau_1}(Y, \eta)) = Ef(Y, \eta)$, and therefore $\theta_{\tau_1}(Y, \eta) \stackrel{d}{=} (Y, \eta)$, which means that (Y, η) is cycle-stationary. Also note that (4) holds in this case by Proposition 11.3.

Next assume that (Y, η) is cycle-stationary with $E t_1(\eta) < \infty$, and define $\mathcal{L}(X, \xi)$ by (4). Then for n and f as before,

$$n \, E\tau_1 \cdot Ef(X, \xi) = E \int_0^{\tau_n} f(\theta_s(Y, \eta)) \, ds,$$

and so for any $t \in \mathbb{R}$,

$$n \, E\tau_1 \cdot Ef(\theta_t(X, \xi)) = E \int_0^{\tau_n} f(\theta_{s+t}(Y, \eta)) \, ds = E \int_t^{\tau_n + t} f(\theta_s(Y, \eta)) \, ds.$$

Hence, by subtraction,

$$|Ef(\theta_t(X, \xi)) - Ef(X, \xi)| \leq \frac{2|t| \, \|f\|}{n \, E\tau_1}.$$

As $n \to \infty$, we get $Ef(\theta_t(X,\xi)) = Ef(X,\xi)$, and so $\theta_t(X,\xi) \stackrel{d}{=} (X,\xi)$, which means that (X,ξ) is stationary.

To see that (X,ξ) and (Y,η) are related by (3), we introduce a possibly unbounded measure space with integration operator $\tilde{E}$ and a random pair $(\tilde{Y},\tilde{\eta})$ satisfying

$$\tilde{E}f(\tilde{Y},\tilde{\eta}) = E\int_0^1 f(\theta_s(X,\xi))\,\xi(ds). \tag{5}$$

Proceeding as in the proof of Proposition 11.3, except that the monotone class argument requires some extra care since $E\tilde{\xi}$ may be infinite, we obtain

$$\tilde{E}\int_0^{t_1(\tilde{\eta})} f(\theta_s(\tilde{Y},\tilde{\eta}))\,ds = Ef(X,\xi) = E\int_0^{t_1(\eta)} f(\theta_s(Y,\eta))\,ds\Big/Et_1(\eta).$$

Replacing $f(x,\mu)$ by $f(\theta_{t_0(\mu)}(x,\mu))$ and noting that $t_0(\theta_s\mu) = -s$ when $\mu\{0\} = 1$ and $s \in [0, t_1(\mu))$, we get

$$\tilde{E}[t_1(\tilde{\eta})f(\tilde{Y},\tilde{\eta})] = E[t_1(\eta)f(Y,\eta)]/Et_1(\eta).$$

Hence, by a suitable substitution,

$$\tilde{E}f(\tilde{Y},\tilde{\eta}) = Ef(Y,\eta)/Et_1(\eta).$$

Inserting this into (5) and dividing by the same formula for $f \equiv 1$, we obtain the required equation. $\square$

When ξ is a simple point process on $\mathbb{R}^d$, we may think of the Palm distribution $Q_{X,\xi}$ as the conditional distribution of (X,ξ), given that $\xi\{0\} = 1$. The interpretation is justified by the following result, which also provides an asymptotic formula for the hitting probabilities of small Borel sets. By $B_n \to 0$ we mean that $\sup\{|s|; s \in B_n\} \to 0$, and we write $\|\cdot\|$ for the total variation norm.

Theorem 11.5 *(local hitting and conditioning, Korolyuk, Ryll-Nardzewski, König, Matthes) Consider a stationary pair (X,ξ) on $\mathbb{R}^d$, where X is a measurable process in S and ξ is a simple point process with $E\tilde{\xi} \in (0,\infty)$. Let $B_1, B_2, \dots \in \mathcal{B}^d$ with $|B_n| > 0$ and $B_n \to 0$, and let f be bounded, measurable, and shift-continuous. On $\{\xi B_n = 1\}$, let σ_n denote the unique point of ξ in B_n. Then*

(i) $P\{\xi B_n = 1\} \sim P\{\xi B_n > 0\} \sim E\xi B_n$;
(ii) $\|P[\theta_{\sigma_n}(X,\xi) \in \cdot | \xi B_n = 1] - Q_{X,\xi}\| \to 0$;
(iii) $E[f(X,\xi)|\xi B_n > 0] \to Q_{X,\xi}f$.

Proof: (i) Since $\eta\{0\} = 1$ a.s., we have $(\theta_s\eta)B_n > 0$ for all $s \in -B_n$. Hence, Proposition 11.3 (ii) yields

$$\frac{P\{\xi B_n > 0\}}{E\tilde{\xi}} = E\int_{V_\eta} 1\{(\theta_s\eta)B_n > 0\}\,ds \geq E|V_\eta \cap (-B_n)|.$$

Dividing by $|B_n|$ and using Fatou's lemma, we obtain

$$\liminf_{n\to\infty} \frac{P\{\xi B_n > 0\}}{E\xi B_n} \geq \liminf_{n\to\infty} \frac{E|V_\eta \cap (-B_n)|}{|B_n|}$$

$$\geq E\liminf_{n\to\infty} \frac{|V_\eta \cap (-B_n)|}{|B_n|} = 1,$$

which implies

$$\liminf_{n\to\infty} \frac{P\{\xi B_n = 1\}}{E\xi B_n} \geq 2\liminf_{n\to\infty} \frac{P\{\xi B_n > 0\}}{E\xi B_n} - 1 \geq 1.$$

The converse relations are obvious since

$$P\{\xi B_n = 1\} \leq P\{\xi B_n > 0\} \leq E\xi B_n.$$

(ii) Introduce on $S^{\mathbb{R}^d} \times \mathcal{N}(\mathbb{R}^d)$ the measures

$$\mu_n = E\int_{B_n} 1\{\theta_s(X,\xi) \in \cdot\}\, \xi(ds),$$

$$\nu_n = P[\theta_{\sigma_n}(X,\xi) \in \cdot;\ \xi B_n = 1],$$

and put $m_n = E\xi B_n$ and $p_n = P\{\xi B_n = 1\}$. By (1) the stated total variation becomes

$$\left\|\frac{\nu_n}{p_n} - \frac{\mu_n}{m_n}\right\| \leq \left\|\frac{\nu_n}{p_n} - \frac{\nu_n}{m_n}\right\| + \left\|\frac{\nu_n}{m_n} - \frac{\mu_n}{m_n}\right\|$$

$$\leq p_n\left|\frac{1}{p_n} - \frac{1}{m_n}\right| + \frac{1}{m_n}|p_n - m_n| = 2\left|1 - \frac{p_n}{m_n}\right|,$$

which tends to 0 in view of (i).

(iii) Here we write

$$|E[f(X,\xi)|\xi B_n > 0] - Q_{X,\xi}f|$$
$$\leq |E[f(X,\xi)|\xi B_n > 0] - E[f(X,\xi)|\xi B_n = 1]|$$
$$+ |E[f(X,\xi) - f(\theta_{\sigma_n}(X,\xi))|\xi B_n = 1]|$$
$$+ |E[f(\theta_{\sigma_n}(X,\xi))|\xi B_n = 1] - Q_{X,\xi}f|.$$

By (i) and (ii) the first and last terms on the right tend to 0 as $n \to \infty$. To estimate the second term, we introduce on $S^{\mathbb{R}^d} \times \mathcal{N}(\mathbb{R}^d)$ the bounded, measurable functions

$$g_\varepsilon(x,\mu) = \sup_{|s|<\varepsilon} |f(\theta_s(x,\mu)) - f(x,\mu)|, \quad \varepsilon > 0,$$

and conclude from (ii) that for large enough n

$$|E[f(X,\xi) - f(\theta_{\sigma_n}(X,\xi))|\xi B_n = 1]|$$
$$\leq E[g_\varepsilon(\theta_{\sigma_n}(X,\xi))|\xi B_n = 1] \to Q_{X,\xi}g_\varepsilon.$$

Since also $Q_{X,\xi}g_\varepsilon \to 0$ by dominated convergence as $\varepsilon \to 0$, the desired convergence follows. □

11. Special Notions of Symmetry and Invariance 209

We turn to a general ergodic theorem for Palm distributions. Given a bounded measure $\nu \neq 0$ on $\mathbb{R}^d$ and a positive or bounded, measurable function f on $S^{\mathbb{R}^d} \times \mathcal{M}(\mathbb{R}^d)$, we introduce the average

$$\bar{f}_\nu(x,\mu) = \int f(\theta_s(x,\mu))\,\nu(ds)/\|\nu\|, \quad x \in S^{\mathbb{R}^d},\ \mu \in \mathcal{M}(\mathbb{R}^d),$$

where x is understood to be a measurable function on $\mathbb{R}^d$, to ensure the existence of the integral. When $\nu = 0$, we take $\bar{f}_\nu = 0$. Let us say that the *weight functions* (probability densities) $g_1, g_2, \ldots$ on $\mathbb{R}^d$ are *asymptotically invariant* if the corresponding property holds for the associated measures $g_n \cdot \lambda^d$. For convenience, we may sometimes write $g \cdot \mu = g\mu$.

Theorem 11.6 *(pointwise averages)* *Consider a stationary and ergodic pair* (X, ξ) *on* $\mathbb{R}^d$, *where* X *is a measurable process in* S *and* ξ *is a random measure with* $\bar{\xi} \in (0, \infty)$ *a.s. Let* $\mathcal{L}(Y, \eta) = Q_{X,\xi}$. *Then for any bounded, measurable function* f *and asymptotically invariant distributions* μ_n *or weight functions* g_n *on* $\mathbb{R}^d$, *we have*

(i) $\bar{f}_{\mu_n}(Y, \eta) \xrightarrow{P} Ef(X, \xi)$;

(ii) $\bar{f}_{g_n\xi}(X, \xi) \xrightarrow{P} Ef(Y, \eta)$.

The same convergence holds a.s. when $\mu_n = 1_{B_n} \cdot \lambda^d$ *or* $g_n = 1_{B_n}$, *respectively, for some bounded, convex sets* $B_1 \subset B_2 \subset \cdots$ *in* $\mathcal{B}^d$ *with* $r(B_n) \to \infty$.

We can give a short and transparent proof by using the general shift coupling in Theorem 10.28. Since the latter result applies directly only when the sample intensity $\bar{\xi}$ is a constant (which holds in particular when ξ is ergodic), we need to replace the Palm distribution $Q_{X,\xi}$ in (1) by a suitably modified version $Q'_{X,\xi}$, given for $f \geq 0$ and $B \in \mathcal{B}^d$ with $|B| \in (0, \infty)$ by

$$Q'_{X,\xi} f = E \int_B f(\theta_s(X,\xi))\,\xi(ds)/\bar{\xi}|B|,$$

whenever $\bar{\xi} \in (0, \infty)$ a.s. If ξ is ergodic, we note that $\bar{\xi} = E\bar{\xi}$ a.s., and therefore $Q'_{X,\xi} = Q_{X,\xi}$. As previously for $Q_{X,\xi}$, it is both suggestive and convenient to introduce a random pair (Z, ζ) with distribution $Q'_{X,\xi}$.

Lemma 11.7 *(shift coupling, Thorisson)* *Consider a stationary pair* (X, ξ) *on* $\mathbb{R}^d$, *where* X *is a measurable process in* S *and* ξ *is a random measure with* $\bar{\xi} \in (0, \infty)$ *a.s. Let* $\mathcal{L}(Z, \zeta) = Q'_{X,\xi}$. *Then there exist some random vectors* σ *and* τ *in* $\mathbb{R}^d$ *such that*

$$(X, \xi) \stackrel{d}{=} \theta_\sigma(Z, \zeta), \qquad (Z, \zeta) \stackrel{d}{=} \theta_\tau(X, \xi).$$

The result suggests that we think of $Q'_{X,\xi}$ as the distribution of (X, ξ) shifted to a "typical" point of ξ. Note that this interpretation fails for $Q_{X,\xi}$ in general.

Proof: Write $\mathcal{I}$ for the shift-invariant σ-field in the measurable path space of (X, ξ), and put $\mathcal{I}_{X,\xi} = (X, \xi)^{-1}\mathcal{I}$. Letting $B = [0,1]^d$ and noting that

$E[\xi B|\mathcal{I}_{X,\xi}] = \bar{\xi}$, we get for any $I \in \mathcal{I}$

$$P\{(Z,\zeta) \in I\} = E\int_B 1_I(\theta_s(X,\xi))\,\xi(ds)/\bar{\xi}$$
$$= E[\xi B/\bar{\xi};\, (X,\xi) \in I] = P\{(X,\xi) \in I\},$$

which shows that $(X,\xi) \stackrel{d}{=} (Z,\zeta)$ on $\mathcal{I}$. Both assertions now follow from Theorem 10.28. □

Proof of Theorem 11.6: (i) By Lemma 11.7 we may assume that $(Y,\eta) = \theta_\tau(X,\xi)$ for some random element τ in $\mathbb{R}^d$. Using Corollary 10.20 (i) and the asymptotic invariance of μ_n, we get

$$|\bar{f}_{\mu_n}(Y,\eta) - Ef(X,\xi)|$$
$$\leq \|\mu_n - \theta_\tau \mu_n\|\,\|f\| + |\bar{f}_{\mu_n}(X,\xi) - Ef(X,\xi)| \stackrel{P}{\to} 0.$$

The a.s. version follows in the same way from Theorem 10.14.

(ii) Let ξ_f be the stationary and ergodic random measure in Lemma 11.2. Applying Corollary 10.20 (ii) to both ξ and ξ_f and using (1), we obtain

$$\bar{f}_{g_n\xi}(X,\xi) = \frac{\xi_f g_n}{\lambda^d g_n}\frac{\lambda^d g_n}{\xi g_n} \stackrel{P}{\to} \frac{\bar{\xi}_f}{\bar{\xi}} = \frac{E\xi_f B}{E\xi B} = Ef(Y,\eta).$$

For the pointwise version, we may use Corollary 10.19 instead. □

Taking expected values in Theorem 11.6, we get for bounded f the formulas

$$E\bar{f}_{\mu_n}(Y,\eta) \to Ef(X,\xi), \qquad E\bar{f}_{g_n\xi}(X,\xi) \to Ef(Y,\eta),$$

which may be interpreted as limit theorems for suitable space averages of the distributions $\mathcal{L}(X,\xi)$ and $\mathcal{L}(Y,\eta)$. We shall prove the less obvious fact that both relations hold uniformly for bounded f. For a striking formulation, we may introduce the possibly defective distributions $\overline{\mathcal{L}}_\mu(X,\xi)$ and $\overline{\mathcal{L}}_{g\xi}(X,\xi)$, given for measurable functions $f \geq 0$ by

$$\overline{\mathcal{L}}_\mu(X,\xi)f = E\bar{f}_\mu(X,\xi), \qquad \overline{\mathcal{L}}_{g\xi}(X,\xi)f = E\bar{f}_{g\xi}(X,\xi).$$

Theorem 11.8 *(distributional averages, Slivnyak, Zähle)* Consider a stationary pair (X,ξ) on $\mathbb{R}^d$, where X is a measurable process in S and ξ is a random measure with $\bar{\xi} \in (0,\infty)$ a.s. Let $\mathcal{L}(Z,\zeta) = Q'_{X,\xi}$. Then for any asymptotically invariant distributions μ_n or weight functions g_n on $\mathbb{R}^d$,

(i) $\|\overline{\mathcal{L}}_{\mu_n}(Z,\zeta) - \mathcal{L}(X,\xi)\| \to 0$;
(ii) $\|\overline{\mathcal{L}}_{g_n\xi}(X,\xi) - \mathcal{L}(Z,\zeta)\| \to 0$.

Proof: (i) By Lemma 11.7 we may assume that $(Z,\zeta) = \theta_\tau(X,\xi)$. Using Fubini's theorem and the stationarity of (X,ξ), we get for any measurable function $f \geq 0$

$$\overline{\mathcal{L}}_{\mu_n}(X,\xi)f = \int Ef(\theta_s(X,\xi))\,\mu_n(ds) = Ef(X,\xi) = \mathcal{L}(X,\xi)f.$$

Hence, by Fubini's theorem and dominated convergence,

$$\begin{aligned}
\|\overline{\mathcal{L}}_{\mu_n}(Z,\zeta) - \mathcal{L}(X,\xi)\| &= \|\overline{\mathcal{L}}_{\mu_n}(\theta_\tau(X,\xi)) - \overline{\mathcal{L}}_{\mu_n}(X,\xi)\| \\
&\leq E\left\|\int 1\{\theta_s(X,\xi) \in \cdot\}(\mu_n - \theta_\tau\mu_n)(ds)\right\| \\
&\leq E\|\mu_n - \theta_\tau\mu_n\| \to 0.
\end{aligned}$$

(ii) Letting $0 \leq f \leq 1$ and defining ξ_f as in Lemma 11.2, we get

$$\xi_f g_n = \int f(\theta_s(X,\xi))\,g_n(s)\,\xi(ds) \leq \xi g_n.$$

Interpreting $\xi_f g_n / \xi g_n$ as 0 when $\xi g_n = 0$, we obtain

$$\begin{aligned}
|\overline{\mathcal{L}}_{g_n\xi}(X,\xi)f - \mathcal{L}(Z,\zeta)| &= |E\bar{f}_{g_n\xi}(X,\xi) - Ef(Z,\zeta)| \\
&\leq E\left|\frac{\xi_f g_n}{\xi g_n} - \frac{\xi_f g_n}{\bar{\xi}}\right| \leq E\left|1 - \frac{\xi g_n}{\bar{\xi}}\right|.
\end{aligned}$$

Here $\xi g_n / \bar{\xi} \xrightarrow{P} 1$ by Corollary 10.20, and moreover

$$E(\xi g_n / \bar{\xi}) = E(E[\xi g_n | \mathcal{I}_{X,\xi}]/\bar{\xi}) = E(\bar{\xi}/\bar{\xi}) = 1.$$

Hence, Proposition 4.12 yields $\xi g_n / \bar{\xi} \to 1$ in L^1, and the assertion follows. $\square$

To motivate our next main topic, we consider a simple limit theorem for multivariate sampling from a stationary process. Here we consider a measurable process X on some index set T, taking values in a space S, and let $\tau = (\tau_1, \tau_2, \ldots)$ be an independent sequence of random elements in T with joint distribution μ. We may then form the associated sampling sequence $\xi = X \circ \tau$ in S^∞, given by

$$\xi = (\xi_1, \xi_2, \ldots) = (X_{\tau_1}, X_{\tau_2}, \ldots)$$

and referred to below as a *sample from X with distribution μ*. The sampling distributions $\mu_1, \mu_2, \ldots$ on T^∞ are said to be *asymptotically invariant* if their projections onto T^k are asymptotically invariant for every $k \in \mathbb{N}$. Recall that $\mathcal{I}_X$ denotes the invariant σ-field of X, and note that the conditional distribution $\eta = P[X_0 \in \cdot | \mathcal{I}_X]$ exists by Theorem 6.3 when S is Borel.

Lemma 11.9 *(asymptotically invariant sampling)* *Let X be a stationary and measurable process on $T = \mathbb{R}$ or $\mathbb{Z}$ with values in a Polish space S, and form $\xi_1, \xi_2, \ldots$ by sampling from X with some asymptotically invariant distributions $\mu_1, \mu_2, \ldots$ on T^∞. Then $\xi_n \xrightarrow{d} \xi$ in S^∞, where $\mathcal{L}(\xi) = E\eta^\infty$ with $\eta = P[X_0 \in \cdot | \mathcal{I}_X]$.*

Proof: Write $\xi = (\xi^k)$ and $\xi_n = (\xi_n^k)$. Fix any asymptotically invariant distributions $\nu_1, \nu_2, \ldots$ on T, and let $f_1, \ldots, f_m$ be measurable functions on S bounded by ± 1. Proceeding as in the proof of Corollary 10.20 (i), we get

$$\left| E\prod_k f_k(\xi_n^k) - E\prod_k f_k(\xi^k)\right|$$
$$\leq E\left|\mu_n \bigotimes_k f_k(X) - \prod_k \eta f_k\right|$$
$$\leq \|\mu_n - \mu_n * \nu_r^m\| + \int E\left|(\nu_r^m * \delta_t)\bigotimes_k f_k(X) - \prod_k \eta f_k\right|\mu_n(dt)$$
$$\leq \int \|\mu_n - \mu_n * \delta_t\|\nu_r^m(dt) + \sum_k \sup_t E|(\nu_r * \delta_t)f_k(X) - \eta f_k|.$$

Using the asymptotic invariance of μ_n and ν_r together with Corollary 10.20 (i) and dominated convergence, we see that the right-hand side tends to 0 as $n \to \infty$ and then $r \to \infty$. The assertion now follows by Theorem 4.29. □

The last result leads immediately to a version of *de Finetti's theorem*, the fact that infinite exchangeable sequences are mixed i.i.d. For a precise statement, consider any finite or infinite random sequence $\xi = (\xi_1, \xi_2, \dots)$ with index set I, and say that ξ is *exchangeable* if

$$(\xi_{k_1}, \xi_{k_2}, \dots) \stackrel{d}{=} (\xi_1, \xi_2, \dots) \tag{6}$$

for any finite permutation $(k_1, k_2, \dots)$ of I. (Here a permutation is said to be *finite* if it affects only finitely many elements). For infinite sequences ξ we also consider the formally weaker property of *spreadability*, where (6) is required for all strictly increasing sequences $k_1 < k_2 < \cdots$. Note that ξ is then stationary and that any sample from ξ with strictly increasing sampling times $\tau_1, \tau_2, \dots$ has the same distribution as ξ. By Lemma 11.9 we conclude that $\mathcal{L}(\xi) = E\eta^\infty$ with $\eta = P[\xi_1 \in \cdot | \mathcal{I}_\xi]$. Below we give a slightly stronger conditional statement. Recall that for any random measure η on a measurable space $(S, \mathcal{S})$, the associated σ-field is generated by the random variables ηB for arbitrary $B \in \mathcal{S}$.

Theorem 11.10 (*exchangeable sequences, de Finetti, Ryll-Nardzewski*) *For any infinite random sequence ξ in a Borel space S, the following conditions are equivalent:*

(i) ξ *is exchangeable;*

(ii) ξ *is spreadable;*

(iii) $P[\xi \in \cdot | \eta] = \eta^\infty$ *a.s. for some random distribution η on S.*

The random measure η is then a.s. unique and equals $P[\xi_1 \in \cdot | \mathcal{I}_\xi]$.

Since η^∞ is a.s. the distribution of an i.i.d. sequence in S based on the measure η, we may state condition (iii) in words by saying that ξ is *conditionally i.i.d.* Taking expectations of both sides in (iii), we obtain the seemingly weaker condition $\mathcal{L}(\xi) = E\eta^\infty$, which says that ξ is *mixed i.i.d.* Now the latter condition implies that ξ is exchangeable, and so, by the stated theorem, the two versions of (iii) are in fact equivalent.

11. Special Notions of Symmetry and Invariance

Proof: Since S is Borel, we may assume that $S = [0,1]$. Letting μ_n be the uniform distribution on the product set $\times_k \{(k-1)n+1, \ldots, kn\}$ and using the spreadability of ξ, we see from Lemma 11.9 that $P\{\xi \in \cdot\} = E\eta^\infty$. More generally, consider any invariant Borel set $I \subset S^\infty$, and note that (6) extends to

$$(1_I(\xi), \xi_{k_1}, \xi_{k_2}, \ldots) \stackrel{d}{=} (1_I(\xi), \xi_1, \xi_2, \ldots), \quad k_1 < k_2 < \cdots .$$

Applying Lemma 11.9 to the sequence of pairs $(\xi_k, 1_I(\xi))$, we get as before $P\{\xi \in \cdot \cap I\} = E[\eta^\infty; \xi \in I]$, and since η is $\mathcal{I}_\xi$-measurable, it follows that $P[\xi \in \cdot | \eta] = \eta^\infty$ a.s..

To see that η is unique, we may use the law of large numbers and Theorem 6.4 to obtain

$$n^{-1} \sum_{k \leq n} 1_B(\xi_k) \to \eta B \quad \text{a.s.,} \quad B \in \mathcal{S}. \qquad \square$$

The statement of Theorem 11.10 is clearly false for finite sequences. To rescue the result in the finite case, we need to replace the inherent i.i.d. sequences by so-called *urn sequences*, generated by successive drawing without replacement from a finite set. For a precise statement, fix any measurable space S, and consider a measure of the form $\mu = \sum_{k \leq n} \delta_{s_k}$ with $s_1, \ldots, s_n \in S$. The associated *factorial measure* $\mu^{(n)}$ on S^n is defined by

$$\mu^{(n)} = \sum_p \delta_{s \circ p},$$

where the summation extends over all permutations $p = (p_1, \ldots, p_n)$ of $1, \ldots, n$, and we write $s \circ p = (s_{p_1}, \ldots, s_{p_n})$. Note that $\mu^{(n)}$ is independent of the order of $s_1, \ldots, s_n$ and is measurable as a function of μ.

Lemma 11.11 *(finite exchangeable sequences)* *Let $\xi_1, \ldots, \xi_n$ be random elements in some measurable space, and put $\xi = (\xi_k)$ and $\eta = \sum_k \delta_{\xi_k}$. Then ξ is exchangeable iff $P[\xi \in \cdot | \eta] = \eta^{(n)}/n!$ a.s.*

Proof: Since η is invariant under permutations of $\xi_1, \ldots, \xi_n$, we note that $(\xi \circ p, \eta) \stackrel{d}{=} (\xi, \eta)$ for any permutation p of $1, \ldots, n$. Now introduce an exchangeable permutation $\pi \perp\!\!\!\perp \xi$ of $1, \ldots, n$. Using Fubini's theorem twice, we get for any measurable sets A and B in appropriate spaces

$$\begin{aligned} P\{\xi \in B, \eta \in A\} &= P\{\xi \circ \pi \in B, \eta \in A\} \\ &= E[P[\xi \circ \pi \in B|\xi]; \eta \in A] \\ &= E[(n!)^{-1}\eta^{(n)} B; \eta \in A]. \end{aligned} \qquad \square$$

Just as for the martingale and Markov properties, even the notions of exchangeability and spreadability may be related to a filtration $\mathcal{F} = (\mathcal{F}_n)$. Thus, a finite or infinite sequence of random elements $\xi = (\xi_1, \xi_2, \ldots)$ is said to be $\mathcal{F}$-*exchangeable* if it is $\mathcal{F}$-adapted and such that, for every $n \geq 0$, the shifted sequence $\theta_n \xi = (\xi_{n+1}, \xi_{n+2}, \ldots)$ is conditionally exchangeable given $\mathcal{F}_n$. For infinite sequences ξ, the definition of $\mathcal{F}$-*spreadability* is similar. (Since both definitions may be stated without reference to regular conditional distributions, no restrictions are needed on S.) When $\mathcal{F}$ is the filtration induced by ξ, the stated properties reduce to the unqualified versions considered earlier.

For an infinite sequence ξ, we define *strong stationarity* or $\mathcal{F}$-*stationarity* by the condition $\theta_\tau \xi \stackrel{d}{=} \xi$ for every finite optional time $\tau \geq 0$. By the *prediction sequence* of ξ we mean the set of conditional distributions

$$\pi_n = P[\theta_n \xi \in \cdot | \mathcal{F}_n], \quad n \in \mathbb{Z}_+. \tag{7}$$

The random probability measures $\pi_0, \pi_1, \ldots$ on S are said to form a *measure-valued martingale* if $(\pi_n B)$ is a real-valued martingale for every measurable set $B \subset S$.

The next result shows that strong stationarity is equivalent to exchangeability; it also exhibits an interesting connection with martingale theory.

Lemma 11.12 *(strong stationarity) Let ξ be an infinite, $\mathcal{F}$-adapted random sequence in a Borel space S, and let π denote the prediction sequence of ξ. Then these conditions are equivalent:*

(i) ξ *is $\mathcal{F}$-exchangeable;*

(ii) ξ *is $\mathcal{F}$-spreadable;*

(iii) ξ *is $\mathcal{F}$-stationary;*

(iv) π *is a measure-valued $\mathcal{F}$-martingale.*

Proof: Conditions (i) and (ii) are equivalent by Theorem 11.10. Assuming (ii), we get a.s. for any $B \in \mathcal{S}^\infty$ and $n \in \mathbb{Z}_+$

$$E[\pi_{n+1} B | \mathcal{F}_n] = P[\theta_{n+1} \xi \in B | \mathcal{F}_n] = P[\theta_n \xi \in B | \mathcal{F}_n] = \pi_n B, \tag{8}$$

which proves (iv). Conversely, (ii) follows by iteration from the second equality in (8), and so (ii) and (iv) are equivalent.

Next we note that (7) extends by Lemma 6.2 to

$$\pi_\tau B = P[\theta_\tau \xi \in B | \mathcal{F}_\tau] \text{ a.s.}, \quad B \in \mathcal{S}^\infty,$$

for any finite optional time τ. By Lemma 7.13 it follows that (iv) is equivalent to

$$P\{\theta_\tau \xi \in B\} = E\pi_\tau B = E\pi_0 B = P\{\xi \in B\}, \quad B \in \mathcal{S}^\infty,$$

which in turn is equivalent to (iii). □

We next aim to show how the exchangeability property extends to a wide class of *random* transformations. For a precise statement, we say that an integer-valued random variable τ is *predictable* with respect to a given filtration $\mathcal{F}$ if the shifted time $\tau - 1$ is $\mathcal{F}$-optional.

11. Special Notions of Symmetry and Invariance

Theorem 11.13 *(predictable sampling)* Let $\xi = (\xi_1, \xi_2, \ldots)$ be a finite or infinite, $\mathcal{F}$-exchangeable random sequence, and let $\tau_1, \ldots, \tau_n$ be a.s. distinct $\mathcal{F}$-predictable times in the index set of ξ. Then

$$(\xi_{\tau_1}, \ldots, \xi_{\tau_n}) \stackrel{d}{=} (\xi_1, \ldots, \xi_n). \tag{9}$$

Of special interest is the case of *optional skipping*, when $\tau_1 < \tau_2 < \cdots$. If $\tau_k \equiv \tau + k$ for some optional time $\tau < \infty$, then (9) reduces to the strong stationarity of Lemma 11.12. In general, we require neither ξ to be infinite nor the τ_k to be increasing.

For both applications and proof, it is useful to introduce the associated *allocation sequence*

$$\alpha_j = \inf\{k; \tau_k = j\}, \quad j \in I,$$

where I is the index set of ξ. Note that any finite value of α_j gives the position of j in the permuted sequence (τ_k). The random times τ_k are clearly predictable iff the α_j form a predictable sequence in the sense of Chapter 7.

Proof of Theorem 11.13: First let ξ be indexed by $I = \{1, \ldots, n\}$, so that $(\tau_1, \ldots, \tau_n)$ and $(\alpha_1, \ldots, \alpha_n)$ are mutually inverse random permutations of I. For each $m \in \{0, \ldots, n\}$, put $\alpha_j^m = \alpha_j$ for all $j \leq m$, and define recursively

$$\alpha_{j+1}^m = \min(I \setminus \{\alpha_1^m, \ldots, \alpha_j^m\}), \quad m \leq j \leq n.$$

Then $(\alpha_1^m, \ldots, \alpha_n^m)$ is a predictable and $\mathcal{F}_{m-1}$-measurable permutation of $1, \ldots, n$. Since also $\alpha_j^m = \alpha_j^{m-1} = \alpha_j$ whenever $j < m$, Theorem 6.4 yields for any bounded measurable functions $f_1, \ldots, f_n$ on S

$$\begin{aligned}
E \prod_j f_{\alpha_j^m}(\xi_j) &= E E\Big[\prod_j f_{\alpha_j^m}(\xi_j) \Big| \mathcal{F}_{m-1}\Big] \\
&= E \prod_{j<m} f_{\alpha_j^m}(\xi_j) E\Big[\prod_{j\geq m} f_{\alpha_j^m}(\xi_j) \Big| \mathcal{F}_{m-1}\Big] \\
&= E \prod_{j<m} f_{\alpha_j^{m-1}}(\xi_j) E\Big[\prod_{j\geq m} f_{\alpha_j^{m-1}}(\xi_j) \Big| \mathcal{F}_{m-1}\Big] \\
&= E \prod_j f_{\alpha_j^{m-1}}(\xi_j).
\end{aligned}$$

Summing over $m \in \{1, \ldots, n\}$ and noting that $\alpha_j^n = \alpha_j$ and $\alpha_j^0 = j$ for all j, we get

$$E \prod_k f_k(\xi_{\tau_k}) = E \prod_j f_{\alpha_j}(\xi_j) = E \prod_k f_k(\xi_k),$$

which extends to (9) by a monotone class argument.

Next assume that $I = \{1, \ldots, m\}$ with $m > n$. We may then extend the sequence (τ_k) to I by recursively defining

$$\tau_{k+1} = \min(I \setminus \{\tau_1, \ldots, \tau_k\}), \quad k \geq n, \tag{10}$$

so that $\tau_1, \ldots, \tau_m$ form a random permutation of I. Using (10), we see by induction that the times $\tau_{n+1}, \ldots, \tau_m$ are again predictable. Hence, the previous case applies, and (9) follows.

Finally, assume that $I = \mathbb{N}$. For each $m \in \mathbb{N}$, we introduce the predictable times
$$\tau_k^m = \tau_k 1\{\tau_k \leq m\} + (m+k)1\{\tau_k > m\}, \quad k = 1, \ldots, n,$$
and conclude from the previous version of (9) that
$$(\xi_{\tau_1^m}, \ldots, \xi_{\tau_n^m}) \stackrel{d}{=} (\xi_1, \ldots, \xi_n). \tag{11}$$
As $m \to \infty$, we have $\tau_k^m \to \tau_k$, and (9) follows from (11) by dominated convergence. $\square$

The last result yields a simple proof of yet another basic property of random walks in $\mathbb{R}$, a striking relation between the first maximum and the number of positive values. The latter result will in turn lead to simple proofs of the arcsine laws in Theorems 13.16 and 14.11.

Corollary 11.14 (*sojourns and maxima, Sparre-Andersen*) *Let $\xi_1, \ldots, \xi_n$ be exchangeable random variables, and put $S_k = \xi_1 + \cdots + \xi_k$. Then*
$$\sum_{k \leq n} 1\{S_k > 0\} \stackrel{d}{=} \min\{k \geq 0; \; S_k = \max_{j \leq n} S_j\}.$$

Proof: Put $\tilde{\xi}_k = \xi_{n-k+1}$ for $k = 1, \ldots, n$, and note that the $\tilde{\xi}_k$ remain exchangeable for the filtration $\mathcal{F}_k = \sigma\{S_n, \tilde{\xi}_1, \ldots, \tilde{\xi}_k\}$, $k = 0, \ldots, n$. Write $\tilde{S}_k = \tilde{\xi}_1 + \cdots + \tilde{\xi}_k$, and introduce the predictable permutation
$$\alpha_k = \sum_{j=0}^{k-1} 1\{\tilde{S}_j < S_n\} + (n-k+1)1\{\tilde{S}_{k-1} \geq S_n\}, \quad k = 1, \ldots, n.$$

Define $\xi_k' = \sum_j \tilde{\xi}_j 1\{\alpha_j = k\}$ for $k = 1, \ldots, n$, and conclude from Theorem 11.13 that $(\xi_k') \stackrel{d}{=} (\xi_k)$. Writing $S_k' = \xi_1' + \cdots + \xi_k'$, we further note that
$$\min\{k \geq 0; \; S_k' = \max_j S_j'\} = \sum_{j=0}^{n-1} 1\{\tilde{S}_j < S_n\} = \sum_{k=1}^{n} 1\{S_k > 0\}. \quad \square$$

Turning to the case of continuous time, we say that a process X in some topological space is *continuous in probability* if $X_s \to X_t$ as $s \to t$. An $\mathbb{R}^d$-valued process X on $\mathbb{R}_+$ is said to be *exchangeable* or *spreadable* if it is continuous in probability with $X_0 = 0$ and such that the increments $X_t - X_s$ over any set of disjoint intervals $(s, t]$ of equal length form an exchangeable or spreadable sequence. Finally, we say that X has *conditionally stationary and independent increments*, given some σ-field $\mathcal{I}$, if the stated property is conditionally true for any finite collection of intervals.

The following continuous-time version of Theorem 11.10 characterizes the exchangeable processes on $\mathbb{R}_+$. We postpone the much harder finite-interval case until Theorem 16.21. The point process case is treated separately by different methods in Theorem 12.12.

Theorem 11.15 *(exchangeable processes on $\mathbb{R}_+$, Bühlmann) Let the process X on $\mathbb{R}_+$ be $\mathbb{R}^d$-valued and continuous in probability with $X_0 = 0$. Then X is spreadable iff it has conditionally stationary and independent increments, given some σ-field $\mathcal{I}$.*

Proof: The sufficiency being obvious, it suffices to show that the stated condition is necessary. Thus, assume that X is spreadable. Then the increments ξ_{nk} over the dyadic intervals $I_{nk} = 2^{-n}(k-1, k]$ are spreadable for fixed n, and so by Theorem 11.10 they are conditionally i.i.d. η_n for some random probability measure η_n on $\mathbb{R}$. Using Corollary 3.12 and the uniqueness in Theorem 11.10, we obtain

$$\eta_n^{*2^{n-m}} = \eta_m \text{ a.s.}, \quad m < n. \tag{12}$$

Thus, for any $m < n$, the increments ξ_{mk} are conditionally i.i.d. η_m, given η_n. Since the σ-fields $\sigma(\eta_n)$ are a.s. nondecreasing by (12), Theorem 7.23 shows that the ξ_{mk} remain conditionally i.i.d. η_m, given $\mathcal{I} \equiv \sigma\{\eta_0, \eta_1, \dots\}$.

Now fix any disjoint intervals $I_1, \dots, I_n$ of equal length with associated increments $\xi_1, \dots, \xi_n$. Here we may approximate by disjoint intervals $I_1^m, \dots, I_n^m$ of equal length with dyadic endpoints. For each m, the associated increments ξ_k^m are conditionally i.i.d., given $\mathcal{I}$. Thus, for any bounded, continuous functions $f_1, \dots, f_n$,

$$E^{\mathcal{I}} \prod_{k \leq n} f_k(\xi_k^m) = \prod_{k \leq n} E^{\mathcal{I}} f_k(\xi_k^m) = \prod_{k \leq n} E^{\mathcal{I}} f_k(\xi_1^m). \tag{13}$$

Since X is continuous in probability, we have $\xi_k^m \xrightarrow{P} \xi_k$ for each k, so (13) extends by dominated convergence to the original variables ξ_k. By suitable approximation and monotone class arguments, we may finally extend the relations to any measurable indicator functions $f_k = 1_{B_k}$. □

We turn to an interesting relationship between the sample intensity $\bar{\xi}$ of a stationary random measure ξ on $\mathbb{R}_+$ and the corresponding maximum over increasing intervals. It is interesting to compare with the more general but less precise maximum inequalities in Proposition 10.10 and Lemmas 10.11 and 10.17. For the need of certain applications, we also consider the case of random measures ξ on $[0,1)$. Here $\bar{\xi} = \xi[0,1)$ by definition, and stationarity is defined as before in terms of the shifts θ_t on $[0,1)$, where $\theta_t s = s + t \pmod{1}$ and correspondingly for sets and measures. Recall that ξ is *singular* if its absolutely continuous component vanishes. This holds in particular for purely atomic measures ξ.

Theorem 11.16 *(ballot theorem)* Let ξ be a stationary and a.s. singular random measure on $\mathbb{R}_+$ or $[0,1)$. Then there exists a $U(0,1)$ random variable $\sigma \perp\!\!\!\perp \mathcal{I}_\xi$ such that

$$\sigma \sup_{t>0} t^{-1}\xi[0,t] = \bar{\xi} \quad a.s. \tag{14}$$

To justify the statement, we note that singularity is a measurable property of a measure μ. Indeed, by Proposition 2.21, it is equivalent that the function $F_t = \mu[0,t]$ be singular. Now it is easy to check that the singularity of F can be described by countably many conditions, each involving the increments of F over finitely many intervals with rational endpoints.

Proof: If ξ is stationary on $[0,1)$, then the periodic continuation $\eta = \sum_{n \leq 0} \theta_n \xi$ is clearly stationary on $\mathbb{R}_+$, and moreover $\mathcal{I}_\eta = \mathcal{I}_\xi$ and $\bar{\eta} = \bar{\xi}$. We may also use the elementary inequality

$$\frac{x_1 + \cdots + x_n}{t_1 + \cdots + t_n} \leq \max_{k \leq n} \frac{x_k}{t_k}, \quad n \in \mathbb{N},$$

valid for arbitrary $x_1, x_2, \cdots \geq 0$ and $t_1, t_2, \cdots > 0$, to see that $\sup_t t^{-1}\eta[0,t] = \sup_t t^{-1}\xi[0,t]$. It is then enough to consider random measures on $\mathbb{R}_+$.

In that case, put $X_t = \xi(0,t]$ and define

$$A_t = \inf_{s \geq t}(s - X_s), \quad \alpha_t = 1\{A_t = t - X_t\}, \quad t \geq 0. \tag{15}$$

Noting that $A_t \leq t - X_t$ and using the monotonicity of X, we get for any $s < t$

$$\begin{aligned} A_s &= \inf_{r \in [s,t)}(r - X_r) \wedge A_t \geq (s - X_t) \wedge A_t \\ &\geq (s - t + A_t) \wedge A_t = s - t + A_t. \end{aligned}$$

If A_0 is finite, then so is A_t for every t, and we obtain by subtraction

$$0 \leq A_t - A_s \leq t - s \quad \text{on } \{A_0 > -\infty\}, \quad s < t. \tag{16}$$

Thus, A is nondecreasing and absolutely continuous on $\{A_0 > -\infty\}$.

Now fix a singular path of X such that A_0 is finite, and let $t \geq 0$ be such that $A_t < t - X_t$. Then $A_t + X_{t\pm} < t$ by monotonicity, and so, by the left and right continuity of A and X, there exists some $\varepsilon > 0$ such that

$$A_s + X_s < s - 2\varepsilon, \quad |s - t| < \varepsilon.$$

Then by (16),

$$s - X_s > A_s + 2\varepsilon > A_t + \varepsilon, \quad |s - t| < \varepsilon,$$

and by (15) it follows that $A_s = A_t$ for $|s-t| < \varepsilon$. In particular, A has derivative $A'_t = 0 = \alpha_t$ at t.

We turn to the complementary set $D = \{t \geq 0; A_t = t - X_t\}$. By Theorem 2.15 both A and X are differentiable a.e., the latter with derivative 0, and we form a set D' by excluding the corresponding null sets. We may also exclude the at most countably many isolated points of D. Then for any $t \in D'$ we may choose some $t_n \to t$ in $D \setminus \{t\}$. By the definition of D,

$$\frac{A_{t_n} - A_t}{t_n - t} = 1 - \frac{X_{t_n} - X_t}{t_n - t}, \quad n \in \mathbb{N},$$

and as $n \to \infty$ we get $A'_t = 1 = \alpha_t$. Combining this with the result in the previous case gives $A' = \alpha$ a.e., and since A is absolutely continuous, we conclude from Theorem 2.15 that

$$A_t - A_0 = \int_0^t \alpha_s ds \quad \text{on } \{A_0 > -\infty\}, \quad t \geq 0. \tag{17}$$

Now recall that $X_t/t \to \bar{\xi}$ a.s. as $t \to \infty$ by Corollary 10.19. When $\bar{\xi} < 1$, we see from (15) that $-\infty < A_t/t \to 1 - \bar{\xi}$ a.s. Also

$$\begin{aligned} A_t + X_t - t &= \inf_{s \geq t}((s-t) - (X_s - X_t)) \\ &= \inf_{s \geq 0}(s - \theta_t \xi(0, s]), \end{aligned}$$

and hence

$$\alpha_t = 1\{\inf_{s \geq 0}(s - \theta_t \xi(0, s]) = 0\}, \quad t \geq 0.$$

Dividing (17) by t and using Corollary 10.9, we get a.s. on $\{\bar{\xi} < 1\}$

$$\begin{aligned} P[\sup_{t>0}(X_t/t) \leq 1 | \mathcal{I}_\xi] &= P[\sup_{t>0}(X_t - t) = 0 | \mathcal{I}_\xi] \\ &= P[A_0 = 0 | \mathcal{I}_\xi] \\ &= E[\alpha_0 | \mathcal{I}_\xi] = 1 - \bar{\xi}. \end{aligned}$$

Replacing ξ by $r\xi$ and taking complements, we obtain more generally

$$P[r \sup_{t>0}(X_t/t) > 1 | \mathcal{I}_\xi] = r\bar{\xi} \wedge 1 \quad \text{a.s.}, \quad r \geq 0, \tag{18}$$

where the result for $r\bar{\xi} \in [1, \infty)$ follows by monotonicity.

When $\bar{\xi} \in (0, \infty)$, we may simply define σ by (14); if instead $\bar{\xi} = 0$ or ∞, we take $\sigma = \vartheta$, where ϑ is $U(0, 1)$ and independent of ξ. Note that (14) remains true in the latter case, since $\xi = 0$ a.s. on $\{\bar{\xi} = 0\}$ and $X_t/t \to \infty$ a.s. on $\{\bar{\xi} = \infty\}$. To verify the distributional claim, we conclude from (18) and Theorem 6.4 that, on $\{\bar{\xi} \in (0, \infty)\}$,

$$P[\sigma < r | \mathcal{I}_\xi] = P[r \sup_t (X_t/t) > \bar{\xi} | \mathcal{I}_\xi] = r \wedge 1 \quad \text{a.s.}, \quad r \geq 0.$$

Since the same relation holds trivially when $\bar{\xi} = 0$ or ∞, we see that σ is conditionally $U(0, 1)$ given $\mathcal{I}_\xi$, which means that σ is $U(0, 1)$ and independent of $\mathcal{I}_\xi$. □

From the last theorem we may easily deduce a corresponding discrete-time result. Here (14) holds only with inequality and will be supplemented by a sharp relation similar to (18). For a stationary sequence $\xi = (\xi_1, \xi_2, \ldots)$ in $\mathbb{R}_+$ with invariant σ-field $\mathcal{I}_\xi$, we define $\bar{\xi} = E[\xi_1 | \mathcal{I}_\xi]$ a.s. On $\{1, \ldots, n\}$ we define stationarity in the obvious way in terms of addition modulo n, and we put $\bar{\xi} = n^{-1} \sum_k \xi_k$.

Corollary 11.17 *(discrete-time ballot theorem)* Let $\xi = (\xi_1, \xi_2, \dots)$ be a finite or infinite, stationary sequence of $\mathbb{R}_+$-valued random variables, and put $S_k = \sum_{j \leq k} \xi_j$. Then there exists a $U(0,1)$ random variable $\sigma \perp\!\!\!\perp \mathcal{I}_\xi$ such that

$$\sigma \sup\nolimits_{k>0} (S_k/k) \leq \bar{\xi} \quad a.s. \tag{19}$$

If the ξ_k are $\mathbb{Z}_+$-valued, we have also

$$P[\sup\nolimits_{k>0} (S_k - k) \geq 0 | \mathcal{I}_\xi] = \bar{\xi} \wedge 1 \quad a.s. \tag{20}$$

Proof: Arguing by periodic continuation as before, we may reduce to the case of infinite sequences ξ. Now let ϑ be $U(0,1)$ and independent of ξ, and define $X_t = S_{[t+\vartheta]}$. Then X has stationary increments, and we note that also $\mathcal{I}_X = \mathcal{I}_\xi$ and $\bar{X} = \bar{\xi}$. By Theorem 11.16 there exists some $U(0,1)$ random variable $\sigma \perp\!\!\!\perp \mathcal{I}_X$ such that a.s.

$$\sup_{k>0} (S_k/k) = \sup_{t>0} (S_{[t]}/t) \leq \sup_{t>0} (X_t/t) = \bar{\xi}/\sigma.$$

If the ξ_k are $\mathbb{Z}_+$-valued, the same result yields a.s.

$$\begin{aligned} P[\sup\nolimits_{k>0} (S_k - k) \geq 0 | \mathcal{I}_\xi] &= P[\sup\nolimits_{t \geq 0} (X_t - t) > 0 | \mathcal{I}_\xi] \\ &= P[\sup\nolimits_{t>0} (X_t/t) > 1 | \mathcal{I}_\xi] \\ &= P[\bar{\xi} > \sigma | \mathcal{I}_\xi] = \bar{\xi} \wedge 1. \end{aligned}$$ $\square$

To state the next result, consider a random element ξ in a countable space S, and put $p_j = P\{\xi = j\}$. Given an arbitrary σ-field $\mathcal{F}$, we define the *information* $I(j)$ and the *conditional information* $I(j|\mathcal{F})$ by

$$I(j) = -\log p_j, \quad I(j|\mathcal{F}) = -\log P[\xi = j|\mathcal{F}], \quad j \in S.$$

For motivation, we note the additivity property

$$I(\xi_1, \dots, \xi_n) = I(\xi_1) + I(\xi_2|\xi_1) + \cdots + I(\xi_n|\xi_1, \dots, \xi_{n-1}), \tag{21}$$

valid for any random elements $\xi_1, \dots, \xi_n$ in S. Next we form the associated *entropy* $H(\xi) = EI(\xi)$ and *conditional entropy* $H(\xi|\mathcal{F}) = EI(\xi|\mathcal{F})$, and note that

$$H(\xi) = EI(\xi) = -\sum\nolimits_j p_j \log p_j.$$

From (21) we see that even H is additive, in the sense that

$$H(\xi_1, \dots, \xi_n) = H(\xi_1) + H(\xi_2|\xi_1) + \cdots + H(\xi_n|\xi_1, \dots, \xi_{n-1}). \tag{22}$$

If the ξ_n form a stationary and ergodic sequence such that $H(\xi_0) < \infty$, we show that the averages of the terms in (21) and (22) converge toward a common limit.

11. Special Notions of Symmetry and Invariance

Theorem 11.18 *(entropy and information, Shannon, McMillan, Breiman, Ionescu Tulcea)* *Let $\xi = (\xi_k)$ be a stationary and ergodic sequence in a countable space S such that $H(\xi_0) < \infty$. Then*
$$n^{-1} I(\xi_1, \ldots, \xi_n) \to H(\xi_0 | \xi_{-1}, \xi_{-2}, \ldots) \quad \text{a.s. and in } L^1.$$

Note that the condition $H(\xi_0) < \infty$ holds automatically when the state space is finite. Our proof will be based on a technical estimate.

Lemma 11.19 *(maximum inequality, Chung, Neveu)* *For any countably valued random variable ξ and discrete filtration $(\mathcal{F}_n)$, we have*
$$E \sup_n I(\xi | \mathcal{F}_n) \le H(\xi) + 1.$$

Proof: Write $p_j = P\{\xi = j\}$ and $\eta = \sup_n I(\xi | \mathcal{F}_n)$. For fixed $r > 0$, we introduce the optional times
$$\tau_j = \inf\{n; \, I(j | \mathcal{F}_n) > r\} = \inf\{n; \, P[\xi = j | \mathcal{F}_n] < e^{-r}\}, \quad j \in S.$$

By Lemma 6.2,
$$\begin{aligned} P\{\eta > r, \xi = j\} &= P\{\tau_j < \infty, \xi = j\} \\ &= E[P[\xi = j | \mathcal{F}_{\tau_j}]; \tau_j < \infty] \\ &\le e^{-r} P\{\tau_j < \infty\} \le e^{-r}. \end{aligned}$$

Since the left-hand side is also bounded by p_j, Lemma 3.4 yields
$$\begin{aligned} E\eta &= \sum_j E[\eta; \xi = j] = \sum_j \int_0^\infty P\{\eta > r, \xi = j\} \, dr \\ &\le \sum_j \int_0^\infty (e^{-r} \wedge p_j) \, dr \\ &= \sum_j p_j (1 - \log p_j) = H(\xi) + 1. \quad \square \end{aligned}$$

Proof of Theorem 11.18 (Breiman): We may assume ξ to be defined on the canonical space S^∞. Then introduce the functions
$$g_k(\xi) = I(\xi_0 | \xi_{-1}, \ldots, \xi_{-k+1}), \quad g(\xi) = I(\xi_0 | \xi_{-1}, \xi_{-2}, \ldots).$$

By (21) we may write the assertion in the form
$$n^{-1} \sum_{k \le n} g_k(\theta^k \xi) \to E g(\xi) \quad \text{a.s. and in } L^1. \tag{23}$$

Here $g_k(\xi) \to g(\xi)$ a.s. by martingale convergence and $E \sup_n g_k(\xi) < \infty$ by Lemma 11.19. Hence, (23) follows by Corollary 10.8. $\square$

Exercises

1. Show that Lemma 11.1 can be strengthened to $\liminf_t t^{-1} \xi[0, t] > 0$ a.s. on $\xi \ne 0$. (*Hint:* Use Corollary 10.19.)

2. Let Ξ be a stationary random set in $\mathbb{R}$. Show that $\sup \Xi = \infty$ a.s. on $\Xi \neq \emptyset$. (*Hint:* Use Lemma 11.1, or prove the result by a similar argument.)

3. For (X, ξ) as in Lemma 11.2, define on the appropriate product space a random measure $\tilde{\xi}$ by $\tilde{\xi}(A \times B) = \int_B 1_A(\theta_s(X, \xi))\xi(ds)$. Show that $\tilde{\xi}$ is again stationary under shifts in $\mathbb{R}^d$.

4. Prove Theorem 11.5 (i) by an elementary argument when the B_n are intervals in $\mathbb{R}^d$. (*Hint:* If an interval I is partitioned for each n into subintervals I_{nj} with $\max_j |I_{nj}| \to 0$, then $\sum_j 1\{\xi I_{nj} = 1\} \to \xi I$ a.s. Now take expected values and use dominated convergence.)

5. In the context of Theorem 11.8, show that $Q_{X,\xi} = Q'_{X,\xi}$ iff $\bar{\xi} = E\bar{\xi} \in (0, \infty)$ a.s. Also give examples where $Q_{X,\xi}$ exists while $Q'_{X,\xi}$ does not, and conversely.

6. Let μ_c be the distribution of the sequence $\tau_1 < \tau_2 < \cdots$, where $\xi = \sum_j \delta_{\tau_j}$ is a stationary Poisson process on $\mathbb{R}_+$ with rate $c > 0$. Show that the μ_c are asymptotically invariant as $c \to 0$.

7. Show by an example that a finite, exchangeable sequence need not be mixed i.i.d.

8. Let the random sequence ξ be conditionally i.i.d. η. Show that ξ is ergodic iff η is a.s. nonrandom.

9. Let ξ and η be random probability measures on some Borel space such that $E\xi^\infty = E\eta^\infty$. Show that $\xi \stackrel{d}{=} \eta$. (*Hint:* Use the law of large numbers.)

10. Let $\xi_1, \xi_2, \ldots$ be spreadable random elements in some Borel space S. Prove the existence of a measurable function $f: [0,1]^2 \to S$ and some i.i.d. $U(0,1)$ random variables $\vartheta_0, \vartheta_1, \ldots$ such that $\xi_n = f(\vartheta_0, \vartheta_n)$ a.s. for all n. (*Hint:* Use Lemma 3.22, Proposition 6.13, and Theorems 6.10 and 11.10.)

11. Let $\xi = (\xi_1, \xi_2, \ldots)$ be an $\mathcal{F}$-spreadable random sequence in some Borel space S. Prove the existence of some random measure η such that, for each $n \in \mathbb{Z}_+$, the sequence $\theta^n \xi$ is conditionally i.i.d. η, given $\mathcal{F}_n$ and η.

12. Let $\xi_1, \ldots, \xi_n$ be exchangeable random variables, fix a Borel set B, and let $\tau_1 < \cdots < \tau_\nu$ be the indices $k \in \{1, \ldots, n\}$ with $\sum_{j<k} \xi_j \in B$. Construct a random vector $(\eta_1, \ldots, \eta_n) \stackrel{d}{=} (\xi_1, \ldots, \xi_n)$ such that $\xi_{\tau_k} = \eta_k$ a.s. for all $k \leq \nu$. (*Hint:* Extend the sequence (τ_k) to $k \in (\nu, n]$, and apply Theorem 11.13.)

13. Prove a version of Corollary 11.14 for the *last* maximum.

14. State and prove a continuous-time version of Lemma 11.12. (If no regularity conditions are imposed on the exchangeable processes of Theorem 11.15, we need to consider optional times taking countably many values.)

15. Anticipating the theory of Lévy processes in Chapter 15, show that any exchangeable process on $\mathbb{R}_+$ as in Theorem 11.15 has a version with rcll paths.

11. Special Notions of Symmetry and Invariance

16. Show by an example that the conclusion of Theorem 11.16 may fail when ξ is not singular.

17. Give an example where the inequality in Corollary 11.17 is a.s. strict. (*Hint:* Examine the proof.)

18. (Bertrand, André) Show that if two candidates A and B in an election get the proportions p and $1-p$ of the votes, then the probability that A will lead throughout the ballot count equals $(2p-1)_+$. (*Hint:* Use Corollary 11.17. Alternatively, use a combinatorial argument based on the reflection principle.)

19. Prove the second claim in Corollary 11.17 by a martingale argument, in the case where $\xi_1, \ldots, \xi_n$ are $\mathbb{Z}_+$-valued and exchangeable. (*Hint:* We may assume that S_n is nonrandom. Then the variables $M_k = S_k/k$ form a reverse martingale, and the result follows by optional sampling.)

20. Prove that the convergence in Theorem 11.18 holds in L^p for arbitrary $p > 0$ when S is finite. (*Hint:* Show as in Lemma 11.19 that $\|\sup_n I(\xi|\mathcal{F}_n)\|_p < \infty$ when ξ is S-valued, and use Corollary 10.8 (ii).)

21. Show that $H(\xi, \eta) \leq H(\xi) + H(\eta)$ for any ξ and η. (*Hint:* Note that $H(\eta|\xi) \leq H(\eta)$ by Jensen's inequality.)

22. Give an example of a stationary Markov chain (ξ_n) such that $H(\xi_1) > 0$ but $H(\xi_1|\xi_0) = 0$.

23. Give an example of a stationary Markov chain (ξ_n) such that $H(\xi_1) = \infty$ but $H(\xi_1|\xi_0) < \infty$. (*Hint:* Choose the state space $\mathbb{Z}_+$, and consider transition probabilities p_{ij} that equal 0 unless $j = i+1$ or $j = 0$.)

Chapter 12

Poisson and Pure Jump-Type Markov Processes

Random measures and point processes; Cox processes, randomization, and thinning; mixed Poisson and binomial processes; independence and symmetry criteria; Markov transition and rate kernels; embedded Markov chains and explosion; compound and pseudo-Poisson processes; ergodic behavior of irreducible chains

Poisson processes and Brownian motion constitute the basic building blocks of modern probability theory. Our first goal in this chapter is to introduce the family of Poisson and related processes. In particular, we construct Poisson processes on bounded sets as mixed binomial processes and derive a variety of Poisson characterizations in terms of independence, symmetry, and renewal properties. A randomization of the underlying intensity measure leads to the richer class of Cox processes. We also consider the related randomizations of general point processes, obtainable through independent motions of the individual point masses. In particular, we will see how the latter type of transformations preserve the Poisson property.

It is usually most convenient to regard Poisson and other point processes on an abstract space as integer-valued random measures. The relevant parts of this chapter may then serve at the same time as an introduction to random measure theory. In particular, Cox processes and randomizations will be used to derive some general uniqueness criteria for simple point processes and diffuse random measures. The notions and results of this chapter form a basis for the corresponding weak convergence theory developed in Chapter 16, where Poisson and Cox processes appear as limits in important special cases.

Our second goal is to continue the theory of Markov processes from Chapter 8 with a detailed study of pure jump-type processes. The evolution of such a process is governed by a rate kernel α, which determines both the rate at which transitions occur and the associated transition probabilities. For bounded α one gets a pseudo-Poisson process, which may be described as a discrete-time Markov chain with transition times given by an independent, homogeneous Poisson process. Of special interest is the case of compound Poisson processes, where the underlying Markov chain is a random walk. In Chapter 19 we shall see how every Feller process can be approximated in a natural way by pseudo-Poisson processes, recognized in that context by the boundedness of their generators. A similar compound Poisson approximation of general Lévy processes is utilized in Chapter 15.

In addition to the already mentioned connections to other topics, we note the fundamental role of Poisson processes for the theory of Lévy processes in Chapter 15 and for excursion theory in Chapter 22. In Chapter 25 the independent-increment characterization of Poisson processes is extended to a criterion in terms of compensators, and we derive some related time-change results. Finally, the ergodic theory for continuous-time Markov chains, developed at the end of this chapter, is analogous to the discrete-time theory of Chapter 8 and will be extended in Chapter 20 to a general class of Feller processes. A related theory for diffusions appears in Chapter 23.

To introduce the basic notions of random measure theory, consider an arbitrary measurable space $(S, \mathcal{S})$. By a random measure on S we mean a σ-finite kernel ξ from the basic probability space $(\Omega, \mathcal{A}, P)$ into S. Here the σ-finiteness means that there exists a partition $B_1, B_2, \ldots \in \mathcal{S}$ of S such that $\xi B_k < \infty$ a.s. for all k. It is often convenient to think of ξ as a random element in the space $\mathcal{M}(S)$ of σ-finite measures on S, endowed with the σ-field generated by the projection maps $\pi_B : \mu \mapsto \mu B$ for arbitrary $B \in \mathcal{S}$. Note that $\xi B = \xi(\cdot, B)$ is a random variable in $[0, \infty]$ for every $B \in \mathcal{S}$. More generally, it is clear by a simple approximation that $\xi f = \int f d\xi$ is a random variable in $[0, \infty]$ for every measurable function $f \geq 0$ on S. The *intensity* of ξ is defined as the measure $E\xi B = E(\xi B)$, $B \in \mathcal{S}$.

We often encounter the situation when S is a topological space with Borel σ-field $\mathcal{S} = \mathcal{B}(S)$. In the special case when S is a locally compact, second countable Hausdorff space (abbreviated as *lcscH*), it is understood that ξ is a.s. finite on the ring $\hat{\mathcal{S}}$ of all relatively compact Borel sets. Equivalently, we assume that $\xi f < \infty$ a.s. for every $f \in C_K^+(S)$, the class of continuous functions $f \geq 0$ on S with compact support. In this case, the σ-field in $\mathcal{M}(S)$ is generated by the projections $\pi_f : \mu \mapsto \mu f$ for all $f \in C_K^+(S)$.

The following elementary result provides the basic uniqueness criteria for random measures. Stronger results are given for simple point processes and diffuse random measures in Theorem 12.8, and related convergence criteria appear in Theorem 16.16.

Lemma 12.1 *(uniqueness for random measures)* *Let ξ and η be random measures on S. Then $\xi \stackrel{d}{=} \eta$ under each of these conditions:*

(i) $(\xi B_1, \ldots, \xi B_n) \stackrel{d}{=} (\eta B_1, \ldots, \eta B_n)$ *for any* $B_1, \ldots, B_n \in \mathcal{S}$, $n \in \mathbb{N}$;

(ii) $\xi f \stackrel{d}{=} \eta f$ *for any measurable function $f \geq 0$ on S.*

If S is lcscH, it suffices in (ii) to consider functions $f \in C_K^+(S)$.

Proof: The sufficiency of (i) is clear from Proposition 3.2. Next we note that (i) follows from (ii), as we apply the latter condition to any positive linear combination $f = \sum_k c_k 1_{B_k}$ and use the Cramér–Wold Corollary 5.5.

Now assume that S is lcscH, and that (ii) holds for all $f \in C_K^+(S)$. Since $C_K^+(S)$ is closed under positive linear combinations, we see as before that

$$(\xi f_1, \ldots, \xi f_n) \stackrel{d}{=} (\eta f_1, \ldots, \eta f_n), \quad f_1, \ldots, f_n \in C_K^+, \, n \in \mathbb{N}.$$

By Theorem 1.1 it follows that $\mathcal{L}(\xi) = \mathcal{L}(\eta)$ on the σ-field $\mathcal{G} = \sigma\{\pi_f; f \in C_K^+\}$, where $\pi_f : \mu \mapsto \mu f$, and it remains to show that $\mathcal{G}$ contains $\mathcal{F} = \sigma\{\pi_B; B \in \mathcal{S}\}$. Then fix any compact set $K \subset S$, and choose some functions $f_n \in C_K^+$ with $f_n \downarrow 1_K$. Since $\mu f_n \downarrow \mu K$ for every $\mu \in \mathcal{M}(S)$, the mapping π_K is $\mathcal{G}$-measurable by Lemma 1.10. Next apply Theorem 1.1 to the Borel subsets of an arbitrary compact set, to see that π_B is $\mathcal{G}$-measurable for any $B \in \hat{\mathcal{S}}$. Hence, $\mathcal{F} \subset \mathcal{G}$. $\square$

By a *point process* on S we mean an integer-valued random measure ξ. In other words, we assume ξB to be a $\overline{\mathbb{Z}}_+$-valued random variable for every $B \in \mathcal{S}$. Alternatively, we may think of ξ as a random element in the space $\mathcal{N}(S) \subset \mathcal{M}(S)$ of all σ-finite, integer-valued measures on S. When S is Borel, we may write $\xi = \sum_{k \leq \kappa} \delta_{\gamma_k}$ for some random elements $\gamma_1, \gamma_2, \ldots$ in S and κ in $\overline{\mathbb{Z}}_+$, and we note that ξ is simple iff the γ_k with $k \leq \kappa$ are distinct. In general, we may eliminate the possible multiplicities to create a simple point process ξ^*, which agrees with the counting measure on the support of ξ. By construction it is clear that ξ^* is a measurable function of ξ.

A random measure ξ on a measurable space S is said to have *independent increments* if the random variables $\xi B_1, \ldots, \xi B_n$ are independent for any disjoint sets $B_1, \ldots, B_n \in \mathcal{S}$. By a *Poisson process* on S with intensity measure $\mu \in \mathcal{M}(S)$ we mean a point process ξ on S with independent increments such that ξB is Poisson with mean μB whenever $\mu B < \infty$. By Lemma 12.1 the stated conditions specify the distribution of ξ, which is then determined by the intensity measure μ. More generally, for any random measure η on S, we say that a point process ξ is a *Cox process directed by* η if it is conditionally Poisson, given η, with $E[\xi|\eta] = \eta$ a.s. In particular, we may take $\eta = \alpha\mu$ for some measure $\mu \in \mathcal{M}(S)$ and random variable $\alpha \geq 0$ to form a *mixed Poisson process* based on μ and α.

We next define a ν-*randomization* ζ of an arbitrary point process ξ on S, where ν is a probability kernel from S to some measurable space T. Assuming first that ξ is nonrandom and equal to $\mu = \sum_k \delta_{s_k}$, we may take $\zeta = \sum_k \delta_{s_k, \gamma_k}$, where the γ_k are independent random elements in T with distributions $\nu(s_k, \cdot)$. Note that the distribution P_μ of ζ depends only on μ. In general, we define a ν-randomization ζ of ξ by the condition $P[\zeta \in \cdot | \xi] = P_\xi$ a.s. In the special case when $T = \{0, 1\}$ and $\nu(s, \{0\}) \equiv p \in [0, 1]$, we refer to the point process $\xi_p = \zeta(\cdot \times \{0\})$ on S as a p-*thinning* of ξ. Another special instance is when $S = \{0\}$, $\xi = \kappa\delta_0$, and $\nu = \mu/\mu T$ for some $\mu \in \mathcal{M}(T)$ with $\mu T \in (0, \infty)$, in which case ζ is called a *mixed binomial (or sample) process* based on μ and κ. Note that ζB is then binomially distributed, conditionally on κ, with parameters νB and κ. If T is Borel, we can write $\zeta = \sum_{k\leq\kappa}\delta_{\gamma_k}$, where the random elements γ_k are i.i.d. ν and independent of κ.

Our first aim is to examine the relationship between the various point processes introduced so far. Here we may simplify the computations by using the *Laplace functional* $\psi_\xi(f) = Ee^{-\xi f}$ of a random measure ξ, defined for any measurable function $f \geq 0$ on the state space S. Note the ψ_ξ determines the distribution $\mathcal{L}(\xi)$ by Lemma 12.1 and the uniqueness theorem for Laplace transforms. The following lemma lists some useful formulas. Recall that a kernel ν between two measurable spaces S and T may be regarded as an operator between the associated function spaces, given by $\nu f(s) = \int \nu(s, dt) f(t)$. For convenience, we write $\hat{\nu}(s, \cdot) = \delta_s \otimes \nu(s, \cdot)$, so that $\mu \otimes \nu = \mu\hat{\nu}$.

Lemma 12.2 *(Laplace functionals) Let $f, g \geq 0$ be measurable.*

(i) *If ξ is a Poisson process with $E\xi = \mu$, then*
$$Ee^{-\xi f} = \exp\{-\mu(1 - e^{-f})\}.$$
Here we may replace f by if when $f\colon S \to \mathbb{R}$ with $\mu(|f| \wedge 1) < \infty$.

(ii) *If ξ is a Cox process directed by η, then*
$$Ee^{-\xi f - \eta g} = E\exp\{-\eta(1 - e^{-f} + g)\}.$$

(iii) *If ζ is a ν-randomization of ξ, then*
$$Ee^{-\zeta f} = E\exp(\xi \log \hat{\nu}e^{-f}).$$

(iv) *If ξ_p is a p-thinning of ξ, then*
$$Ee^{-\xi_p f - \xi g} = E\exp\{-\xi(g - \log\{1 - p(1 - e^{-f})\})\}.$$

(v) *If ξ is a mixed binomial process based on μ and κ, then*
$$Ee^{-\xi f} = E(\mu e^{-f}/\mu S)^\kappa.$$

Proof: (i) If α is a Poisson random variable with mean m, then clearly
$$Ee^{-c\alpha} = e^{-m}\sum_{k\geq 0}(me^{-c})^k/k! = \exp\{-m(1 - e^{-c})\}, \quad c \in \mathbb{C}.$$

Now let $f = \sum_{k\leq m} c_k 1_{B_k}$, where $c_k \in \mathbb{C}$ and the sets $B_k \in \mathcal{S}$ are disjoint with $\mu B_k < \infty$. Then
$$\begin{aligned}
Ee^{-\xi f} &= E\exp\left\{-\sum_k c_k \xi B_k\right\} = \prod_k Ee^{-c_k \xi B_k} \\
&= \prod_k \exp\{-\mu B_k(1 - e^{-c_k})\} \\
&= \exp\left\{-\sum_k \mu B_k(1 - e^{-c_k})\right\} \\
&= \exp\{-\mu(1 - e^{-f})\}.
\end{aligned}$$

For general $f \geq 0$, we may choose some simple functions $f_n \geq 0$ with $f_n \uparrow f$ and conclude by monotone convergence that $\xi f_n \to \xi f$ and $\mu(1 - e^{-f_n}) \to \mu(1 - e^{-f})$. The asserted formula then follows by dominated convergence from the version for f_n.

Now assume that $\mu(|f| \wedge 1) < \infty$. Replacing f by $c|f|$ in the previous formula and letting $c \downarrow 0$, we get by dominated convergence $P\{\xi|f| < \infty\} = e^0 = 1$, or $\xi|f| < \infty$ a.s. Next choose some simple functions $f_n \to f$ with $|f_n| \leq |f|$ and $\mu|f_n| < \infty$, and note that $|1 - e^{-if_n}| \leq |f| \wedge 2$ by Lemma 5.14. By dominated convergence we obtain $\xi f_n \to \xi f$ and $\mu(1 - e^{-if_n}) \to \mu(1 - e^{-if})$. The extended formula now follows from the version for f_n.

(ii) By (i) we have
$$\begin{aligned} Ee^{-\xi f - \eta g} &= Ee^{-\eta g} E[e^{-\xi f}|\eta] \\ &= Ee^{-\eta g} \exp\{-\eta(1 - e^{-f})\} \\ &= E\exp\{-\eta(1 - e^{-f} + g)\}. \end{aligned}$$

(iii) First assume that $\xi = \sum_k \delta_{s_k}$ is nonrandom. Introducing some independent random elements γ_k in T with distributions $\nu(s_k, \cdot)$, we get
$$\begin{aligned} Ee^{-\zeta f} &= E\exp\left\{-\sum_k f(s_k, \gamma_k)\right\} \\ &= \prod_k Ee^{-f(s_k, \gamma_k)} = \prod_k \hat{\nu} e^{-f}(s_k) \\ &= \exp\sum_k \log \hat{\nu} e^{-f}(s_k) = \exp \xi \log \hat{\nu} e^{-f}. \end{aligned}$$

Hence, in general,
$$Ee^{-\zeta f} = EE[e^{-\zeta f}|\xi] = E \exp \xi \log \hat{\nu} e^{-f}.$$

(iv) Apply (iii), or use the same method of proof.

(v) We may assume that $\xi = \sum_{k \leq \kappa} \delta_{\gamma_k}$, where $\gamma_1, \gamma_2, \ldots$ are i.i.d. and independent of κ with distribution $\mu/\mu S$. Using Fubini's theorem, we get
$$\begin{aligned} Ee^{-\xi f} &= E\exp\left\{-\sum_{k \leq \kappa} f(\gamma_k)\right\} = E\prod_{k \leq \kappa} Ee^{-f(\gamma_k)} \\ &= E\prod_{k \leq \kappa}(\mu e^{-f}/\mu S) = E(\mu e^{-f}/\mu S)^\kappa. \quad \Box \end{aligned}$$

It is now easy to prove that the Poisson property is preserved under randomizations. Here is a more general result.

Proposition 12.3 *(preservation laws)* For any measurable spaces S, T, and U, consider some probability kernels $\mu: S \to T$ and $\nu: S \times T \to U$.

(i) If ξ is a Cox process on S directed by η and $\zeta \perp\!\!\!\perp_\xi \eta$ is a μ-randomization of ξ, then ζ is a Cox process directed by $\eta \otimes \mu$.

(ii) If η is a μ-randomization of ξ and ζ is a ν-randomization of η, then ζ is a $\mu \otimes \nu$-randomization of ξ.

Note that the conditional independence in (i) holds automatically when ζ is constructed from ξ by independent randomization, as in Lemma 6.9.

Proof: (i) Using Proposition 6.6 and Lemma 12.2 (ii) and (iii), we get for any measurable functions $f, g \geq 0$

$$\begin{aligned}
Ee^{-\xi f - \eta \hat{\mu} g} &= Ee^{-\eta \hat{\mu} g} E[e^{-\xi f} | \xi, \eta] \\
&= E \exp\{\xi \log \hat{\mu} e^{-f} - \eta \hat{\mu} g\} \\
&= E \exp\{-\eta(1 - \hat{\mu} e^{-f} + \hat{\mu} g)\} \\
&= E \exp\{-\eta \hat{\mu}(1 - e^{-f} + g)\}.
\end{aligned}$$

The result now follows by Lemmas 12.1 (ii) and 12.2 (ii).

(ii) By Lemma 12.2 (iii),

$$\begin{aligned}
Ee^{-\xi f} &= E \exp\{\eta \log \hat{\nu} e^{-f}\} \\
&= E \exp\{\xi \log \hat{\mu} \hat{\nu} e^{-f}\} \\
&= E \exp\{\xi \log(\mu \otimes \nu)\hat{e}^{-f}\}. \qquad \square
\end{aligned}$$

We continue with a basic relationship between Poisson and binomial processes. The result leads to an easy construction of the general Poisson process in Theorem 12.7. The significance of mixed Poisson and binomial processes is further clarified by Theorem 12.12 below.

Theorem 12.4 *(mixed Poisson and binomial processes)* Consider a point process ξ and a σ-finite measure μ on a common space $(S, \mathcal{S})$, and let $B_1, B_2, \ldots \in \mathcal{S}$ with $B_n \uparrow S$. Then ξ is a mixed Poisson or binomial process based on μ iff the same property holds on B_n for every $n \in \mathbb{N}$.

Proof: First assume that ξ is a mixed Poisson process based on μ and α. Then the same property holds for the restriction to any set $B \in \mathcal{S}$ with $\mu B = \infty$. If instead $\mu B \in (0, \infty)$, let η be a mixed binomial process based on $1_B \cdot \mu$ and κ, where κ is conditionally Poisson with mean $\alpha \mu B$. By Lemma 12.2 (ii) and (v) we have for any measurable function $f: S \to \mathbb{R}_+$ supported by B

$$\begin{aligned}
Ee^{-\eta f} &= E\big(\mu[e^{-f}; B]/\mu B\big)^{\kappa} \\
&= E \exp\big(-\alpha \mu B(1 - \mu[e^{-f}; B]/\mu B)\big) \\
&= E \exp\big(-\alpha \mu[1 - e^{-f}; B]\big) \\
&= E \exp\big(-\alpha \mu(1 - e^{-f})\big) = Ee^{-\xi f}.
\end{aligned}$$

Thus, $\xi \stackrel{d}{=} \eta$ on B, as required.

Next let ξ be a mixed binomial process on S based on μ and κ, and fix any $B \in \mathcal{S}$ with $\mu B > 0$. Let κ_p be a p-thinning of κ, where $p = \mu B/\mu S$, and consider a mixed binomial process η based on $1_B \cdot \mu$ and κ_p. Using Lemma 12.2 (iv) and (v), we get for any measurable function $f \geq 0$ supported by B

$$Ee^{-\eta f} = E(\mu[e^{-f}; B]/\mu B)^{\kappa p}$$

$$= E\left\{1 - \frac{\mu B}{\mu S}\left(1 - \frac{\mu[e^{-f}; B]}{\mu B}\right)\right\}^{\kappa}$$

$$= E(1 - \mu(1 - e^{-f})/\mu S)^{\kappa}$$

$$= E(\mu e^{-f}/\mu S)^{\kappa} = Ee^{-\xi f}.$$

Again it follows that $\xi \stackrel{d}{=} \eta$ on B.

To prove the converse assertion, we may clearly assume that $\mu B_n \in (0, \infty)$ for all n, so that $1_{B_n} \cdot \xi$ is a mixed binomial process based on $1_{B_n} \cdot \mu$ and ξB_n. If $f \geq 0$ is supported by B_m, then by Lemma 12.2 we have for $n \geq m$

$$Ee^{-\xi f} = E\left(\frac{\mu[e^{-f}; B_n]}{\mu B_n}\right)^{\xi B_n} = E\left(1 - \frac{\mu(1 - e^{-f})}{\mu B_n}\right)^{\xi B_n}. \tag{1}$$

If $\mu S < \infty$, then as $n \to \infty$ we get by dominated convergence

$$Ee^{-\xi f} = E\left(1 - \frac{\mu(1 - e^{-f})}{\mu S}\right)^{\xi S} = E\left(\frac{\mu e^{-f}}{\mu S}\right)^{\xi S}.$$

Taking $f = \varepsilon 1_{B_m}$ and letting $\varepsilon \to 0$, we see in particular that $\xi S < \infty$ a.s. The relation extends by dominated convergence to arbitrary $f \geq 0$, and so by Lemma 12.2 we conclude that ξ is a mixed binomial process based on μ and ξS.

If instead $\mu S = \infty$, Theorem 5.19 shows that $\xi B_n/\mu B_n \stackrel{d}{\to} \alpha$ in $[0, \infty]$ along some subsequence $N' \subset \mathbb{N}$, where $0 \leq \alpha \leq \infty$. By Theorem 4.30 we may choose some $\alpha_n \stackrel{d}{=} \xi B_n/\mu B_n$ such that $\alpha_n \to \alpha$ a.s. along N', and so by dominated convergence in (1)

$$Ee^{-\xi f} = E \exp(-\alpha\mu(1 - e^{-f})).$$

As before, we see that $\alpha < \infty$ a.s., and by monotone and dominated convergence we may extend the relation to arbitrary $f \geq 0$. Hence, Lemma 12.2 shows that ξ is a mixed Poisson process based on μ and α. □

The last result leads in particular to a criterion for a Poisson process to be simple. Recall that a measure μ on S is said to be *diffuse* if $\mu\{s\} = 0$ for all $s \in S$.

Corollary 12.5 (*simplicity and diffuseness*) *Let ξ be a Cox process directed by some random measure η, both defined on a Borel space S. Then ξ is a.s. simple iff η is a.s. diffuse.*

Proof: It is enough to establish the corresponding property for mixed binomial processes. Then let $\gamma_1, \gamma_2, \ldots$ be i.i.d. with distribution μ. By

Fubini's theorem

$$P\{\gamma_i = \gamma_j\} = \int \mu\{s\}\mu(ds) = \sum_s (\mu\{s\})^2, \quad i \neq j,$$

and so the γ_j are a.s. distinct iff μ is diffuse. □

The following uniqueness assertion will play a crucial role in a subsequent proof.

Lemma 12.6 *(uniqueness for Cox processes and thinnings)* *Fix a* $p \in (0,1)$.

(i) *For any Cox processes ξ and ξ' directed by η and η', we have $\xi \stackrel{d}{=} \xi'$ iff $\eta \stackrel{d}{=} \eta'$.*

(ii) *For any p-thinnings ξ_p and ξ'_p of ξ and ξ', we have $\xi_p \stackrel{d}{=} \xi'_p$ iff $\xi \stackrel{d}{=} \xi'$.*

Proof: We prove only (i), the argument for (ii) being similar. By Lemma 12.2 (ii) we have for any measurable function $g \geq 0$ on S

$$Ee^{-\xi g} = E\exp\{-\eta(1 - e^{-g})\} = Ee^{-\eta f},$$

where $f = 1 - e^{-g}$, and similarly for ξ' and η'. Assuming $\xi \stackrel{d}{=} \xi'$, we conclude that $Ee^{-\eta f} = Ee^{-\eta' f}$ for any measurable function $f: S \to [0,1)$. Then also

$$Ee^{-t\eta f} = Ee^{-t\eta' f}, \quad t \in [0,1],$$

and since both sides are analytic for $\Re t > 0$, the relation extends to all $t \geq 0$. Hence, $Ee^{-\eta f} = Ee^{-\eta' f}$ for all bounded, measurable functions $f \geq 0$, and Lemma 12.1 (ii) yields $\eta \stackrel{d}{=} \eta'$. □

We proceed to establish the existence of a Poisson process with arbitrary intensity measure on a general measurable space. More generally, we can prove the existence of arbitrary Cox processes and randomizations, which also covers the cases of thinnings and mixed binomial processes.

Theorem 12.7 *(existence)* *Fix any measurable spaces S and T, and allow suitable extensions of the basic probability space.*

(i) *For any random measure η on S, there exists a Cox process ξ directed by η.*

(ii) *For any point process ξ on S and probability kernel $\nu: S \to T$, there exists a ν-randomization ζ of ξ.*

Proof: (i) First assume that $\eta = \mu$ is nonrandom with $\mu S \in (0, \infty)$. By Corollary 6.18 we may choose a Poisson distributed random variable κ with $E\kappa = \mu S$ and an independent sequence of i.i.d. random elements $\gamma_1, \gamma_2, \ldots$ in S with distribution $\mu/\mu S$. By Theorem 12.4 the random measure $\xi = \sum_{j \leq \kappa} \delta_{\gamma_j}$ is then Poisson with intensity μ.

Next let $\mu S = \infty$. Since μ is σ-finite, we may split S into disjoint subsets $B_1, B_2, \cdots \in S$ such that $\mu B_k \in (0, \infty)$ for each k. As before, there exists for every k a Poisson process ξ_k on S with intensity $\mu_k = 1_{B_k} \cdot \mu$, and by

Corollary 6.18 we may choose the ξ_k to be independent. Writing $\xi = \sum_k \xi_k$ and using Lemma 12.2 (i), we get for any measurable function $f \geq 0$ on S

$$\begin{aligned} Ee^{-\xi f} &= \prod_k Ee^{-\xi_k f} = \prod_k \exp\{-\mu_k(1 - e^{-f})\} \\ &= \exp\left\{-\sum_k \mu_k(1 - e^{-f})\right\} \\ &= \exp\{-\mu(1 - e^{-f})\}. \end{aligned}$$

Using Lemmas 12.1 (ii) and 12.2 (i), we conclude that ξ is a Poisson process with intensity μ.

Now let ξ_μ be a Poisson process with intensity μ. Then for any numbers $m_1, \ldots, m_n \in \mathbb{Z}_+$ and disjoint sets $B_1, \ldots, B_n \in \mathcal{S}$, we have

$$P\bigcap_{k \leq n} \{\xi_\mu B_k = m_k\} = \prod_{k \leq n} e^{-\mu B_k}(\mu B_k)^{m_k}/m_k!,$$

which is a measurable function of μ. (Here the expression on the right is understood to be 0 when $\mu B_k = \infty$.) The measurability extends to arbitrary sets $B_k \in \mathcal{S}$, since the general probability on the left is a finite sum of such products. Now the sets on the left form a π-system generating the σ-field in $\mathcal{N}(S)$, and so by Theorem 1.1 we conclude that $P_\mu = \mathcal{L}(\xi_\mu)$ is a probability kernel from $\mathcal{M}(S)$ to $\mathcal{N}(S)$. But then Lemma 6.9 ensures the existence, for any random measure η on S, of a Cox process ξ directed by η.

(ii) First let $\mu = \sum_k \delta_{s_k}$ be nonrandom in $\mathcal{N}(S)$. By Corollary 6.18 there exist some independent random elements γ_k in T with distributions $\nu(s_k, \cdot)$, and we note that $\zeta_\mu = \sum_k \delta_{s_k, \gamma_k}$ is a ν-randomization of μ. Letting $B_1, \ldots, B_n \in \mathcal{S} \times \mathcal{T}$ and $s_1, \ldots, s_n \in (0,1)$, we get by Lemma 12.2 (iii)

$$\begin{aligned} E\prod_k s_k^{\zeta_\mu B_k} &= E \exp \zeta_\mu \sum_k 1_{B_k} \log s_k \\ &= \exp \mu \log \hat{\nu} \exp \sum_k 1_{B_k} \log s_k \\ &= \exp \mu \log \hat{\nu} \prod_k s_k^{1_{B_k}}. \end{aligned}$$

Using Lemma 1.41 (i) twice, we see that $\hat{\nu} \prod_k s_k^{1_{B_k}}$ is a measurable function on S for fixed $s_1, \ldots, s_n$, and hence that the right-hand side is a measurable function of μ. Differentiating m_k times with respect to s_k for each k and taking $s_1 = \cdots = s_n = 0$, we conclude that the probability $P\bigcap_k \{\zeta_\mu B_k = m_k\}$ is a measurable function of μ for any $m_1, \ldots, m_n \in \mathbb{Z}_+$. As before, it follows that $P_\mu = \mathcal{L}(\zeta_\mu)$ is a probability kernel from $\mathcal{N}(S)$ to $\mathcal{N}(S \times T)$, and the general result follows by Lemma 6.9. $\square$

We may use Cox transformations and thinnings to derive some general uniqueness criteria for simple point processes and diffuse random measures, improving the elementary statements in Lemma 12.1. Related convergence criteria are given in Proposition 16.17 and Theorems 16.28 and 16.29.

Theorem 12.8 *(one-dimensional uniqueness criteria) Let $(S, \mathcal{S})$ be Borel.*

(i) *For any simple point processes ξ and η on S, we have $\xi \stackrel{d}{=} \eta$ iff $P\{\xi B = 0\} = P\{\eta B = 0\}$ for all $B \in \mathcal{S}$.*

(ii) *Let ξ and η be simple point processes or diffuse random measures on S, and fix any $c > 0$. Then $\xi \stackrel{d}{=} \eta$ iff $Ee^{-c\xi B} = Ee^{-c\eta B}$ for all $B \in \mathcal{S}$.*

(iii) *Let ξ be a simple point process or diffuse random measure on S, and let η be an arbitrary random measure on S. Then $\xi \stackrel{d}{=} \eta$ iff $\xi B \stackrel{d}{=} \eta B$ for all $B \in \mathcal{S}$.*

Proof: We may clearly assume that $S = (0, 1]$.

(i) Let $\mathcal{C}$ denote the class of sets $\{\mu; \mu B = 0\}$ with $B \in \mathcal{S}$, and note that $\mathcal{C}$ is a π-system since

$$\{\mu B = 0\} \cap \{\mu C = 0\} = \{\mu(B \cup C) = 0\}, \quad B, C \in \mathcal{S}.$$

By Theorem 1.1 it follows that $\xi \stackrel{d}{=} \eta$ on $\sigma(\mathcal{C})$. Furthermore, writing $I_{nj} = 2^{-n}(j-1, j]$ for $n \in \mathbb{N}$ and $j = 1, \ldots, 2^n$, we have

$$\mu^* B = \lim_{n \to \infty} \sum_j (\mu(B \cap I_{nj}) \wedge 1), \quad \mu \in \mathcal{N}(S), \ B \in \mathcal{S},$$

which shows that the mapping $\mu \mapsto \mu^*$ is $\sigma(\mathcal{C})$-measurable. Since ξ and η are simple, we conclude that $\xi = \xi^* \stackrel{d}{=} \eta^* = \eta$.

(ii) First let ξ and η be diffuse. By Theorem 12.7 we may choose some Cox processes $\tilde{\xi}$ and $\tilde{\eta}$ directed by $c\xi$ and $c\eta$. Conditioning on ξ or η, respectively, we obtain

$$P\{\tilde{\xi} B = 0\} = Ee^{-c\xi B} = Ee^{-c\eta B} = P\{\tilde{\eta} B = 0\}, \quad B \in \mathcal{S}. \tag{2}$$

Since $\tilde{\xi}$ and $\tilde{\eta}$ are a.s. simple by Corollary 12.5, assertion (i) yields $\tilde{\xi} \stackrel{d}{=} \tilde{\eta}$, and so $\xi \stackrel{d}{=} \eta$ by Lemma 12.6. If ξ and η are instead simple point processes, then (2) holds by Lemma 12.2 (iv) when $\tilde{\xi}$ and $\tilde{\eta}$ are p-thinnings of ξ and η with $p = 1 - e^{-c}$, and the proof may be completed as before.

(iii) First let ξ be a simple point process. Fix any $B \in \mathcal{S}$ such that $\eta B < \infty$ a.s. Defining I_{nj} as before, we note that $\eta(B \cap I_{nj}) \in \mathbb{Z}_+$ outside a fixed null set. It follows easily that $1_B \cdot \eta$ is a.s. integer valued, and so even η is a.s. a point process. Noting that

$$P\{\eta^* B = 0\} = P\{\eta B = 0\} = P\{\xi B = 0\}, \quad B \in \mathcal{S},$$

we conclude from (i) that $\xi \stackrel{d}{=} \eta^*$. In particular, $\eta B \stackrel{d}{=} \xi B \stackrel{d}{=} \eta^* B$ for all B, and so $\eta^* = \eta$ a.s.

Next assume that ξ is a.s. diffuse. Letting $\tilde{\xi}$ and $\tilde{\eta}$ be Cox processes directed by ξ and η, we note that $\tilde{\xi} B \stackrel{d}{=} \tilde{\eta} B$ for every $B \in \mathcal{S}$. Since $\tilde{\xi}$ is a.s. simple by Proposition 12.5, it follows as before that $\tilde{\xi} \stackrel{d}{=} \tilde{\eta}$, and so $\xi \stackrel{d}{=} \eta$ by Lemma 12.6. $\square$

As an easy consequence, we get the following characterization of Poisson processes. To simplify the statement, we may allow a Poisson random variable to have infinite mean, hence to be a.s. infinite.

Corollary 12.9 (*one-dimensional Poisson criterion, Rényi*) *Let ξ be a random measure on a Borel space S such that $\xi\{s\} = 0$ a.s. for all $s \in S$. Then ξ is a Poisson process iff ξB is Poisson for every $B \in \mathcal{S}$, in which case $E\xi$ is σ-finite and diffuse.*

Proof: Assume the stated condition. Then $\mu = E\xi$ is clearly σ-finite and diffuse, and by Theorem 12.7 there exists a Poisson process η on S with intensity μ. Then $\eta B \stackrel{d}{=} \xi B$ for all $B \in \mathcal{S}$, and since η is a.s. simple by Corollary 12.5, we conclude from Theorem 12.8 that $\xi \stackrel{d}{=} \eta$. □

Much of the previous theory can be extended to the case of marks. Given any measurable spaces $(S, \mathcal{S})$ and $(K, \mathcal{K})$, we define a *K-marked point process on S* as a point process ξ on $S \times K$ in the usual sense satisfying $\xi(\{s\} \times K) \leq 1$ identically and such that the projections $\xi(\cdot \times K_j)$ are σ-finite point processes on S for some measurable partition $K_1, K_2, \ldots$ of K.

We say that ξ has *independent increments* if the point processes $\xi(B_1 \times \cdot)$, $\ldots, \xi(B_n \times \cdot)$ on K are independent for any disjoint sets $B_1, \ldots, B_n \in \mathcal{S}$. We also say that ξ is a *Poisson process* if ξ is Poisson in the usual sense on the product space $S \times K$. The following result characterizes Poisson processes in terms of the independence property. The result plays a crucial role in Chapters 15 and 22. A related characterization in terms of compensators is given in Corollary 25.25.

Theorem 12.10 (*independence criterion for Poisson, Erlang, Lévy*) *Let ξ be a K-marked point process on a Borel space S such that $\xi(\{s\} \times K) = 0$ a.s. for all $s \in S$. Then ξ is Poisson iff it has independent increments, in which case $E\xi$ is σ-finite with diffuse projections onto S.*

Proof: We may assume that $S = (0, 1]$. Fix any set $B \in \mathcal{S} \otimes \mathcal{K}$ with $\xi B < \infty$ a.s., and note that the projection $\eta = (1_B \cdot \xi)(\cdot \times K)$ is a simple point process on S with independent increments such that $\eta\{s\} = 0$ a.s. for all $s \in S$. Introduce the dyadic intervals $I_{nj} = 2^{-n}(j-1, j]$, and note that $\max_j \eta I_{nj} \vee 1 \to 1$ a.s.

Next fix any $\varepsilon > 0$. By dominated convergence, every point $s \in [0, 1]$ has an open neighborhood G^s such that $P\{\eta G^s > 0\} < \varepsilon$, and by compactness we may cover $[0, 1]$ by finitely many such sets $G_1, \ldots, G_m$. Choosing n so large that every interval I_{nj} lies in one of the G_k, we get $\max_j P\{\eta I_{nj} > 0\} < \varepsilon$. This shows that the variables ηI_{nj} form a null array.

Now apply Theorem 5.7 to see that the random variable $\xi B = \eta S = \sum_j \eta I_{nj}$ is Poisson. Since B was arbitrary, Corollary 12.9 then shows that ξ is a Poisson process on $S \times K$. The last assertion is now obvious. □

The last theorem yields in particular a representation of random measures with independent increments. A version for general processes on $\mathbb{R}_+$ will be proved in Theorem 15.4.

Corollary 12.11 *(independent increments)* Let ξ be a random measure on a Borel space S such that $\xi\{s\} = 0$ a.s. for all s. Then ξ has independent increments iff a.s.

$$\xi B = \alpha B + \int_0^\infty x\, \eta(B \times dx), \quad B \in \mathcal{S}, \tag{3}$$

for some nonrandom measure α on S and some Poisson process η on $S \times (0, \infty)$. Furthermore, $\xi B < \infty$ a.s. for some $B \in \mathcal{S}$ iff $\alpha B < \infty$ and

$$\int_0^\infty (x \wedge 1)\, E\eta(B \times dx) < \infty. \tag{4}$$

Proof: Introduce on $S \times (0, \infty)$ the point process $\eta = \sum_s \delta_{s,\xi\{s\}}$, where the required measurability follows by a simple approximation. Noting that η has independent S-increments, and also that

$$\eta(\{s\} \times (0, \infty)) = 1\{\xi\{s\} > 0\} \le 1, \quad s \in S,$$

we conclude from Theorem 12.10 that η is a Poisson process. Subtracting the atomic part from ξ, we get a diffuse random measure α satisfying (3), and we note that α has again independent increments. Hence, α is a.s. nonrandom by Theorem 5.11. Next, Lemma 12.2 (i) yields for any $B \in \mathcal{S}$ and $r > 0$

$$-\log E \exp\left\{-r \int_0^\infty x\, \eta(B \times dx)\right\} = \int_0^\infty (1 - e^{-rx})\, E\eta(B \times dx).$$

As $r \to 0$, it follows by dominated convergence that $\int_0^\infty x\, \eta(B \times dx) < \infty$ a.s. iff (4) holds. $\square$

We proceed to characterize the mixed Poisson and binomial processes by a natural symmetry condition. Related results for more general processes appear in Theorems 11.15 and 16.21. Given a random measure ξ and a diffuse measure μ on S, we say that ξ is μ-*symmetric* if $\xi \circ f^{-1} \stackrel{d}{=} \xi$ for every μ-preserving mapping f on S.

Theorem 12.12 *(symmetric point processes)* Consider a simple point process ξ and a diffuse, σ-finite measure μ on a Borel space S. Then ξ is μ-symmetric iff it is a mixed Poisson or binomial process based on μ.

Proof: By Theorem 12.4 and scaling we may assume that $\mu S = 1$. By the symmetry of ξ there exists a function φ on $[0, 1]$ such that $P\{\xi B = 0\} = \varphi(\mu B)$ for all B, and by Theorem 12.8 (i) it is enough to show that φ has the desired form. For notational convenience, we may then assume that μ equals Lebesgue measure on $(0, 1]$, the general case being similar. Then introduce for suitable $j, n \in \mathbb{N}$ the intervals $I_{nj} = n^{-1}(j-1, j]$, and put $\xi_{nj} = \xi I_{nj} \wedge 1$. Writing $\kappa_n = \sum_j \xi_{nj}$, we get by symmetry

$$\varphi(k/n) = E \prod_{j=0}^{k-1} \frac{n - \kappa_n - j}{n - j}, \quad 0 \leq k \leq n.$$

As $n \to \infty$, we have $\kappa_n \to \kappa = \xi(0,1]$, and so for $k/n \to t \in (0,1)$

$$\log \prod_{j=0}^{k-1} \frac{n - \kappa_n - j}{n - j} = \sum_{r=n-k+1}^{n} \log\left(1 - \frac{\kappa_n}{r}\right)$$

$$\sim -\kappa \sum_{r=n-k+1}^{n} r^{-1} \sim -\kappa \int_{n-k}^{n} x^{-1} dx$$

$$= \kappa \log(1 - k/n) \to \kappa \log(1 - t).$$

Hence, the product on the left tends to $(1-t)^\kappa$, and so by dominated convergence we get $\varphi(t) = E(1-t)^\kappa$ for rational $t \in (0,1)$, which extends by monotonicity to all real $t \in [0,1]$. This clearly agrees with the result for a mixed binomial process on $(0,1]$ with κ points. □

Integrals with respect to Poisson processes occur frequently in applications. The next result gives criteria for the existence of the integrals ξf, $(\xi - \xi')f$, and $(\xi - \mu)f$, where ξ and ξ' are independent Poisson processes with a common intensity measure μ. In each case the integral may be defined as a limit in probability of elementary integrals ξf_n, $(\xi - \xi')f_n$, or $(\xi - \mu)f_n$, respectively, where the f_n are bounded with compact support and such that $|f_n| \leq |f|$ and $f_n \to f$. We say that the integral of f exists if the appropriate limit exists and is independent of the choice of approximating functions f_n.

Lemma 12.13 *(Poisson integrals)* *Let ξ and ξ' be independent Poisson processes on S with the same intensity measure μ. Then for any measurable function f on S, we have*

(i) ξf *exists iff* $\mu(|f| \wedge 1) < \infty$;

(ii) $(\xi - \xi')f$ *exists iff* $\mu(f^2 \wedge 1) < \infty$;

(iii) $(\xi - \mu)f$ *exists iff* $\mu(f^2 \wedge |f|) < \infty$.

In each case, it is also equivalent that the corresponding set of approximating elementary integrals is tight.

Proof: (i) If $\xi|f| < \infty$ a.s., then $\mu(|f| \wedge 1) < \infty$ by Lemma 12.2. The converse implication was established in the proof of the same lemma.

(ii) First consider a deterministic counting measure $\nu = \sum_k \delta_{s_k}$, and define $\tilde{\nu} = \sum_k \vartheta_k \delta_{s_k}$, where $\vartheta_1, \vartheta_2, \ldots$ are i.i.d. random variables with $P\{\vartheta_k = \pm 1\} = \frac{1}{2}$. By Theorem 4.17, the series $\tilde{\nu} f$ converges a.s. iff $\nu f^2 < \infty$, and otherwise $|\tilde{\nu} f_n| \xrightarrow{P} \infty$ for any bounded approximations $f_n = 1_{B_n} f$ with $B_n \in \hat{S}$. The result extends by conditioning to arbitrary point processes ν and their symmetric randomizations $\tilde{\nu}$. Now Proposition

12.3 exhibits $\xi - \xi'$ as such a randomization of the Poisson process $\xi + \xi'$, and by part (i) we have $(\xi + \xi')f^2 < \infty$ a.s. iff $\mu(f^2 \wedge 1) < \infty$.

(iii) Write $f = g + h$, where $g = f1\{|f| \leq 1\}$ and $h = f1\{|f| > 1\}$. First assume that $\mu g^2 + \mu|h| = \mu(f^2 \wedge |f|) < \infty$. Since clearly $E(\xi f - \mu f)^2 = \mu f^2$, the integral $(\xi - \mu)g$ exists. Furthermore, ξh exists by part (i). Hence, even $(\xi - \mu)f = (\xi - \mu)g + \xi h - \mu h$ exists.

Conversely, assume that $(\xi - \mu)f$ exists. Then so does $(\xi - \xi')f$, and by part (ii) we get $\mu g^2 + \mu\{h \neq 0\} = \mu(f^2 \wedge 1) < \infty$. The existence of $(\xi - \mu)g$ now follows by the direct assertion, and trivially even ξh exists. Thus, the existence of $\mu h = (\xi - \mu)g + \xi h - (\xi - \mu)f$ follows, and so $\mu|h| < \infty$. □

A Poisson process ξ on $\mathbb{R}_+$ is said to be *time-homogeneous* with rate $c \geq 0$ if $E\xi = c\lambda$. In that case Proposition 8.5 shows that $N_t = \xi[0, t]$, $t \geq 0$, is a space- and time-homogeneous Markov process. We now introduce a more general class of Markov processes.

Say that a process X in some measurable space $(S, \mathcal{S})$ is of *pure jump type* if its paths are a.s. right-continuous and constant apart from isolated jumps. In that case we may denote the jump times of X by $\tau_1, \tau_2, \ldots$, with the understanding that $\tau_n = \infty$ if there are fewer than n jumps. By Lemma 7.3 and a simple approximation, the times τ_n are optional with respect to the right-continuous filtration $\mathcal{F} = (\mathcal{F}_t)$ induced by X. For convenience we may choose X to be the identity mapping on the canonical path space Ω. When X is Markov, the distribution with initial state x is denoted by P_x, and we note that the mapping $x \mapsto P_x$ is a kernel from $(S, \mathcal{S})$ to $(\Omega, \mathcal{F}_\infty)$.

We begin our study of pure jump-type Markov processes by proving an extension of the elementary strong Markov property in Proposition 8.9. A further extension appears as Theorem 19.17.

Theorem 12.14 (strong Markov property, Doob) *A pure jump-type Markov process satisfies the strong Markov property at every optional time.*

Proof: For any optional time τ, we may choose some optional times $\sigma_n \geq \tau + 2^{-n}$ taking countably many values such that $\sigma_n \to \tau$ a.s. By Proposition 8.9 we get, for any $A \in \mathcal{F}_\tau \cap \{\tau < \infty\}$ and $B \in \mathcal{F}_\infty$,

$$P[\theta_{\sigma_n} X \in B; A] = E[P_{X_{\sigma_n}} B; A]. \tag{5}$$

By the right-continuity of X, we have $P\{X_{\sigma_n} \neq X_\tau\} \to 0$. If B depends on finitely many coordinates, it is also clear that

$$P(\{\theta_{\sigma_n} X \in B\} \triangle \{\theta_\tau X \in B\}) \to 0, \quad n \to \infty.$$

Hence, (5) remains true for such sets B with σ_n replaced by τ, and the relation extends to the general case by a monotone class argument. □

We shall now see how the homogeneous Poisson processes may be characterized as special renewal processes. Recall that a random variable γ is said to be *exponentially distributed* with rate $c > 0$ if $P\{\gamma > t\} = e^{-ct}$ for all $t \geq 0$. In this case, clearly $E\gamma = c^{-1}$.

Proposition 12.15 *(Poisson and renewal processes) Let ξ be a simple point process on $\mathbb{R}_+$ with atoms at $\tau_1 < \tau_2 < \cdots$, and put $\tau_0 = 0$. Then ξ is homogeneous Poisson with rate $c > 0$ iff the differences $\tau_n - \tau_{n-1}$ are i.i.d. and exponentially distributed with mean c^{-1}.*

Proof: First assume that ξ is Poisson with rate c. Then $N_t = \xi[0,t]$ is a space- and time-homogeneous pure jump-type Markov process. By Lemma 7.6 and Theorem 12.14, the strong Markov property holds at each τ_n, and by Theorem 8.10 we get

$$\tau_1 \stackrel{d}{=} \tau_{n+1} - \tau_n \perp\!\!\!\perp (\tau_1, \ldots, \tau_n), \quad n \in \mathbb{N}.$$

Thus, the variables $\tau_n - \tau_{n-1}$ are i.i.d., and it remains to note that

$$P\{\tau_1 > t\} = P\{\xi[0,t] = 0\} = e^{-c}.$$

Conversely, assume that $\tau_1, \tau_2, \ldots$ have the stated properties. Consider a homogeneous Poisson process η with rate c and with atoms at $\sigma_1 < \sigma_2 < \cdots$, and conclude from the necessity part that $(\sigma_n) \stackrel{d}{=} (\tau_n)$. Hence,

$$\xi = \sum_n \delta_{\tau_n} \stackrel{d}{=} \sum_n \delta_{\sigma_n} = \eta. \qquad \square$$

We proceed to examine the structure of a general pure jump-type Markov process. Here the first and crucial step is to describe the distributions associated with the first jump. Say that a state $x \in S$ is *absorbing* if $P_x\{X \equiv x\} = 1$ or, equivalently, if $P_x\{\tau_1 = \infty\} = 1$.

Lemma 12.16 *(first jump) If x is nonabsorbing, then under P_x the time τ_1 until the first jump is exponentially distributed and independent of $\theta_{\tau_1} X$.*

Proof: Put $\tau_1 = \tau$. Using the Markov property at fixed times, we get for any $s, t \geq 0$

$$P_x\{\tau > s + t\} = P_x\{\tau > s, \tau \circ \theta_s > t\} = P_x\{\tau > s\} P_x\{\tau > t\}.$$

The only nonincreasing solutions to this Cauchy equation are of the form $P_x\{\tau > t\} = e^{-ct}$ with $c \in [0, \infty]$. Since x is nonabsorbing and $\tau > 0$ a.s., we have $c \in (0, \infty)$, and so τ is exponentially distributed with parameter c.

By the Markov property at fixed times, we further get for any $B \in \mathcal{F}_\infty$

$$P_x\{\tau > t, \theta_\tau X \in B\} = P_x\{\tau > t, (\theta_\tau X) \circ \theta_t \in B\}$$
$$= P_x\{\tau > t\} P_x\{\theta_\tau X \in B\},$$

which shows that $\tau \perp\!\!\!\perp \theta_\tau X$. $\qquad \square$

Writing $X_\infty = x$ when X is eventually absorbed at x, we may define the *rate function* c and *jump transition kernel* μ by

$$c(x) = (E_x \tau_1)^{-1}, \quad \mu(x, B) = P_x\{X_{\tau_1} \in B\}, \quad x \in S, B \in \mathcal{S}.$$

It is often convenient to combine c and μ into a *rate kernel* $\alpha(x, B) = c(x)\mu(x, B)$ or $\alpha = c\mu$, where the required measurability is clear from that for the kernel (P_x). Note that μ may be reconstructed from α, if we add the requirement that $\mu(x,\cdot) = \delta_x$ when $\alpha(x,\cdot) = 0$, conforming with our convention for absorbing states. This ensures that μ is a measurable function of α.

The following theorem gives an explicit representation of the process in terms of a discrete-time Markov chain and a sequence of exponentially distributed random variables. The result shows in particular that the distributions P_x are uniquely determined by the rate kernel α. As usual, we assume the existence of required randomization variables.

Theorem 12.17 *(embedded Markov chain) Let X be a pure jump-type Markov process with rate kernel $\alpha = c\mu$. Then there exist a Markov process Y on $\mathbb{Z}_+$ with transition kernel μ and an independent sequence of i.i.d., exponentially distributed random variables $\gamma_1, \gamma_2, \ldots$ with mean 1 such that a.s.*

$$X_t = Y_n, \quad t \in [\tau_n, \tau_{n+1}), \quad n \in \mathbb{Z}_+, \tag{6}$$

where

$$\tau_n = \sum_{k=1}^{n} \frac{\gamma_k}{c(Y_{k-1})}, \quad n \in \mathbb{Z}_+. \tag{7}$$

Proof: To satisfy (6), put $\tau_0 = 0$, and define $Y_n = X_{\tau_n}$ for $n \in \mathbb{Z}_+$. Introduce some i.i.d. exponentially distributed random variables $\gamma_1', \gamma_2', \ldots \perp\!\!\!\perp X$ with mean 1, and define for $n \in \mathbb{N}$

$$\gamma_n = (\tau_n - \tau_{n-1})c(Y_n)1\{\tau_{n-1} < \infty\} + \gamma_n'1\{c(Y_n) = 0\}.$$

By Lemma 12.16, we get for any $t \geq 0$, $B \in \mathcal{S}$, and $x \in S$ with $c(x) > 0$

$$P_x\{\gamma_1 > t, Y_1 \in B\} = P_x\{\tau_1 c(x) > t, Y_1 \in B\} = e^{-t}\mu(x, B),$$

and this clearly remains true when $c(x) = 0$. By the strong Markov property we obtain for every n, a.s. on $\{\tau_n < \infty\}$,

$$P_x[\gamma_{n+1} > t, Y_{n+1} \in B|\mathcal{F}_{\tau_n}] = P_{Y_n}\{\gamma_1 > t, Y_1 \in B\} = e^{-t}\mu(Y_n, B). \tag{8}$$

The strong Markov property also gives $\tau_{n+1} < \infty$ a.s. on the set $\{\tau_n < \infty, c(Y_n) > 0\}$. Arguing recursively, we get $\{c(Y_n) = 0\} = \{\tau_{n+1} = \infty\}$ a.s., and (7) follows. Using the same relation, it is also easy to check that (8) remains a.s. true on $\{\tau_n = \infty\}$, and in both cases we may clearly replace $\mathcal{F}_{\tau_n}$ by $\mathcal{G}_n = \mathcal{F}_{\tau_n} \vee \sigma\{\gamma_1', \ldots, \gamma_n'\}$. Thus, the pairs (γ_n, Y_n) form a discrete-time Markov process with the desired transition kernel. By Proposition 8.2, the latter property together with the initial distribution determine uniquely the joint distribution of Y and (γ_n). $\square$

In applications the rate kernel α is normally given, and one needs to know whether a corresponding Markov process X exists. As before we may write $\alpha(x, B) = c(x)\mu(x, B)$ for a suitable choice of rate function $c: S \to \mathbb{R}_+$ and transition kernel μ on S, where $\mu(x, \cdot) = \delta_x$ when $c(x) = 0$ and otherwise $\mu(x, \{x\}) = 0$. If X does exist, it clearly may be constructed as in Theorem

12.17. The construction fails when $\zeta \equiv \sup_n \tau_n < \infty$, in which case an *explosion* is said to occur at time ζ.

Theorem 12.18 *(synthesis)* *For any kernel $\alpha = c\mu$ on S with $\alpha(x, \{x\}) \equiv 0$, consider a Markov chain Y with transition kernel μ and some i.i.d., exponentially distributed random variables $\gamma_1, \gamma_2, \ldots \perp\!\!\!\perp Y$ with mean 1. Assume that $\sum_n \gamma_n/c(Y_{n-1}) = \infty$ a.s. under every initial distribution for Y. Then (6) and (7) define a pure jump-type Markov process with rate kernel α.*

Proof: Let P_x be the distribution of the sequences $Y = (Y_n)$ and $\Gamma = (\gamma_n)$ when $Y_0 = x$. For convenience, we may regard (Y, Γ) as the identity mapping on the canonical space $\Omega = S^\infty \times \mathbb{R}_+^\infty$. Construct X from (Y, Γ) as in (6) and (7), with $X_t = s_0$ arbitrary for $t \geq \sup_n \tau_n$, and introduce the filtrations $\mathcal{G} = (\mathcal{G}_n)$ induced by (Y, γ) and $\mathcal{F} = (\mathcal{F}_t)$ induced by X. It suffices to prove the Markov property $P_x[\theta_t X \in \cdot | \mathcal{F}_t] = P_{X_t}\{X \in \cdot\}$, since the rate kernel may then be identified via Theorem 12.17.

Then fix any $t \geq 0$ and $n \in \mathbb{Z}_+$, and define

$$\kappa = \sup\{k; \tau_k \leq t\}, \qquad \beta = (t - \tau_n)c(Y_n).$$

Put $T^m(Y, \Gamma) = \{(Y_k, \gamma_{k+1}); k \geq m\}$, $(Y', \Gamma') = T^{n+1}(Y, \Gamma)$, and $\gamma' = \gamma_{n+1}$. Since clearly

$$\mathcal{F}_t = \mathcal{G}_n \vee \sigma\{\gamma' > \beta\} \text{ on } \{\kappa = n\},$$

it is enough by Lemma 6.2 to prove that

$$P_x[(Y', \Gamma') \in \cdot, \gamma' - \beta > r | \mathcal{G}_n, \gamma' > \beta] = P_{Y_n}\{T(Y, \Gamma) \in \cdot, \gamma_1 > r\}.$$

Now $(Y', \Gamma') \perp\!\!\!\perp_{\mathcal{G}_n} (\gamma', \beta)$ because $\gamma' \perp\!\!\!\perp (\mathcal{G}_n, Y', \Gamma')$, and so the left-hand side equals

$$\frac{P_x[(Y', \Gamma') \in \cdot, \gamma' - \beta > r | \mathcal{G}_n]}{P_x[\gamma' > \beta | \mathcal{G}_n]}$$

$$= P_x[(Y', \Gamma') \in \cdot | \mathcal{G}_n] \frac{P_x[\gamma' - \beta > r | \mathcal{G}_n]}{P_x[\gamma' > \beta | \mathcal{G}_n]} = (P_{Y_n} \circ T^{-1})e^{-r},$$

as required. □

To complete the picture, we need a convenient criterion for nonexplosion.

Proposition 12.19 *(explosion)* *For any rate kernel α and initial state x, let (Y_n) and (τ_n) be such as in Theorem 12.17. Then a.s.*

$$\tau_n \to \infty \quad \text{iff} \quad \sum_n \{c(Y_n)\}^{-1} = \infty. \qquad (9)$$

In particular, $\tau_n \to \infty$ a.s. when x is recurrent for (Y_n).

Proof: Write $\beta_n = \{c(Y_{n-1})\}^{-1}$. Noting that $Ee^{-u\gamma_n} = (1+u)^{-1}$ for all $u \geq 0$, we get by (7) and Fubini's theorem

$$E[e^{-u\zeta}|Y] = \prod_n (1 + u\beta_n)^{-1} = \exp\left\{-\sum_n \log(1 + u\beta_n)\right\} \quad \text{a.s.} \qquad (10)$$

Since $\frac{1}{2}(r\wedge 1) \le \log(1+r) \le r$ for all $r > 0$, the series on the right converges for every $u > 0$ iff $\sum_n \beta_n < \infty$. Letting $u \to 0$ in (10), we get by dominated convergence

$$P[\zeta < \infty | Y] = 1\left\{\sum_n \beta_n < \infty\right\} \text{ a.s.,}$$

which implies (9). If x is visited infinitely often, then the series $\sum_n \beta_n$ has infinitely many terms $c_x^{-1} > 0$, and the last assertion follows. □

By a *pseudo-Poisson* process in some measurable space S we mean a process of the form $X = Y \circ N$ a.s., where Y is a discrete-time Markov process in S and N is an independent homogeneous Poisson process. Letting μ be the transition kernel of Y and writing c for the constant rate of N, we may construct a kernel

$$\alpha(x, B) = c\mu(x, B \setminus \{x\}), \quad x \in S,\ B \in \mathcal{B}(S), \quad (11)$$

which is measurable since $\mu(x, \{x\})$ is a measurable function of x. The next result characterizes pseudo-Poisson processes in terms of the rate kernel.

Proposition 12.20 (*pseudo-Poisson processes*) *A process X in some Borel space S is pseudo-Poisson iff it is pure jump-type Markov with a bounded rate function. Specifically, if $X = Y \circ N$ a.s. for some Markov chain Y with transition kernel μ and an independent Poisson process N with constant rate c, then X has the rate kernel in* (11).

Proof: Assume that $X = Y \circ N$ with Y and N as stated. Letting $\tau_1, \tau_2, \ldots$ be the jump times of N and writing $\mathcal{F}$ for the filtration induced by the pair (X, N), it may be seen as in Theorem 12.18 that X is $\mathcal{F}$-Markov. To identify the rate kernel α, fix any initial state x, and note that the first jump of X occurs at the first time τ_n when Y_n leaves x. For each transition of Y, this happens with probability $p_x = \mu(x, \{x\}^c)$. By Proposition 12.3 the time until first jump is then exponentially distributed with parameter cp_x. If $p_x > 0$, we further note that the location of X after the first jump has distribution $\mu(x, \cdot \setminus \{x\})/p_x$. Thus, α is given by (11).

Conversely, let X be a pure jump-type Markov process with uniformly bounded rate kernel $\alpha \ne 0$. Put $r_x = \alpha(x, S)$ and $c = \sup_x r_x$, and note that the kernel

$$\mu(x, \cdot) = c^{-1}\{\alpha(x, \cdot) + (c - r_x)\delta_x\}, \quad x \in S,$$

satisfies (11). Thus, if $X' = Y' \circ N'$ is a pseudo-Poisson process based on μ and c, then X' is again Markov with rate kernel α, and so $X \stackrel{d}{=} X'$. Hence, Corollary 6.11 yields $X = Y \circ N$ a.s. for some pair $(Y, N) \stackrel{d}{=} (Y', N')$. □

If the underlying Markov chain Y is a random walk in some measurable Abelian group S, then $X = Y \circ N$ is called a *compound Poisson process*. In this case $X - X_0 \perp\!\!\!\perp X_0$, the jump sizes are i.i.d., and the jump times are given by an independent homogeneous Poisson process. Thus, the distribution of $X - X_0$ is determined by the *characteristic measure* $\nu = c\mu$, where c

is the rate of the jump time process and μ is the common distribution of the jumps. A kernel α on S is said to be *homogeneous* if $\alpha(x, B) = \alpha(0, B - x)$ for all x and B. Let us also say that a process X in S has *independent increments* if $X_t - X_s \perp\!\!\!\perp \{X_r; r \le s\}$ for any $s < t$.

The next result characterizes compound Poisson processes in two ways, analytically in terms of the rate kernel and probabilistically in terms of the increments of the process.

Corollary 12.21 (*compound Poisson processes*) *For any pure jump-type process X in some measurable Abelian group, these conditions are equivalent:*

(i) *X is Markov with homogeneous rate kernel;*

(ii) *X has independent increments;*

(iii) *X is compound Poisson.*

Proof: If a pure jump-type Markov process is space-homogeneous, then its rate kernel is clearly homogeneous; the converse follows from the representation in Theorem 12.17. Thus, (i) and (ii) are equivalent by Proposition 8.5. Next Theorem 12.17 shows that (i) implies (iii), and the converse follows by Theorem 12.18. □

Our next aim is to derive a combined differential and integral equation for the transition kernels μ_t. An abstract version of this result appears in Theorem 19.6. For any measurable and suitably integrable function $f: S \to \mathbb{R}$, we define

$$T_t f(x) = \int f(y) \mu_t(x, dy) = E_x f(X_t), \quad x \in S, \ t \ge 0.$$

Theorem 12.22 (*backward equation, Kolmogorov*) *Let α be the rate kernel of a pure jump-type Markov process on S, and fix any bounded, measurable function $f: S \to \mathbb{R}$. Then $T_t f(x)$ is continuously differentiable in t for fixed x, and we have*

$$\frac{\partial}{\partial t} T_t f(x) = \int \alpha(x, dy)\{T_t f(y) - T_t f(x)\}, \quad t \ge 0, \ x \in S. \tag{12}$$

Proof: Put $\tau = \tau_1$, and let $x \in S$ and $t \ge 0$. By the strong Markov property at $\sigma = \tau \wedge t$ and Theorem 6.4,

$$\begin{aligned}
T_t f(x) &= E_x f(X_t) = E_x f((\theta_\sigma X)_{t-\sigma}) = E_x T_{t-\sigma} f(X_\sigma) \\
&= f(x) P_x\{\tau > t\} + E_x[T_{t-\tau} f(X_\tau); \tau \le t] \\
&= f(x) e^{-t c_x} + \int_0^t e^{-s c_x} ds \int \alpha(x, dy) T_{t-s} f(y),
\end{aligned}$$

and so

$$e^{t c_x} T_t f(x) = f(x) + \int_0^t e^{s c_x} ds \int \alpha(x, dy) T_s f(y). \tag{13}$$

12. Poisson and Pure Jump-Type Markov Processes

Here the use of the disintegration theorem is justified by the fact that $X(\omega, t)$ is product measurable on $\Omega \times \mathbb{R}_+$ because of the right-continuity of the paths.

From (13) we note that $T_t f(x)$ is continuous in t for each x, and so by dominated convergence the inner integral on the right is continuous in s. Hence, $T_t f(x)$ is continuously differentiable in t, and (12) follows by an easy computation. □

The next result relates the invariant distributions of a pure jump-type Markov process to those of the embedded Markov chain.

Proposition 12.23 *(invariance)* *Let the processes X and Y be related as in Theorem 12.17, and fix a probability measure ν on S with $\int c\, d\nu < \infty$. Then ν is invariant for X iff $c \cdot \nu$ is invariant for Y.*

Proof: By Theorem 12.22 and Fubini's theorem, we have for any bounded measurable function $f: S \to \mathbb{R}$

$$E_\nu f(X_t) = \int f(x)\nu(dx) + \int_0^t ds \int \nu(dx) \int \alpha(x, dy)\{T_s f(y) - T_s f(x)\}.$$

Thus, ν is invariant for X iff the second term on the right is identically zero. Now (12) shows that $T_t f(x)$ is continuous in t, and by dominated convergence this is also true for the integral

$$I_t = \int \nu(dx) \int \alpha(x, dy)\{T_t f(y) - T_t f(x)\}, \quad t \geq 0.$$

Thus, the condition becomes $I_t \equiv 0$. Since f is arbitrary, it is enough to take $t = 0$. Our condition then reduces to $(\nu\alpha)f \equiv \nu(cf)$ or $(c \cdot \nu)\mu = c \cdot \nu$, which means that $c \cdot \nu$ is invariant for Y. □

By a *continuous-time Markov chain* we mean a pure jump-type Markov process on a countable state space S. Here the kernels μ_t may be specified by the set of *transition functions* $p_{ij}^t = \mu_t(i, \{j\})$. The connectivity properties are simpler than in discrete time, and the notion of periodicity has no counterpart in the continuous-time theory.

Lemma 12.24 *(positivity)* *For any $i, j \in S$, we have either $p_{ij}^t > 0$ for all $t > 0$, or $p_{ij}^t = 0$ for all $t \geq 0$. In particular, $p_{ii}^t > 0$ for all t and i.*

Proof: Let $q = (q_{ij})$ be the transition matrix of the embedded Markov chain Y in Theorem 12.17. If $q_{ij}^n = P_i\{Y_n = j\} = 0$ for all $n \geq 0$, then clearly $1\{X_t \neq j\} \equiv 1$ a.s. P_i, and so $p_{ij}^t = 0$ for all $t \geq 0$. If instead $q_{ij}^n > 0$ for some $n \geq 0$, there exist some states $i = i_0, i_1, \ldots, i_n = j$ with $q_{i_{k-1}, i_k} > 0$ for $k = 1, \ldots, n$. Noting that the distribution of $(\gamma_1, \ldots, \gamma_{n+1})$ has positive density $\prod_{k \leq n+1} e^{-x_k} > 0$ on $\mathbb{R}_+^{n+1}$, we obtain for any $t > 0$

$$p_{ij}^t \geq P\left\{\sum_{k=1}^n \frac{\gamma_k}{c_{i_{k-1}}} \leq t < \sum_{k=1}^{n+1} \frac{\gamma_k}{c_{i_{k-1}}}\right\} \prod_{k=1}^n q_{i_{k-1}, i_k} > 0.$$

Since $p_{ii}^0 = q_{ii}^0 = 1$, we get in particular $p_{ii}^t > 0$ for all $t \geq 0$. □

A continuous-time Markov chain is said to be *irreducible* if $p_{ij}^t > 0$ for all $i, j \in S$ and $t > 0$. Note that this holds iff the associated discrete-time process Y in Theorem 12.17 is irreducible. In that case clearly $\sup\{t > 0; X_t = j\} < \infty$ iff $\sup\{n > 0; Y_n = j\} < \infty$. Thus, when Y is recurrent, the sets $\{t; X_t = j\}$ are a.s. unbounded under P_i for all $i \in S$; otherwise, they are a.s. bounded. The two possibilities are again referred to as *recurrence* and *transience*, respectively.

The basic ergodic Theorem 8.18 for discrete-time Markov chains has an analogous version in continuous time. Further extensions are considered in Chapter 20.

Theorem 12.25 (*ergodic behavior*) *For any irreducible, continuous-time Markov chain in S, exactly one of these cases occurs:*

(i) *There exists a unique invariant distribution ν, the latter satisfies $\nu_i > 0$ for all $i \in S$, and for any distribution μ on S,*

$$\lim_{t \to \infty} \| P_\mu \circ \theta_t^{-1} - P_\nu \| = 0. \tag{14}$$

(ii) *No invariant distribution exists, and $p_{ij}^t \to 0$ for all $i, j \in S$.*

Proof: By Lemma 12.24 the discrete-time chain X_{nh}, $n \in \mathbb{Z}_+$, is irreducible and aperiodic. Assume that (X_{nh}) is positive recurrent for some $h > 0$, say with invariant distribution ν. Then the chain $(X_{nh'})$ is positive recurrent for every h' of the form $2^{-m}h$, and by the uniqueness in Theorem 8.18 it has the same invariant distribution. Since the paths are right-continuous, we may conclude by a simple approximation that ν is invariant even for the original process X.

For any distribution μ on S we have

$$\| P_\mu \circ \theta_t^{-1} - P_\nu \| = \left\| \sum_i \mu_i \sum_j (p_{ij}^t - \nu_j) P_j \right\| \leq \sum_i \mu_i \sum_j |p_{ij}^t - \nu_j|.$$

Thus, (14) follows by dominated convergence if we can show that the inner sum on the right tends to zero. This is clear if we put $n = \lfloor t/h \rfloor$ and $r = t - nh$ and note that by Theorem 8.18

$$\sum_k |p_{ik}^t - \nu_k| \leq \sum_j \sum_k |p_{ij}^{nh} - \nu_j| p_{jk}^r = \sum_j |p_{ij}^{nh} - \nu_j| \to 0.$$

It remains to consider the case when (X_{nh}) is null recurrent or transient for every $h > 0$. Fixing any $i, k \in S$ and writing $n = \lfloor t/h \rfloor$ and $r = t - nh$ as before, we get

$$p_{ik}^t = \sum_j p_{ij}^r p_{jk}^{nh} \leq p_{ik}^{nh} + \sum_{j \neq i} p_{ij}^r = p_{ik}^{nh} + (1 - p_{ii}^r),$$

which tends to zero as $t \to \infty$ and then $h \to 0$, due to Theorem 8.18 and the continuity of p_{ii}^t. □

As in discrete time, we note that condition (ii) of the last theorem holds for any transient Markov chain, whereas a recurrent chain may satisfy either condition. Recurrent chains satisfying (i) and (ii) are again referred to as *positive recurrent* and *null-recurrent*, respectively. It is interesting to note that X may be positive recurrent even when the embedded, discrete-time chain Y is null-recurrent, and vice versa. On the other hand, X clearly has the same ergodic properties as the discrete-time processes (X_{nh}), $h > 0$.

Let us next introduce the *first exit* and *recurrence times*

$$\gamma_j = \inf\{t > 0;\ X_t \neq j\}, \qquad \tau_j = \inf\{t > \gamma_j;\ X_t = j\}.$$

As in Theorem 8.22 for the discrete-time case, we may express the asymptotic transition probabilities in terms of the mean recurrence times $E_j \tau_j$. To avoid trivial exceptions, we confine our attention to nonabsorbing states.

Theorem 12.26 *(mean recurrence times)* *For any continuous-time Markov chain in S and states $i, j \in S$ with j nonabsorbing, we have*

$$\lim_{t \to \infty} p_{ij}^t = \frac{P_i\{\tau_j < \infty\}}{c_j E_j \tau_j}. \tag{15}$$

Proof: It is enough to take $i = j$, since the general statement will then follow as in the proof of Theorem 8.22. If j is transient, then $1\{X_t = j\} \to 0$ a.s. P_j, and so by dominated convergence $p_{jj}^t = P_j\{X_t = j\} \to 0$. This agrees with (15), since in this case $P_j\{\tau_j = \infty\} > 0$. Turning to the recurrent case, let S_j denote the class of states i accessible from j. Then S_j is clearly irreducible, and so p_{jj}^t converges by Theorem 12.25.

To identify the limit, define

$$L_t^j = \lambda\{s \leq t;\ X_s = j\} = \int_0^t 1\{X_s = j\} ds, \quad t \geq 0,$$

and let τ_j^n denote the instant of nth return to j. Letting $m, n \to \infty$ with $|m - n| \leq 1$, and using the strong Markov property and the law of large numbers, we get a.s. P_j

$$\frac{L^j(\tau_j^m)}{\tau_j^n} = \frac{L^j(\tau_j^m)}{m} \cdot \frac{n}{\tau_j^n} \cdot \frac{m}{n} \to \frac{E_j \gamma_j}{E_j \tau_j} = \frac{1}{c_j E_j \tau_j}.$$

By the monotonicity of L^j, it follows that $t^{-1} L_t^j \to (c_j E_j \tau_j)^{-1}$ a.s. Hence, by Fubini's theorem and dominated convergence,

$$\frac{1}{t} \int_0^t p_{jj}^s ds = \frac{E_j L_t^j}{t} \to \frac{1}{c_j E_j \tau_j},$$

and (15) follows. □

Exercises

1. Let ξ be a point process on a Borel space S. Show that $\xi = \sum_k \delta_{\tau_k}$ for some random elements τ_k in $S \cup \{\Delta\}$, where $\Delta \notin S$ is arbitrary. Extend the result to general random measures. (*Hint:* We may assume that $S = \mathbb{R}_+$.)

2. Show that two random measures ξ and η are independent iff $Ee^{-\xi f - \eta g} = Ee^{-\xi f} Ee^{-\eta g}$ for all measurable $f, g \geq 0$. Also, in case of simple point processes, prove the equivalence of $P\{\xi B + \eta C = 0\} = P\{\xi B = 0\} P\{\eta C = 0\}$ for any $B, C \in \mathcal{S}$. (*Hint:* Regard (ξ, η) as a random measure on $2S$.)

3. Let $\xi_1, \xi_2, \ldots$ be independent Poisson processes with intensity measures $\mu_1, \mu_2, \ldots$ such that the measure $\mu = \sum_k \mu_k$ is σ-finite. Show that $\xi = \sum_k \xi_k$ is again Poisson with intensity measure μ.

4. Show that the classes of mixed Poisson and binomial processes are preserved under randomization.

5. Let ξ be a Cox process on S directed by some random measure η, and let f be a measurable mapping into some space T such that $\eta \circ f^{-1}$ is a.s. σ-finite. Prove directly from definitions that $\xi \circ f^{-1}$ is a Cox process on T directed by $\eta \circ f^{-1}$. Derive a corresponding result for p-thinnings. Also show how the result follows from Proposition 12.3.

6. Consider a p-thinning η of ξ and a q-thinning ζ of η with $\zeta \perp\!\!\!\perp_\eta \xi$. Show that ζ is a pq-thinning of ξ.

7. Let ξ be a Cox process directed by η or a p-thinning of η with $p \in (0,1)$, and fix two disjoint sets $B, C \in \mathcal{S}$. Show that $\xi B \perp\!\!\!\perp \xi C$ iff $\eta B \perp\!\!\!\perp \eta C$. (*Hint:* Compute the Laplace transforms. The *if* assertions can also be obtained from Proposition 6.8.)

8. Use Lemma 12.2 to derive expressions for $P\{\xi B = 0\}$ when ξ is a Cox process directed by η, a μ-randomization of η, or a p-thinning of η. (*Hint:* Note that $Ee^{-t\xi B} \to P\{\xi B = 0\}$ as $t \to 0$.)

9. Let ξ be a p-thinning of η, where $p \in (0,1)$. Show that ξ and η are simultaneously Cox. (*Hint:* Use Lemma 12.6.)

10. (Fichtner) For a fixed $p \in (0,1)$, let η be a p-thinning of a point process ξ on S. Show that ξ is Poisson iff $\eta \perp\!\!\!\perp \xi - \eta$. (*Hint:* Extend by iteration to arbitrary p. Then a uniform randomization of ξ on $S \times [0,1]$ has independent increments in the second variable, and the result follows by Theorem 18.3.)

11. Use Theorem 12.8 to give a simplified proof of Theorem 12.4 in the case when ξ is simple.

12. Derive Theorem 12.4 from Theorem 12.12. (*Hint:* Note that ξ is symmetric on S iff it is symmetric on B_n for every n. If ξ is simple, the assertion follows immediately from Theorem 12.12. Otherwise, apply the same result to a uniform randomization on $S \times [0,1]$.)

13. For ξ as Theorem 12.12, show that $P\{\xi B = 0\} = \varphi(\mu B)$ for some completely monotone function φ. Conclude from the Hausdorff–Bernstein characterization and Theorem 12.8 that ξ is a mixed Poisson or binomial process based on μ.

14. Show that the distribution of a simple point process ξ on $\mathbb{R}$ is *not* determined, in general, by the distributions of ξI for all intervals I. (*Hint:* If ξ is restricted to $\{1, \ldots, n\}$, then the distributions of all ξI give $\sum_{k \leq n} k(n - k + 1) \leq n^3$ linear relations between the $2^n - 1$ parameters.)

15. Show that the distribution of a point process is *not* determined, in general, by the one-dimensional distributions. (*Hint:* If ξ is restricted to $\{0, 1\}$ with $\xi\{0\} \vee \xi\{1\} \leq n$, then the one-dimensional distributions give $4n$ linear relations between the $n(n+2)$ parameters.)

16. Show that Lemma 12.1 remains valid with $B_1, \ldots, B_n$ restricted to an arbitrary preseparating class $\mathcal{C}$, as defined in Chapter 16 or Appendix A2. Also show that Theorem 12.8 holds with B restricted to a separating class. (*Hint:* Extend to the case when $\mathcal{C} = \{B \in \mathcal{S};\ (\xi + \eta)\partial B = 0\ \text{a.s.}\}$. Then use monotone class arguments for sets in S and in $\mathcal{M}(S)$.)

17. Show that Theorem 12.10 fails in general without the condition $\xi(\{s\} \times K) = 0$ a.s. for all s.

18. Give an example of a non-Poisson point process ξ on S such that ξB is Poisson for every $B \in \mathcal{S}$. (*Hint:* It suffices to take $S = \{0, 1\}$.)

19. Extend Corollary 12.11 to the case when $p_s = P\{\xi\{s\} > 0\}$ may be positive. (*Hint:* By Fatou's lemma, $p_s > 0$ for at most countably many s.)

20. Prove Theorem 12.13 (i) and (iii) by means of characteristic functions.

21. Let ξ and η be independent Poisson processes on S with $E\xi = E\eta = \mu$, and let $f_1, f_2, \ldots : S \to \mathbb{R}$ be measurable with $\infty > \mu(f_n^2 \wedge 1) \to \infty$. Show that $|(\xi - \eta)f_n| \xrightarrow{P} \infty$. (*Hint:* Consider the symmetrization $\tilde{\nu}$ of a fixed measure $\nu \in \mathcal{N}(S)$ with $\nu f_n^2 \to \infty$, and argue along subsequences as in the proof of Theorem 4.17.)

22. For any pure jump-type Markov process on S, show that $P_x\{\tau_2 \leq t\} = o(t)$ for all $x \in S$. Also note that the bound can be sharpened to $O(t^2)$ if the rate function is bounded, but not in general. (*Hint:* Use Lemma 12.16 and dominated convergence.)

23. Show that any transient, discrete-time Markov chain Y can be embedded into an exploding (resp., nonexploding) continuous-time chain X. (*Hint:* Use Propositions 8.12 and 12.19.)

24. In Corollary 12.21, use the measurability of the mapping $X = Y \circ N$ to deduce the implication (iii) $\Rightarrow$ (i) from its converse. (*Hint:* Proceed as in the proof of Proposition 12.15.) Also use Proposition 12.3 to show that (iii) implies (ii), and prove the converse by means of Theorem 12.10.

25. Consider a pure jump-type Markov process on $(S, \mathcal{S})$ with transition kernels μ_t and rate kernel α. Show for any $x \in S$ and $B \in \mathcal{S}$ that $\alpha(x, B) = \dot{\mu}_0(x, B \setminus \{x\})$. (*Hint:* Take $f = 1_{B \setminus \{x\}}$ in Theorem 12.22, and use dominated convergence.)

26. Use Theorem 12.22 to derive a system of differential equations for the transition functions $p_{ij}(t)$ of a continuous-time Markov chain. (*Hint:* Take $f(i) = \delta_{ij}$ for fixed j.)

27. Give an example of a positive recurrent, continuous-time Markov chain such that the embedded discrete-time chain is null-recurrent, and vice versa. (*Hint:* Use Proposition 12.23.)

28. Establish Theorem 12.25 by a direct argument, mimicking the proof of Theorem 8.18.

Chapter 13

Gaussian Processes and Brownian Motion

Symmetries of Gaussian distribution; existence and path properties of Brownian motion; strong Markov and reflection properties; arcsine and uniform laws; law of the iterated logarithm; Wiener integrals and isonormal Gaussian processes; multiple Wiener–Itô integrals; chaos expansion of Brownian functionals

The main purpose of this chapter is to initiate the study of Brownian motion, arguably the single most important object in modern probability theory. Indeed, we shall see in Chapters 14 and 16 how the Gaussian limit theorems of Chapter 5 can be extended to approximations of broad classes of random walks and discrete-time martingales by a Brownian motion. In Chapter 18 we show how every continuous local martingale may be represented in terms of Brownian motion through a suitable random time-change. Similarly, the results of Chapters 21 and 23 demonstrate how large classes of diffusion processes may be constructed from Brownian motion by various pathwise transformations. Finally, a close relationship between Brownian motion and classical potential theory is uncovered in Chapters 24 and 25.

The easiest construction of Brownian motion is via a so-called isonormal Gaussian process on $L^2(\mathbb{R}_+)$, whose existence is a consequence of the characteristic spherical symmetry of the multivariate Gaussian distributions. Among the many important properties of Brownian motion, this chapter covers the Hölder continuity and existence of quadratic variation, the strong Markov and reflection properties, the three arcsine laws, and the law of the iterated logarithm.

The values of an isonormal Gaussian process on $L^2(\mathbb{R}_+)$ may be identified with integrals of L^2-functions with respect to the associated Brownian motion. Many processes of interest have representations in terms of such integrals, and in particular we shall consider spectral and moving average representations of stationary Gaussian processes. More generally, we shall introduce the multiple Wiener–Itô integrals $I_n f$ of functions $f \in L^2(\mathbb{R}_+^n)$ and establish the fundamental chaos expansion of Brownian L^2-functionals.

The present material is related to practically every other chapter in the book. Thus, we refer to Chapter 5 for the definition of Gaussian distributions and the basic Gaussian limit theorem, to Chapter 6 for the transfer theorem, to Chapter 7 for properties of martingales and optional times, to Chapter 8 for basic facts about Markov processes, to Chapter 9 for similarities with random walks, to Chapter 11 for some basic symmetry results, and to Chapter 12 for analogies with the Poisson process.

Our study of Brownian motion per se is continued in Chapter 18 with the basic recurrence or transience dichotomy, some further invariance properties, and a representation of Brownian martingales. Brownian local time and additive functionals are studied in Chapter 22. In Chapter 24 we consider some basic properties of Brownian hitting distributions, and in Chapter 25 we examine the relationship between excessive functions and additive functionals of Brownian motion. A further discussion of multiple integrals and chaos expansions appears in Chapter 18.

To begin with some basic definitions, we say that a process X on some parameter space T is *Gaussian* if the random variable $c_1 X_{t_1} + \cdots + c_n X_{t_n}$ is Gaussian for any choice of $n \in \mathbb{N}$, $t_1, \ldots, t_n \in T$, and $c_1, \ldots, c_n \in \mathbb{R}$. This holds in particular if the X_t are independent Gaussian random variables. A Gaussian process X is said to be *centered* if $EX_t = 0$ for all $t \in T$. Let us also say that the processes X^i on T_i, $i \in I$, are *jointly Gaussian* if the combined process $X = \{X_t^i; t \in T_i, i \in I\}$ is Gaussian. The latter condition is certainly fulfilled if the processes X^i are independent and Gaussian.

The following simple facts clarify the fundamental role of the covariance function. As usual, we assume all distributions to be defined on the σ-fields generated by the evaluation maps.

Lemma 13.1 *(covariance function)*

(i) *The distribution of a Gaussian process X on T is determined by the functions EX_t and $\text{cov}(X_s, X_t)$, $s, t \in T$.*

(ii) *The jointly Gaussian processes X^i on T_i, $i \in I$, are independent iff $\text{cov}(X_s^i, X_t^j) = 0$ for all $s \in T_i$ and $t \in T_j$, $i \neq j$ in I.*

Proof: (i) Let X and Y be Gaussian processes on T with the same means and covariances. Then the random variables $c_1 X_{t_1} + \cdots + c_n X_{t_n}$ and $c_1 Y_{t_1} + \cdots + c_n Y_{t_n}$ have the same mean and variance for any $c_1, \ldots, c_n \in \mathbb{R}$ and $t_1, \ldots, t_n \in T$, $n \in \mathbb{N}$, and since both variables are Gaussian, their distributions must agree. By the Cramér–Wold theorem it follows that $(X_{t_1}, \ldots, X_{t_n}) \stackrel{d}{=} (Y_{t_1}, \ldots, Y_{t_n})$ for any $t_1, \ldots, t_n \in T$, $n \in \mathbb{N}$, and so $X \stackrel{d}{=} Y$ by Proposition 3.2.

(ii) Assume the stated condition. To prove the asserted independence, we may assume I to be finite. Introduce some independent processes Y^i, $i \in I$, with the same distributions as the X^i, and note that the combined processes $X = (X^i)$ and $Y = (Y^i)$ have the same means and covariances. Hence, the joint distributions agree by part (i). In particular, the independence between the processes Y^i implies the corresponding property for the processes X^i. □

The following result characterizes the Gaussian distributions by a simple symmetry property.

Proposition 13.2 *(spherical symmetry, Maxwell)* *Let $\xi_1, \ldots, \xi_d$ be independent random variables, where $d \geq 2$. Then the distribution of $(\xi_1, \ldots, \xi_d)$ is spherically symmetric iff the ξ_i are i.i.d. centered Gaussian.*

Proof: Let φ denote the common characteristic function of $\xi_1, \ldots, \xi_d$, and assume the stated condition. In particular, $-\xi_1 \stackrel{d}{=} \xi_1$, and so φ is real valued and symmetric. Noting that $s\xi_1 + t\xi_2 \stackrel{d}{=} \xi_1 \sqrt{s^2 + t^2}$, we obtain the functional equation $\varphi(s)\varphi(t) = \varphi(\sqrt{s^2 + t^2})$, and so by iteration $\varphi^n(t) = \varphi(t\sqrt{n})$ for all n. Thus, for rational t^2 we have $\varphi(t) = e^{at^2}$ for some constant a, and by continuity this extends to all $t \in \mathbb{R}$. Finally, we have $a \leq 0$ since $|\varphi| \leq 1$.

Conversely, let $\xi_1, \ldots, \xi_d$ be i.i.d. centered Gaussian, and assume that $(\eta_1, \ldots, \eta_d) = T(\xi_1, \ldots, \xi_d)$ for some orthogonal transformation T. Then both random vectors are Gaussian, and we may easily verify that $\text{cov}(\eta_i, \eta_j) = \text{cov}(\xi_i, \xi_j)$ for all i and j. Hence, the two distributions agree by Lemma 13.1. □

In infinite dimensions, the Gaussian property is essentially a consequence of the rotational symmetry alone, without any assumption of independence.

Theorem 13.3 *(unitary invariance, Schoenberg, Freedman)* *For any infinite sequence of random variables $\xi_1, \xi_2, \ldots$, the distribution of $(\xi_1, \ldots, \xi_n)$ is spherically symmetric for every $n \geq 1$ iff the ξ_k are conditionally i.i.d. $N(0, \sigma^2)$, given some random variable $\sigma^2 \geq 0$.*

Proof: The ξ_n are clearly exchangeable, and so by Theorem 11.10 there exists a random probability measure μ such that the ξ_n are conditionally μ-i.i.d. given μ. By the law of large numbers,

$$\mu B = \lim_{n \to \infty} n^{-1} \sum_{k \leq n} 1\{\xi_k \in B\} \text{ a.s.,} \quad B \in \mathcal{B},$$

and in particular μ is a.s. $\{\xi_3, \xi_4, \ldots\}$-measurable. Now the spherical symmetry implies that, for any orthogonal transformation T on $\mathbb{R}^2$,

$$P[(\xi_1, \xi_2) \in B | \xi_3, \ldots, \xi_n] = P[T(\xi_1, \xi_2) \in B | \xi_3, \ldots, \xi_n], \quad B \in \mathcal{B}(\mathbb{R}^2).$$

As $n \to \infty$, we get $\mu^2 = \mu^2 \circ T^{-1}$ a.s. Considering a countable dense set of mappings T, it is clear that the exceptional null set can be chosen to be independent of T. Thus, μ^2 is a.s. spherically symmetric, and so μ is a.s. centered Gaussian by Proposition 13.2. It remains to take $\sigma^2 = \int x^2 \mu(dx)$. □

Now fix a separable Hilbert space H. By an *isonormal Gaussian process* on H we mean a centered Gaussian process ηh, $h \in H$, such that $E(\eta h \, \eta k) = \langle h, k \rangle$, the inner product of h and k. To construct such a process η, we may introduce an orthonormal basis (ONB) $e_1, e_2, \cdots \in H$, and let $\xi_1, \xi_2, \ldots$ be independent $N(0, 1)$ random variables. For any element

$h = \sum_i b_i e_i$, we define $\eta h = \sum_i b_i \xi_i$, where the series converges a.s. and in L^2 since $\sum_i b_i^2 < \infty$. The process η is clearly centered Gaussian. It is also linear, in the sense that $\eta(ah + bk) = a\eta h + b\eta k$ a.s. for all $h, k \in H$ and $a, b \in \mathbb{R}$. Assuming $k = \sum_i c_i e_i$, we may compute

$$E(\eta h\, \eta k) = \sum_{i,j} b_i c_j E(\xi_i \xi_j) = \sum_i b_i c_i = \langle h, k \rangle.$$

By Lemma 13.1 the stated conditions uniquely determine the distribution of η. In particular, the symmetry in Proposition 13.2 extends to a distributional invariance of η under any unitary transformation on H.

The following result shows how the Gaussian distribution arises naturally in the context of processes with independent increments. It is interesting to compare with the similar Poisson characterization in Theorem 12.10.

Theorem 13.4 *(independence and Gaussian property, Lévy)* Let X be a continuous process in $\mathbb{R}^d$ with independent increments and $X_0 = 0$. Then X is Gaussian, and there exist some continuous functions b in $\mathbb{R}^d$ and a in $\mathbb{R}^{d^2}$, the latter with nonnegative definite increments, such that $X_t - X_s$ is $N(b_t - b_s, a_t - a_s)$ for all $s < t$.

Proof: Fix any $s < t$ in $\mathbb{R}_+$ and $u \in \mathbb{R}^d$. For every $n \in \mathbb{N}$ we may divide the interval $[s, t]$ into n subintervals of equal length, and we denote the corresponding increments of uX by $\xi_{n1}, \ldots, \xi_{nn}$. By the continuity of X we have $\max_j |\xi_{nj}| \to 0$ a.s., and so Theorem 5.15 shows that $u(X_t - X_s) = \sum_j \xi_{nj}$ is a Gaussian random variable. Since X has independent increments, it follows that X is Gaussian. Writing $b_t = EX_t$ and $a_t = \text{cov}(X_t)$, we get $E(X_t - X_s) = EX_t - EX_s = b_t - b_s$, and so by independence

$$0 \leq \text{cov}(X_t - X_s) = \text{cov}(X_t) - \text{cov}(X_s) = a_t - a_s, \quad s < t.$$

The continuity of X yields $X_s \xrightarrow{d} X_t$ as $s \to t$, and so $b_s \to b_t$ and $a_s \to a_t$. Thus, both functions are continuous. $\square$

If the process X in Theorem 13.4 has stationary, independent increments, then the mean and covariance functions are clearly linear. The simplest choice in one dimension is to take $b = 0$ and $a_t = t$, so that $X_t - X_s$ is $N(0, t - s)$ for all $s < t$. The next result shows that the corresponding process exists; it also gives an estimate of the local modulus of continuity. More precise rates of continuity are obtained in Theorem 13.18 and Lemma 14.7.

Theorem 13.5 *(existence of Brownian motion, Wiener)* There exists a continuous Gaussian process B in $\mathbb{R}$ with stationary independent increments and $B_0 = 0$ such that B_t is $N(0, t)$ for every $t \geq 0$. Furthermore, B is a.s. locally Hölder continuous with exponent c for any $c \in (0, \frac{1}{2})$.

Proof: Let η be an isonormal Gaussian process on $L^2(\mathbb{R}_+, \lambda)$, and define $B_t = \eta 1_{[0,t]}$, $t \geq 0$. Since indicator functions of disjoint intervals are orthogonal, the increments of the process B are uncorrelated and hence independent. Furthermore, we have $\|1_{(s,t]}\|^2 = t - s$ for any $s \leq t$, and so $B_t - B_s$ is $N(0, t-s)$. For any $s \leq t$ we get

$$B_t - B_s \stackrel{d}{=} B_{t-s} \stackrel{d}{=} (t-s)^{1/2} B_1, \tag{1}$$

whence,

$$E|B_t - B_s|^c = (t-s)^{c/2} E|B_1|^c < \infty, \quad c > 0.$$

The asserted Hölder continuity now follows by Theorem 3.23. □

A process B as in Theorem 13.5 is called a *(standard) Brownian motion* or a *Wiener process*. By a Brownian motion in $\mathbb{R}^d$ we mean a process $B_t = (B_t^1, \ldots, B_t^d)$, where $B^1, \ldots, B^d$ are independent, one-dimensional Brownian motions. From Proposition 13.2 we note that the distribution of B is invariant under orthogonal transformations of $\mathbb{R}^d$. It is also clear that any continuous process X in $\mathbb{R}^d$ with stationary independent increments and $X_0 = 0$ can be written as $X_t = bt + \sigma B_t$ for some vector b and matrix σ.

From Brownian motion we may construct other important Gaussian processes. For example, a *Brownian bridge* may be defined as a process on $[0, 1]$ with the same distribution as $X_t = B_t - tB_1$, $t \in [0, 1]$. An easy computation shows that X has covariance function $r_{s,t} = s(1 - t)$, $0 \leq s \leq t \leq 1$.

The Brownian motion and bridge have many nice symmetry properties. For example, if B is a Brownian motion, then so is $-B$ as well as the process $c^{-1} B(c^2 t)$ for any $c > 0$. The latter transformation is especially useful and is often referred to as a *Brownian scaling*. We also note that, for each $u > 0$, the processes $B_{u \pm t} - B_u$ are Brownian motions on $\mathbb{R}_+$ and $[0, u]$, respectively. If B is instead a Brownian bridge, then so are the processes $-B_t$ and B_{1-t}.

The following result gives some less obvious invariance properties. Further, possibly random mappings that preserve the distribution of a Brownian motion or bridge are exhibited in Theorem 13.11, Lemma 13.14, and Proposition 18.9.

Lemma 13.6 *(scaling and inversion)* *If B is a Brownian motion, then so is the process $tB_{1/t}$, whereas $(1-t)B_{t/(1-t)}$ and $tB_{(1-t)/t}$ are Brownian bridges. If B is instead a Brownian bridge, then the processes $(1+t)B_{t/(1+t)}$ and $(1+t)B_{1/(1+t)}$ are Brownian motions.*

Proof: Since all processes are centered Gaussian, it suffices by Lemma 13.1 to verify that they have the desired covariance functions. This is clear from the expressions $s \wedge t$ and $(s \wedge t)(1 - s \vee t)$ for the covariance functions of the Brownian motion and bridge. □

From Proposition 8.5 together with Theorem 13.4 we note that any space- and time-homogeneous, continuous Markov process in $\mathbb{R}^d$ has the form $\sigma B_t + tb + c$, where B is a Brownian motion in $\mathbb{R}^d$, σ is a $d \times d$ matrix, and

b and c are vectors in $\mathbb{R}^d$. The next result gives a general characterization of Gaussian Markov processes. Here we use the convention $0/0 = 0$.

Proposition 13.7 *(Gaussian Markov processes) Let X be a Gaussian process on some index set $T \subset \mathbb{R}$, and define $r_{s,t} = \mathrm{cov}(X_s, X_t)$. Then X is Markov iff*

$$r_{s,u} = r_{s,t} r_{t,u}/r_{t,t}, \quad s \leq t \leq u \text{ in } T. \tag{2}$$

If X is further stationary and defined on $\mathbb{R}$, then $r_{s,t} = ae^{-b|s-t|}$ for some constants $a \geq 0$ and $b \in [0, \infty]$.

Proof: Subtracting the means if necessary, we may assume that $EX_t \equiv 0$. Now fix any times $t \leq u$ in T, and choose $a \in \mathbb{R}$ such that $X'_u \equiv X_u - aX_t \perp X_t$. Then $a = r_{t,u}/r_{t,t}$ when $r_{t,t} \neq 0$, and if $r_{t,t} = 0$, we may take $a = 0$. By Lemma 13.1 we get $X'_u \perp\!\!\!\perp X_t$.

First assume that X is Markov, and let $s \leq t$ be arbitrary. Then $X_s \perp\!\!\!\perp_{X_t} X_u$, and so $X_s \perp\!\!\!\perp_{X_t} X'_u$. Since also $X_t \perp\!\!\!\perp X'_u$ by the choice of a, Proposition 6.8 yields $X_s \perp\!\!\!\perp X'_u$. Hence, $r_{s,u} = ar_{s,t}$, and (2) follows as we insert the expression for a. Conversely, (2) implies $X_s \perp X'_u$ for all $s \leq t$, and so $\mathcal{F}_t \perp\!\!\!\perp X'_u$ by Lemma 13.1, where $\mathcal{F}_t = \sigma\{X_s; s \leq t\}$. By Proposition 6.8 it follows that $\mathcal{F}_t \perp\!\!\!\perp_{X_t} X_u$, which is the required Markov property of X at t.

If X is stationary, then $r_{s,t} = r_{|s-t|,0} = r_{|s-t|}$, and (2) reduces to the Cauchy equation $r_0 r_{s+t} = r_s r_t$, $s, t \geq 0$, which admits the only bounded solutions $r_t = ae^{-bt}$. □

A continuous, centered Gaussian process on $\mathbb{R}$ with covariance function $r_t = \frac{1}{2}e^{-|t|}$ is called a stationary *Ornstein–Uhlenbeck process*. Such a process Y can be expressed in terms of a Brownian motion B as $Y_t = e^{-t}B(\frac{1}{2}e^{2t})$, $t \in \mathbb{R}$. The last result shows that the Ornstein–Uhlenbeck process is essentially the only stationary Gaussian process that is also a Markov process.

We will now study some basic sample path properties of Brownian motion.

Lemma 13.8 *(level sets) If B is a Brownian motion or bridge, then*

$$\lambda\{t; B_t = u\} = 0 \text{ a.s.}, \quad u \in \mathbb{R}.$$

Proof: Introduce the processes $X_t^n = B_{[nt]/n}$, $t \in \mathbb{R}_+$ or $[0, 1]$, $n \in \mathbb{N}$, and note that $X_t^n \to B^t$ for every t. Since each process X^n is product measurable on $\Omega \times \mathbb{R}_+$ or $\Omega \times [0, 1]$, the same thing is true for B. Now use Fubini's theorem to conclude that

$$E\lambda\{t; B_t = u\} = \int P\{B_t = u\}dt = 0, \quad u \in \mathbb{R}. \qquad \square$$

The next result shows that Brownian motion has locally finite quadratic variation. An extension to general continuous semimartingales is obtained in Proposition 17.17.

13. Gaussian Processes and Brownian Motion

Theorem 13.9 *(quadratic variation, Lévy)* Let B be a Brownian motion, and fix any $t > 0$ and a sequence of partitions $0 = t_{n,0} < t_{n,1} < \cdots < t_{n,k_n} = t$, $n \in \mathbb{N}$, such that $h_n \equiv \max_k(t_{n,k} - t_{n,k-1}) \to 0$. Then

$$\zeta_n \equiv \sum_k (B_{t_{n,k}} - B_{t_{n,k-1}})^2 \to t \quad \text{in } L^2. \tag{3}$$

If the partitions are nested, then also $\zeta_n \to t$ a.s.

Proof (Doob): To prove (3), we may use the scaling property $B_t - B_s \stackrel{d}{=} |t-s|^{1/2} B_1$ to obtain

$$
\begin{aligned}
E\zeta_n &= \sum_k E(B_{t_{n,k}} - B_{t_{n,k-1}})^2 \\
&= \sum_k (t_{n,k} - t_{n,k-1}) E B_1^2 = t, \\
\text{var}(\zeta_n) &= \sum_k \text{var}(B_{t_{n,k}} - B_{t_{n,k-1}})^2 \\
&= \sum_k (t_{n,k} - t_{n,k-1})^2 \text{var}(B_1^2) \le h_n t E B_1^4 \to 0.
\end{aligned}
$$

For nested partitions we may prove the a.s. convergence by showing that the sequence (ζ_n) is a reverse martingale, that is,

$$E[\zeta_{n-1} - \zeta_n | \zeta_n, \zeta_{n+1}, \ldots] = 0 \text{ a.s.}, \quad n \in \mathbb{N}. \tag{4}$$

Inserting intermediate partitions if necessary, we may assume that $k_n = n$ for all n. In that case there exist some numbers $t_1, t_2, \cdots \in [0,t]$ such that the nth partition has division points $t_1, \ldots, t_n$. To verify (4) for a fixed n, we may further introduce an auxiliary random variable $\vartheta \perp\!\!\!\perp B$ with $P\{\vartheta = \pm 1\} = \frac{1}{2}$, and replace B by the Brownian motion

$$B'_s = B_{s \wedge t_n} + \vartheta(B_s - B_{s \wedge t_n}), \quad s \ge 0.$$

Since B' has the same sums $\zeta_n, \zeta_{n+1}, \ldots$ as B whereas $\zeta_{n-1} - \zeta_n$ is replaced by $\vartheta(\zeta_n - \zeta_{n-1})$, it is enough to show that $E[\vartheta(\zeta_n - \zeta_{n-1})|\zeta_n, \zeta_{n+1}, \ldots]$ = 0 a.s. This is clear from the choice of ϑ if we first condition on $\zeta_{n-1}, \zeta_n, \ldots$. $\square$

The last result implies that B has locally unbounded variation. This explains why the stochastic integral $\int V dB$ cannot be defined as an ordinary Stieltjes integral and a more sophisticated approach is required in Chapter 17.

Corollary 13.10 *(linear variation)* *Brownian motion has a.s. unbounded variation on every interval $[s,t]$ with $s < t$.*

Proof: The quadratic variation vanishes for any continuous function of bounded variation on $[s,t]$. $\square$

From Proposition 8.5 we note that Brownian motion B is a space-homogeneous Markov process with respect to its induced filtration. If the Markov property holds for some more general filtration $\mathcal{F} = (\mathcal{F}_t)$ —that is, if B is adapted to $\mathcal{F}$ and such that the process $B'_t = B_{s+t} - B_s$ is independent of

$\mathcal{F}_s$ for each $s \geq 0$ —we say that B is a Brownian motion with respect to $\mathcal{F}$, or an $\mathcal{F}$-Brownian motion. In particular, we may take $\mathcal{F}_t = \mathcal{G}_t \vee \mathcal{N}$, $t \geq 0$, where $\mathcal{G}$ is the filtration induced by B and $\mathcal{N} = \sigma\{N \subset A;\ A \in \mathcal{A},\ PA = 0\}$. With this construction, $\mathcal{F}$ becomes right-continuous by Corollary 7.25.

The Markov property of B will now be extended to suitable optional times. A more general version of this result appears in Theorem 19.17. As in Chapter 7, we write $\mathcal{F}_t^+ = \mathcal{F}_{t+}$.

Theorem 13.11 *(strong Markov property, Hunt) For any $\mathcal{F}$-Brownian motion B in $\mathbb{R}^d$ and a.s. finite $\mathcal{F}^+$-optional time τ, the process $B'_t = B_{\tau+t} - B_\tau$, $t \geq 0$, is again a Brownian motion independent of $\mathcal{F}_\tau^+$.*

Proof: As in Lemma 7.4, we may choose some optional times $\tau_n \to \tau$ that take countably many values and satisfy $\tau_n \geq \tau + 2^{-n}$. Then $\mathcal{F}_\tau^+ \subset \bigcap_n \mathcal{F}_{\tau_n}$ by Lemmas 7.1 and 7.3, and so by Proposition 8.9 and Theorem 8.10 each process $B_t^n = B_{\tau_n+t} - B_{\tau_n}$, $t \geq 0$, is a Brownian motion independent of $\mathcal{F}_\tau^+$. The continuity of B yields $B_t^n \to B'_t$ a.s. for every t. By dominated convergence we then obtain, for any $A \in \mathcal{F}_\tau^+$ and $t_1, \ldots, t_k \in \mathbb{R}_+$, $k \in \mathbb{N}$, and for bounded continuous functions $f: \mathbb{R}^k \to \mathbb{R}$,

$$E[f(B'_{t_1}, \ldots, B'_{t_k}); A] = Ef(B_{t_1}, \ldots, B_{t_k}) \cdot PA.$$

The general relation $P[B' \in \cdot, A] = P\{B \in \cdot\} \cdot PA$ now follows by a straightforward extension argument. □

If B is a Brownian motion in $\mathbb{R}^d$, then a process with the same distribution as $|B|$ is called a *Bessel process* of order d. More general Bessel processes may be obtained as solutions to suitable SDEs. The next result shows that $|B|$ inherits the strong Markov property from B.

Corollary 13.12 *(Bessel processes) If B is an $\mathcal{F}$-Brownian motion in $\mathbb{R}^d$, then $|B|$ is a strong $\mathcal{F}^+$-Markov process.*

Proof: By Theorem 13.11 it is enough to show that $|B + x| \stackrel{d}{=} |B + y|$ whenever $|x| = |y|$. We may then choose an orthogonal transformation T on $\mathbb{R}^d$ with $Tx = y$, and note that

$$|B + x| = |T(B + x)| = |TB + y| \stackrel{d}{=} |B + y|.$$ □

We shall use the strong Markov property to derive the distribution of the maximum of Brownian motion up to a fixed time. A stronger result is obtained in Corollary 22.3.

Proposition 13.13 *(maximum process, Bachelier) Let B be a Brownian motion in $\mathbb{R}$, and define $M_t = \sup_{s \leq t} B_s$, $t \geq 0$. Then*

$$M_t \stackrel{d}{=} M_t - B_t \stackrel{d}{=} |B_t|, \quad t \geq 0.$$

For the proof we need the following continuous-time counterpart to Lemma 9.10.

Lemma 13.14 *(reflection principle) Consider a Brownian motion B and an associated optional time τ. Then B has the same distribution as the reflected process*
$$\tilde{B}_t = B_{t\wedge\tau} - (B_t - B_{t\wedge\tau}), \quad t \geq 0.$$

Proof: It is enough to compare the distributions up to a fixed time t, and so we may assume that $\tau < \infty$. Define $B_t^\tau = B_{\tau \wedge t}$ and $B_t' = B_{\tau+t} - B_\tau$. By Theorem 13.11 the process B' is a Brownian motion independent of (τ, B^τ). Since, moreover, $-B' \stackrel{d}{=} B'$, we get $(\tau, B^\tau, B') \stackrel{d}{=} (\tau, B^\tau, -B')$. It remains to note that
$$B_t = B_t^\tau + B'_{(t-\tau)_+}, \quad \tilde{B}_t = B_t^\tau - B'_{(t-\tau)_+}, \quad t \geq 0. \qquad \Box$$

Proof of Proposition 13.13: By scaling it suffices to take $t = 1$. Applying Lemma 13.14 with $\tau = \inf\{t;\, B_t = x\}$ gives
$$P\{M_1 \geq x,\, B_1 \leq y\} = P\{\tilde{B}_1 \geq 2x - y\}, \quad x \geq y \vee 0.$$
By differentiation it follows that the pair (M_1, B_1) has probability density $-2\varphi'(2x - y)$, where φ denotes the standard normal density. Changing variables, we may conclude that $(M_1, M_1 - B_1)$ has density $-2\varphi'(x+y)$, $x, y \geq 0$. In particular, both M_1 and $M_1 - B_1$ have density $2\varphi(x)$, $x \geq 0$. $\qquad \Box$

To prepare for the next main result, we shall derive another elementary sample path property.

Lemma 13.15 *(local extremes) The local maxima and minima of a Brownian motion or bridge are a.s. distinct.*

Proof: Let B be a Brownian motion, and fix any intervals $I = [a, b]$ and $J = [c, d]$ with $b < c$. Write
$$\sup_{t\in J} B_t - \sup_{t\in I} B_t = \sup_{t\in J}(B_t - B_c) + (B_c - B_b) - \sup_{t\in I}(B_t - B_b).$$
Here the second term on the right has a diffuse distribution, and by independence the same thing is true for the whole expression. In particular, the difference on the left is a.s. nonzero. Since I and J are arbitrary, this proves the result for local maxima. The case of local minima and the mixed case are similar.

The result for the Brownian bridge B° follows from that for Brownian motion, since the distributions of the two processes are equivalent (mutually absolutely continuous) on any interval $[0, t]$ with $t < 1$. To see this, construct from B and B° the corresponding "bridges"
$$X_s = B_s - \frac{s}{t}B_t, \quad Y_s = B_s^\circ - \frac{s}{t}B_t^\circ, \quad s \in [0, t],$$
and check that $B_t \perp\!\!\!\perp X \stackrel{d}{=} Y \perp\!\!\!\perp B_t^\circ$. The stated equivalence now follows from the fact that $N(0, t) \sim N(0, t(1-t))$ when $t \in [0, 1)$. $\qquad \Box$

The next result involves the *arcsine law*, which may be defined as the distribution of $\xi = \sin^2 \alpha$ when α is $U(0, 2\pi)$. The name comes from the fact that

$$P\{\xi \le t\} = P\{|\sin \alpha| \le \sqrt{t}\} = \frac{2}{\pi} \arcsin \sqrt{t}, \quad t \in [0, 1].$$

Note that the arcsine distribution is symmetric around $\frac{1}{2}$, since

$$\xi = \sin^2 \alpha \stackrel{d}{=} \cos^2 \alpha = 1 - \sin^2 \alpha = 1 - \xi.$$

The following celebrated result exhibits three interesting functionals of Brownian motion, all of which are arcsine distributed.

Theorem 13.16 (arcsine laws, Lévy) *Let B be a Brownian motion on $[0, 1]$ with maximum M_1. Then these random variables are all arcsine distributed:*

$$\tau_1 = \lambda\{t; B_t > 0\}, \quad \tau_2 = \inf\{t; B_t = M_1\}, \quad \tau_3 = \sup\{t; B_t = 0\}.$$

It is interesting to compare the relations $\tau_1 \stackrel{d}{=} \tau_2 \stackrel{d}{=} \tau_3$ with the discrete-time versions obtained in Theorem 9.11 and Corollary 11.14. In Theorems 14.11 and 15.21, the arcsine laws are extended by approximation to appropriate random walks and Lévy processes.

Proof: To see that $\tau_1 \stackrel{d}{=} \tau_2$, let $n \in \mathbb{N}$, and note that by Corollary 11.14

$$n^{-1} \sum_{k \le n} 1\{B_{k/n} > 0\} \stackrel{d}{=} n^{-1} \min\{k \ge 0; B_{k/n} = \max_{j \le n} B_{j/n}\}.$$

By Lemma 13.15 the right-hand side tends a.s. to τ_2 as $n \to \infty$. To see that the left-hand side converges to τ_1, we may conclude from Lemma 13.8 that

$$\lambda\{t \in [0, 1]; B_t > 0\} + \lambda\{t \in [0, 1]; B_t < 0\} = 1 \quad \text{a.s.}$$

It remains to note that, for any open set $G \subset [0, 1]$,

$$\liminf_{n \to \infty} n^{-1} \sum_{k \le n} 1_G(k/n) \ge \lambda G.$$

In case of τ_2, fix any $t \in [0, 1]$, let ξ and η be independent $N(0, 1)$, and let α be $U(0, 2\pi)$. Using Proposition 13.13 and the circular symmetry of the distribution of (ξ, η), we get

$$\begin{aligned}
P\{\tau_2 \le t\} &= P\{\sup_{s \le t}(B_s - B_t) \ge \sup_{s \ge t}(B_s - B_t)\} \\
&= P\{|B_t| \ge |B_1 - B_t|\} = P\{t\xi^2 \ge (1-t)\eta^2\} \\
&= P\left\{\frac{\eta^2}{\xi^2 + \eta^2} \le t\right\} = P\{\sin^2 \alpha \le t\}.
\end{aligned}$$

In case of τ_3, we may write

$$\begin{aligned}
P\{\tau_3 < t\} &= P\{\sup_{s \ge t} B_s < 0\} + P\{\inf_{s \ge t} B_s > 0\} \\
&= 2P\{\sup_{s \ge t}(B_s - B_t) < -B_t\} = 2P\{|B_1 - B_t| < B_t\} \\
&= P\{|B_1 - B_t| < |B_t|\} = P\{\tau_2 \le t\}.
\end{aligned}$$

□

The first two arcsine laws have the following counterparts for the Brownian bridge.

Theorem 13.17 *(uniform laws)* Let B be a Brownian bridge with maximum M_1. Then these random variables are both $U(0,1)$:
$$\tau_1 = \lambda\{t;\ B_t > 0\}, \qquad \tau_2 = \inf\{t;\ B_t = M_1\}.$$

Proof: The relation $\tau_1 \stackrel{d}{=} \tau_2$ may be proved in the same way as for Brownian motion. To see that τ_2 is $U(0,1)$, write $(x) = x - [x]$, and consider for each $u \in [0,1]$ the process $B^u_t = B_{(u+t)} - B_u$, $t \in [0,1]$. It is easy to check that $B^u \stackrel{d}{=} B$ for each u, and further that the maximum of B^u occurs at $(\tau_2 - u)$. By Fubini's theorem we hence obtain for any $t \in [0,1]$
$$P\{\tau_2 \le t\} = \int_0^1 P\{(\tau_2 - u) \le t\}\,du = E\,\lambda\{u;\ (\tau_2 - u) \le t\} = t. \qquad \Box$$

From Theorem 13.5 we note that $t^{-c}B_t \to 0$ a.s. as $t \to 0$ for any $c \in [0, \tfrac{1}{2})$. The following classical result gives the exact growth rate of Brownian motion at 0 and ∞. Extensions to random walks and renewal processes are obtained in Corollaries 14.8 and 14.14. A functional version appears in Theorem 27.18.

Theorem 13.18 *(laws of the iterated logarithm, Khinchin)* For a Brownian motion B in $\mathbb{R}$, we have a.s.
$$\limsup_{t \to 0} \frac{B_t}{\sqrt{2t\log\log(1/t)}} = \limsup_{t \to \infty} \frac{B_t}{\sqrt{2t\log\log t}} = 1.$$

Proof: The Brownian inversion $\tilde{B}_t = tB_{1/t}$ of Lemma 13.6 converts the two formulas into one another, so it is enough to prove the result for $t \to \infty$. Then we note that as $u \to \infty$
$$\int_u^\infty e^{-x^2/2}dx \sim u^{-1}\int_u^\infty xe^{-x^2/2}dx = u^{-1}e^{-u^2/2}.$$

By Proposition 13.13 we hence obtain, uniformly in $t > 0$,
$$P\{M_t > ut^{1/2}\} = 2P\{B_t > ut^{1/2}\} \sim (2/\pi)^{1/2}u^{-1}e^{-u^2/2},$$

where $M_t = \sup_{s \le t} B_s$. Writing $h_t = (2t\log\log t)^{1/2}$, we get for any $r > 1$ and $c > 0$
$$P\{M(r^n) > ch(r^{n-1})\} \lesssim n^{-c^2/r}(\log n)^{-1/2}, \quad n \in \mathbb{N}.$$

Fixing $c > 1$ and choosing $r < c^2$, it follows by the Borel–Cantelli lemma that
$$P\{\limsup_{t \to \infty}(B_t/h_t) > c\} \le P\{M(r^n) > ch(r^{n-1})\ \text{i.o.}\} = 0,$$

which shows that $\limsup_{t \to \infty}(B_t/h_t) \le 1$ a.s.

To prove the reverse inequality, we may write
$$P\{B(r^n) - B(r^{n-1}) > ch(r^n)\} \gtrsim n^{-c^2r/(r-1)}(\log n)^{-1/2}, \quad n \in \mathbb{N}.$$

Taking $c = \{(r-1)/r\}^{1/2}$, we get by the Borel–Cantelli lemma

$$\limsup_{t\to\infty} \frac{B_t - B_{t/r}}{h_t} \geq \limsup_{n\to\infty} \frac{B(r^n) - B(r^{n-1})}{h(r^n)} \geq \left(\frac{r-1}{r}\right)^{1/2} \quad \text{a.s.}$$

The upper bound obtained earlier yields $\limsup_{t\to\infty}(-B_{t/r}/h_t) \leq r^{-1/2}$, and combining the two estimates gives

$$\limsup_{t\to\infty} \frac{B_t}{h_t} \geq (1 - r^{-1})^{1/2} - r^{-1/2} \quad \text{a.s.}$$

Here we may finally let $r \to \infty$ to obtain $\limsup_{t\to\infty}(B_t/h_t) \geq 1$ a.s. □

In the proof of Theorem 13.5 we constructed a Brownian motion B from an isonormal Gaussian process η on $L^2(\mathbb{R}_+, \lambda)$ such that $B_t = \eta 1_{[0,t]}$ a.s. for all $t \geq 0$. If instead we are starting from a Brownian motion B on $\mathbb{R}_+$, the existence of an associated isonormal Gaussian process η may be inferred from Theorem 6.10. Since every function $h \in L^2(\mathbb{R}_+, \lambda)$ can be approximated by simple step functions, as in the proof of Lemma 1.35, we note that the random variables ηh are a.s. unique. We shall see how they can also be constructed directly from B as suitable *Wiener integrals* $\int h\,dB$. As already noted, the latter fail to exist in the pathwise Stieltjes sense, and so a different approach is needed.

As a first step, we may consider the class $\mathcal{S}$ of simple step functions of the form

$$h_t = \sum_{j \leq n} a_j 1_{(t_{j-1}, t_j]}(t), \quad t \geq 0,$$

where $n \in \mathbb{Z}_+$, $0 = t_0 < \cdots < t_n$, and $a_1, \ldots, a_n \in \mathbb{R}$. For such integrands h, we may define the integral in the obvious way as

$$\eta h = \int_0^\infty h_t\,dB_t = Bh = \sum_{j \leq n} a_j(B_{t_j} - B_{t_{j-1}}).$$

Here ηh is clearly centered Gaussian with variance

$$E(\eta h)^2 = \sum_{j \leq n} a_j^2(t_j - t_{j-1}) = \int_0^\infty h_t^2\,dt = \|h\|^2,$$

where $\|h\|$ denotes the norm in $L^2(\mathbb{R}_+, \lambda)$. Thus, the integration $h \mapsto \eta h = \int h\,dB$ defines a linear isometry from $\mathcal{S} \subset L^2(\mathbb{R}_+, \lambda)$ into $L^2(\Omega, P)$.

Since $\mathcal{S}$ is dense in $L^2(\mathbb{R}_+, \lambda)$, we may extend the integral by continuity to a linear isometry $h \mapsto \eta h = \int h\,dB$ from $L^2(\lambda)$ to $L^2(P)$. Here ηh is again centered Gaussian for every $h \in L^2(\lambda)$, and by linearity the whole process $h \mapsto \eta h$ is then Gaussian. By a polarization argument it is also clear that the integration preserves inner products, in the sense that

$$E(\eta h\,\eta k) = \int_0^\infty h_t k_t\,dt = \langle h, k \rangle, \quad h, k \in L^2(\lambda).$$

We shall consider two general ways of representing stationary Gaussian processes in terms of Wiener integrals ηh. Here a complex notation is convenient. By a *complex-valued, isonormal Gaussian process* on a (real) Hilbert space H we mean a process $\zeta = \xi + i\eta$ on H such that ξ and η are independent, real-valued, isonormal Gaussian processes on H. For any $f = g + ih$ with $g, h \in H$, we define $\zeta f = \xi g - \eta h + i(\xi h + \eta g)$.

Now let X be a stationary, centered Gaussian process on $\mathbb{R}$ with covariance function $r_t = E X_s X_{s+t}$, $s, t \in \mathbb{R}$. We know that r is nonnegative definite, and it is further continuous whenever X is continuous in probability. In that case Bochner's theorem yields a unique *spectral representation*

$$r_t = \int_{-\infty}^{\infty} e^{itx} \mu(dx), \quad t \in \mathbb{R},$$

where the *spectral measure* μ is a bounded, symmetric measure on $\mathbb{R}$.

The following result gives a similar spectral representation of the process X itself. By a different argument, the result extends to suitable non-Gaussian processes. As usual, we assume that the basic probability space is rich enough to support the required randomization variables.

Proposition 13.19 *(spectral representation, Stone, Cramér)* Let X be an L^2-continuous, stationary, centered Gaussian process on $\mathbb{R}$ with spectral measure μ. Then there exists a complex, isonormal Gaussian process ζ on $L^2(\mu)$ such that

$$X_t = \Re \int_{-\infty}^{\infty} e^{itx} d\zeta_x \quad a.s., \quad t \in \mathbb{R}. \tag{5}$$

Proof: Denoting the right-hand side of (5) by Y, we may compute

$$E Y_s Y_t = E \int (\cos sx \, d\xi_x - \sin sx \, d\eta_x) \int (\cos tx \, d\xi_x - \sin tx \, d\eta_x)$$

$$= \int (\cos sx \cos tx - \sin sx \sin tx) \mu(dx)$$

$$= \int \cos(s-t)x \, \mu(dx) = \int e^{i(s-t)x} \mu(dx) = r_{s-t}.$$

Since both X and Y are centered Gaussian, Lemma 13.1 shows that $Y \stackrel{d}{=} X$. Now both X and ζ are continuous and defined on the separable spaces $L^2(X)$ and $L^2(\mu)$, and so they may be regarded as random elements in suitable Polish spaces. The a.s. representation in (5) then follows by Theorem 6.10. □

Another useful representation may be obtained under suitable regularity conditions on the spectral measure μ.

Proposition 13.20 *(moving average representation)* Let X be an L^2-continuous, stationary, centered Gaussian process on $\mathbb{R}$ with absolutely continuous spectral measure μ. Then there exist an isonormal Gaussian process η on $L^2(\mathbb{R}, \lambda)$ and a function $f \in L^2(\lambda)$ such that

$$X_t = \int_{-\infty}^{\infty} f_{t-s} d\eta_s \quad a.s., \quad t \in \mathbb{R}. \tag{6}$$

Proof: Fix a symmetric density $g \geq 0$ of μ, and define $h = g^{1/2}$. Then $h \in L^2(\lambda)$, and we may introduce the Fourier transform in the sense of Plancherel,

$$f_s = \hat{h}_s = (2\pi)^{-1/2} \lim_{a \to \infty} \int_{-a}^{a} e^{isx} h_x dx, \quad s \in \mathbb{R}, \tag{7}$$

which is again real valued and square integrable. For each $t \in \mathbb{R}$ the function $k_x = e^{-itx} h_x$ has Fourier transform $\hat{k}_s = f_{s-t}$, and so by Parseval's relation

$$r_t = \int_{-\infty}^{\infty} e^{itx} h_x^2 dx = \int_{-\infty}^{\infty} h_x \bar{k}_x dx = \int_{-\infty}^{\infty} f_s f_{s-t} ds. \tag{8}$$

Now consider any isonormal Gaussian process η on $L^2(\lambda)$. For f as in (7), we may define a process Y on $\mathbb{R}$ by the right-hand side of (6). Using (8), we get $EY_s Y_{s+t} = r_t$ for arbitrary $s, t \in \mathbb{R}$, and so $Y \stackrel{d}{=} X$ by Lemma 13.1. Again an appeal to Theorem 6.10 yields the desired a.s. representation of X. $\square$

For an example, we may consider a moving average representation of the stationary Ornstein–Uhlenbeck process. Then introduce an isonormal Gaussian process η on $L^2(\mathbb{R}, \lambda)$ and define

$$X_t = \int_{-\infty}^{t} e^{s-t} d\eta_s, \quad t \geq 0.$$

The process X is clearly centered Gaussian, and we get

$$r_{s,t} = E X_s X_t = \int_{-\infty}^{s \wedge t} e^{u-s} e^{u-t} du = \tfrac{1}{2} e^{-|s-t|}, \quad s, t \in \mathbb{R},$$

as desired. The Markov property of X follows most easily from the fact that

$$X_t = e^{s-t} X_s + \int_s^t e^{u-t} d\eta_u, \quad s \leq t.$$

We proceed to introduce multiple integrals $I_n = \eta^{\otimes n}$ with respect to an isonormal Gaussian process η on a separable (infinite-dimensional) Hilbert space H. Without loss of generality, we may take H to be of the form $L^2(S, \mu)$. Then $H^{\otimes n}$ can be identified with $L^2(S^n, \mu^{\otimes n})$, where $\mu^{\otimes n}$ denotes the n-fold product measure $\mu \otimes \cdots \otimes \mu$, and the tensor product $\bigotimes_{k \leq n} h_k = h_1 \otimes \cdots \otimes h_n$ of the elements $h_1, \ldots, h_n \in H$ is equivalent to the function $h_1(t_1) \cdots h_n(t_n)$ on S^n. Recall that for any ONB $e_1, e_2, \ldots$ in H, the tensor products $\bigotimes_{j \leq n} e_{k_j}$ with arbitrary $k_1, \ldots, k_n \in \mathbb{N}$ form an ONB in $H^{\otimes n}$.

We may now state the basic existence and uniqueness result for the integrals I_n.

13. Gaussian Processes and Brownian Motion

Theorem 13.21 *(multiple stochastic integrals, Wiener, Itô)* *Let η be an isonormal Gaussian process on some separable Hilbert space H. Then for every $n \in \mathbb{N}$ there exists a unique continuous linear mapping $I_n : H^{\otimes n} \to L^2(P)$ such that a.s.*

$$I_n \bigotimes_{k \leq n} h_k = \prod_{k \leq n} \eta h_k, \quad h_1, \ldots, h_n \in H \text{ orthogonal.}$$

Here the *uniqueness* means that $I_n h$ is a.s. unique for every h, and the *linearity* means that $I_n(af + bg) = aI_n f + bI_n g$ a.s. for any $a, b \in \mathbb{R}$ and $f, g \in H^{\otimes n}$. Note in particular that $I_1 h = \eta h$ a.s. For consistency, we define I_0 as the identity mapping on $\mathbb{R}$.

For the proof we may clearly assume that $H = L^2([0,1], \lambda)$. Let $\mathcal{E}_n$ denote the class of *elementary* functions of the form

$$f = \sum_{j \leq m} c_j \bigotimes_{k \leq n} 1_{A_j^k}, \tag{9}$$

where the sets $A_j^1, \ldots, A_j^n \in \mathcal{B}[0,1]$ are disjoint for each $j \in \{1, \ldots, m\}$. The indicator functions $1_{A_j^k}$ are then orthogonal for fixed j, and we need to take

$$I_n f = \sum_{j \leq m} c_j \prod_{k \leq n} \eta A_j^k, \tag{10}$$

where $\eta A = \eta 1_A$. From the linearity in each factor it is clear that the value of $I_n f$ is independent of the choice of representation (9) for f.

To extend the definition of I_n to the entire space $L^2(\mathbb{R}_+^n, \lambda^{\otimes n})$, we need two lemmas. For any function f on $\mathbb{R}_+^n$, we introduce the *symmetrization*

$$\tilde{f}(t_1, \ldots, t_n) = (n!)^{-1} \sum_p f(t_{p_1}, \ldots, t_{p_n}), \quad t_1, \ldots, t_n \in \mathbb{R}_+,$$

where the summation extends over all permutations p of $\{1, \ldots, n\}$. The following result gives the basic L^2-structure, which later carries over to the general integrals.

Lemma 13.22 *(isometry)* *The elementary integrals $I_n f$ in (10) are orthogonal for different n and satisfy*

$$E(I_n f)^2 = n! \|\tilde{f}\|^2 \leq n! \|f\|^2, \quad f \in \mathcal{E}_n. \tag{11}$$

Proof: The second relation in (11) follows from Minkowski's inequality. To prove the remaining assertions, we may first reduce to the case when all sets A_j^k are chosen from some fixed collection of disjoint sets $B_1, B_2, \ldots$. For any finite index sets $J \neq K$ in $\mathbb{N}$, we note that

$$E \prod_{j \in J} \eta B_j \prod_{k \in K} \eta B_k = \prod_{j \in J \cap K} E(\eta B_j)^2 \prod_{j \in J \triangle K} E \eta B_j = 0.$$

This proves the asserted orthogonality. Since clearly $\langle f, g \rangle = 0$ when f and g involve different index sets, it also reduces the proof of the isometry in (11) to the case when all terms in f involve the same sets $B_1, \ldots, B_n$, though in possibly different order. Since $I_n f = I_n \tilde{f}$, we may further assume that $f = \bigotimes_k 1_{B_k}$. But then

$$E(I_n f)^2 = \prod_k E(\eta B_k)^2 = \prod_k \lambda B_k = \|f\|^2 = n! \|\tilde{f}\|^2,$$

where the last relation holds since, in the present case, the permutations of f are orthogonal. □

To extend the integral, we need to show that the elementary functions are dense in $L^2(\lambda^{\otimes n})$.

Lemma 13.23 *(approximation)* *The set $\mathcal{E}_n$ is dense in $L^2(\lambda^{\otimes n})$.*

Proof: By a standard argument based on monotone convergence and a monotone class argument, any function $f \in L^2(\lambda^{\otimes n})$ can be approximated by linear combinations of products $\bigotimes_{k \le n} 1_{A_k}$, and so it is enough to approximate functions f of the latter type. Then divide $[0,1]$ for each m into 2^m intervals B_{mj} of length 2^{-m}, and define

$$f_m = f \sum_{j_1, \ldots, j_n} \bigotimes_{k \le n} 1_{B_{m,j_k}}, \qquad (12)$$

where the summation extends over all collections of *distinct* indices $j_1, \ldots, j_n \in \{1, \ldots, 2^m\}$. Here $f_m \in \mathcal{E}_n$ for each m, and the sum in (12) tends to 1 a.e. $\lambda^{\otimes n}$. Thus, by dominated convergence $f_m \to f$ in $L^2(\lambda^{\otimes n})$. □

By the last two lemmas, I_n is defined as a uniformly continuous mapping on a dense subset of $L^2(\lambda^{\otimes n})$, and so it extends by continuity to all of $L^2(\lambda^{\otimes n})$, with preservation of both the linearity and the norm relations in (11). To complete the proof of Theorem 13.21, it remains to show that $I_n \bigotimes_{k \le n} h_k = \prod_k \eta h_k$ for any orthogonal functions $h_1, \ldots, h_n \in L^2(\lambda)$. This is an immediate consequence of the following lemma, where for any $f \in L^2(\lambda^{\otimes n})$ and $g \in L^2(\lambda)$ we write

$$(f \otimes_1 g)(t_1, \ldots, t_{n-1}) = \int f(t_1, \ldots, t_n) g(t_n) dt_n.$$

Lemma 13.24 *(recursion)* *For any $f \in L^2(\lambda^{\otimes n})$ and $g \in L^2(\lambda)$ with $n \in \mathbb{N}$, we have*

$$I_{n+1}(f \otimes g) = I_n f \cdot \eta g - n I_{n-1}(\tilde{f} \otimes_1 g). \qquad (13)$$

Proof: By Fubini's theorem and the Cauchy–Buniakowski inequality,

$$\|f \otimes g\| = \|f\| \|g\|, \qquad \|\tilde{f} \otimes_1 g\| \le \|\tilde{f}\| \|g\| \le \|f\| \|g\|.$$

Hence, the two sides of (13) are continuous in probability in both f and g, and it is enough to prove the formula for $f \in \mathcal{E}_n$ and $g \in \mathcal{E}_1$. By the linearity of each side we may next reduce to the case when $f = \bigotimes_{k \le n} 1_{A_k}$ and $g = 1_A$, where $A_1, \ldots, A_n$ are disjoint and either $A \cap \bigcup_k A_k = \emptyset$ or

$A = A_1$. In the former case we have $\tilde{f} \otimes_1 g = 0$, so (13) is immediate from the definitions. In the latter case, (13) becomes

$$I_{n+1}(A^2 \times A_2 \times \cdots \times A_n) = \{(\eta A)^2 - \lambda A\}\eta A_2 \cdots \eta A_n. \tag{14}$$

Approximating 1_{A^2} as in Lemma 13.23 by functions $f_m \in \mathcal{E}_2$ with support in A^2, it is clear that the left-hand side equals $I_2 A^2 \, \eta A_2 \cdots \eta A_n$. This reduces the proof of (14) to the two-dimensional version $I_2 A^2 = (\eta A)^2 - \lambda A$. To prove the latter, we may divide A for each m into 2^m subsets B_{mj} of measure $\leq 2^{-m}$, and note as in Theorem 13.9 and Lemma 13.23 that

$$(\eta A)^2 = \sum_i (\eta B_{mi})^2 + \sum_{i \neq j} \eta B_{mi} \eta B_{mj} \to \lambda A + I_2 A^2 \text{ in } L^2. \qquad \square$$

The last lemma will be used to derive an explicit representation of the integrals I_n in terms of the *Hermite polynomials* $p_0, p_1, \ldots$. The latter are defined as orthogonal polynomials of degrees $0, 1, \ldots$ with respect to the standard Gaussian distribution on $\mathbb{R}$. This condition determines each p_n up to a normalization, which we choose for convenience such that the leading coefficient becomes 1. The first few polynomials are then

$$p_0(x) = 1, \quad p_1(x) = x, \quad p_2(x) = x^2 - 1, \quad p_3(x) = x^3 - 3x, \quad \ldots .$$

Theorem 13.25 *(orthogonal representation, Itô)* *On a separable Hilbert space H, let η be an isonormal Gaussian process with associated multiple Wiener–Itô integrals $I_1, I_2, \ldots$. Then for any orthonormal elements $e_1, \ldots, e_m \in H$ and integers $n_1, \ldots, n_m \geq 1$ with sum n, we have*

$$I_n \bigotimes_{j \leq m} e_j^{\otimes n_j} = \prod_{j \leq m} p_{n_j}(\eta e_j).$$

Using the linearity of I_n and writing $\hat{h} = h/\|h\|$, we see that the stated formula is equivalent to the factorization

$$I_n \bigotimes_{j \leq m} h_j^{\otimes n_j} = \prod_{j \leq m} I_{n_j} h_j^{\otimes n_j}, \quad h_1, \ldots, h_k \in H \text{ orthogonal}, \tag{15}$$

together with the representation of the individual factors

$$I_n h^{\otimes n} = \|h\|^n p_n(\eta \hat{h}), \quad h \in H \setminus \{0\}. \tag{16}$$

Proof: We prove (15) by induction on n. Then assume the relation to hold for all integrals up to order n, fix any orthonormal elements $h, h_1, \ldots, h_m \in H$ and integers $k, n_1, \ldots, n_m \in \mathbb{N}$ with sum $n+1$, and write $f = \bigotimes_{j \leq m} h_j^{\otimes n_j}$. By Lemma 13.24 and the induction hypothesis,

$$\begin{aligned}
I_{n+1}(f \otimes h^{\otimes k}) &= I_n(f \otimes h^{\otimes(k-1)}) \cdot \eta h - (k-1)I_{n-1}(f \otimes h^{\otimes(k-2)}) \\
&= (I_{n-k+1}f)\left\{I_{k-1}h^{\otimes(k-1)} \cdot \eta h - (k-1)I_{k-2}h^{\otimes(k-2)}\right\} \\
&= I_{n-k+1}f \cdot I_k h^{\otimes k}.
\end{aligned}$$

Using the induction hypothesis again, we obtain the desired extension to I_{n+1}.

It remains to prove (16) for an arbitrary element $h \in H$ with $\|h\| = 1$. Then conclude from Lemma 13.24 that

$$I_{n+1}h^{\otimes(n+1)} = I_n h^{\otimes n} \cdot \eta h - n I_{n-1} h^{\otimes(n-1)}, \quad n \in \mathbb{N}.$$

Since $I_0 1 = 1$ and $I_1 h = \eta h$, we see by induction that $I_n h^{\otimes n}$ is a polynomial in ηh of degree n and with leading coefficient 1. By the definition of Hermite polynomials, it remains to show that the integrals $I_n h^{\otimes n}$ for different n are orthogonal, which holds by Lemma 13.22. □

Given an isonormal Gaussian process η on some separable Hilbert space H, we introduce the space $L^2(\eta) = L^2(\Omega, \sigma\{\eta\}, P)$ of η-measurable random variables ξ with $E\xi^2 < \infty$. The nth *polynomial chaos* $\mathcal{P}_n$ is defined as the closed linear subspace generated by all polynomials of degree $\leq n$ in the random variables ηh, $h \in H$. We also introduce for every $n \in \mathbb{Z}_+$ the nth *homogeneous chaos* $\mathcal{H}_n$, consisting of all integrals $I_n f$, $f \in H^{\otimes n}$.

The relationship between the mentioned spaces is clarified by the following result. As usual, we write $\oplus$ and $\ominus$ for direct sums and orthogonal complements, respectively.

Theorem 13.26 *(chaos expansion, Wiener)* On a separable Hilbert space H, let η be an isonormal Gaussian process with associated polynomial and homogeneous chaoses $\mathcal{P}_n$ and $\mathcal{H}_n$, respectively. Then the $\mathcal{H}_n$ are orthogonal, closed, linear subspaces of $L^2(\eta)$, satisfying

$$\mathcal{P}_n = \bigoplus_{k=0}^{n} \mathcal{H}_k, \quad n \in \mathbb{Z}_+; \qquad L^2(\eta) = \bigoplus_{n=0}^{\infty} \mathcal{H}_n. \tag{17}$$

Furthermore, every $\xi \in L^2(\eta)$ has a unique a.s. representation $\xi = \sum_n I_n f_n$ with symmetric elements $f_n \in H^{\otimes n}$, $n \geq 0$.

In particular, we note that $\mathcal{H}_0 = \mathcal{P}_0 = \mathbb{R}$ and

$$\mathcal{H}_n = \mathcal{P}_n \ominus \mathcal{P}_{n-1}, \quad n \in \mathbb{N}.$$

Proof: The properties in Lemma 13.22 extend to arbitrary integrands, and so the spaces $\mathcal{H}_n$ are mutually orthogonal, closed, linear subspaces of $L^2(\eta)$. From Lemma 13.23 or Theorem 13.25 we see that also $\mathcal{H}_n \subset \mathcal{P}_n$. Conversely, let ξ be an nth-degree polynomial in the variables ηh. We may then choose some orthonormal elements $e_1, \ldots, e_m \in H$ such that ξ is an nth-degree polynomial in $\eta e_1, \ldots, \eta e_m$. Since any power $(\eta e_j)^k$ is a linear combination of the variables $p_0(\eta e_j), \ldots, p_k(\eta e_j)$, Theorem 13.25 shows that ξ is a linear combination of multiple integrals $I_k f$ with $k \leq n$, which means that $\xi \in \bigoplus_{k \leq n} \mathcal{H}_k$. This proves the first relation in (17).

To prove the second relation, let $\xi \in L^2(\eta) \ominus \bigoplus_n \mathcal{H}_n$. In particular, $\xi \perp (\eta h)^n$ for every $h \in H$ and $n \in \mathbb{Z}_+$. Since $\sum_n |\eta h|^n/n! = e^{|\eta h|} \in L^2$, the series $e^{i\eta h} = \sum_n (i\eta h)^n/n!$ converges in L^2, and we get $\xi \perp e^{i\eta h}$ for every $h \in H$. By the linearity of the integral ηh, we hence obtain for any

$h_1, \ldots, h_n \in H$, $n \in \mathbb{N}$,

$$E\left[\xi \exp \sum_{k \leq n} i u_k \eta h_k\right] = 0, \quad u_1, \ldots, u_n \in \mathbb{R}.$$

Applying the uniqueness theorem for characteristic functions to the distributions of $(\eta h_1, \ldots, \eta h_n)$ under the bounded measures $\mu^{\pm} = E[\xi^{\pm}; \cdot]$, we may conclude that

$$E[\xi;\, (\eta h_1, \ldots, \eta h_n) \in B] = 0, \quad B \in \mathcal{B}(\mathbb{R}^n).$$

By a monotone class argument, this extends to $E[\xi; A] = 0$ for arbitrary $A \in \sigma\{\eta\}$, and since ξ is η-measurable, it follows that $\xi = E[\xi|\eta] = 0$ a.s. The proof of (17) is then complete.

In particular, any element $\xi \in L^2(\eta)$ has an orthogonal expansion

$$\xi = \sum_{n \geq 0} I_n f_n = \sum_{n \geq 0} I_n \tilde{f}_n,$$

for some elements $f_n \in H^{\otimes n}$ with symmetric versions $\tilde{f}_n$, $n \in \mathbb{Z}_+$. Now assume that also $\xi = \sum_n I_n g_n$. Projecting onto $\mathcal{H}_n$ and using the linearity of I_n, we get $I_n(g_n - f_n) = 0$. By the isometry in (11) it follows that $\|\tilde{g}_n - \tilde{f}_n\| = 0$, and so $\tilde{g}_n = \tilde{f}_n$. □

Exercises

1. Let $\xi_1, \ldots, \xi_n$ be i.i.d. $N(m, \sigma^2)$. Show that the random variables $\bar{\xi} = n^{-1} \sum_k \xi_k$ and $s^2 = (n-1)^{-1} \sum_k (\xi_k - \bar{\xi})^2$ are independent and that $(n-1)s^2 \stackrel{d}{=} \sum_{k<n}(\xi_k - m)^2$. (*Hint:* Use the symmetry in Proposition 13.2, and no calculations.)

2. For a Brownian motion B, put $t_{nk} = k2^{-n}$, and define $\xi_{0,k} = B_k - B_{k-1}$ and $\xi_{nk} = B_{t_{n,2k-1}} - \frac{1}{2}(B_{t_{n-1,k-1}} + B_{t_{n-1,k}})$, $k, n \geq 1$. Show that the ξ_{nk} are independent Gaussian. Use this fact to construct a Brownian motion from a sequence of i.i.d. $N(0,1)$ random variables.

3. Let B be a Brownian motion on $[0,1]$, and define $X_t = B_t - tB_1$. Show that $X \perp\!\!\!\perp B_1$. Use this fact to express the conditional distribution of B, given B_1, in terms of a Brownian bridge.

4. Combine the transformations in Lemma 13.6 with the Brownian scaling $c^{-1}B(c^2 t)$ to construct a family of transformations preserving the distribution of a Brownian bridge.

5. Show that the Brownian bridge is an inhomogeneous Markov process. (*Hint:* Use the transformations in Lemma 13.6 or verify the condition in Proposition 13.7.)

6. Let $B = (B^1, B^2)$ be a Brownian motion in $\mathbb{R}^2$, and consider some times t_{nk} as in Theorem 13.9. Show that $\sum_k (B^1_{t_{n,k}} - B^1_{t_{n,k-1}})(B^2_{t_{n,k}} - B^2_{t_{n,k-1}}) \to 0$ in L^2 or a.s., respectively. (*Hint:* Reduce to the case of the quadratic variation.)

7. Use Theorem 7.27 to construct an rcll version B of Brownian motion. Then show as in Theorem 13.9 that B has quadratic variation $[B]_t \equiv t$, and conclude that B is a.s. continuous.

8. For a Brownian motion B, show that $\inf\{t > 0;\ B_t > 0\} = 0$ a.s. (*Hint:* Conclude from Kolmogorov's 0–1 law that the stated event has probability 0 or 1. Alternatively, use Theorem 13.18.)

9. For a Brownian motion B, define $\tau_a = \inf\{t > 0;\ B_t = a\}$. Compute the density of the distribution of τ_a for $a \neq 0$, and show that $E\tau_a = \infty$. (*Hint:* Use Proposition 13.13.)

10. For a Brownian motion B, show that $Z_t = \exp(cB_t - \frac{1}{2}c^2 t)$ is a martingale for every c. Use optional sampling to compute the Laplace transform of τ_a above, and compare with the preceding result.

11. (Paley, Wiener, and Zygmund) Show that Brownian motion B is a.s. nowhere Lipschitz continuous, and hence nowhere differentiable. (*Hint:* If B is Lipschitz at $t < 1$, there exist some $K, \delta > 0$ such that $|B_r - B_s| \leq 2hK$ for all $r, s \in (t - h, t + h)$ with $h < \delta$. Apply this to three consecutive n-dyadic intervals (r, s) around t.)

12. Refine the preceding argument to show that B is a.s. nowhere Hölder continuous with exponent $c > \frac{1}{2}$.

13. Show that the local maxima of a Brownian motion are a.s. dense in $\mathbb{R}$ and that the corresponding times are a.s. dense in $\mathbb{R}_+$. (*Hint:* Use the preceding result.)

14. Show by a direct argument that $\limsup_t t^{-1/2} B_t = \infty$ a.s. as $t \to 0$ and ∞, where B is a Brownian motion. (*Hint:* Use Kolmogorov's 0–1 law.)

15. Show that the law of the iterated logarithm for Brownian motion at 0 remains valid for the Brownian bridge.

16. Show for a Brownian motion B in $\mathbb{R}^d$ that the process $|B|$ satisfies the law of the iterated logarithm at 0 and ∞.

17. Let $\xi_1, \xi_2, \ldots$ be i.i.d. $N(0, 1)$. Show that $\limsup_n (2 \log n)^{-1/2} \xi_n = 1$ a.s.

18. For a Brownian motion B, show that $M_t = t^{-1} B_t$ is a reverse martingale, and conclude that $t^{-1} B_t \to 0$ a.s. and in L^p, $p > 0$, as $t \to \infty$. (*Hint:* The limit is degenerate by Kolmogorov's 0–1 law.) Deduce the same result from Theorem 10.9.

19. For a Brownian bridge B, show that $M_t = (1 - t)^{-1} B_t$ is a martingale on $[0, 1)$. Check that M is not L^1-bounded.

20. Let I_n be the n-fold Wiener–Itô integral w.r.t. Brownian motion B on $\mathbb{R}_+$. Show that the process $M_t = I_n(1_{[0,t]^n})$ is a martingale. Express M in terms of B, and compute the expression for $n = 1, 2, 3$. (*Hint:* Use Theorem 13.25.)

21. Let $\eta_1, \ldots, \eta_n$ be independent, isonormal Gaussian processes on a separable Hilbert space H. Show that there exists a unique continuous linear mapping $\bigotimes_k \eta_k$ from $H^{\otimes n}$ to $L^2(P)$ such that $\bigotimes_k \eta_k \bigotimes_k h_k = \prod_k \eta_k h_k$ a.s. for all $h_1, \ldots, h_n \in H$. Also show that $\bigotimes_k \eta_k$ is an isometry.

Chapter 14

Skorohod Embedding and Invariance Principles

Embedding of random variables; approximation of random walks; functional central limit theorem; laws of the iterated logarithm; arcsine laws; approximation of renewal processes; empirical distribution functions; embedding and approximation of martingales

In Chapter 5 we used analytic methods to derive criteria for a sum of independent random variables to be approximately Gaussian. Though this may remain the easiest approach to the classical limit theorems, the results are best understood when viewed as consequences of some general approximation theorems for random processes. The aim of this chapter is to develop a purely probabilistic technique, the so-called Skorohod embedding, for deriving such functional limit theorems.

In the simplest setting, we may consider a random walk (S_n) based on some i.i.d. random variables ξ_k with mean 0 and variance 1. In this case there exist a Brownian motion B and some optional times $\tau_1 \leq \tau_2 \leq \cdots$ such that $S_n = B_{\tau_n}$ a.s. for every n. For applications it is essential to choose the τ_n such that the differences $\Delta\tau_n$ are again i.i.d. with mean one. The step process $S_{[t]}$ will then be close to the path of B, and many results for Brownian motion carry over, at least approximately, to the random walk. In particular, the procedure yields versions for random walks of the arcsine laws and the law of the iterated logarithm.

From the statements for random walks, similar results may be deduced rather easily for various related processes. In particular, we shall derive a functional central limit theorem and a law of the iterated logarithm for renewal processes, and we shall also see how suitably normalized versions of the empirical distribution functions from an i.i.d. sample can be approximated by a Brownian bridge. For an extension in another direction, we shall obtain a version of the Skorohod embedding for general L^2-martingales and show how any suitably time-changed martingale with small jumps can be approximated by a Brownian motion.

The present exposition depends in many ways on material from previous chapters. Thus, we rely on the basic theory of Brownian motion, as set forth in Chapter 13. We also make frequent use of ideas and results from Chapter 7 on martingales and optional times. Finally, occasional references

14. Skorohod Embedding and Invariance Principles

are made to Chapter 4 for empirical distributions, to Chapter 6 for the transfer theorem, to Chapter 9 for random walks and renewal processes, and to Chapter 12 for the Poisson process.

More general approximations and functional limit theorems are obtained by different methods in Chapters 15, 16, and 19. We also note the close relationship between the present approximation result for martingales with small jumps and the time-change results for continuous local martingales in Chapter 18.

To clarify the basic ideas, we begin with a detailed discussion of the classical Skorohod embedding for random walks. The main result in this context is the following.

Theorem 14.1 *(embedding of random walk, Skorohod)* *Let* $\xi_1, \xi_2, \ldots$ *be i.i.d. random variables with mean 0, and put* $S_n = \xi_1 + \cdots + \xi_n$. *Then there exists a filtered probability space with a Brownian motion B and some optional times* $0 = \tau_0 \leq \tau_1 \leq \ldots$ *such that* $(B_{\tau_n}) \stackrel{d}{=} (S_n)$ *and the differences* $\Delta \tau_n = \tau_n - \tau_{n-1}$ *are i.i.d. with* $E\Delta \tau_n = E\xi_1^2$ *and* $E(\Delta \tau_n)^2 \leq 4E\xi_1^4$.

Here the moment requirements on the differences $\Delta \tau_n$ are crucial for applications. Without those conditions the statement would be trivially true, since we could then choose $B \perp\!\!\!\perp (\xi_n)$ and define the τ_n recursively by $\tau_n = \inf\{t \geq \tau_{n-1}; B_t = S_n\}$. In that case $E\tau_n = \infty$ unless $\xi_1 = 0$ a.s.

The proof of Theorem 14.1 is based on a sequence of lemmas. First we exhibit some martingales associated with Brownian motion.

Lemma 14.2 *(Brownian martingales)* *For a Brownian motion B, the processes* B_t, $B_t^2 - t$, *and* $B_t^4 - 6tB_t^2 + 3t^2$ *are all martingales.*

Proof: Note that $EB_t = EB_t^3 = 0$, $EB_t^2 = t$, and $EB_t^4 = 3t^2$. Write $\mathcal{F}$ for the filtration induced by B, let $0 \leq s \leq t$, and recall that the process $\tilde{B}_t = B_{s+t} - B_s$ is again a Brownian motion independent of $\mathcal{F}_s$. Hence,

$$E[B_t^2|\mathcal{F}_s] = E[B_s^2 + 2B_s\tilde{B}_{t-s} + \tilde{B}_{t-s}^2|\mathcal{F}_s] = B_s^2 + t - s.$$

Moreover,

$$\begin{aligned} E[B_t^4|\mathcal{F}_s] &= E[B_s^4 + 4B_s^3\tilde{B}_{t-s} + 6B_s^2\tilde{B}_{t-s}^2 + 4B_s\tilde{B}_{t-s}^3 + \tilde{B}_{t-s}^4|\mathcal{F}_s] \\ &= B_s^4 + 6(t-s)B_s^2 + 3(t-s)^2, \end{aligned}$$

and so

$$E[B_t^4 - 6tB_t^2|\mathcal{F}_s] = B_s^4 - 6sB_s^2 + 3(s^2 - t^2). \qquad \square$$

By optional sampling, we may deduce some useful formulas.

Lemma 14.3 *(moment relations)* *Consider a Brownian motion B and an optional time τ such that B^τ is bounded. Then*

$$EB_\tau = 0, \qquad E\tau = EB_\tau^2, \qquad E\tau^2 \leq 4EB_\tau^4. \qquad (1)$$

Proof: By optional stopping and Lemma 14.2, we get for any $t \geq 0$

$$EB_{\tau \wedge t} = 0, \qquad E(\tau \wedge t) = EB_{\tau \wedge t}^2, \qquad (2)$$

$$3E(\tau \wedge t)^2 + EB^4_{\tau \wedge t} = 6E(\tau \wedge t)B^2_{\tau \wedge t}. \tag{3}$$

The first two relations in (1) follow from (2) by dominated and monotone convergence as $t \to \infty$. In particular, we have $E\tau < \infty$. We may then take limits even in (3) and conclude by dominated and monotone convergence together with the Cauchy–Buniakovsky inequality that

$$3E\tau^2 + EB^4_\tau = 6E\tau B^2_\tau \le 6(E\tau^2 EB^4_\tau)^{1/2}.$$

Writing $r = (E\tau^2/EB^4_\tau)^{1/2}$, we get $3r^2 + 1 \le 6r$. Thus, $3(r-1)^2 \le 2$, and finally, $r \le 1 + (2/3)^{1/2} < 2$. □

The next result shows how an arbitrary distribution with mean 0 can be expressed as a mixture of centered two-point distributions. For any $a \le 0 \le b$, let $\nu_{a,b}$ denote the unique probability measure on $\{a,b\}$ with mean 0. Clearly, $\nu_{a,b} = \delta_0$ when $ab = 0$, and otherwise

$$\nu_{a,b} = \frac{b\delta_a - a\delta_b}{b-a}, \quad a < 0 < b.$$

It is easy to verify that ν is a probability kernel from $\mathbb{R}_- \times \mathbb{R}_+$ to $\mathbb{R}$. For mappings between two measure spaces, measurability is defined in terms of the σ-fields generated by all evaluation maps $\pi_B : \mu \mapsto \mu B$, where B is an arbitrary set in the underlying σ-field.

Lemma 14.4 *(randomization) For any distribution μ on $\mathbb{R}$ with mean zero, there exists a distribution $\tilde{\mu}$ on $\mathbb{R}_- \times \mathbb{R}_+$ with $\mu = \int \tilde{\mu}(dx\,dy)\nu_{x,y}$, and we can choose $\tilde{\mu}$ to be a measurable function of μ.*

Proof (Chung): Let $\mu_\pm$ denote the restrictions of μ to $\mathbb{R}_\pm \setminus \{0\}$, define $l(x) \equiv x$, and put $c = \int l\,d\mu_+ = -\int l\,d\mu_-$. For any measurable function $f : \mathbb{R} \to \mathbb{R}_+$ with $f(0) = 0$, we get

$$c\int f\,d\mu = \int l\,d\mu_+ \int f\,d\mu_- - \int l\,d\mu_- \int f\,d\mu_+$$

$$= \int\int (y-x)\mu_-(dx)\mu_+(dy) \int f\,d\nu_{x,y},$$

and so we may take

$$\tilde{\mu}(dx\,dy) = \mu\{0\}\delta_{0,0}(dx\,dy) + c^{-1}(y-x)\mu_-(dx)\mu_+(dy).$$

The measurability of the mapping $\mu \mapsto \tilde{\mu}$ is clear by a monotone class argument, once we note that $\tilde{\mu}(A \times B)$ is a measurable function of μ for arbitrary $A, B \in \mathcal{B}(\mathbb{R})$. □

The embedding in Theorem 14.1 will now be constructed recursively, beginning with the first random variable ξ_1.

Lemma 14.5 *(embedding of random variable)* *For any probability measure μ on $\mathbb{R}$ with mean 0, consider a random pair (α, β) with distribution $\tilde{\mu}$ as in Lemma 14.4, and let B be an independent Brownian motion. Then the time $\tau = \inf\{t \geq 0;\ B_t \in \{\alpha, \beta\}\}$ is optional for the filtration $\mathcal{F}_t = \sigma\{\alpha, \beta;\ B_s,\ s \leq t\}$, and we have*

$$\mathcal{L}(B_\tau) = \mu, \qquad E\tau = \int x^2 \mu(dx), \qquad E\tau^2 \leq 4 \int x^4 \mu(dx).$$

Proof: The process B is clearly an $\mathcal{F}$-Brownian motion, and τ is $\mathcal{F}$-optional as in Lemma 7.6 (ii). Using Lemma 14.3 and Fubini's theorem gives

$$\begin{aligned}
\mathcal{L}(B_\tau) &= E\,P[B_\tau \in \cdot\,|\,\alpha, \beta] = E\nu_{\alpha,\beta} = \mu, \\
E\tau &= E\,E[\tau\,|\,\alpha, \beta] = E\int x^2 \nu_{\alpha,\beta}(dx) = \int x^2 \mu(dx), \\
E\tau^2 &= E\,E[\tau^2\,|\,\alpha, \beta] \leq 4E\int x^4 \nu_{\alpha,\beta}(dx) = 4\int x^4 \mu(dx). \qquad \square
\end{aligned}$$

Proof of Theorem 14.1: Let μ be the common distribution of the ξ_n. Introduce a Brownian motion B and some independent i.i.d. pairs (α_n, β_n), $n \in \mathbb{N}$, with the distribution $\tilde{\mu}$ of Lemma 14.4. Define recursively the random times $0 = \tau_0 \leq \tau_1 \leq \cdots$ by

$$\tau_n = \inf\{t \geq \tau_{n-1};\ B_t - B_{\tau_{n-1}} \in \{\alpha_n, \beta_n\}\}, \quad n \in \mathbb{N}.$$

Here each τ_n is clearly optional for the filtration $\mathcal{F}_t = \sigma\{\alpha_k, \beta_k,\ k \geq 1;\ B^t\}$, $t \geq 0$, and B is an $\mathcal{F}$-Brownian motion. By the strong Markov property at τ_n, the process $B_t^{(n)} = B_{\tau_n + t} - B_{\tau_n}$ is then a Brownian motion independent of $\mathcal{G}_n = \sigma\{\tau_k, B_{\tau_k};\ k \leq n\}$. Since moreover $(\alpha_{n+1}, \beta_{n+1}) \perp\!\!\!\perp (B^{(n)}, \mathcal{G}_n)$, we obtain $(\alpha_{n+1}, \beta_{n+1}, B^{(n)}) \perp\!\!\!\perp \mathcal{G}_n$, and so the pairs $(\Delta\tau_n, \Delta B_{\tau_n})$ are i.i.d. The remaining assertions now follow by Lemma 14.5. $\square$

The last theorem enables us to approximate the entire random walk by a Brownian motion. As before, we assume the underlying probability space to be rich enough to support the required randomization variables.

Theorem 14.6 *(approximation of random walk, Skorohod, Strassen)* *Let $\xi_1, \xi_2, \ldots$ be i.i.d. random variables with mean 0 and variance 1, and write $S_n = \xi_1 + \cdots + \xi_n$. Then there exists a Brownian motion B such that*

$$t^{-1/2} \sup_{s \leq t} |S_{[s]} - B_s| \xrightarrow{P} 0, \quad t \to \infty, \tag{4}$$

$$\lim_{t \to \infty} \frac{S_{[t]} - B_t}{\sqrt{2t \log \log t}} = 0 \quad a.s. \tag{5}$$

The proof of (5) requires the following estimate.

Lemma 14.7 *(rate of continuity)* For a Brownian motion B in $\mathbb{R}$, we have
$$\lim_{r\downarrow 1}\limsup_{t\to\infty}\sup_{t\le u\le rt}\frac{|B_u-B_t|}{\sqrt{2t\log\log t}}=0 \quad a.s.$$

Proof: Write $h(t)=(2t\log\log t)^{1/2}$. It is enough to show that
$$\lim_{r\downarrow 1}\limsup_{n\to\infty}\sup_{r^n\le t\le r^{n+1}}\frac{|B_t-B_{r^n}|}{h(r^n)}=0 \quad a.s. \tag{6}$$

Proceeding as in the proof of Theorem 13.18, we get as $n\to\infty$ for fixed $r>1$ and $c>0$
$$P\left\{\sup_{t\in[r^n,r^{n+1}]}|B_t-B_{r^n}|>ch(r^n)\right\} \le P\{B(r^n(r-1))>ch(r^n)\}$$
$$\lesssim n^{-c^2/(r-1)}(\log n)^{-1/2}.$$

(As before, $a\lesssim b$ means that $a\le cb$ for some constant $c>0$.) If $c^2>r-1$, it is clear from the Borel–Cantelli lemma that the lim sup in (6) is a.s. bounded by c, and the relation follows as we let $r\to 1$. □

For the main proof, we need to introduce the *modulus of continuity*
$$w(f,t,h)=\sup_{r,s\le t,\,|r-s|\le h}|f_r-f_s|, \quad t,h>0.$$

Proof of Theorem 14.6: By Theorems 6.10 and 14.1 we may choose a Brownian motion B and some optional times $0\equiv\tau_0\le\tau_1\le\cdots$ such that $S_n=B_{\tau_n}$ a.s. for all n, and the differences $\tau_n-\tau_{n-1}$ are i.i.d. with mean 1. Then $\tau_n/n\to 1$ a.s. by the law of large numbers, and so $\tau_{[t]}/t\to 1$ a.s. Relation (5) now follows by Lemma 14.7.

Next define
$$\delta_t=\sup_{s\le t}|\tau_{[s]}-s|, \quad t\ge 0,$$
and note that the a.s. convergence $\tau_n/n\to 1$ implies $\delta_t/t\to 0$ a.s. Fix any $t,h,\varepsilon>0$, and conclude by the scaling property of B that
$$P\left\{t^{-1/2}\sup_{s\le t}|B_{\tau_{[s]}}-B_s|>\varepsilon\right\}$$
$$\le P\{w(B,t+th,th)>\varepsilon t^{1/2}\}+P\{\delta_t>th\}$$
$$= P\{w(B,1+h,h)>\varepsilon\}+P\{t^{-1}\delta_t>h\}.$$

Here the right-hand side tends to zero as $t\to\infty$ and then $h\to 0$, and (4) follows. □

As an immediate application of the last theorem, we may extend the law of the iterated logarithm to suitable random walks.

Corollary 14.8 *(law of the iterated logarithm, Hartman and Wintner)*
Let $\xi_1, \xi_2, \ldots$ be i.i.d. random variables with mean 0 and variance 1, and define $S_n = \xi_1 + \cdots + \xi_n$. Then
$$\limsup_{n \to \infty} \frac{S_n}{\sqrt{2n \log \log n}} = 1 \quad a.s.$$

Proof: Combine Theorems 13.18 and 14.6. □

To derive a weak convergence result, let $D[0,1]$ denote the space of all functions on $[0,1]$ that are right-continuous with left-hand limits (rcll). For our present needs, it is convenient to equip $D[0,1]$ with the norm $\|x\| = \sup_t |x_t|$ and the σ-field $\mathcal{D}$ generated by all evaluation maps $\pi_t : x \mapsto x_t$. The norm is clearly $\mathcal{D}$-measurable, and so the same thing is true for the open balls $B_{x,r} = \{y; \|x - y\| < r\}$, $x \in D[0,1]$, $r > 0$. (However, $\mathcal{D}$ is strictly smaller than the Borel σ-field induced by the norm.)

Given a process X with paths in $D[0,1]$ and a mapping $f: D[0,1] \to \mathbb{R}$, we say that f is a.s. continuous at X if $X \notin D_f$ a.s., where D_f is the set of functions $x \in D[0,1]$ where f is discontinuous. (The measurability of D_f is irrelevant here, provided that we interpret the condition in the sense of inner measure.)

We may now state a functional version of the classical central limit theorem.

Theorem 14.9 *(functional central limit theorem, Donsker)* Let $\xi_1, \xi_2, \ldots$ be i.i.d. random variables with mean 0 and variance 1, and define
$$X_t^n = n^{-1/2} \sum_{k \leq nt} \xi_k, \quad t \in [0,1], \, n \in \mathbb{N}.$$
Consider a Brownian motion B on $[0,1]$, and let $f : D[0,1] \to \mathbb{R}$ be measurable and a.s. continuous at B. Then $f(X^n) \xrightarrow{d} f(B)$.

The result follows immediately from Theorem 14.6 together with the following lemma.

Lemma 14.10 *(approximation and convergence)* Let $X_1, X_2, \ldots$ and $Y_1, Y_2, \ldots$ be rcll processes on $[0,1]$ with $Y_n \xrightarrow{d} Y_1 \equiv Y$ for all n and $\|X_n - Y_n\| \xrightarrow{P} 0$, and let $f: D[0,1] \to \mathbb{R}$ be measurable and a.s. continuous at Y. Then $f(X_n) \xrightarrow{d} f(Y)$.

Proof: Put $T = \mathbb{Q} \cap [0,1]$. By Theorem 6.10 there exist some processes X'_n on T such that $(X'_n, Y) \xrightarrow{d} (X_n, Y_n)$ on T for all n. Then each X'_n is a.s. bounded and has finitely many upcrossings of any nondegenerate interval, and so the process $\tilde{X}_n(t) = X'_n(t+)$ exists a.s. with paths in $D[0,1]$. From the right continuity of paths it is also clear that $(\tilde{X}_n, Y) \xrightarrow{d} (X_n, Y_n)$ on $[0,1]$ for every n.

To obtain the desired convergence, we note that $\|\tilde{X}_n - Y\| \xrightarrow{d} \|X_n - Y_n\| \xrightarrow{P} 0$, and hence $f(X_n) \xrightarrow{d} f(\tilde{X}_n) \xrightarrow{P} f(Y)$ as in Lemma 4.3. □

In particular, we may recover the central limit theorem in Proposition 5.9 by taking $f(x) = x_1$ in Theorem 14.9. We may also obtain results that go beyond the classical theory, such as for the choice $f(x) = \sup_t |x_t|$. As a less obvious application, we shall see how the arcsine laws of Theorem 13.16 can be extended to suitable random walks. Recall that a random variable ξ is said to be *arcsine distributed* if $\xi \stackrel{d}{=} \sin^2 \alpha$, where α is $U(0, 2\pi)$.

Theorem 14.11 *(arcsine laws, Erdös and Kac, Sparre-Andersen)* Let (S_n) be a random walk based on some distribution μ with mean 0 and variance 1, and define for $n \in \mathbb{N}$

$$\begin{aligned}
\tau_n^1 &= n^{-1} \sum_{k \leq n} 1\{S_k > 0\}, \\
\tau_n^2 &= n^{-1} \min\{k \geq 0;\ S_k = \max_{j \leq n} S_j\}, \\
\tau_n^3 &= n^{-1} \max\{k \leq n;\ S_k S_n \leq 0\}.
\end{aligned}$$

Then $\tau_n^i \stackrel{d}{\to} \tau$ for $i = 1, 2, 3$, where τ is arcsine distributed. The results for $i = 1, 2$ remain valid for any nondegenerate, symmetric distribution μ.

For the proof, we consider on $D[0, 1]$ the functionals

$$\begin{aligned}
f_1(x) &= \lambda\{t \in [0, 1];\ x_t > 0\}, \\
f_2(x) &= \inf\{t \in [0, 1];\ x_t \vee x_{t-} = \sup_{s \leq 1} x_s\}, \\
f_3(x) &= \sup\{t \in [0, 1];\ x_t x_1 \leq 0\}.
\end{aligned}$$

The following result is elementary.

Lemma 14.12 *(continuity of functionals)* *The functionals f_i are measurable. Furthermore, f_1 is continuous at x iff $\lambda\{t;\ x_t = 0\} = 0$, f_2 is continuous at x iff $x_t \vee x_{t-}$ has a unique maximum, and f_3 is continuous at x if 0 is not a local extreme of x_t or x_{t-} on $(0, 1]$.*

Proof of Theorem 14.11: Clearly, $\tau_n^i = f_i(X^n)$ for $n \in \mathbb{N}$ and $i = 1, 2, 3$, where

$$X_t^n = n^{-1/2} S_{[nt]}, \quad t \in [0, 1],\ n \in \mathbb{N}.$$

To prove the first assertion, it suffices by Theorems 13.16 and 14.9 to show that each f_i is a.s. continuous at B. Thus, we need to verify that B a.s. satisfies the conditions in Lemma 14.12. For f_1 this is obvious, since by Fubini's theorem

$$E\lambda\{t \leq 1;\ B_t = 0\} = \int_0^1 P\{B_t = 0\} dt = 0.$$

The conditions for f_2 and f_3 follow easily from Lemma 13.15.

To prove the last assertion, it is enough to consider τ_n^1, since τ_n^2 has the same distribution by Corollary 11.14. Then introduce an independent Brownian motion B and define

$$\sigma_n^\varepsilon = n^{-1} \sum_{k \leq n} 1\{\varepsilon B_k + (1 - \varepsilon) S_k > 0\}, \quad n \in \mathbb{N},\ \varepsilon \in (0, 1].$$

By the first assertion together with Theorem 9.11 and Corollary 11.14, we have $\sigma_n^\varepsilon \stackrel{d}{=} \sigma_n^1 \stackrel{d}{\to} \tau$. Since $P\{S_n = 0\} \to 0$, e.g. by Theorem 4.17, we also note that

$$\limsup_{\varepsilon \to 0} |\sigma_n^\varepsilon - \tau_n^1| \leq n^{-1} \sum_{k \leq n} 1\{S_k = 0\} \stackrel{P}{\to} 0.$$

Hence, we may choose some constants $\varepsilon_n \to 0$ with $\sigma_n^{\varepsilon_n} - \tau_n^1 \stackrel{P}{\to} 0$, and by Theorem 4.28 we get $\tau_n^1 \stackrel{d}{\to} \tau$. □

Theorem 14.9 is often referred to as an *invariance principle*, because the limiting distribution of $f(X^n)$ is the same for all i.i.d. sequences (ξ_k) with mean 0 and variance 1. This fact is often useful for applications, since a direct computation may be possible for some special choice of distribution, such as for $P\{\xi_k = \pm 1\} = \frac{1}{2}$.

The approximation Theorem 14.6 yields a corresponding result for renewal processes, regarded here as nondecreasing step processes.

Theorem 14.13 *(approximation of renewal processes)* Let N be a renewal process based on some distribution μ with mean 1 and variance $\sigma^2 \in (0, \infty)$. Then there exists a Brownian motion B such that

$$t^{-1/2} \sup_{s \leq t} |N_s - s - \sigma B_s| \stackrel{P}{\to} 0, \quad t \to \infty, \tag{7}$$

$$\lim_{t \to \infty} \frac{N_t - t - \sigma B_t}{\sqrt{2t \log \log t}} = 0 \quad a.s. \tag{8}$$

Proof: Let $\tau_0, \tau_1, \ldots$ be the renewal times of N, and introduce the random walk $S_n = n - \tau_n + \tau_0$, $n \in \mathbb{Z}_+$. Choosing a Brownian motion B as in Theorem 14.6, we get

$$\lim_{n \to \infty} \frac{N_{\tau_n} - \tau_n - \sigma B_n}{\sqrt{2n \log \log n}} = \lim_{n \to \infty} \frac{S_n - \sigma B_n}{\sqrt{2n \log \log n}} = 0 \quad a.s.$$

Since $\tau_n \sim n$ a.s. by the law of large numbers, we may replace n in the denominator by τ_n, and by Lemma 14.7 we may further replace B_n by B_{τ_n}. Hence,

$$\frac{N_t - t - \sigma B_t}{\sqrt{2t \log \log t}} \to 0 \quad \text{a.s. along } (\tau_n).$$

Invoking Lemma 14.7, we see that (8) will follow if we can only show that

$$\frac{\tau_{n+1} - \tau_n}{\sqrt{2\tau_n \log \log \tau_n}} \to 0 \quad a.s.$$

This may be seen most easily from Theorem 14.6.

From Theorem 14.6 we see that also

$$n^{-1/2} \sup_{k \leq n} |N_{\tau_k} - \tau_k - \sigma B_k| = n^{-1/2} \sup_{k \leq n} |S_k - \tau_0 - \sigma B_k| \stackrel{P}{\to} 0,$$

and by Brownian scaling,

$$n^{-1/2} w(B, n, 1) \stackrel{d}{=} w(B, 1, n^{-1}) \to 0.$$

To get (7), it is then enough to show that
$$n^{-1/2}\sup_{k\leq n}|\tau_k - \tau_{k-1} - 1| = n^{-1/2}\sup_{k\leq n}|S_k - S_{k-1}| \xrightarrow{P} 0,$$
which is again clear from Theorem 14.6. □

We may now proceed as in Corollary 14.8 and Theorem 14.9 to deduce an associated law of the iterated logarithm and a weak convergence result.

Corollary 14.14 (*limits of renewal processes*) *Let N be a renewal process based on some distribution μ with mean 1 and variance $\sigma^2 < \infty$. Then*
$$\limsup_{t\to\infty} \frac{\pm(N_t - t)}{\sqrt{2t\log\log t}} = \sigma \quad a.s.$$
If B is a Brownian motion and
$$X_t^r = \frac{N_{rt} - rt}{\sigma r^{1/2}}, \quad t \in [0,1],\ r > 0,$$
then also $f(X^r) \xrightarrow{d} f(B)$ as $r \to \infty$ for any measurable function $f: D[0,1] \to \mathbb{R}$ that is a.s. continuous at B.

The weak convergence part of the last corollary yields a similar result for the empirical distribution functions associated with a sequence of i.i.d. random variables. In this case the asymptotic behavior can be expressed in terms of a Brownian bridge.

Theorem 14.15 (*approximation of empirical distribution functions*) *Let $\xi_1, \xi_2, \ldots$ be i.i.d. random variables with distribution function F and empirical distribution functions $\hat{F}_1, \hat{F}_2, \ldots$. Then there exist some Brownian bridges $B^1, B^2, \ldots$ such that*
$$\sup_x \left| n^{1/2}\left(\hat{F}_n(x) - F(x)\right) - B^n \circ F(x)\right| \xrightarrow{P} 0, \quad n \to \infty. \tag{9}$$

Proof: Arguing as in the proof of Proposition 4.24, we may reduce the discussion to the case when the ξ_n are $U(0,1)$, and $F(t) \equiv t$ on $[0,1]$. Then clearly
$$n^{1/2}(\hat{F}_n(t) - F(t)) = n^{-1/2}\sum_{k\leq n}(1\{\xi_k \leq t\} - t), \quad t \in [0,1].$$

Now introduce for each n an independent Poisson random variable κ_n with mean n, and conclude from Proposition 12.4 that $N_t^n = \sum_{k\leq \kappa_n} 1\{\xi_k \leq t\}$ is a homogeneous Poisson process on $[0,1]$ with rate n. By Theorem 14.13 there exist some Brownian motions W^n on $[0,1]$ with
$$\sup_{t\leq 1}\left|n^{-1/2}(N_t^n - nt) - W_t^n\right| \xrightarrow{P} 0.$$

For the associated Brownian bridges $B_t^n = W_t^n - tW_1^n$, we get
$$\sup_{t\leq 1}\left|n^{-1/2}(N_t^n - tN_1^n) - B_t^n\right| \xrightarrow{P} 0.$$

To deduce (9), it is enough to show that

$$n^{-1/2} \sup_{t \leq 1} \left| \sum_{k \leq |\kappa_n - n|} (1\{\xi_k \leq t\} - t) \right| \xrightarrow{P} 0. \tag{10}$$

Here $|\kappa_n - n| \xrightarrow{P} \infty$, e.g. by Proposition 5.9, and so (10) holds by Proposition 4.24 with $n^{1/2}$ replaced by $|\kappa_n - n|$. It remains to note that $n^{-1/2}|\kappa_n - n|$ is tight, since $E(\kappa_n - n)^2 = n$. □

Our next aim is to establish martingale versions of the Skorohod embedding Theorem 14.1 and the associated approximation Theorem 14.6.

Theorem 14.16 *(embedding of martingales)* *Let (M_n) be a martingale with $M_0 = 0$ and induced filtration $(\mathcal{G}_n)$. Then there exist a Brownian motion B and some associated optional times $0 = \tau_0 \leq \tau_1 \leq \cdots$ such that $M_n = B_{\tau_n}$ a.s. for all n and*

$$E[\Delta\tau_n | \mathcal{F}_{n-1}] = E[(\Delta M_n)^2 | \mathcal{G}_{n-1}], \tag{11}$$

$$E[(\Delta\tau_n)^2 | \mathcal{F}_{n-1}] \leq 4E[(\Delta M_n)^4 | \mathcal{G}_{n-1}], \tag{12}$$

where $(\mathcal{F}_n)$ denotes the filtration induced by the pairs (M_n, τ_n).

Proof: Let $\mu_1, \mu_2, \ldots$ be probability kernels satisfying

$$P[\Delta M_n \in \cdot | \mathcal{G}_{n-1}] = \mu_n(M_1, \ldots, M_{n-1}; \cdot) \text{ a.s.}, \quad n \in \mathbb{N}. \tag{13}$$

Since the M_n form a martingale, we may assume that $\mu_n(x; \cdot)$ has mean 0 for all $x \in \mathbb{R}^{n-1}$. Define the associated measures $\tilde{\mu}_n(x; \cdot)$ on $\mathbb{R}^2$ as in Lemma 14.4, and conclude from the measurability part of the lemma that $\tilde{\mu}_n$ is a probability kernel from $\mathbb{R}^{n-1}$ to $\mathbb{R}^2$. Next choose some measurable functions $f_n: \mathbb{R}^n \to \mathbb{R}^2$ as in Lemma 3.22 such that $f_n(x, \vartheta)$ has distribution $\tilde{\mu}_n(x, \cdot)$ when ϑ is $U(0, 1)$.

Now fix any Brownian motion B' and some independent i.i.d. $U(0, 1)$ random variables $\vartheta_1, \vartheta_2, \ldots$. Take $\tau'_0 = 0$, and recursively define the random variables α_n, β_n, and τ'_n, $n \in \mathbb{N}$, through the relations

$$(\alpha_n, \beta_n) = f_n(B'_{\tau'_1}, \ldots, B'_{\tau'_{n-1}}, \vartheta_n), \tag{14}$$

$$\tau'_n = \inf\left\{t \geq \tau'_{n-1}; B'_t - B'_{\tau'_{n-1}} \in \{\alpha_n, \beta_n\}\right\}. \tag{15}$$

Since B' is a Brownian motion for the filtration $\mathcal{B}_t = \sigma\{(B')^t, (\vartheta_n)\}$, $t \geq 0$, and each τ'_n is $\mathcal{B}$-optional, the strong Markov property shows that $B_t^{(n)} = B'_{\tau'_n + t} - B'_{\tau'_n}$ is again a Brownian motion independent of $\mathcal{F}'_n = \sigma\{\tau'_k, B'_{\tau'_k}; k \leq n\}$. Since also $\vartheta_{n+1} \perp\!\!\!\perp (B^{(n)}, \mathcal{F}'_n)$, we have $(B^{(n)}, \vartheta_{n+1}) \perp\!\!\!\perp \mathcal{F}'_n$. Writing $\mathcal{G}'_n = \sigma\{B'_{\tau'_k}; k \leq n\}$, it follows easily that

$$(\Delta\tau'_{n+1}, \Delta B'_{\tau'_{n+1}}) \perp\!\!\!\perp_{\mathcal{G}'_n} \mathcal{F}'_n. \tag{16}$$

By (14) and Theorem 6.4 we have

$$P[(\alpha_n, \beta_n) \in \cdot | \mathcal{G}'_{n-1}] = \tilde{\mu}_n(B'_{\tau'_1}, \ldots, B'_{\tau'_{n-1}}; \cdot). \tag{17}$$

Since also $B^{(n-1)} \perp\!\!\!\perp (\alpha_n, \beta_n, \mathcal{G}'_{n-1})$, we have $B^{(n-1)} \perp\!\!\!\perp_{\mathcal{G}'_{n-1}} (\alpha_n, \beta_n)$, and $B^{(n-1)}$ is conditionally a Brownian motion. Applying Lemma 14.5 to the conditional distributions given $\mathcal{G}'_{n-1}$, we get by (15), (16), and (17)

$$P[\Delta B'_{\tau'_n} \in \cdot | \mathcal{G}'_{n-1}] = \mu_n(B'_{\tau'_1}, \ldots, B'_{\tau'_{n-1}}; \cdot), \quad (18)$$

$$E[\Delta \tau'_n | \mathcal{F}'_{n-1}] = E[\Delta \tau'_n | \mathcal{G}'_{n-1}] = E[(\Delta B'_{\tau'_n})^2 | \mathcal{G}'_{n-1}], \quad (19)$$

$$E[(\Delta \tau'_n)^2 | \mathcal{F}'_{n-1}] = E[(\Delta \tau'_n)^2 | \mathcal{G}'_{n-1}] \leq 4 E[(\Delta B'_{\tau'_n})^4 | \mathcal{G}'_{n-1}]. \quad (20)$$

Comparing (13) and (18) gives $(B'_{\tau'_n}) \stackrel{d}{=} (M_n)$. By Theorem 6.10 we may then choose a Brownian motion B with associated optional times $\tau_1, \tau_2, \ldots$ such that

$$\{B, (M_n), (\tau_n)\} \stackrel{d}{=} \{B', (B'_{\tau'_n}), (\tau'_n)\}.$$

All a.s. relations between the objects on the right, including also their conditional expectations given any induced σ-fields, remain valid for the objects on the left. In particular, $M_n = B_{\tau_n}$ a.s. for all n, and relations (19) and (20) imply the corresponding formulas (11) and (12). □

We may use the last theorem to show how martingales with small jumps can be approximated by a Brownian motion. For martingales M on $\mathbb{Z}_+$, we then introduce the *quadratic variation* $[M]$ and *predictable quadratic variation* $\langle M \rangle$, given by

$$[M]_n = \sum_{k \leq n} (\Delta M_k)^2, \quad \langle M \rangle_n = \sum_{k \leq n} E[(\Delta M_k)^2 | \mathcal{F}_{k-1}].$$

Continuous-time versions of those processes are considered in Chapters 17 and 26.

Theorem 14.17 *(approximation of martingales with small jumps)* For each $n \in \mathbb{N}$, let M^n be an $\mathcal{F}^n$-martingale on $\mathbb{Z}_+$ with $M_0^n = 0$ and $|\Delta M_k^n| \leq 1$, and assume that $\sup_k |\Delta M_k^n| \stackrel{P}{\to} 0$. Define

$$X_t^n = \sum_k \Delta M_k^n \mathbf{1}\{[M^n]_k \leq t\}, \quad t \in [0,1], \ n \in \mathbb{N},$$

and put $\zeta_n = [M^n]_\infty$. Then $(X^n - B^n)^*_{\zeta_n \wedge 1} \stackrel{P}{\to} 0$ for some Brownian motions B^n. This remains true with $[M^n]$ replaced by $\langle M^n \rangle$, and we may also replace the condition $\sup_k |\Delta M_k^n| \stackrel{P}{\to} 0$ by

$$\sum_k P[|\Delta M_k^n| > \varepsilon | \mathcal{F}_{k-1}^n] \stackrel{P}{\to} 0, \quad \varepsilon > 0. \quad (21)$$

For the proof, we need to show that the time scales given by the sequences (τ_k^n), $[M^n]$, and $\langle M^n \rangle$ are asymptotically equivalent.

14. Skorohod Embedding and Invariance Principles

Lemma 14.18 *(time-scale comparison) Assume in Theorem 14.17 that $M_k^n = B^n(\tau_k^n)$ a.s. for some Brownian motions B^n and associated optional times τ_k^n as in Theorem 14.16. Put $\kappa_t^n = \inf\{k; [M^n]_k > t\}$. Then as $n \to \infty$ for fixed $t > 0$, we have*

$$\sup_{k \leq \kappa_t^n} (|\tau_k^n - [M^n]_k| \vee |[M^n]_k - \langle M^n \rangle_k|) \xrightarrow{P} 0. \tag{22}$$

Proof: By optional stopping, we may assume that $[M^n]$ is uniformly bounded and take the supremum in (22) over all k. To handle the second difference in (22), we note that $D_n = [M^n] - \langle M^n \rangle$ is a martingale for each n. Using the martingale property, Proposition 7.16, and dominated convergence, we get

$$
\begin{aligned}
E(D^n)^{*2} &\leq \sup_k E(D_k^n)^2 = \sum_k E(\Delta D_k^n)^2 \\
&= \sum_k EE[(\Delta D_k^n)^2 | \mathcal{F}_{k-1}^n] \\
&\leq \sum_k EE[(\Delta [M^n]_k)^2 | \mathcal{F}_{k-1}^n] \\
&= E \sum_k (\Delta M_k^n)^4 \leq E \sup_k (\Delta M_k^n)^2 \to 0,
\end{aligned}
$$

and so $(D^n)^* \xrightarrow{P} 0$. This clearly remains true if each sequence $\langle M^n \rangle$ is defined in terms of the filtration $\mathcal{G}^n$ induced by M^n.

To complete the proof of (22), it is enough to show, for the latter versions of $\langle M^n \rangle$, that $(\tau^n - \langle M^n \rangle)^* \xrightarrow{P} 0$. Then let $\mathcal{T}^n$ denote the filtration induced by the pairs (M_k^n, τ_k^n), $k \in \mathbb{N}$, and conclude from (11) that

$$\langle M^n \rangle_m = \sum_{k \leq m} E[\Delta \tau_k^n | \mathcal{T}_{k-1}^n], \quad m, n \in \mathbb{N}.$$

Hence, $\tilde{D}^n = \tau^n - \langle M^n \rangle$ is a $\mathcal{T}^n$-martingale. Using (11) and (12), we then get as before

$$
\begin{aligned}
E(\tilde{D}^n)^{*2} &\leq \sup_k E(\tilde{D}_k^n)^2 = \sum_k EE[(\Delta \tilde{D}_k^n)^2 | \mathcal{T}_{k-1}^n] \\
&\leq \sum_k EE[(\Delta \tau_k^n)^2 | \mathcal{T}_{k-1}^n] \\
&\leq \sum_k EE[(\Delta M_k^n)^4 | \mathcal{G}_{k-1}^n] \\
&= E \sum_k (\Delta M_k^n)^4 \leq E \sup_k (\Delta M_k^n)^2 \to 0. \quad \square
\end{aligned}
$$

The sufficiency of (21) is a consequence of the following simple estimate.

Lemma 14.19 *(Dvoretzky) For any filtration $\mathcal{F}$ on $\mathbb{Z}_+$ and sets $A_n \in \mathcal{F}_n$, $n \in \mathbb{N}$, we have*

$$P \bigcup_n A_n \leq P\left\{ \sum_n P[A_n | \mathcal{F}_{n-1}] > \varepsilon \right\} + \varepsilon, \quad \varepsilon > 0.$$

Proof: Write $\xi_n = 1_{A_n}$ and $\hat{\xi}_n = P[A_n|\mathcal{F}_{n-1}]$, fix any $\varepsilon > 0$, and define $\tau = \inf\{n;\ \hat{\xi}_1 + \cdots + \hat{\xi}_n > \varepsilon\}$. Then $\{\tau \leq n\} \in \mathcal{F}_{n-1}$ for each n, and so

$$E \sum_{n < \tau} \xi_n = \sum_n E[\xi_n;\ \tau > n] = \sum_n E[\hat{\xi}_n;\ \tau > n] = E \sum_{n < \tau} \hat{\xi}_n \leq \varepsilon.$$

Hence,

$$P \bigcup_n A_n \leq P\{\tau < \infty\} + E \sum_{n < \tau} \xi_n \leq P\left\{\sum_n \hat{\xi}_n > \varepsilon\right\} + \varepsilon. \qquad \Box$$

Proof of Theorem 14.17: To prove the result for the time-scales $[M^n]$, we may reduce by optional stopping to the case when $[M^n] \leq 2$ for all n. For each n we may choose some Brownian motion B^n and associated optional times τ_k^n as in Theorem 14.16. Then

$$(X^n - B^n)^*_{\zeta_n \wedge 1} \leq w(B^n, 1 + \delta_n, \delta_n), \quad n \in \mathbb{N},$$

where

$$\delta_n = \sup_k\{|\tau_k^n - [M^n]_k| + (\Delta M_k^n)^2\},$$

and so

$$E[(X^n - B^n)^*_{\zeta_n \wedge 1} \wedge 1] \leq E[w(B^n, 1 + h, h) \wedge 1] + P\{\delta_n > h\}.$$

Since $\delta_n \xrightarrow{P} 0$ by Lemma 14.18, the right-hand side tends to zero as $n \to \infty$ and then $h \to 0$, and the assertion follows.

In the case of the time scales $\langle M^n \rangle$, define $\kappa_n = \inf\{k;\ [M^n] > 2\}$. Then $[M^n]_{\kappa_n} - \langle M^n \rangle_{\kappa_n} \xrightarrow{P} 0$ by Lemma 14.18, and so $P\{\langle M^n \rangle_{\kappa_n} < 1,\ \kappa_n < \infty\} \to 0$. We may then reduce by optional stopping to the case when $[M^n] \leq 3$. The proof may now be completed as before. $\qquad \Box$

Though the Skorohod embedding has no natural extension to higher dimensions, one can still obtain useful multidimensional approximations by applying the previous results to each component separately. To illustrate the method, we proceed to show how suitable random walks in $\mathbb{R}^d$ can be approximated by continuous processes with stationary, independent increments. Extensions to more general limits are obtained by different methods in Corollary 15.20 and Theorem 16.14.

Theorem 14.20 *(approximation of random walks in $\mathbb{R}^d$)* Let $S^1, S^2, \ldots$ be random walks in $\mathbb{R}^d$ such that $\mathcal{L}(S^n_{m_n}) \xrightarrow{w} N(0, \sigma\sigma')$ for some $d \times d$ matrix σ and integers $m_n \to \infty$. Then there exist some Brownian motions $B^1, B^2, \ldots$ in $\mathbb{R}^d$ such that the processes $X^n_t = S^n_{[m_n t]}$ satisfy $(X^n - \sigma B_n)^*_t \xrightarrow{P} 0$ for all $t \geq 0$.

Proof: By Theorem 5.15 we have

$$\max_{k \leq m_n t} |\Delta S^n_k| \xrightarrow{P} 0, \quad t \geq 0,$$

and so we may assume that $|\Delta S_k^n| \leq 1$ for all n and k. Subtracting the means, we may further assume that $ES_k^n \equiv 0$. Applying Theorem 14.17 in each coordinate, we get $w(X^n, t, h) \xrightarrow{P} 0$ as $n \to \infty$ and then $h \to 0$. Furthermore, $w(\sigma B, t, h) \to 0$ a.s. as $h \to 0$.

Using Theorem 5.15 in both directions gives $X_{t_n}^n \xrightarrow{d} \sigma B_t$ as $t_n \to t$. By independence it follows that $(X_{t_1}^n, \ldots, X_{t_m}^n) \xrightarrow{d} \sigma(B_{t_1}, \ldots, B_{t_m})$ for all $n \in \mathbb{N}$ and $t_1, \ldots, t_n \geq 0$, and so $X^n \xrightarrow{d} \sigma B$ on $\mathbb{Q}_+$ by Theorem 4.29. By Theorem 4.30 or, more conveniently, by Corollary 6.12 and Theorem A2.2, there exist some rcll processes $Y^n \stackrel{d}{=} X^n$ with $Y_t^n \to \sigma B_t$ a.s. for all $t \in \mathbb{Q}_+$. For any $t, h > 0$ we have

$$E[(Y^n - \sigma B)_t^* \wedge 1] \leq E\left[\max_{j \leq t/h} |Y_{jh}^n - \sigma B_{jh}| \wedge 1\right]$$
$$+ E[w(Y^n, t, h) \wedge 1] + E[w(\sigma B, t, h) \wedge 1].$$

Multiplying by e^{-t}, integrating over $t > 0$, and letting $n \to \infty$ and then $h \to 0$ along $\mathbb{Q}_+$, we get by dominated convergence

$$\int_0^\infty e^{-t} E[(Y^n - \sigma B)_t^* \wedge 1] dt \to 0.$$

Hence, by monotonicity, the last integrand tends to zero as $n \to \infty$, and so $(Y^n - \sigma B)_t^* \xrightarrow{P} 0$ for each $t > 0$. It remains to use Theorem 6.10. □

Exercises

1. Proceed as in Lemma 14.2 to construct Brownian martingales with leading terms B_t^3 and B_t^5. Use multiple Wiener–Itô integrals to give an alternative proof of the lemma, and find for every $n \in \mathbb{N}$ a martingale with leading term B_t^n. (*Hint:* Use Theorem 13.25.)

2. Given a Brownian motion B and an associated optional time $\tau < \infty$, show that $E\tau \geq EB_\tau^2$. (*Hint:* Truncate τ and use Fatou's lemma.)

3. For S_n as in Corollary 14.8, show that sequence of random variables $(2n \log \log n)^{-1/2} S_n$, $n \geq 3$, is a.s. relatively compact with set of limit points equal to $[-1, 1]$. (*Hint:* Prove the corresponding property for Brownian motion, and use Theorem 14.6.)

4. Let $\xi_1, \xi_2, \ldots$ be i.i.d. random vectors in $\mathbb{R}^d$ with mean 0 and covariances δ_{ij}. Show that the conclusion of Corollary 14.8 holds with S_n replaced by $|S_n|$. More precisely, show that the sequence $(2n \log \log n)^{-1/2} S_n$, $n \geq 3$, is relatively compact in $\mathbb{R}^d$, and that the set of limit points is contained in the closed unit ball. (*Hint:* Apply Corollary 14.8 to the projections $u \cdot S_n$ for arbitrary $u \in \mathbb{R}^d$ with $|u| = 1$.)

5. In Theorem 13.18, show that for any $c \in (0,1)$ there exists a sequence $t_n \to \infty$ such that the limsup along (t_n) equals c a.s. Conclude that the set of limit points in the preceding exercise agrees with the closed unit ball in $\mathbb{R}^d$.

6. Condition (21) clearly follows from $\sum_k E[|\Delta M_k^n| \wedge 1 | \mathcal{F}_{n-1}^n] \xrightarrow{P} 0$. Show by an example that the latter condition is strictly stronger. (*Hint:* Consider a sequence of random walks.)

7. Specialize Lemma 14.18 to random walks, and give a direct proof in this case.

8. In the special case of random walks, show that condition (21) is also necessary. (*Hint:* Use Theorem 5.15.)

9. Specialize Theorem 14.17 to a sequence of random walks in $\mathbb{R}$, and derive a corresponding extension of Theorem 14.9. Then derive a functional version of Theorem 5.12.

10. Specialize further to the case of successive renormalizations of a single random walk S_n. Then derive a limit theorem for the values at $t = 1$, and compare with Proposition 5.9.

11. In the second arcsine law of Theorem 14.11, show that the first maximum on $[0, 1]$ can be replaced by the last one. Conclude that the associated times σ_n and τ_n satisfy $\tau_n - \sigma_n \xrightarrow{P} 0$. (*Hint:* Use the corresponding result for Brownian motion. Alternatively, use the symmetry of (S_n) and of the arcsine distribution.)

12. Extend Theorem 14.11 to an arbitrary sequence of symmetric random walks satisfying a Lindeberg condition. Also extend the results for τ_n^1 and τ_n^2 to sequences of random walks based on diffuse, symmetric distributions. Finally, show that the result for τ_n^3 may fail in the latter case. (*Hint:* Consider the n^{-1}-increments of a compound Poisson process based on the uniform distribution on $[-1, 1]$, perturbed by a small diffusion term $\varepsilon_n B$, where B is an independent Brownian motion.)

13. In the context of Theorem 14.20, show that for any Brownian motion B there exist some processes $Y^n \stackrel{d}{=} X^n$ such that $(Y^n - \sigma B)_t^* \to 0$ a.s. for all $t \geq 0$. Prove a corresponding version of Theorem 14.17. (*Hint:* Use Theorem 4.30 or Corollary 6.12.)

Chapter 15

Independent Increments
and Infinite Divisibility

Regularity and integral representation; Lévy processes and subordinators; stable processes and first-passage times; infinitely divisible distributions; characteristics and convergence criteria; approximation of Lévy processes and random walks; limit theorems for null arrays; convergence of extremes

In Chapters 12 and 13 we saw how Poisson processes and Brownian motion arise as special processes with independent increments. Our present aim is to study more general processes of this type. Under a mild regularity assumption, we shall derive a general representation of independent-increment processes in terms of a Gaussian component and a jump component, where the latter is expressible as a suitably compensated Poisson integral. Of special importance is the time-homogeneous case of so-called Lévy processes, which admit a description in terms of a characteristic triple (a, b, ν), where a is the diffusion rate, b is the drift coefficient, and ν is the Lévy measure that determines the rates for jumps of different sizes.

In the same way that Brownian motion is the basic example of both a a diffusion process and a continuous martingale, the general Lévy processes constitute the fundamental cases of both Markov processes and general semimartingales. As a motivation for the general weak convergence theory of Chapter 16, we shall further see how Lévy processes serve as the natural approximations to random walks. In particular, such approximations may be used to extend two of the arcsine laws for Brownian motion to general symmetric Lévy processes. Increasing Lévy processes, even called subordinators, play a basic role in Chapter 22, where they appear in representations of local time and regenerative sets.

The distributions of Lévy processes at fixed times coincide with the infinitely divisible laws, which also arise as the most general limit laws in the classical limit theorems for null arrays. The special cases of convergence toward Poisson and Gaussian limits were considered in Chapter 5, and now we shall be able to characterize the convergence toward an arbitrary infinitely divisible law. Though characteristic functions will still be needed occasionally as a technical tool, the present treatment is more probabilistic in flavor and involves as crucial steps a centering at truncated means followed by a compound Poisson approximation.

To resume our discussion of general independent-increment processes, say that a process X in $\mathbb{R}^d$ is *continuous in probability* if $X_s \xrightarrow{P} X_t$ whenever $s \to t$. Let us further say that a function f on $\mathbb{R}_+$ or $[0,1]$ is *right-continuous with left-hand limits* (abbreviated as *rcll*) if the right- and left-hand limits $f_{t\pm}$ exist and are finite and if, moreover, $f_{t+} \equiv f_t$. A process X is said to be rcll if its paths have this property. In that case only jump discontinuities may occur, and we say that X has a *fixed jump* at some time $t > 0$ if $P\{X_t \neq X_{t-}\} > 0$.

The following result gives the basic regularity properties of independent-increment processes. A similar result for Feller processes is obtained by different methods in Theorem 19.15.

Theorem 15.1 *(regularization, Lévy)* *If a process X in $\mathbb{R}^d$ is continuous in probability and has independent increments, then X has an rcll version without fixed jumps.*

For the proof we shall use a martingale argument based on the characteristic functions

$$\varphi_{s,t}(u) = E\exp\{iu(X_t - X_s)\}, \quad u \in \mathbb{R}^d,\ 0 \leq s \leq t.$$

Note that $\varphi_{r,s}\varphi_{s,t} = \varphi_{r,t}$ for any $r \leq s \leq t$, and put $\varphi_{0,t} = \varphi_t$. In order to construct associated martingales, we need to know that $\varphi_{s,t} \neq 0$.

Lemma 15.2 *(zeros)* *For any $u \in \mathbb{R}^d$ and $s \leq t$, we have $\varphi_{s,t}(u) \neq 0$.*

Proof: Fix any $u \in \mathbb{R}^d$ and $s \leq t$. Since X is continuous in probability, there exists for any $r \geq 0$ some $h > 0$ such that $\varphi_{r,r'}(u) \neq 0$ whenever $|r - r'| < h$. By compactness we may then choose finitely many division point $s = t_0 < t_1 < \cdots < t_n = t$ such that $\varphi_{t_{k-1},t_k}(u) \neq 0$ for all k, and by the independence of the increments we get $\varphi_{s,t}(u) = \prod_k \varphi_{t_{k-1},t_k}(u) \neq 0$. □

We also need the following deterministic convergence criterion.

Lemma 15.3 *(complex exponentials)* *Fix any $a_1, a_2, \cdots \in \mathbb{R}^d$. Then a_n converges iff e^{iua_n} converges for almost every $u \in \mathbb{R}^d$.*

Proof: Assume the stated condition. Fix a nondegenerate Gaussian random vector η in $\mathbb{R}^d$, and note that $\exp\{it\eta(a_m - a_n)\} \to 1$ a.s. as $m, n \to \infty$ for fixed $t \in \mathbb{R}$. By dominated convergence the characteristic function of $\eta(a_m - a_n)$ tends to 1, and so $\eta(a_m - a_n) \xrightarrow{P} 0$ by Theorem 5.3, which implies $a_m - a_n \to 0$. Thus, (a_n) is Cauchy and therefore convergent. □

Proof of Theorem 15.1: We may clearly assume that $X_0 = 0$. By Lemma 15.2 we may define

$$M_t^u = \frac{e^{iuX_t}}{\varphi_t(u)}, \quad t \geq 0,\ u \in \mathbb{R}^d,$$

which is clearly a martingale in t for each u. Letting $\Omega_u \subset \Omega$ denote the set where e^{iuX_t} has limits from the left and right along $\mathbb{Q}_+$ at every $t \geq 0$, we see from Theorem 7.18 that $P\Omega_u = 1$.

Restating the definition of Ω_u in terms of upcrossings, we note that the set $A = \{(u,\omega); \omega \in \Omega_u\}$ is product measurable in $\mathbb{R}^d \times \Omega$. Writing $A_\omega = \{u \in \mathbb{R}^d; \omega \in \Omega_u\}$, it follows by Fubini's theorem that the set $\Omega' = \{\omega; \lambda^d A_\omega^c = 0\}$ has probability 1. If $\omega \in \Omega'$, we have $u \in A_\omega$ for almost every $u \in \mathbb{R}^d$, and so Lemma 15.3 shows that X itself has finite right- and left-hand limits along $\mathbb{Q}_+$. Now define $\tilde{X}_t = X_{t+}$ on Ω' and $\tilde{X} = 0$ on Ω'^c, and note that $\tilde{X}$ is rcll everywhere. Further note that $\tilde{X}$ is a version of X, since $X_{t+h} \xrightarrow{P} X_t$ as $h \to 0$ for fixed t by hypothesis. For the same reason, $\tilde{X}$ has no fixed jumps. □

We proceed to state the general representation theorem. Given any Poisson process η with intensity measure $\nu = E\eta$, we recall from Theorem 12.13 that the integral $(\eta - \nu)f = \int f(x)(\eta - \nu)(dx)$ exists in the sense of approximation in probability iff $\nu(f^2 \wedge |f|) < \infty$.

Theorem 15.4 *(independent-increment processes, Lévy, Itô)* Let X be an rcll process in $\mathbb{R}^d$ with $X_0 = 0$. Then X has independent increments and no fixed jumps iff, a.s. for each $t \geq 0$,

$$X_t = m_t + G_t + \int_0^t\!\!\int_{|x|\leq 1} x\,(\eta - E\eta)(ds\,dx) + \int_0^t\!\!\int_{|x|>1} x\,\eta(ds\,dx), \quad (1)$$

for some continuous function m with $m_0 = 0$, some continuous centered Gaussian process G with independent increments and $G_0 = 0$, and some independent Poisson process η on $(0,\infty) \times (\mathbb{R}^d \setminus \{0\})$ with

$$\int_0^t\!\!\int (|x|^2 \wedge 1) E\eta(ds\,dx) < \infty, \quad t > 0. \quad (2)$$

In the special case when X is real and nondecreasing, (1) simplifies to

$$X_t = a_t + \int_0^t\!\!\int_0^\infty x\,\eta(ds\,dx), \quad t \geq 0, \quad (3)$$

for some nondecreasing continuous function a with $a_0 = 0$ and some Poisson process η on $(0,\infty)^2$ with

$$\int_0^t\!\!\int_0^\infty (x \wedge 1) E\eta(ds\,dx) < \infty, \quad t > 0. \quad (4)$$

Both representations are a.s. unique, and all functions m or a and processes G and η with the stated properties may occur.

We begin the proof by analyzing the jump structure of X. Let us then introduce the random measure

$$\eta = \sum_t \delta_{t,\Delta X_t} = \sum_t 1\{(t, \Delta X_t) \in \cdot\}, \quad (5)$$

where the summation extends over all times $t > 0$ with $\Delta X_t \equiv X_t - X_{t-} \neq 0$. We say that η is *locally X-measurable* if, for any $s < t$, the measure $\eta((s,t] \times \cdot)$ is a measurable function of the process $X_r - X_s$, $r \in [s,t]$.

Lemma 15.5 *(Poisson process of jumps) Let X be an rcll process in $\mathbb{R}^d$ with independent increments and no fixed jumps. Then η in (5) is a locally X-measurable Poisson process on $(0,\infty) \times (\mathbb{R}^d \setminus \{0\})$ satisfying (2). If X is further real-valued and nondecreasing, then η is supported by $(0,\infty)^2$ and satisfies (4).*

Proof (beginning): Fix any times $s < t$, and consider a sequence of partitions $s = t_{n,0} < \cdots < t_{n,n}$ with $\max_k(t_{n,k} - t_{n,k-1}) \to 0$. For any continuous function f on $\mathbb{R}^d$ that vanishes in a neighborhood of 0, we have

$$\sum_k f(X_{t_{n,k}} - X_{t_{n,k-1}}) \to \int f(x)\eta((s,t] \times dx),$$

which implies the measurability of the integrals on the right. By a simple approximation we may conclude that $\eta((s,t] \times B)$ is measurable for every compact set $B \subset \mathbb{R}^d \setminus \{0\}$. The measurability extends by a monotone class argument to all random variables ηA with A included in some fixed bounded rectangle $[0,t] \times B$, and the further extension to arbitrary Borel sets is immediate.

Since X has independent increments and no fixed jumps, the same properties hold for η, which is then Poisson by Theorem 12.10. If X is real-valued and nondecreasing, then (4) holds by Theorem 12.13. □

The proof of (2) requires a further lemma, which is also needed for the main proof.

Lemma 15.6 *(orthogonality and independence) Let X and Y be rcll processes in $\mathbb{R}^d$ with $X_0 = Y_0 = 0$ such that (X,Y) has independent increments and no fixed jumps. Assume also that Y is a.s. a step process and that $\Delta X \cdot \Delta Y = 0$ a.s. Then $X \perp\!\!\!\perp Y$.*

Proof: Define η as in (5) in terms of Y, and note as before that η is locally Y-measurable whereas Y is locally η-measurable. By a simple transformation of η we may reduce to the case when Y has bounded jumps. Since η is Poisson, Y then has integrable variation on every finite interval. By Corollary 3.7 we need to show that $(X_{t_1}, \ldots, X_{t_n}) \perp\!\!\!\perp (Y_{t_1}, \ldots, Y_{t_n})$ for any $t_1 < \cdots < t_n$, and by Lemma 3.8 it suffices to show for all $s < t$ that $X_t - X_s \perp\!\!\!\perp Y_t - Y_s$. Without loss of generality, we may take $s = 0$ and $t = 1$.

Then fix any $u, v \in \mathbb{R}^d$, and introduce the locally bounded martingales

$$M_t = \frac{e^{iuX_t}}{Ee^{iuX_t}}, \quad N_t = \frac{e^{ivY_t}}{Ee^{ivY_t}}, \quad t \geq 0.$$

Note that N again has integrable variation on $[0,1]$. For $n \in \mathbb{N}$, we get by the martingale property and dominated convergence

$$E M_1 N_1 - 1 = E \sum_{k \leq n} (M_{k/n} - M_{(k-1)/n})(N_{k/n} - N_{(k-1)/n})$$

$$= E \int_0^1 (M_{[sn+1-]/n} - M_{[sn-]/n}) \, dN_s$$

$$\to E \int_0^1 \Delta M_s dN_s = E \sum_{s \leq 1} \Delta M_s \Delta N_s = 0.$$

Thus, $E M_1 N_1 = 1$, and so

$$E e^{iu X_1 + iv Y_1} = E e^{iu X_1} E e^{iv Y_1}, \quad u, v \in \mathbb{R}^d.$$

The asserted independence $X_1 \perp\!\!\!\perp Y_1$ now follows by the uniqueness theorem for characteristic functions. $\square$

End of proof of Lemma 15.5: It remains to prove (2). Then define $\eta_t = \eta([0,t] \times \cdot)$, and note that $\eta_t\{x; |x| > \varepsilon\} < \infty$ a.s. for all $t, \varepsilon > 0$ because X is rcll. Since η is Poisson, the same relations hold for the measures $E\eta_t$, and so it suffices to prove that

$$\int_{|x| \leq 1} |x|^2 E\eta_t(dx) < \infty, \quad t > 0. \tag{6}$$

Then introduce for each $\varepsilon > 0$ the process

$$X_t^\varepsilon = \sum_{s \leq t} \Delta X_s 1\{|\Delta X_s| > \varepsilon\} = \int_{|x| > \varepsilon} x \eta_t(dx), \quad t \geq 0,$$

and note that $X^\varepsilon \perp\!\!\!\perp X - X^\varepsilon$ by Lemma 15.6. By Lemmas 12.2 (i) and 15.2 we get for any $\varepsilon, t > 0$ and $u \in \mathbb{R}^d \setminus \{0\}$

$$0 < |E e^{iu X_t}| \leq |E e^{iu X_t^\varepsilon}| = \left| E \exp \int_{|x| > \varepsilon} iux \, \eta_t(dx) \right|$$

$$= \left| \exp \int_{|x| > \varepsilon} (e^{iux} - 1) E\eta_t(dx) \right| = \exp \int_{|x| > \varepsilon} (\cos ux - 1) E\eta_t(dx).$$

Letting $\varepsilon \to 0$ gives

$$\int_{|ux| \leq 1} |ux|^2 E\eta_t(dx) \lesssim \int (1 - \cos ux) E\eta_t(dx) < \infty,$$

and (6) follows since u is arbitrary. $\square$

Proof of Theorem 15.4: In the nondecreasing case, we may subtract the jump component to obtain a continuous, nondecreasing process Y with independent increments, and from Theorem 5.11 it is clear that Y is a.s. nonrandom. Thus, in this case we get a representation as in (3).

In the general case, introduce for each $\varepsilon \in [0,1]$ the martingale

$$M_t^\varepsilon = \int_0^t \int_{|x| \in (\varepsilon, 1]} x\, (\eta - E\eta)(ds\, dx), \quad t \geq 0.$$

Put $M_t = M_t^0$, and let J_t denote the last term in (1). By Proposition 7.16 we have $E(M^\varepsilon - M^0)_t^{*2} \to 0$ for each t. Thus, $M + J$ has a.s. the same jumps as X, and so the process $Y = X - M - J$ is a.s. continuous. Since η is locally X-measurable, the same thing is true for Y. Theorem 13.4 then shows that Y is Gaussian with continuous mean and covariance functions. Subtracting the means m_t yields a continuous, centered Gaussian process G, and by Lemma 15.6 we get $G \perp\!\!\!\perp (M^\varepsilon + J)$ for every $\varepsilon > 0$. The independence extends to M by Lemma 3.6, and so $G \perp\!\!\!\perp \eta$.

The uniqueness of η is clear from (5), and G is then determined by subtraction. From Theorem 12.13 it is further seen that the integrals in (1) and (3) exist for any Poisson process η with the stated properties, and we note that the resulting process has independent increments. $\square$

We may now specialize to the time-homogeneous case, when the distribution of $X_{t+h} - X_t$ depends only on h. An rcll process X in $\mathbb{R}^d$ with stationary independent increments and $X_0 = 0$ is called a *Lévy process*. If X is also real and nonnegative, it is often called a *subordinator*.

Corollary 15.7 (*Lévy processes and subordinators*) *An rcll process X in $\mathbb{R}^d$ is Lévy iff (1) holds with $m_t \equiv bt$, $G_t \equiv \sigma B_t$, and $E\eta = \lambda \otimes \nu$ for some $b \in \mathbb{R}^d$, some $d \times d$-matrix σ, some measure ν on $\mathbb{R}^d \setminus \{0\}$ with $\int (|x|^2 \wedge 1)\nu(dx) < \infty$, and some Brownian motion $B \perp\!\!\!\perp \eta$ in $\mathbb{R}^d$. Furthermore, X is a subordinator iff (3) holds with $a_t \equiv at$ and $E\eta = \lambda \otimes \nu$ for some $a \geq 0$ and some measure ν on $(0, \infty)$ with $\int (x \wedge 1)\nu(dx) < \infty$. The triple $(\sigma\sigma', b, \nu)$ or pair (a, ν) is then determined by $\mathcal{L}(X)$, and any a, b, σ, and ν with the stated properties may occur.*

The measure ν above is called the *Lévy measure* of X, and the quantities $\sigma\sigma', b$, and ν or a and ν are referred to collectively as the *characteristics* of X.

Proof: The stationarity of the increments excludes the possibility of fixed jumps, and so X has a representation as in Theorem 15.4. The stationarity also implies that $E\eta$ is time invariant. Thus, Theorem 2.6 yields $E\eta = \lambda \otimes \nu$ for some measure ν on $\mathbb{R}^d \setminus \{0\}$ or $(0, \infty)$. The stated conditions on ν are immediate from (2) and (4). Finally, Theorem 13.4 gives the form of the continuous component. Formula (5) shows that η is a measurable function of X, and so ν is uniquely determined by $\mathcal{L}(X)$. The uniqueness of the remaining characteristics then follows by subtraction. $\square$

From the representations in Theorem 15.4 we may easily deduce the following so-called *Lévy-Khinchin formulas* for the associated characteristic functions or Laplace transforms. Here we write u' for the transpose of u.

15. Independent Increments and Infinite Divisibility

Corollary 15.8 *(characteristic exponents, Kolmogorov, Lévy)* Let X be a Lévy process in $\mathbb{R}^d$ with characteristics (a, b, ν). Then $Ee^{iu'X_t} = e^{t\psi_u}$ for all $t \geq 0$ and $u \in \mathbb{R}^d$, where

$$\psi_u = iu'b - \tfrac{1}{2}u'au + \int (e^{iu'x} - 1 - iu'x 1\{|x| \leq 1\})\nu(dx), \quad u \in \mathbb{R}^d. \tag{7}$$

If X is a subordinator with characteristics (a, ν), then also $Ee^{-uX_t} = e^{-t\chi_u}$ for all $t, u \geq 0$, where

$$\chi_u = ua + \int (1 - e^{-ux})\nu(dx), \quad u \geq 0. \tag{8}$$

In both cases, the characteristics are determined by $\mathcal{L}(X_1)$.

Proof: Formula (8) follows immediately from (3) and Lemma 12.2 (i). Similarly, (7) is obtained from (1) by the same lemma when ν is bounded, and the general case then follows by dominated convergence.

To prove the last assertion, we note that ψ is the unique continuous function with $\psi_0 = 0$ satisfying $e^{\psi_u} = Ee^{iuX_1}$. By the uniqueness theorem for characteristic functions and the independence of the increments, ψ determines all finite-dimensional distributions of X, and so the uniqueness of the characteristics follows from the uniqueness in Corollary 15.7. □

From Proposition 8.5 we note that a Lévy process X is Markov for the induced filtration $\mathcal{G} = (\mathcal{G}_t)$ with translation-invariant transition kernels $\mu_t(x, B) = \mu_t(B-x) = P\{X_t \in B-x\}$. More generally, given any filtration $\mathcal{F}$, we say that X is Lévy with respect to $\mathcal{F}$, or simply $\mathcal{F}$-Lévy, if X is adapted to $\mathcal{F}$ and such that $(X_t - X_s) \perp\!\!\!\perp \mathcal{F}_s$ for all $s < t$. In particular, we may take $\mathcal{F}_t = \mathcal{G}_t \vee \mathcal{N}$, $t \geq 0$, where $\mathcal{N} = \sigma\{N \subset A; A \in \mathcal{A}, PA = 0\}$. Note that the latter filtration is right-continuous by Corollary 7.25. Just as for Brownian motion in Theorem 13.11, we further see that any process X which is $\mathcal{F}$-Lévy for some right-continuous, complete filtration $\mathcal{F}$ is a strong Markov process, in the sense that the process $X' = \theta_\tau X - X_\tau$ satisfies $X \stackrel{d}{=} X' \perp\!\!\!\perp \mathcal{F}_\tau$ for any finite optional time τ.

We turn to a brief discussion of some basic symmetry properties. A process X on $\mathbb{R}_+$ is said to be *self-similar* if for any $r > 0$ there exists some $s = h(r) > 0$ such that the process X_{rt}, $t \geq 0$, has the same distribution as sX. Excluding the trivial case when $X_t = 0$ a.s. for all $t > 0$, it is clear that h satisfies the Cauchy equation $h(xy) = h(x)h(y)$. If X is right-continuous, then h is continuous, and the only solutions are of the form $h(x) = x^\alpha$ for some $\alpha \in \mathbb{R}$.

Let us now return to the context of Lévy processes. We say that such a process X is *strictly stable* if it is self-similar, and *weakly stable* if it is self-similar apart from a centering, so that for each $r > 0$ the process (X_{rt}) has the same distribution as $(sX_t + bt)$ for suitable s and b. In the latter case, the corresponding symmetrized process is strictly stable, and so s is again of the form r^α. In both cases it is clear that $\alpha > 0$. We may then introduce the *index* $p = \alpha^{-1}$ and say that X is strictly or weakly p-stable.

The terminology carries over to random variables or vectors with the same distribution as X_1.

Proposition 15.9 *(stable Lévy processes) Let X be a nondegenerate Lévy process in $\mathbb{R}$ with characteristics (a, b, ν). Then X is weakly p-stable for some $p > 0$ iff either of these conditions holds:*

(i) $p = 2$ and $\nu = 0$;

(ii) $p \in (0, 2)$, $a = 0$, and $\nu(dx) = c_\pm |x|^{-p-1} dx$ on $\mathbb{R}_\pm$ for some $c_\pm \geq 0$.

For subordinators, weak p-stability is equivalent to the condition

(iii) $p \in (0, 1)$ and $\nu(dx) = c x^{-p-1} dx$ on $(0, \infty)$ for some $c > 0$.

Proof: Writing $S_r : x \mapsto rx$ for any $r > 0$, we note that the processes $X(r^p t)$ and rX have characteristics $r^p(a, b, \nu)$ and $(r^2 a, rb, \nu \circ S_r^{-1})$, respectively. Since the latter are determined by the distributions, it follows that X is weakly p-stable iff $r^p a = r^2 a$ and $r^p \nu = \nu \circ S_r^{-1}$ for all $r > 0$. In particular, $a = 0$ when $p \neq 2$. Writing $F(x) = \nu[x, \infty)$ or $\nu(-\infty, -x]$, we also note that $r^p F(rx) = F(x)$ for all $r, x > 0$, and so $F(x) = x^{-p} F(1)$, which yields the stated form of the density. The condition $\int (x^2 \wedge 1) \nu(dx) < \infty$ implies $p \in (0, 2)$ when $\nu \neq 0$. If $X \geq 0$, we have the stronger condition $\int (x \wedge 1) \nu(dx) < \infty$, so in this case $p < 1$. □

If X is weakly p-stable for some $p \neq 1$, it can be made strictly p-stable by a suitable centering. In particular, a weakly p-stable subordinator is strictly stable iff the drift component vanishes. In the latter case we simply say that X is *stable*.

The next result shows how stable subordinators may arise naturally even in the study of continuous processes. Given a Brownian motion B in $\mathbb{R}$, we introduce the maximum process $M_t = \sup_{s \leq t} B_s$ and its right-continuous inverse

$$T_r = \inf\{t \geq 0;\ M_t > r\} = \inf\{t \geq 0;\ B_t > r\}, \quad r \geq 0. \tag{9}$$

Theorem 15.10 *(first-passage times, Lévy) For a Brownian motion B, the process T in (9) is a $\frac{1}{2}$-stable subordinator with Lévy measure*

$$\nu(dx) = (2\pi)^{-1/2} x^{-3/2} dx, \quad x > 0.$$

Proof: By Lemma 7.6, the random times T_r are optional with respect to the right-continuous filtration $\mathcal{F}$ induced by B. By the strong Markov property of B, the process $\theta_r T - T_r$ is then independent of $\mathcal{F}_{T_r}$ with the same distribution as T. Since T is further adapted to the filtration $(\mathcal{F}_{T_r})$, it follows that T has stationary independent increments and hence is a subordinator.

To see that T is $\frac{1}{2}$-stable, fix any $c > 0$, put $\tilde{B}_t = c^{-1} B(c^2 t)$, and define $\tilde{T}_r = \inf\{t \geq 0;\ \tilde{B}_t > r\}$. Then

$$T_{cr} = \inf\{t \geq 0;\ B_t > cr\} = c^2 \inf\{t \geq 0;\ \tilde{B}_t > r\} = c^2 \tilde{T}_r.$$

By Proposition 15.9 the Lévy measure of T has a density of the form $ax^{-3/2}$, $x > 0$, and it remains to identify a. Then note that the process
$$X_t = \exp(uB_t - \tfrac{1}{2}u^2 t), \quad t \geq 0,$$
is a martingale for any $u \in \mathbb{R}$. In particular, $EX_{\tau_r \wedge t} = 1$ for any $r, t \geq 0$, and since clearly $B_{\tau_r} = r$, we get by dominated convergence
$$E \exp(-\tfrac{1}{2}u^2 T_r) = e^{-ur}, \quad u, r \geq 0.$$
Taking $u = \sqrt{2}$ and comparing with Corollary 15.8, we obtain
$$\frac{\sqrt{2}}{a} = \int_0^\infty (1 - e^{-x}) x^{-3/2} dx = 2 \int_0^\infty e^{-x} x^{-1/2} dx = 2\sqrt{\pi},$$
which shows that $a = (2\pi)^{-1/2}$. □

If we add a negative drift to a Brownian motion, the associated maximum process M becomes bounded, and so $T = M^{-1}$ terminates by a jump to infinity. For such occasions, it is useful to consider subordinators with possibly infinite jumps. By a *generalized subordinator* we mean a process of the form $X_t \equiv Y_t + \infty \cdot 1\{t \geq \zeta\}$ a.s., where Y is an ordinary subordinator and ζ is an independent, exponentially distributed random variable. In this case we say that X is obtained from Y by *exponential killing*. The representation in Theorem 15.4 remains valid in the generalized case, except that ν may now have positive mass at ∞.

The following characterization is needed in Chapter 22.

Lemma 15.11 *(generalized subordinators) Let X be a nondecreasing and right-continuous process in $[0, \infty]$ with $X_0 = 0$, and let $\mathcal{F}$ denote the filtration induced by X. Then X is a generalized subordinator iff*
$$P[X_{s+t} - X_s \in \cdot | \mathcal{F}_s] = P\{X_t \in \cdot\} \quad \text{a.s. on } \{X_s < \infty\}, \quad s, t > 0. \tag{10}$$

Proof: Writing $\zeta = \inf\{t; X_t = \infty\}$, we get from (10) the Cauchy equation
$$P\{\zeta > s + t\} = P\{\zeta > s\} P\{\zeta > t\}, \quad s, t \geq 0, \tag{11}$$
which shows that ζ is exponentially distributed with mean $m \in (0, \infty]$. Next define $\mu_t = P[X_t \in \cdot | X_t < \infty]$, $t \geq 0$, and conclude from (10) and (11) that the μ_t form a semigroup under convolution. By Theorem 8.4 there exists a corresponding process Y with stationary, independent increments. From the right-continuity of X, it follows that Y is continuous in probability. Hence, Y has a version that is a subordinator. Now choose $\tilde{\zeta} \stackrel{d}{=} \zeta$ with $\tilde{\zeta} \!\perp\!\!\!\perp Y$, and let $\tilde{X}$ denote the process Y killed at $\tilde{\zeta}$. Comparing with (10), we note that $\tilde{X} \stackrel{d}{=} X$. By Theorem 6.10 we may assume that even $X = \tilde{X}$ a.s., which means that X is a generalized subordinator. The converse assertion is obvious. □

The next result provides the basic link between Lévy processes and triangular arrays. A random vector ξ or its distribution is said to be *in-*

finitely divisible if for every $n \in \mathbb{N}$ there exist some i.i.d. random vectors $\xi_{n1}, \ldots, \xi_{nn}$ with $\sum_k \xi_{nk} \stackrel{d}{=} \xi$. By an *i.i.d. array* we mean a triangular array of random vectors ξ_{nj}, $j \leq m_n$, where the ξ_{nj} are i.i.d. for each n and $m_n \to \infty$.

Theorem 15.12 *(Lévy processes and infinite divisibility)* *For any random vector ξ in $\mathbb{R}^d$, these conditions are equivalent:*

(i) ξ *is infinitely divisible;*

(ii) $\sum_j \xi_{nj} \stackrel{d}{\to} \xi$ *for some i.i.d. array* (ξ_{nj});

(iii) $\xi \stackrel{d}{=} X_1$ *for some Lévy process X in $\mathbb{R}^d$.*

Under those conditions, $\mathcal{L}(X)$ is determined by $\mathcal{L}(\xi) = \mathcal{L}(X_1)$.

A simple lemma is needed for the proof.

Lemma 15.13 *(individual terms)* *If the ξ_{nj} are such as in Theorem 15.12 (ii), then $\xi_{n1} \stackrel{P}{\to} 0$.*

Proof: Let μ and μ_n denote the distributions of ξ and ξ_{nj}, respectively. Choose $r > 0$ so small that $\hat{\mu} \neq 0$ on $[-r, r]$, and write $\hat{\mu} = e^\psi$ on this interval, where $\psi: [-r, r] \to \mathbb{C}$ is continuous with $\psi(0) = 0$. Since the convergence $\hat{\mu}_n^{m_n} \to \hat{\mu}$ is uniform on bounded intervals, it follows that $\hat{\mu}_n \neq 0$ on $[-r, r]$ for sufficiently large n. Thus, we may write $\hat{\mu}_n(u) = e^{\psi_n(u)}$ for $|u| \leq r$, where $m_n \psi_n \to \psi$ on $[-r, r]$. Then $\psi_n \to 0$ on the same interval, and therefore $\hat{\mu}_n \to 1$. Now let $\varepsilon \leq r^{-1}$, and note as in Lemma 5.1 that

$$\int_{-r}^r (1 - \hat{\mu}_n(u))du = 2r \int \left(1 - \frac{\sin rx}{rx}\right) \mu_n(dx)$$

$$\geq 2r\left(1 - \frac{\sin r\varepsilon}{r\varepsilon}\right) \mu_n\{|x| \geq \varepsilon\}.$$

As $n \to \infty$, the left-hand side tends to 0 by dominated convergence, and we get $\mu_n \stackrel{w}{\to} \delta_0$. □

Proof of Theorem 15.12: Trivially (iii) $\Rightarrow$ (i) $\Rightarrow$ (ii). Now let ξ_{nj}, $j \leq m_n$, be an i.i.d. array satisfying (ii), put $\mu_n = \mathcal{L}(\xi_{nj})$, and fix any $k \in \mathbb{N}$. By Lemma 15.13 we may assume that k divides each m_n and write $\sum_j \xi_{nj} = \eta_{n1} + \cdots + \eta_{nk}$, where the η_{nj} are i.i.d. with distribution $\mu_n^{*(m_n/k)}$. For any $u \in \mathbb{R}^d$ and $r > 0$ we have

$$(P\{u\eta_{n1} > r\})^k = P\{\min_{j \leq k} u\eta_{nj} > r\} \leq P\left\{\sum_{j \leq k} u\eta_{nj} > kr\right\},$$

and so the tightness of $\sum_j \eta_{nj}$ carries over to the sequence (η_{n1}). By Proposition 5.21 we may extract a weakly convergent subsequence, say with limiting distribution ν_k. Since $\sum_j \eta_{nj} \stackrel{d}{\to} \xi$, it follows by Theorem 5.3 that ξ has distribution ν_k^{*k}. Thus, (ii) $\Rightarrow$ (i).

Next assume (i), so that $\mathcal{L}(\xi) \equiv \mu = \mu_n^{*n}$ for each n. By Lemma 15.13 we get $\hat{\mu}_n \to 1$ uniformly on bounded intervals, and so $\hat{\mu} \neq 0$. We may then write $\hat{\mu} = e^{\psi}$ and $\hat{\mu}_n = e^{\psi_n}$ for some continuous functions ψ and ψ_n with $\psi(0) = \psi_n(0) = 0$, and we get $\psi = n\psi_n$ for each n. Hence, $e^{t\psi}$ is a characteristic function for every $t \in \mathbb{Q}_+$, and then also for $t \in \mathbb{R}_+$ by Theorem 5.22. By Theorem 6.16 there exists a process X with stationary independent increments such that X_t has characteristic function $e^{t\psi}$ for every t. Here X is continuous in probability, and so by Theorem 15.1 it has an rcll version, which is the desired Lévy process. Thus, (i) $\Rightarrow$ (iii). The last assertion is clear from Corollary 15.8. □

Justified by the one-to-one correspondence between infinitely divisible distributions μ and their characteristics (a, b, ν) or (a, ν), we may write $\mu = \text{id}(a, b, \nu)$ or $\mu = \text{id}(a, \nu)$, respectively. The last result shows that the class of infinitely divisible laws is closed under weak convergence, and we proceed to derive explicit convergence criteria. Then define for each $h > 0$

$$a^h = a + \int_{|x| \leq h} xx'\nu(dx), \qquad b^h = b - \int_{h < |x| \leq 1} x\nu(dx),$$

where $\int_{h < |x| \leq 1} = -\int_{1 < |x| \leq h}$ when $h > 1$. In the positive case, we define instead $a^h = a + \int_{x \leq h} x\nu(dx)$. Let $\overline{\mathbb{R}^d}$ denote the one-point compactification of $\mathbb{R}^d$.

Theorem 15.14 *(convergence of infinitely divisible distributions)*

(i) *Let $\mu = \text{id}(a, b, \nu)$ and $\mu_n = \text{id}(a_n, b_n, \nu_n)$ on $\mathbb{R}^d$, and fix any $h > 0$ with $\nu\{|x| = h\} = 0$. Then $\mu_n \xrightarrow{w} \mu$ iff $a_n^h \to a^h$, $b_n^h \to b^h$, and $\nu_n \xrightarrow{v} \nu$ on $\overline{\mathbb{R}^d} \setminus \{0\}$.*

(ii) *Let $\mu = \text{id}(a, \nu)$ and $\mu_n = \text{id}(a, \nu)$ on $\mathbb{R}_+$, and fix any $h > 0$ with $\nu\{h\} = 0$. Then $\mu_n \xrightarrow{w} \mu$ iff $a_n^h \to a^h$ and $\nu_n \xrightarrow{v} \nu$ on $(0, \infty]$.*

For the proof, we consider first the one-dimensional case, which allows some important simplifications. Thus, (7) may then be written as

$$\psi_u = icu + \int \left(e^{iux} - 1 - \frac{iux}{1+x^2}\right)\frac{1+x^2}{x^2}\tilde{\nu}(dx), \qquad (12)$$

where

$$\tilde{\nu}(dx) = \sigma^2\delta_0(dx) + \frac{x^2}{1+x^2}\nu(dx), \qquad (13)$$

$$c = b + \int \left(\frac{x}{1+x^2} - x1\{|x| \leq 1\}\right)\nu(dx), \qquad (14)$$

and the integrand in (12) is defined by continuity as $-u^2/2$ when $x = 0$. For infinitely divisible distributions on $\mathbb{R}_+$, we may instead introduce the measure

$$\tilde{\nu}(dx) = a\delta_0 + (1 - e^{-x})\nu(dx). \qquad (15)$$

The associated distributions μ are denoted by $\text{Id}(c, \tilde{\nu})$ and $\text{Id}(\tilde{\nu})$, respectively.

Lemma 15.15 *(one-dimensional convergence criteria)*

(i) *Let* $\mu = \mathrm{Id}(c, \tilde\nu)$ *and* $\mu_n = \mathrm{Id}(c_n, \tilde\nu_n)$ *on* $\mathbb{R}$. *Then* $\mu_n \xrightarrow{w} \mu$ *iff* $c_n \to c$ *and* $\tilde\nu_n \xrightarrow{w} \tilde\nu$.

(ii) *Let* $\mu = \mathrm{Id}(\tilde\nu)$ *and* $\mu_n = \mathrm{Id}(\tilde\nu_n)$ *on* $\mathbb{R}_+$. *Then* $\mu_n \xrightarrow{w} \mu$ *iff* $\tilde\nu_n \xrightarrow{w} \tilde\nu$.

Proof: (i) Defining ψ and ψ_n as in (12), we may write $\hat\mu = e^\psi$ and $\hat\mu_n = e^{\psi_n}$. If $c_n \to c$ and $\tilde\nu_n \xrightarrow{w} \tilde\nu$, then $\psi_n \to \psi$ by the boundedness and continuity of the integrand in (12), and so $\hat\mu_n \to \hat\mu$, which implies $\mu_n \xrightarrow{w} \mu$ by Theorem 5.3. Conversely, $\mu_n \xrightarrow{w} \mu$ implies $\hat\mu_n \to \hat\mu$, uniformly on bounded intervals, and we get $\psi_n \to \psi$ in the same sense. Now define

$$\chi(u) = \int_{-1}^{1} (\psi(u) - \psi(u+s))\, ds = 2\int e^{iux}\left(1 - \frac{\sin x}{x}\right)\frac{1+x^2}{x^2}\tilde\nu(dx),$$

and similarly for χ_n, where the interchange of integrations is justified by Fubini's theorem. Then $\chi_n \to \chi$, and so by Theorem 5.3

$$\left(1 - \frac{\sin x}{x}\right)\frac{1+x^2}{x^2}\tilde\nu_n(dx) \xrightarrow{w} \left(1 - \frac{\sin x}{x}\right)\frac{1+x^2}{x^2}\tilde\nu(dx).$$

Since the integrand is continuous and bounded away from 0, it follows that $\tilde\nu_n \xrightarrow{w} \tilde\nu$. This implies convergence of the integral in (12), and by subtraction $c_n \to c$.

(ii) This may be proved directly by the same method, where we note that the functions in (8) satisfy $\chi(u+1) - \chi(u) = \int e^{-ux}\tilde\nu(dx)$. □

Proof of Theorem 15.14: For any finite measures m_n and m on $\mathbb{R}$ we note that $m_n \xrightarrow{w} m$ iff $m_n \xrightarrow{v} m$ on $\mathbb{R} \setminus \{0\}$ and $m_n(-h, h) \to m(-h, h)$ for some $h > 0$ with $m\{\pm h\} = 0$. Thus, for distributions μ and μ_n on $\mathbb{R}$, we have $\tilde\nu_n \xrightarrow{w} \tilde\nu$ iff $\nu_n \xrightarrow{v} \nu$ on $\mathbb{R} \setminus \{0\}$ and $a_n^h \to a^h$ for any $h > 0$ with $\nu\{\pm h\} = 0$. Similarly, $\tilde\nu_n \xrightarrow{w} \tilde\nu$ holds for distributions μ and μ_n on $\mathbb{R}_+$ iff $\nu_n \xrightarrow{v} \nu$ on $(0, \infty]$ and $a_n^h \to a^h$ for all $h > 0$ with $\nu\{h\} = 0$. Thus, (ii) follows immediately from Lemma 15.15. To obtain (i) from the same lemma when $d = 1$, it remains to notice that the conditions $b_n^h \to b^h$ and $c_n \to c$ are equivalent when $\tilde\nu_n \xrightarrow{w} \tilde\nu$ and $\nu\{\pm h\} = 0$, since $|x - x(1+x^2)^{-1}| \leq |x|^3$.

Turning to the proof of (i) when $d > 1$, let us first assume that $\nu_n \xrightarrow{v} \nu$ on $\overline{\mathbb{R}}^d \setminus \{0\}$ and that $a_n^h \to a^h$ and $b_n^h \to b^h$ for some $h > 0$ with $\nu\{|x| = h\} = 0$. To prove $\mu_n \xrightarrow{w} \mu$, it suffices by Corollary 5.5 to show that, for any one-dimensional projection $\pi_u : x \mapsto u'x$ with $u \neq 0$, $\mu_n \circ \pi_u^{-1} \xrightarrow{w} \mu \circ \pi_u^{-1}$. Then fix any $k > 0$ with $\nu\{|u'x| = k\} = 0$, and note that $\mu \circ \pi_u^{-1}$ has the associated characteristics $\nu^u = \nu \circ \pi_u^{-1}$ and

$$a^{u,k} = u'a^h u + \int (u'x)^2 \{1_{(0,k]}(|u'x|) - 1_{(0,h]}(|x|)\}\nu(dx),$$

$$b^{u,k} = u'b^h + \int u'x\{1_{(1,k]}(|u'x|) - 1_{(1,h]}(|x|)\}\nu(dx).$$

Let $a_n^{u,k}$, $b_n^{u,k}$, and ν_n^u denote the corresponding characteristics of $\mu_n \circ \pi_u^{-1}$. Then $\nu_n^u \xrightarrow{v} \nu^u$ on $\overline{\mathbb{R}} \setminus \{0\}$, and furthermore $a_n^{u,k} \to a^{u,k}$ and $b_n^{u,k} \to b^{u,k}$. The desired convergence now follows from the one-dimensional result.

Conversely, assume that $\mu_n \xrightarrow{w} \mu$. Then $\mu_n \circ \pi_u^{-1} \xrightarrow{w} \mu \circ \pi_u^{-1}$ for every $u \neq 0$, and the one-dimensional result yields $\nu_n^u \xrightarrow{v} \nu^u$ on $\overline{\mathbb{R}} \setminus \{0\}$ as well as $a_n^{u,k} \to a^{u,k}$ and $b_n^{u,k} \to b^{u,k}$ for any $k > 0$ with $\nu\{|u'x| = k\} = 0$. In particular, the sequence $(\nu_n K)$ is bounded for every compact set $K \subset \overline{\mathbb{R}}^d \setminus \{0\}$, and so the sequences $(u'a_n^h u)$ and $(u'b_n^h)$ are bounded for any $u \neq 0$ and $h > 0$. In follows easily that (a_n^h) and (b_n^h) are bounded for every $h > 0$, and therefore all three sequences are relatively compact.

Given any subsequence $N' \subset \mathbb{N}$, we have $\nu_n \xrightarrow{v} \nu'$ along a further subsequence $N'' \subset N'$ for some measure ν' satisfying $\int (|x|^2 \wedge 1)\nu'(dx) < \infty$. Fixing any $h > 0$ with $\nu'\{|x| = h\} = 0$, we may choose a still further subsequence N''' such that even a_n^h and b_n^h converge toward some limits a' and b'. The direct assertion then yields $\mu_n \xrightarrow{w} \mu'$ along N''', where μ' is infinitely divisible with characteristics determined by (a', b', ν'). Since $\mu' = \mu$, we get $\nu' = \nu$, $a' = a^h$, and $b' = b^h$. Thus, the convergence remains valid along the original sequence. $\square$

By a simple approximation, we may now derive explicit criteria for the convergence $\sum_j \xi_{nj} \xrightarrow{d} \xi$ in Theorem 15.12. Note that the compound Poisson distribution with characteristic measure $\mu = \mathcal{L}(\xi)$ is given by $\tilde{\mu} = \mathrm{id}(0, b, \mu)$, where $b = E[\xi; |\xi| \leq 1]$. For any array of random vectors ξ_{nj}, we may introduce an *associated compound Poisson array*, consisting of rowwise independent compound Poisson random vectors $\tilde{\xi}_{nj}$ with characteristic measures $\mathcal{L}(\xi_{nj})$.

Corollary 15.16 *(i.i.d. arrays)* *Consider in* $\mathbb{R}^d$ *an i.i.d. array* (ξ_{nj}) *and an associated compound Poisson array* $(\tilde{\xi}_{nj})$, *and let* ξ *be* $\mathrm{id}(a, b, \nu)$. *Then* $\sum_j \xi_{nj} \xrightarrow{d} \xi$ *iff* $\sum_j \tilde{\xi}_{nj} \xrightarrow{d} \xi$. *For any* $h > 0$ *with* $\nu\{|x| = h\} = 0$, *it is also equivalent that*

(i) $m_n \mathcal{L}(\xi_{n1}) \xrightarrow{v} \nu$ *on* $\overline{\mathbb{R}}^d \setminus \{0\}$;

(ii) $m_n E[\xi_{n1}\xi'_{n1}; |\xi_{n1}| \leq h] \to a^h$;

(iii) $m_n E[\xi_{n1}; |\xi_{n1}| \leq h] \to b^h$.

Proof: Let $\mu = \mathcal{L}(\xi)$ and write $\hat{\mu} = e^\psi$, where ψ is continuous with $\psi(0) = 0$. If $\mu_n^{*m_n} \xrightarrow{w} \mu$, then $\hat{\mu}_n^{m_n} \to \hat{\mu}$ uniformly on compacts. Thus, on any bounded set B we may write $\hat{\mu}_n = e^{\psi_n}$ for large enough n, where the ψ_n are continuous with $m_n \psi_n \to \psi$ uniformly on B. Hence, $m_n(e^{\psi_n} - 1) \to \psi$, and so $\tilde{\mu}_n^{*m_n} \xrightarrow{w} \mu$. The proof in the other direction is similar. Since $\tilde{\mu}_n^{*m_n}$ is $\mathrm{id}(0, b_n, m_n \mu_n)$ with $b_n = m_n \int_{|x| \leq 1} x \mu_n(dx)$, the last assertion follows by Theorem 15.14. $\square$

The weak convergence of infinitely divisible laws extends to a pathwise approximation property for the corresponding Lévy processes.

Theorem 15.17 *(approximation of Lévy processes, Skorohod)* Let $X, X^1, X^2, \ldots$ be Lévy processes in $\mathbb{R}^d$ with $X_1^n \xrightarrow{d} X_1$. Then there exist some processes $\tilde{X}^n \stackrel{d}{=} X^n$ such that $(\tilde{X}^n - X)_t^* \xrightarrow{P} 0$ for all $t \geq 0$.

Before proving the general result, we consider two special cases.

Lemma 15.18 *(compound Poisson case)* The conclusion of Theorem 15.17 holds when $X, X^1, X^2, \ldots$ are compound Poisson with characteristic measures $\nu, \nu_1, \nu_2, \ldots$ satisfying $\nu_n \xrightarrow{w} \nu$.

Proof: Allowing positive mass at the origin, we may assume that ν and the ν_n have the same total mass, which may then be reduced to 1 through a suitable scaling. If $\xi_1, \xi_2, \ldots$ and $\xi_1^n, \xi_2^n, \ldots$ are associated i.i.d. sequences, then $(\xi_1^n, \xi_2^n, \ldots) \xrightarrow{d} (\xi_1, \xi_2, \ldots)$ by Theorem 4.29, and by Theorem 4.30 we may assume that the convergence holds a.s. Letting N be an independent unit-rate Poisson process, and defining $X_t = \sum_{j \leq N_t} \xi_j$ and $X_t^n = \sum_{j \leq N_t} \xi_j^n$, it follows that $(X^n - X)_t^* \to 0$ a.s. for each $t \geq 0$. □

Lemma 15.19 *(case of small jumps)* The conclusion of Theorem 15.17 holds when $EX^n \equiv 0$ and $1 \geq (\Delta X^n)_1^* \xrightarrow{P} 0$.

Proof: Since $(\Delta X^n)_1^* \xrightarrow{P} 0$, we may choose some constants $h_n \to 0$ with $m_n = h_n^{-1} \in \mathbb{N}$ such that $w(X^n, 1, h_n) \xrightarrow{P} 0$. By the stationarity of the increments, it follows that $w(X^n, t, h_n) \xrightarrow{P} 0$ for all $t \geq 0$. Next, Theorem 15.14 shows that X is centered Gaussian. Thus, there exist as in Theorem 14.20 some processes $Y^n \stackrel{d}{=} (X_{[m_n t]h_n}^n)$ with $(Y^n - X)_t^* \xrightarrow{P} 0$ for all $t \geq 0$. By Corollary 6.11 we may further choose some processes $\tilde{X}^n \stackrel{d}{=} X^n$ with $Y^n \equiv \tilde{X}_{[m_n t]h_n}^n$ a.s. Then, as $n \to \infty$ for fixed $t \geq 0$,

$$E[(\tilde{X}^n - X)_t^* \wedge 1] \leq E[(Y^n - X)_t^* \wedge 1] + E[w(X^n, t, h_n) \wedge 1] \to 0. \quad □$$

Proof of Theorem 15.17: The asserted convergence is clearly equivalent to $\rho(\tilde{X}^n, X) \to 0$, where ρ denotes the metric

$$\rho(X, Y) = \int_0^\infty e^{-t} E[(X - Y)_t^* \wedge 1] dt.$$

For any $h > 0$ we may write $X = L^h + M^h + J^h$ and $X^n = L^{n,h} + M^{n,h} + J^{n,h}$ with $L_t^h \equiv b^h t$ and $L_t^{n,h} \equiv b_n^h t$, where M^h and $M^{n,h}$ are martingales containing the Gaussian components and all centered jumps of size $\leq h$, and the processes J^h and $J^{n,h}$ are formed by all remaining jumps. Write B for the Gaussian component of X, and note that $\rho(M^h, B) \to 0$ as $h \to 0$ by Proposition 7.16.

For any $h > 0$ with $\nu\{|x| = h\} = 0$, it is clear from Theorem 15.14 that $b_n^h \to b^h$ and $\nu_n^h \xrightarrow{w} \nu^h$, where ν^h and ν_n^h denote the restrictions of ν and ν_n, respectively, to the set $\{|x| > h\}$. The same theorem yields $a_n^h \to a$ as $n \to \infty$ and then $h \to 0$, and so under those conditions $M_1^{n,h} \xrightarrow{d} B_1$.

Now fix any $\varepsilon > 0$. By Lemma 15.19 there exist some constants $h, r > 0$ and processes $\tilde{M}^{n,h} \stackrel{d}{=} M^{n,h}$ such that $\rho(M^h, B) \leq \varepsilon$ and $\rho(\tilde{M}^{n,h}, B) \leq \varepsilon$ for all $n > r$. Furthermore, if $\nu\{|x| = h\} = 0$, there exist by Lemma 15.18 some number $r' \geq r$ and processes $\tilde{J}^{n,h} \stackrel{d}{=} J^{n,h}$ independent of $\tilde{M}^{n,h}$ such that $\rho(\tilde{J}^h, \tilde{J}^{n,h}) \leq \varepsilon$ for all $n > r'$. We may finally choose $r'' \geq r'$ so large that $\rho(L^h, L^{n,h}) \leq \varepsilon$ for all $n > r''$. The processes $\tilde{X}^n \equiv L^{n,h} + \tilde{M}^{n,h} + \tilde{J}^{n,h} \stackrel{d}{=} X^n$ then satisfy $\rho(X, \tilde{X}^n) \leq 4\varepsilon$ for all $n > r''$. □

Combining Theorem 15.17 with Corollary 15.16, we get a similar approximation theorem for random walks, which extends the result for Gaussian limits in Theorem 14.20. A slightly weaker result is obtained by different methods in Theorem 16.14.

Corollary 15.20 *(approximation of random walks)* Consider in $\mathbb{R}^d$ a Lévy process X and some random walks $S^1, S^2, \ldots$ such that $S^n_{m_n} \stackrel{d}{\to} X_1$ for some integers $m_n \to \infty$, and let N be an independent unit-rate Poisson process. Then there exist some processes $X^n \stackrel{d}{=} (S^n \circ N_{m_n t})$ such that $(X^n - X)^*_t \stackrel{P}{\to} 0$ for all $t \geq 0$.

In particular, we may use this result to extend the first two arcsine laws in Theorem 13.16 to symmetric Lévy processes.

Theorem 15.21 *(arcsine laws)* Let X be a symmetric Lévy process in $\mathbb{R}$ with $X_1 \neq 0$ a.s. Then these random variables are arcsine distributed:

$$\tau_1 = \lambda\{t \leq 1; X_t > 0\}, \quad \tau_2 = \inf\{t \geq 0; X_t \vee X_{t-} = \sup_{s \leq 1} X_s\}. \quad (16)$$

The purpose of the condition $X_1 \neq 0$ a.s. is to exclude the degenerate case of pure jump-type processes.

Lemma 15.22 *(diffuseness, Doeblin)* A measure $\mu = \mathrm{id}(a, b, \nu)$ in $\mathbb{R}^d$ is diffuse iff $a \neq 0$ or $\nu \mathbb{R}^d = \infty$.

Proof: If $a = 0$ and $\nu \mathbb{R}^d < \infty$, then μ is compound Poisson apart from a shift, and so it is clearly not diffuse. When either condition fails, then it does so for at least one coordinate projection, and we may take $d = 1$. If $a > 0$, the diffuseness is obvious by Lemma 1.28. Next assume that ν is unbounded, say with $\nu(0, \infty) = \infty$. For each $n \in \mathbb{N}$ we may then write $\nu = \nu_n + \nu'_n$, where ν'_n is supported by $(0, n^{-1})$ and has total mass $\log 2$. For μ we get a corresponding decomposition $\mu_n * \mu'_n$, where μ'_n is compound Poisson with Lévy measure ν'_n and $\mu'_n\{0\} = \frac{1}{2}$. For any $x \in \mathbb{R}$ and $\varepsilon > 0$ we get

$$\begin{aligned}\mu\{x\} &\leq \mu_n\{x\}\mu'_n\{0\} + \mu_n[x - \varepsilon, x]\mu'_n(0, \varepsilon] + \mu'_n(\varepsilon, \infty) \\ &\leq \tfrac{1}{2}\mu_n[x - \varepsilon, x] + \mu'_n(\varepsilon, \infty).\end{aligned}$$

Letting $n \to \infty$ and then $\varepsilon \to 0$, and noting that $\mu'_n \stackrel{w}{\to} \delta_0$ and $\mu_n \stackrel{w}{\to} \mu$, we get $\mu\{x\} \leq \tfrac{1}{2}\mu\{x\}$ by Theorem 4.25, and so $\mu\{x\} = 0$. □

Proof of Theorem 15.21: Introduce the random walk $S_k^n = X_{k/n}$, let N be an independent unit-rate Poisson process, and define $X_t^n = S^n \circ N_{nt}$. By Corollary 15.20 there exist some processes $\tilde{X}^n \stackrel{d}{=} X^n$ with $(\tilde{X}^n - X)_1^* \stackrel{P}{\to} 0$. Define τ_1^n and τ_2^n as in (16) in terms of X^n, and conclude from Lemmas 14.12 and 15.22 that $\tau_i^n \stackrel{d}{\to} \tau_i$ for $i = 1, 2$.

Now define
$$\sigma_1^n = N_n^{-1} \sum_{k \leq N_n} 1\{S_k^n > 0\}, \quad \sigma_2^n = N_n^{-1} \min\{k;\ S_k^n = \max_{j \leq N_n} S_j^n\}.$$

Since $t^{-1} N_t \to 1$ a.s. by the law of large numbers, we have $\sup_{t \leq 1} |n^{-1} N_{nt} - t| \to 0$ a.s., and so $\sigma_2^n - \tau_2^n \to 0$ a.s. Applying the same law to the sequence of holding times in N, we further note that $\sigma_1^n - \tau_1^n \stackrel{P}{\to} 0$. Hence, $\sigma_i^n \stackrel{d}{\to} \tau_i$ for $i = 1, 2$. Now $\sigma_1^n \stackrel{d}{=} \sigma_2^n$ by Corollary 11.14, and by Theorem 14.11 we have $\sigma_2^n \stackrel{d}{\to} \sin^2 \alpha$ where α is $U(0, 2\pi)$. Hence, $\tau_1 \stackrel{d}{=} \tau_2 \stackrel{d}{=} \sin^2 \alpha$. □

The preceding results will now be used to complete the classical limit theory for sums of independent random variables begun in Chapter 5. Recall that a *null array* in $\mathbb{R}^d$ is defined as a family of random vectors ξ_{nj}, $j = 1, \ldots, m_n$, $n \in \mathbb{N}$, such that the ξ_{nj} are independent for each n and satisfy $\sup_j E[|\xi_{nj}| \wedge 1] \to 0$. Our first goal is to extend Theorem 5.11, by giving the basic connection between sums with positive and symmetric terms. Here we write p_2 for the mapping $x \mapsto x^2$.

Proposition 15.23 *(positive and symmetric terms)* Let (ξ_{nj}) be a null array of symmetric random variables, and let ξ and η be infinitely divisible with characteristics $(a, 0, \nu)$ and $(a, \nu \circ p_2^{-1})$, respectively, where ν is symmetric and $a \geq 0$. Then $\sum_j \xi_{nj} \stackrel{d}{\to} \xi$ iff $\sum_j \xi_{nj}^2 \stackrel{d}{\to} \eta$.

Again the proof may be based on a simple compound Poisson approximation. Here $\xi_n \stackrel{d}{\sim} \eta_n$ means that $\xi_n \stackrel{d}{\to} \xi$ iff $\eta_n \stackrel{d}{\to} \xi$ for any ξ.

Lemma 15.24 *(approximation)* Let (ξ_{nj}) be a null array of positive or symmetric random variables, and let $(\tilde{\xi}_{nj})$ be an associated compound Poisson array. Then $\sum_j \xi_{nj} \stackrel{d}{\sim} \sum_j \tilde{\xi}_{nj}$.

Proof: Write $\mu = \mathcal{L}(\xi)$ and $\mu_{nj} = \mathcal{L}(\xi_{nj})$. In the symmetric case we need to show that
$$\prod_j \hat{\mu}_{nj} \to \hat{\mu} \quad \Longleftrightarrow \quad \prod_j \exp(\hat{\mu}_{nj} - 1) \to \hat{\mu},$$
which is immediate from Lemmas 5.6 and 5.8. In the nonnegative case, a similar argument applies to the Laplace transforms. □

Proof of Proposition 15.23: Define $\mu_{nj} = \mathcal{L}(\xi_{nj})$, and fix any $h > 0$ with $\nu\{|x| = h\} = 0$. By Theorem 15.14 (i) and Lemma 15.24 we have

$\sum_j \xi_{nj} \xrightarrow{d} \xi$ iff

$$\sum_j \mu_{nj} \xrightarrow{v} \nu \text{ on } \mathbb{R} \setminus \{0\},$$

$$\sum_j E[\xi_{nj}^2; |\xi_{nj}| \leq h] \to a + \int_{|x| \leq h} x^2 \nu(dx),$$

whereas $\sum_j \xi_{nj}^2 \xrightarrow{d} \eta$ iff

$$\sum_j \mu_{nj} \circ p_2^{-1} \xrightarrow{v} \nu \circ p_2^{-1} \text{ on } (0, \infty],$$

$$\sum_j E[\xi_{nj}^2; \xi_{nj}^2 \leq h^2] \to a + \int_{y \leq h^2} y(\nu \circ p_2^{-1})(dy).$$

The two sets of conditions are equivalent by Lemma 1.22. □

The limit problem for general null arrays is more delicate, since a compound Poisson approximation as in Corollary 15.16 or Lemma 15.24 applies only after a careful centering, as specified by the following key result.

Theorem 15.25 (*compound Poisson approximation*) *Let* (ξ_{nj}) *be a null array of random vectors in* $\mathbb{R}^d$, *and fix any* $h > 0$. *Define* $\eta_{nj} = \xi_{nj} - b_{nj}$, *where* $b_{nj} = E[\xi_{nj}; |\xi_{nj}| \leq h]$, *and let* $(\tilde{\eta}_{nj})$ *be an associated compound Poisson array. Then*

$$\sum_j \xi_{nj} \stackrel{d}{\sim} \sum_j (\tilde{\eta}_{nj} + b_{nj}). \tag{17}$$

A technical estimate is needed for the proof.

Lemma 15.26 (*uniform summability*) *Let the random vectors* $\eta_{nj} = \xi_{nj} - b_{nj}$ *in Theorem 15.25 have characteristic functions* φ_{nj}. *Then either condition in* (17) *implies*

$$\limsup_{n \to \infty} \sum_j |1 - \varphi_{nj}(u)| < \infty, \quad u \in \mathbb{R}^d.$$

Proof: By the definitions of b_{nj}, η_{nj}, and φ_{nj}, we have

$$1 - \varphi_{nj}(u) = E\left[1 - e^{iu'\eta_{nj}} + iu'\eta_{nj}1\{|\xi_{nj}| \leq h\}\right] - iu'b_{nj}P\{|\xi_{nj}| > h\}.$$

Putting

$$a_n = \sum_j E[\eta_{nj}\eta'_{nj}; |\xi_{nj}| \leq h], \quad p_n = \sum_j P\{|\xi_{nj}| > h\},$$

and using Lemma 5.14, we get

$$\sum_j |1 - \varphi_{nj}(u)| \leq \tfrac{1}{2} u' a_n u + (2 + |u|) p_n.$$

Hence, it is enough to show that $(u'a_n u)$ and (p_n) are bounded.

Assuming the second condition in (17), the desired boundedness follows easily from Theorem 15.14, together with the fact that $\max_j |b_{nj}| \to 0$. If instead $\sum_j \xi_{nj} \xrightarrow{d} \xi$, we may introduce an independent copy (ξ'_{nj}) of the array (ξ_{nj}) and apply Theorem 15.14 and Lemma 15.24 to the symmetric random variables $\zeta_{nj}^u = u'\xi_{nj} - u'\xi'_{nj}$. For any $h' > 0$, this gives

$$\limsup_{n\to\infty} \sum_j P\{|\zeta_{nj}^u| > h'\} < \infty, \tag{18}$$

$$\limsup_{n\to\infty} \sum_j E[(\zeta_{nj}^u)^2; |\zeta_{nj}^u| \leq h'] < \infty. \tag{19}$$

The boundedness of p_n follows from (18) and Lemma 4.19. Next we note that (19) remains true with the condition $|\zeta_{nj}^u| \leq h'$ replaced by $|\xi_{nj}| \vee |\xi'_{nj}| \leq h$. Furthermore, by the independence of ξ_{nj} and ξ'_{nj},

$$\tfrac{1}{2}\sum_j E[(\zeta_{nj}^u)^2; |\xi_{nj}| \vee |\xi'_{nj}| \leq h]$$
$$= \sum_j E[(u'\eta_{nj})^2; |\xi_{nj}| \leq h]P\{|\xi_{nj}| \leq h\} - \sum_j (E[u'\eta_{nj}; |\xi_{nj}| \leq h])^2$$
$$\geq u'a_n u \min_j P\{|\xi_{nj}| \leq h\} - \sum_j (u'b_{nj} P\{|\xi_{nj}| > h\})^2.$$

Here the last sum is bounded by $p_n \max_j (u'b_{nj})^2 \to 0$, and the minimum on the right tends to 1. The boundedness of $(u'a_n u)$ now follows by (19). □

Proof of Theorem 15.25: By Lemma 5.13 it is enough to show that $\sum_j |\varphi_{nj}(u) - \exp\{\varphi_{nj}(u) - 1\}| \to 0$, where φ_{nj} denotes the characteristic function of η_{nj}. This is clear from Taylor's formula, together with Lemmas 5.6 and 15.26. □

In particular, we may now identify the possible limits.

Corollary 15.27 *(limit laws, Feller, Khinchin)* *Let (ξ_{nj}) be a null array of random vectors in $\mathbb{R}^d$ such that $\sum_j \xi_{nj} \xrightarrow{d} \xi$ for some random vector ξ. Then ξ is infinitely divisible.*

Proof: The random vectors $\tilde{\eta}_{nj}$ in Theorem 15.25 are infinitely divisible, so the same thing is true for the sums $\sum_j (\tilde{\eta}_{nj} - b_{nj})$. The infinite divisibility of ξ then follows by Theorem 15.12. □

We may further combine Theorems 15.14 and 15.25 to obtain explicit convergence criteria for general null arrays. The present result generalizes Theorem 5.15 for Gaussian limits and Corollary 15.16 for i.i.d. arrays. For convenience, we write $\text{cov}[\xi; A]$ for the covariance matrix of the random vector $1_A \xi$.

Theorem 15.28 *(convergence criteria for null arrays, Doeblin, Gnedenko)* Let (ξ_{nj}) be a null array of random vectors in $\mathbb{R}^d$, let ξ be $\mathrm{id}(a,b,\nu)$, and fix any $h > 0$ with $\nu\{|x| = h\} = 0$. Then $\sum_j \xi_{nj} \xrightarrow{d} \xi$ iff these conditions hold:

(i) $\sum_j \mathcal{L}(\xi_{nj}) \xrightarrow{v} \nu$ on $\overline{\mathbb{R}^d} \setminus \{0\}$;

(ii) $\sum_j \mathrm{cov}[\xi_{nj}; |\xi_{nj}| \leq h] \to a^h$;

(iii) $\sum_j E[\xi_{nj}; |\xi_{nj}| \leq h] \to b^h$.

Proof: Define $a_{nj} = \mathrm{cov}[\xi_{nj}; |\xi_{nj}| \leq h]$ and $b_{nj} = E[\xi_{nj}; |\xi_{nj}| \leq h]$. By Theorems 15.14 and 15.25 the convergence $\sum_j \xi_{nj} \xrightarrow{d} \xi$ is equivalent to the conditions

(i') $\sum_j \mathcal{L}(\eta_{nj}) \xrightarrow{v} \nu$ on $\overline{\mathbb{R}^d} \setminus \{0\}$,

(ii') $\sum_j E[\eta_{nj} \eta'_{nj}; |\eta_{nj}| \leq h] \to a^h$,

(iii') $\sum_j (b_{nj} + E[\eta_{nj}; |\eta_{nj}| \leq h]) \to b^h$.

Here (i) and (i') are equivalent since $\max_j |b_{nj}| \to 0$. Using (i) and the facts that $\max_j |b_{nj}| \to 0$ and $\nu\{|x| = h\} = 0$, it is further clear that the sets $\{|\eta_{nj}| \leq h\}$ in (ii') and (iii') can be replaced by $\{|\xi_{nj}| \leq h\}$. To prove the equivalence of (ii) and (ii'), it is then enough to note that, in view of (i),

$$\left\|\sum_j \{a_{nj} - E[\eta_{nj}\eta'_{nj}; |\xi_{nj}| \leq h]\}\right\| \leq \left\|\sum_j b_{nj} b'_{nj} P\{|\xi_{nj}| > h\}\right\|$$
$$\lesssim \max_j |b_{nj}|^2 \sum_j P\{|\xi_{nj}| > h\} \to 0.$$

Similarly, (iii) and (iii') are equivalent because

$$\left|\sum_j E[\eta_{nj}; |\xi_{nj}| \leq h]\right| = \left|\sum_j b_{nj} P\{|\xi_{nj}| > h\}\right|$$
$$\leq \max_j |b_{nj}| \sum_j P\{|\xi_{nj}| > h\} \to 0. \qquad \square$$

In the one-dimensional case, we give two probabilistic interpretations of the first condition in Theorem 15.28, one of which involves the row-wise extremes. For random measures η and η_n on $\mathbb{R}\setminus\{0\}$, the convergence $\eta_n \xrightarrow{d} \eta$ on $\overline{\mathbb{R}} \setminus \{0\}$ is defined by the condition $\eta_n f \xrightarrow{d} \eta f$ for all $f \in C_K^+(\overline{\mathbb{R}} \setminus \{0\})$.

Theorem 15.29 *(sums and extremes)* Let (ξ_{nj}) be a null array of random variables with distributions μ_{nj}, and define $\eta_n = \sum_j \delta_{\xi_{nj}}$ and $\alpha_n^\pm = \max_j(\pm \xi_{nj})$, $n \in \mathbb{N}$. Fix a Lévy measure ν on $\mathbb{R} \setminus \{0\}$, let η be a Poisson process on $\mathbb{R} \setminus \{0\}$ with $E\eta = \nu$, and put $\alpha^\pm = \sup\{x \geq 0;\, \eta\{\pm x\} > 0\}$. Then these conditions are equivalent:

(i) $\sum_j \mu_{nj} \xrightarrow{v} \nu$ on $\overline{\mathbb{R}} \setminus \{0\}$;

(ii) $\eta_n \xrightarrow{d} \eta$ on $\overline{\mathbb{R}} \setminus \{0\}$;

(iii) $\alpha_n^\pm \xrightarrow{d} \alpha^\pm$.

The equivalence of (i) and (ii) is an immediate consequence of Theorem 16.18 in the next chapter. Here we give a direct elementary proof.

Proof: Condition (i) holds iff

$$\sum_j \mu_{nj}(x, \infty) \to \nu(x, \infty), \qquad \sum_j \mu_{nj}(-\infty, -x) \to \nu(-\infty, -x), \qquad (20)$$

for all $x > 0$ with $\nu\{\pm x\} = 0$. By Lemma 5.8, the first condition in (20) is equivalent to

$$P\{\alpha_n^+ \leq x\} = \prod_j (1 - P\{\xi_{nj} > x\}) \to e^{-\nu(x,\infty)} = P\{\alpha^+ \leq x\},$$

which holds for all continuity points $x > 0$ iff $\alpha_n^+ \xrightarrow{d} \alpha^+$. Similarly, the second condition in (20) holds iff $\alpha_n^- \xrightarrow{d} \alpha^-$. Thus, (i) and (iii) are equivalent.

To show that (i) implies (ii), we may write the latter condition in the form

$$\sum_j f(\xi_{nj}) \xrightarrow{d} \eta f, \qquad f \in C_K^+(\mathbb{R} \setminus \{0\}). \qquad (21)$$

Here the variables $f(\xi_{nj})$ form a null array with distributions $\mu_{nj} \circ f^{-1}$, and ηf is compound Poisson with characteristic measure $\nu \circ f^{-1}$. Thus, Theorem 15.14 (ii) shows that (21) is equivalent to the conditions

$$\sum_j \mu_{nj} \circ f^{-1} \xrightarrow{v} \nu \circ f^{-1} \quad \text{on } (0, \infty], \qquad (22)$$

$$\lim_{\varepsilon \to 0} \limsup_{n \to \infty} \sum_j \int_{f(x) \leq \varepsilon} f(x) \mu_{nj}(dx) = 0. \qquad (23)$$

Now (22) follows immediately from (i). To deduce (23), it suffices to note that the sum on the left is bounded by $\sum_j \mu_{nj}(f \wedge \varepsilon) \to \nu(f \wedge \varepsilon)$.

Finally, assume (ii). By a simple approximation, $\eta_n(x, \infty) \xrightarrow{d} \eta(x, \infty)$ for any $x > 0$ with $\nu\{x\} = 0$. In particular, for such an x,

$$P\{\alpha_n^+ \leq x\} = P\{\eta_n(x, \infty) = 0\} \to P\{\eta(x, \infty) = 0\} = P\{\alpha^+ \leq x\},$$

and so $\alpha_n^+ \xrightarrow{d} \alpha^+$. Similarly, $\alpha_n^- \xrightarrow{d} \alpha^-$, which proves (iii). □

Exercises

1. Show that a Lévy process X in $\mathbb{R}$ is a subordinator iff $X_1 \geq 0$ a.s.

2. Show that the Cauchy distribution $\mu(dx) = \pi^{-1}(1 + x^2)^{-1} dx$ is strictly 1-stable, and determine the corresponding Lévy measure ν. (*Hint:* Check that $\hat\mu(u) = e^{-|u|}$. By symmetry, $\nu(dx) = cx^{-2} dx$ for some $c > 0$, and it remains to determine c.)

3. Let X be a weakly p-stable Lévy process. If $p \neq 1$, show that the process $X_t - ct$ is strictly p-stable for a suitable constant c. Note that the centering fails for $p = 1$.

4. Extend Proposition 15.23 to null arrays of spherically symmetric random vectors in $\mathbb{R}^d$.

5. Show by an example that Theorem 15.25 fails without the centering at truncated means. (*Hint:* Without the centering, condition (ii) of Theorem 15.28 becomes $\sum_j E[\xi_{nj}\xi'_{nj}; |\xi_{nj}| \leq h] \to a^h$.)

6. Deduce Theorems 5.7 and 5.11 from Theorem 15.14 and Lemma 15.24.

7. For a Lévy process X of effective dimension $d \geq 3$, show that $|X_t| \to \infty$ a.s. as $t \to \infty$. (*Hint:* Define $\tau = \inf\{t; |X_t| > 1\}$, and iterate to form a random walk (S_n). Show that the latter has the same effective dimension as X, and use Theorem 9.8.)

8. Let X be a Lévy process in $\mathbb{R}$, and fix any $p \in (0, 2)$. Show that $t^{-1/p} X_t$ converges a.s. iff $E|X_1|^p < \infty$ and either $p \leq 1$ or $EX_1 = 0$. (*Hint:* Define a random walk (S_n) as before, show that S_1 satisfies the same moment condition as X_1, and apply Theorem 4.23.)

9. If ξ is id(a, b, ν) and $p > 0$, show that $E|\xi|^p < \infty$ iff $\int_{|x|>1} |x|^p \nu(dx) < \infty$. (*Hint:* If ν has bounded support, then $E|\xi|^p < \infty$ for all p. It is then enough to consider compound Poisson distributions, for which the result is elementary.)

10. Show by a direct argument that a $\mathbb{Z}_+$-valued random variable ξ is infinitely divisible (on $\mathbb{Z}_+$) iff $-\log Es^\xi = \sum_k (1 - s^k)\nu_k$, $s \in (0, 1]$, for some unique, bounded measure $\nu = (\nu_k)$ on $\mathbb{N}$. (*Hint:* Assuming $\mathcal{L}(\xi) = \mu_n^{*n}$, use the inequality $1 - x \leq e^{-x}$ to show that the sequence $(n\mu_n)$ is tight on $\mathbb{N}$. Then $n\mu_n \xrightarrow{w} \nu$ along a subsequence for some bounded measure ν on $\mathbb{N}$. Finally note that $-\log(1 - x) \sim x$ as $x \to 0$. For the uniqueness, take differences and use the uniqueness theorem for power series.)

11. Show by a direct argument that a random variable $\xi \geq 0$ is infinitely divisible iff $-\log Ee^{-u\xi} = ua + \int(1 - e^{-ux})\nu(dx)$, $u \geq 0$, for some unique constant $a \geq 0$ and measure ν on $(0, \infty)$ with $\int(|x| \wedge 1) < \infty$. (*Hint:* If $\mathcal{L}(\xi) = \mu_n^{*n}$, note that the measures $\chi_n(dx) = n(1 - e^{-x})\mu_n(dx)$ are tight on $\mathbb{R}_+$. Then $\chi_n \xrightarrow{w} \chi$ along a subsequence, and we may write $\chi(dx) = a\delta_0(dx) + (1 - e^{-x})\nu(dx)$. The desired representation now follows as before. To get the uniqueness, take differences and use the uniqueness theorem for Laplace transforms.)

12. Show by a direct argument that a random variable ξ is infinitely divisible iff $\psi_u = \log Ee^{iu\xi}$ exists and is given by (7) for some unique constants $a \geq 0$ and b and measure ν on $\mathbb{R} \setminus \{0\}$ with $\int(x^2 \wedge 1)\nu(dx) < \infty$. (*Hint:* Proceed as in Lemma 15.15.)

13. Given a semigroup of infinitely divisible distributions μ_t, show that there exists a process X on $\mathbb{R}_+$ with stationary, independent increments and $\mathcal{L}(X_t) = \mu_t$ for all $t \geq 0$. Starting from a suitable Poisson process and an independent Brownian motion, construct a Lévy process Y with the same property. Conclude that X has a version with rcll paths and a similar representation as Y. (*Hint:* Use Lemma 3.24 and Theorems 6.10 and 6.16.)

Chapter 16

Convergence of Random Processes, Measures, and Sets

Relative compactness and tightness; uniform topology on $C(K,S)$; Skorohod's J_1-topology; equicontinuity and tightness; convergence of random measures; superposition and thinning; exchangeable sequences and processes; simple point processes and random closed sets

The basic notions of weak or distributional convergence were introduced in Chapter 4, and in Chapter 5 we studied the special case of distributions on Euclidean spaces. The purpose of this chapter is to develop the general weak convergence theory into a powerful tool that applies to a wide range of set, measure, and function spaces. In particular, some functional limit theorems derived in the last two chapters by cumbersome embedding and approximation techniques will then be accessible by straightforward compactness arguments.

The key result is Prohorov's theorem, which gives the basic connection between tightness and relative distributional compactness. This result will enable us to convert some classical compactness criteria into convenient probabilistic versions. In particular, we shall see how the Arzelà–Ascoli theorem yields a corresponding criterion for distributional compactness of continuous processes. Similarly, an optional equicontinuity condition will be shown to guarantee the appropriate compactness for processes that are right-continuous with left-hand limits (rcll). We shall also derive some general criteria for convergence in distribution of random measures and sets, with special attention to the point process case.

The general criteria will be applied to some interesting concrete situations. In addition to some already familiar results from Chapters 14 and 13, we shall obtain a general functional limit theorem for sampling from finite populations and derive convergence criteria for superpositions and thinnings of point processes. Further applications appear in subsequent chapters, such as a general approximation result for Markov chains in Chapter 19 and a method for constructing weak solutions to SDEs in Chapter 21.

Beginning with the case of continuous processes, let us fix two metric spaces (K,d) and (S,ρ), where K is compact and S is separable and complete, and consider the space $C(K,S)$ of continuous functions from K to

S, endowed with the uniform metric $\hat{\rho}(x,y) = \sup_{t \in K} \rho(x_t, y_t)$. For each $t \in K$ we may introduce the evaluation map $\pi_t : x \mapsto x_t$ from $C(K,S)$ to S. The following result shows that the random elements in $C(K,S)$ are precisely the continuous S-valued processes on K.

Lemma 16.1 *(Borel sets and evaluations)* $\mathcal{B}(C(K,S)) = \sigma\{\pi_t; \, t \in K\}$.

Proof: The maps π_t are continuous, hence Borel measurable, and so the generated σ-field $\mathcal{C}$ is contained in $\mathcal{B}(C(K,S))$. To prove the reverse relation, we need to show that any open subset $G \subset C(K,S)$ lies in $\mathcal{C}$. From the Arzelà–Ascoli Theorem A2.1 we note that $C(K,S)$ is σ-compact and hence separable. Thus, G is a countable union of open balls $B_{x,r} = \{y \in C(K,S); \, \hat{\rho}(x,y) < r\}$, and it suffices to prove that the latter lie in $\mathcal{C}$. But this is clear, since for any countable dense set $D \subset K$,

$$\overline{B}_{x,r} = \bigcap_{t \in D} \{y \in C(K,S); \, \rho(x_t, y_t) \leq r\}. \qquad \Box$$

If X and X^n are random processes on K, we write $X^n \xrightarrow{fd} X$ for convergence of the finite-dimensional distributions, in the sense that

$$(X^n_{t_1}, \ldots, X^n_{t_k}) \xrightarrow{d} (X_{t_1}, \ldots, X_{t_k}), \quad t_1, \ldots, t_k \in K, \, k \in \mathbb{N}. \qquad (1)$$

Though by Proposition 3.2 the distribution of a random process is determined by the family of finite-dimensional distributions, condition (1) is insufficient in general for the convergence $X^n \xrightarrow{d} X$ in $C(K,S)$. This is already clear when the processes are nonrandom, since pointwise convergence of a sequence of functions need not be uniform. To overcome this difficulty, we may add a compactness condition. Recall that a sequence of random elements $\xi_1, \xi_2, \ldots$ is said to be *relatively compact in distribution* if every subsequence has a further subsequence that converges in distribution.

Lemma 16.2 *(weak convergence via compactness)* *Let $X, X_1, X_2, \ldots$ be random elements in $C(K,S)$. Then $X_n \xrightarrow{d} X$ iff $X_n \xrightarrow{fd} X$ and (X_n) is relatively compact in distribution.*

Proof: If $X_n \xrightarrow{d} X$, then $X_n \xrightarrow{fd} X$ by Theorem 4.27, and (X_n) is trivially relatively compact in distribution. Now assume instead that (X_n) satisfies the two conditions. If $X_n \not\xrightarrow{d} X$, we may choose a bounded continuous function $f : C(K,S) \to \mathbb{R}$ and an $\varepsilon > 0$ such that $|Ef(X_n) - Ef(X)| > \varepsilon$ along some subsequence $N' \subset \mathbb{N}$. By the relative compactness we may choose a further subsequence N'' and a process Y such that $X_n \xrightarrow{d} Y$ along N''. But then $X_n \xrightarrow{fd} Y$ along N'', and since also $X_n \xrightarrow{fd} X$, Proposition 3.2 yields $X \stackrel{d}{=} Y$. Thus, $X_n \xrightarrow{d} X$ along N'', and so $Ef(X_n) \to Ef(X)$ along the same sequence, a contradiction. We conclude that $X_n \xrightarrow{d} X$. $\quad\Box$

The last result shows the importance of finding tractable conditions for a random sequence $\xi_1, \xi_2, \ldots$ in a metric space S to be relatively compact.

Generalizing a notion from Chapter 4, we say that (ξ_n) is *tight* if

$$\sup_K \liminf_{n\to\infty} P\{\xi_n \in K\} = 1, \qquad (2)$$

where the supremum extends over all compact subsets $K \subset S$.

We may now state the key result of weak convergence theory, the equivalence between tightness and relative compactness for random elements in sufficiently regular metric spaces. A version for Euclidean spaces was obtained in Proposition 5.21.

Theorem 16.3 *(tightness and relative compactness, Prohorov) For any sequence of random elements $\xi_1, \xi_2, \ldots$ in a metric space S, tightness implies relative compactness in distribution, and the two conditions are equivalent when S is separable and complete.*

In particular, we note that when S is separable and complete, a single random element ξ in S is *tight*, in the sense that $\sup_K P\{\xi \in K\} = 1$. In that case we may clearly replace the "lim inf" in (2) by "inf."

For the proof of Theorem 16.3 we need a simple lemma. Recall from Lemma 1.6 that a random element in a subspace of a metric space S may also be regarded as a random element in S.

Lemma 16.4 *(preservation of tightness) Tightness is preserved by continuous mappings. In particular, if (ξ_n) is a tight sequence of random elements in a subspace A of some metric space S, then (ξ_n) remains tight when regarded as a sequence in S.*

Proof: Compactness is preserved by continuous mappings. This applies in particular to the natural embedding $I: A \to S$. □

Proof of Theorem 16.3 (Varadarajan): For $S = \mathbb{R}^d$ the result was proved in Proposition 5.21. Turning to the case when $S = \mathbb{R}^\infty$, consider a tight sequence of random elements $\xi^n = (\xi_1^n, \xi_2^n, \ldots)$ in $\mathbb{R}^\infty$. Writing $\eta_k^n = (\xi_1^n, \ldots, \xi_k^n)$, we conclude from Lemma 16.4 that the sequence $(\eta_k^n;\, n \in \mathbb{N})$ is tight in $\mathbb{R}^k$ for each $k \in \mathbb{N}$. Given any subsequence $N' \subset \mathbb{N}$, we may then use a diagonal argument to extract a further subsequence N'' such that $\eta_k^n \xrightarrow{d}$ some η_k as $n \to \infty$ along N'' for fixed $k \in \mathbb{N}$. The sequence $(\mathcal{L}(\eta_k))$ is projective by the continuity of the coordinate projections, and so by Theorem 6.14 there exists a random sequence $\xi = (\xi_1, \xi_2, \ldots)$ such that $(\xi_1, \ldots, \xi_k) \stackrel{d}{=} \eta_k$ for each k. But then $\xi^n \xrightarrow{fd} \xi$ along N'', and so Theorem 4.29 yields $\xi^n \xrightarrow{d} \xi$ along the same sequence.

Next assume that $S \subset \mathbb{R}^\infty$. If (ξ_n) is tight in S, then by Lemma 16.4 it remains tight as a sequence in $\mathbb{R}^\infty$. Hence, for any sequence $N' \subset \mathbb{N}$ there exist a further subsequence N'' and some random element ξ such that $\xi_n \xrightarrow{d} \xi$ in $\mathbb{R}^\infty$ along N''. To show that the convergence remains valid in S, it suffices by Lemma 4.26 to verify that $\xi \in S$ a.s. Then choose some compact sets $K_m \subset S$ with $\liminf_n P\{\xi_n \in K_m\} \geq 1 - 2^{-m}$ for each $m \in \mathbb{N}$.

Since the K_m remain closed in $\mathbb{R}^\infty$, Theorem 4.25 yields
$$P\{\xi \in K_m\} \geq \limsup_{n \in N''} P\{\xi_n \in K_m\} \geq \liminf_{n \to \infty} P\{\xi_n \in K_m\} \geq 1 - 2^{-m},$$
and so $\xi \in \bigcup_m K_m \subset S$ a.s.

Now assume that S is σ-compact. In particular, it is then separable and therefore homeomorphic to a subset $A \subset \mathbb{R}^\infty$. By Lemma 16.4 the tightness of (ξ_n) carries over to the image sequence $(\tilde{\xi}_n)$ in A, and by Lemma 4.26 the possible relative compactness of $(\tilde{\xi}_n)$ implies the same property for (ξ_n). This reduces the discussion to the previous case.

Now turn to the general case. If (ξ_n) is tight, there exist some compact sets $K_m \subset S$ with $\liminf_n P\{\xi_n \in K_m\} \geq 1 - 2^{-m}$. In particular, $P\{\xi_n \in A\} \to 1$, where $A = \bigcup_m K_m$, and so we may choose some random elements η_n in A with $P\{\xi_n = \eta_n\} \to 1$. Here (η_n) is again tight, even as a sequence in A, and since A is σ-compact, the previous argument shows that (η_n) is relatively compact as a sequence in A. By Lemma 4.26 it remains relatively compact in S, and by Theorem 4.28 the relative compactness carries over to (ξ_n).

To prove the converse assertion, let S be separable and complete, and assume that (ξ_n) is relatively compact. For any $r > 0$ we may cover S by some open balls $B_1, B_2, \ldots$ of radius r. Writing $G_k = B_1 \cup \cdots \cup B_k$, we claim that
$$\lim_{k \to \infty} \inf_n P\{\xi_n \in G_k\} = 1. \tag{3}$$

Indeed, we may otherwise choose some integers $n_k \uparrow \infty$ with $\sup_k P\{\xi_{n_k} \in G_k\} = c < 1$. By the relative compactness we have $\xi_{n_k} \xrightarrow{d} \xi$ along a subsequence $N' \subset \mathbb{N}$ for a suitable ξ, and so
$$P\{\xi \in G_m\} \leq \liminf_{k \in N'} P\{\xi_{n_k} \in G_m\} \leq c < 1, \quad m \in \mathbb{N},$$
which leads as $m \to \infty$ to the absurdity $1 < 1$. Thus, (3) must be true.

Now take $r = m^{-1}$ and write G_k^m for the corresponding sets G_k. For any $\varepsilon > 0$ there exist by (3) some $k_1, k_1, \cdots \in \mathbb{N}$ with
$$\inf_n P\{\xi_n \in G_{k_m}^m\} \geq 1 - \varepsilon 2^{-m}, \quad m \in \mathbb{N}.$$

Writing $A = \bigcap_m G_{k_m}^m$, we get $\inf_n P\{\xi_n \in A\} \geq 1 - \varepsilon$. Also, note that $\bar{A}$ is complete and totally bounded, hence compact. Thus, (ξ_n) is tight. □

In order to apply the last theorem, we need convenient criteria for tightness. Beginning with the space $C(K, S)$, we may convert the classical Arzelà–Ascoli compactness criterion into a condition for tightness. Then introduce the *modulus of continuity*
$$w(x, h) = \sup\{\rho(x_s, x_t); d(s, t) \leq h\}, \quad x \in C(K, S), \; h > 0.$$

The function $w(x, h)$ is clearly continuous for fixed $h > 0$ and hence a measurable function of x.

Theorem 16.5 *(tightness in $C(K,S)$, Prohorov)* *For any metric spaces K and S, where K is compact and S is separable and complete, let $X, X_1, X_2, \ldots$ be random elements in $C(K,S)$. Then $X_n \xrightarrow{d} X$ iff $X_n \xrightarrow{fd} X$ and*

$$\lim_{h \to 0} \limsup_{n \to \infty} E[w(X_n, h) \wedge 1] = 0. \tag{4}$$

Proof: Since $C(K,S)$ is separable and complete, Theorem 16.3 shows that tightness and relative compactness are equivalent for (X_n). By Lemma 16.2 it is then enough to show that, under the condition $X_n \xrightarrow{fd} X$, the tightness of (X^n) is equivalent to (4).

First let (X_n) be tight. For any $\varepsilon > 0$ we may then choose a compact set $B \subset C(K,S)$ such that $\limsup_n P\{X_n \in B^c\} < \varepsilon$. By the Arzelà–Ascoli Theorem A2.1 we may next choose $h > 0$ so small that $w(x, h) \leq \varepsilon$ for all $x \in B$. But then $\limsup_n P\{w(X_n, h) > \varepsilon\} < \varepsilon$, and (4) follows since ε was arbitrary.

Next assume that (4) holds and $X_n \xrightarrow{fd} X$. Since each X_n is continuous, $w(X_n, h) \to 0$ a.s. as $h \to 0$ for fixed n, so the "lim sup" in (4) may be replaced by "sup." For any $\varepsilon > 0$ we may then choose $h_1, h_2, \ldots > 0$ so small that

$$\sup_n P\{w(X_n, h_k) > 2^{-k}\} \leq 2^{-k-1}\varepsilon, \quad k \in \mathbb{N}. \tag{5}$$

Letting $t_1, t_2, \ldots$ be dense in K, we may further choose some compact sets $C_1, C_2, \cdots \subset S$ such that

$$\sup_n P\{X_n(t_k) \in C_k^c\} \leq 2^{-k-1}\varepsilon, \quad k \in \mathbb{N}. \tag{6}$$

Now define

$$B = \bigcap_k \{x \in C(K,S);\ x(t_k) \in C_k,\ w(x, h_k) \leq 2^{-k}\}.$$

Then $\overline{B}$ is compact by the Arzelà–Ascoli Theorem A2.1, and from (5) and (6) we get $\sup_n P\{X_n \in B^c\} \leq \varepsilon$. Thus, (X_n) is tight. $\square$

One often needs to replace the compact parameter space K by some more general index set T. Here we assume T to be locally compact, second-countable, and Hausdorff (abbreviated as *lcscH*) and endow the space $C(T,S)$ of continuous functions from T to S with the topology of uniform convergence on compacts. As before, the Borel σ-field in $C(T,S)$ is generated by the evaluation maps π_t, and so the random elements in $C(T,S)$ are precisely the continuous processes on T taking values in S. The following result characterizes convergence in distribution of such processes.

Proposition 16.6 *(locally compact parameter space)* Let $X, X^1, X^2, \ldots$ be random elements in $C(T, S)$, where S is a metric space and T is lcscH. Then $X^n \overset{d}{\to} X$ iff convergence holds for the restrictions to any compact subset $K \subset T$.

Proof: The necessity is obvious from Theorem 4.27, since the restriction map $\pi_K : C(T, S) \to C(K, S)$ is continuous for any compact set $K \subset T$. To prove the sufficiency, we may choose some compact sets $K_1 \subset K_2 \subset \cdots \subset T$ with $K_j^\circ \uparrow T$, and let $X_i, X_i^1, X_i^2, \ldots$ denote the restrictions of the processes $X, X^1, X^2, \ldots$ to K_i. By hypothesis we have $X_i^n \overset{d}{\to} X_i$ for every i, and so Theorem 4.29 yields $(X_1^n, X_2^n, \ldots) \overset{d}{\to} (X_1, X_2, \ldots)$. Now $\pi = (\pi_{K_1}, \pi_{K_2}, \ldots)$ is a homeomorphism from $C(T, S)$ onto its range in $\times_j C(K_j, S)$, and so $X^n \overset{d}{\to} X$ by Lemma 4.26 and Theorem 4.27. □

For a simple illustration, we may prove a version of Donsker's Theorem 14.9. Since Theorem 16.5 applies only to processes with continuous paths, we need to replace the original step processes by their linearly interpolated versions

$$X_t^n = n^{-1/2} \left\{ \sum_{k \leq nt} \xi_k + (nt - [nt]) \xi_{[nt]+1} \right\}, \quad t \geq 0,\ n \in \mathbb{N}. \tag{7}$$

Corollary 16.7 *(functional central limit theorem, Donsker)* Let $\xi_1, \xi_2, \ldots$ be i.i.d. random variables with mean 0 and variance 1, define $X^1, X^2, \ldots$ by (7), and let B denote a Brownian motion on $\mathbb{R}_+$. Then $X^n \overset{d}{\to} B$ in $C(\mathbb{R}_+)$.

The following simple estimate may be used to verify the tightness.

Lemma 16.8 *(maximum inequality, Ottaviani)* Let $\xi_1, \xi_2, \ldots$ be i.i.d. random variables with mean 0 and variance 1, and put $S_n = \sum_{j \leq n} \xi_j$. Then

$$P\{S_n^* \geq 2r\sqrt{n}\} \leq \frac{P\{|S_n| \geq r\sqrt{n}\}}{1 - r^{-2}}, \quad r > 1,\ n \in \mathbb{N}.$$

Proof: Put $c = r\sqrt{n}$, and define $\tau = \inf\{k \in \mathbb{N};\ |S_k| \geq 2c\}$. By the strong Markov property at τ and Theorem 6.4,

$$\begin{aligned} P\{|S_n| \geq c\} &\geq P\{|S_n| \geq c,\ S_n^* \geq 2c\} \\ &\geq P\{\tau \leq n,\ |S_n - S_\tau| \leq c\} \\ &\geq P\{S_n^* \geq 2c\} \min_{k \leq n} P\{|S_k| \leq c\}, \end{aligned}$$

and by Chebyshev's inequality,

$$\min_{k \leq n} P\{|S_k| \leq c\} \geq \min_{k \leq n}(1 - kc^{-2}) \geq (1 - nc^{-2}) = 1 - r^{-2}. \quad \square$$

Proof of Corollary 16.7: By Proposition 16.6 it is enough to prove the convergence on $[0, 1]$. Clearly, $X_n \overset{fd}{\to} X$ by Proposition 5.9 and Corollary 5.5. Combining the former result with Lemma 16.8, we further get the

rough estimate
$$\lim_{r\to\infty} r^2 \limsup_{n\to\infty} P\{S_n^* \geq r\sqrt{n}\} = 0,$$
which implies
$$\lim_{h\to 0} h^{-1} \limsup_{n\to\infty} \sup_t P\left\{\sup_{0\leq r\leq h} |X_{t+r}^n - X_t^n| > \varepsilon\right\} = 0.$$

Now (4) follows easily, as we divide $[0,1]$ into subintervals of length $\leq h$. □

Next we show how the Kolmogorov–Chentsov criterion in Theorem 3.23 may be converted into a sufficient condition for tightness in $C(\mathbb{R}^d, S)$. An important application appears in Theorem 21.9.

Corollary 16.9 *(moments and tightness) Let $X^1, X^2, \ldots$ be continuous processes on $\mathbb{R}^d$ with values in a separable, complete metric space (S, ρ). Assume that (X_0^n) is tight in S and that for suitable constants $a, b > 0$,*
$$E\{\rho(X_s^n, X_t^n)\}^a \lesssim |s-t|^{d+b}, \quad s, t \in \mathbb{R}^d, \ n \in \mathbb{N}, \tag{8}$$
uniformly in n. Then (X^n) is tight in $C(\mathbb{R}^d, S)$, and for any $c \in (0, b/a)$ the limiting processes are a.s. locally Hölder continuous with exponent c.

Proof: For each process X^n we define the associated quantities ξ_{nk} as in the proof of Theorem 3.23, and we get $E\xi_{nk}^a \lesssim 2^{-kb}$. Hence, Lemma 1.29 yields for $m, n \in \mathbb{N}$
$$\|w(X_n, 2^{-m})\|_a^{a\wedge 1} \lesssim \sum_{k\geq m} \|\xi_{nk}\|_a^{a\wedge 1} \lesssim \sum_{k\geq m} 2^{-kb/(a\vee 1)} \lesssim 2^{-mb/(a\vee 1)},$$
which implies (4). Condition (8) extends by Lemma 4.11 to any limiting process X, and the last assertion then follows by Theorem 3.23. □

Let us now fix a separable, complete metric space S, and consider random processes with paths in $D(\mathbb{R}_+, S)$, the space of rcll functions $f: \mathbb{R}_+ \to S$. We endow $D(\mathbb{R}_+, S)$ with the *Skorohod J_1-topology*, whose basic properties are summarized in Appendix A2. Note in particular that the path space is again Polish and that compactness may be characterized in terms of a modified modulus of continuity $\tilde{w}$, as defined in Theorem A2.2.

The following result gives a criterion for weak convergence in $D(\mathbb{R}_+, S)$, similar to Theorem 16.5 for $C(K, S)$.

Theorem 16.10 *(tightness in $D(\mathbb{R}_+, S)$, Skorohod, Prohorov) For any separable, complete metric space S, let $X, X_1, X_2, \ldots$ be random elements in $D(\mathbb{R}_+, S)$. Then $X_n \xrightarrow{d} X$ iff $X_n \xrightarrow{fd} X$ on some dense subset of $T = \{t \geq 0; \Delta X_t = 0 \text{ a.s.}\}$ and*
$$\lim_{h\to 0} \limsup_{n\to\infty} E[\tilde{w}(X_n, t, h) \wedge 1] = 0, \quad t > 0. \tag{9}$$

Proof: Since π_t is continuous at every path $x \in D(\mathbb{R}_+, S)$ with $\Delta x_t = 0$, $X_n \xrightarrow{d} X$ implies $X_n \xrightarrow{fd} X$ on T by Theorem 4.27. Now use Theorem A2.2 and proceed as in the proof of Theorem 16.5. □

Tightness in $D(\mathbb{R}_+, S)$ is often verified most easily by means of the following sufficient condition. Given a process X, we say that a random time is X-*optional* if it is optional with respect to the filtration induced by X.

Theorem 16.11 *(optional equicontinuity and tightness, Aldous)* *For any metric space (S, ρ), let $X^1, X^2, \ldots$ be random elements in $D(\mathbb{R}_+, S)$. Then (9) holds if for any bounded sequence of X^n-optional times τ_n and any positive constants $h_n \to 0$,*

$$\rho(X^n_{\tau_n}, X^n_{\tau_n+h_n}) \xrightarrow{P} 0, \quad n \to \infty. \tag{10}$$

The proof will be based on two lemmas, where the first one is a restatement of condition (10).

Lemma 16.12 *The condition in Theorem 16.11 is equivalent to*

$$\lim_{h \to 0} \limsup_{n \to \infty} \sup_{\sigma, \tau} E[\rho(X^n_\sigma, X^n_\tau) \wedge 1] = 0, \quad t > 0, \tag{11}$$

where the supremum extends over all X^n-optional times $\sigma, \tau \leq t$ with $\sigma \leq \tau \leq \sigma + h$.

Proof: Replacing ρ by $\rho \wedge 1$ if necessary, we may assume that $\rho \leq 1$. The condition in Theorem 16.11 is then equivalent to

$$\lim_{\delta \to 0} \limsup_{n \to \infty} \sup_{\tau \leq t} \sup_{h \in [0, \delta]} E\rho(X^n_\tau, X^n_{\tau+h}) = 0, \quad t > 0,$$

where the first supremum extends over all X^n-optional times $\tau \leq t$. To deduce (11), assume that $0 \leq \tau - \sigma \leq \delta$. Then $[\tau, \tau + \delta] \subset [\sigma, \sigma + 2\delta]$, and so by the triangle inequality and a simple substitution,

$$\delta \rho(X_\sigma, X_\tau) \leq \int_0^\delta \{\rho(X_\sigma, X_{\tau+h}) + \rho(X_\tau, X_{\tau+h})\} dh$$

$$\leq \int_0^{2\delta} \rho(X_\sigma, X_{\sigma+h}) dh + \int_0^\delta \rho(X_\tau, X_{\tau+h}) dh.$$

Thus,

$$\sup_{\sigma, \tau} E\rho(X_\sigma, X_\tau) \leq 3 \sup_\tau \sup_{h \in [0, 2\delta]} E\rho(X_\tau, X_{\tau+h}),$$

where the suprema extend over all optional times $\tau \leq t$ and $\sigma \in [\tau - \delta, \tau]$. $\square$

We also need the following elementary estimate.

Lemma 16.13 *Let $\xi_1, \ldots, \xi_n \geq 0$ be random variables with sum S_n. Then*
$$Ee^{-S_n} \leq e^{-nc} + \max_{k \leq n} P\{\xi_k < c\}, \quad c > 0.$$

Proof: Let p denote the maximum on the right. By the Hölder and Chebyshev inequalities we get
$$Ee^{-S_n} = E\prod_k e^{-\xi_k} \leq \prod_k (Ee^{-n\xi_k})^{1/n} \leq \left\{(e^{-nc} + p)^{1/n}\right\}^n = e^{-nc} + p. \quad \Box$$

Proof of Theorem 16.11: Again we may assume that $\rho \leq 1$, and by suitable approximation we may extend condition (11) to weakly optional times σ and τ. For each $n \in \mathbb{N}$ and $\varepsilon > 0$, we recursively define the weakly X^n-optional times
$$\sigma_{k+1}^n = \inf\{s > \sigma_k^n; \rho(X_{\sigma_k^n}^n, X_s^n) > \varepsilon\}, \quad k \in \mathbb{Z}_+,$$
starting with $\sigma_0^n = 0$. Note that for $m \in \mathbb{N}$ and $t, h > 0$,
$$\tilde{w}(X^n, t, h) \leq 2\varepsilon + \sum_{k < m} 1\{\sigma_{k+1}^n - \sigma_k^n < h, \sigma_k^n < t\} + 1\{\sigma_m^n < t\}. \quad (12)$$

Now let $\nu_n(t, h)$ denote the supremum in (11). By Chebyshev's inequality and a simple truncation,
$$P\{\sigma_{k+1}^n - \sigma_k^n < h, \sigma_k^n < t\} \leq \varepsilon^{-1}\nu_n(t+h, h), \quad k \in \mathbb{N}, \, t, h > 0, \quad (13)$$
and so by (11) and (12),
$$\lim_{h \to 0} \limsup_{n \to \infty} E\tilde{w}(X^n, t, h) \leq 2\varepsilon + \limsup_{n \to \infty} P\{\sigma_m^n < t\}. \quad (14)$$

Next we conclude from (13) and Lemma 16.13 that, for any $c > 0$,
$$P\{\sigma_m^n < t\} \leq e^t E[e^{-\sigma_m^n}; \sigma_m^n < t] \leq e^t\{e^{-mc} + \varepsilon^{-1}\nu_n(t+c, c)\}.$$

By (11) the right-hand side tends to 0 as $m, n \to \infty$ and then $c \to 0$. Hence, the last term in (14) tends to 0 as $m \to \infty$, and (9) follows since ε is arbitrary. $\quad \Box$

We may illustrate the use of Theorem 16.11 by proving an extension of Corollary 16.7. A more precise result is obtained by different methods in Corollary 15.20. An extension to Markov chains appears in Theorem 19.28.

Theorem 16.14 *(approximation of random walks, Skorohod) Let $S^1, S^2, \ldots$ be random walks in $\mathbb{R}^d$ such that $S_{m_n}^n \xrightarrow{d} X_1$ for some Lévy process X and some integers $m_n \to \infty$. Then the processes $X_t^n = S_{[m_n t]}^n$ satisfy $X^n \xrightarrow{d} X$ in $D(\mathbb{R}_+, \mathbb{R}^d)$.*

Proof: By Corollary 15.16 we have $X^n \xrightarrow{fd} X$, and so by Theorem 16.11 it is enough to show that $|X_{\tau_n + h_n}^n - X_{\tau_n}^n| \xrightarrow{P} 0$ for any finite optional times τ_n and constants $h_n \to 0$. By the strong Markov property of S^n, or alternatively by Theorem 11.13, we may reduce to the case when $\tau_n = 0$ for all n. Thus, it suffices to show that $X_{h_n}^n \xrightarrow{P} 0$ as $h_n \to 0$, which again may be seen from Corollary 15.16. □

For the remainder of this chapter, we assume that S is lcscH with Borel σ-field $\mathcal{S}$. Write $\hat{\mathcal{S}}$ for the class of relatively compact sets in $\mathcal{S}$. Let $\mathcal{M}(S)$ denote the space of locally finite measures on S, endowed with the *vague topology* induced by the mappings $\pi_f : \mu \mapsto \mu f = \int f d\mu$, $f \in C_K^+$. The basic properties of this topology are summarized in Theorem A2.3. Note in particular that $\mathcal{M}(S)$ is Polish and that the random elements in $\mathcal{M}(S)$ are precisely the random measures on S. Similarly, the point processes on S are random elements in the vaguely closed subspace $\mathcal{N}(S)$, consisting of all integer-valued measures in $\mathcal{M}(S)$.

We begin with the basic tightness criterion.

Lemma 16.15 *(tightness of random measures, Prohorov)* Let $\xi_1, \xi_2, \ldots$ *be random measures on some lcscH space S. Then the sequence (ξ_n) is relatively compact in distribution iff $(\xi_n B)$ is tight in $\mathbb{R}_+$ for every $B \in \hat{\mathcal{S}}$.*

Proof: By Theorems 16.3 and A2.3 the notions of relative compactness and tightness are equivalent for (ξ_n). If (ξ_n) is tight, then so is $(\xi_n f)$ for every $f \in C_K^+$ by Lemma 16.4, and hence $(\xi_n B)$ is tight for all $B \in \hat{\mathcal{S}}$. Conversely, assume the latter condition. Choose an open cover $G_1, G_2, \cdots \in \hat{\mathcal{S}}$ of S, fix any $\varepsilon > 0$, and let $r_1, r_2, \cdots > 0$ be large enough that

$$\sup_n P\{\xi_n G_k > r_k\} < \varepsilon 2^{-k}, \quad k \in \mathbb{N}. \tag{15}$$

Then the set $A = \bigcap_k \{\mu;\ \mu G_k \leq r_k\}$ is relatively compact by Theorem A2.3 (ii), and (15) yields $\inf_n P\{\xi_n \in A\} > 1 - \varepsilon$. Thus, (ξ_n) is tight. □

We may now derive some general convergence criteria for random measures, corresponding to the uniqueness results in Lemma 12.1 and Theorem 12.8. Define $\hat{\mathcal{S}}_\xi = \{B \in \hat{\mathcal{S}};\ \xi \partial B = 0 \text{ a.s.}\}$.

Theorem 16.16 *(convergence of random measures)* Let $\xi, \xi_1, \xi_2, \ldots$ *be random measures on an lcscH space S. Then these conditions are equivalent:*

(i) $\xi_n \xrightarrow{d} \xi$;

(ii) $\xi_n f \xrightarrow{d} \xi f$ for all $f \in C_K^+$;

(iii) $(\xi_n B_1, \ldots, \xi_n B_k) \xrightarrow{d} (\xi B_1, \ldots, \xi B_k)$ for all $B_1, \ldots, B_k \in \hat{\mathcal{S}}_\xi$, $k \in \mathbb{N}$.

If ξ is a simple point process or a diffuse random measure, it is also equivalent that

(iv) $\xi_n B \xrightarrow{d} \xi B$ for all $B \in \hat{\mathcal{S}}_\xi$.

Proof: By Theorems 4.27 and A2.3 (iii), condition (i) implies both (ii) and (iii). Conversely, Lemma 16.15 shows that (ξ_n) is relatively compact in distribution under both (ii) and (iii). Arguing as in the proof of Lemma

16.2, it remains to show for any random measures ξ and η on S that $\xi \stackrel{d}{=} \eta$ if $\xi f \stackrel{d}{=} \eta f$ for all $f \in C_K^+$, or if

$$(\xi B_1, \ldots, \xi B_k) \stackrel{d}{=} (\eta B_1, \ldots, \eta B_k), \quad B_1, \ldots, B_k \in \hat{S}_{\xi+\eta}, \ k \in \mathbb{N}. \quad (16)$$

In the former case, this holds by Lemma 12.1; in the latter case it follows by a monotone class argument from Theorem A2.3 (iv). The last assertion is obtained in a similar way from a suitable version of Theorem 12.8 (iii). $\square$

Weaker conditions are required for convergence to a simple point process, as suggested by Theorem 12.8. The following conditions are only sufficient, and a precise criterion is given in Theorem 16.29.

Here a class $\mathcal{U} \subset \hat{S}$ is said to be *separating* if, for any compact and open sets K and G with $K \subset G$, there exists some $U \in \mathcal{U}$ with $K \subset U \subset G$. Furthermore, we say that $\mathcal{I} \subset \hat{S}$ is *preseparating* if the finite unions of sets in $\mathcal{I}$ form a separating class. Applying Lemma A2.6 to the function $h(B) = Ee^{-\xi B}$, we note that the class $\hat{S}_\xi$ is separating for any random measure ξ. For Euclidean spaces S, a preseparating class typically consist of rectangular boxes, whereas the corresponding finite unions form a separating class.

Proposition 16.17 *(convergence of point processes)* Let $\xi, \xi_1, \xi_2, \ldots$ be point processes on an lcscH space S, where ξ is simple, and fix a separating class $\mathcal{U} \subset \hat{S}$. Then $\xi_n \stackrel{d}{\to} \xi$ under these conditions:
 (i) $P\{\xi_n U = 0\} \to P\{\xi U = 0\}$ for all $U \in \mathcal{U}$;
 (ii) $\limsup_n E\xi_n K \leq E\xi K < \infty$ for all compact sets $K \subset S$.

Proof: First note that both (i) and (ii) extend by suitable approximation to sets in $\hat{S}_\xi$. By the usual compactness argument together with Lemma 4.11, it is enough to prove that a point process η is distributed as ξ whenever

$$P\{\eta B = 0\} = P\{\xi B = 0\}, \quad E\eta B \leq E\xi B, \quad B \in \hat{S}_{\xi+\eta}.$$

Here the first relation yields $\eta^* \stackrel{d}{=} \xi$ as in Theorem 12.8 (i). From the second relation we then obtain $E\eta B \leq E\eta^* B$ for all $B \in \hat{S}_\xi$, which shows that η is a.s. simple. $\square$

We may illustrate the use of Theorem 16.16 by showing how Poisson and Cox processes may arise as limits under superposition or thinning. Say that the random measures ξ_{nj}, $n, j \in \mathbb{N}$, form a *null array* if they are independent for fixed n and such that, for every $B \in \hat{S}$, the random variables $\xi_{nj} B$ form a null array in the sense of Chapter 5. The following result is a point process version of Theorem 5.7.

Theorem 16.18 *(convergence of superpositions, Grigelionis) Let (ξ_{nj}) be a null array of point processes on an lcscH space S, and consider a Poisson process ξ on S with $E\xi = \mu$. Then $\sum_j \xi_{nj} \xrightarrow{d} \xi$ iff these conditions hold:*

(i) $\sum_j P\{\xi_{nj} B > 0\} \to \mu B$ *for all* $B \in \hat{S}_\mu$;

(ii) $\sum_j P\{\xi_{nj} B > 1\} \to 0$ *for all* $B \in \hat{S}$.

Proof: If $\sum_j \xi_{nj} \xrightarrow{d} \xi$, then $\sum_j \xi_{nj} B \xrightarrow{d} \xi B$ for all $B \in \hat{S}_\mu$ by Theorem 16.16, which implies (i) and (ii) by Theorem 5.7. Conversely, assume (i) and (ii). To prove that $\sum_j \xi_{nj} \xrightarrow{d} \xi$, we may restrict our attention to an arbitrary compact set $C \in \hat{S}_\mu$. For notational convenience, we may also assume that S itself is compact. Now define $\eta_{nj} = \xi_{nj} 1\{\xi_{nj} S \leq 1\}$, and note that (i) and (ii) remain true for the array (η_{nj}). Moreover, $\sum_j \eta_{nj} \xrightarrow{d} \xi$ implies $\sum_j \xi_{nj} \xrightarrow{d} \xi$ by Theorem 4.28. This reduces the discussion to the case when $\xi_{nj} S \leq 1$ for all n and j.

Now define $\mu_{nj} = E\xi_{nj}$. By (i) we get

$$\sum_j \mu_{nj} B = \sum_j E\xi_{nj} B = \sum_j P\{\xi_{nj} B > 0\} \to \mu B, \quad B \in \hat{S}_\mu,$$

and so $\sum_j \mu_{nj} \xrightarrow{w} \mu$ by Theorem 4.25. Noting that $m(1 - e^{-f}) = 1 - e^{-mf}$ when $m = \delta_x$ or 0 and writing $\xi_n = \sum_j \xi_{nj}$, we get by Lemmas 5.8 and 12.2 (i)

$$\begin{aligned}
Ee^{-\xi_n f} &= \prod_j Ee^{-\xi_{nj} f} = \prod_j E\{1 - \xi_{nj}(1 - e^{-f})\} \\
&= \prod_j \{1 - \mu_{nj}(1 - e^{-f})\} \sim \exp\left\{-\sum_j \mu_{nj}(1 - e^{-f})\right\} \\
&\to \exp(-\mu(1 - e^{-f})) = Ee^{-\xi f}. \quad \square
\end{aligned}$$

We may next establish a basic limit theorem for independent thinnings of point processes.

Theorem 16.19 *(convergence of thinnings) For every $n \in \mathbb{N}$, let ξ_n be a p_n-thinning of some point process η_n on S, where S is lcscH and $p_n \to 0$. Then $\xi_n \xrightarrow{d}$ some ξ iff $p_n \eta_n \xrightarrow{d}$ some η, in which case ξ is distributed as a Cox process directed by η.*

Proof: For any $f \in C_K^+$, we get by Lemma 12.2

$$E^{-\xi_n f} = E\exp(\eta_n \log\{1 - p_n(1 - e^{-f})\}).$$

Noting that $px \leq -\log(1 - px) \leq -x\log(1 - p)$ for any $p, x \in [0, 1)$ and writing $p'_n = -\log(1 - p_n)$, we obtain

$$E\exp\{-p'_n \eta_n(1 - e^{-f})\} \leq Ee^{-\xi_n f} \leq E\exp\{-p_n \eta_n(1 - e^{-f})\}. \quad (17)$$

If $p_n \eta_n \xrightarrow{d} \eta$, then even $p'_n \eta_n \xrightarrow{d} \eta$, and so by Lemma 12.2

$$Ee^{-\xi_n f} \to E\exp\{-\eta(1 - e^{-f})\} = Ee^{-\xi f},$$

where ξ is a Cox process directed by η. Hence, $\xi_n \xrightarrow{d} \xi$.

Conversely, assume that $\xi_n \xrightarrow{d} \xi$. Fix any $g \in C_K^+$ and let $0 \leq t < \|g\|^{-1}$. Applying (17) with $f = -\log(1 - tg)$, we get

$$\liminf_{n \to \infty} E \exp\{-tp_n\eta_n g\} \geq E \exp\{\xi \log(1 - tg)\}.$$

Here the right-hand side tends to 1 as $t \to 0$, and so by Lemmas 5.2 and 16.15 the sequence $(p_n\eta_n)$ is tight. For any subsequence $N' \subset \mathbb{N}$, we may then choose a further subsequence N'' such that $p_n\eta_n \xrightarrow{d}$ some η along N''. By the direct assertion, ξ is then distributed as a Cox process directed by η, which by Lemma 12.6 determines the distribution of η. Hence, $\eta_n \xrightarrow{d} \eta$ remains true along the original sequence. □

The last result leads in particular to an interesting characterization of Cox processes.

Corollary 16.20 *(Cox processes and thinnings, Mecke) Let ξ be a point process on S. Then ξ is Cox iff for every $p \in (0, 1)$ there exists a point process ξ_p such that ξ is distributed as a p-thinning of ξ_p.*

Proof: If ξ and ξ_p are Cox processes directed by η and η/p, respectively, then Proposition 12.3 shows that ξ is distributed as a p-thinning of ξ_p. Conversely, assuming the stated condition for every $p \in (0, 1)$, we note that ξ is Cox by Theorem 16.19. □

The previous theory will now be used to derive a general limit theorem for sums of exchangeable random variables. The result applies in particular to sequences obtained by sampling without replacement from a finite population. It is also general enough to contain a version of Donsker's theorem. The appropriate function space in this case is $D([0, 1], \mathbb{R}) = D[0, 1]$, to which the results for $D(\mathbb{R}_+)$ apply with obvious modifications.

For motivation, we begin with a description of the possible limits, which are precisely the exchangeable processes on $[0, 1]$. Here we say that a process X on $[0, 1]$ is *exchangeable* if it is continuous in probability with $X_0 = 0$ and has exchangeable increments over any set of disjoint intervals of equal length. The following result is a finite-interval version of Theorem 11.15.

Theorem 16.21 *(exchangeable processes on $[0, 1]$) A process X on $[0, 1]$ is exchangeable iff it has a version with representation*

$$X_t = \alpha t + \sigma B_t + \sum_j \beta_j(1\{\tau_j \leq t\} - t), \quad t \in [0, 1], \tag{18}$$

for some Brownian bridge B, some independent i.i.d. $U(0, 1)$ random variables $\tau_1, \tau_2, \ldots$, and some independent set of coefficients α, σ, and $\beta_1, \beta_2, \ldots$ such that $\sum_j \beta_j^2 < \infty$ a.s. In that case, the sum in (18) converges in probability, uniformly on $[0, 1]$, toward an rcll limit.

In particular, we note that a simple point process on $[0, 1]$ is symmetric with respect to Lebesgue measure λ iff it is a mixed binomial process based on λ, in agreement with Theorem 12.12. Combining the present result with

Theorem 11.15, we also see that a continuous process X on $\mathbb{R}_+$ or $[0,1]$ with $X_0 = 0$ is exchangeable iff it can be written in the form $X_t = \alpha t + \sigma B_t$, where B is a Brownian motion or bridge, respectively, and (α, σ) is an independent pair of random variables.

We first examine the convergence of the series in (18).

Lemma 16.22 *(convergence of series)* *For any $t \in (0,1)$, the series in (18) converges a.s. iff $\sum_j \beta_j^2 < \infty$ a.s. In that case, it converges in probability with respect to the uniform metric on $[0,1]$, and the sum has a version in $D[0,1]$.*

Proof: For both assertions, we may assume that the coefficients β_j are nonrandom. Then for fixed $t \in (0,1)$, the terms are independent and bounded with mean 0 and variance $\beta_j^2 t(1-t)$, and so by Theorem 4.18 the series converges iff $\sum_j \beta_j^2 < \infty$.

To prove the second assertion, let X^n denote the nth partial sum in (18), and note that the processes $M_t^n = X_t^n/(1-t)$ are L^2-martingales on $[0,1)$ with respect to the filtration induced by the processes $1\{\tau_j \leq t\}$. By Doob's inequality we have for any $m < n$ and $t \in [0,1)$

$$E(X^n - X^m)_t^{*2} \leq E(M^n - M^m)_t^{*2} \leq 4E(M_t^n - M_t^m)^2$$
$$= 4(1-t)^{-2} E(X_t^n - X_t^m)^2$$
$$\leq 4t(1-t)^{-1} \sum_{j>m} \beta_j^2,$$

which tends to 0 as $m \to \infty$ for fixed t. Hence, $(X^n - X)_t^* \to 0$ a.s. along a subsequence for some process X, and then also $(X^n - X)_t^* \xrightarrow{P} 0$ along $\mathbb{N}$. By symmetry we have also $(\tilde{X}^n - \tilde{X})_t^* \xrightarrow{P} 0$ for the reflected processes $\tilde{X}_t = X_{1-t-}$ and $\tilde{X}_t^n = X_{1-t-}^n$, and so by combination $(X^n - X)_1^* \xrightarrow{P} 0$. The last assertion now follows from the fact that X^n is rcll for every n. $\square$

We plan to prove Theorem 16.21 together with the following approximation result. Here we consider for every $n \in \mathbb{N}$ some exchangeable random variables ξ_{nj}, $j \leq m_n$, where $m_n \to \infty$, and introduce the summation processes

$$X_t^n = \sum_{j \leq m_n t} \xi_{nj}, \quad t \in [0,1], \; n \in \mathbb{N}. \tag{19}$$

Our aim is to show that the X^n can be approximated by exchangeable processes as in (18). The convergence criteria will be stated in terms of the random variables and measures

$$\alpha_n = \sum_j \xi_{nj}, \quad \kappa_n = \sum_j \xi_{nj}^2 \delta_{\xi_{nj}}, \quad n \in \mathbb{N}, \tag{20}$$

$$\kappa = \sigma^2 \delta_0 + \sum_j \beta_j^2 \delta_{\beta_j}. \tag{21}$$

Theorem 16.23 *(approximation of exchangeable sums)* *For every $n \in \mathbb{N}$, consider some exchangeable random variables ξ_{nj}, $j \leq m_n$, and define X^n, α_n, and κ_n by (19) and (20). Assume $m_n \to \infty$. Then $X^n \xrightarrow{d}$ some X in $D[0,1]$ iff $(\alpha_n, \kappa_n) \xrightarrow{d}$ some (α, κ) in $\mathbb{R} \times \mathcal{M}(\overline{\mathbb{R}})$, in which case X and (α, κ) are related by (18) and (21).*

Our proof is based on three auxiliary results. We begin with a simple randomization lemma, which will enable us to reduce the proof to the case of non-random coefficients. Recall that if ν is a measure on S and μ is a kernel from S to T, then $\nu\mu$ denotes the measure $\int \mu(s, \cdot) \nu(ds)$ on T. For any measurable function $f : T \to \mathbb{R}_+$, we define the measurable function μf on S by $\mu f(s) = \int \mu(s, dt) f(t)$.

Lemma 16.24 *(randomization)* *For any metric spaces S and T, let $\nu, \nu_1, \nu_2, \ldots$ be probability measures on S with $\nu_n \xrightarrow{w} \nu$, and let $\mu, \mu_1, \mu_2, \ldots$ be probability kernels from S to T such that $s_n \to s$ in S implies $\mu_n(s_n, \cdot) \xrightarrow{w} \mu(s, \cdot)$. Then $\nu_n \mu_n \xrightarrow{w} \nu\mu$.*

Proof: Fix any bounded, continuous function f on T. Then $\mu_n f(s_n) \to \mu f(s)$ as $s_n \to s$, and so by Theorem 4.27

$$(\nu_n \mu_n) f = \nu_n(\mu_n f) \to \nu(\mu f) = (\nu\mu) f. \qquad \Box$$

To establish tightness of the random measures κ_n, we need the following conditional hyper-contractivity criterion.

Lemma 16.25 *(hyper-contraction and tightness)* *Let the random variables $\xi_1, \xi_2, \ldots \geq 0$ and σ-fields $\mathcal{F}_1, \mathcal{F}_2, \ldots$ be such that, for some $a > 0$,*

$$E[\xi_n^2 | \mathcal{F}_n] \leq a(E[\xi_n | \mathcal{F}_n])^2 < \infty \text{ a.s.}, \quad n \in \mathbb{N}.$$

Then if (ξ_n) is tight, so is the sequence $\eta_n = E[\xi_n | \mathcal{F}_n]$, $n \in \mathbb{N}$.

Proof: By Lemma 4.9 we need to show that $c_n \eta_n \xrightarrow{P} 0$ whenever $0 \leq c_n \to 0$. Then conclude from Lemma 4.1 that, for any $r \in (0,1)$ and $\varepsilon > 0$,

$$0 < (1-r)^2 a^{-1} \leq P[\xi_n \geq r\eta_n | \mathcal{F}_n] \leq P[c_n \xi_n \geq r\varepsilon | \mathcal{F}_n] + 1\{c_n \eta_n < \varepsilon\}.$$

Here the first term on the right tends in probability to 0 since $c_n \xi_n \xrightarrow{P} 0$ by Lemma 4.9. Hence, $1\{c_n \eta_n < \varepsilon\} \xrightarrow{P} 1$, which means that $P\{c_n \eta_n \geq \varepsilon\} \to 0$. Since ε is arbitrary, we get $c_n \eta_n \xrightarrow{P} 0$. $\qquad \Box$

Since the summation processes in (19) will be approximated by exchangeable processes, as in Theorem 16.21, we finally need a convergence criterion for the latter. This result also has some independent interest.

Proposition 16.26 *(convergence of exchangeable processes) Let the processes X^n and pairs (α_n, κ_n) be related as in (18) and (21). Then $X^n \xrightarrow{d}$ some X in $D[0,1]$ iff $(\alpha_n, \kappa_n) \xrightarrow{d}$ some (α, κ) in $\mathbb{R} \times \mathcal{M}(\overline{\mathbb{R}})$, in which case even X and (α, κ) are related by (18) and (21).*

Proof: First let $(\alpha_n, \kappa_n) \xrightarrow{d} (\alpha, \kappa)$. To prove $X^n \xrightarrow{d} X$ for the corresponding processes in (18), it suffices by Lemma 16.24 to assume that all the α_n and κ_n are nonrandom. Thus, we may restrict our attention to processes X^n with constant coefficients α_n, σ_n, and β_{nj}, $j \in \mathbb{N}$.

To prove that $X^n \xrightarrow{fd} X$, we begin with four special cases. First we note that if $\alpha_n \to \alpha$, then trivially $\alpha_n t \to \alpha t$ uniformly on $[0,1]$. Similarly, $\sigma_n \to \sigma$ implies $\sigma_n B \to \sigma B$ in the same sense. Next we consider the case when $\alpha_n = \sigma_n = 0$ and $\beta_{n,m+1} = \beta_{n,m+2} = \cdots = 0$ for some fixed $m \in \mathbb{N}$. Here we may assume that even $\alpha = \sigma = 0$ and $\beta_{m+1} = \beta_{m+2} = \cdots = 0$, and that moreover $\beta_{nj} \to \beta_j$ for all j. The convergence $X^n \to X$ is then obvious. Finally, we may assume that $\alpha_n = \sigma_n = 0$ and $\alpha = \beta_1 = \beta_2 = \cdots = 0$. Then $\max_j |\beta_{nj}| \to 0$, and for any $s \leq t$ we have

$$E(X^n_s X^n_t) = s(1-t) \sum_j \beta_{nj}^2 \to s(1-t)\sigma^2 = E(X_s X_t). \qquad (22)$$

In this case, $X^n \xrightarrow{fd} X$ by Theorem 5.12 and Corollary 5.5. By independence we may combine the four special cases to obtain $X^n \xrightarrow{fd} X$ whenever $\beta_j = 0$ for all but finitely many j. From here on, it is easy to extend to the general case by means of Theorem 4.28, where the required uniform error estimate may be obtained as in (22).

To strengthen the convergence to $X^n \xrightarrow{d} X$ in $D[0,1]$, it is enough to verify the tightness criterion in Theorem 16.11. Thus, for any X^n-optional times τ_n and positive constants $h_n \to 0$ with $\tau_n + h_n \leq 1$, we need to show that $X^n_{\tau_n + h_n} - X^n_{\tau_n} \xrightarrow{P} 0$. By Theorem 11.13 and a simple approximation, it is equivalent that $X^n_{h_n} \xrightarrow{P} 0$, which is clear since

$$E(X^n_{h_n})^2 = h_n^2 \alpha_n^2 + h_n(1-h_n)\kappa_n \mathbb{R} \to 0.$$

To obtain the reverse implication, we assume that $X^n \xrightarrow{d} X$ in $D[0,1]$ for some process X. Since $\alpha_n = X_1^n \xrightarrow{d} X_1$, the sequence (α_n) is tight. Next define for $n \in \mathbb{N}$

$$\eta_n = 2X^n_{1/2} - X^n_1 = 2\sigma_n B_{1/2} + 2\sum_j \beta_{nj}(1\{\tau_j \leq \tfrac{1}{2}\} - \tfrac{1}{2}).$$

Then

$$E[\eta_n^2 | \kappa_n] = \sigma_n^2 + \sum_j \beta_{nj}^2 = \kappa_n \mathbb{R},$$

$$E[\eta_n^4 | \kappa_n] = 3\left\{\sigma_n^2 + \sum_j \beta_{nj}^2\right\}^2 - 2\sum_j \beta_{nj}^4 \leq 3(\kappa_n \mathbb{R})^2.$$

Since (η_n) is tight, Lemmas 16.15 and 16.25 show that even (κ_n) is tight, and so the same thing is true for the sequence of pairs (α_n, κ_n).

The tightness implies relative compactness in distribution, and so every subsequence contains a further subsequence that converges in $\mathbb{R} \times \mathcal{M}(\overline{\mathbb{R}})$ toward some random pair (α, κ). Since the measures in (21) form a vaguely closed subset of $\mathcal{M}(\overline{\mathbb{R}})$, the limit κ has the same form for suitable σ and $\beta_1, \beta_2, \ldots$. The direct assertion then yields $X^n \overset{d}{\to} Y$ with Y as in (18), and therefore $X \overset{d}{=} Y$. Now the coefficients in (18) are measurable functions of Y, and so the distribution of (α, κ) is uniquely determined by that of X. Thus, the limiting distribution is independent of subsequence, and the convergence $(\alpha_n, \kappa_n) \overset{d}{\to} (\alpha, \kappa)$ remains valid along $\mathbb{N}$. We may finally use Corollary 6.11 to transfer the representation (18) to the original process X. □

Proof of Theorem 16.23: Let $\tau_1, \tau_2, \ldots$ be i.i.d. $U(0,1)$ and independent of all ξ_{nj}, and define
$$Y_t^n = \sum_j \xi_{nj} 1\{\tau_j \le t\} = \alpha_n t + \sum_j \xi_{nj}(1\{\tau_j \le t\} - t), \quad t \in [0,1].$$
Writing $\tilde{\xi}_{nk}$ for the kth jump from the left of Y^n (including possible 0 jumps when $\xi_{nj} = 0$), we note that $(\tilde{\xi}_{nj}) \overset{d}{=} (\xi_{nj})$ by exchangeability. Thus, $\tilde{X}^n \overset{d}{=} X^n$, where $\tilde{X}_t^n = \sum_{j \le m_n t} \tilde{\xi}_{nj}$. Furthermore, $d(\tilde{X}^n, Y^n) \to 0$ a.s. by Proposition 4.24, where d is the metric in Theorem A2.2. Hence, by Theorem 4.28 it is equivalent to replace X^n by Y^n. But then the assertion follows by Proposition 16.26. □

Using similar compactness arguments, we may finally prove the main representation theorem for exchangeable processes on $[0,1]$.

Proof of Theorem 16.21: The sufficiency part being obvious, it is enough to prove the necessity. Thus, assume that X has exchangeable increments. Introduce the step processes
$$X_t^n = X(2^{-n}[2^n t]), \quad t \in [0,1], \ n \in \mathbb{N},$$
define κ_n as in (20) in terms of the jump sizes of X^n, and put $\alpha_n \equiv X_1$. If the sequence (κ_n) is tight, then $(\alpha_n, \kappa_n) \overset{d}{\to} (\alpha, \kappa)$ along some subsequence, and by Theorem 16.23 we get $X^n \overset{d}{\to} Y$ along the same subsequence, where Y can be represented as in (18). In particular, $X^n \overset{fd}{\to} Y$, and so the finite-dimensional distributions of X and Y agree for dyadic times. The agreement extends to arbitrary times, since both processes are continuous in probability. By Lemma 3.24 it follows that X has a version in $D[0,1]$, and by Corollary 6.11 we obtain the desired representation.

To prove the required tightness of (κ_n), denote the increments in X^n by ξ_{nj}, put $\zeta_{nj} = \xi_{nj} - 2^{-n}\alpha_n$, and note that
$$\kappa_n \mathbb{R} = \sum_j \xi_{nj}^2 = \sum_j \zeta_{nj}^2 + 2^{-n}\alpha_n^2. \tag{23}$$

Writing $\eta_n = 2X_{1/2}^n - X_1^n = 2X_{1/2} - X_1$ and noting that $\sum_j \zeta_{nj} = 0$, we get the elementary estimates

$$E[\eta_n^4 | \kappa_n] \lesssim \sum_j \zeta_{nj}^4 + \sum_{i \neq j} \zeta_{ni}^2 \zeta_{nj}^2 = \left\{ \sum_j \zeta_{nj}^2 \right\}^2 \lesssim (E[\eta_n^2 | \kappa_n])^2.$$

Since η_n is independent of n, the sequence of sums $\sum_j \zeta_{nj}^2$ is tight by Lemma 16.25, and the tightness of (κ_n) follows by (23). □

For measure-valued processes X^n with rcll paths, we show that tightness can be characterized in terms of the real-valued projections $X_t^n f = \int f(s) X_t^n(ds)$, $f \in C_K^+$.

Theorem 16.27 (*measure-valued processes*) *Let $X^1, X^2, \ldots$ be random elements in $D(\mathbb{R}_+, \mathcal{M}(S))$, where S is lcscH. Then (X^n) is tight iff $(X^n f)$ is tight in $D(\mathbb{R}_+, \mathbb{R}_+)$ for every $f \in C_K^+(S)$.*

Proof: Assume that $(X^n f)$ is tight for every $f \in C_K^+$, and fix any $\varepsilon > 0$. Let $f_1, f_2, \ldots$ be such as in Theorem A2.4, and choose some compact sets $B_1, B_2, \cdots \subset D(\mathbb{R}_+, \mathbb{R}_+)$ with

$$P\{X^n f_k \in B_k\} \geq 1 - \varepsilon 2^{-k}, \quad k, n \in \mathbb{N}. \tag{24}$$

Then $A = \bigcap_k \{\mu; \mu f_k \in B_k\}$ is relatively compact in $D(\mathbb{R}_+, \mathcal{M}(S))$, and (24) yields $P\{X^n \in A\} \geq 1 - \varepsilon$. □

We turn our attention to random sets. Then fix an lcscH space S, and let $\mathcal{F}$, $\mathcal{G}$, and $\mathcal{K}$ denote the classes of closed, open, and compact subsets, respectively. We endow $\mathcal{F}$ with the so-called *Fell topology*, generated by the sets $\{F; F \cap G \neq \emptyset\}$ and $\{F; F \cap K = \emptyset\}$ for arbitrary $G \in \mathcal{G}$ and $K \in \mathcal{K}$. Some basic properties of this topology are summarized in Theorem A2.5. In particular, $\mathcal{F}$ is compact and metrizable, and $\{F; F \cap B = \emptyset\}$ is universally measurable for every $B \in \hat{\mathcal{S}}$.

By a *random closed set* in S we mean a random element φ in $\mathcal{F}$. In this context we often write $\varphi \cap B = \varphi B$, and we note that the probabilities $P\{\varphi B = \emptyset\}$ are well defined. For any random closed set φ, we introduce the class

$$\hat{\mathcal{S}}_\varphi = \left\{ B \in \hat{\mathcal{S}}; P\{\varphi B^\circ = \emptyset\} = P\{\varphi \overline{B} = \emptyset\} \right\},$$

which is separating by Lemma A2.6. We may now state the basic convergence criterion for random sets. It is interesting to note the formal agreement with the first condition in Proposition 16.17.

Theorem 16.28 *(convergence of random sets, Norberg)* Let $\varphi, \varphi_1, \varphi_2, \ldots$ be random closed sets in an lcscH space S. Then $\varphi_n \xrightarrow{d} \varphi$ iff
$$P\{\varphi_n U = \emptyset\} \to P\{\varphi U = \emptyset\}, \quad U \in \mathcal{U}, \tag{25}$$
for some separating class $\mathcal{U} \subset \hat{\mathcal{S}}$, in which case we may take $\mathcal{U} = \hat{\mathcal{S}}_\varphi$.

Proof: Write $h(B) = P\{\varphi B \neq \emptyset\}$ and $h_n(B) = P\{\varphi_n B \neq \emptyset\}$. If $\varphi_n \xrightarrow{d} \varphi$, then by Theorem 4.25,
$$h(B^\circ) \leq \liminf_{n \to \infty} h_n(B) \leq \limsup_{n \to \infty} h_n(B) \leq h(\overline{B}), \quad B \in \hat{\mathcal{S}},$$
and so for any $B \in \hat{\mathcal{S}}_\varphi$ we get $h_n(B) \to h(B)$.

Next assume that (25) holds for some separating class $\mathcal{U}$. Fix any $B \in \hat{\mathcal{S}}_\varphi$, and conclude from (25) that, for any $U, V \in \mathcal{U}$ with $U \subset B \subset V$,
$$h(U) \leq \liminf_{n \to \infty} h_n(B) \leq \limsup_{n \to \infty} h_n(B) \leq h(V).$$

Since $\mathcal{U}$ is separating, we may let $U \uparrow B^\circ$ to get $\{\varphi U \neq \emptyset\} \uparrow \{\varphi B^\circ \neq \emptyset\}$ and hence $h(U) \uparrow h(B^\circ) = h(B)$. Next choose some sets $V \in \mathcal{U}$ with $\overline{V} \downarrow B$, and conclude by the finite intersection property that $\{\varphi \overline{V} \neq \emptyset\} \downarrow \{\varphi \overline{B} \neq \emptyset\}$, which gives $h(V) \downarrow h(\overline{B}) = h(B)$. Thus, $h_n(B) \to h(B)$, and so (25) remains true for $\mathcal{U} = \hat{\mathcal{S}}_\varphi$.

Since $\mathcal{F}$ is compact, the sequence $\{\varphi_n\}$ is relatively compact by Theorem 16.3. Thus, for any subsequence $N' \subset \mathbb{N}$, we have $\varphi_n \xrightarrow{d} \psi$ along a further subsequence for some random closed set ψ. By the direct statement together with (25) we get
$$P\{\varphi B = \emptyset\} = P\{\psi B = \emptyset\}, \quad B \in \hat{\mathcal{S}}_\varphi \cap \hat{\mathcal{S}}_\psi. \tag{26}$$

Since $\hat{\mathcal{S}}_\varphi \cap \hat{\mathcal{S}}_\psi$ is separating by Lemma A2.6, we may approximate as before to extend (26) to arbitrary compact sets B. The class of sets $\{F; F \cap K = \emptyset\}$ with K compact is clearly a π-system, and so a monotone class argument gives $\varphi \stackrel{d}{=} \psi$. Since N' is arbitrary, we obtain $\varphi_n \xrightarrow{d} \varphi$ along $\mathbb{N}$. $\square$

Simple point processes allow the dual descriptions as integer-valued random measures or locally finite random sets. The corresponding notions of convergence are different, and we proceed to clarify their relationship. Since the mapping $\mu \mapsto \operatorname{supp} \mu$ is continuous on $\mathcal{N}(S)$, we note that $\xi_n \xrightarrow{d} \xi$ implies $\operatorname{supp} \xi_n \xrightarrow{d} \operatorname{supp} \xi$. Conversely, assuming the intensity measures $E\xi$ and $E\xi_n$ to be locally finite, we see from Proposition 16.17 and Theorem 16.28 that $\xi_n \xrightarrow{d} \xi$ whenever $\operatorname{supp} \xi_n \xrightarrow{d} \operatorname{supp} \xi$ and $E\xi_n \xrightarrow{v} E\xi$. The next result gives a general criterion.

Theorem 16.29 *(supports of point processes)* *Let $\xi, \xi_1, \xi_2, \ldots$ be point processes on an lcscH space S, where ξ is simple, and fix a preseparating class $\mathcal{I} \subset \hat{\mathcal{S}}_\xi$. Then $\xi_n \xrightarrow{d} \xi$ iff $\operatorname{supp} \xi_n \xrightarrow{d} \operatorname{supp} \xi$ and*

$$\limsup_{n\to\infty} P\{\xi_n I > 1\} \leq P\{\xi I > 1\}, \quad I \in \mathcal{I}. \tag{27}$$

Proof: By Corollary 6.12 we may assume that $\operatorname{supp}\xi_n \xrightarrow{f} \operatorname{supp}\xi$ a.s., and since ξ is simple we get by Proposition A2.8

$$\limsup_{n\to\infty}(\xi_n B \wedge 1) \leq \xi B \leq \liminf_{n\to\infty} \xi_n B \quad \text{a.s.}, \quad B \in \mathcal{B}_\xi. \tag{28}$$

Next we have for any $a, b \in \mathbb{Z}_+$

$$\begin{aligned}\{b \leq a \leq 1\}^c &= \{a > 1\} \cup \{a < b \wedge 2\} \\ &= \{b > 1\} \cup \{a = 0, b = 1\} \cup \{a > 1 \geq b\},\end{aligned}$$

where all unions are disjoint. Substituting $a = \xi I$ and $b = \xi_n I$, we get by (27) and (28)

$$\lim_{n\to\infty} P\{\xi I < \xi_n I \wedge 2\} = 0, \quad I \in \mathcal{I}. \tag{29}$$

Next let $B \subset I \in \mathcal{I}$ and $B' = I \setminus B$, and note that

$$\begin{aligned}\{\xi_n B > \xi B\} &\subset \{\xi_n I > \xi I\} \cup \{\xi_n B' < \xi B'\} \\ &\subset \{\xi_n I \wedge 2 > \xi I\} \cup \{\xi I > 1\} \cup \{\xi_n B' < \xi B'\}. \tag{30}\end{aligned}$$

More generally, assume that $B \in \mathcal{B}_\xi$ is covered by $I_1, \ldots, I_m \in \mathcal{I}$. It may then be partitioned into sets $B_k \in \mathcal{B}_\xi \cap I_k$, $k = 1, \ldots, m$, and by (28), (29), and (30) we get

$$\limsup_{n\to\infty} P\{\xi_n B > \xi B\} \leq P\bigcup_k \{\xi I_k > 1\}. \tag{31}$$

Now let $B \in \mathcal{B}_\xi$ and $K \in \mathcal{K}$ with $\overline{B} \subset K^\circ$. Fix a metric d in S and let $\varepsilon > 0$. Since $\mathcal{I}$ is preseparating, we may choose some $I_1, \ldots, I_m \in \mathcal{I}$ with d-diameters $< \varepsilon$ such that $B \subset \bigcup_k I_k \subset K$. Letting ρ_K denote the minimum d-distance between points in $K \cap \operatorname{supp}\xi$, it follows that the right-hand side of (31) is bounded by $P\{\rho_K < \varepsilon\}$. Since $\rho_K > 0$ a.s. and $\varepsilon > 0$ is arbitrary, we get $P\{\xi_n B > \xi B\} \to 0$. In view of the second relation in (28), we obtain $\xi_n B \xrightarrow{P} \xi B$. Thus, $\xi_n \xrightarrow{d} \xi$ by Theorem 16.16. □

Exercises

1. For any metric space (S, ρ), show that if $x^n \to x$ in $D(\mathbb{R}_+, S)$ with x continuous, then $\sup_{s \leq t} \rho(x_s^n, x_s) \to 0$ for every $t \geq 0$. (*Hint:* Note that x is uniformly continuous on every interval $[0, t]$.)

2. For any separable and complete metric space (S, ρ), show that if $X^n \xrightarrow{d} X$ in $D(\mathbb{R}_+, S)$ with X continuous, there exist some processes $Y^n \stackrel{d}{=} X^n$ such that $\sup_{s \le t} \rho(Y_s^n, X_s) \to 0$ a.s. for every $t \ge 0$. (*Hint:* Use the preceding result together with Theorems 4.30 and A2.2.)

3. Give an example where $x_n \to x$ and $y_n \to y$ in $D(\mathbb{R}_+, \mathbb{R})$ and yet $(x_n, y_n) \not\to (x, y)$ in $D(\mathbb{R}_+, \mathbb{R}^2)$.

4. Let X and X^n be random elements in $D(\mathbb{R}_+, \mathbb{R}^d)$ with $X^n \xrightarrow{fd} X$ and such that $uX^n \xrightarrow{d} uX$ in $D(\mathbb{R}_+, \mathbb{R})$ for every $u \in \mathbb{R}^d$. Show that $X^n \xrightarrow{d} X$. (*Hint:* Proceed as in Theorems 16.27 and A2.4.)

5. Let f be a continuous mapping between two metric spaces S and T, where S is separable and complete. Show that if $X^n \xrightarrow{d} X$ in $D(\mathbb{R}_+, S)$, then $f(X^n) \xrightarrow{d} f(X)$ in $D(\mathbb{R}_+, T)$. (*Hint:* By Theorem 4.27 it suffices to show that $x_n \to x$ in $D(\mathbb{R}_+, S)$ implies $f(x_n) \to f(x)$ in $D(\mathbb{R}_+, T)$. Since $A = \{x, x_1, x_2, \ldots\}$ is relatively compact in $D(\mathbb{R}_+, S)$, Theorem A2.2 shows that $U_t = \bigcup_{s \le t} \pi_s A$ is relatively compact in S for every $t > 0$. Hence, f is uniformly continuous on each U_t.)

6. Show by an example that the condition in Theorem 16.11 is not necessary for tightness. (*Hint:* Consider nonrandom processes X_n.)

7. In Theorem 16.11, show that it is enough to consider optional times taking finitely many values. (*Hint:* Approximate from the right and use the right-continuity of the paths.)

8. Let the process X on $\mathbb{R}_+$ be continuous in probability with values in a separable and complete metric space (S, ρ). Assume that $\rho(X_{\tau_n}, X_{\tau_n + h_n}) \xrightarrow{P} 0$ for any bounded optional times τ_n and constants $h_n \to 0$. Show that X has an rcll version. (*Hint:* Approximate by suitable step processes and use Theorems 16.10 and 16.11.)

9. Extend Corollary 16.7 to random vectors in $\mathbb{R}^d$.

10. Let $X, X^1, X^2, \ldots$ be Lévy processes in $\mathbb{R}^d$. Show that $X^n \xrightarrow{d} X$ in $D(\mathbb{R}_+, \mathbb{R}^d)$ iff $X_1^n \xrightarrow{d} X_1$ in $\mathbb{R}^d$. Compare with Theorem 15.17.

11. Show that conditions (iii) and (iv) of Theorem 16.16 remain sufficient if we replace $\hat{S}_\xi$ by an arbitrary separating class. (*Hint:* Restate the conditions in terms of Laplace transforms, and extend to $\hat{S}_\xi$ by a suitable approximation.)

12. Deduce Theorem 16.18 from Theorem 5.7. (*Hint:* First assume that μ is diffuse and use Theorem 16.17. Then extend to the general case by a suitable randomization.)

13. Strengthen the conclusion in Theorem 16.19 to $(\xi_n, p_n \eta_n) \xrightarrow{d} (\xi, \eta)$, where ξ is a Cox process directed by η.

14. For any lcscH space S, let $\xi, \xi_1, \xi_2, \ldots$ be Cox processes on S directed by $\eta, \eta_1, \eta_2, \ldots$. Show that $\xi_n \xrightarrow{d} \xi$ iff $\eta_n \xrightarrow{d} \eta$. Prove the corresponding result for p-thinnings with a fixed $p \in (0, 1)$.

15. Let $\eta, \eta_1, \eta_2, \ldots$ be λ-randomizations of some point processes $\xi, \xi_1, \xi_2, \ldots$ on an lcscH space S. Show that $\xi_n \xrightarrow{d} \xi$ iff $\eta_n \xrightarrow{d} \eta$.

16. Specialize Theorem 16.23 to suitably normalized sequences of i.i.d. random variables, and compare with Corollary 16.7.

17. Let X be a continuous process on $I = \mathbb{R}_+$ or $[0,1]$ with $X_0 = 0$. Show that X is exchangeable iff a.s. $X_t = \alpha t + \sigma B_t$, $t \in I$, for some Brownian motion or bridge B and some independent pair of random variables α and $\sigma \geq 0$. Also show that α and σ are a.s. unique. (*Hint:* For the last assertion, use the laws of large numbers and the iterated logarithm.)

18. Characterize the Lévy processes on $[0,1]$ as special exchangeable processes, in terms of the coefficients in Theorem 16.21.

19. Show that a process X on $\mathbb{R}_+$ is exchangeable iff it has a version that is conditionally Lévy with random characteristics (α, β, ν). (*Hint:* Theorem 16.21 shows that X has an rcll version. By Theorem 11.15 it is then conditionally Lévy, given some σ-field $\mathcal{I}$. Finally, the characteristics (α, β, ν) of X are $\mathcal{I}$-measurable by the law of large numbers.)

20. Let X be an rcll, exchangeable process on $\mathbb{R}_+$. Show directly from Corollary 16.20 and Theorem 16.21 that the point process of jump sizes on $[0, 1]$ is Cox. Also conclude from Theorem 16.19 and the law of large numbers that the point process of jump times and sizes is Cox with directing random measure of the form $\nu \otimes \lambda$.

21. For an lcscH space S, let $\mathcal{U} \subset \hat{\mathcal{S}}$ be separating. Show that if $K \subset G$ with K compact and G open, there exists some $U \in \mathcal{U}$ with $K \subset U^\circ \subset \overline{U} \subset G$. (*Hint:* First choose $B, C \in \hat{\mathcal{S}}$ with $K \subset B^\circ \subset \overline{B} \subset C^\circ \subset \overline{C} \subset G$.)

Chapter 17

Stochastic Integrals and Quadratic Variation

Continuous local martingales and semimartingales; quadratic variation and covariation; existence and basic properties of the integral; integration by parts and Itô's formula; Fisk–Stratonovich integral; approximation and uniqueness; random time-change; dependence on parameter

This chapter introduces the basic notions of stochastic calculus in the special case of continuous integrators. As a first major task, we shall construct the quadratic variation $[M]$ of a continuous local martingale M, using an elementary approximation and completeness argument. The processes M and $[M]$ will be related by some useful continuity and norm relations, most importantly by the powerful BDG inequalities.

Given the quadratic variation $[M]$, we may next construct the stochastic integral $\int V dM$ for suitable progressive processes V, using a simple Hilbert space argument. Combining with the ordinary Stieltjes integral $\int V dA$ for processes A of locally finite variation, we may finally extend the integral to arbitrary continuous semimartingales $X = M + A$. The continuity properties of quadratic variation carry over to the stochastic integral, and in conjunction with the obvious linearity they characterize the integration.

The key result for applications is Itô's formula, which shows how semimartingales are transformed under smooth mappings. The present substitution rule differs from the corresponding result for Stieltjes integrals, but the two formulas can be brought into agreement by a suitable modification of the integral. We conclude the chapter with some special topics of importance for applications, such as the transformation of stochastic integrals under a random time-change, and the integration of processes depending on a parameter.

The present material may be regarded as continuing the martingale theory from Chapter 7. Though no results for Brownian motion are used explicitly in this chapter, the existence of the Brownian quadratic variation in Chapter 13 may serve as a motivation. We shall also need the representation and measurability of limits obtained in Chapter 4. The stochastic calculus developed in this chapter plays an important role throughout the remainder of this book, especially in Chapters 18, 21, 22, and 23. In Chapter 26 the theory is extended to possibly discontinuous semimartingales.

Throughout the chapter we let $\mathcal{F} = (\mathcal{F}_t)$ be a right-continuous and complete filtration on $\mathbb{R}_+$. A process M is said to be a *local martingale* if it is adapted to $\mathcal{F}$ and such that the stopped and shifted processes $M^{\tau_n} - M_0$ are true martingales for suitable optional times $\tau_n \uparrow \infty$. By a similar *localization* we may define local L^2-martingales, locally bounded martingales, locally integrable processes, and so on. The associated optional times τ_n are said to form a *localizing sequence*.

Any continuous local martingale may clearly be reduced by localization to a sequence of bounded, continuous martingales. Conversely, we see by dominated convergence that every bounded local martingale is a true martingale. The following useful result may be less obvious.

Lemma 17.1 *(localization)* Fix any optional times $\tau_n \uparrow \infty$. Then a process M is a local martingale iff M^{τ_n} has this property for every n.

Proof: If M is a local martingale with localizing sequence (σ_n), and if τ is an arbitrary optional time, then the processes $(M^\tau)^{\sigma_n} = (M^{\sigma_n})^\tau$ are true martingales. Thus, M^τ is again a local martingale with localizing sequence (σ_n).

Conversely, assume that each process M^{τ_n} is a local martingale with localizing sequence (σ_k^n). Since $\sigma_k^n \to \infty$ a.s. for each n, we may choose some indices k_n with

$$P\{\sigma_{k_n}^n < \tau_n \wedge n\} \leq 2^{-n}, \quad n \in \mathbb{N}.$$

Writing $\tau_n' = \tau_n \wedge \sigma_{k_n}^n$, we get $\tau_n' \to \infty$ a.s. by the Borel–Cantelli lemma, and so the optional times $\tau_n'' = \inf_{m \geq n} \tau_m'$ satisfy $\tau_n'' \uparrow \infty$ a.s. It remains to note that the processes $M^{\tau_n''} = (M^{\tau_n})^{\tau_n''}$ are true martingales. $\square$

The next result shows that every continuous martingale of finite variation is a.s. constant. An extension appears as Lemma 25.11.

Proposition 17.2 *(finite-variation martingales)* If M is a continuous local martingale of locally finite variation, then $M = M_0$ a.s.

Proof: By localization we may reduce to the case when $M_0 = 0$ and M has bounded variation. In fact, let V_t denote the total variation of M on the interval $[0,t]$, and note that V is continuous and adapted. For each $n \in \mathbb{N}$ we may then introduce the optional time $\tau_n = \inf\{t \geq 0;\ V_t = n\}$, and we note that $M^{\tau_n} - M_0$ is a continuous martingale with total variation bounded by n. Note also that $\tau_n \to \infty$ and that if $M^{\tau_n} = M_0$ a.s. for each n, then even $M = M_0$ a.s.

In the reduced case, fix any $t > 0$, write $t_{n,k} = kt/n$, and conclude from the continuity of M that a.s.

$$\zeta_n \equiv \sum_{k \leq n} (M_{t_{n,k}} - M_{t_{n,k-1}})^2 \leq V_t \max_{k \leq n} |M_{t_{n,k}} - M_{t_{n,k-1}}| \to 0.$$

Since $\zeta_n \leq V_t^2$, which is bounded by a constant, it follows by the martingale property and dominated convergence that $EM_t^2 = E\zeta_n \to 0$, and so $M_t = 0$ a.s. for each $t > 0$. □

Our construction of stochastic integrals depends on the quadratic variation and covariation processes, which then need to be constructed first. Here we use a direct approach that has the further advantage of giving some insight into the nature of the basic integration-by-parts formula of Theorem 17.16. An alternative but less elementary approach would be to use the Doob–Meyer decomposition in Chapter 25.

The construction utilizes *predictable step processes* of the form

$$V_t = \sum_k \xi_k 1\{t > \tau_k\} = \sum_k \eta_k 1_{(\tau_k, \tau_{k+1}]}(t), \quad t \geq 0, \qquad (1)$$

where the τ_n are optional times with $\tau_n \uparrow \infty$ a.s., and the ξ_k and η_k are $\mathcal{F}_{\tau_k}$-measurable random variables for each $k \in \mathbb{N}$. For any process X we may introduce the elementary integral process $V \cdot X$, given as in Chapter 7 by

$$(V \cdot X)_t \equiv \int_0^t V\, dX = \sum_k \xi_k (X_t - X_t^{\tau_k}) = \sum_k \eta_k (X_{\tau_{k+1}}^t - X_{\tau_k}^t), \qquad (2)$$

where the series converge since they have only finitely many nonzero terms. Note that $(V \cdot X)_0 = 0$ and that $V \cdot X$ inherits the possible continuity properties of X. It is further useful to note that $V \cdot X = V \cdot (X - X_0)$. The following simple estimate will be needed later.

Lemma 17.3 (L^2-bound) *For any continuous L^2-martingale M with $M_0 = 0$ and predictable step process V with $|V| \leq 1$, the process $V \cdot M$ is again an L^2-martingale, and $E(V \cdot M)_t^2 \leq EM_t^2$.*

Proof: First assume that the sum in (1) has only finitely many nonzero terms. Then Corollary 7.14 shows that $V \cdot M$ is a martingale, and the L^2-bound follows by the computation

$$E(V \cdot M)_t^2 = E \sum_k \eta_k^2 (M_{\tau_{k+1}}^t - M_{\tau_k}^t)^2$$
$$\leq E \sum_k (M_{\tau_{k+1}}^t - M_{\tau_k}^t)^2 = EM_t^2.$$

The estimate extends to the general case by Fatou's lemma, and the martingale property then extends by uniform integrability. □

Let us now introduce the space $\mathcal{M}^2$ of all L^2-bounded, continuous martingales M with $M_0 = 0$, and equip $\mathcal{M}^2$ with the norm $\|M\| = \|M_\infty\|_2$. Recall that $\|M^*\|_2 \leq 2\|M\|$ by Proposition 7.16.

Lemma 17.4 (completeness) *The space $\mathcal{M}^2$ is a Hilbert space.*

Proof: Fix any Cauchy sequence $M^1, M^2, \ldots$ in $\mathcal{M}^2$. The sequence (M_∞^n) is then Cauchy in L^2 and thus converges toward some element $\xi \in L^2$.

Introduce the L^2-martingale $M_t = E[\xi|\mathcal{F}_t]$, $t \geq 0$, and note that $M_\infty = \xi$ a.s. since ξ is $\mathcal{F}_\infty$-measurable. Hence,

$$\|(M^n - M)^*\|_2 \leq 2\|M^n - M\| = 2\|M_\infty^n - M_\infty\|_2 \to 0,$$

and so $\|M^n - M\| \to 0$. Moreover, $(M^n - M)^* \to 0$ a.s. along some subsequence, which shows that M is a.s. continuous with $M_0 = 0$. □

We are now ready to prove the existence of the *quadratic variation* and *covariation* processes $[M]$ and $[M, N]$. Extensions to possibly discontinuous processes are considered in Chapter 26.

Theorem 17.5 *(covariation)* *For any continuous local martingales M and N, there exists an a.s. unique continuous process $[M, N]$ of locally finite variation and with $[M, N]_0 = 0$ such that $MN - [M, N]$ is a local martingale. The form $[M, N]$ is a.s. symmetric and bilinear with $[M, N] = [M - M_0, N - N_0]$ a.s. Furthermore, $[M] = [M, M]$ is a.s. nondecreasing, and for any optional time τ,*

$$[M^\tau, N] = [M^\tau, N^\tau] = [M, N]^\tau \quad a.s.$$

Proof: The a.s. uniqueness of $[M, N]$ follows from Proposition 17.2, and the symmetry and bilinearity are immediate consequences. If $[M, N]$ exists with the stated properties and τ is an optional time, then by Lemma 17.1 the process $M^\tau N^\tau - [M, N]^\tau$ is a local martingale, and so is the process $M^\tau(N - N^\tau)$ by Corollary 7.14. Hence, even $M^\tau N - [M, N]^\tau$ is a local martingale, and so $[M^\tau, N] = [M^\tau, N^\tau] = [M, N]^\tau$ a.s. Furthermore,

$$MN - (M - M_0)(N - N_0) = M_0 N_0 + M_0(N - N_0) + N_0(M - M_0)$$

is a local martingale, and so $[M - M_0, N - N_0] = [M, N]$ a.s. whenever either side exists. If both $[M + N]$ and $[M - N]$ exist, then

$$4MN - ([M + N] - [M - N])$$
$$= ((M + N)^2 - [M + N]) - ((M - N)^2 - [M - N])$$

is a local martingale, and so we may take $[M, N] = ([M+N]-[M-N])/4$. It is then enough to prove the existence of $[M]$ when $M_0 = 0$.

First assume that M is bounded. For each $n \in \mathbb{N}$, let $\tau_0^n = 0$ and define recursively

$$\tau_{k+1}^n = \inf\{t > \tau_k^n; |M_t - M_{\tau_k^n}| = 2^{-n}\}, \quad k \geq 0.$$

Clearly, $\tau_k^n \to \infty$ as $k \to \infty$ for fixed n. Introduce the processes

$$V_t^n = \sum_k M_{\tau_k^n} 1\{t \in (\tau_k^n, \tau_{k+1}^n]\}, \quad Q_t^n = \sum_k (M_{t \wedge \tau_k^n} - M_{t \wedge \tau_{k-1}^n})^2.$$

The V^n are bounded predictable step processes, and we note that

$$M_t^2 = 2(V^n \cdot M)_t + Q_t^n, \quad t \geq 0. \tag{3}$$

By Lemma 17.3 the integrals $V^n \cdot M$ are continuous L^2-martingales, and since $|V^n - M| \leq 2^n$ for each n, we have

$$\|V^m \cdot M - V^n \cdot M\| = \|(V^m - V^n) \cdot M\| \leq 2^{-m+1}\|M\|, \quad m \leq n.$$

Hence, by Lemma 17.4 there exists some continuous martingale N such that $(V^n \cdot M - N)^* \xrightarrow{P} 0$. The process $[M] = M^2 - 2N$ is again continuous, and by (3) we have

$$(Q^n - [M])^* = 2(N - V^n \cdot M)^* \xrightarrow{P} 0.$$

In particular, $[M]$ is a.s. nondecreasing on the random time set $T = \{\tau_k^n; n, k \in \mathbb{N}\}$, and the monotonicity extends by continuity to the closure $\overline{T}$. Also note that $[M]$ is constant on each interval in $\overline{T}^c$, since this is true for M and hence also for every Q^n. Thus, $[M]$ is a.s. nondecreasing.

Turning to the unbounded case, we define $\tau_n = \inf\{t > 0; |M_t| = n\}$, $n \in \mathbb{N}$. The processes $[M^{\tau_n}]$ exist as before, and we note that $[M^{\tau_m}]^{\tau_m} = [M^{\tau_n}]^{\tau_m}$ a.s. for all $m < n$. Hence, $[M^{\tau_m}] = [M^{\tau_n}]$ a.s. on $[0, \tau_m]$, and since $\tau_n \to \infty$ there exists a nondecreasing, continuous, and adapted process $[M]$ such that $[M] = [M^{\tau_n}]$ a.s. on $[0, \tau_n]$ for each n. Here $(M^{\tau_n})^2 - [M]^{\tau_n}$ is a local martingale for each n, and so $M^2 - [M]$ is a local martingale by Lemma 17.1. □

We proceed to establish a basic continuity property.

Proposition 17.6 (*continuity*) *For any continuous local martingales M_n starting at 0, we have $M_n^* \xrightarrow{P} 0$ iff $[M_n]_\infty \xrightarrow{P} 0$.*

Proof: First let $M_n^* \xrightarrow{P} 0$. Fix any $\varepsilon > 0$, and define $\tau_n = \inf\{t \geq 0; |M_n(t)| > \varepsilon\}$, $n \in \mathbb{N}$. Write $N_n = M_n^2 - [M_n]$, and note that $N_n^{\tau_n}$ is a true martingale on $\mathbb{R}_+$. In particular, $E[M_n]_{\tau_n} \leq \varepsilon^2$, and so by Chebyshev's inequality

$$P\{[M_n]_\infty > \varepsilon\} \leq P\{\tau_n < \infty\} + \varepsilon^{-1} E[M_n]_{\tau_n} \leq P\{M_n^* > \varepsilon\} + \varepsilon.$$

Here the right-hand side tends to zero as $n \to \infty$ and then $\varepsilon \to 0$, which shows that $[M_n]_\infty \xrightarrow{P} 0$.

The proof in the other direction is similar, except that we need to use a localization argument together with Fatou's lemma to see that a continuous local martingale M with $M_0 = 0$ and $E[M]_\infty < \infty$ is necessarily L^2-bounded. □

Next we prove a pair of basic norm inequalities involving the quadratic variation, known as the *BDG inequalities*. Partial extensions to discontinuous martingales are established in Theorem 26.12.

Theorem 17.7 (*norm inequalities, Burkholder, Millar, Gundy, Novikov*) *There exist some constants $c_p \in (0, \infty)$, $p > 0$, such that for any continuous local martingale M with $M_0 = 0$,*

$$c_p^{-1} E[M]_\infty^{p/2} \leq EM^{*p} \leq c_p E[M]_\infty^{p/2}, \quad p > 0.$$

Proof: By optional stopping we may assume that M and $[M]$ are bounded. Write $M' = M - M^\tau$ with $\tau = \inf\{t; M_t^2 = r\}$ and define

$N = (M')^2 - [M']$. By Corollary 7.30 we have for any $r > 0$ and $c \in (0, 2^{-p})$

$$\begin{aligned}
P\{M^{*2} \geq 4r\} - P\{[M]_\infty \geq cr\} &\leq P\{M^{*2} \geq 4r, [M]_\infty < cr\} \\
&\leq P\{N > -cr, \sup_t N_t > r - cr\} \\
&\leq cP\{N^* > 0\} \leq cP\{M^{*2} \geq r\}.
\end{aligned}$$

Multiplying by $(p/2)r^{p/2-1}$ and integrating over $\mathbb{R}_+$, we get by Lemma 3.4

$$2^{-p} EM^{*p} - c^{-p/2} E[M]_\infty^{p/2} \leq cEM^{*p},$$

and the right-hand inequality follows with $c_p = c^{-p/2}/(2^{-p} - c)$.

Next let N be as before with $\tau = \inf\{t; [M]_t = r\}$, and write for any $r > 0$ and $c \in (0, 2^{-p/2-2})$

$$\begin{aligned}
P\{[M]_\infty \geq 2r\} - P\{M^{*2} \geq cr\} &\leq P\{[M]_\infty \geq 2r, M^{*2} < cr\} \\
&\leq P\{N < 4cr, \inf_t N_t < 4cr - r\} \\
&\leq 4cP\{[M]_\infty \geq r\}.
\end{aligned}$$

Integrating as before yields

$$2^{-p/2} E[M]_\infty^{p/2} - c^{-p/2} EM^{*p} \leq 4cE[M]_\infty^{p/2},$$

and the left-hand inequality follows with $c_p = c^{-p/2}/(2^{-p/2} - 4c)$. □

It is often important to decide whether a local martingale is in fact a true martingale. The last proposition yields a useful criterion.

Corollary 17.8 *(uniform integrability)* *Let M be a continuous local martingale satisfying $E(|M_0| + [M]_\infty^{1/2}) < \infty$. Then M is a uniformly integrable martingale.*

Proof: By Theorem 17.7 we have $EM^* < \infty$, and the martingale property follows by dominated convergence. □

The basic properties of $[M, N]$ suggest that we think of the covariation process as an inner product. A further justification is given by the following useful Cauchy–Buniakovsky-type inequalities.

Proposition 17.9 *(Cauchy-type inequalities, Courrège)* *For any continuous local martingales M and N, we have a.s.*

$$|[M, N]| \leq \int |d[M, N]| \leq [M]^{1/2}[N]^{1/2}. \tag{4}$$

More generally, we have a.s. for any measurable processes U and V

$$\int_0^t |UV d[M, N]| \leq (U^2 \cdot [M])_t^{1/2} (V^2 \cdot [N])_t^{1/2}, \quad t \geq 0.$$

Proof: Using the positivity and bilinearity of the covariation, we get a.s. for any $a, b \in \mathbb{R}$ and $t > 0$

$$0 \leq [aM + bN]_t = a^2[M]_t + 2ab[M, N]_t + b^2[N]_t.$$

By continuity we can choose a common exceptional null set for all a and b, and so $[M,N]_t^2 \leq [M]_t[N]_t$ a.s. Applying this inequality to the processes $M - M^s$ and $N - N^s$ for any $s < t$, we obtain a.s.

$$|[M,N]_t - [M,N]_s| \leq ([M]_t - [M]_s)^{1/2}([N]_t - [N]_s)^{1/2}, \qquad (5)$$

and by continuity we may again choose a common null set. Now let $0 = t_0 < t_1 < \cdots < t_n = t$ be arbitrary, and conclude from (5) and the classical Cauchy–Buniakovsky inequality that

$$|[M,N]_t| \leq \sum_k |[M,N]_{t_k} - [M,N]_{t_{k-1}}| \leq [M]_t^{1/2}[N]_t^{1/2}.$$

To get (4), it remains to take the supremum over all partitions of $[0,t]$.

Next write $d\mu = d[M]$, $d\nu = d[N]$, and $d\rho = |d[M,N]|$, and conclude from (4) that $(\rho I)^2 \leq \mu I \, \nu I$ a.s. for every interval I. By continuity we may choose the exceptional null set A to be independent of I. Letting $G \subset \mathbb{R}_+$ be open with connected components I_k and using the Cauchy–Buniakovsky inequality, we get on A^c

$$\rho G = \sum_k \rho I_k \leq \sum_k (\mu I_k \nu I_k)^{1/2} \leq \left\{\sum_j \mu I_j \sum_k \nu I_k\right\}^{1/2} = (\mu G \, \nu G)^{1/2}.$$

By Lemma 1.34 the last relation extends to any $B \in \mathcal{B}(\mathbb{R}_+)$.

Now fix any simple measurable functions $f = \sum_k a_k 1_{B_k}$ and $g = \sum_k b_k 1_{B_k}$. Using the Cauchy–Buniakovsky inequality again, we obtain on A^c

$$\rho|fg| \leq \sum_k |a_k b_k| \rho B_k \leq \sum_k |a_k b_k|(\mu B_k \nu B_k)^{1/2}$$
$$\leq \left\{\sum_j a_j^2 \mu B_j \sum_k b_k^2 \nu B_k\right\}^{1/2} \leq (\mu f^2 \nu g^2)^{1/2},$$

which extends by monotone convergence to any measurable functions f and g on $\mathbb{R}_+$. In particular, in view of Lemma 1.33, we may take $f(t) = U_t(\omega)$ and $g(t) = V_t(\omega)$ for fixed $\omega \in A^c$. $\square$

Let $\mathcal{E}$ denote the class of bounded, predictable step processes with jumps at finitely many fixed times. To motivate the construction of general stochastic integrals and for subsequent needs, we shall establish a basic identity for elementary integrals.

Lemma 17.10 *(covariation of elementary integrals) For any continuous local martingales M, N and processes $U, V \in \mathcal{E}$, the integrals $U \cdot M$ and $V \cdot N$ are again continuous local martingales, and we have*

$$[U \cdot M, V \cdot N] = (UV) \cdot [M,N] \quad \text{a.s.} \qquad (6)$$

Proof: We may clearly take $M_0 = N_0 = 0$. The first assertion follows by localization from Lemma 17.3. To prove (6), let $U_t = \sum_{k \leq n} \xi_k 1_{(t_k, t_{k+1}]}(t)$, where ξ_k is bounded and $\mathcal{F}_{t_k}$-measurable for each k. By localization we may assume M, N, and $[M,N]$ to be bounded, so that M, N, and $MN - [M,N]$

are martingales on $\bar{\mathbb{R}}_+$. Then

$$\begin{aligned}
E(U \cdot M)_\infty N_\infty &= E\sum_j \xi_j(M_{t_{j+1}} - M_{t_j})\sum_k(N_{t_{k+1}} - N_{t_k})\\
&= E\sum_k \xi_k(M_{t_{k+1}}N_{t_{k+1}} - M_{t_k}N_{t_k})\\
&= E\sum_k \xi_k([M,N]_{t_{k+1}} - [M,N]_{t_k})\\
&= E(U \cdot [M,N])_\infty.
\end{aligned}$$

Replacing M and N by M^τ and N^τ for an arbitrary optional time τ, we get

$$E(U \cdot M)_\tau N_\tau = E(U \cdot M^\tau)_\infty N^\tau_\infty = E(U \cdot [M^\tau, N^\tau])_\infty = E(U \cdot [M,N])_\tau.$$

By Lemma 7.13 the process $(U \cdot M)N - U \cdot [M,N]$ is then a martingale, and so $[U \cdot M, N] = U \cdot [M,N]$ a.s. The general formula follows by iteration. □

In order to extend the stochastic integral $V \cdot M$ to more general processes V, it is convenient to take (6) as the characteristic property. Given a continuous local martingale M, let $L(M)$ denote the class of all progressive processes V such that $(V^2 \cdot [M])_t < \infty$ a.s. for every $t > 0$.

Theorem 17.11 (*stochastic integral, Itô, Kunita and Watanabe*) *For any continuous local martingale M and process $V \in L(M)$, there exists an a.s. unique continuous local martingale $V \cdot M$ with $(V \cdot M)_0 = 0$ such that $[V \cdot M, N] = V \cdot [M,N]$ a.s. for every continuous local martingale N.*

Proof: To prove the uniqueness, let M' and M'' be continuous local martingales with $M'_0 = M''_0 = 0$ such that $[M', N] = [M'', N] = V \cdot [M, N]$ a.s. for all continuous local martingales N. By linearity we get $[M' - M'', N] = 0$ a.s. Taking $N = M' - M''$ gives $[M' - M''] = 0$ a.s. But then $(M' - M'')^2$ is a local martingale starting at 0, and it easily follows that $M' = M''$ a.s.

To prove the existence, we may first assume that $\|V\|_M^2 = E(V^2 \cdot [M])_\infty < \infty$. Since V is measurable, we get by Proposition 17.9 and the Cauchy–Buniakovsky inequality

$$|E(V \cdot [M,N])_\infty| \le \|V\|_M \|N\|, \quad N \in \mathcal{M}^2.$$

The mapping $N \mapsto E(V \cdot [M,N])_\infty$ is then a continuous linear functional on $\mathcal{M}^2$, and so by Lemma 17.4 there exists an element $V \cdot M \in \mathcal{M}^2$ with

$$E(V \cdot [M,N])_\infty = E(V \cdot M)_\infty N_\infty, \quad N \in \mathcal{M}^2.$$

Now replace N by N^τ for an arbitrary optional time τ. By Theorem 17.5 and optional sampling we get

$$\begin{aligned}
E(V \cdot [M,N])_\tau &= E(V \cdot [M,N]^\tau)_\infty = E(V \cdot [M, N^\tau])_\infty\\
&= E(V \cdot M)_\infty N_\tau = E(V \cdot M)_\tau N_\tau.
\end{aligned}$$

Since V is progressive, it follows by Lemma 7.13 that $V \cdot [M,N] - (V \cdot M)N$ is a martingale, which means that $[V \cdot M, N] = V \cdot [M,N]$ a.s. The last relation extends by localization to arbitrary continuous local martingales N.

In the general case, define $\tau_n = \inf\{t > 0; (V^2 \cdot [M])_t = n\}$. By the previous argument there exist some continuous local martingales $V \cdot M^{\tau_n}$ such that, for any continuous local martingale N,

$$[V \cdot M^{\tau_n}, N] = V \cdot [M^{\tau_n}, N] \text{ a.s., } n \in \mathbb{N}. \tag{7}$$

For $m < n$ it follows that $(V \cdot M^{\tau_n})^{\tau_m}$ satisfies the corresponding relation with $[M^{\tau_m}, N]$, and so $(V \cdot M^{\tau_n})^{\tau_m} = V \cdot M^{\tau_m}$ a.s. Hence, there exists a continuous process $V \cdot M$ with $(V \cdot M)^{\tau_n} = V \cdot M^{\tau_n}$ a.s. for all n, and Lemma 17.1 shows that $V \cdot M$ is again a local martingale. Finally, (7) yields $[V \cdot M, N] = V \cdot [M, N]$ a.s. on $[0, \tau_n]$ for each n, and so the same relation holds on $\mathbb{R}_+$. □

By Lemma 17.10 we note that the stochastic integral $V \cdot M$ of the last theorem extends the previously defined elementary integral. It is also clear that $V \cdot M$ is a.s. bilinear in the pair (V, M) and satisfies the following basic continuity property.

Lemma 17.12 *(continuity)* *For any continuous local martingales M_n and processes $V_n \in L(M_n)$, we have $(V_n \cdot M_n)^* \xrightarrow{P} 0$ iff $(V_n^2 \cdot [M_n])_\infty \xrightarrow{P} 0$.*

Proof: Recall that $[V_n \cdot M_n] = V_n^2 \cdot [M_n]$ and use Proposition 17.6. □

Before continuing the study of stochastic integrals, it is convenient to extend the definition to a larger class of integrators. A process X is said to be a *continuous semimartingale* if it can be written as a sum $M + A$, where M is a continuous local martingale and A is a continuous, adapted process of locally finite variation and with $A_0 = 0$. By Proposition 17.2 the decomposition $X = M + A$ is then a.s. unique, and it is often referred to as the *canonical decomposition* of X. By a continuous semimartingale in $\mathbb{R}^d$ we mean a process $X = (X^1, \ldots, X^d)$ such that the component processes X^k are one-dimensional continuous semimartingales.

Let $L(A)$ denote the class of progressive processes V such that the process $(V \cdot A)_t = \int_0^t V dA$ exists in the sense of ordinary Stieltjes integration. For any continuous semimartingale $X = M + A$ we may write $L(X) = L(M) \cap L(A)$, and we define the integral of a process $V \in L(X)$ as the sum $V \cdot X = V \cdot M + V \cdot A$. Note that $V \cdot X$ is again a continuous semimartingale with canonical decomposition $V \cdot M + V \cdot A$. For progressive processes V, it is further clear that $V \in L(X)$ iff $V^2 \in L([M])$ and $V \in L(A)$.

From Lemma 17.12 we may easily deduce the following stochastic version of the dominated convergence theorem.

Corollary 17.13 *(dominated convergence)* *For any continuous semimartingale X, let $U, V, V_1, V_2, \cdots \in L(X)$ with $|V_n| \leq U$ and $V_n \to V$. Then $(V_n \cdot X - V \cdot X)_t^* \xrightarrow{P} 0$, $t \geq 0$.*

Proof: Assume that $X = M + A$. Since $U \in L(X)$, we have $U^2 \in L([M])$ and $U \in L(A)$. Hence, by dominated convergence for ordinary Stieltjes integrals, $((V_n - V)^2 \cdot [M])_t \to 0$ and $(V_n \cdot A - V \cdot A)_t^* \to 0$ a.s. By Lemma 17.12 the former convergence implies $(V_n \cdot M - V \cdot M)_t^* \overset{P}{\to} 0$, and the assertion follows. □

The next result extends the elementary chain rule of Lemma 1.23 to stochastic integrals.

Proposition 17.14 *(chain rule)* *Consider a continuous semimartingale X and two progressive processes U and V, where $V \in L(X)$. Then $U \in L(V \cdot X)$ iff $UV \in L(X)$, in which case $U \cdot (V \cdot X) = (UV) \cdot X$ a.s.*

Proof: Let $M+A$ be the canonical decomposition of X. Then $U \in L(V \cdot X)$ iff $U^2 \in L([V \cdot M])$ and $U \in L(V \cdot A)$, whereas $UV \in L(X)$ iff $(UV)^2 \in L([M])$ and $UV \in L(A)$. Since $[V \cdot M] = V^2 \cdot [M]$, the two pairs of conditions are equivalent.

The formula $U \cdot (V \cdot A) = (UV) \cdot A$ is elementary. To see that even $U \cdot (V \cdot M) = (UV) \cdot M$ a.s., let N be an arbitrary continuous local martingale, and note that

$$[(UV) \cdot M, N] = (UV) \cdot [M, N] = U \cdot (V \cdot [M, N])$$
$$= U \cdot [V \cdot M, N] = [U \cdot (V \cdot M), N]. \qquad \square$$

The next result shows how the stochastic integral behaves under optional stopping.

Proposition 17.15 *(optional stopping)* *For any continuous semimartingale X, process $V \in L(X)$, and optional time τ, we have a.s.*

$$(V \cdot X)^\tau = V \cdot X^\tau = (V1_{[0,\tau]}) \cdot X.$$

Proof: The relations being obvious for ordinary Stieltjes integrals, we may assume that $X = M$ is a continuous local martingale. Then $(V \cdot M)^\tau$ is a continuous local martingale starting at 0, and we have

$$[(V \cdot M)^\tau, N] = [V \cdot M, N^\tau] = V \cdot [M, N^\tau] = V \cdot [M^\tau, N]$$
$$= V \cdot [M, N]^\tau = (V1_{[0,\tau]}) \cdot [M, N].$$

Thus, $(V \cdot M)^\tau$ satisfies the conditions characterizing the integrals $V \cdot M^\tau$ and $(V1_{[0,\tau]}) \cdot M$. □

We may extend the definitions of quadratic variation and covariation to arbitrary continuous semimartingales X and Y with canonical decompositions $M + A$ and $N + B$, respectively, by putting $[X] = [M]$ and $[X, Y] = [M, N]$. As a key step toward the development of a stochastic calculus, we show how the covariation process can be expressed in terms of stochastic integrals. In the martingale case, the result is implicit in the proof of Theorem 17.5.

Theorem 17.16 *(integration by parts)* *For any continuous semimartingales X and Y, we have a.s.*

$$XY = X_0 Y_0 + X \cdot Y + Y \cdot X + [X, Y]. \tag{8}$$

Proof: We may take $X = Y$, since the general result will then follow by polarization. First let $X = M \in \mathcal{M}^2$, and define V^n and Q^n as in the proof of Theorem 17.5. Then $V^n \to M$ and $|V_t^n| \leq M_t^* < \infty$, and so Corollary 17.13 yields $(V^n \cdot M)_t \xrightarrow{P} (M \cdot M)_t$ for each $t \geq 0$. Thus, (8) follows in this case as we let $n \to \infty$ in the relation $M^2 = V^n \cdot M + Q^n$, and it extends by localization to general continuous local martingales M with $M_0 = 0$. If instead $X = A$, formula (8) reduces to $A^2 = 2A \cdot A$, which holds by Fubini's theorem.

Turning to the general case, we may assume that $X_0 = 0$, since the formula for general X_0 will then follow by an easy computation from the result for $X - X_0$. In this case (8) reduces to $X^2 = 2X \cdot X + [M]$. Subtracting the formulas for M^2 and A^2, it remains to prove that $AM = A \cdot M + M \cdot A$ a.s. Then fix any $t > 0$, and introduce the processes

$$A_s^n = A_{(k-1)t/n}, \quad M_s^n = M_{kt/n}, \quad s \in t(k-1, k]/n, \ k, n \in \mathbb{N},$$

which satisfy

$$A_t M_t = (A^n \cdot M)_t + (M^n \cdot A)_t, \quad n \in \mathbb{N}.$$

Here $(A^n \cdot M)_t \xrightarrow{P} (A \cdot M)_t$ by Corollary 17.13 and $(M^n \cdot A)_t \to (M \cdot A)_t$ by dominated convergence for ordinary Stieltjes integrals. □

The terms quadratic variation and covariation are justified by the following result, which extends Theorem 13.9 for Brownian motion.

Proposition 17.17 *(approximation, Fisk)* *Let X and Y be continuous semimartingales, fix any $t > 0$, and consider for every $n \in \mathbb{N}$ a partition $0 = t_{n,0} < t_{n,1} < \cdots < t_{n,k_n} = t$ such that $\max_k(t_{n,k} - t_{n,k-1}) \to 0$. Then*

$$\zeta_n \equiv \sum_k (X_{t_{n,k}} - X_{t_{n,k-1}})(Y_{t_{n,k}} - Y_{t_{n,k-1}}) \xrightarrow{P} [X, Y]_t. \tag{9}$$

Proof: We may clearly assume that $X_0 = Y_0 = 0$. Introduce the predictable step processes

$$X_s^n = X_{t_{n,k-1}}, \quad Y_s^n = Y_{t_{n,k-1}}, \quad s \in (t_{n,k-1}, t_{n,k}], \ k, n \in \mathbb{N},$$

and note that

$$X_t Y_t = (X^n \cdot Y)_t + (Y^n \cdot X)_t + \zeta_n, \quad n \in \mathbb{N}.$$

Since $X^n \to X$ and $Y^n \to Y$, and also $(X^n)_t^* \leq X_t^* < \infty$ and $(Y^n)_t^* \leq X_t^* < \infty$, we get by Corollary 17.13 and Theorem 17.16

$$\zeta_n \xrightarrow{P} X_t Y_t - (X \cdot Y)_t - (Y \cdot X)_t = [X, Y]_t. \quad \Box$$

We proceed to prove a version of *Itô's formula*, arguably the most important formula in modern probability. The result shows that the class of continuous semimartingales is preserved under smooth mappings; it also exhibits the canonical decomposition of the image process in terms of the components of the original process. Extended versions appear in Corollaries 17.19 and 17.20 as well as in Theorems 22.5 and 26.7.

Let $C^k = C^k(\mathbb{R}^d)$ denote the class of k times continuously differentiable functions on $\mathbb{R}^d$. When $f \in C^2$, we write f'_i and f''_{ij} for the first- and second-order partial derivatives of f. Here and below, summation over repeated indices is understood.

Theorem 17.18 *(substitution rule, Itô)* For any continuous semimartingale X in $\mathbb{R}^d$ and function $f \in C^2(\mathbb{R}^d)$, we have a.s.

$$f(X) = f(X_0) + f'_i(X) \cdot X^i + \tfrac{1}{2} f''_{ij}(X) \cdot [X^i, X^j]. \tag{10}$$

The result is often written in differential form as

$$df(X) = f'_i(X)\, dX^i + \tfrac{1}{2} f''_{ij}(X)\, d[X^i, X^j].$$

It is suggestive to think of Itô's formula as a second-order Taylor expansion

$$df(X) = f'_i(X)\, dX^i + \tfrac{1}{2} f''_{ij}(X)\, dX^i dX^j,$$

where the second-order differential $dX^i dX^j$ is interpreted as $d[X^i, X^j]$.

If X has canonical decomposition $M + A$, we get the corresponding decomposition of $f(X)$ by substituting $M^i + A^i$ for X^i on the right of (10). When $M = 0$, the last term vanishes, and (10) reduces to the familiar substitution rule for ordinary Stieltjes integrals. In general, the appearance of this *Itô correction term* shows that the Itô integral does not obey the rules of ordinary calculus.

Proof of Theorem 17.18: For notational convenience we may assume that $d = 1$, the general case being similar. Then fix a one-dimensional, continuous semimartingale X, and let $\mathcal{C}$ denote the class of functions $f \in C^2$ satisfying (10), now appearing in the form

$$f(X) = f(X_0) + f'(X) \cdot X + \tfrac{1}{2} f''(X) \cdot [X]. \tag{11}$$

The class $\mathcal{C}$ is clearly a linear subspace of C^2 containing the functions $f(x) \equiv 1$ and $f(x) \equiv x$. We shall prove that $\mathcal{C}$ is closed under multiplication and hence contains all polynomials.

To see this, assume that (11) holds for both f and g. Then $F = f(X)$ and $G = g(X)$ are continuous semimartingales, and so, by the definition of the integral together with Proposition 17.14 and Theorem 17.16, we have

$$\begin{aligned}
(fg)(X) &- (fg)(X_0) \\
&= FG - F_0 G_0 = F \cdot G + G \cdot F + [F, G] \\
&= F \cdot (g'(X) \cdot X + \tfrac{1}{2} g''(X) \cdot [X]) \\
&\quad + G \cdot (f'(X) \cdot X + \tfrac{1}{2} f''(X) \cdot [X]) + [f'(X) \cdot X, g'(X) \cdot X] \\
&= (fg' + f'g)(X) \cdot X + \tfrac{1}{2}(fg'' + 2f'g' + f''g)(X) \cdot [X] \\
&= (fg)'(X) \cdot X + \tfrac{1}{2}(fg)''(X) \cdot [X].
\end{aligned}$$

Now let $f \in C^2$ be arbitrary. By Weierstrass' approximation theorem, we may choose some polynomials $p_1, p_2, \ldots$ such that $\sup_{|x| \le c} |p_n(x) - f''(x)| \to 0$ for every $c > 0$. Integrating the p_n twice yields polynomials f_n satisfying

$$\sup_{|x| \le c} (|f_n(x) - f(x)| \vee |f_n'(x) - f'(x)| \vee |f_n''(x) - f''(x)|) \to 0, \quad c > 0.$$

In particular, $f_n(X_t) \to f(X_t)$ for each $t > 0$. Letting $M + A$ be the canonical decomposition of X and using dominated convergence for ordinary Stieltjes integrals, we get for any $t \ge 0$

$$(f_n'(X) \cdot A + \tfrac{1}{2} f_n''(X) \cdot [X])_t \to (f'(X) \cdot A + \tfrac{1}{2} f''(X) \cdot [X])_t.$$

Similarly, $(f_n'(X) - f'(X))^2 \cdot [M]_t \to 0$ for all t, and so by Lemma 17.12

$$(f_n'(X) \cdot M)_t \xrightarrow{P} (f'(X) \cdot M)_t, \quad t \ge 0.$$

Thus, equation (11) for the polynomials f_n extends in the limit to the same formula for f. □

We sometimes need a local version of the last theorem, involving stochastic integrals up to the time ζ_D when X first leaves a given domain $D \subset \mathbb{R}^d$. If X is continuous and adapted, then ζ_D is clearly *predictable*, in the sense of being *announced* by some optional times $\tau_n \uparrow \zeta_D$ such that $\tau_n < \zeta_D$ a.s. on $\{\zeta_D > 0\}$ for all n. In fact, writing ρ for the Euclidean metric in $\mathbb{R}^d$, we may choose

$$\tau_n = \inf\{t \in [0, n]; \rho(X_t, D^c) \le n^{-1}\}, \quad n \in \mathbb{N}. \tag{12}$$

We say that X is a semimartingale on $[0, \zeta_D)$ if the stopped process X^{τ_n} is a semimartingale in the usual sense for every $n \in \mathbb{N}$. In that case, we may define the covariation processes $[X^i, X^j]$ on the interval $[0, \zeta_D)$ by requiring $[X^i, X^j]^{\tau_n} = [(X^i)^{\tau_n}, (X^j)^{\tau_n}]$ a.s. for every n. Stochastic integrals with respect to $X^1, \ldots, X^d$ are defined on $[0, \zeta_D)$ in a similar way.

Corollary 17.19 *(local Itô-formula) For any domain $D \subset \mathbb{R}^d$, let X be a continuous semimartingale on $[0, \zeta_D)$. Then (10) holds a.s. on $[0, \zeta_D)$ for every $f \in C^2(D)$.*

Proof: Choose some functions $f_n \in C^2(\mathbb{R}^d)$ with $f_n(x) = f(x)$ when $\rho(x, D^c) \ge n^{-1}$. Applying Theorem 17.18 to $f_n(X^{\tau_n})$ with τ_n as in (12), we get (10) on $[0, \tau_n]$. Since n was arbitrary, the result extends to $[0, \zeta_D)$. □

By a *complex-valued, continuous semimartingale* we mean a process of the form $Z = X + iY$, where X and Y are real continuous semimartingales. The bilinearity of the covariation process suggests that we define the quadratic variation of Z as

$$[Z] = [Z, Z] = [X + iY, X + iY] = [X] + 2i[X, Y] - [Y].$$

Let $L(Z)$ denote the class of processes $W = U + iV$ with $U, V \in L(X) \cap L(Y)$. For such a process W, we define the integral by
$$W \cdot Z = (U + iV) \cdot (X + iY)$$
$$= U \cdot X - V \cdot Y + i(U \cdot Y + V \cdot X).$$

Corollary 17.20 *(conformal mapping) Let f be an analytic function on some domain $D \subset \mathbb{C}$. Then (10) holds for any D-valued, continuous semimartingale Z.*

Proof: Writing $f(x + iy) = g(x, y) + ih(x, y)$ for any $x + iy \in D$, we get
$$g_1' + ih_1' = f', \qquad g_2' + ih_2' = if',$$
and so by iteration
$$g_{11}'' + ih_{11}'' = f'', \quad g_{12}'' + ih_{12}'' = if'', \quad g_{22}'' + ih_{22}'' = -f''.$$
Equation (10) now follows for $Z = X + iY$, as we apply Corollary 17.19 to the semimartingale (X, Y) and the functions g and h. $\square$

We also consider a modification of the Itô integral that does obey the rules of ordinary calculus. Assuming both X and Y to be continuous semimartingales, we define the *Fisk-Stratonovich integral* by
$$\int_0^t X \circ dY = (X \cdot Y)_t + \tfrac{1}{2}[X, Y]_t, \quad t \geq 0, \tag{13}$$
or in differential form $X \circ dY = X dY + \tfrac{1}{2} d[X, Y]$, where the first term on the right is an ordinary Itô integral.

Corollary 17.21 *(modified substitution rule, Fisk, Stratonovich) For any continuous semimartingale X in $\mathbb{R}^d$ and function $f \in C^3(\mathbb{R}^d)$, we have a.s.*
$$f(X_t) = f(X_0) + \int_0^t f_i'(X) \circ dX^i, \quad t \geq 0.$$

Proof: By Itô's formula,
$$f_i'(X) = f_i'(X_0) + f_{ij}''(X) \cdot X^j + \tfrac{1}{2} f_{ijk}'''(X) \cdot [X^j, X^k].$$
Using Itô's formula again, together with (6) and (13), we get
$$\int_0^t f_i'(X) \circ dX^i = f_i'(X) \cdot X^i + \tfrac{1}{2}[f_i'(X), X^i]$$
$$= f_i'(X) \cdot X^i + \tfrac{1}{2} f_{ij}''(X) \cdot [X^j, X^i] = f(X) - f(X_0). \square$$

Unfortunately, the more convenient substitution rule of Corollary 17.21 comes at a high price: The new integral does not preserve the martingale property, and it requires even the integrand to be a continuous semimartingale. It is the latter restriction that forces us to impose stronger regularity conditions on the function f in the substitution rule.

Our next task is to establish a basic uniqueness property, justifying our reference to the process $V \cdot M$ in Theorem 17.11 as an integral.

Theorem 17.22 *(uniqueness)* *The integral $V \cdot M$ in Theorem 17.11 is the a.s. unique linear extension of the elementary stochastic integral such that, for any $t > 0$, the convergence $(V_n^2 \cdot [M])_t \xrightarrow{P} 0$ implies $(V_n \cdot M)_t^* \xrightarrow{P} 0$.*

The statement follows immediately from Lemmas 17.10 and 17.12, together with the following approximation of progressive processes by predictable step processes.

Lemma 17.23 *(approximation)* *For any continuous semimartingale $X = M + A$ and process $V \in L(X)$, there exist some processes $V_1, V_2, \cdots \in \mathcal{E}$ such that a.s. $((V_n - V)^2 \cdot [M])_t \to 0$ and $((V_n - V) \cdot A)_t^* \to 0$ for every $t > 0$.*

Proof: It is enough to take $t = 1$, since we can then combine the processes V_n for disjoint finite intervals to construct an approximating sequence on $\mathbb{R}_+$. Furthermore, it suffices to consider approximations in the sense of convergence in probability, since the a.s. versions will then follow for a suitable subsequence. This allows us to perform the construction in steps, first approximating V by bounded and progressive processes V', next approximating each V' by continuous and adapted processes V'', and finally approximating each V'' by predictable step processes V'''.

Here the first and last steps are elementary, so we may concentrate on the second step. Then let V be bounded. We need to construct some continuous, adapted processes V_n such that $((V_n-V)^2 \cdot [M])_1 \to 0$ and $((V_n-V)\cdot A)_1^* \to 0$ a.s. Since the V_n can be taken to be uniformly bounded, we may replace the former condition by $(|V_n - V| \cdot [M])_1 \to 0$ a.s. Thus, it is enough to establish the approximation $(|V_n - V| \cdot A)_1 \to 0$ in the case when A is a nondecreasing, continuous, adapted process with $A_0 = 0$. Replacing A_t by $A_t + t$ if necessary, we may even assume that A is strictly increasing.

To construct the required approximations, we may introduce the inverse process $T_s = \sup\{t \geq 0;\, A_t \leq s\}$, and define

$$V_t^h = h^{-1} \int_{T(A_t - h)}^t V\, dA = h^{-1} \int_{(A_t - h)_+}^{A_t} V(T_s)\, ds, \quad t, h > 0.$$

By Theorem 2.15 we have $V^h \circ T \to V \circ T$ as $h \to 0$, a.e. on $[0, A_1]$. Thus, by dominated convergence,

$$\int_0^1 |V^h - V|\, dA = \int_0^{A_1} |V^h(T_s) - V(T_s)|\, ds \to 0.$$

The processes V^h are clearly continuous. To prove that they are also adapted, we note that the process $T(A_t - h)$ is adapted for every $h > 0$ by the definition of T. Since V is progressive, it is further seen that $V \cdot A$ is adapted and hence progressive. The adaptedness of $(V \cdot A)_{T(A_t - h)}$ now follows by composition. $\square$

Though the class $L(X)$ of stochastic integrands is sufficient for most purposes, it is sometimes useful to allow the integration of slightly more general processes. Given any continuous semimartingale $X = M + A$, let $\hat{L}(X)$ denote the class of product-measurable processes V such that $(V - \tilde{V}) \cdot [M] = 0$ and $(V - \tilde{V}) \cdot A = 0$ a.s. for some process $\tilde{V} \in L(X)$. For $V \in \hat{L}(X)$ we define $V \cdot X = \tilde{V} \cdot X$ a.s. The extension clearly enjoys all the previously established properties of stochastic integration.

It is often important to see how semimartingales, covariation processes, and stochastic integrals are transformed by a random time-change. Let us then consider a nondecreasing, right-continuous family of finite optional times τ_s, $s \geq 0$, here referred to as a *finite random time-change* τ. If even $\mathcal{F}$ is right-continuous, then by Lemma 7.3 the same thing is true for the *induced filtration* $\mathcal{G}_s = \mathcal{F}_{\tau_s}$, $s \geq 0$. A process X is said to be τ-*continuous* if it is a.s. continuous on $\mathbb{R}_+$ and constant on every interval $[\tau_{s-}, \tau_s]$, $s \geq 0$, where $\tau_{0-} = X_{0-} = 0$ by convention.

Theorem 17.24 *(random time-change, Kazamaki)* Let τ be a finite random time-change with induced filtration $\mathcal{G}$, and let $X = M + A$ be a τ-continuous $\mathcal{F}$-semimartingale. Then $X \circ \tau$ is a continuous $\mathcal{G}$-semimartingale with canonical decomposition $M \circ \tau + A \circ \tau$ and such that $[X \circ \tau] = [X] \circ \tau$ a.s. Furthermore, $V \in L(X)$ implies $V \circ \tau \in \hat{L}(X \circ \tau)$ and
$$(V \circ \tau) \cdot (X \circ \tau) = (V \cdot X) \circ \tau \quad a.s. \tag{14}$$

Proof: It is easy to check that the time-change $X \mapsto X \circ \tau$ preserves continuity, adaptedness, monotonicity, and the local martingale property. In particular, $X \circ \tau$ is then a continuous $\mathcal{G}$-semimartingale with canonical decomposition $M \circ \tau + A \circ \tau$. Since $M^2 - [M]$ is a continuous local martingale, the same thing is true for the time-changed process $M^2 \circ \tau - [M] \circ \tau$, and so
$$[X \circ \tau] = [M \circ \tau] = [M] \circ \tau = [X] \circ \tau \quad \text{a.s.}$$

If $V \in L(X)$, we also note that $V \circ \tau$ is product-measurable, since this is true for both V and τ.

Fixing any $t \geq 0$ and using the τ-continuity of X, we get
$$(1_{[0,t]} \circ \tau) \cdot (X \circ \tau) = 1_{[0,\tau_t^{-1}]} \cdot (X \circ \tau) = (X \circ \tau)^{\tau_t^{-1}} = (1_{[0,t]} \cdot X) \circ \tau,$$

which proves (14) when $V = 1_{[0,t]}$. If X has locally finite variation, the result extends by a monotone class argument and monotone convergence to arbitrary $V \in L(X)$. In general, Lemma 17.23 yields the existence of some continuous, adapted processes $V_1, V_2, \ldots$ such that $\int (V_n - V)^2 d[M] \to 0$ and $\int |(V_n - V) dA| \to 0$ a.s. By (14) the corresponding properties hold for the time-changed processes, and since the processes $V_n \circ \tau$ are right-continuous and adapted, hence progressive, we obtain $V \circ \tau \in \hat{L}(X \circ \tau)$.

Now assume instead that the approximating processes $V_1, V_2, \ldots$ are predictable step processes. The previous calculation then shows that (14) holds for each V_n, and by Lemma 17.12 the relation extends to V. $\square$

Let us next consider stochastic integrals of processes depending on a parameter. Given any measurable space $(S, \mathcal{S})$, we say that a process V on

$S \times \mathbb{R}_+$ is *progressive* if its restriction to $S \times [0, t]$ is $\mathcal{S} \otimes \mathcal{B}_t \otimes \mathcal{F}_t$-measurable for every $t \geq 0$, where $\mathcal{B}_t = \mathcal{B}([0, t])$. A simple version of the following result will be useful in Chapter 18.

Theorem 17.25 *(dependence on parameter, Doléans, Stricker and Yor)* *Let X be a continuous semimartingale, fix a measurable space S, and consider a progressive process $V_s(t)$, $s \in S$, $t \geq 0$, such that $V_s \in L(X)$ for every $s \in S$. Then the process $Y_s(t) = (V_s \cdot X)_t$ has a version that is progressive on $S \times \mathbb{R}_+$ and a.s. continuous for each $s \in S$.*

Proof: Let $M + A$ be the canonical decomposition of X. Assume the existence of some progressive processes V_s^n on $S \times \mathbb{R}_+$ such that, for any $t \geq 0$ and $s \in S$,

$$((V_s^n - V_s)^2 \cdot [M])_t \xrightarrow{P} 0, \qquad ((V_s^n - V_s) \cdot A)_t^* \xrightarrow{P} 0.$$

Then Lemma 17.12 yields $(V_s^n \cdot X - V_s \cdot X)_t^* \xrightarrow{P} 0$ for every s and t. Proceeding as in the proof of Proposition 4.31, we may choose a subsequence $(n_k(s)) \subset \mathbb{N}$, depending measurably on s, such that the same convergence holds a.s. along $(n_k(s))$ for any s and t. Define $Y_{s,t} = \limsup_k (V_s^{n_k} \cdot X)_t$ whenever this is finite, and put $Y_{s,t} = 0$ otherwise. If we can choose versions of the processes $(V_s^n \cdot X)_t$ that are progressive on $S \times \mathbb{R}_+$ and a.s. continuous for each s, then $Y_{s,t}$ is clearly a version of the process $(V_s \cdot X)_t$ with the same properties. This argument will now be applied in three steps.

First we reduce to the case of bounded and progressive integrands by taking $V^n = V 1\{|V| \leq n\}$. Next we apply the transformation in the proof of Lemma 17.23, to reduce to the case of continuous and progressive integrands. In the final step, we approximate any continuous, progressive process V by the predictable step processes $V_s^n(t) = V_s(2^{-n}[2^n t])$. Here the integrals $V_s^n \cdot X$ are elementary, and the desired continuity and measurability are obvious by inspection. □

We turn to the related topic of functional representations. To motivate the problem, note that the construction of the stochastic integral $V \cdot X$ depends in a subtle way on the underlying probability measure P and filtration $\mathcal{F}$. Thus, we cannot expect any universal representation $F(V, X)$ of the integral process $V \cdot X$. In view of Proposition 4.31, one might still hope for a modified representation $F(\mu, V, X)$, where μ denotes the distribution of (V, X). Even this could be too optimistic, however, since the canonical decomposition of X may also depend on $\mathcal{F}$.

Dictated by our needs in Chapter 21, we restrict our attention to a very special situation, which is still general enough to cover most applications of interest. Fixing any progressive functions σ_j^i and b^i of suitable dimension, defined on the path space $C(\mathbb{R}_+, \mathbb{R}^d)$, we may consider an arbitrary adapted process X satisfying the stochastic differential equation

$$dX_t^i = \sigma_j^i(t, X) dB_t^j + b^i(t, X) dt, \qquad (15)$$

where B is a Brownian motion in $\mathbb{R}^r$. A detailed discussion of such equations is given in Chapter 21. For the moment, we need only the simple fact from Lemma 21.1 that the coefficients $\sigma_j^i(t, X)$ and $b^i(t, X)$ are again progressive. Write $a^{ij} = \sigma_k^i \sigma_k^j$.

Proposition 17.26 *(functional representation)* For any progressive functions σ, b, and f of suitable dimension, there exists a measurable mapping

$$F: \mathcal{P}(C(\mathbb{R}_+, \mathbb{R}^d)) \times C(\mathbb{R}_+, \mathbb{R}^d) \to C(\mathbb{R}_+, \mathbb{R}) \qquad (16)$$

such that, whenever X is a solution to (15) with $\mathcal{L}(X) = \mu$ and $f^i(X) \in L(X^i)$ for all i, we have $f^i(X) \cdot X^i = F(\mu, X)$ a.s.

Proof: From (15) we note that X is a semimartingale with covariation processes $[X^i, X^j] = a^{ij}(X) \cdot \lambda$ and drift components $b^i(X) \cdot \lambda$. Hence, $f^i(X) \in L(X^i)$ for all i iff the processes $(f^i)^2 a^{ii}(X)$ and $f^i b^i(X)$ are a.s. Lebesgue integrable. Note that this holds in particular when f is bounded. Now assume that $f_1, f_2, \ldots$ are progressive with

$$(f_n^i - f^i)^2 a^{ii}(X) \cdot \lambda \to 0, \qquad |(f_n^i - f^i) b^i(X)| \cdot \lambda \to 0. \qquad (17)$$

Then $(f_n^i(X) \cdot X^i - f^i(X) \cdot X^i)_t^* \xrightarrow{P} 0$ for every $t \geq 0$ by Lemma 17.12. Thus, if $f_n^i(X) \cdot X^i = F_n(\mu, X)$ a.s. for some measurable mappings F_n as in (16), then Proposition 4.31 yields a similar representation for the limit $f^i(X) \cdot X^i$.

As in the preceding proof, we may apply this argument in three steps, reducing first to the case when f is bounded, next to the case of continuous f, and finally to the case when f is a predictable step function. Here the first and last steps are again elementary. For the second step, we may now use the simpler approximation

$$f_n(t, x) = n \int_{(t-n^{-1})_+}^t f(s, x) ds, \quad t \geq 0, \ n \in \mathbb{N}, \ x \in C(\mathbb{R}_+, \mathbb{R}^d).$$

By Theorem 2.15 we have $f_n(t, x) \to f(t, x)$ a.e. in t for each $x \in C(\mathbb{R}_+, \mathbb{R}^d)$, and (17) follows by dominated convergence. $\square$

Exercises

1. Show that if M is a local martingale and ξ is an $\mathcal{F}_0$-measurable random variable, then the process $N_t = \xi M_t$ is again a local martingale.

2. Use Fatou's lemma to show that every local martingale $M \geq 0$ with $EM_0 < \infty$ is a supermartingale. Also show by an example that M may fail to be a martingale. (*Hint:* Let $M_t = X_{t/(1-t)_+}$, where X is a Brownian motion starting at 1, stopped when it reaches 0.)

3. Fix a continuous local martingale M. Show that M and $[M]$ have a.s. the same intervals of constancy. (*Hint:* For any $r \in \mathbb{Q}_+$, put $\tau = \inf\{t > r;\ [M]_t > [M]_r\}$. Then M^τ is a continuous local martingale on $[r,\infty)$ with quadratic variation 0, so M^τ is a.s. constant on $[s,\tau]$. Use a similar argument in the other direction.)

4. For any continuous local martingales M_n starting at 0 and associated optional times τ_n, show that $(M_n)^*_{\tau_n} \xrightarrow{P} 0$ iff $[M_n]_{\tau_n} \xrightarrow{P} 0$. State the corresponding result for stochastic integrals.

5. Show that there exist some continuous semimartingales $X_1, X_2, \ldots$ such that $X_n^* \xrightarrow{P} 0$ and yet $[X_n]_t \not\xrightarrow{P} 0$ for all $t > 0$. (*Hint:* Let B be a Brownian motion stopped at time 1, put $A_{k2^{-n}} = B_{(k-1)+2^{-n}}$, and interpolate linearly. Define $X^n = B - A^n$.)

6. Consider a Brownian motion B and an optional time τ. Show that $EB_\tau = 0$ when $E\tau^{1/2} < \infty$ and that $EB_\tau^2 = E\tau$ when $E\tau < \infty$. (*Hint:* Use optional sampling and Theorem 17.7.)

7. Deduce the first inequality in Proposition 17.9 from Proposition 17.17 and the classical Cauchy–Buniakovsky inequality.

8. Prove for any continuous semimartingales X and Y that $[X+Y]^{1/2} \leq [X]^{1/2} + [Y]^{1/2}$ a.s.

9. (Kunita and Watanabe) Let M and N be continuous local martingales, and fix any $p, q, r > 0$ with $p^{-1} + q^{-1} = r^{-1}$. Show that $\|[M,N]_t\|_{2r}^2 \leq \|[M]_t\|_p \|[N]_t\|_q$ for all $t > 0$.

10. Let M, N be continuous local martingales with $M_0 = N_0 = 0$. Show that $M \perp\!\!\!\perp N$ implies $[M,N] \equiv 0$ a.s. Also show by an example that the converse is false. (*Hint:* Let $M = U \cdot B$ and $N = V \cdot B$ for a Brownian motion B and suitable $U, V \in L(B)$.)

11. Fix a continuous semimartingale X, and let $U, V \in L(X)$ with $U = V$ a.s. on some set $A \in \mathcal{F}_0$. Show that $U \cdot X = V \cdot X$ a.s. on A. (*Hint:* Use Proposition 17.15.)

12. Fix a continuous local martingale M, and let $U, U_1, U_2, \ldots$ and $V, V_1, V_2, \ldots \in L(M)$ with $|U_n| \leq V_n$, $U_n \to U$, $V_n \to V$, and $((V_n - V) \cdot M)_t^* \xrightarrow{P} 0$ for all $t > 0$. Show that $(U_n \cdot M)_t \xrightarrow{P} (U \cdot M)_t$ for all t. (*Hint:* Write $(U_n - U)^2 \leq 2(V_n - V)^2 + 8V^2$, and use Theorem 1.21 and Lemmas 4.2 and 17.12.)

13. Let B be a Brownian bridge. Show that $X_t = B_{t \wedge 1}$ is a semimartingale on $\mathbb{R}_+$ w.r.t. the induced filtration. (*Hint:* Note that $M_t = (1-t)^{-1} B_t$ is a martingale on $[0,1)$, integrate by parts, and check that the compensator has finite variation.)

14. Show by an example that the canonical decomposition of a continuous semimartingale may depend on the filtration. (*Hint:* Let B be Brownian motion with induced filtration $\mathcal{F}$, put $\mathcal{G}_t = \mathcal{F}_t \vee \sigma(B_1)$, and use the preceding result.)

15. Show by stochastic calculus that $t^{-p}B_t \to 0$ a.s. as $t \to \infty$, where B is a Brownian motion and $p > \frac{1}{2}$. (*Hint:* Integrate by parts to find the canonical decomposition. Compare with the L^1-limit.)

16. Extend Theorem 17.16 to a product of n semimartingales.

17. Consider a Brownian bridge X and a bounded, progressive process V with $\int_0^1 V_t dt = 0$ a.s. Show that $E\int_0^1 VdX = 0$. (*Hint:* Integrate by parts to get $\int_0^1 VdX = \int_0^1 (V-U)dB$, where B is a Brownian motion and $U_t = (1-t)^{-1}\int_t^1 V_s ds$.)

18. Show that Proposition 17.17 remains valid for any finite optional times t and t_{nk} satisfying $\max_k(t_{nk} - t_{n,k-1}) \xrightarrow{P} 0$.

19. Let M be a continuous local martingale. Find the canonical decomposition of $|M|^p$ when $p \geq 2$, and deduce for such a p the second relation in Theorem 17.7. (*Hint:* Use Theorem 17.18. For the last part, use Hölder's inequality.)

20. Let M be a continuous local martingale with $M_0 = 0$ and $[M]_\infty \leq 1$. Show for any $r \geq 0$ that $P\{\sup_t M_t \geq r\} \leq e^{-r^2/2}$. (*Hint:* Consider the supermartingale $Z = \exp(cM - c^2[M]/2)$ for a suitable $c > 0$.)

21. Let X and Y be continuous semimartingales. Fix a $t > 0$ and a sequence of partitions (t_{nk}) of $[0, t]$ with $\max_k(t_{nk} - t_{n,k-1}) \to 0$. Show that $\frac{1}{2}\sum_k(Y_{t_{nk}} + Y_{t_{n,k-1}})(X_{t_{nk}} - X_{t_{n,k-1}}) \xrightarrow{P} (Y \circ X)_t$. (*Hint:* Use Corollary 17.13 and Proposition 17.17.)

22. Show that the Fisk–Stratonovich integral satisfies the chain rule $U \circ (V \circ X) = (UV) \circ X$. (*Hint:* Reduce to Itô integrals and use Theorems 17.11 and 17.16 and Proposition 17.14.)

23. A process is *predictable* if it is measurable with respect to the σ-field in $\mathbb{R}_+ \times \Omega$ induced by all predictable step processes. Show that every predictable process is progressive. Conversely, given a progressive process X and a constant $h > 0$, show that the process $Y_t = X_{(t-h)_+}$ is predictable.

24. Given a progressive process V and a nondecreasing, continuous, adapted process A, show that there exists some predictable process $\tilde{V}$ with $|V - \tilde{V}| \cdot A = 0$ a.s. (*Hint:* Use Lemma 17.23.)

25. Given the preceding statement, deduce Lemma 17.23. (*Hint:* Begin with predictable V, using a monotone class argument.)

26. Construct the stochastic integral $V \cdot M$ by approximation from elementary integrals, using Lemmas 17.10 and 17.23. Show that the resulting integral satisfies the relation in Theorem 17.11. (*Hint:* First let $M \in \mathcal{M}^2$ and $E(V^2 \cdot [M])_\infty < \infty$, and extend by localization.)

27. Let $(V, B) \stackrel{d}{=} (\tilde{V}, \tilde{B})$, where B and $\tilde{B}$ are Brownian motions on possibly different filtered probability spaces and $V \in L(B)$, $\tilde{V} \in L(\tilde{B})$. Show that $(V, B, V \cdot B) \stackrel{d}{=} (\tilde{V}, \tilde{B}, \tilde{V} \cdot \tilde{B})$. (*Hint:* Argue as in the proof of Proposition 17.26.)

28. Let X be a continuous $\mathcal{F}$-semimartingale. Show that X remains a semimartingale conditionally on $\mathcal{F}_0$, and that the conditional quadratic variation agrees with $[X]$. Also show that if $V \in L(X)$, where $V = \sigma(Y)$ for some continuous process Y and measurable function σ, then V remains conditionally X-integrable, and the conditional integral agrees with $V \cdot X$. (*Hint:* Conditioning on $\mathcal{F}_0$ preserves martingales.)

Chapter 18

Continuous Martingales and Brownian Motion

Real and complex exponential martingales; martingale characterization of Brownian motion; random time-change of martingales; integral representation of martingales; iterated and multiple integrals; change of measure and Girsanov's theorem; Cameron–Martin theorem; Wald's identity and Novikov's condition

This chapter deals with a wide range of applications of the stochastic calculus, the principal tools of which were introduced in the preceding chapter. A recurrent theme is the notion of exponential martingales, which appear in both a real and a complex variety. Exploring the latter yields an effortless approach to Lévy's celebrated martingale characterization of Brownian motion as well as to the basic random time-change reduction of isotropic continuous local martingales to a Brownian motion. By applying the latter result to suitable compositions of Brownian motion with harmonic or analytic functions, we may deduce some important information about Brownian motion in $\mathbb{R}^d$. Similar methods can be used to analyze a variety of other transformations that lead to Gaussian processes.

As a further application of the exponential martingales, we shall derive stochastic integral representations of Brownian functionals and martingales and examine their relationship to the chaos expansions obtained by different methods in Chapter 13. In this context, we show how the previously introduced multiple Wiener–Itô integrals can be expressed as iterated single Itô integrals. A similar problem, of crucial importance for Chapter 21, is to represent a continuous local martingale with absolutely continuous covariation processes in terms of stochastic integrals with respect to a suitable Brownian motion.

Our last main topic is to examine the transformations induced by an absolutely continuous change of probability measure. The density process turns out to be a real exponential martingale, and any continuous local martingale in the original setting will remain a martingale under the new measure, apart from an additional drift term. The observation is useful for applications, where it is often employed to remove the drift from a given semimartingale. The appropriate change of measure then depends on the process, and it becomes important to derive effective criteria for a proposed exponential process to be a true martingale.

Our present exposition may be regarded as a continuation of the discussion of martingales and Brownian motion from Chapters 7 and 13, respectively. Changes of time and measure are both important for the theory of stochastic differential equations, as developed in Chapters 21 and 23. The time-change results for continuous martingales have a counterpart for point processes explored in Chapter 25, where general Poisson processes play a role similar to that of the Gaussian processes here. The results about changes of measure are extended in Chapter 26 to the context of possibly discontinuous semimartingales.

To elaborate on the new ideas, we begin with an introduction of complex exponential martingales. It is instructive to compare them with the real versions appearing in Lemma 18.21.

Lemma 18.1 *(complex exponential martingales)* Let M be a real continuous local martingale with $M_0 = 0$. Then
$$Z_t = \exp(iM_t + \tfrac{1}{2}[M]_t), \quad t \geq 0,$$
is a complex local martingale satisfying $Z_t = 1 + i(Z \cdot M)_t$ a.s.

Proof: Applying Corollary 17.20 to the complex-valued semimartingale $X_t = iM_t + \tfrac{1}{2}[M]_t$ and the entire function $f(z) = e^z$, we get
$$dZ_t = Z_t(dX_t + \tfrac{1}{2}d[X]_t) = Z_t(idM_t + \tfrac{1}{2}d[M]_t - \tfrac{1}{2}d[M]_t) = iZ_t dM_t. \quad \square$$

The next result gives the basic connection between continuous martingales and Gaussian processes. For any subset K of a Hilbert space, we write $\hat{K}$ for the closed linear subspace generated by K.

Lemma 18.2 *(isometries and Gaussian processes)* Given a subset K of a Hilbert space H, consider for each $h \in K$ a continuous local $\mathcal{F}$-martingale M^h with $M_0^h = 0$ such that
$$[M^h, M^k]_\infty = \langle h, k \rangle \quad \text{a.s.}, \quad h, k \in K. \tag{1}$$
Then there exists an isonormal Gaussian process $\eta \perp\!\!\!\perp \mathcal{F}_0$ on $\hat{K}$ such that $M_\infty^h = \eta h$ a.s. for all $h \in K$.

Proof: Fix any linear combination $N_t = u_1 M_t^{h_1} + \cdots + u_n M_t^{h_n}$, and conclude from (1) that
$$[N]_\infty = \sum_{j,k} u_j u_k [M^{h_j}, M^{h_k}]_\infty = \sum_{j,k} u_j u_k \langle h_j, h_k \rangle = \|h\|^2,$$
where $h = u_1 h_1 + \cdots + u_n h_n$. The process $Z = \exp(iN + \tfrac{1}{2}[N])$ is a.s. bounded, and so by Lemma 18.1 it is a uniformly integrable martingale. Writing $\xi = N_\infty$, we hence obtain for any $A \in \mathcal{F}_0$
$$PA = E[Z_\infty; A] = E[\exp(iN_\infty + \tfrac{1}{2}[N]_\infty); A] = E[e^{i\xi}; A]e^{\|h\|^2/2}.$$

Since $u_1, \ldots, u_n$ were arbitrary, we conclude from the uniqueness theorem for characteristic functions that the random vector $(M_\infty^{h_1}, \ldots, M_\infty^{h_n})$ is independent of $\mathcal{F}_0$ and centered Gaussian with covariances $\langle h_j, h_k\rangle$. It is now easy to construct a process η with the stated properties. □

As a first application, we may establish the following basic martingale characterization of Brownian motion.

Theorem 18.3 *(characterization of Brownian motion, Lévy)* Let $B = (B^1, \ldots, B^d)$ be a process in $\mathbb{R}^d$ with $B_0 = 0$. Then B is an $\mathcal{F}$-Brownian motion iff it is a continuous local $\mathcal{F}$-martingale with $[B^i, B^j]_t \equiv \delta_{ij} t$ a.s.

Proof: For fixed $s < t$, we may apply Lemma 18.2 to the continuous local martingales $M_r^i = B_{r\wedge t}^i - B_{r\wedge s}^i$, $r \geq s$, $i = 1, \ldots, d$, to see that the differences $B_t^i - B_s^i$ are i.i.d. $N(0, t - s)$ and independent of $\mathcal{F}_s$. □

The last theorem suggests the possibility of transforming an arbitrary continuous local martingale M into a Brownian motion through a suitable random time-change. The proposed result is indeed true and admits a natural extension to higher dimensions; for convenience, we consider directly the version in $\mathbb{R}^d$. A continuous local martingale $M = (M^1, \ldots, M^d)$ is said to be *isotropic* if a.s. $[M^i] = [M^j]$ and $[M^i, M^j] = 0$ for all $i \neq j$. Note in particular that this holds for Brownian motion in $\mathbb{R}^d$. When M is a continuous local martingale in $\mathbb{C}$, the condition is clearly equivalent to $[M] = 0$ a.s., or $[\Re M] = [\Im M]$ and $[\Re M, \Im M] = 0$ a.s. For isotropic processes M, we refer to $[M^1] = \cdots = [M^d]$ or $[\Re M] = [\Im M]$ as the *rate process* of M.

The proof is straightforward when $[M]_\infty = \infty$ a.s., but in general it requires a rather subtle extension of the filtered probability space. To simplify our statements, we assume the existence of any requested randomization variables. This can always be achieved, as in the elementary context of Chapter 6, by passing from the original setup $(\Omega, \mathcal{A}, \mathcal{F}, P)$ to the product space $(\hat{\Omega}, \hat{\mathcal{A}}, \hat{\mathcal{F}}, \hat{P})$, where $\hat{\Omega} = \Omega \times [0, 1]$, $\hat{\mathcal{A}} = \mathcal{A} \otimes \mathcal{B}$, $\hat{\mathcal{F}}_t = \mathcal{F}_t \times [0, 1]$, and $\hat{P} = P \otimes \lambda$. Given two filtrations $\mathcal{F}$ and $\mathcal{G}$ on Ω, we say that $\mathcal{G}$ is a *standard extension* of $\mathcal{F}$ if $\mathcal{F}_t \subset \mathcal{G}_t \perp\!\!\!\perp_{\mathcal{F}_t} \mathcal{F}$ for all $t \geq 0$. This is precisely the condition needed to ensure that all adaptedness and conditioning properties will be preserved. The notion is still flexible enough to admit a variety of useful constructions.

Theorem 18.4 *(time-change reduction, Dambis, Dubins and Schwarz)* Let M be an isotropic continuous local $\mathcal{F}$-martingale in $\mathbb{R}^d$ with $M_0 = 0$, and define
$$\tau_s = \inf\{t \geq 0;\ [M^1]_t > s\}, \quad \mathcal{G}_s = \mathcal{F}_{\tau_s}, \quad s \geq 0.$$
Then there exists in $\mathbb{R}^d$ a Brownian motion B with respect to a standard extension of $\mathcal{G}$, such that a.s. $B = M \circ \tau$ on $[0, [M^1]_\infty)$ and $M = B \circ [M^1]$.

Proof: We may take $d = 1$, the proof in higher dimensions being similar. Introduce a Brownian motion $X \perp\!\!\!\perp \mathcal{F}$ with induced filtration $\mathcal{X}$, and put $\hat{\mathcal{G}}_t = \mathcal{G}_t \vee \mathcal{X}_t$. Since $\mathcal{G} \perp\!\!\!\perp \mathcal{X}$, it is clear that $\hat{\mathcal{G}}$ is a standard extension of both

$\mathcal{G}$ and $\mathcal{X}$. In particular, X remains a Brownian motion under $\hat{\mathcal{G}}$. Now define

$$B_s = M_{\tau_s} + \int_0^s 1\{\tau_r = \infty\} dX_r, \quad s \geq 0. \qquad (2)$$

Since M is τ-continuous by Proposition 17.6, Theorem 17.24 shows that the first term $M \circ \tau$ is a continuous $\mathcal{G}$-martingale, hence also a $\hat{\mathcal{G}}$-martingale, with quadratic variation

$$[M \circ \tau]_s = [M]_{\tau_s} = s \wedge [M]_\infty, \quad s \geq 0.$$

The second term in (2) has quadratic variation $s - s \wedge [M]_\infty$, and the covariation vanishes since $M \circ \tau \perp\!\!\!\perp X$. Thus, $[B]_s = s$ a.s., and so Theorem 18.3 shows that B is a $\hat{\mathcal{G}}$-Brownian motion. Finally, $B_s = M_{\tau_s}$ for $s < [M]_\infty$, which implies $M = B \circ [M]$ a.s. by the τ-continuity of M. □

In two dimensions, isotropic martingales arise naturally through the composition of a complex Brownian motion B with an arbitrary (possibly multi-valued) analytic function f. For a general continuous process X, we may clearly choose a continuous evolution of $f(X)$, as long as X avoids the possible singularities of f. Similar results are available for harmonic functions, which is especially useful in dimensions $d \geq 3$, when no analytic functions exist.

Theorem 18.5 *(harmonic and analytic maps, Lévy)*

(i) *Let M be an isotropic, continuous local martingale in $\mathbb{R}^d$, and fix an harmonic function f such that M a.s. avoids the sigularities of f. Then $f(M)$ is a local martingale with $[f(M)] = |\nabla f(M)|^2 \cdot [M^1]$.*

(ii) *Let M be a complex, isotropic, continuous local martingale, and fix an analytic function f such that M a.s. avoids the singularities of f. Then $f(M)$ is again an isotropic local martingale, and $[\Re f(M)] = |f'(M)|^2 \cdot [\Re M]$. If B is a Brownian motion and $f' \not\equiv 0$, then $[\Re f(B)]$ is a.s. unbounded and strictly increasing.*

Proof: (i) Using the isotropy of M, we get by Corollary 17.19

$$f(M) = f(M_0) + f'_i \cdot M^i + \tfrac{1}{2} \Delta f(M) \cdot [M^1].$$

Here the last term vanishes since f is harmonic, and so $f(M)$ is a local martingale. From the isotropy of M we further obtain

$$[f(M)] = \sum_i [f'_i(M) \cdot M^i] = \sum_i (f'_i(M))^2 \cdot [M^1] = |\nabla f(M)|^2 \cdot [M^1].$$

(ii) Since f is analytic, we get by Corollary 17.20

$$f(M) = f(M_0) + f'(M) \cdot M + \tfrac{1}{2} f''(M) \cdot [M]. \qquad (3)$$

Here the last term vanishes since M is isotropic. The same property also yields

$$[f(M)] = [f'(M) \cdot M] = (f'(M))^2 \cdot [M] = 0,$$

and so $f(M)$ is again isotropic. Finally, writing $M = X + iY$ and $f'(M) = U + iV$, we get

$$[\Re f(M)] = [U \cdot X - V \cdot Y] = (U^2 + V^2) \cdot [X] = |f'(M)|^2 \cdot [\Re M].$$

If f' is not identically 0, it has at most countably many zeros. Hence, by Fubini's theorem

$$E\lambda\{t \geq 0; f'(B_t) = 0\} = \int_0^\infty P\{f'(B_t) = 0\}dt = 0,$$

and so $[\Re f(B)] = |f'(B)|^2 \cdot \lambda$ is a.s. strictly increasing. To see that it is also a.s. unbounded, we note that $f(B)$ converges a.s. on the set $\{[\Re f(B)] < \infty\}$. However, $f(B)$ diverges a.s. since f is nonconstant and the random walk $B_0, B_1, \ldots$ is recurrent by Theorem 9.2. $\square$

Combining the last two results, we may derive two basic properties of Brownian motion in $\mathbb{R}^d$, namely the polarity of singleton sets when $d \geq 2$ and the transience when $d \geq 3$. Note that the latter property is a continuous-time counterpart of Theorem 9.8 for random walks. Both properties play important roles for the potential theory developed in Chapter 24. Define $\tau_a = \inf\{t > 0; B_t = a\}$.

Theorem 18.6 (*point polarity and transience, Lévy, Kakutani*) *For a Brownian motion B in $\mathbb{R}^d$, we have the following:*

(i) *If $d \geq 2$, then $\tau_a = \infty$ a.s. for all $a \in \mathbb{R}^d$.*

(ii) *If $d \geq 3$, then $|B_t| \to \infty$ a.s. as $t \to \infty$.*

Proof: (i) Here we may clearly take $d = 2$, so we may let B be a complex Brownian motion. Applying Theorem 18.5 (ii) to the entire function e^z, it is seen that $M = e^B$ is a conformal local martingale with unbounded rate $[\Re M]$. By Theorem 18.4 we have $M - 1 = X \circ [\Re M]$ a.s. for some Brownian motion X, and since $M \neq 0$ it follows that X a.s. avoids -1. Hence, $\tau_{-1} = \infty$ a.s., and by the scaling and rotational symmetries of B we get $\tau_a = \infty$ a.s. for every $a \neq 0$. To extend the result to $a = 0$, we may conclude from the Markov property at $h > 0$ that

$$P_0\{\tau_0 \circ \theta_h < \infty\} = E_0 P_{B_h}\{\tau_0 < \infty\} = 0, \quad h > 0.$$

As $h \to 0$, we get $P_0\{\tau_0 < \infty\} = 0$, and so $\tau_0 = \infty$ a.s.

(ii) Here we may take $d = 3$. For any $a \neq 0$ we have $\tau_a = \infty$ a.s. by claim (i), and so by Theorem 18.5 (i) the process $M = |B - a|^{-1}$ is a continuous local martingale. By Fatou's lemma M is then an L^1-bounded supermartingale, and so by Theorem 7.18 it converges a.s. toward some random variable ξ. Since $M_t \overset{d}{\to} 0$ we have $\xi = 0$ a.s. $\square$

Combining part (i) of the last result with Theorem 19.11, we note that a complex, isotropic continuous local martingale avoids every fixed point outside the origin. Thus, Theorem 18.5 (ii) applies to any analytic function f with only isolated singularities. Since f is allowed to be multi-valued, the result applies even to functions with essential singularities, such as to $f(z) = \log(1+z)$. For a simple application, we may consider the windings of planar Brownian motion around a fixed point.

Corollary 18.7 *(skew-product representation, Galmarino)* *Let B be a complex Brownian motion starting at 1, and choose a continuous version of $V = \arg B$ with $V_0 = 0$. Then $V_t \equiv Y \circ (|B|^{-2} \cdot \lambda)_t$ a.s. for some real Brownian motion $Y \perp\!\!\!\perp |B|$.*

Proof: Applying Theorem 18.5 (ii) with $f(z) = \log(1+z)$, we note that $M_t = \log|B_t| + iV_t$ is an isotropic martingale with rate $[\Re M] = |B|^{-2} \cdot \lambda$. Hence, by Theorem 18.4 there exists some complex Brownian motion $Z = X + iY$ with $M = Z \circ [\Re M]$ a.s., and the assertion follows. □

For a nonisotropic continuous local martingale M in $\mathbb{R}^d$, there is no single random time-change that will reduce the process to a Brownian motion. However, we may transform each component M^i separately, as in Theorem 18.4, to obtain a collection of one-dimensional Brownian motions $B^1, \ldots, B^d$. If the latter processes happen to be independent, they may clearly be combined into a d-dimensional Brownian motion $B = (B^1, \ldots, B^d)$. It is remarkable that the required independence arises automatically whenever the original components M^i are *strongly orthogonal*, in the sense that $[M^i, M^j] = 0$ a.s. for all $i \neq j$.

Proposition 18.8 *(orthogonality and independence, Knight)* *Let $M^1, M^2, \ldots$ be strongly orthogonal, continuous local martingales starting at 0. Then there exist some independent Brownian motions $B^1, B^2, \ldots$ such that $M^k = B^k \circ [M^k]$ a.s. for every k.*

Proof: When $[M^k]_\infty = \infty$ a.s. for all k, the result is an easy consequence of Lemma 18.2. In general, we may introduce a sequence of independent Brownian motions $X^1, X^2, \ldots \perp\!\!\!\perp \mathcal{F}$ with induced filtration $\mathcal{X}$. Define

$$B^k_s = M^k(\tau^k_s) + X^k((s - [M^k]_\infty)_+), \quad s \geq 0, \ k \in \mathbb{N},$$

write $\psi_t = -\log(1-t)_+$, and put $\mathcal{G}_t = \mathcal{F}_{\psi_t} + \mathcal{X}_{(t-1)_+}$, $t \geq 0$. To check that $B^1, B^2, \ldots$ have the desired joint distribution, we may clearly assume the $[M^k]$ to be bounded. Then the processes $N^k_t = M^k_{\psi_t} + X^k_{(t-1)_+}$ are strongly orthogonal, continuous $\mathcal{G}$-martingales with quadratic variations $[N^k]_t = [M^k]_{\psi_t} + (t-1)_+$, and we note that $B^k_s = N^k_{\sigma^k_s}$, where $\sigma^k_s = \inf\{t \geq 0; [N^k]_t > s\}$. The assertion now follows from the result for $[M^k]_\infty = \infty$ a.s. □

As a further application of Lemma 18.2, we consider a simple continuous-time version of Theorem 11.13. Given a continuous semimartingale X on $I = \mathbb{R}_+$ or $[0,1)$ and a progressive process T on I that takes values in $\bar{I} = [0,\infty]$ or $[0,1]$, respectively, we may define

$$(X \circ T^{-1})_t = \int_I 1\{T_s \leq t\} dX_s, \quad t \in I,$$

as long as the integrals on the right exist. For motivation, we note that if ξ is a random measure on I with "distribution function" $X_t = \xi[0,t]$, $t \in I$, then $X \circ T^{-1}$ is the distribution function of the transformed measure $\xi \circ T^{-1}$.

Proposition 18.9 *(measure-preserving progressive maps)* Let B be a Brownian motion or bridge on $I = \mathbb{R}_+$ or $[0,1]$, respectively, and let T be a progressive process on I such that $\lambda \circ T^{-1} = \lambda$ a.s. Then $B \circ T^{-1} \stackrel{d}{=} B$.

Proof: The result for $I = \mathbb{R}_+$ is an immediate consequence of Lemma 18.2, and so we may assume that B is a Brownian bridge on $[0,1]$. Then $M_t = B_t/(1-t)$ is a martingale on $[0,1)$, and therefore B is a semimartingale on the same interval. Integrating by parts gives

$$dB_t = (1-t)dM_t - M_t dt \equiv dX_t - M_t dt. \tag{4}$$

Thus, $[X]_t = [B]_t = t$ a.s. for all t, and X is a Brownian motion by Theorem 18.3.

Now let V be a bounded, progressive process on $[0,1]$ such that the integral $\overline{V} = \int_0^1 V_t dt$ is a.s. nonrandom. Integrating by parts, we get for any $u \in [0,1)$

$$\int_0^u V_t M_t dt = M_u \int_0^u V_t dt - \int_0^u dM_t \int_0^t V_s ds$$
$$= \int_0^u dM_t \int_t^1 V_s ds - M_u \int_u^1 V_t dt.$$

As $u \to 1$, we have $(1-u)M_u = B_u \to 0$, and so the last term tends to 0. Hence, by dominated convergence and (4),

$$\int_0^1 V_t dB_t = \int_0^1 V_t dX_t - \int_0^1 V_t M_t dt = \int_0^1 (V_t - \overline{V}_t) dX_t,$$

where $\overline{V}_t = (1-t)^{-1}\int_t^1 V_s ds$. If U is another bounded, progressive process, we get by a simple calculation

$$\int_0^1 (U_t - \overline{U}_t)(V_t - \overline{V}_t) dt = \int_0^1 U_t V_t dt - \overline{U}\,\overline{V}.$$

For $U_r = 1\{T_r \leq s\}$ and $V_r = 1\{T_r \leq t\}$, the right-hand side becomes $s \wedge t - st = E(B_s B_t)$, and the assertion follows by Lemma 18.2. $\square$

We turn to a basic representation of martingales with respect to a Brownian filtration.

Theorem 18.10 *(Brownian martingales) Let $\mathcal{F}$ be the complete filtration induced by a Brownian motion $B = (B^1, \ldots, B^d)$ in $\mathbb{R}^d$. Then any local $\mathcal{F}$-martingale M is a.s. continuous, and there exist some $(P \times \lambda)$-a.e. unique processes $V^1, \ldots, V^d \in L(B^1)$ such that*

$$M = M_0 + \sum_{k \leq d} V^k \cdot B^k \quad a.s. \tag{5}$$

The statement is essentially equivalent to the following representation of Brownian functionals, which we prove first.

Lemma 18.11 *(Brownian functionals, Itô) Let $B = (B^1, \ldots, B^d)$ be a Brownian motion in $\mathbb{R}^d$, and fix any B-measurable random variable $\xi \in L^2$ with $E\xi = 0$. Then there exist some $(P \times \lambda)$-a.e. unique processes $V^1, \ldots, V^d \in L(B^1)$ such that $\xi = \sum_k (V^k \cdot B^k)_\infty$ a.s.*

Proof (Dellacherie): Let H denote the Hilbert space of B-measurable random variables $\xi \in L^2$ with $E\xi = 0$, and write K for the subspace of elements ξ admitting the desired representation $\sum_k (V^k \cdot B^k)_\infty$. For such a ξ we get $E\xi^2 = E\sum_k ((V^k)^2 \cdot \lambda)_\infty$, which implies the asserted uniqueness. By the obvious completeness of $L(B^1)$, it is further seen from the same formula that K is closed. To obtain $K = H$, we need to show that any $\xi \in H \ominus K$ vanishes a.s.

Then fix any nonrandom functions $u^1, \ldots, u^d \in L^2(\mathbb{R})$. Put $M = \sum_k u^k \cdot B^k$, and define the process Z as in Lemma 18.1. Then $Z - 1 = iZ \cdot M = i\sum_k (Zu^k) \cdot B^k$ by Proposition 17.14, and so $\xi \perp (Z_\infty - 1)$, or $E\xi \exp\{i\sum_k (u^k \cdot B^k)_\infty\} = 0$. Specializing to step functions u^k and using the uniqueness theorem for characteristic functions, we get

$$E[\xi; (B_{t_1}, \ldots, B_{t_n}) \in C] = 0, \quad t_1, \ldots, t_n \in \mathbb{R}_+, \ C \in \mathcal{B}^n, \ n \in \mathbb{N}.$$

By a monotone class argument this extends to $E[\xi; A] = 0$ for arbitrary $A \in \mathcal{F}_\infty$, and so $\xi = E[\xi|\mathcal{F}_\infty] = 0$ a.s. □

Proof of Theorem 18.10: We may clearly take $M_0 = 0$, and by suitable localization we may assume that M is uniformly integrable. Then M_∞ exists in $L^1(\mathcal{F}_\infty)$ and may be approximated in L^1 by some random variables $\xi_1, \xi_2, \cdots \in L^2(\mathcal{F}_\infty)$. The martingales $M^n_t = E[\xi_n|\mathcal{F}_t]$ are a.s. continuous by Lemma 18.11, and by Proposition 7.15 we get, for any $\varepsilon > 0$,

$$P\{(\Delta M)^* > 2\varepsilon\} \leq P\{(M^n - M)^* > \varepsilon\} \leq \varepsilon^{-1} E|\xi_n - M_\infty| \to 0.$$

Hence, $(\Delta M)^* = 0$ a.s., and so M is a.s. continuous. The remaining assertions now follow by localization from Lemma 18.11. □

Our next theorem deals with the converse problem of finding a Brownian motion B satisfying (5) when the representing processes V^k are given. The result plays a crucial role in Chapter 21.

Theorem 18.12 *(integral representation, Doob) Let M be a continuous local $\mathcal{F}$-martingale in $\mathbb{R}^d$ with $M_0 = 0$ such that $[M^i, M^j] = V_k^i V_k^j \cdot \lambda$ a.s. for some $\mathcal{F}$-progressive processes V_k^i, $1 \le i \le d$, $1 \le k \le n$. Then there exists in $\mathbb{R}^d$ a Brownian motion B with respect to a standard extension of $\mathcal{F}$ such that $M^i = V_k^i \cdot B^k$ a.s. for all i.*

Proof: For any $t \ge 0$, let N_t and R_t be the null and range spaces of the matrix V_t, and write $N_t^\perp$ and $R_t^\perp$ for their orthogonal complements. Denote the corresponding orthogonal projections by π_{N_t}, π_{R_t}, $\pi_{N_t^\perp}$, and $\pi_{R_t^\perp}$, respectively. Note that V_t is a bijection from $N_t^\perp$ to R_t, and write V_t^{-1} for the inverse mapping from R_t to $N_t^\perp$. All these mappings are clearly Borel-measurable functions of V_t, and hence again progressive.

Now introduce a Brownian motion $X \perp\!\!\!\perp \mathcal{F}$ in $\mathbb{R}^n$ with induced filtration $\mathcal{X}$, and note that $\mathcal{G}_t = \mathcal{F}_t \vee \mathcal{X}_t$, $t \ge 0$, is a standard extension of both $\mathcal{F}$ and $\mathcal{X}$. Thus, V remains $\mathcal{G}$-progressive, and the martingale properties of M and X are still valid for $\mathcal{G}$. Consider in $\mathbb{R}^n$ the local $\mathcal{G}$-martingale

$$B = V^{-1}\pi_R \cdot M + \pi_N \cdot X.$$

The covariation matrix of B has density

$$(V^{-1}\pi_R)VV'(V^{-1}\pi_R)' + \pi_N \pi_N' = \pi_{N^\perp}\pi_{N^\perp}' + \pi_N \pi_N' = \pi_{N^\perp} + \pi_N = I,$$

and so Theorem 18.3 shows that B is a Brownian motion. Furthermore, the process $\pi_{R^\perp} \cdot M = 0$ vanishes a.s. since its covariation matrix has density $\pi_{R^\perp} V V' \pi_{R^\perp}' = 0$. Hence, by Proposition 17.14,

$$V \cdot B = VV^{-1}\pi_R \cdot M + V\pi_N \cdot Y = \pi_R \cdot M = (\pi_R + \pi_{R^\perp}) \cdot M = M. \quad \square$$

We may next prove a Fubini-type theorem, which shows how the multiple Wiener–Itô integrals defined in Chapter 13 can be expressed in terms of iterated Itô integrals. Then introduce for each $n \in \mathbb{N}$ the simplex

$$\Delta_n = \{(t_1, \ldots, t_n) \in \mathbb{R}_+^n;\ t_1 < \cdots < t_n\}.$$

Given a function $f \in L^2(\mathbb{R}_+^n, \lambda^n)$, we write $\hat{f} = n!\tilde{f}1_{\Delta_n}$, where $\tilde{f}$ denotes the symmetrization of f defined in Chapter 13.

Theorem 18.13 *(multiple and iterated integrals) Consider a Brownian motion B in $\mathbb{R}$ with associated multiple Wiener–Itô integrals I_n, and fix any $f \in L^2(\mathbb{R}_+^n)$. Then*

$$I_n f = \int dB_{t_n} \int dB_{t_{n-1}} \cdots \int \hat{f}(t_1, \ldots, t_n) dB_{t_1} \quad a.s. \tag{6}$$

Though a formal verification is easy, the existence of the integrals on the right depends in a subtle way on the possibility of choosing suitable versions in each step. The existence of such versions is implicitly regarded as part of the assertion.

Proof: We shall prove by induction that the iterated integral

$$V^k_{t_{k+1},\ldots,t_n} = \int dB_{t_k} \int dB_{t_{k-1}} \cdots \int \hat{f}(t_1,\ldots,t_n) dB_{t_1}$$

exists for almost all $t_{k+1},\ldots,t_n$, and that V^k has a version supported by Δ_{n-k} that is progressive as a process in t_{k+1} with parameters $t_{k+2},\ldots,t_n$. Furthermore, we shall establish the relation

$$E\left(V^k_{t_{k+1},\ldots,t_n}\right)^2 = \int \cdots \int \{\hat{f}(t_1,\ldots,t_n)\}^2 dt_1 \cdots dt_k. \tag{7}$$

This allows us, in the next step, to define $V^{k+1}_{t_{k+2},\ldots,t_n}$ for almost all $t_{k+2},\ldots,t_n$.

The integral $V^0 = \hat{f}$ clearly has the stated properties. Now assume that a version of the integral $V^{k-1}_{t_k,\ldots,t_n}$ has been constructed with the desired properties. For any $t_{k+1},\ldots,t_n$ such that (7) is finite, Theorem 17.25 shows that the process

$$X^k_{t,t_{k+1},\ldots,t_n} = \int_0^t V^{k-1}_{t_k,\ldots,t_n} dB_{t_k}, \quad t \geq 0,$$

has a progressive version that is a.s. continuous in t for fixed $t_{k+1},\ldots,t_n$. By Proposition 17.15 we obtain

$$V^k_{t_{k+1},\ldots,t_n} = X^k_{t_{k+1},t_{k+1},\ldots,t_n} \quad \text{a.s.}, \quad t_{k+1},\ldots,t_n \geq 0,$$

and the progressivity clearly carries over to V^k, regarded as a process in t_{k+1} with parameters $t_{k+2},\ldots,t_n$. Since V^{k-1} is supported by Δ_{n-k+1}, we may choose X^k to be supported by $\mathbb{R}_+ \times \Delta_{n-k}$, which ensures V^k to be supported by Δ_{n-k}. Finally, equation (7) for V^{k-1} yields

$$E\left(V^k_{t_{k+1},\ldots,t_n}\right)^2 = E\int \left(V^{k-1}_{t_k,\ldots,t_n}\right)^2 dt_k$$
$$= \int \cdots \int \{\hat{f}(t_1,\ldots,t_n)\}^2 dt_1 \cdots dt_k.$$

To prove (6), we note that the right-hand side is linear and L^2-continuous in f. Furthermore, the two sides agree for indicator functions of rectangular boxes in Δ_n. The relation extends by a monotone class argument to arbitrary indicator functions in Δ_n, and the further extension to $L^2(\Delta_n)$ is immediate. It remains to note that $I_n f = I_n \tilde{f} = I_n \hat{f}$ for any $f \in L^2(\mathbb{R}^n_+)$. □

Our previous developments have provided two entirely different representations of Brownian functionals with zero mean and finite variance, namely the chaos expansion in Theorem 13.26 and the stochastic integral representation in Lemma 18.11. We proceed to examine how the two formulas are related. For any function $f \in L^2(\mathbb{R}^n_+)$, we define $f_t(t_1,\ldots,t_{n-1}) = f(t_1,\ldots,t_{n-1},t)$ and write $I_{n-1}f(t) = I_{n-1}f_t$ when $\|f_t\| < \infty$.

Proposition 18.14 *(chaos and integral representations) Fix a Brownian motion B in $\mathbb{R}$, and let ξ be a B-measurable random variable with chaos expansion $\sum_{n\geq 1} I_n f_n$. Then $\xi = (V \cdot B)_\infty$ a.s., where*

$$V_t = \sum_{n\geq 1} I_{n-1}\hat{f}_n(t), \quad t \geq 0.$$

Proof: For any $m \in \mathbb{N}$ we get, as in the last proof,

$$\int dt \sum_{n\geq m} E\{I_{n-1}\hat{f}_n(t)\}^2 = \sum_{n\geq m} \|\hat{f}_n\|^2 = \sum_{n\geq m} E(I_n f_n)^2 < \infty. \quad (8)$$

Since integrals $I_n f$ with different n are orthogonal, it follows that the series for V_t converges in L^2 for almost every $t \geq 0$. On the exceptional set we may redefine V_t to be 0. As before, we may choose progressive versions of the integrals $I_{n-1}\hat{f}_n(t)$, and from the proof of Corollary 4.32 it is clear that even the sum V can be chosen to be progressive. Applying (8) with $m = 1$, we then obtain $V \in L(B)$.

Using Theorem 18.13, we get by a formal calculation

$$\xi = \sum_{n\geq 1} I_n f_n = \sum_{n\geq 1} \int I_{n-1}\hat{f}_n(t)dB_t = \int dB_t \sum_{n\geq 1} I_{n-1}\hat{f}_n(t) = \int V_t dB_t.$$

To justify the interchange of integration and summation, we may use (8) and conclude as $m \to \infty$ that

$$E\left\{\int dB_t \sum_{n\geq m} I_{n-1}\hat{f}_n(t)\right\}^2 = \int dt \sum_{n\geq m} E\{I_{n-1}\hat{f}_n(t)\}^2$$
$$= \sum_{n\geq m} E(I_n f_n)^2 \to 0. \quad \Box$$

Let us now consider two different probability measures P and Q on the same measurable space $(\Omega, \mathcal{A})$, equipped with a right-continuous and P-complete filtration $(\mathcal{F}_t)$. If $Q \ll P$ on $\mathcal{F}_t$, we denote the corresponding density by Z_t, so that $Q = Z_t \cdot P$ on $\mathcal{F}_t$. Since the martingale property depends on the choice of probability measure, we need to distinguish between P-martingales and Q-martingales. Integration with respect to P is denoted by E as usual, and we write $\mathcal{F}_\infty = \bigvee_t \mathcal{F}_t$.

Lemma 18.15 *(absolute continuity) Let $Q = Z_t \cdot P$ on $\mathcal{F}_t$ for all $t \geq 0$. Then Z is a P-martingale, and it is further uniformly integrable iff $Q \ll P$ on $\mathcal{F}_\infty$. More generally, an adapted process X is a Q-martingale iff XZ is a P-martingale.*

Proof: For any adapted process X, we note that X_t is Q-integrable iff $X_t Z_t$ is P-integrable. If this holds for all t, we may write the Q-martingale property of X as

$$\int_A X_s dQ = \int_A X_t dQ, \quad A \in \mathcal{F}_s, \; s < t.$$

By the definition of Z, it is equivalent that
$$E[X_s Z_s; A] = E[X_t Z_t; A], \quad A \in \mathcal{F}_s, \; s < t,$$
which means that XZ is a P-martingale. This proves the last assertion, and the first statement follows as we take $X_t \equiv 1$.

Next assume that Z is uniformly P-integrable, say with L^1-limit Z_∞. For any $t < u$ and $A \in \mathcal{F}_t$ we have $QA = E[Z_u; A]$. As $u \to \infty$, it follows that $QA = E[Z_\infty; A]$, which extends by a monotone class argument to arbitrary $A \in \mathcal{F}_\infty$. Thus, $Q = Z_\infty \cdot P$ on $\mathcal{F}_\infty$. Conversely, if $Q = \xi \cdot P$ on $\mathcal{F}_\infty$, then $E\xi = 1$, and the P-martingale $M_t = E[\xi|\mathcal{F}_t]$ satisfies $Q = M_t \cdot P$ on $\mathcal{F}_t$ for each t. But then $Z_t = M_t$ a.s. for each t, and Z is uniformly P-integrable with limit ξ. □

By the last lemma and Theorem 7.27, we may henceforth assume that the density process Z is rcll. The basic properties may then be extended to optional times and local martingales as follows.

Lemma 18.16 *(localization) Let $Q = Z_t \cdot P$ on $\mathcal{F}_t$ for all $t \geq 0$. Then for any optional time τ, we have*
$$Q = Z_\tau \cdot P \quad \text{on} \quad \mathcal{F}_\tau \cap \{\tau < \infty\}. \tag{9}$$
Furthermore, an adapted rcll process X is a local Q-martingale iff XZ is a local P-martingale.

Proof: By optional sampling,
$$QA = E[Z_{\tau \wedge t}; A], \quad A \in \mathcal{F}_{\tau \wedge t}, \; t \geq 0,$$
and so
$$Q[A; \tau \leq t] = E[Z_\tau; A \cap \{\tau \leq t\}], \quad A \in \mathcal{F}_\tau, \; t \geq 0.$$
Equation (9) now follows by monotone convergence as $t \to \infty$.

To prove the last assertion, it is enough to show for any optional time τ that X^τ is a Q-martingale iff $(XZ)^\tau$ is a P-martingale. This may be seen as before if we note that $Q = Z_t^\tau \cdot P$ on $\mathcal{F}_{\tau \wedge t}$ for each t. □

We also need the following positivity property.

Lemma 18.17 *(positivity) For every $t > 0$ we have $\inf_{s \leq t} Z_s > 0$ a.s. Q.*

Proof: By Lemma 7.31 it is enough to show for each $t > 0$ that $Z_t > 0$ a.s. Q. This is clear from the fact that $Q\{Z_t = 0\} = E[Z_t; Z_t = 0] = 0$. □

In typical applications, the measure Q is not given at the outset but needs to be constructed from the martingale Z. This requires some regularity conditions on the underlying probability space.

Lemma 18.18 *(existence)* For any Polish space S, let P be a probability measure on $\Omega = D(\mathbb{R}_+, S)$, endowed with the right-continuous and complete induced filtration $\mathcal{F}$. Consider an $\mathcal{F}$-martingale $Z \geq 0$ with $Z_0 = 1$. Then there exists a probability measure Q on Ω with $Q = Z_t \cdot P$ on $\mathcal{F}_t$ for all $t \geq 0$.

Proof: For each $t \geq 0$, we may introduce the probability measure $Q_t = Z_t \cdot P$ on $\mathcal{F}_t$, which may be regarded as a measure on $D([0,t], S)$. Since the spaces $D([0,t], S)$ are Polish for the Skorohod topology, Corollary 6.15 ensures the existence of some probability measure Q on $D(\mathbb{R}_+, S)$ with projections Q_t. It is easy to verify that Q has the stated properties. □

The following basic result shows how the drift term of a continuous semimartingale is transformed under a change of measure with a continuous density Z. An extension appears in Theorem 26.9.

Theorem 18.19 *(transformation of drift, Girsanov, van Schuppen and Wong)* Let $Q = Z_t \cdot P$ on $\mathcal{F}_t$ for each $t \geq 0$, where Z is a.s. continuous. Then for any continuous local P-martingale M, the process $\tilde{M} = M - Z^{-1} \cdot [M, Z]$ is a local Q-martingale.

Proof: First assume that Z^{-1} is bounded on the support of $[M]$. Then $\tilde{M}$ is a continuous P-semimartingale, and we get by Proposition 17.14 and an integration by parts

$$\begin{aligned}
\tilde{M}Z - (\tilde{M}Z)_0 &= \tilde{M} \cdot Z + Z \cdot \tilde{M} + [\tilde{M}, Z] \\
&= \tilde{M} \cdot Z + Z \cdot M - [M, Z] + [\tilde{M}, Z] \\
&= \tilde{M} \cdot Z + Z \cdot M,
\end{aligned}$$

which shows that $\tilde{M}Z$ is a local P-martingale. Hence, $\tilde{M}$ is a local Q-martingale by Lemma 18.16.

For general M, we may define $\tau_n = \inf\{t \geq 0;\ Z_t < 1/n\}$ and conclude as before that $\tilde{M}^{\tau_n}$ is a local Q-martingale for each $n \in \mathbb{N}$. Since $\tau_n \to \infty$ a.s. Q by Lemma 18.17, it follows by Lemma 17.1 that $\tilde{M}$ is a local Q-martingale. □

The next result shows how the basic notions of stochastic calculus are preserved under a change of measure. Here $[X]_P$ denotes the quadratic variation of X under the probability measure P. We further write $L_P(X)$ for the class of X-integrable processes V under P, and let $(V \cdot X)_P$ be the corresponding stochastic integral.

Proposition 18.20 *(preservation laws)* Let $Q = Z_t \cdot P$ on $\mathcal{F}_t$ for each $t \geq 0$, where Z is continuous. Then any continuous P-semimartingale X is also a Q-semimartingale, and $[X]_P = [X]_Q$ a.s. Q. Furthermore, $L_P(X) \subset L_Q(X)$, and for any $V \in L_P(X)$ we have $(V \cdot X)_P = (V \cdot X)_Q$ a.s. Q. Finally, any continuous local P-martingale M satisfies $(V \cdot M)^{\sim} = V \cdot \tilde{M}$ a.s. Q whenever either side exists.

Proof: Consider a continuous P-semimartingale $X = M+A$, where M is a continuous local P-martingale and A is a process of locally finite variation. Under Q we may write $X = \tilde{M}+Z^{-1}\cdot[M,Z]+A$, where $\tilde{M}$ is the continuous local Q-martingale of Theorem 18.19, and we note that $Z^{-1}\cdot[M,Z]$ has locally finite variation since $Z > 0$ a.s. Q by Lemma 18.17. Thus, X is also a Q-semimartingale. The statement for $[X]$ is now clear from Proposition 17.17.

Now assume that $V \in L_P(X)$. Then $V^2 \in L_P([X])$ and $V \in L_P(A)$, so the same relations hold under Q, and we get $V \in L_Q(\tilde{M} + A)$. Thus, to get $V \in L_Q(X)$, it remains to show that $V \in L_Q(Z^{-1}[M,Z])$. Since $Z > 0$ under Q, it is equivalent to show that $V \in L_Q([M,Z])$. But this is clear by Proposition 17.9, since $[M,Z]_Q = [\tilde{M},Z]_Q$ and $V \in L_Q(\tilde{M})$.

To prove the last assertion, we note as before that $L_Q(M) = L_Q(\tilde{M})$. If V belongs to either class, then by Proposition 17.14 we get under Q the a.s. relations

$$(V \cdot M)^{\sim} = V \cdot M - Z^{-1} \cdot [V \cdot M, Z]$$
$$= V \cdot M - VZ^{-1} \cdot [M,Z] = V \cdot \tilde{M}. \qquad \square$$

In particular, we note that if B is a P-Brownian motion in $\mathbb{R}^d$, then $\tilde{B}$ is a Q-Brownian motion by Theorem 18.3, since both processes are continuous martingales with the same covariation process.

The preceding theory simplifies when P and Q are equivalent on each $\mathcal{F}_t$, since in that case $Z > 0$ a.s. P by Lemma 18.17. If Z is also continuous, it may be expressed as an exponential martingale. More general processes of this type are considered in Theorem 26.8.

Lemma 18.21 *(real exponential martingales) A continuous process $Z > 0$ is a local martingale iff it has an a.s. representation*

$$Z_t = \mathcal{E}(M)_t \equiv \exp(M_t - \tfrac{1}{2}[M]_t), \quad t \geq 0, \qquad (10)$$

for some continuous local martingale M. In that case M is a.s. unique, and for any continuous local martingale N we have $[M,N] = Z^{-1}\cdot[Z,N]$.

Proof: If M is a continuous local martingale, then so is $\mathcal{E}(M)$ by Itô's formula. Conversely, assume that $Z > 0$ is a continuous local martingale. Then by Corollary 17.19,

$$\log Z - \log Z_0 = Z^{-1}\cdot Z - \tfrac{1}{2}Z^{-2}\cdot[Z] = Z^{-1}\cdot Z - \tfrac{1}{2}[Z^{-1}\cdot Z],$$

and (10) follows with $M = \log Z_0 + Z^{-1}\cdot Z$. The last assertion is clear from this expression, and the uniqueness of M follows from Proposition 17.2. $\qquad \square$

We shall now see how Theorem 18.19 can be used to eliminate the drift of a continuous semimartingale, and we begin with the simple case of Brownian motion B with a deterministic drift. Here we need the fact that $\mathcal{E}(B)$ is a true martingale, as can be seen most easily by a direct computation. By $P \sim Q$ we mean that $P \ll Q$ and $Q \ll P$. Write L^2_{loc} for the class of functions $f: \mathbb{R}_+ \to \mathbb{R}^d$ such that $|f|^2$ is locally Lebesgue integrable. For any $f \in L^2_{\text{loc}}$ we define $f \cdot \lambda = (f^1 \cdot \lambda, \ldots, f^d \cdot \lambda)$, where the components on the right are ordinary Lebesgue integrals.

Theorem 18.22 *(shifted Brownian motion, Cameron and Martin)* *Let $\mathcal{F}$ be the complete filtration induced by canonical Brownian motion B in $\mathbb{R}^d$, fix a continuous function $h: \mathbb{R}_+ \to \mathbb{R}^d$ with $h_0 = 0$, and write P_h for the distribution of $B + h$. Then $P_h \sim P_0$ on $\mathcal{F}_t$ for all $t \geq 0$ iff $h = f \cdot \lambda$ for some $f \in L^2_{\text{loc}}$, in which case $P_h = \mathcal{E}(f \cdot B)_t \cdot P_0$.*

Proof: If $P_h \sim P_0$ on each $\mathcal{F}_t$, then by Lemmas 18.15 and 18.17 there exists some P_0-martingale $Z > 0$ such that $P_h = Z_t \cdot P_0$ on $\mathcal{F}_t$ for all $t \geq 0$. Theorem 18.10 shows that Z is a.s. continuous, and by Lemma 18.21 it can then be written as $\mathcal{E}(M)$ for some continuous local P_0-martingale M. Using Theorem 18.10 again, we note that $M = V \cdot B = \sum_i V^i \cdot B^i$ a.s. for some processes $V^i \in L(B^1)$, and in particular $V \in L^2_{\text{loc}}$ a.s.

By Theorem 18.19 the process $\tilde{B} = B - [B, M] = B - V \cdot \lambda$ is a P_h-Brownian motion, and so, under P_h, the canonical process B has two semimartingale decompositions, namely
$$B = \tilde{B} + V \cdot \lambda = (B - h) + h.$$

By Proposition 17.2 the decomposition is a.s. unique, and so $V \cdot \lambda = h$ a.s. Thus, $h = f \cdot \lambda$ for some nonrandom function $f \in L^2_{\text{loc}}$, and furthermore $\lambda\{t \geq 0;\ V_t \neq f_t\} = 0$ a.s., which implies $M = V \cdot B = f \cdot B$ a.s.

Conversely, assume that $h = f \cdot \lambda$ for some $f \in L^2_{\text{loc}}$. Since $M = f \cdot B$ is a time-changed Brownian motion under P_0, the process $Z = \mathcal{E}(M)$ is a P_0-martingale, and by Lemma 18.18 there exists a probability measure Q on $C(\mathbb{R}_+, \mathbb{R}^d)$ such that $Q = Z_t \cdot P_0$ on $\mathcal{F}_t$ for all $t \geq 0$. Moreover, Theorem 18.19 shows that $\tilde{B} = B - [B, M] = B - h$ is a Q-Brownian motion, which means that $Q = P_h$. In particular, $P_h \sim P_0$ on each $\mathcal{F}_t$. $\square$

In more general cases, Theorem 18.19 and Lemma 18.21 suggest that we might try to remove the drift of a semimartingale through a change of measure of the form $Q = \mathcal{E}(M)_t \cdot P$ on $\mathcal{F}_t$ for each $t \geq 0$, where M is a continuous local martingale with $M_0 = 0$. By Lemma 18.15 it is then necessary for $Z = \mathcal{E}(M)$ to be a true martingale. This is ensured by the following condition.

Theorem 18.23 *(uniform integrability, Novikov)* *Let M be a continuous local martingale with $M_0 = 0$ such that $Ee^{[M]_\infty/2} < \infty$. Then $\mathcal{E}(M)$ is a uniformly integrable martingale.*

The result will first be proved in a special case.

Lemma 18.24 *(Wald's identity)* *If B is a real Brownian motion and τ is an optional time with $Ee^{\tau/2} < \infty$, then $E \exp(B_\tau - \frac{1}{2}\tau) = 1$.*

Proof: We first consider the special optional times
$$\tau_b = \inf\{t \geq 0;\ B_t = t - b\}, \quad b > 0.$$

Since the τ_b remain optional with respect to the right-continuous, induced filtration, we may assume B to be canonical Brownian motion with associated distribution $P = P_0$. Defining $h_t \equiv t$ and $Z = \mathcal{E}(B)$, we see from Theorem 18.22 that $P_h = Z_t \cdot P$ on $\mathcal{F}_t$ for all $t \geq 0$. Since $\tau_b < \infty$ a.s. under both P and P_h, Lemma 18.16 yields

$$E \exp(B_{\tau_b} - \tfrac{1}{2}\tau_b) = EZ_{\tau_b} = E[Z_{\tau_b};\, \tau_b < \infty] = P_h\{\tau_b < \infty\} = 1.$$

In the general case, the stopped process $M_t \equiv Z_{t \wedge \tau_b}$ is a positive martingale, and Fatou's lemma shows that M is also a supermartingale on $[0, \infty]$. Since, moreover, $EM_\infty = EZ_{\tau_b} = 1 = EM_0$, it is clear from the Doob decomposition that M is a true martingale on $[0, \infty]$. Hence, by optional sampling,

$$1 = EM_\tau = EZ_{\tau \wedge \tau_b} = E[Z_\tau;\, \tau \leq \tau_b] + E[Z_{\tau_b};\, \tau > \tau_b]. \tag{11}$$

By the definition of τ_b and the hypothesis on τ, we get as $b \to \infty$

$$E[Z_{\tau_b};\, \tau > \tau_b] = e^{-b} E[e^{\tau_b/2};\, \tau > \tau_b] \leq e^{-b} E e^{\tau/2} \to 0,$$

and so the last term in (11) tends to zero. Since, moreover, $\tau_b \to \infty$, the first term on the right tends to EZ_τ by monotone convergence, and the desired relation $EZ_\tau = 1$ follows. □

Proof of Theorem 18.23: Since $\mathcal{E}(M)$ is always a supermartingale on $[0, \infty]$, it is enough, under the stated condition, to show that $E\mathcal{E}(M)_\infty = 1$. We may then use Theorem 18.4 and Proposition 7.9 to reduce to the statement of Lemma 18.24. □

In particular, we obtain the following classical result for Brownian motion.

Corollary 18.25 *(removal of drift, Girsanov)* *Consider in $\mathbb{R}^d$ a Brownian motion B and a progressive process V with $E \exp\{\tfrac{1}{2}(|V|^2 \cdot \lambda)_\infty\} < \infty$. Then $Q = \mathcal{E}(V' \cdot B)_\infty \cdot P$ is a probability measure, and $\tilde{B} = B - V \cdot \lambda$ is a Q-Brownian motion.*

Proof: Combine Theorems 18.19 and 18.23. □

Exercises

1. Assume in Theorem 18.4 that $[M]_\infty = \infty$ a.s. Show that M is τ-continuous in the sense of Theorem 17.24, and use Theorem 18.3 to conclude that $B = M \circ \tau$ is a Brownian motion. Also show for any $V \in L(M)$ that $(V \circ \tau) \cdot B = (V \cdot M) \circ \tau$ a.s.

2. If B is a real Brownian motion and $V \in L(B)$, then $X = V \cdot B$ is a time-changed Brownian motion. Express the required time-change τ in terms of V, and verify that X is τ-continuous.

3. Let M be a real continuous local martingale. Show that M converges a.s. on the set $\{\sup_t M_t < \infty\}$. (*Hint:* Use Theorem 18.4.)

4. Let $\mathcal{F}$ and $\mathcal{G}$ be filtrations on a common probability space $(\Omega, \mathcal{A}, P)$. Show that $\mathcal{G}$ is a standard extension of $\mathcal{F}$ iff every $\mathcal{F}$-martingale is also a $\mathcal{G}$-martingale. (*Hint:* Consider martingales of the form $M_t = E[\xi|\mathcal{F}_t]$, where $\xi \in L^1(\mathcal{F}_\infty)$. Here M_t is $\mathcal{G}_t$-measurable for all ξ iff $\mathcal{F}_t \subset \mathcal{G}_t$, and then $M_t = E[\xi|\mathcal{G}_t]$ a.s. for all ξ iff $\mathcal{F}\perp\!\!\!\perp_{\mathcal{F}_t} \mathcal{G}_t$ by Proposition 6.6.)

5. Let $\mathcal{F}$ and $\mathcal{G}$ be right-continuous filtrations such that $\mathcal{G}$ is a standard extension of $\mathcal{F}$, and let τ be an $\mathcal{F}$-optional time. Show that $\mathcal{F}_\tau \subset \mathcal{G}_\tau \perp\!\!\!\perp_{\mathcal{F}_\tau} \mathcal{F}$. (*Hint:* Apply optional sampling to the uniformly integrable martingale $M_t = E[\xi|\mathcal{F}_t]$ for any $\xi \in L^1(\mathcal{F}_\infty)$.)

6. Let M be a nontrivial isotropic continuous local martingale in $\mathbb{R}^d$, and fix an affine transformation f on $\mathbb{R}^d$. Show that even $f(M)$ is isotropic iff f is conformal (i.e., the composition of a rigid motion with a change of scale).

7. Deduce Theorem 18.6 (ii) from Theorem 9.8. (*Hint:* Define $\tau = \inf\{t; |B_t| = 1\}$, and iterate the construction to form a random walk in $\mathbb{R}^d$ with steps of size 1.)

8. Deduce Theorem 18.3 for $d = 1$ from Theorem 14.17. (*Hint:* Proceed as above to construct a discrete-time martingale with jumps of size h. Let $h \to 0$, and use a version of Proposition 17.17.)

9. Consider a real Brownian motion B and a family of progressive processes $V^t \in L(B)$, $t \geq 0$. Give necessary and sufficient conditions on the V^t for the existence of a Brownian motion B', such that $B'_t = (V^t \cdot B)_\infty$ a.s. for each t. Verify the conditions in the case of Proposition 18.9.

10. Extend Proposition 18.9 to any continuous, $\mathcal{F}$-exchangeable process X on $\mathbb{R}_+$ or $[0, 1]$. (*Hint:* Recall that $X_t = \alpha t + \sigma B_t$ for some Brownian motion or bridge B and some independent pair of random variables α and $\sigma \geq 0$. Note that X remains exchangeable for the filtration $\mathcal{G}_t = \mathcal{F}_t \vee \sigma\{\alpha, \sigma\}$. Hence, so is B, and we may apply Proposition 18.9.)

11. Use Proposition 18.9 to give direct proofs of the relation $\tau_1 \stackrel{d}{=} \tau_2$ in Theorems 13.16 and 13.17. (*Hint:* Imitate the proof of Theorem 11.14.)

12. For a Brownian motion B and optional time $\tau < \infty$, show that $E \exp(B_\tau - \tfrac{1}{2}\tau) \leq 1$ where the inequality may be strict. (*Hint:* Truncate and use Fatou's lemma. Note that $t - 2B_t \to \infty$ by the law of large numbers.)

Chapter 19

Feller Processes and Semigroups

Semigroups, resolvents, and generators; closure and core; Hille–Yosida theorem; existence and regularization; strong Markov property; characteristic operator; diffusions and elliptic operators; convergence and approximation

Our aim in this chapter is to continue the general discussion of continuous-time Markov processes initiated in Chapter 8. We have already seen several important examples of such processes, such as the pure jump-type processes in Chapter 12, Brownian motion in Chapters 13 and 18, and the general Lévy processes in Chapter 15. The present treatment will be supplemented by detailed studies of ergodic properties in Chapter 20, of diffusions in Chapters 21 and 23, and of excursions and additive functionals in Chapters 22 and 25.

The crucial new idea is to regard the transition kernels as operators T_t on an appropriate function space. The Chapman–Kolmogorov relation then turns into the semigroup property $T_s T_t = T_{s+t}$, which suggests a formal representation $T_t = e^{tA}$ in terms of a generator A. Under suitable regularity conditions—the so-called Feller properties—it is indeed possible to define a generator A that describes the infinitesimal evolution of the underlying process X. Under further hypotheses, X will be shown to have continuous paths iff A is (an extension of) an elliptic differential operator. In general, the powerful Hille–Yosida theorem provides the precise conditions for the existence of a Feller process corresponding to a given operator A.

Using the basic regularity theorem for submartingales from Chapter 7, it will be shown that every Feller process has a version that is right-continuous with left-hand limits (rcll). Given this fundamental result, it is straightforward to extend the strong Markov property to arbitrary Feller processes. We shall also explore some profound connections with martingale theory. Finally, we shall establish a general continuity theorem for Feller processes and deduce a corresponding approximation of discrete-time Markov chains by diffusions and other continuous-time Markov processes. The proofs of the latter results will require some weak convergence theory from Chapter 16.

To clarify the connection between transition kernels and operators, let μ be an arbitrary probability kernel on some measurable space $(S, \mathcal{S})$. We may then introduce an associated *transition operator* T, given by

$$Tf(x) = (Tf)(x) = \int \mu(x, dy)f(y), \quad x \in S, \tag{1}$$

where $f \colon S \to \mathbb{R}$ is assumed to be measurable and either bounded or nonnegative. Approximating f by simple functions, we see that by monotone convergence Tf is again a measurable function on S. It is also clear that T is a *positive contraction operator*, in the sense that $0 \le f \le 1$ implies $0 \le Tf \le 1$. A special role is played by the identity operator I, which corresponds to the kernel $\mu(x, \cdot) \equiv \delta_x$. The importance of transition operators for the study of Markov processes is due to the following simple fact.

Lemma 19.1 *(transition kernels and operators)* *The probability kernels μ_t, $t \ge 0$, satisfy the Chapman–Kolmogorov relation iff the corresponding transition operators T_t have the semigroup property*

$$T_{s+t} = T_s T_t, \quad s, t \ge 0. \tag{2}$$

Proof: For any $B \in \mathcal{S}$ we have $T_{s+t} 1_B(x) = \mu_{s+t}(x, B)$ and

$$\begin{aligned}(T_s T_t) 1_B(x) &= T_s(T_t 1_B)(x) = \int \mu_s(x, dy)(T_t 1_B)(y) \\ &= \int \mu_s(x, dy)\mu_t(y, B) = (\mu_s \mu_t)(x, B).\end{aligned}$$

Thus, the Chapman–Kolmogorov relation is equivalent to $T_{s+t} 1_B = (T_s T_t) 1_B$ for any $B \in \mathcal{S}$. The latter relation extends to (2) by linearity and monotone convergence. □

By analogy with the situation for the Cauchy equation, one might hope to represent the semigroup in the form $T_t = e^{tA}$, $t \ge 0$, for a suitable *generator* A. For the formula to make sense, the operator A must be suitably bounded, so that the exponential function can be defined through a Taylor expansion. We shall consider a simple case when such a representation exists.

Proposition 19.2 *(pseudo-Poisson processes)* *Let (T_t) be the transition semigroup of a pure jump-type Markov process in S with bounded rate kernel α. Then $T_t = e^{tA}$ for all $t \ge 0$, where for any bounded measurable function $f \colon S \to \mathbb{R}$,*

$$Af(x) = \int (f(y) - f(x))\alpha(x, dy), \quad x \in S.$$

Proof: Choose a probability kernel μ and a constant $c \ge 0$ such that $\alpha(x, B) \equiv c\mu(x, B \setminus \{x\})$. From Proposition 12.20 we see that the process is pseudo-Poisson of the form $X = Y \circ N$, where Y is a discrete-time Markov chain with transition kernel μ, and N is an independent Poisson process with fixed rate c. Letting T denote the transition operator associated with

μ, we get for any $t \geq 0$ and f as stated,

$$\begin{aligned} T_t f(x) &= E_x f(X_t) = \sum_{n \geq 0} E_x[f(Y_n); N_t = n] \\ &= \sum_{n \geq 0} P\{N_t = n\} E_x f(Y_n) \\ &= \sum_{n \geq 0} e^{-ct} \frac{(ct)^n}{n!} T^n f(x) = e^{ct(T-I)} f(x). \end{aligned}$$

Hence, $T_t = e^{tA}$ holds for $t \geq 0$ with

$$\begin{aligned} Af(x) &= c(T - I)f(x) = c \int (f(y) - f(x)) \mu(x, dy) \\ &= \int (f(y) - f(x)) \alpha(x, dy). \end{aligned} \qquad \square$$

For the further analysis, we assume S to be a locally compact, separable metric space, and we write $C_0 = C_0(S)$ for the class of continuous functions $f : S \to \mathbb{R}$ with $f(x) \to 0$ as $x \to \infty$. We can make C_0 into a Banach space by introducing the norm $\|f\| = \sup_x |f(x)|$. A semigroup of positive contraction operators T_t on C_0 is called a *Feller semigroup* if it has the additional regularity properties

(F_1) $T_t C_0 \subset C_0$, $t \geq 0$,

(F_2) $T_t f(x) \to f(x)$ as $t \to 0$, $f \in C_0$, $x \in S$.

In Theorem 19.6 we show that (F_1) and (F_2) together with the semigroup property imply the *strong continuity*

(F_3) $T_t f \to f$ as $t \to 0$, $f \in C_0$.

For motivation, we proceed to clarify the probabilistic significance of those conditions. Then assume for simplicity that S is compact, and also that (T_t) is *conservative* in the sense that $T_t 1 = 1$ for all t. For every initial state x, we may then introduce an associated Markov process X_t^x, $t \geq 0$, with transition operators T_t.

Lemma 19.3 *(Feller properties)* Let (T_t) be a conservative transition semigroup on a compact metric space (S, ρ). Then

(F_1) holds iff $X_t^x \xrightarrow{d} X_t^y$ as $x \to y$ for fixed $t \geq 0$;
(F_2) holds iff $X_t^x \xrightarrow{P} x$ as $t \to 0$ for fixed x;
(F_3) holds iff $\sup_x E_x[\rho(X_s, X_t) \wedge 1] \to 0$ as $s - t \to 0$.

Proof: The first two statements are obvious, so we shall prove only the third one. Then choose a dense sequence $f_1, f_2, \ldots$ in $C = C(S)$. By the compactness of S we note that $x_n \to x$ in S iff $f_k(x_n) \to f_k(x)$ for each k. Thus, ρ is topologically equivalent to the metric

$$\rho'(x, y) = \sum_k 2^{-k} (|f_k(x) - f_k(y)| \wedge 1), \quad x, y \in S.$$

Since S is compact, the identity mapping on S is uniformly continuous with respect to ρ and ρ', and so we may assume that $\rho = \rho'$.

Next we note that, for any $f \in C$, $x \in S$, and $t, h \geq 0$,

$$\begin{aligned} E_x(f(X_t) - f(X_{t+h}))^2 &= E_x(f^2 - 2fT_h f - T_h f^2)(X_t) \\ &\leq \|f^2 - 2fT_h f + T_h f^2\| \\ &\leq 2\|f\|\,\|f - T_h f\| + \|f^2 - T_h f^2\|. \end{aligned}$$

Assuming (F$_3$), we get $\sup_x E_x|f_k(X_s) - f_k(X_t)| \to 0$ as $s - t \to 0$ for fixed k, and so by dominated convergence $\sup_x E_x \rho(X_s, X_t) \to 0$. Conversely, the latter condition yields $T_h f_k \to f_k$ for each k, which implies (F$_3$). $\square$

Our aim is now to construct the generator of an arbitrary Feller semigroup (T_t) on C_0. In general, there is no bounded linear operator A satisfying $T_t = e^{tA}$, and we need to look for a suitable substitute. For motivation, we note that if p is a real-valued function on $\mathbb{R}_+$ with representation $p_t = e^{ta}$, then a can be recovered from p by either differentiation or integration:

$$t^{-1}(p_t - 1) \to a \quad \text{as } t \to 0;$$

$$\int_0^\infty e^{-\lambda t} p_t dt = (\lambda - a)^{-1}, \quad \lambda > 0.$$

Motivated by the latter formula, we introduce for each $\lambda > 0$ the associated *resolvent* or *potential* R_λ, defined as the Laplace transform

$$R_\lambda f = \int_0^\infty e^{-\lambda t}(T_t f) dt, \quad f \in C_0.$$

Note that the integral exists, since $T_t f(x)$ is bounded and right-continuous in $t \geq 0$ for fixed $x \in S$.

Theorem 19.4 *(resolvents and generator)* *Let (T_t) be a Feller semigroup on C_0 with resolvents R_λ, $\lambda > 0$. Then the operators λR_λ are injective contractions on C_0 such that $\lambda R_\lambda \to I$ strongly as $\lambda \to \infty$. Furthermore, the range $\mathcal{D} = R_\lambda C_0$ is independent of λ and dense in C_0, and there exists an operator A on C_0 with domain $\mathcal{D}$ such that $R_\lambda^{-1} = \lambda - A$ on $\mathcal{D}$ for every $\lambda > 0$. Finally, A commutes on $\mathcal{D}$ with every T_t.*

Proof: If $f \in C_0$, then (F$_1$) yields $T_t f \in C_0$ for every t, and so by dominated convergence we have even $R_\lambda f \in C_0$. To prove the stated contraction property, we write for any $f \in C_0$

$$\|\lambda R_\lambda f\| \leq \lambda \int_0^\infty e^{-\lambda t} \|T_t f\| dt \leq \lambda \|f\| \int_0^\infty e^{-\lambda t} dt = \|f\|.$$

A simple computation yields the *resolvent equation*

$$R_\lambda - R_\mu = (\mu - \lambda) R_\lambda R_\mu, \quad \lambda, \mu > 0, \tag{3}$$

which shows that the operators R_λ commute and have a common range $\mathcal{D}$. If $f = R_1 g$ with $g \in C_0$, we get by (3) and as $\lambda \to \infty$

$$\|\lambda R_\lambda f - f\| = \|(\lambda R_\lambda - I)R_1 g\| = \|(R_1 - I)R_\lambda g\|$$
$$\leq \lambda^{-1}\|R_1 - I\|\,\|g\| \to 0.$$

The convergence extends by a simple approximation to the closure of $\mathcal{D}$.

Now introduce the one-point compactification $\hat{S} = S \cup \{\Delta\}$ of S, and extend any $f \in C_0$ to $\hat{C} = C(\hat{S})$ by putting $f(\Delta) = 0$. If $\overline{\mathcal{D}} \neq C_0$, then by the Hahn–Banach theorem there exists a bounded linear functional $\varphi \neq 0$ on $\hat{C}$ such that $\varphi R_1 f = 0$ for all $f \in C_0$. By Riesz's representation Theorem 2.22 we may extend φ to a bounded, signed measure on $\hat{S}$. Letting $f \in C_0$ and using (F$_2$), we get by dominated convergence as $\lambda \to \infty$

$$0 = \lambda \varphi R_\lambda f = \int \varphi(dx) \int_0^\infty \lambda e^{-\lambda t} T_t f(x) dt$$
$$= \int \varphi(dx) \int_0^\infty e^{-s} T_{s/\lambda} f(x) dt \to \varphi f,$$

and so $\varphi \equiv 0$. The contradiction shows that $\mathcal{D}$ is dense in C_0.

To see that the operators R_λ are injective, let $f \in C_0$ with $R_{\lambda_0} f = 0$ for some $\lambda_0 > 0$. Then (3) yields $R_\lambda f = 0$ for every $\lambda > 0$, and since $\lambda R_\lambda f \to f$ as $\lambda \to \infty$, we get $f = 0$. Hence, the inverses R_λ^{-1} exist on $\mathcal{D}$. Multiplying (3) by R_λ^{-1} from the left and by R_μ^{-1} from the right, we get on $\mathcal{D}$ the relation $R_\mu^{-1} - R_\lambda^{-1} = \mu - \lambda$. Thus, the operator $A = \lambda - R_\lambda^{-1}$ on $\mathcal{D}$ is independent of λ.

To prove the final assertion, we note that T_t and R_λ commute for any $t, \lambda > 0$, and write

$$T_t(\lambda - A)R_\lambda = T_t = (\lambda - A)R_\lambda T_t = (\lambda - A)T_t R_\lambda. \qquad \square$$

The operator A in Theorem 19.4 is called the *generator* of the semigroup (T_t). If we want to emphasize the role of the domain $\mathcal{D}$, we say that (T_t) has generator $(A, \mathcal{D})$. The term is justified by the following lemma.

Lemma 19.5 *(uniqueness) A Feller semigroup is uniquely determined by its generator.*

Proof: The operator A determines $R_\lambda = (\lambda - A)^{-1}$ for all $\lambda > 0$. By the uniqueness theorem for Laplace transforms, it then determines the measure $\mu(dt) = T_t f(x) dt$ on $\mathbb{R}_+$ for any $f \in C_0$ and $x \in S$. Since the density $T_t f(x)$ is right-continuous in t for fixed x, the assertion follows. $\qquad \square$

We now aim to show that any Feller semigroup is strongly continuous and to derive abstract versions of Kolmogorov's forward and backward equations.

Theorem 19.6 *(strong continuity, forward and backward equations) Let (T_t) be a Feller semigroup with generator $(A, \mathcal{D})$. Then (T_t) is strongly continuous and satisfies*

$$T_t f - f = \int_0^t T_s A f \, ds, \quad f \in \mathcal{D}, \; t \geq 0. \tag{4}$$

Furthermore, $T_t f$ is differentiable at 0 iff $f \in \mathcal{D}$, in which case

$$\frac{d}{dt}(T_t f) = T_t A f = A T_t f, \quad t \geq 0. \tag{5}$$

To prove this result, we introduce the so-called *Yosida approximation*

$$A^\lambda = \lambda A R_\lambda = \lambda(\lambda R_\lambda - I), \quad \lambda > 0, \tag{6}$$

and the associated semigroup $T_t^\lambda = e^{tA^\lambda}$, $t \geq 0$. The latter is clearly the transition semigroup of a pseudo-Poisson process with rate λ based on the transition operator λR_λ.

Lemma 19.7 *(Yosida approximation) For any $f \in \mathcal{D}$, we have*

$$\|T_t f - T_t^\lambda f\| \leq t \|A f - A^\lambda f\|, \quad t, \lambda > 0, \tag{7}$$

and $A^\lambda f \to A f$ as $\lambda \to \infty$. Furthermore, $T_t^\lambda f \to T_t f$ as $\lambda \to \infty$ for each $f \in C_0$, uniformly for bounded $t \geq 0$.

Proof: By Theorem 19.4 we have $A^\lambda f = \lambda R_\lambda A f \to A f$ for any $f \in \mathcal{D}$. For fixed $\lambda > 0$ it is further clear that $h^{-1}(T_h^\lambda - I) \to A^\lambda$ in the norm topology as $h \to 0$. Now for any commuting contraction operators B and C,

$$\begin{aligned} \|B^n f - C^n f\| &\leq \|B^{n-1} + B^{n-2}C + \cdots + C^{n-1}\| \, \|B f - C f\| \\ &\leq n \|B f - C f\|. \end{aligned}$$

Fixing any $f \in C_0$ and $t, \lambda, \mu > 0$, we hence obtain as $h = t/n \to 0$

$$\begin{aligned} \|T_t^\lambda f - T_t^\mu f\| &\leq n \|T_h^\lambda f - T_h^\mu f\| \\ &= t \left\| \frac{T_h^\lambda f - f}{h} - \frac{T_h^\mu f - f}{h} \right\| \to t \|A^\lambda f - A^\mu f\|. \end{aligned}$$

For $f \in \mathcal{D}$ it follows that $T_t^\lambda f$ is Cauchy convergent as $\lambda \to \infty$ for fixed t, and since $\mathcal{D}$ is dense in C_0, the same property holds for arbitrary $f \in C_0$. Denoting the limit by $\tilde{T}_t f$, we get in particular

$$\left\|T_t^\lambda f - \tilde{T}_t f\right\| \leq t \|A^\lambda f - A f\|, \quad f \in \mathcal{D}, \; t \geq 0. \tag{8}$$

Thus, for each $f \in \mathcal{D}$ we have $T_t^\lambda f \to \tilde{T}_t f$ as $\lambda \to \infty$, uniformly for bounded t, which again extends to all $f \in C_0$.

To identify $\tilde{T}_t$, we may use the resolvent equation (3) to obtain, for any $f \in C_0$ and $\lambda, \mu > 0$,

$$\int_0^\infty e^{-\lambda t} T_t^\mu \mu R_\mu f \, dt = (\lambda - A^\mu)^{-1} \mu R_\mu f = \frac{\mu}{\lambda + \mu} R_\nu f, \tag{9}$$

where $\nu = \lambda\mu(\lambda+\mu)^{-1}$. As $\mu \to \infty$, we have $\nu \to \lambda$, and so $R_\nu f \to R_\lambda$. Furthermore,

$$\|T_t^\mu \mu R_\mu f - \tilde{T}_t f\| \le \|\mu R_\mu f - f\| + \|T_t^\mu f - \tilde{T}_t f\| \to 0,$$

so from (9) we get by dominated convergence $\int e^{-\lambda t} \tilde{T}_t f \, dt = R_\lambda f$. Hence, the semigroups (T_t) and $(\tilde{T}_t)$ have the same resolvent operators R_λ, and so they agree by Lemma 19.5. In particular, (7) then follows from (8). □

Proof of Theorem 19.6: The semigroup (T_t^λ) is clearly norm continuous in t for each $\lambda > 0$, and so the strong continuity of (T_t) follows by Lemma 19.7 as $\lambda \to \infty$. Furthermore, we note that $h^{-1}(T_h^\lambda - I) \to A^\lambda$ as $h \downarrow 0$. Using the semigroup relation and continuity, we obtain more generally

$$\frac{d}{dt} T_t^\lambda = A^\lambda T_t^\lambda = T_t^\lambda A^\lambda, \quad t \ge 0,$$

which implies

$$T_t^\lambda f - f = \int_0^t T_s^\lambda A^\lambda f \, ds, \quad f \in C_0, \, t \ge 0. \tag{10}$$

If $f \in \mathcal{D}$, then by Lemma 19.7 we get as $\lambda \to \infty$

$$\|T_s^\lambda A^\lambda f - T_s A f\| \le \|A^\lambda f - A f\| + \|T_s^\lambda A f - T_s A f\| \to 0,$$

uniformly for bounded s, and so (4) follows from (10) as $\lambda \to \infty$. By the strong continuity of T_t we may differentiate (4) to get the first relation in (5). The second relation holds by Theorem 19.4.

Conversely, assume that $h^{-1}(T_h f - f) \to g$ for some pair of functions $f, g \in C_0$. As $h \to 0$, we get

$$AR_\lambda f \leftarrow \frac{T_h - I}{h} R_\lambda f = R_\lambda \frac{T_h f - f}{h} \to R_\lambda g,$$

and so

$$f = (\lambda - A)R_\lambda f = \lambda R_\lambda f - A R_\lambda f = R_\lambda(\lambda f - g) \in \mathcal{D}. \quad \square$$

In applications, the domain of a generator A is often hard to identify or too large to be convenient for computations. It is then useful to restrict A to a suitable subdomain. An operator A with domain $\mathcal{D}$ on some Banach space B is said to be *closed* if its graph $G = \{(f, Af); f \in \mathcal{D}\}$ is a closed subset of B^2. In general, we say that A is *closable* if the closure $\overline{G}$ is the graph of a single-valued operator $\overline{A}$, the so-called *closure* of A. Note that A is closable iff the conditions $\mathcal{D} \ni f_n \to 0$ and $Af_n \to g$ imply $g = 0$.

When A is closed, a *core* for A is defined as a linear subspace $D \subset \mathcal{D}$ such that the restriction $A|_D$ has closure A. In this case, A is clearly uniquely determined by $A|_D$. We shall give some conditions ensuring that $D \subset \mathcal{D}$ is a core when A is the generator of a Feller semigroup (T_t) on C_0.

Lemma 19.8 *(closure and cores)* The generator $(A, \mathcal{D})$ of a Feller semigroup is closed, and for any $\lambda > 0$ a subspace $D \subset \mathcal{D}$ is a core for A iff $(\lambda - A)D$ is dense in C_0.

Proof: Assume that $f_1, f_2, \cdots \in \mathcal{D}$ with $f_n \to f$ and $Af_n \to g$. Then $(I - A)f_n \to f - g$, and since R_1 is bounded, it follows that $f_n \to R_1(f - g)$. Hence, $f = R_1(f - g) \in \mathcal{D}$, and we have $(I - A)f = f - g$, or $g = Af$. Thus, A is closed.

If D is a core for A, then for any $g \in C_0$ and $\lambda > 0$ there exist some $f_1, f_2, \cdots \in D$ with $f_n \to R_\lambda g$ and $Af_n \to AR_\lambda g$, and we get $(\lambda - A)f_n \to (\lambda - A)R_\lambda g = g$. Thus, $(\lambda - A)D$ is dense in C_0.

Conversely, assume that $(\lambda - A)D$ is dense in C_0. To show that D is a core, fix any $f \in \mathcal{D}$. By hypothesis we may choose some $f_1, f_2, \cdots \in D$ with
$$g_n \equiv (\lambda - A)f_n \to (\lambda - A)f \equiv g.$$
Since R_λ is bounded, we obtain $f_n = R_\lambda g_n \to R_\lambda g = f$, and thus
$$Af_n = \lambda f_n - g_n \to \lambda f - g = Af. \qquad \Box$$

A subspace $D \subset C_0$ is said to be *invariant* under (T_t) if $T_t D \subset D$ for all $t \geq 0$. In particular, we note that, for any subset $B \subset C_0$, the linear span of $\bigcup_t T_t B$ is an invariant subspace of C_0.

Proposition 19.9 *(invariance and cores, Watanabe)* If $(A, \mathcal{D})$ is the generator of a Feller semigroup, then any dense, invariant subspace $D \subset \mathcal{D}$ is a core for A.

Proof: By the strong continuity of (T_t) we note that R_1 can be approximated in the strong topology by some finite linear combinations $L_1, L_2, \ldots$ of the operators T_t. Now fix any $f \in D$, and define $g_n = L_n f$. Noting that A and L_n commute on $\mathcal{D}$ by Theorem 19.4, we get
$$(I - A)g_n = (I - A)L_n f = L_n(I - A)f \to R_1(I - A)f = f.$$
Since $g_n \in D$ and D is dense in C_0, it follows that $(I - A)D$ is dense in C_0. Hence, D is a core by Lemma 19.8. $\qquad \Box$

The Lévy processes in $\mathbb{R}^d$ are the archetypes of Feller processes, and we proceed to identify their generators. Let C_0^∞ denote the class of all infinitely differentiable functions f on $\mathbb{R}^d$ such that f and all its derivatives belong to $C_0 = C_0(\mathbb{R}^d)$.

Theorem 19.10 *(Lévy processes)* Let T_t, $t \geq 0$, be the transition operators of a Lévy process in $\mathbb{R}^d$ with characteristics (a, b, ν). Then (T_t) is a Feller semigroup, and C_0^∞ is a core for the associated generator A. Moreover, we have for any $f \in C_0^\infty$ and $x \in \mathbb{R}^d$
$$Af(x) = \tfrac{1}{2}\sum_{i,j} a_{ij} f''_{ij}(x) + \sum_i b_i f'_i(x)$$
$$+ \int \left\{ f(x+y) - f(x) - \sum_i y_i f'_i(x) 1\{|y| \leq 1\} \right\} \nu(dy). \quad (11)$$

In particular, a standard Brownian motion in $\mathbb{R}^d$ has generator $\frac{1}{2}\Delta$, and the uniform motion with velocity $b \in \mathbb{R}^d$ has generator $b\nabla$, both on the core C_0^∞. Here Δ and ∇ denote the Laplace and gradient operators, respectively. Also note that the generator of the jump component has the same form as for the pseudo-Poisson processes in Proposition 19.2, apart from the compensation for small jumps by a linear drift term.

Proof of Theorem 19.10: As $t \to 0$, we have $\mu_t^{*[t^{-1}]} \xrightarrow{w} \mu_1$. Thus, Corollary 15.20 yields $\mu_t/t \xrightarrow{v} \nu$ on $\overline{\mathbb{R}^d \setminus \{0\}}$ and

$$a^{t,h} \equiv t^{-1} \int_{|x| \leq h} xx' \mu_t(dx) \to a^h, \quad b^{t,h} \equiv t^{-1} \int_{|x| \leq h} x\mu_t(dx) \to b^h, \quad (12)$$

provided that $h > 0$ satisfies $\nu\{|x| = h\} = 0$. Now fix any $f \in C_0^\infty$, and write

$$t^{-1}(T_t f(x) - f(x)) = t^{-1} \int (f(x+y) - f(x))\mu_t(dy)$$

$$= t^{-1} \int_{|y| \leq h} \left\{ f(x+y) - f(x) - \sum_i y_i f_i'(x) - \tfrac{1}{2} \sum_{i,j} y_i y_j f_{ij}''(x) \right\} \mu_t(dy)$$

$$+ t^{-1} \int_{|y| > h} (f(x+y) - f(x))\mu_t(dy) + \sum_i b_i^{t,h} f_i'(x) + \tfrac{1}{2} \sum_{i,j} a_{ij}^{t,h} f_{ij}''(x).$$

As $t \to 0$, the last three terms approach the expression in (11), though with a_{ij} replaced by a_{ij}^h and with the integral taken over $\{|x| > h\}$. To establish the required convergence, it is then enough to show that the first term on the right tends to zero as $h \to 0$, uniformly for small $t > 0$. But this is clear from (12), since the integrand is of the order $h|y|^2$ by Taylor's formula. From the uniform boundedness of the derivatives of f, we also see that the convergence is uniform in x. Thus, $C_0^\infty \subset \mathcal{D}$ by Theorem 19.6, and (11) holds on C_0^∞.

It remains to show that C_0^∞ is a core for A. Since C_0^∞ is dense in C_0, it suffices by Proposition 19.9 to show that it is also invariant under (T_t). Then note that, by dominated convergence, the differentiation operators commute with each T_t, and use condition (F_1). $\square$

We proceed to characterize the linear operators A on C_0 whose closures $\bar{A}$ are generators of a Feller semigroups.

Theorem 19.11 *(characterization of generators, Hille, Yosida) Let A be a linear operator on C_0 with domain $\mathcal{D}$. Then A is closable and its closure $\bar{A}$ is the generator of a Feller semigroup on C_0 iff these conditions hold:*

(i) *$\mathcal{D}$ is dense in C_0;*

(ii) *the range of $\lambda_0 - A$ is dense in C_0 for some $\lambda_0 > 0$;*

(iii) *if $f \vee 0 \leq f(x)$ for some $f \in \mathcal{D}$ and $x \in S$, then $Af(x) \leq 0$.*

Condition (iii) is known as the *positive-maximum principle*.

Proof: First assume that $\bar{A}$ is the generator of a Feller semigroup (T_t). Then (i) and (ii) hold by Theorem 19.4. To prove (iii), let $f \in \mathcal{D}$ and $x \in S$ with $f^+ = f \vee 0 \leq f(x)$. Then

$$T_t f(x) \leq T_t f^+(x) \leq \|T_t f^+\| \leq \|f^+\| = f(x), \quad t \geq 0,$$

and so $h^{-1}(T_h f - f)(x) \leq 0$. As $h \to 0$, we get $Af(x) \leq 0$.

Conversely, assume that A satisfies (i), (ii), and (iii). Let $f \in \mathcal{D}$ be arbitrary, choose $x \in S$ with $|f(x)| = \|f\|$, and put $g = f \operatorname{sgn} f(x)$. Then $g \in \mathcal{D}$ with $g^+ \leq g(x)$, and so (iii) yields $Ag(x) \leq 0$. Thus, we get for any $\lambda > 0$

$$\|(\lambda - A)f\| \geq \lambda g(x) - Ag(x) \geq \lambda g(x) = \lambda \|f\|. \tag{13}$$

To show that A is closable, let $f_1, f_2, \cdots \in \mathcal{D}$ with $f_n \to 0$ and $Af_n \to g$. By (i) we may choose $g_1, g_2, \cdots \in \mathcal{D}$ with $g_n \to g$, and by (13) we have

$$\|(\lambda - A)(g_m + \lambda f_n)\| \geq \lambda \|g_m + \lambda f_n\|, \quad m, n \in \mathbb{N}, \ \lambda > 0.$$

As $n \to \infty$, we get $\|(\lambda - A)g_m - \lambda g\| \geq \lambda \|g_m\|$. Here we may divide by λ and let $\lambda \to \infty$ to obtain $\|g_m - g\| \geq \|g_m\|$, which yields $\|g\| = 0$ as $m \to \infty$. Thus, A is closable, and from (13) we note that the closure $\bar{A}$ satisfies

$$\|(\lambda - \bar{A})f\| \geq \lambda \|f\|, \quad \lambda > 0, \ f \in \operatorname{dom}(\bar{A}). \tag{14}$$

Now assume that $\lambda_n \to \lambda > 0$ and $(\lambda_n - \bar{A})f_n \to g$ for some $f_1, f_2, \cdots \in \operatorname{dom}(\bar{A})$. By (14) the sequence (f_n) is then Cauchy, say with limit $f \in C_0$. By the definition of $\bar{A}$ we get $(\lambda - \bar{A})f = g$, and so g belongs to the range of $\lambda - \bar{A}$. Letting Λ denote the set of constants $\lambda > 0$ such that $\lambda - \bar{A}$ has range C_0, it follows in particular that Λ is closed. If we can show that Λ is open as well, then by (ii) we have $\Lambda = (0, \infty)$.

Then fix any $\lambda \in \Lambda$, and conclude from (14) that $\lambda - \bar{A}$ has a bounded inverse R_λ with norm $\|R_\lambda\| \leq \lambda^{-1}$. For any $\mu > 0$ with $|\lambda - \mu|\|R_\lambda\| < 1$, we may form the bounded linear operator

$$\tilde{R}_\mu = \sum_{n \geq 0} (\lambda - \mu)^n R_\lambda^{n+1},$$

and we note that

$$(\mu - \bar{A})\tilde{R}_\mu = (\lambda - \bar{A})\tilde{R}_\mu - (\lambda - \mu)\tilde{R}_\mu = I.$$

In particular, $\mu \in \Lambda$, which shows that $\lambda \in \Lambda^\circ$.

We may next establish the resolvent equation (3). Then start from the identity $(\lambda - \bar{A})R_\lambda = (\mu - \bar{A})R_\mu = I$. By a simple rearrangement,

$$(\lambda - \bar{A})(R_\lambda - R_\mu) = (\mu - \lambda)R_\mu,$$

and (3) follows as we multiply from the left by R_λ. In particular, (3) shows that the operators R_λ and R_μ commute for any $\lambda, \mu > 0$.

Since $R_\lambda(\lambda - \bar{A}) = I$ on $\operatorname{dom}(\bar{A})$ and $\|R_\lambda\| \leq \lambda^{-1}$, we have for any $f \in \operatorname{dom}(\bar{A})$ as $\lambda \to \infty$

$$\|\lambda R_\lambda f - f\| = \|R_\lambda \bar{A}f\| \leq \lambda^{-1}\|\bar{A}f\| \to 0.$$

From (i) and the contractivity of λR_λ, it follows easily that $\lambda R_\lambda \to I$ in the strong topology. Now define A^λ as in (6) and let $T_t^\lambda = e^{tA^\lambda}$. As in the proof of Lemma 19.7, we get $T_t^\lambda f \to T_t f$ for each $f \in C_0$, uniformly for bounded t, where the T_t form a strongly continuous family of contraction operators on C_0 such that $\int e^{-\lambda t} T_t dt = R_\lambda$ for all $\lambda > 0$. To deduce the semigroup property, fix any $f \in C_0$ and $s, t \geq 0$, and note that as $\lambda \to \infty$

$$(T_{s+t} - T_s T_t)f = (T_{s+t} - T_{s+t}^\lambda)f + T_s^\lambda(T_t^\lambda - T_t)f + (T_s^\lambda - T_s)T_t f \to 0.$$

The positivity of the operators T_t will follow immediately, if we can show that R_λ is positive for each $\lambda > 0$. Then fix any function $g \geq 0$ in C_0, and put $f = R_\lambda g$, so that $g = (\lambda - \bar{A})f$. By the definition of $\bar{A}$, there exist some $f_1, f_2, \ldots \in \mathcal{D}$ with $f_n \to f$ and $Af_n \to \bar{A}f$. If $\inf_x f(x) < 0$, we have $\inf_x f_n(x) < 0$ for all sufficiently large n, and we may choose some $x_n \in S$ with $f_n(x_n) \leq f_n \wedge 0$. By (iii) we have $Af_n(x_n) \geq 0$, and so

$$\inf_x (\lambda - A) f_n(x) \leq (\lambda - A) f_n(x_n)$$
$$\leq \lambda f_n(x_n) = \lambda \inf_x f_n(x).$$

As $n \to \infty$, we get the contradiction

$$0 \leq \inf_x g(x) = \inf_x (\lambda - \bar{A}) f(x) \leq \lambda \inf_x f(x) < 0.$$

It remains to show that $\bar{A}$ is the generator of the semigroup (T_t). But this is clear from the fact that the operators $\lambda - \bar{A}$ are inverses to the resolvent operators R_λ. □

From the proof we note that any operator A on C_0 satisfying the positive maximum principle in (iii) must be *dissipative*, in the sense that $\|(\lambda - A)f\| \geq \lambda \|f\|$ for all $f \in \text{dom}(A)$ and $\lambda > 0$. This leads to the following simple observation, which will be needed later.

Lemma 19.12 *(maximality)* Let $(A, \mathcal{D})$ be the generator of a Feller semigroup on C_0, and assume that A extends to a linear operator $(A', \mathcal{D}')$ satisfying the positive-maximum principle. Then $\mathcal{D}' = \mathcal{D}$.

Proof: Fix any $f \in \mathcal{D}'$, and put $g = (I - A')f$. Since A' is dissipative and $(I - A)R_1 = I$ on C_0, we get

$$\|f - R_1 g\| \leq \|(I - A')(f - R_1 g)\| = \|g - (I - A)R_1 g\| = 0,$$

and so $f = R_1 g \in \mathcal{D}$. □

Our next aim is to show how a nice Markov process can be associated with every Feller semigroup (T_t). In order for the corresponding transition kernels μ_t to have total mass 1, we need the operators T_t to be *conservative*, in the sense that $\sup_{f \leq 1} T_t f(x) = 1$ for all $x \in S$. This can be achieved by a suitable extension.

Let us then introduce an auxiliary state $\Delta \notin S$ and form the compactified space $\hat{S} = S \cup \{\Delta\}$, where Δ is regarded as the *point at infinity* when S is noncompact, and otherwise as isolated from S. Note that any function

$f \in C_0$ has a continuous extension to $\hat{S}$, obtained by putting $f(\Delta) = 0$. We may now extend the original semigroup on C_0 to a conservative semigroup on the space $\hat{C} = C(\hat{S})$.

Lemma 19.13 *(compactification)* *Any Feller semigroup (T_t) on C_0 admits an extension to a conservative Feller semigroup $(\hat{T}_t)$ on $\hat{C}$, given by*

$$\hat{T}_t f = f(\Delta) + T_t \{f - f(\Delta)\}, \quad t \geq 0, \ f \in \hat{C}.$$

Proof: It is straightforward to verify that $(\hat{T}_t)$ is a strongly continuous semigroup on $\hat{C}$. To show that the operators $\hat{T}_t$ are positive, fix any $f \in \hat{C}$ with $f \geq 0$, and note that $g \equiv f(\Delta) - f \in C_0$ with $g \leq f(\Delta)$. Hence,

$$T_t g \leq T_t g^+ \leq \|T_t g^+\| \leq \|g^+\| \leq f(\Delta),$$

and so $\hat{T}_t f = f(\Delta) - T_t g \geq 0$. The contraction and conservation properties now follow from the fact that $\hat{T}_t 1 = 1$. □

Our next step is to construct an associated semigroup of Markov transition kernels μ_t on $\hat{S}$, satisfying

$$T_t f(x) = \int f(y) \mu_t(x, dy), \quad f \in C_0. \tag{15}$$

We say that a state $x \in \hat{S}$ is *absorbing* for (μ_t) if $\mu_t(x, \{x\}) = 1$ for each $t \geq 0$.

Proposition 19.14 *(existence)* *For any Feller semigroup (T_t) on C_0, there exists a unique semigroup of Markov transition kernels μ_t on $\hat{S}$ satisfying (15) and such that Δ is absorbing for (μ_t).*

Proof: For fixed $x \in S$ and $t \geq 0$, the mapping $f \mapsto \hat{T}_t f(x)$ is a positive linear functional on $\hat{C}$ with norm 1, so by Riesz's representation Theorem 2.22 there exist some probability measures $\mu_t(x, \cdot)$ on $\hat{S}$ satisfying

$$\hat{T}_t f(x) = \int f(y) \mu_t(x, dy), \quad f \in \hat{C}, \ x \in \hat{S}, \ t \geq 0. \tag{16}$$

The measurability of the right-hand side is clear by continuity. By a standard approximation followed by a monotone class argument, we then obtain the desired measurability of $\mu_t(x, B)$ for any $t \geq 0$ and Borel set $B \subset \hat{S}$. The Chapman–Kolmogorov relation holds on $\hat{S}$ by Lemma 19.1. Relation (15) is a special case of (16), and from (16) we further get

$$\int f(y) \mu_t(\Delta, dy) = \hat{T}_t f(\Delta) = f(\Delta) = 0, \quad f \in C_0,$$

which shows that Δ is absorbing. The uniqueness of (μ_t) is a consequence of the last two properties. □

For any probability measure ν on $\hat{S}$, there exists by Theorem 8.4 a Markov process X^ν in $\hat{S}$ with initial distribution ν and transition kernels μ_t. As before, we denote the distribution of X^ν by P_ν and write E_ν for the corresponding integration operator. When $\nu = \delta_x$, we often prefer the simpler forms P_x and E_x, respectively. We may now extend Theorem 15.1 to a basic regularization theorem for Feller processes. Given a process X, we say that Δ is *absorbing* for $X^\pm$ if $X_t = \Delta$ or $X_{t-} = \Delta$ implies $X_u = \Delta$ for all $u \geq t$.

Theorem 19.15 *(regularization, Kinney)* Let X be a Feller process in $\hat{S}$ with arbitrary initial distribution ν. Then X has an rcll version $\tilde{X}$ in $\hat{S}$ such that Δ is absorbing for $\tilde{X}^\pm$. If (T_t) is conservative and ν is restricted to S, we can choose $\tilde{X}$ to be rcll even in S.

The idea of the proof is to construct a sufficiently rich class of supermartingales, to which the regularity theorems of Chapter 7 can be applied. Let C_0^+ denote the class of nonnegative functions in C_0.

Lemma 19.16 *(resolvents and excessive functions)* If $f \in C_0^+$, then the process $Y_t = e^{-t} R_1 f(X_t)$, $t \geq 0$, is a supermartingale under P_ν for every ν.

Proof: Writing $(\mathcal{G}_t)$ for the filtration induced by X, we get for any $t, h \geq 0$

$$\begin{aligned}
E[Y_{t+h}|\mathcal{G}_t] &= E[e^{-t-h} R_1 f(X_{t+h})|\mathcal{G}_t] = e^{-t-h} T_h R_1 f(X_t) \\
&= e^{-t-h} \int_0^\infty e^{-s} T_{s+h} f(X_t) ds \\
&= e^{-t} \int_h^\infty e^{-s} T_s f(X_t) ds \leq Y_t.
\end{aligned}$$
$\square$

Proof of Theorem 19.15: By Lemma 19.16 and Theorem 7.27, the process $f(X_t)$ has a.s. right- and left-hand limits along $\mathbb{Q}_+$ for any $f \in \mathcal{D} \equiv \text{dom}(A)$. Since $\mathcal{D}$ is dense in C_0, the stated property holds for every $f \in C_0$. By the separability of C_0 we may choose the exceptional null set N to be independent of f. If $x_1, x_2, \cdots \in \hat{S}$ are such that $f(x_n)$ converges for every $f \in C_0$, then the compactness of $\hat{S}$ ensures that x_n converges in the topology of $\hat{S}$. Thus, on N^c the process X itself has right- and left-hand limits $X_{t\pm}$ along $\mathbb{Q}_+$; on N we may redefine X to be 0. Then clearly $\tilde{X}_t = X_{t+}$ is rcll. It remains to show that $\tilde{X}$ is a version of X or, equivalently, that $X_{t+} = X_t$ a.s. for each $t \geq 0$. But this follows from the fact that $X_{t+h} \xrightarrow{P} X_t$ as $h \downarrow 0$ by Lemma 19.3 and dominated convergence.

Now fix any $f \in C_0$ with $f > 0$ on S, and note from the strong continuity of (T_t) that even $R_1 f > 0$ on S. Applying Lemma 7.31 to the supermartingale $Y_t = e^{-t} R_1 f(\tilde{X}_t)$, we conclude that $X \equiv \Delta$ a.s. on the interval $[\zeta, \infty)$, where $\zeta = \inf\{t \geq 0;\ \Delta \in \{\tilde{X}_t, \tilde{X}_{t-}\}\}$. Discarding the exceptional null set, we can make this hold identically. If (T_t) is conservative and ν is restricted to S, then $\tilde{X}_t \in S$ a.s. for every $t \geq 0$. Thus, $\zeta > t$ a.s. for all t, and hence $\zeta = \infty$ a.s. Again we may assume that this holds identically. Then $\tilde{X}_t$ and $\tilde{X}_{t-}$ take values in S, and the stated regularity properties remain valid in S.
$\square$

In view of the last theorem, we may choose Ω to be the space of all $\hat{S}$-valued rcll functions such that the state Δ is absorbing, and let X be the canonical process on Ω. Processes with different initial distributions ν are then distinguished by their distributions P_ν on Ω. Thus, under P_ν the process X is Markov with initial distribution ν and transition kernels μ_t, and X has all the regularity properties stated in Theorem 19.15. In particular, $X \equiv \Delta$ on the interval $[\zeta, \infty)$, where ζ denotes the *terminal time*

$$\zeta = \inf\{t \geq 0;\ X_t = \Delta \text{ or } X_{t-} = \Delta\}.$$

We take $(\mathcal{F}_t)$ to be the right-continuous filtration generated by X, and put $\mathcal{A} = \mathcal{F}_\infty = \bigvee_t \mathcal{F}_t$. The *shift operators* θ_t on Ω are defined as before by

$$(\theta_t \omega)_s = \omega_{s+t}, \quad s, t \geq 0.$$

The process X with associated distributions P_ν, filtration $\mathcal{F} = (\mathcal{F}_t)$, and shift operators θ_t is called the *canonical Feller process* with semigroup (T_t).

We are now ready to state a general version of the strong Markov property. The result extends the special versions obtained in Proposition 8.9 and Theorems 12.14 and 13.11. A further instant of this property appears in Theorem 21.11.

Theorem 19.17 *(strong Markov property, Dynkin and Yushkevich, Blumenthal)* *For any canonical Feller process X, initial distribution ν, optional time τ, and random variable $\xi \geq 0$, we have*

$$E_\nu[\xi \circ \theta_\tau | \mathcal{F}_\tau] = E_{X_\tau} \xi \quad \text{a.s. } P_\nu \text{ on } \{\tau < \infty\}.$$

Proof: By Lemmas 6.2 and 7.1 we may assume that $\tau < \infty$. Let $\mathcal{G}$ denote the filtration induced by X. Then Lemma 7.4 shows that the times $\tau_n = 2^{-n}[2^n \tau + 1]$ are $\mathcal{G}$-optional, and by Lemma 7.3 we have $\mathcal{F}_\tau \subset \mathcal{G}_{\tau_n}$ for all n. Thus, Proposition 8.9 yields

$$E_\nu[\xi \circ \theta_{\tau_n}; A] = E_\nu[E_{X_{\tau_n}} \xi; A], \quad A \in \mathcal{F}_\tau,\ n \in \mathbb{N}. \tag{17}$$

To extend the relation to τ, we first assume that $\xi = \prod_{k \leq m} f_k(X_{t_k})$ for some $f_1, \ldots, f_m \in C_0$ and $t_1 < \cdots < t_m$. Then $\xi \circ \theta_{\tau_n} \to \xi \circ \theta_\tau$ by the right-continuity of X and the continuity of $f_1, \ldots, f_m$. Writing $h_k = t_k - t_{k-1}$ with $t_0 = 0$, it is also clear from the first Feller property and the right-continuity of X that

$$\begin{aligned}
E_{X_{\tau_n}} \xi &= T_{h_1}(f_1 T_{h_2} \cdots (f_{m-1} T_{h_m} f_m) \cdots)(X_{\tau_n}) \\
&\to T_{h_1}(f_1 T_{h_2} \cdots (f_{m-1} T_{h_m} f_m) \cdots)(X_\tau) = E_{X_\tau} \xi.
\end{aligned}$$

Thus, (17) extends to τ by dominated convergence on both sides. We may finally use standard approximation and monotone class arguments to extend the result to arbitrary ξ. □

As a simple application, we get the following useful zero–one law.

Corollary 19.18 *(Blumenthal's 0-1 law)* *For any canonical Feller process, we have*
$$P_x A = 0 \text{ or } 1, \quad x \in S, \ A \in \mathcal{F}_0.$$

Proof: Taking $\tau = 0$ in Theorem 19.17, we get for any $x \in S$ and $A \in \mathcal{F}_0$
$$1_A = P_x[A|\mathcal{F}_0] = P_{X_0} A = P_x A \text{ a.s. } P_x. \qquad \square$$

To appreciate the last result, recall that $\mathcal{F}_0 = \mathcal{F}_{0+}$. In particular, we note that $P_x\{\tau = 0\} = 0$ or 1 for any state $x \in S$ and $\mathcal{F}$-optional time τ.

The strong Markov property is often used in the following extended form.

Corollary 19.19 *(optional projection)* *For any canonical Feller process X, nondecreasing adapted process Y, and random variable $\xi \geq 0$, we have*
$$E_x \int_0^\infty (E_{X_t}\xi)\, dY_t = E_x \int_0^\infty (\xi \circ \theta_t)\, dY_t, \quad x \in S.$$

Proof: We may assume that $Y_0 = 0$. Introduce the right-continuous inverse
$$\tau_s = \inf\{t \geq 0;\ Y_t > s\}, \quad s \geq 0,$$
and note that the times τ_s are optional by Lemma 7.6. By Theorem 19.17 we have
$$E_x[E_{X_{\tau_s}}\xi;\ \tau_s < \infty] = E_x[E_x[\xi \circ \theta_{\tau_s}|\mathcal{F}_{\tau_s}];\ \tau_s < \infty]$$
$$= E_x[\xi \circ \theta_{\tau_s};\ \tau_s < \infty].$$
Since $\tau_s < \infty$ iff $s < Y_\infty$, we get by integration
$$E_x \int_0^{Y_\infty} (E_{X_{\tau_s}}\xi)\, ds = E_x \int_0^{Y_\infty} (\xi \circ \theta_{\tau_s})\, ds,$$
and the asserted formula follows by Lemma 1.22. $\qquad \square$

Our next aim is to show that any martingale on the canonical space of a Feller process X is a.s. continuous outside the discontinuity set of X. For Brownian motion, the result was already noted as a consequence of the integral representation in Theorem 18.10.

Theorem 19.20 *(discontinuity sets)* *Let X be a canonical Feller process with arbitrary initial distribution ν, and let M be a local P_ν-martingale. Then*
$$\{t > 0;\ \Delta M_t \neq 0\} \subset \{t > 0;\ X_{t-} \neq X_t\} \text{ a.s.} \tag{18}$$

Proof (Chung and Walsh): By localization we may reduce to the case when M is uniformly integrable and hence of the form $M_t = E[\xi|\mathcal{F}_t]$ for some $\xi \in L^1$. Let $\mathcal{C}$ denote the class of random variables $\xi \in L^1$ such that the corresponding M satisfies (18). Then $\mathcal{C}$ is a linear subspace of L^1. It is further closed, since if $M_t^n = E[\xi_n|\mathcal{F}_t]$ with $\|\xi_n\|_1 \to 0$, then
$$P\{\sup_t |M_t^n| > \varepsilon\} \leq \varepsilon^{-1} E|\xi_n| \to 0, \quad \varepsilon > 0,$$

and so $\sup_t |M_t^n| \xrightarrow{P} 0$.

Now let $\xi = \prod_{k \leq n} f_k(X_{t_k})$ for some $f_1, \ldots, f_n \in C_0$ and $t_1 < \cdots < t_n$. Writing $h_k = t_k - t_{k-1}$, we note that

$$M_t = \prod_{k \leq m} f_k(X_{t_k}) T_{t_{m+1}-t} g_{m+1}(X_t), \quad t \in [t_m, t_{m+1}], \tag{19}$$

where

$$g_k = f_k T_{h_{k+1}}(f_{k+1} T_{h_{k+2}}(\cdots T_{h_n} f_n) \cdots), \quad k = 1, \ldots, n,$$

with the obvious conventions for $t < t_1$ and $t > t_n$. Since $T_t g(x)$ is jointly continuous in (t, x) for each $g \in C_0$, equation (19) defines a right-continuous version of M satisfying (18), and so $\xi \in \mathcal{C}$. By a simple approximation it follows that $\mathcal{C}$ contains all indicator functions of sets $\bigcap_{k \leq n} \{X_{t_k} \in G_k\}$ with $G_1, \ldots, G_n$ open. The result extends by a monotone class argument to any X-measurable indicator function ξ, and a routine argument yields the final extension to L_1. □

A basic role in the theory is played by the processes

$$M_t^f = f(X_t) - f(X_0) - \int_0^t Af(X_s) ds, \quad t \geq 0, \ f \in \mathcal{D}.$$

Lemma 19.21 *(Dynkin's formula) The processes M^f are martingales under any initial distribution ν for X. In particular, we have for any bounded optional time τ*

$$E_x f(X_\tau) = f(x) + E_x \int_0^\tau Af(X_s) ds, \quad x \in S, \ f \in \mathcal{D}. \tag{20}$$

Proof: For any $t, h \geq 0$, we have

$$M_{t+h}^f - M_t^f = f(X_{t+h}) - f(X_t) - \int_t^{t+h} Af(X_s) ds = M_h^f \circ \theta_t,$$

and so by the Markov property at t and Theorem 19.6

$$E_\nu[M_{t+h}^f | \mathcal{F}_t] - M_t^f = E_\nu[M_h^f \circ \theta_t | \mathcal{F}_t] = E_{X_t} M_h^f = 0.$$

Thus, M^f is a martingale, and (20) follows by optional sampling. □

As a preparation for the next major result, we introduce the optional times

$$\tau_h = \inf\{t \geq 0; \rho(X_t, X_0) > h\}, \quad h > 0,$$

where ρ denotes the metric in S. Note that a state x is *absorbing* iff $\tau_h = \infty$ a.s. P_x for every $h > 0$.

Lemma 19.22 *(escape times) For any nonabsorbing state $x \in S$, we have $E_x \tau_h < \infty$ for all sufficiently small $h > 0$.*

Proof: If x is not absorbing, then $\mu_t(x, B_x^\varepsilon) < p < 1$ for some $t, \varepsilon > 0$, where $B_x^\varepsilon = \{y; \rho(x, y) \leq \varepsilon\}$. By Lemma 19.3 and Theorem 4.25 we may choose $h \in (0, \varepsilon]$ so small that

$$\mu_t(y, B_x^h) \leq \mu_t(y, B_x^\varepsilon) \leq p, \quad y \in B_x^h.$$

Then Proposition 8.2 yields

$$P_x\{\tau_h \geq nt\} \leq P_x \bigcap_{k \leq n} \{X_{kt} \in B_x^h\} \leq p^n, \quad n \in \mathbb{Z}_+,$$

and so by Lemma 3.4

$$E_x \tau_h = \int_0^\infty P\{\tau_h \geq s\} ds \leq t \sum_{n \geq 0} P\{\tau_h \geq nt\} = t \sum_{n \geq 0} p^n = \frac{t}{1-p} < \infty. \quad \Box$$

We turn to a probabilistic description of the generator and its domain. Say that A is *maximal* within a class of linear operators if it extends every member of the class.

Theorem 19.23 *(characteristic operator, Dynkin)* Let $(A, \mathcal{D})$ be the generator of a Feller process. Then for any $f \in \mathcal{D}$ we have $Af(x) = 0$ if x is absorbing, and otherwise

$$Af(x) = \lim_{h \to 0} \frac{E_x f(X_{\tau_h}) - f(x)}{E_x \tau_h}. \tag{21}$$

Furthermore, A is the maximal operator on C_0 with those properties.

Proof: Fix any $f \in \mathcal{D}$. If x is absorbing, then $T_t f(x) = f(x)$ for all $t \geq 0$, and so $Af(x) = 0$. For a nonabsorbing x, we get instead by Lemma 19.21

$$E_x f(X_{\tau_h \wedge t}) - f(x) = E_x \int_0^{\tau_h \wedge t} Af(X_s) ds, \quad t, h > 0. \tag{22}$$

By Lemma 19.22 we have $E\tau_h < \infty$ for sufficiently small $h > 0$, and so (22) extends by dominated convergence to $t = \infty$. Relation (21) now follows from the continuity of Af, together with the fact that $\rho(X_s, x) \leq h$ for all $s < \tau_h$. Since the positive maximum principle holds for any extension of A with the stated properties, the last assertion follows by Lemma 19.12. $\quad \Box$

In the special case when $S = \mathbb{R}^d$, let C_K^∞ denote the class of infinitely differentiable functions on $\mathbb{R}^d$ with bounded support. An operator $(A, \mathcal{D})$ with $\mathcal{D} \supset C_K^\infty$ is said to be *local* on C_K^∞ if $Af(x) = 0$ whenever f vanishes in some neighborhood of x. For any generator with this property, we note that the positive-maximum principle implies a *local positive-maximum principle*, asserting that if $f \in C_K^\infty$ has a local maximum ≥ 0 at some point x, then $Af(x) \leq 0$.

The following result gives the basic connection between diffusion processes and elliptic differential operators. This connection is explored further in Chapters 21 and 24.

Theorem 19.24 *(Feller diffusions and elliptic operators, Dynkin)* Let $(A, \mathcal{D})$ be the generator of a Feller process X in $\mathbb{R}^d$, and assume that $C_K^\infty \subset \mathcal{D}$. Then X is continuous on $[0, \zeta)$, a.s. P_ν for every ν, iff A is local on C_K^∞. In that case there exist some functions $a_{ij}, b_i, c \in C(\mathbb{R}^d)$, where $c \geq 0$ and the a_{ij} form a symmetric, nonnegative definite matrix, such that for any $f \in C_K^\infty$ and $x \in \mathbb{R}_+$,

$$Af(x) = \tfrac{1}{2}\sum_{i,j} a_{ij}(x) f''_{ij}(x) + \sum_i b_i(x) f'_i(x) - c(x) f(x). \tag{23}$$

In the situation described by this result, we may choose Ω to consist of all paths that are continuous on $[0, \zeta)$. The resulting Markov process is referred to as a *canonical Feller diffusion*.

Proof: If X is continuous on $[0, \zeta)$, then A is local by Theorem 19.23. Conversely, assume that A is local on C_K^∞. Fix any $x \in \mathbb{R}^d$ and $0 < h < m$, and choose $f \in C_K^\infty$ with $f \geq 0$ and support $\{y;\ h \leq |y-x| \leq m\}$. Then $Af(y) = 0$ for all $y \in B_x^h$, and so Lemma 19.21 shows that $f(X_{t \wedge \tau_h})$ is a martingale under P_x. By dominated convergence we get $E_x f(X_{\tau_h}) = 0$, and since m was arbitrary,

$$P_x\{|X_{\tau_h} - x| \leq h \text{ or } X_{\tau_h} = \Delta\} = 1, \quad x \in \mathbb{R}^d,\ h > 0.$$

Applying the Markov property at fixed times, we obtain for any initial distribution ν

$$P_\nu \bigcap_{t \in \mathbb{Q}_+} \theta_t^{-1}\{|X_{\tau_h} - X_0| \leq h \text{ or } X_{\tau_h} = \Delta\} = 1, \quad h > 0,$$

which implies

$$P_\nu\{\sup_{t < \zeta} |\Delta X_t| \leq h\} = 1, \quad h > 0.$$

Hence, X is continuous on $[0, \zeta)$ a.s. P_μ.

To show that (23) holds for suitable a_{ij}, b_i, and c, we choose for every $x \in \mathbb{R}^d$ some functions $f_0^x, f_i^x, f_{ij}^x \in C_K^\infty$ such that, for any y close to x,

$$f_0^x(y) = 1, \quad f_i^x(y) = y_i - x_i, \quad f_{ij}^x(y) = (y_i - x_i)(y_j - x_j).$$

Putting

$$c(x) = -Af_0^x(x), \quad b_i(x) = Af_i^x(x), \quad a_{ij}(x) = Af_{ij}^x(x),$$

we note that (23) holds locally for any function $f \in C_K^\infty$ that agrees near x with a second-degree polynomial. In particular, we may choose $f_0(y) = 1$, $f_i(y) = y_i$, and $f_{ij}(y) = y_i y_j$ near x to obtain

$$\begin{aligned} Af_0(x) &= -c(x), \\ Af_i(x) &= b_i(x) - x_i c(x), \\ Af_{ij}(x) &= a_{ij}(x) + x_i b_j(x) + x_j b_i(x) - x_i x_j c(x). \end{aligned}$$

This shows that c, b_i, and $a_{ij} = a_{ji}$ are continuous.

Applying the local positive-maximum principle to f_0^x gives $c(x) \geq 0$. By the same principle applied to the function

$$f = -\left\{\sum_i u_i f_i^x\right\}^2 = -\sum_{ij} u_i u_j f_{ij}^x,$$

we get $\sum_{ij} u_i u_j a_{ij}(x) \geq 0$, which shows that (a_{ij}) is nonnegative definite. Finally, we consider any function $f \in C_K^\infty$ with a second-order Taylor expansion $\tilde{f}$ around x. Here each function

$$g_\pm^\varepsilon(y) = \pm(f(y) - \tilde{f}(y)) - \varepsilon|x - y|^2, \quad \varepsilon > 0,$$

has a local maximum 0 at x, and so

$$A g_\pm^\varepsilon(x) = \pm(Af(x) - A\tilde{f}(x)) - \varepsilon \sum_i a_{ii}(x) \leq 0, \quad \varepsilon > 0.$$

Letting $\varepsilon \to 0$ gives $Af(x) = A\tilde{f}(x)$, which shows that (23) is generally true. □

We consider next a basic convergence theorem for Feller processes, essentially generalizing the result for Lévy processes in Theorem 15.17.

Theorem 19.25 *(convergence, Trotter, Sova, Kurtz, Mackevičius)* Let X and X^n be Feller processes in S with semigroups (T_t) and $(T_{n,t})$ and generators $(A, \mathcal{D})$ and $(A_n, \mathcal{D}_n)$, respectively. Fix a core D for A. Then these conditions are equivalent:

(i) If $f \in D$, there exist some $f_n \in \mathcal{D}_n$ with $f_n \to f$ and $A_n f_n \to Af$.

(ii) $T_{n,t} \to T_t$ strongly for each $t > 0$.

(iii) $T_{n,t} f \to T_t f$ for every $f \in C_0$, uniformly for bounded $t > 0$.

(iv) If $X_0^n \xrightarrow{d} X_0$ in S, then $X^n \xrightarrow{d} X$ in $D(\mathbb{R}_+, \hat{S})$.

For the proof we need two lemmas, the first of which extends Lemma 19.7.

Lemma 19.26 *(norm inequality)* Let (T_t) and (T_t') be Feller semigroups with generators $(A, \mathcal{D})$ and $(A', \mathcal{D}')$, respectively, where A' is bounded. Then

$$\|T_t f - T_t' f\| \leq \int_0^t \|(A - A')T_s f\| \, ds, \quad f \in \mathcal{D}, \, t \geq 0. \tag{24}$$

Proof: Fix any $f \in \mathcal{D}$ and $t > 0$. Since (T_s') is norm continuous, we get by Theorem 19.6

$$\frac{\partial}{\partial s}(T_{t-s}' T_s f) = T_{t-s}'(A - A')T_s f, \quad 0 \leq s \leq t.$$

Here the right-hand side is continuous in s, because of the strong continuity of (T_s), the boundedness of A', the commutativity of A and T_s, and the norm continuity of (T_s'). Hence,

$$T_t f - T_t' f = \int_0^t \frac{\partial}{\partial s}(T_{t-s}' T_s f) \, ds = \int_0^t T_{t-s}'(A - A')T_s f \, ds,$$

and (24) follows by the contractivity of T'_{t-s}. □

We may next establish a continuity property for the Yosida approximations A^λ and A_n^λ of A and A_n, respectively.

Lemma 19.27 *(continuity of Yosida approximation)* Let $(A, \mathcal{D})$ and $(A_n, \mathcal{D}_n)$ be the generators of some Feller semigroups satisfying condition (i) of Theorem 19.25. Then $A_n^\lambda \to A^\lambda$ strongly for every $\lambda > 0$.

Proof: By Lemma 19.8 it suffices to show that $A_n^\lambda f \to A^\lambda f$ for every $f \in (\lambda - A)\mathcal{D}$. Then define $g \equiv R^\lambda f \in \mathcal{D}$. By (i) we may choose some $g_n \in \mathcal{D}_n$ with $g_n \to g$ and $A_n g_n \to Ag$. Then $f_n \equiv (\lambda - A_n)g_n \to (\lambda - A)g = f$, and so

$$\begin{aligned}
\|A_n^\lambda f - A^\lambda f\| &= \lambda^2 \|R_n^\lambda f - R^\lambda f\| \\
&\leq \lambda^2 \|R_n^\lambda (f - f_n)\| + \lambda^2 \|R_n^\lambda f_n - R^\lambda f\| \\
&\leq \lambda \|f - f_n\| + \lambda^2 \|g_n - g\| \to 0.
\end{aligned}$$
□

Proof of Theorem 19.25: First we show that (i) implies (iii). Since D is dense in C_0, it is enough to verify (iii) for $f \in \mathcal{D}$. Then choose some functions f_n as in (i), and conclude by Lemmas 19.7 and 19.26 that, for any $n \in \mathbb{N}$ and $t, \lambda > 0$,

$$\begin{aligned}
\|T_{n,t} f - T_t f\| &\leq \|T_{n,t}(f - f_n)\| + \|(T_{n,t} - T_{n,t}^\lambda) f_n\| + \|T_{n,t}^\lambda (f_n - f)\| \\
&\quad + \|(T_{n,t}^\lambda - T_t^\lambda) f\| + \|(T_t^\lambda - T_t) f\| \\
&\leq 2\|f_n - f\| + t\|(A^\lambda - A) f\| + t\|(A_n - A_n^\lambda) f_n\| \\
&\quad + \int_0^t \|(A_n^\lambda - A^\lambda) T_s^\lambda f\| \, ds.
\end{aligned} \quad (25)$$

By Lemma 19.27 and dominated convergence, the last term tends to zero as $n \to \infty$. For the third term on the right, we get

$$\begin{aligned}
\|(A_n - A_n^\lambda) f_n\| &\leq \|A_n f_n - A f\| + \|(A - A^\lambda) f\| \\
&\quad + \|(A^\lambda - A_n^\lambda) f\| + \|A_n^\lambda (f - f_n)\|,
\end{aligned}$$

which tends to $\|(A - A^\lambda) f\|$ by the same lemma. Hence, by (25)

$$\limsup_{n \to \infty} \sup_{t \leq u} \|T_{n,t} f - T_t f\| \leq 2u\|(A^\lambda - A)f\|, \quad u, \lambda > 0,$$

and the desired convergence follows by Lemma 19.7 as we let $\lambda \to \infty$.

Conversely, (iii) trivially implies (ii), and so the equivalence of (i)–(iii) will follow if we can show that (ii) implies (i). Then fix any $f \in \mathcal{D}$ and $\lambda > 0$, and define $g = (\lambda - A)f$ and $f_n = R_n^\lambda g$. Assuming (ii), we get by dominated convergence $f_n \to R^\lambda g = f$. Since $(\lambda - A_n)f_n = g = (\lambda - A)f$, we also note that $A_n f_n \to Af$. Thus, even (i) holds.

It remains to show that conditions (i)–(iii) are equivalent to (iv). For convenience, we may then assume that S is compact and the semigroups (T_t) and $(T_{n,t})$ are conservative. First assume (iv). We may establish (ii) by showing that, for any $f \in C$ and $t > 0$, we have $T_t^n f(x_n) \to T_t f(x)$

whenever $x_n \to x$ in S. Then assume that $X_0 = x$ and $X_0^n = x_n$. By Lemma 19.3 the process X is a.s. continuous at t. Thus, (iv) yields $X_t^n \xrightarrow{d} X_t$, and the desired convergence follows.

Conversely, assume conditions (i)–(iii), and let $X_0^n \xrightarrow{d} X_0$. To obtain $X^n \xrightarrow{fd} X$, it is enough to show that, for any $f_0, \ldots, f_m \in C$ and $0 = t_0 < t_1 \cdots t_m$,

$$\lim_{n\to\infty} E \prod_{k\le m} f_k(X_{t_k}^n) = E \prod_{k \le m} f_k(X_{t_k}). \qquad (26)$$

This holds by hypothesis when $m = 0$. Proceeding by induction, we may use the Markov property to rewrite (26) in the form

$$E \prod_{k<m} f_k(X_{t_k}^n) \cdot T_{h_m}^n f_m(X_{t_{m-1}}^n) \to E \prod_{k<m} f_k(X_{t_k}) \cdot T_{h_m} f_m(X_{t_{m-1}}), \qquad (27)$$

where $h_m = t_m - t_{m-1}$. Since (ii) implies $T_{h_m}^n f_m \to T_{h_m} f_m$, it is equivalent to prove (27) with $T_{h_m}^n$ replaced by T_{h_m}. The resulting condition is of the form (26) with m replaced by $m - 1$. This completes the induction and shows that $X^n \xrightarrow{fd} X$.

To strengthen the conclusion to $X^n \xrightarrow{d} X$, it suffices by Theorems 16.10 and 16.11 to show that $\rho(X_{\tau_n}^n, X_{\tau_n+h_n}^n) \xrightarrow{P} 0$ for any finite optional times τ_n and positive constants $h_n \to 0$. By the strong Markov property we may prove instead that $\rho(X_0^n, X_{h_n}^n) \xrightarrow{P} 0$ for any initial distributions ν_n. In view of the compactness of S and Theorem 16.3, we may then assume that $\nu_n \xrightarrow{w} \nu$ for some probability measure ν. Fixing any $f, g \in C$ and noting that $T_{h_n}^n g \to g$ by (iii), we get

$$Ef(X_0^n) g(X_{h_n}^n) = E f T_{h_n}^n g(X_0^n) \to E fg(X_0),$$

where $\mathcal{L}(X_0) = \nu$. Then $(X_0^n, X_{h_n}^n) \xrightarrow{d} (X_0, X_0)$ as before, and in particular $\rho(X_0^n, X_{h_n}^n) \xrightarrow{d} \rho(X_0, X_0) = 0$. This completes the proof of (iv). □

From the last theorem and its proof we may easily deduce a similar approximation property for discrete-time Markov chains. The result extends the approximations for random walks obtained in Corollary 15.20 and Theorem 16.14.

Theorem 19.28 (*approximation of Markov chains*) *Let $Y^1, Y^2, \ldots$ be discrete-time Markov chains in S with transition operators $U_1, U_2, \ldots$, and consider a Feller process X in S with semigroup (T_t) and generator A. Fix a core D for A, and assume that $0 < h_n \to 0$. Then conditions (i)–(iv) of Theorem 19.25 remain equivalent for the operators and processes*

$$A_n = h_n^{-1}(U_n - I), \qquad T_{n,t} = U_n^{[t/h_n]}, \qquad X_t^n = Y_{[t/h_n]}^n.$$

Proof: Let N be an independent, unit-rate Poisson process, and note that the processes $\tilde{X}_t^n = Y^n \circ N_{t/h_n}$ are pseudo-Poisson with generators A_n. Theorem 19.25 shows that (i) is equivalent to (iv) with X^n replaced by $\tilde{X}^n$. By the strong law of large numbers for N together with Theorem 4.28, we also see that (iv) holds simultaneously for the processes X^n and $\tilde{X}^n$. Thus, (i) and (iv) are equivalent.

Since X is a.s. continuous at fixed times, condition (iv) yields $X_{t_n}^n \xrightarrow{d} X_t$ whenever $t_n \to t$ and the processes X^n and X start at fixed points $x_n \to x$ in $\hat{S}$. Hence, $T_{n,t_n} f(x_n) \to T_t f(x)$ for any $f \in \hat{C}$, and (iii) follows. Since (iii) trivially implies (ii), it remains to show that (ii) implies (i).

Arguing as in the preceding proof, we then need to show that $\tilde{R}_n^\lambda g \to R^\lambda g$ for any $\lambda > 0$ and $g \in C_0$, where $\tilde{R}_n^\lambda = (\lambda - A_n)^{-1}$. Now (ii) yields $R_n^\lambda g \to R^\lambda g$, where $R_n^\lambda = \int e^{-\lambda t} T_{n,t} dt$, and so it suffices to prove that $(R_n^\lambda - \tilde{R}_n^\lambda) g \to 0$. Then note that

$$\lambda R_n^\lambda g - \lambda \tilde{R}_n^\lambda g = Eg(Y_{\kappa_n - 1}^n) - Eg(Y_{\tilde{\kappa}_n - 1}^n),$$

where the random variables κ_n and $\tilde{\kappa}_n$ are independent of Y^n and geometrically distributed with parameters $p_n = 1 - e^{-\lambda h_n}$ and $\tilde{p}_n = \lambda h_n (1 + \lambda h_n)^{-1}$, respectively. Since $p_n \sim \tilde{p}_n$, we have $\|\mathcal{L}(\kappa_n) - \mathcal{L}(\tilde{\kappa}_n)\| \to 0$, and the desired convergence follows by Fubini's theorem. $\square$

Exercises

1. Examine how the proofs of Theorems 19.4 and 19.6 can be simplified if we assume (F$_3$) instead of the weaker condition (F$_2$).

2. Consider a pseudo-Poisson process X on S with rate kernel α. Give conditions ensuring X to be Feller.

3. Verify the resolvent equation (3), and conclude that the range of R_λ is independent of λ.

4. Show that a Feller semigroup (T_t) is uniquely determined by the resolvent operator R_λ for a fixed $\lambda > 0$. Interpret the result probabilistically in terms of an independent, exponentially distributed random variable with mean λ^{-1}. (*Hint:* Use Theorem 19.4 and Lemma 19.5.)

5. Consider a discrete-time Markov process in S with transition operator T, and let τ be an independent random variable with a fixed geometric distribution. Show that T is uniquely determined by $E_x f(X_\tau)$ for arbitrary $x \in S$ and $f \geq 0$. (*Hint:* Apply the preceding result to the associated pseudo-Poisson process.)

6. Give a probabilistic description of the Yosida approximation T_t^λ in terms the original process X and *two* independent Poisson processes with rate λ.

7. Given a Feller diffusion semigroup, write the second differential equation in Theorem 19.6, for suitable f, as a PDE for the function $T_t f(x)$ on $\mathbb{R}_+ \times \mathbb{R}^d$. Also show that the backward equation of Theorem 12.22 is a special case of the same equation.

8. Consider a Feller process X and an independent subordinator T. Show that $Y_t = X(T_t)$ is again Markov, and that Y is Lévy whenever this is true for X. If both T and X are stable, then so is Y. Find the relationship between the transition semigroups, respectively between the indices of stability.

9. Consider a Feller process X and an independent renewal process $\tau_0, \tau_1, \ldots$. Show that $Y_n = X_{\tau_n}$ is a discrete-time Markov process, and express its transition kernel in terms of the transition semigroup of X. Also show that $Y_t = X(\tau_{[t]})$ may fail to be Markov, even when (τ_n) is Poisson.

10. Let X and Y be independent Feller processes in S and T with generators A and B. Show that (X, Y) is a Feller process in $S \times T$ with generator extending $\tilde{A} + \tilde{B}$, where $\tilde{A}$ and $\tilde{B}$ denote the natural extensions of A and B to $C_0(S \times T)$.

11. Consider in S a Feller process with generator A and a pseudo-Poisson process with generator B. Construct a Markov process with generator $A + B$.

12. Use Theorem 19.23 to show that the generator of Brownian motion in $\mathbb{R}$ extends $A = \frac{1}{2}\Delta$ on the set D of functions $f \in C_0^2$ with $Af \in C_0$.

13. Let R_λ be the λ-resolvent of Brownian motion in $\mathbb{R}$. For any $f \in C_0$, put $h = R_\lambda f$, and show by direct computation that $\lambda h - \frac{1}{2} h'' = f$. Conclude by Theorem 19.4 that $\frac{1}{2}\Delta$ with domain D, defined as above, extends the generator A. Thus, $A = \frac{1}{2}\Delta$ by the preceding exercise or by Lemma 19.12.

14. Show that if A is a bounded generator on C_0, then the associated Markov process is pseudo-Poisson. (*Hint:* Note as in Theorem 19.11 that A satisfies the positive-maximum principle. Next use Riesz' representation theorem to express A in terms of bounded kernels, and show that A has the form of Proposition 19.2.)

15. Let the processes X^n and X be such as in Theorem 16.14. Show that $X_t^n \xrightarrow{d} X_t$ for all $t > 0$ implies $X^n \xrightarrow{d} X$ in $D(\mathbb{R}_+, \mathbb{R}^d)$, and compare with the stated theorem. Also prove a corrsponding result for a sequence of Lévy processes X^n. (*Hint:* Use Theorems 19.28 and 19.25, respectively.)

Chapter 20

Ergodic Properties of Markov Processes

transition and contraction operators; ratio ergodic theorem; space-time invariance and tail triviality; mixing and convergence in total variation; Harris recurrence and transience; existence and uniqueness of invariant measure; distributional and pathwise limits

In Chapters 8 and 12 we have seen, under suitable regularity conditions, how the transition probabilities of a discrete- or continuous-time Markov chain converge in total variation toward a unique invariant distribution. Here our main purpose is to study the asymptotic behavior of more general Markov processes and their associated transition kernels. A wide range of powerful tools will then come into play.

We first extend the basic ergodic theorem of Chapter 10 to suitable contraction operators on an arbitrary measure space and establish a general operator version of the ratio ergodic theorem. The relevance of those results for the study of Markov processes is due to the fact that the transition operators are positive $L^1 - L^\infty$-contractions with respect to any invariant measure λ on the state space S. The mentioned results cover both the positive recurrent case, where $\lambda S < \infty$, and the null-recurrent case, where $\lambda S = \infty$. Even more remarkably, the same ergodic theorems apply to both the transition probabilities and the sample paths, in each case giving conclusive information about the asymptotic behavior.

Next we prove for an arbitrary Markov process that a certain strong ergodicity condition is equivalent to the triviality of the tail σ-field, the constancy of all bounded, space-time invariant functions, and a uniform mixing condition. We also consider a similar result where all four conditions are replaced by suitably averaged versions. For both sets of equivalences, one gets very simple and transparent proofs by applying the general coupling results of Chapter 10.

In order to apply the mentioned theorems to specific Markov processes, one needs to find regularity conditions ensuring the existence of an invariant measure or the triviality of the tail σ-field. Here we consider a general class of Feller processes which satisfy either a strong recurrence or a uniform transience condition. In the former case, we prove the existence of an invariant measure, required for the application of the mentioned ergodic theorems, and show that the space-time invariant functions are constant, which implies the mentioned strong ergodicity. Our proofs of the latter results depend on some potential theoretic tools related to those developed in Chapter 19.

To begin with the technical developments, we consider a Markov transition operator T on an arbitrary measurable space $(S, \mathcal{S})$. Note that T is *positive*, in the sense that $f \geq 0$ implies $Tf \geq 0$, and also that $T1 = 1$. As before, we write P_x for the distribution of a Markov process on $\mathbb{Z}_+$ with transition operator T starting at $x \in S$. More generally, we define $P_\mu = \int_S P_x \mu(dx)$ for any measure μ on S. A measure λ on S is said to be *invariant* if $\lambda T f = \lambda f$ for any measurable function $f \geq 0$. Writing θ for the shift on the path space S^∞, we define the associated operator $\tilde{\theta}$ by $\tilde{\theta} f = f \circ \theta$.

For any $p \geq 1$, we say that an operator T on some measure space $(S, \mathcal{S}, \mu)$ is an L^p-*contraction* if $\|Tf\|_p \leq \|f\|_p$ for every $f \in L^p$. By an $L^1 - L^\infty$-*contraction* we mean an operator that is an L^p-contraction for every $p \in [1, \infty]$. The following result shows the relevance of the mentioned notions for the theory of Markov processes.

Lemma 20.1 *(Markov processes and contractions)* Let T be a Markov transition operator on $(S, \mathcal{S})$ with invariant measure λ. Then

(i) T is a positive $L^1 - L^\infty$-contraction on (S, λ);

(ii) $\tilde{\theta}$ is a positive $L^1 - L^\infty$-contraction on (S^∞, P_λ).

Proof: (i) Applying Jensen's inequality to the transition kernel $\mu(x, B) = T 1_B(x)$ and using the invariance of λ, we get for any $p \in [1, \infty)$ and $f \in L^p$

$$\|Tf\|_p^p = \lambda |\mu f|^p \leq \lambda \mu |f|^p = \lambda |f|^p = \|f\|_p^p.$$

The result for $p = \infty$ is obvious.

(ii) Proceeding as in Lemma 8.11, we see that θ is a measure-preserving transformation on (S^∞, P_λ). Hence, for any measurable function $f \geq 0$ on S^∞ and constant $p \geq 1$, we have

$$P_\lambda |\tilde{\theta} f|^p = P_\lambda |f \circ \theta|^p = (P_\lambda \circ \theta^{-1})|f|^p = P_\lambda |f|^p.$$

The contraction property for $p = \infty$ is again obvious. □

We shall see how some crucial results of Chapter 10 carry over to the context of positive $L^1 - L^\infty$-contractions on an arbitrary measure space. First we consider an operator version of Birkhoff's ergodic theorem. To simplify our writing, we introduce the operators $S_n = \sum_{k<n} T^k$, $A_n = S_n/n$, and $Mf = \sup_n A_n f$. Say that f is T-*invariant* if $Tf = f$.

Theorem 20.2 *(operator ergodic theorem, Hopf, Dunford and Schwartz)*
Let T be a positive $L^1 - L^\infty$-contraction on a measure space $(S, \mathcal{S}, \mu)$. Then $A_n f$ converges a.e. for every $f \in L^1$ toward a T-invariant function $Af \in L^1$.

For the proof, we need to extend the inequalities in Lemmas 10.7 and 10.11 and in Proposition 10.10 (i) to an operator setting.

Lemma 20.3 *(maximum inequalities)* For any positive L^1-contraction T on a measure space $(S, \mathcal{S}, \mu)$, we have

(i) $\mu[f; Mf > 0] \geq 0$, $f \in L^1$.

If T is even an $L^1 - L^\infty$-contraction, then also

(ii) $r\mu\{Mf > 2r\} \leq \mu[f; f > r]$, $f \in L^1$, $r > 0$;

(iii) $\|Mf\|_p \leq \|f\|_p$, $f \in L^p$, $p > 1$.

Proof: (i) For any $f \in L^1$ we write $M_n f = S_1 f \vee \cdots \vee S_n f$ and conclude by positivity that
$$S_k f = f + T S_{k-1} f \leq f + T(M_n f)_+, \quad k = 1, \ldots, n.$$
Hence, $M_n f \leq f + T(M_n f)_+$ for all n, and so by positivity and contractivity
$$\begin{aligned} \mu[f; M_n f > 0] &\geq \mu[M_n f - T(M_n f)_+; M_n f > 0] \\ &\geq \mu[(M_n f)_+ - T(M_n f)_+] \\ &= \|(M_n f)_+\|_1 - \|T(M_n f)_+\|_1 \geq 0. \end{aligned}$$
As before, it remains to let $n \to \infty$.

(ii) Put $f_r = f 1\{f > r\}$. By the L^∞-contractivity and positivity of A_n,
$$A_n f - 2r \leq A_n(f - 2r) \leq A_n(f_r - r), \quad n \in \mathbb{N},$$
which implies $Mf - 2r \leq M(f_r - r)$. Hence, by part (i),
$$\begin{aligned} r\mu\{Mf > 2r\} &\leq r\mu\{M(f_r - r) > 0\} \\ &\leq \mu[f_r; M(f_r - r) > 0] \\ &\leq \mu f_r = \mu[f; f > r]. \end{aligned}$$

(iii) Here the earlier proof applies with only notational changes. □

Proof of Theorem 20.2: Fix any $f \in L^1$. By dominated convergence, we may approximate f in L^1 by functions $\tilde{f} \in L^1 \cap L^\infty \subset L^2$. By Lemma 10.18, we may next approximate $\tilde{f}$ in L^2 by functions of the form $\hat{f} + (g - Tg)$, where $\hat{f}, g \in L^2$ and $T\hat{f} = \hat{f}$. Finally, we may approximate g in L^2 by functions $\tilde{g} \in L^1 \cap L^\infty$. Since T contracts L^2, the functions $\tilde{g} - T\tilde{g}$ will then approximate $g - Tg$ in L^2. Combining the three approximations, we have for any $\varepsilon > 0$
$$f = f_\varepsilon + (g_\varepsilon - Tg_\varepsilon) + h_\varepsilon + k_\varepsilon, \tag{1}$$
where $f_\varepsilon \in L^2$ with $Tf_\varepsilon = f_\varepsilon$, $g_\varepsilon \in L^1 \cap L^\infty$, and $\|h_\varepsilon\|_2 \vee \|k_\varepsilon\|_1 < \varepsilon$.

Since f_ε is invariant, we have $A_n f_\varepsilon \equiv f_\varepsilon$. Next we note that

$$\|A_n(g_\varepsilon - Tg_\varepsilon)\|_\infty = n^{-1}\|g_\varepsilon - T^n g_\varepsilon\|_\infty \leq 2n^{-1}\|g_\varepsilon\|_\infty \to 0. \quad (2)$$

Hence,

$$\limsup_{n\to\infty} A_n f \leq f_\varepsilon + M h_\varepsilon + M k_\varepsilon < \infty \text{ a.e.,}$$

and similarly for $\liminf_n A_n f$. Combining the two estimates gives

$$(\limsup_n - \liminf_n) A_n f \leq 2M|h_\varepsilon| + 2M|k_\varepsilon|.$$

Now Lemma 20.3 yields for any $\varepsilon, r > 0$

$$\||M|h_\varepsilon|\|_2 \leq \|h_\varepsilon\|_2 < \varepsilon,$$
$$\mu\{M|k_\varepsilon| > 2r\} \leq r^{-1}\|k_\varepsilon\|_1 < \varepsilon/r,$$

and so $M|h_\varepsilon| + M|k_\varepsilon| \to 0$ a.e. as $\varepsilon \to 0$ along a suitable sequence. Thus, $A_n f$ converges a.e. toward some limit Af.

To see that Af is T-invariant, we note that by (1) and (2) the a.e. limits Ah_ε and Ak_ε exist and satisfy $TAf - Af = (TA - A)(h_\varepsilon + k_\varepsilon)$. By the contraction property and Fatou's lemma, the right-hand side tends to 0 a.e. as $\varepsilon \to 0$ along some sequence, and we get $TAf = Af$ a.e. $\square$

A problem with the last theorem is that the limit Af may be 0, in which case the a.s. convergence $A_n f \to Af$ gives little information about the asymptotic behavior of $A_n f$. For example, this happens when $\mu S = \infty$ and T is the operator induced by a μ-preserving and ergodic transformation θ on S. Then Af is a constant, and the condition $Af \in L^1$ implies $Af = 0$. To get around this difficulty, we may instead compare the asymptotic behavior of $S_n f$ with that of $S_n g$ for a suitable reference function $g \in L^1$. This idea leads to a far-reaching and powerful extension of Birkhoff's theorem.

Theorem 20.4 *(ratio ergodic theorem, Chacon and Ornstein) Let T be a positive L^1-contraction on a measure space $(S, \mathcal{S}, \mu)$, and fix any $f \in L^1$ and $g \in L^1_+$. Then $S_n f / S_n g$ converges a.e. on the set $\{S_\infty g > 0\}$.*

Our proof will be based on three lemmas.

Lemma 20.5 *(individual terms) $T^n f / S_{n+1} g \to 0$ a.e. on $\{S_\infty g > 0\}$.*

Proof: We may assume that $f \geq 0$. Fix any $\varepsilon > 0$, and define

$$h_n = T^n f - \varepsilon S_{n+1} g, \quad A_n = \{h_n > 0\}, \quad n \geq 0.$$

By positivity,

$$h_n = Th_{n-1} - \varepsilon g \leq Th_{n-1}^+ - \varepsilon g, \quad n \geq 1.$$

Examining the cases A_n and A_n^c separately, we conclude that

$$h_n^+ \leq Th_{n-1}^+ - \varepsilon 1_{A_n} g, \quad n \geq 1,$$

and so by contractivity

$$\varepsilon\mu[g; A_n] \leq \mu(Th_{n-1}^+) - \mu h_n^+ \leq \mu h_{n-1}^+ - \mu h_n^+.$$

Summing over n gives

$$\varepsilon\mu \sum\nolimits_{n\geq 1} 1_{A_n} g \leq \mu h_0^+ = \mu(f - \varepsilon g) \leq \mu f < \infty,$$

which implies $\mu[g;\, A_n \text{ i.o.}] = 0$ and hence $\limsup_n (T^n f / S_{n+1} g) \leq \varepsilon$ a.e. on $\{g > 0\}$. Since ε was arbitrary, we obtain $T^n f / S_{n+1} g \to 0$ a.e. on $\{g > 0\}$. Applying this result to the functions $T^m f$ and $T^m g$ gives the same convergence on $\{S_{m-1} g = 0 < S_m g\}$ for arbitrary $m > 1$. □

To state the next result, we introduce the nonlinear *filling operator* U on L^1, given by $Uh = Th_+ - h_-$. It is suggestive to think of the sequence $U^n h$ as resulting from successive attempts to fill a hole h_-, by mapping in each step only the matter that has not yet fallen into the hole. We also define $M_n h = S_1 h \vee \cdots \vee S_n h$.

Lemma 20.6 *(filling operator)* *For any $h \in L^1$ and $n \in \mathbb{N}$, we have $U^{n-1} h \geq 0$ on $\{M_n h > 0\}$.*

Proof: Writing $h_k = h_+ + (Uh)_+ + \cdots + (U^k h)_+$, we claim that

$$h_k \geq S_{k+1} h, \quad k \geq 0. \tag{3}$$

This holds for $k = 0$ since $h_+ = h + h_- \geq h$. Proceeding by induction, we assume (3) to be true for $k = m \geq 0$. Using the induction hypothesis and the definitions of S_k, h_k, and U, we get for $m+1$

$$\begin{aligned}
S_{m+2} h &= h + T S_{m+1} h \leq h + T h_m = h + \sum\nolimits_{k \leq m} T(U^k h)_+ \\
&= h + \sum\nolimits_{k \leq m} \left(U^{k+1} h + (U^k h)_- \right) \\
&= h + \sum\nolimits_{k \leq m} \left((U^{k+1} h)_+ - (U^{k+1} h)_- + (U^k h)_- \right) \\
&= h + h_{m+1} - h_+ + h_- - (U^{m+1} h)_- \leq h_{m+1}.
\end{aligned}$$

This completes the proof of (3).

If $M_n h > 0$ at some point in S, then $S_k h > 0$ for some $k \leq n$, and so by (3) we have $h_k > 0$ for some $k < n$. But then $(U^k h)_+ > 0$, and therefore $(U^k h)_- = 0$ for the same k. Since $(U^k h)_-$ is nonincreasing, it follows that $(U^{n-1} h)_- = 0$, and hence $U^{n-1} h \geq 0$. □

To state our third and crucial lemma, we write $g \in \mathcal{T}_1(f)$ for a given $f \in L_+^1$ if there exists a decomposition $f = f_1 + f_2$ with $f_1, f_2 \in L_+^1$ such that $g = Tf_1 + f_2$. In particular, we note that $f, g \in L_+^1$ implies $U(f - g) = f' - g$ for some $f' \in \mathcal{T}(f)$. The classes $\mathcal{T}_n(f)$ are defined recursively by $\mathcal{T}_{n+1}(f) = \mathcal{T}_1(\mathcal{T}_n(f))$ and we put $\mathcal{T}(f) = \bigcup_n \mathcal{T}_n(f)$. We may now introduce the functionals

$$\psi_B f = \sup\{\mu[g; B];\, g \in \mathcal{T}(f)\}, \quad f \in L_+^1, \; B \in \mathcal{S}.$$

Lemma 20.7 *(filling functionals)* Let $f, g \in L_+^1$ and $B \in \mathcal{S}$. Then
$$B \subset \{\limsup_n S_n(f-g) > 0\} \implies \psi_B f \geq \psi_B g.$$

Proof: Fix any $g' \in \mathcal{T}(g)$ and $c > 1$. First we show that
$$\{\limsup_n S_n(f-g) > 0\} \subset \{\limsup_n S_n(cf - g') > 0\} \quad \text{a.e.} \tag{4}$$

We may then assume that $g' \in \mathcal{T}_1(g)$, since the general result then follows by iteration in finitely many steps. Letting $g' = r + Ts$ for some $r, s \in L_+^1$ with $r + s = g$, we obtain
$$\begin{aligned} S_n(cf - g') &= \sum_{k<n} T^k(cf - r - Ts) \\ &= S_n(f - g) + (c-1)S_n f + s - T^n s. \end{aligned}$$

Since $T^n s / S_n f = T^{n-1} Ts / S_n f \to 0$ a.e. on $\{S_\infty f > 0\}$ by Lemma 20.5, we conclude that eventually $S_n(cf - g') \geq S_n(f - g)$ a.e. on the same set, and (4) follows.

Combining the given hypothesis with (4), we obtain the a.e. relation $B \subset \{M(cf - g') > 0\}$. Now Lemma 20.6 yields
$$U^{n-1}(cf - g') \geq 0 \quad \text{on } B_n \equiv B \cap \{M_n(cf - g') > 0\}, \quad n \in \mathbb{N}.$$

Since $B_n \uparrow B$ a.e. and $U^{n-1}(cf - g') = f' - g'$ for some $f' \in \mathcal{T}(cf)$, we get
$$\begin{aligned} 0 &\leq \mu[U^{n-1}(cf - g'); B_n] \\ &= \mu[f'; B_n] - \mu[g'; B_n] \\ &\leq \psi_B(cf) - \mu[g'; B_n] \\ &\to c\psi_B f - \mu[g'; B], \end{aligned}$$

and so $c\psi_B f \geq \mu[g'; B]$. It remains to let $c \to 1$ and take the supremum over $g' \in \mathcal{T}(g)$. $\square$

Proof of Theorem 20.4: We may assume that $f \geq 0$. On $\{S_\infty g > 0\}$, put
$$\alpha = \liminf_n (S_n f / S_n g) \leq \limsup_n (S_n f / S_n g) = \beta,$$

and define $\alpha = \beta = 0$ otherwise. Since $S_n g$ is nondecreasing, we have for any $c > 0$
$$\begin{aligned} \{\beta > c\} &\subset \{\limsup_n (S_n(f - cg)/S_n g) > 0, \, S_\infty g > 0\} \\ &\subset \{\limsup_n S_n(f - cg) > 0\}. \end{aligned}$$

Writing $B = \{\beta = \infty, \, S_\infty g > 0\}$, we see from Lemma 20.7 that
$$c\psi_B g = \psi_B(cg) \leq \psi_B f \leq \mu f < \infty,$$

and as $c \to \infty$ we get $\psi_B g = 0$. But then $\mu[T^n g; B] = 0$ for all $n \geq 0$, and therefore $\mu[S_\infty g; B] = 0$. Since $S_\infty g > 0$ on B, we obtain $\mu B = 0$, which means that $\beta < \infty$ a.e.

Now define $C = \{\alpha < a < b < \beta\}$ for fixed $b > a > 0$. As before,
$$C \subset \{\limsup_n S_n(f - bg) \,\wedge\, \limsup_n S_n(ag - f) > 0\},$$

and so by Lemma 20.7
$$b\psi_C g = \psi_C(bg) \le \psi_C f \le \psi_C(ag) = a\psi_C g < \infty,$$
which implies $\psi_C g = 0$, and therefore $\mu C = 0$. Hence,
$$\mu\{\alpha < \beta\} \le \sum_{a<b} \mu\{\alpha < a < b < \beta\} = 0,$$
where the summation extends over all rational $a < b$, and so $\alpha = \beta$ a.e., which proves the asserted convergence. □

We illustrate the use of the last theorem by considering a striking application to discrete-time Markov processes. Given such a process X on S and a measurable function f on S^∞, we define $S_n f = \sum_{k<n} f(\theta_k X)$.

Corollary 20.8 *(ratio limit theorem)* *Given a discrete-time Markov process in S with invariant measure λ, we have for any $f \in L^1(P_\lambda)$ and $g \in L^1_+(P_\lambda)$*

(i) $S_n f / S_n g$ *converges a.e. P_λ on $\{y \in S^\infty; S_\infty g(y) > 0\}$;*

(ii) $E_x S_n f / E_x S_n g$ *converges a.e. λ on $\{x \in S; E_x S_\infty g > 0\}$.*

Proof: (i) By Lemma 20.1 (ii) we may apply Theorem 20.4 to the $L^1 - L^\infty$-contraction $\tilde{\theta}$ on (S^∞, P_λ) induced by the shift θ, and the result follows.
(ii) Writing $\tilde{f}(x) = E_x f(X)$ and using the Markov property at k, we get
$$E_x f(\theta_k X) = E_x E_{X_k} f(X) = E_x \tilde{f}(X_k) = T^k \tilde{f}(x),$$
and so
$$E_x S_n f = \sum_{k<n} E_x f(\theta_k X) = \sum_{k<n} T^k \tilde{f}(x) = S_n \tilde{f}(x),$$
where $S_n = T^0 + \cdots + T^{n-1}$ on the right. We also note that
$$\lambda \tilde{f} = \int \tilde{f}(x) \lambda(dx) = \int E_x f(X) \lambda(dx) = E_\lambda f(X) = P_\lambda f.$$

Now Lemma 20.1 (i) shows that T is a positive $L^1 - L^\infty$-contraction on (S, λ). By Theorem 20.4 we conclude that $S_n \tilde{f}(x)/S_n \tilde{g}(x)$ converges a.e. λ on the set $\{x \in S; S_\infty \tilde{g}(x) > 0\}$, which translates immediately into the asserted statement. □

Now consider a conservative, continuous-time Markov process on an arbitrary state space $(S, \mathcal{S})$ with distributions P_x and associated expectation operators E_x. On the canonical path space $\Omega = S^{\mathbb{R}_+}$ we introduce the shift operators θ_t and filtration $\mathcal{F} = (\mathcal{F}_t)$. A bounded function $f: S \to \mathbb{R}$ is said to be *invariant* or *harmonic* if it is measurable and such that
$$f(x) = T_t f(x) = E_x f(X_t), \quad x \in S, \ t \ge 0.$$
More generally, we say that a bounded function $f: S \times \mathbb{R}_+ \to \mathbb{R}$ is *space-time invariant* or *harmonic* if it is measurable and satisfies
$$f(x, s) = E_x f(X_t, s+t), \quad x \in S, \ s, t \ge 0. \tag{5}$$

For motivation, we note that f is then invariant for the associated *space-time process* $\tilde{X}_t = (X_t, s + t)$ in $S \times \mathbb{R}_+$, where the second component is deterministic apart from the possibly random initial value $s \geq 0$. Note that $\tilde{X}$ is again a time-homogeneous Markov process with transition operators $\tilde{T}_t f(x, s) = E_x f(X_t, s + t)$. We need the following useful martingale connection.

Lemma 20.9 *(space-time invariance)* *A bounded, measurable function f: $S \times \mathbb{R}_+ \to \mathbb{R}$ is space-time invariant iff the process $M_t = f(X_t, s + t)$ is a P_μ-martingale for any μ and $s \geq 0$.*

Proof: Assume that f is space-time invariant. Letting $s, t, h \geq 0$ and using the Markov property of X, we get

$$\begin{aligned} E_\mu[M_{t+h}|\mathcal{F}_t] &= E_\mu[f(X_{t+h}, s + t + h)|\mathcal{F}_t] \\ &= E_{X_t} f(X_h, s + t + h) \\ &= f(X_t, s + t) = M_t, \end{aligned}$$

which shows that M is a P_μ-martingale. Conversely, the martingale property of M for $\mu = \delta_x$ yields

$$E_x f(X_t, s + t) = E_x M_t = E_x M_0 = f(x, s) \text{ a.s.}, \quad x \in S, \; s, t \geq 0,$$

which means that f is space-time invariant. □

The *tail σ-field* on Ω is defined as $\mathcal{T} = \bigcap_t \mathcal{T}_t$, where $\mathcal{T}_t = \sigma(\theta_t) = \sigma\{X_s;\, s \geq t\}$. A σ-field $\mathcal{G}$ on Ω is said to be P_μ-*trivial* if $P_\mu A = 0$ or 1 for every $A \in \mathcal{G}$. We write $P_\mu^B = P_\mu[\,\cdot\,|B]$ and say that P_μ is *mixing* if

$$\lim_{t \to \infty} \|P_\mu \circ \theta_t^{-1} - P_\mu^B \circ \theta_t^{-1}\| = 0, \quad B \in \mathcal{F}_\infty \text{ with } P_\mu B > 0.$$

The following key result defines the notion of *strong ergodicity*, as opposed to the weak ergodicity of Theorem 20.11.

Theorem 20.10 *(strong ergodicity, Orey)* *For any conservative, discrete- or continuous-time Markov semigroup with distributions P_μ, these conditions are equivalent:*

(i) *the tail σ-field $\mathcal{T}$ is P_μ-trivial for every μ;*
(ii) *P_μ is mixing for every μ;*
(iii) *every bounded, space-time invariant function is a constant;*
(iv) *$\|P_\mu \circ \theta_t^{-1} - P_\nu \circ \theta_t^{-1}\| \to 0$ as $t \to \infty$ for any μ and ν.*

First proof: By Theorem 10.27 (i) we note that (ii) and (iv) are equivalent to the conditions
 (ii') $P_\mu = P_\mu^B$ on $\mathcal{T}$ for any μ and B;
 (iv') $P_\mu = P_\nu$ on $\mathcal{T}$ for any μ and ν.
We may then prove that (ii') ⇔ (i) ⇒ (iv') and (iv) ⇒ (iii) ⇒ (i).

(i) ⇔ (ii'): If $P_\mu A = 0$ or 1, then clearly also $P_\mu^B A = 0$ or 1, which shows that (i) ⇒ (ii'). Conversely, let $A \in \mathcal{T}$ be arbitrary with $P_\mu A > 0$. Taking $B = A$ in (ii') gives $P_\mu A = (P_\mu A)^2$, which implies $P_\mu A = 1$.

(i) ⇒ (iv'): Applying (i) to the distribution $\frac{1}{2}(\mu+\nu)$ gives $P_\mu A + P_\nu A = 0$ or 2 for every $A \in \mathcal{T}$, which implies $P_\mu A = P_\nu A = 0$ or 1.

(iv) ⇒ (iii): Let f be bounded and space-time invariant. Using (iv) with $\mu = \delta_x$ and $\nu = \delta_y$ gives

$$|f(x,s) - f(y,s)| = |E_x f(X_t, s+t) - E_y f(X_t, s+t)|$$
$$\leq \|f\| \, \|P_x \circ \theta_t^{-1} - P_y \circ \theta_t^{-1}\| \to 0,$$

which shows that $f(x,s) = f(s)$ is independent of x. But then $f(s) = f(s+t)$ by (5), and so f is a constant.

(iii) ⇒ (i): Fix any $A \in \mathcal{T}$. Since $A \in \mathcal{T}_t = \sigma(\theta_t)$ for every $t \geq 0$, we have $A = \theta_t^{-1} A_t$ for some sets $A_t \in \mathcal{F}_\infty$, and we note that A_t is unique since θ_t is surjective. For any $s, t \geq 0$,

$$\theta_t^{-1}\theta_s^{-1} A_{s+t} = \theta_{s+t}^{-1} A_{s+t} = A = \theta_t^{-1} A_t, \quad s,t \geq 0,$$

and so $\theta_s^{-1} A_{s+t} = A_t$. Putting $f(x,t) = P_x A_t$ and using the Markov property at time s, we get

$$E_x f(X_s, s+t) = E_x P_{X_s} A_{s+t} = E_x P_x[\theta_s^{-1} A_{s+t} | \mathcal{F}_s] = P_x A_t = f(x,t).$$

Thus, f is space-time invariant and therefore equal to a constant $c \in [0,1]$. By the Markov property at t and martingale convergence as $t \to \infty$, we have a.s.

$$c = f(X_t, t) = P_{X_t} A_t = P_\mu[\theta_t^{-1} A_t | \mathcal{F}_t] = P_\mu[A | \mathcal{F}_t] \to 1_A,$$

which implies $P_\mu A = c \in \{0,1\}$. This shows that $\mathcal{T}$ is P_μ-trivial. □

Second proof: We can avoid using the rather deep Theorem 10.27 by giving direct proofs of the implications (i) ⇒ (ii) ⇒ (iv).

(i) ⇒ (ii): Assuming (i), we get by reverse martingale convergence

$$\|P_\mu(\cdot \cap B) - P_\mu(\cdot) P_\mu(B)\|_{\mathcal{T}_t} = \|E_\mu[P_\mu[B|\mathcal{T}_t] - P_\mu B; \cdot]\|_{\mathcal{T}_t}$$
$$\leq E_\mu |P_\mu[B|\mathcal{T}_t] - P_\mu B| \to 0.$$

(ii) ⇒ (iv): Let $\mu' - \nu'$ be the Hahn decomposition of $\mu - \nu$ and choose $B \in \mathcal{S}$ with $\mu' B^c = \nu B = 0$. Writing $\chi = \mu' + \nu'$ and $A = \{X_0 \in B\}$, we get by (ii)

$$\|P_\mu \circ \theta_t^{-1} - P_\nu \circ \theta_t^{-1}\| = \|\mu'\| \, \|P_\chi^A \circ \theta_t^{-1} - P_\chi^{A^c} \circ \theta_t^{-1}\| \to 0. \quad □$$

The *invariant* σ-field $\mathcal{I}$ on Ω consists of all events $A \subset \Omega$ such that $\theta_t^{-1} A = A$ for all $t \geq 0$. Note that a random variable ξ on Ω is $\mathcal{I}$-measurable iff $\xi \circ \theta_t = \xi$ for all $t \geq 0$. The invariant σ-field $\mathcal{I}$ is clearly contained in the tail σ-field $\mathcal{T}$. We say that P_μ is *weakly mixing* if

$$\lim_{t \to \infty} \left\| \int_0^1 (P_\mu - P_\mu^B) \circ \theta_{st}^{-1} ds \right\| = 0, \quad B \in \mathcal{F}_\infty \text{ with } P_\mu B > 0,$$

where it is understood that $\theta_s = \theta_{[s]}$ when the time scale is discrete. We may now state the weak counterpart of Theorem 20.10.

Theorem 20.11 *(weak ergodicity) For any measurable, conservative, discrete- or continuous-time Markov semigroup with distributions P_μ, these conditions are equivalent:*

(i) *the invariant σ-field $\mathcal{I}$ is P_μ-trivial for every μ;*

(ii) *P_μ is weakly mixing for every μ;*

(iii) *every bounded, invariant function is a constant;*

(iv) *$\|\int_0^1 (P_\mu - P_\nu) \circ \theta_{st}^{-1} ds\| \to 0$ as $t \to \infty$ for any μ and ν.*

Proof: By Theorem 10.27 (ii) we note that (ii) and (iv) are equivalent to the conditions

(ii') $P_\mu = P_\mu^B$ on $\mathcal{I}$ for any μ and B;

(iv') $P_\mu = P_\nu$ on $\mathcal{I}$ for any μ and ν.

Here the implications (ii') $\Leftrightarrow$ (i) $\Rightarrow$ (iv') may be established as before, and so it is enough to show that (iv) $\Rightarrow$ (iii) $\Rightarrow$ (i).

(iv) $\Rightarrow$ (iii): Let f be bounded and invariant. Then $f(x) = E_x f(X_t) = T_t f(x)$, and therefore $f(x) = \int_0^1 T_{st} f(x) \, ds$. Using (iv) gives

$$|f(x) - f(y)| = \left| \int_0^1 (T_{st} f(x) - T_{st} f(y)) \, ds \right|$$

$$\leq \|f\| \left\| \int_0^1 (P_x - P_y) \circ \theta_{st}^{-1} ds \right\| \to 0,$$

which shows that f is a constant.

(iii) $\Rightarrow$ (i): Fix any $A \in \mathcal{I}$, and define $f(x) = P_x A$. Using the Markov property at t and the invariance of A, we get

$$E_x f(X_t) = E_x P_{X_t} A = E_x P_x [\theta_t^{-1} A | \mathcal{F}_t] = P_x \theta_t^{-1} A = P_x A = f(x),$$

which shows that f is invariant. By (iii) it follows that f equals a constant $c \in [0,1]$. Hence, by the Markov property and martingale convergence, we have a.s.

$$c = f(X_t) = P_{X_t} A = P_\mu [\theta_t^{-1} A | \mathcal{F}_t] = P_\mu [A | \mathcal{F}_t] \to 1_A,$$

which implies $P_\mu A = c \in \{0,1\}$. Thus, $\mathcal{I}$ is P_μ-trivial. $\square$

Let us now specialize to the case of conservative Feller processes X with distributions P_x, defined on an lcscH (locally compact, second countable Hausdorff) space S with Borel σ-field $\mathcal{S}$. We say that the process is *regular* if there exist a locally finite measure ρ on S and a continuous function $(x, y, t) \mapsto p_t(x, y) > 0$ on $S^2 \times (0, \infty)$ such that

$$P_x\{X_t \in B\} = \int_B p_t(x, y) \rho(dy), \quad x \in S, \ B \in \mathcal{S}, \ t > 0.$$

Note that the *supporting measure* ρ is then unique, up to an equivalence, and that $\mathrm{supp}(\rho) = S$ by the Feller property. A Feller process is said to be

Harris recurrent if it is regular with a supporting measure ρ satisfying

$$\int_0^\infty 1_B(X_t)\,dt = \infty \text{ a.s. } P_x, \quad x \in S,\ B \in \mathcal{S} \text{ with } \rho B > 0. \qquad (6)$$

Theorem 20.12 *(Harris recurrence and ergodicity, Orey)* *Any Harris recurrent Feller process is strongly ergodic.*

Proof: By Theorem 20.10 it suffices to prove that any bounded, space-time invariant function $f : S \times \mathbb{R}_+ \to \mathbb{R}$ is a constant. First we show for fixed $x \in S$ that $f(x,t)$ is independent of t. Then assume instead that $f(x,h) \neq f(x,0)$ for some $h > 0$, say $f(x,h) > f(x,0)$. Recall from Lemma 20.9 that $M_t^s = f(X_t, s+t)$ is a P_y-martingale for any $y \in S$ and $s \geq 0$. In particular, the limit M_∞^s exists a.s. along $h\mathbb{Q}$, and we get a.s. P_x

$$E_x[M_\infty^h - M_\infty^0 | \mathcal{F}_0] = M_0^h - M_0^0 = f(x,h) - f(x,0) > 0,$$

which implies $P_x\{M_\infty^h > M_\infty^0\} > 0$. We may then choose some constants $a < b$ such that

$$P_x\{M_\infty^0 < a < b < M_\infty^h\} > 0. \qquad (7)$$

We also note that

$$M_t^{s+h} \circ \theta_s = f(X_{s+t}, s+t+h) = M_{s+t}^h, \quad s,t,h \geq 0. \qquad (8)$$

With s and t restricted to $h\mathbb{Q}_+$, we define

$$g(y,s) = P_y \bigcap_{t \geq 0}\{M_t^s \leq a < b \leq M_t^{s+h}\}, \quad y \in S,\ s \geq 0.$$

Using the Markov property at s and (8), we get a.s. P_x for any $r \leq s$

$$\begin{aligned}
g(X_s, s) &= P_{X_s} \bigcap_{t \geq 0}\{M_t^s \leq a < b \leq M_t^{s+h}\} \\
&= P_x^{\mathcal{F}_s} \bigcap_{t \geq 0}\{M_t^s \circ \theta_s \leq a < b \leq M_t^{s+h} \circ \theta_s\} \\
&= P_x^{\mathcal{F}_s} \bigcap_{t \geq s}\{M_t^0 \leq a < b \leq M_t^h\} \\
&\geq P_x^{\mathcal{F}_s} \bigcap_{t \geq r}\{M_t^0 \leq a < b \leq M_t^h\}.
\end{aligned}$$

By martingale convergence, we get a.s. as $s \to \infty$ along $h\mathbb{Q}$ and then $r \to \infty$

$$\begin{aligned}
\liminf_{s \to \infty} g(X_s, s) &\geq \liminf_{t \to \infty} 1\{M_t^0 \leq a < b \leq M_t^h\} \\
&\geq 1\{M_\infty^0 < a < b < M_\infty^h\},
\end{aligned}$$

and so by (7)

$$P_x\{g(X_s, s) \to 1\} \geq P_x\{M_\infty^0 < a < b < M_\infty^h\} > 0. \qquad (9)$$

Now fix any nonempty, bounded, open set $B \subset S$. Using (6) and the right-continuity of X, we note that $\limsup_s 1_B(X_s) = 1$ a.s. P_x, and so in view of (9)

$$P_x\{\limsup_s 1_B(X_s)\,g(X_s, s) = 1\} > 0. \qquad (10)$$

Furthermore, we have by regularity

$$p_h(u,v) \wedge p_{2h}(u,v) \geq \varepsilon > 0, \quad u,v \in B, \tag{11}$$

for some $\varepsilon > 0$. By (10) we may choose some $y \in B$ and $s \geq 0$ such that $g(y,s) > 1 - \frac{1}{2}\varepsilon\rho B$. Define for $i = 1, 2$

$$B_i = B \setminus \{u \in S;\ f(u, s+ih) \leq a < b \leq f(u, s+(i+1)h)\}.$$

Using (11), the definitions of B_i, M^s, and g, and the properties of y and s, we get

$$\begin{aligned}
\varepsilon\rho B_i &\leq P_y\{X_{ih} \in B_i\} \\
&\leq 1 - P_y\{f(X_{ih}, s+ih) \leq a < b \leq f(X_{ih}, s+(i+1)h)\} \\
&= 1 - P_y\{M_{ih}^s \leq a < b \leq M_{ih}^{s+h}\} \\
&\leq 1 - g(y,s) < \tfrac{1}{2}\varepsilon\rho B.
\end{aligned}$$

Thus, $\rho B_1 + \rho B_2 < \rho B$, and there exists some $u \in B \setminus (B_1 \cup B_2)$. But this yields the contradiction $a < b \leq f(u, s+2h) \leq a$, which shows that $f(x,t) = f(x)$ is indeed independent of t.

To see that $f(x)$ is also independent of x, we assume that instead

$$\rho\{x;\ f(x) \leq a\} \wedge \rho\{x;\ f(x) \geq b\} > 0$$

for some $a < b$. Then by (6) the martingale $M_t = f(X_t)$ satisfies

$$\int_0^\infty 1\{M_t \leq a\}\,dt = \int_0^\infty 1\{M_t \geq b\}\,dt = \infty \quad \text{a.s.} \tag{12}$$

Writing $\tilde{M}$ for the right-continuous version of M, which exists by Theorem 7.27 (ii), we get by Fubini's theorem for any $x \in S$

$$\begin{aligned}
\sup_{u>0} E_x \left| \int_0^u 1\{M_t \leq a\}\,dt - \int_0^u 1\{\tilde{M}_t \leq a\}\,dt \right| \\
\leq \int_0^\infty E_x \left| 1\{M_t \leq a\} - 1\{\tilde{M}_t \leq a\} \right| dt \\
\leq \int_0^\infty P_x\{M_t \neq \tilde{M}_t\}\,dt = 0,
\end{aligned}$$

and similarly for the events $M_t \geq b$ and $\tilde{M}_t \geq b$. Thus, the integrals on the left agree a.s. P_x for all x, and so (12) remains true with M replaced by $\tilde{M}$. In particular,

$$\liminf_{t\to\infty} \tilde{M}_t \leq a < b \leq \limsup_{t\to\infty} \tilde{M}_t \quad \text{a.s. } P_x.$$

But this is impossible, since $\tilde{M}$ is a bounded, right-continuous martingale and therefore converges a.s. The contradiction shows that $f(x) = c$ a.e. ρ for some constant $c \in \mathbb{R}$. Then for any $t > 0$,

$$f(x) = E_x f(X_t) = \int f(y)\,p_t(x,y)\,\rho(dy) = c, \quad x \in S,$$

and so $f(x)$ is indeed independent of x. □

Our further analysis of regular Feller processes requires some potential theory. For any measurable functions $f, h \geq 0$ on S, we define the *h-potential of $U_h f$ of f* by

$$U_h f(x) = E_x \int_0^\infty e^{-A_t^h} f(X_t)\, dt, \quad x \in S,$$

where A^h denotes the elementary additive functional

$$A_t^h = \int_0^t h(X_s)\, ds, \quad t \geq 0.$$

When h is a constant $a \geq 0$, we note that $U_h = U_a$ agrees with the resolvent operator R_a of the semigroup (T_t), in which case

$$U_a f(x) = E_x \int_0^\infty e^{-at} f(X_t)\, dt = \int_0^\infty e^{-at} T_t f(x)\, dt, \quad x \in S.$$

The classical resolvent equation extends to general h-potentials as follows.

Lemma 20.13 *(resolvent equation)* *Let $f \geq 0$ and $h \geq k \geq 0$ be measurable functions on S, and assume that h is bounded. Then $U_h h \leq 1$, and*

$$U_k f = U_h f + U_h(h-k) U_k f = U_h f + U_k(h-k) U_h f.$$

Proof: For convenience, we define $F = f(X)$, $H = h(X)$, and $K = k(X)$. By Itô's formula for continuous functions of bounded variation,

$$e^{-A_t^h} = 1 - \int_0^t e^{-A_s^h} H_s\, ds, \quad t \geq 0, \tag{13}$$

which implies $U_h h \leq 1$. We may also conclude from the Markov property of X that a.s.

$$\begin{aligned} U_k f(X_t) &= E_{X_t} \int_0^\infty e^{-A_s^k} F_s\, ds \\ &= E_x^{\mathcal{F}_t} \int_0^\infty e^{-A_s^k \circ \theta_t} F_{s+t}\, ds \\ &= E_x^{\mathcal{F}_t} \int_t^\infty e^{-A_u^k + A_t^k} F_u\, du. \tag{14} \end{aligned}$$

Using (13) and (14) together with Fubini's theorem, we get

$$\begin{aligned}
U_h(h-k)U_k f(x) &= E_x \int_0^\infty e^{-A_t^h}(H_t - K_t)\,dt \int_t^\infty e^{-A_u^k + A_t^k} F_u\,du \\
&= E_x \int_0^\infty e^{-A_u^k} F_u\,du \int_0^u e^{-A_t^h + A_t^k}(H_t - K_t)\,dt \\
&= E_x \int_0^\infty e^{-A_u^k} F_u \left(1 - e^{-A_u^h + A_u^k}\right) du \\
&= E_x \int_0^\infty \left(e^{-A_u^k} - e^{-A_u^h}\right) F_u\,du \\
&= U_k f(x) - U_h f(x).
\end{aligned}$$

A similar calculation gives the same expression for $U_k(h-k)U_h f(x)$. □

For a simple application of the resolvent equation, we show that any bounded potential function $U_h f$ is continuous.

Lemma 20.14 *(boundedness and continuity)* *For any regular Feller process on S, let $f, h \geq 0$ be bounded, measurable functions on S such that $U_h f$ is bounded. Then $U_h f$ is continuous.*

Proof: Using Fatou's lemma and the continuity of $p_t(\cdot, y)$, we get for any time $t > 0$ and sequence $x_n \to x$ in S

$$\begin{aligned}
\liminf_{n\to\infty} T_t f(x_n) &= \liminf_{n\to\infty} \int p_t(x_n, y) f(y)\, \rho(dy) \\
&\geq \int p_t(x, y) f(y)\, \rho(dy) = T_t f(x).
\end{aligned}$$

If $f \leq c$, the same relation applies to the function $T_t(c-f) = c - T_t f$, and by combination it follows that $T_t f$ is continuous. By dominated convergence, $U_a f$ is then continuous for every $a > 0$.

Now assume that $h \leq a$. Applying the previous result to the bounded, measurable function $(a-h)U_h f \geq 0$, we conclude that even $U_a(a-h)U_h f$ is continuous. The continuity of $U_h f$ now follows from Lemma 20.13 with h and k replaced by a and h. □

We proceed with some useful estimates.

Lemma 20.15 *(lower bounds)* *For any regular Feller process on S, there exist some continuous functions $h, k : S \to (0, 1]$ such that, for every measurable function $f \geq 0$ on S,*

(i) $U_2 f(x) \geq \rho(kf)\, h(x)$;
(ii) $U_h f(x) \geq \rho(kf)\, U_h h(x)$.

Proof: Fix any compact sets $K \subset S$ and $T \subset (0, \infty)$ with $\rho K > 0$ and $\lambda T > 0$. Define

$$u_a^T(x, y) = \int_T e^{-at} p_t(x, y)\,dt, \quad x, y \in S,$$

and note that for any measurable function $f \geq 0$ on S,
$$U_a f(x) \geq \int u_a^T(x,y) f(y) \rho(dy), \quad x \in S.$$

Using Lemma 20.13 for the constants 4 and 2 gives
$$\begin{aligned} U_2 f(x) &= U_4 f(x) + 2U_4 U_2 f(x) \\ &\geq 2 \int_K u_4^T(x,y) \rho(dy) \int u_2^T(y,z) f(z) \rho(dz), \end{aligned}$$

and (i) follows with
$$h(x) = 2 \int_K u_4^T(x,y) \rho(dy) \wedge 1, \qquad k(x) = \inf_{y \in K} u_2^T(y,x) \wedge 1.$$

To deduce (ii), we may combine (i) with Lemma 20.13 for the functions 2 and h to obtain
$$\begin{aligned} U_h f(x) &\geq U_h(2-h) U_2 f(x) \\ &\geq U_h U_2 f(x) \geq U_h h(x) \rho(kf). \end{aligned}$$

The continuity of h is clear by dominated convergence. For the same reason, the function u_2^T is jointly continuous on S^2. Since K is compact, the functions $u_2^T(y,\cdot)$, $y \in K$, are then equicontinuous on S, which yields the required continuity of k. Finally, the relation $h > 0$ is obvious, whereas $k > 0$ holds by the compactness of K. □

Fixing a function h as in Lemma 20.15, we introduce the kernel
$$Q_x B = U_h(h 1_B)(x), \quad x \in S, \ B \in \mathcal{S}, \tag{15}$$
and note that $Q_x S = U_h h(x) \leq 1$ by Lemma 20.13.

Lemma 20.16 *(convergence dichotomy)* Let Q be given by (15) in terms of some function h as in Lemma 20.15.
  (i) If $U_h h \not\equiv 1$, then $\|Q^n S\| \leq r^{n-1}$, $n \in \mathbb{N}$, for some $r \in (0,1)$.
  (ii) If $U_h h \equiv 1$, then $\|Q_x^n - \nu\| \to 0$, $x \in S$, for some Q-invariant probability measure $\nu \sim \rho$, and every σ-finite, Q-invariant measure on S is proportional to ν.

Proof: (i) Choose k as in Lemma 20.15, fix any $a \in S$ with $U_h h(a) < 1$, and define
$$r = 1 - h(a) \rho(hk(1 - U_h h)).$$
Note that $\rho(hk(1 - U_h h)) > 0$ since $U_h h$ is continuous by Lemma 20.14. Using Lemma 20.15 (i), we obtain
$$0 < 1 - r \leq h(a) \rho k \leq U_2 1(a) = \tfrac{1}{2}.$$
Next we see from Lemma 20.15 (ii) that
$$(1-r) U_h h = h(a) \rho(hk(1 - U_h h)) U_h h \leq U_h h(1 - U_h h).$$

Hence,
$$Q^2 S = U_h h U_h h \leq r U_h h = rQS,$$
and so by iteration
$$Q^n S \leq r^{n-1} QS \leq r^{n-1}, \quad n \in \mathbb{N}.$$

(ii) Introduce a measure $\tilde{\rho} = hk \cdot \rho$ on S. Since $U_h h = 1$, we get by Lemma 20.15 (ii)
$$\tilde{\rho} B = \rho(hk1_B) \leq U_h(h1_B)(x) = Q_x B, \quad B \in \mathcal{S}. \tag{16}$$
Regarding $\tilde{\rho}$ as a kernel, we have for any $x, y \in S$ and $m, n \in \mathbb{Z}_+$
$$(Q_x^m - Q_y^n)\tilde{\rho}^k = (Q_x^m S - Q_y^n S)\tilde{\rho}^k = 0. \tag{17}$$
Iterating (17) and using (16), we get as $n \to \infty$
$$\begin{aligned}
\|Q_x^n - Q_y^{n+k}\| &= \|(\delta_x - Q_y^k)Q^n\| \\
&= \|(\delta_x - Q_y^k)(Q - \tilde{\rho})^n\| \\
&\leq \|\delta_x - Q_y^k\| \sup_z \|Q_z - \tilde{\rho}\|^n \\
&\leq 2(1 - \tilde{\rho} S)^n \to 0.
\end{aligned}$$

Hence, $\sup_x \|Q_x^n - \nu\| \to 0$ for some set function ν on S, and it is easy to see that ν is a Q-invariant probability measure. By Fubini's theorem we note that $Q_x \ll \rho$ for all x, and so $\nu = \nu Q \ll \rho$. Conversely, Lemma 20.15 (ii) yields
$$\nu B = \nu Q(B) = \nu(U_h(h1_B)) \geq \rho(hk1_B),$$
which shows that even $\rho \ll \nu$.

Now consider any σ-finite, Q-invariant measure μ on S. By Fatou's lemma, we get for any $B \in \mathcal{S}$
$$\mu B = \liminf_{n \to \infty} \mu Q^n B \geq \mu \nu B = \mu S \nu B.$$
Choosing B such that $\mu B < \infty$ and $\nu B > 0$, we obtain $\mu S < \infty$. We may then conclude by dominated convergence that $\mu = \mu Q^n \to \mu \nu = \mu S \nu$, which proves the asserted uniqueness of ν. $\square$

We are now ready to prove the basic recurrence dichotomy for regular Feller processes. Write $U = U_0$ and say that X is *uniformly transient* if
$$\|U 1_K\| = \sup_x E_x \int_0^\infty 1_K(X_t) \, dt < \infty, \quad K \subset S \text{ compact}.$$

Theorem 20.17 *(recurrence dichotomy)* *A regular Feller process is either Harris recurrent or uniformly transient.*

Proof: Choose h and k as in Lemma 20.15 and Q, r, and ν as in Lemma 20.16. First assume that $U_h h \not\equiv 1$. Letting $a \in (0, \|h\|)$, we note that $ah \leq (h \wedge a)\|h\|$, and hence
$$a U_{h \wedge a} h \leq U_{h \wedge a}(h \wedge a)\|h\| \leq \|h\|. \tag{18}$$

Furthermore, Lemma 20.13 yields
$$U_{h\wedge a}h \leq U_h h + U_h h U_{h\wedge a}h = Q(1 + U_{h\wedge a}h).$$
Iterating this relation and using Lemma 20.16 (i) and (18), we get
$$\begin{aligned} U_{h\wedge a}h &\leq \sum_{l\leq n} Q^l 1 + Q^n U_{h\wedge a}h \\ &\leq \sum_{l\leq n} r^{l-1} + r^{n-1}\|U_{h\wedge a}h\| \\ &\leq (1-r)^{-1} + r^{n-1}\|h\|/a. \end{aligned}$$
Letting $n \to \infty$ and then $a \to 0$, we conclude by dominated and monotone convergence that $Uh \leq (1-r)^{-1}$. Now fix any compact set $K \subset S$. Since $b \equiv \inf_K h > 0$, we get
$$U1_K(x) \leq b^{-1} Uh(x) \leq b^{-1}(1-r)^{-1} < \infty, \quad x \in S,$$
which shows that X is uniformly transient.

Now assume instead that $U_h h \equiv 1$. Fix any measurable function f on S with $0 \leq f \leq h$ and $\rho f > 0$, and put $g = 1 - U_f f$. By Lemma 20.13 we get
$$\begin{aligned} g &= 1 - U_f f = U_h h - U_h f - U_h(h-f) U_f f \\ &= U_h(h-f)(1 - U_f f) \\ &= U_h(h-f)g \leq U_h hg = Qg. \end{aligned} \quad (19)$$
Iterating this relation and using Lemma 20.16 (ii), we obtain $g \leq Q^n g \to \nu g$, where $\nu \sim \rho$ is the unique Q-invariant distribution on S. Inserting this into (19) gives $g \leq U_h(h-f)\nu g$, and so by Lemma 20.15 (ii)
$$\nu g \leq \nu(U_h(h-f))\nu g \leq (1 - \rho(kf))\nu g.$$
Since $\rho(kf) > 0$, we obtain $\nu g = 0$, and so $U_f f = 1 - g = 1$ a.e. $\nu \sim \rho$. Recalling that $U_f f$ is continuous by Lemma 20.14 and $\operatorname{supp}\rho = S$, we obtain $U_f f \equiv 1$. Taking expected values in (13), we conclude that $A_\infty^f = \infty$ a.s. P_x for every $x \in S$. Now fix any compact set $K \subset S$ with $\rho K > 0$. Since $b \equiv \inf_K h > 0$, we may choose $f = b1_K$, and the desired Harris recurrence follows. □

A measure λ on S is said to be *invariant* for the semigroup (T_t) if $\lambda(T_t f) = \lambda f$ for all $t > 0$ and every measurable function $f \geq 0$ on S. In the Harris recurrent case, the existence of an invariant measure λ can be inferred from Lemma 20.16.

Theorem 20.18 *(invariant measure, Harris, Watanabe) Any Harris recurrent Feller process on S with supporting measure ρ has a locally finite, invariant measure $\lambda \sim \rho$, and every σ-finite, invariant measure agrees with λ up to a normalization.*

To prepare for the proof, we first express the required invariance in terms of the resolvent operators.

Lemma 20.19 *(invariance equivalence)* Let (T_t) be a Feller semigroup on S with resolvent (U_a), and fix any locally finite measure λ on S and constant $c > 0$. Then λ is (T_t)-invariant iff it is aU_a-invariant for every $a \geq c$.

Proof: If λ is (T_t)-invariant, then Fubini's theorem yields for any measurable function $f \geq 0$ and constant $a > 0$

$$\lambda(U_a f) = \int_0^\infty e^{-at} \lambda(T_t f)\, dt = \int_0^\infty e^{-at} \lambda f\, dt = \lambda f / a, \tag{20}$$

which shows that λ is aU_a-invariant.

Conversely, assume that λ is aU_a-invariant for every $a \geq c$. Then for any measurable function $f \geq 0$ on S with $\lambda f < \infty$, the integrals in (20) agree for all $a \geq c$. Hence, by Theorem 5.3 the measures $\lambda(T_t f)e^{-ct}dt$ and $\lambda f e^{-ct} dt$ agree on $\mathbb{R}_+$, which implies $\lambda(T_t f) = \lambda f$ for almost every $t \geq 0$. By the semigroup property and Fubini's theorem we then obtain for any $t \geq 0$

$$\begin{aligned}
\lambda(T_t f) &= c\lambda U_c(T_t f) = c\lambda \int_0^\infty e^{-cs} T_s T_t f\, ds \\
&= c \int_0^\infty e^{-cs} \lambda(T_{s+t} f)\, ds \\
&= c \int_0^\infty e^{-cs} \lambda f\, ds = \lambda f,
\end{aligned}$$

which shows that λ is (T_t)-invariant. $\square$

Proof of Theorem 20.18: Let h, Q, and ν be such as in Lemmas 20.15 and 20.16, and put $\lambda = h^{-1} \cdot \nu$. Using the definition of λ (twice), the Q-invariance of ν (three times), and Lemma 20.13, we get for any constant $a \geq \|h\|$ and bounded, measurable function $f \geq 0$ on S

$$\begin{aligned}
a\lambda U_a f &= a\nu(h^{-1} U_a f) = a\nu U_h U_a f \\
&= \nu(U_h f - U_a f + U_h h U_a f) \\
&= \nu U_h f = \nu(h^{-1} f) = \lambda f,
\end{aligned}$$

which shows that λ is aU_a-invariant for every such a. By Lemma 20.19 it follows that λ is also (T_t)-invariant.

To prove the asserted uniqueness, consider any σ-finite, (T_t)-invariant measure λ' on S. By Lemma 20.19, λ' is even aU_a-invariant for every $a \geq \|h\|$. Now define $\nu' = h \cdot \lambda'$. Letting $f \geq 0$ be bounded and measurable on S and using Lemma 20.13, we get as before

$$\begin{aligned}
\nu' U_h(hf) &= \lambda'(hU_h(hf)) = a\lambda' U_a h U_h(hf) \\
&= a\lambda'(U_a(hf) - U_h(hf) + aU_a U_h(hf)) \\
&= a\lambda' U_a(hf) = \lambda'(hf) = \nu' f,
\end{aligned}$$

which shows that ν' is Q-invariant. Hence, the uniqueness part of Lemma 20.16 (ii) yields $\nu' = c\nu$ for some constant $c \geq 0$, which implies $\lambda' = c\lambda$. □

A Harris recurrent Feller process is said to be *positive recurrent* if the invariant measure λ is bounded and *null-recurrent* otherwise. In the former case, we may assume that λ is a probability measure on S. For any process X in S, the divergence $X_t \to \infty$ a.s. or $X_t \xrightarrow{P} \infty$ means that $1_K(X_t) \to 0$ in the same sense for every compact set $K \subset S$.

Theorem 20.20 *(distributional limits)* *For any regular Feller process X and distribution μ on S, the following holds as $t \to \infty$:*

(i) *If X is positive recurrent with invariant distribution λ and $A \in \mathcal{F}_\infty$ with $P_\mu A > 0$, then $\|P_\mu^A \circ \theta_t^{-1} - P_\lambda\| \to 0$.*

(ii) *If X is null-recurrent or transient, then $X_t \xrightarrow{P_\mu} \infty$.*

Proof: (i) Since $P_\lambda \circ \theta_t^{-1} = P_\lambda$ by Lemma 8.11, the assertion follows from Theorem 20.12 together with properties (ii) and (iv) of Theorem 20.10.

(ii) (*null-recurrent case*): For any compact set $K \subset S$ and constant $\varepsilon > 0$, we define
$$B_t = \{x \in S;\ T_t 1_K(x) \geq \mu T_t 1_K - \varepsilon\}, \quad t > 0,$$
and note that, for any invariant measure λ,
$$(\mu T_t 1_K - \varepsilon) \lambda B_t \leq \lambda(T_t 1_K) = \lambda K < \infty. \tag{21}$$

Since $\mu T_t 1_K - T_t 1_K(x) \to 0$ for all $x \in S$ by Theorem 20.12, we have $\liminf_t B_t = S$, and so $\lambda B_t \to \infty$ by Fatou's lemma. Hence, (21) yields $\limsup_t \mu T_t 1_K \leq \varepsilon$, and since ε was arbitrary, we obtain $P_\mu\{X_t \in K\} = \mu T_t 1_K \to 0$.

(ii) (*transient case*): Fix any compact set $K \subset S$ with $\rho K > 0$, and conclude from the uniform transience of X that $U 1_K$ is bounded. Hence, by the Markov property at t and dominated convergence,
$$E_\mu U 1_K(X_t) = E_\mu E_{X_t} \int_0^\infty 1_K(X_s)\, ds = E_\mu \int_t^\infty 1_K(X_s)\, ds \to 0,$$

which shows that $U 1_K(X_t) \xrightarrow{P_\mu} 0$. Since $U 1_K$ is strictly positive and also continuous by Lemma 20.14, we conclude that $X_t \xrightarrow{P_\mu} \infty$. □

We complete our discussion of regular Feller processes with a pathwise limit theorem. Recall that "almost surely" means a.s. P_μ for every initial distribution μ on S.

Theorem 20.21 *(pathwise limits)* *For any regular Feller process X on S, the following holds as $t \to \infty$:*

(i) *If X is positive recurrent with invariant distribution λ, then*
$$t^{-1} \int_0^t f(\theta_s X)\, ds \to E_\lambda f(X) \quad a.s., \quad f \text{ bounded, measurable.}$$

(ii) *If X is null-recurrent, then*
$$t^{-1} \int_0^t 1_K(X_s)\, ds \to 0 \quad a.s., \quad K \subset S \text{ compact.}$$

(iii) *If X is transient, then $X_t \to \infty$ a.s.*

Proof: (i) From Lemma 8.11 and Theorems 20.10 (i) and 20.12 we note that P_λ is stationary and ergodic, and so the assertion holds a.s. P_λ by Corollary 10.9. Since the stated convergence is a tail event and $P_\mu = P_\lambda$ on $\mathcal{T}$ for any μ, the general result follows.

(ii) Since P_λ is shift-invariant with $P_\lambda\{X_s \in K\} = \lambda K < \infty$, the left-hand side converges a.e. P_λ by Theorem 20.2. From Theorems 20.10 and 20.12 we see that the limit is a.e. a constant $c \geq 0$. Using Fatou's lemma and Fubini's theorem gives
$$E_\lambda c \leq \liminf_{t \to \infty} t^{-1} \int_0^t P_\lambda\{X_s \in K\}\, ds = \lambda K < \infty,$$
which implies $c = 0$ since $\|P_\lambda\| = \|\lambda\| = \infty$. The general result follows from the fact that $P_\mu = P_\nu$ on $\mathcal{T}$ for any distributions μ and ν.

(iii) Fix any compact set $K \subset S$ with $\rho K > 0$, and conclude from the Markov property at $t > 0$ that a.s. P_μ
$$U1_K(X_t) = E_{X_t} \int_0^\infty 1_K(X_r)\, dr = E_\mu^{\mathcal{F}_t} \int_t^\infty 1_K(X_r)\, dr.$$
Using the chain rule for conditional expectations, we get for any $s < t$
$$E_\mu[U1_K(X_t)|\mathcal{F}_s] = E_\mu^{\mathcal{F}_s} \int_t^\infty 1_K(X_r)\, dr$$
$$\leq E_\mu^{\mathcal{F}_s} \int_s^\infty 1_K(X_r)\, dr = U1_K(X_s),$$
which shows that $U1_K(X_t)$ is a supermartingale. Since it is also nonnegative and right-continuous, it converges a.s. P_μ as $t \to \infty$, and the limit equals 0 a.s. since $U1_K(X_t) \xrightarrow{P_\mu} 0$ by the preceding proof. Since $U1_K$ is strictly positive and continuous, it follows that $X_t \to \infty$ a.s. P_μ. □

Exercises

1. Given a measure space $(S, \mathcal{S}, \mu)$, let T be a positive, linear operator on $L^1 \cap L^\infty$. Show that if T is both an L^1-contraction and an L^∞-contraction, then it is also an L^p-contraction for every $p \in [1, \infty]$. (*Hint:* Prove a Hölder-type inequality for T.)

2. Extend Lemma 10.3 to arbitrary transition operators T on a measurable space $(S, \mathcal{S})$. In other words, letting $\mathcal{I}$ denote the class of sets $B \in \mathcal{S}$ with $T1_B = 1_B$, show that an $\mathcal{S}$-measurable function $f \geq 0$ is T-invariant iff it is $\mathcal{I}$-measurable.

3. Prove a continuous-time version of Theorem 20.2 for measurable semigroups of positive $L^1 - L^\infty$-contraction. (*Hint:* Interpolate in the discrete-time result.)

4. Let (T_t) be a measurable, discrete- or continuous-time semigroup of positive $L^1 - L^\infty$-contractions on $(S, \mathcal{S}, \nu)$, let $\mu_1, \mu_2, \ldots$ be asymptotically invariant distributions on $\mathbb{Z}_+$ or $\mathbb{R}_+$, and define $A_n = \int T_t \mu_n(dt)$. Show that $A_n f \xrightarrow{\nu} Af$ for any $f \in L^1(\lambda)$, where $\xrightarrow{\nu}$ denotes convergence in measure. (*Hint:* Proceed as in Theorem 20.2, using the contractivity together with Minkowski's and Chebyshev's inequalities to estimate the remainder terms.)

5. Prove a continuous-time version of Theorem 20.4. (*Hint:* Use Lemma 20.5 to interpolate in the discrete-time result.)

6. Derive Theorem 10.6 from Theorem 20.4. (*Hint:* Take $g \equiv 1$, and proceed as in Corollary 10.9 to identify the limit.)

7. Show that when $f \geq 0$, the limit in Theorem 20.4 is strictly positive on the set $\{S_\infty f \wedge S_\infty g > 0\}$.

8. Show that the limit in Theorem 20.4 is invariant, at least when T is induced by a measure-preserving map on S.

9. Derive Lemma 20.3 (i) from Lemma 20.6. (*Hint:* Note that if $g \in \mathcal{T}(f)$ with $f \in L^1_+$, then $\mu g \leq \mu f$. Conclude that for any $h \in L^1$, $\mu[h; M_n h > 0] \geq \mu[U^{n-1}h; M_n h > 0] \geq 0$.)

10. Show that Brownian motion X in $\mathbb{R}^d$ is regular and strongly ergodic for every $d \in \mathbb{N}$ with an invariant measure that is unique up to a constant factor. Also show that X is Harris recurrent for $d = 1, 2$, uniformly transient for $d \geq 3$.

11. Let X be a Markov process with associated space-time process $\tilde{X}$. Show that X is strongly ergodic in the sense of Theorem 20.10 iff $\tilde{X}$ is weakly ergodic in the sense of Theorem 20.11. (*Hint:* Note that a function is space-time invariant for X iff it is invariant for $\tilde{X}$.)

12. For a Harris recurrent process on $\mathbb{R}_+$ or $\mathbb{Z}_+$, every tail event is clearly a.s. invariant. Show by an example that the statement may fail in the transient case.

13. State and prove discrete-time versions of Theorems 20.12, 20.17, and 20.18. (*Hint:* The continuous-time arguments apply with obvious changes.)

14. Derive discrete-time versions of Theorems 20.17 and 20.18 from the corresponding continuous-time results.

15. Show that a regular Markov process may be weakly but not strongly ergodic. (*Hint:* For any strongly ergodic process, the associated space-time process has the stated property. For a less trivial example, consider a suitable supercritical branching process.)

16. Give examples of nonregular Markov processes with no invariant measure, with exactly one (up to a normalization), and with more than one.

17. Show that a discrete-time Markov process X and the corresponding pseudo-Poisson process Y have the same invariant measures. Furthermore, regularity of X implies that Y is regular, but not conversely.

Chapter 21

Stochastic Differential Equations and Martingale Problems

Linear equations and Ornstein–Uhlenbeck processes; strong existence, uniqueness, and nonexplosion criteria; weak solutions and local martingale problems; well-posedness and measurability; pathwise uniqueness and functional solution; weak existence and continuity; transformation of SDEs; strong Markov and Feller properties

In this chapter we shall study classical stochastic differential equations (SDEs) driven by a Brownian motion and clarify the connection with the associated local martingale problems. Originally, the mentioned equations were devised to provide a pathwise construction of diffusions and more general continuous semimartingales. They have later turned out to be useful in a wide range of applications, where they may provide models for a diversity of dynamical systems with random perturbations. The coefficients determine a possibly time-dependent elliptic operator A as in Theorem 19.24, which suggests the associated martingale problem of finding a process X such that the processes M^f in Lemma 19.21 become martingales. It turns out to be essentially equivalent for X to be a weak solutions to the given SDE, as will be seen from the fundamental Theorem 21.7.

The theory of SDEs utilizes the basic notions and ideas of stochastic calculus, as developed in Chapters 17 and 18. Occasional references will be made to other chapters, such as to Chapter 6 for conditional independence, to Chapter 7 for martingale theory, to Chapter 16 for weak convergence, and to Chapter 19 for Feller processes. Some further aspects of the theory are displayed at the beginning of Chapter 23 as well as in Theorems 24.2, 26.8, and 27.14.

The SDEs studied in this chapter are typically of the form

$$dX_t^i = \sigma_j^i(t, X)dB_t^j + b^i(t, X)dt, \tag{1}$$

or more explicitly,

$$X_t^i = X_0^i + \sum_j \int_0^t \sigma_j^i(s, X)dB_s^j + \int_0^t b^i(s, X)ds, \quad t \geq 0. \tag{2}$$

Here $B = (B^1, \ldots, B^r)$ is a Brownian motion in $\mathbb{R}^r$ with respect to some filtration $\mathcal{F}$, and the solution $X = (X^1, \ldots, X^d)$ is a continuous

$\mathcal{F}$-semimartingale in $\mathbb{R}^d$. Furthermore, the coefficients σ and b are progressive functions of suitable dimension, defined on the canonical path space $C(\mathbb{R}_+, \mathbb{R}^d)$ equipped with the induced filtration $\mathcal{G}_t = \sigma\{w_s;\ s \leq t\}$, $t \geq 0$. For convenience, we shall often refer to (1) as *equation* (σ, b).

For the integrals in (2) to exist in the sense of Itô and Lebesgue integration, X must fulfill the integrability conditions

$$\int_0^t (|a^{ij}(s, X)| + |b^i(s, X)|) ds < \infty \text{ a.s.}, \quad t \geq 0, \tag{3}$$

where $a^{ij} = \sigma_k^i \sigma_k^j$ or $a = \sigma\sigma'$, and the bars denote any norms in the spaces of $d \times d$-matrices and d-vectors, respectively. For the existence and adaptedness of the right-hand side, it is also necessary that the integrands in (2) be progressive. This is ensured by the following result.

Lemma 21.1 *(progressive functions)* Let the function f on $\mathbb{R}_+ \times C(\mathbb{R}_+, \mathbb{R}^d)$ be progressive for the induced filtration $\mathcal{G}$ on $C(\mathbb{R}_+, \mathbb{R}^d)$, and let X be a continuous, $\mathcal{F}$-adapted process in $\mathbb{R}^d$. Then the process $Y_t = f(t, X)$ is $\mathcal{F}$-progressive.

Proof: Fix any $t \geq 0$. Since X is adapted, we note that $\pi_s(X) = X_s$ is $\mathcal{F}_t$-measurable for every $s \leq t$, where $\pi_s(w) = w_s$ on $C(\mathbb{R}_+, \mathbb{R}^d)$. Since $\mathcal{G}_t = \sigma\{\pi_s;\ s \leq t\}$, Lemma 1.4 shows that X is $\mathcal{F}_t/\mathcal{G}_t$-measurable. Hence, by Lemma 1.8 the mapping $\varphi(s, \omega) = (s, X(\omega))$ is $\mathcal{B}_t \otimes \mathcal{F}_t/\mathcal{B}_t \otimes \mathcal{G}_t$-measurable from $[0, t] \times \Omega$ to $[0, t] \times C(\mathbb{R}_+, \mathbb{R}^d)$, where $\mathcal{B}_t = \mathcal{B}[0, t]$. Also note that f is $\mathcal{B}_t \otimes \mathcal{G}_t$-measurable on $[0, t] \times C(\mathbb{R}_+, \mathbb{R}^d)$ since f is progressive. By Lemma 1.7 we conclude that $Y = f \circ \varphi$ is $\mathcal{B}_t \otimes \mathcal{F}_t/\mathcal{B}$-measurable on $[0, t] \times \Omega$. $\square$

Equation (2) exhibits the solution process X as an $\mathbb{R}^d$-valued semimartingale with drift components $b^i(X) \cdot \lambda$ and covariation processes $[X^i, X^j] = a^{ij}(X) \cdot \lambda$, where $a^{ij}(w) = a^{ij}(\cdot, w)$ and $b^i(w) = b^i(\cdot, w)$. It is natural to regard the densities $a(t, X)$ and $b(t, X)$ as *local characteristics* of X at time t. Of special interest is the *diffusion case*, where σ and b have the form

$$\sigma(t, w) = \sigma(w_t), \quad b(t, w) = b(w_t), \quad t \geq 0,\ w \in C(\mathbb{R}_+, \mathbb{R}^d), \tag{4}$$

for some measurable functions on $\mathbb{R}^d$. In that case, the local characteristics at time t depend only on the current position X_t of the process, and the progressivity holds automatically.

We shall distinguish between strong and weak solutions to an SDE (σ, b). For the former, the filtered probability space $(\Omega, \mathcal{F}, P)$ is regarded as given, along with an $\mathcal{F}$-Brownian motion B and an $\mathcal{F}_0$-measurable random vector ξ. A *strong solution* is then defined as an adapted process X with $X_0 = \xi$ a.s. satisfying (1). In case of a *weak solution*, only the initial distribution μ is given, and the solution consists of the triple $(\Omega, \mathcal{F}, P)$ together with an $\mathcal{F}$-Brownian motion B and an adapted process X with $P \circ X_0^{-1} = \mu$ satisfying (1).

This leads to different notions of existence and uniqueness for a given equation (σ, b). Thus, *weak existence* is said to hold for the initial distribution μ if there is a corresponding weak solution $(\Omega, \mathcal{F}, P, B, X)$. By contrast, *strong existence* for the given μ means that there is a strong solution X for *every* basic triple $(\mathcal{F}, B, \xi)$ such that ξ has distribution μ. We further say that *uniqueness in law* holds for the initial distribution μ if the corresponding weak solutions X have the same distribution. Finally, we say that *pathwise uniqueness* holds for the initial distribution μ if, for any two solutions X and Y on a common filtered probability space with a given Brownian motion B such that $X_0 = Y_0$ a.s. with distribution μ, we have $X = Y$ a.s.

One of the simplest SDEs is the *Langevin equation*
$$dX_t = dB_t - X_t dt, \tag{5}$$
which is of great importance for both theory and applications. Integrating by parts, we get from (5) the equation
$$d(e^t X_t) = e^t dX_t + e^t X_t dt = e^t dB_t,$$
which admits the explicit solution
$$X_t = e^{-t} X_0 + \int_0^t e^{-(t-s)} dB_s, \quad t \geq 0, \tag{6}$$
recognized as an *Ornstein-Uhlenbeck process*. Conversely, the process in (6) is easily seen to satisfy (5). We further note that $\theta_t X \xrightarrow{d} Y$ as $t \to \infty$, where Y denotes the stationary version of the process considered in Chapter 13. We can also get the stationary version directly from (6), by choosing X_0 to be $N(0, \frac{1}{2})$ and independent of B.

We turn to a more general class of equations that can be solved explicitly. A further extension appears in Theorem 26.8.

Proposition 21.2 *(linear equations)* *Let U and V be continuous semimartingales, and put $Z = \exp(V - V_0 - \frac{1}{2}[V])$. Then the equation $dX = dU + X dV$ has the unique solution*
$$X = Z\{X_0 + Z^{-1} \cdot (U - [U, V])\}. \tag{7}$$

Proof: Define $Y = X/Z$. Integrating by parts and noting that $dZ = Z dV$, we get
$$dU = dX - X dV = Y dZ + Z dY + d[Y, Z] - X dV = Z dY + d[Y, Z]. \tag{8}$$
In particular,
$$[U, V] = Z \cdot [Y, V] = [Y, Z]. \tag{9}$$
Substituting (9) into (8) yields $Z dY = dU - d[U, V]$, which implies $dY = Z^{-1} d(U - [U, V])$. To get (7), it remains to integrate from 0 to t and note that $Y_0 = X_0$. Since all steps are reversible, the same argument shows that (7) is indeed a solution. □

21. Stochastic Differential Equations and Martingale Problems

Though most SDEs have no explicit solution, we may still derive general conditions for strong existence, pathwise uniqueness, and continuous dependence on the initial conditions, by imitating the classical Picard iteration for ordinary differential equations. Recall that the relation $\lesssim$ denotes inequality up to a constant factor.

Theorem 21.3 *(strong solutions and stochastic flows, Itô)* Let σ and b be bounded, progressive functions satisfying a Lipschitz condition

$$(\sigma(w) - \sigma(w'))_t^* + (b(w) - b(w'))_t^* \lesssim (w - w')_t^*, \quad t \geq 0, \tag{10}$$

and fix a Brownian motion B in $\mathbb{R}^r$ with associated complete filtration $\mathcal{F}$. Then there exists a jointly continuous process $X = (X_t^x)$ on $\mathbb{R}_+ \times \mathbb{R}^d$ such that, for any $\mathcal{F}_0$-measurable random vector ξ in $\mathbb{R}^d$, equation (σ, b) has the a.s. unique solution X^ξ starting at ξ.

For one-dimensional diffusion equations, a stronger result is established in Theorem 23.3. The solution process $X = (X_t^x)$ on $\mathbb{R}_+ \times \mathbb{R}^d$ is called the *stochastic flow* generated by B. Our proof is based on two lemmas, and we begin with an elementary estimate.

Lemma 21.4 *(Gronwall)* Let f be a continuous function on $\mathbb{R}_+$ such that

$$f(t) \leq a + b \int_0^t f(s) ds, \quad t \geq 0, \tag{11}$$

for some $a, b \geq 0$. Then $f(t) \leq ae^{bt}$ for all $t \geq 0$.

Proof: We may write (11) as

$$\frac{d}{dt}\left\{e^{-bt}\int_0^t f(s)ds\right\} \leq ae^{-bt}, \quad t \geq 0.$$

It remains to integrate over $[0, t]$ and combine with (11). □

To state the next result, let $S(X)$ denote the process defined by the right-hand side of (2).

Lemma 21.5 *(local contraction)* Let σ and b be bounded, progressive functions satisfying (10), and fix any $p \geq 2$. Then there exists a nondecreasing function $c \geq 0$ on $\mathbb{R}_+$ such that, for any continuous adapted processes X and Y in $\mathbb{R}^d$,

$$E(S(X) - S(Y))_t^{*p} \leq 2E|X_0 - Y_0|^p + c_t \int_0^t E(X - Y)_s^{*p} ds, \quad t \geq 0.$$

Proof: By Theorem 17.7, condition (10), and Jensen's inequality,

$$\begin{aligned}
E(S(X) - S(Y))_t^{*p} &- 2E|X_0 - Y_0|^p \\
&\lesssim E((\sigma(X) - \sigma(Y)) \cdot B)_t^{*p} + E((b(X) - b(Y)) \cdot \lambda)_t^{*p} \\
&\lesssim E(|\sigma(X) - \sigma(Y)|^2 \cdot \lambda)_t^{p/2} + E(|b(X) - b(Y)| \cdot \lambda)_t^p \\
&\lesssim E\left|\int_0^t (X-Y)_s^{*2} ds\right|^{p/2} + E\left|\int_0^t (X-Y)_s^* ds\right|^p \\
&\leq (t^{p/2-1} + t^{p-1}) \int_0^t E(X-Y)_s^{*p} ds. \qquad \square
\end{aligned}$$

Proof of Theorem 21.3: To prove the existence, fix any $\mathcal{F}_0$-measurable random vector ξ in $\mathbb{R}^d$, put $X_t^0 \equiv \xi$, and define recursively $X^n = S(X^{n-1})$ for $n \geq 1$. Since σ and b are bounded, we have $E(X^1 - X^0)_t^{*2} < \infty$, and by Lemma 21.5

$$E(X^{n+1} - X^n)_t^{*2} \leq c_t \int_0^t E(X^n - X^{n-1})_s^{*2} ds, \quad t \geq 0, \ n \geq 1.$$

Hence, by induction,

$$E(X^{n+1} - X^n)_t^{*2} \leq \frac{c_t^n t^n}{n!} E(X^1 - \xi)_t^{*2} < \infty, \quad t, n \geq 0.$$

For any $k \in \mathbb{N}$, we get

$$\begin{aligned}
\|\sup_{n \geq k}(X^n - X^k)_t^*\|_2 &\leq \sum_{n \geq k} \|(X^{n+1} - X^n)_t^*\|_2 \\
&\leq \|(X^1 - \xi)_t^*\|_2 \sum_{n \geq k} (c_t^n t^n / n!)^{1/2} < \infty.
\end{aligned}$$

Thus, by Lemma 4.6 there exists a continuous adapted process X with $X_0 = \xi$ such that $(X^n - X)_t^* \to 0$ a.s. and in L^2 for each $t \geq 0$. To see that X solves equation (σ, b), we may use Lemma 21.5 to obtain

$$E(X^n - S(X))_t^{*2} \leq c_t \int_0^t E(X^{n-1} - X)_s^{*2} ds, \quad t \geq 0.$$

As $n \to \infty$, we get $E(X - S(X))_t^{*2} = 0$ for all t, which implies $X = S(X)$ a.s.

Now consider any two solutions X and Y with $|X_0 - Y_0| \leq \varepsilon$ a.s. By Lemma 21.5 we get for any $p \geq 2$

$$E(X - Y)_t^{*p} \leq 2\varepsilon^p + c_t \int_0^t E(X-Y)_s^{*p} ds, \quad t \geq 0,$$

and by Lemma 21.4 it follows that

$$E(X - Y)_t^{*p} \leq 2\varepsilon^p e^{c_t t}, \quad t \geq 0. \qquad (12)$$

If $X_0 = Y_0$ a.s., we may take $\varepsilon = 0$ and conclude that $X = Y$ a.s., which proves the asserted uniqueness. Letting X^x denote the solution X with

$X_0 = x$ a.s., we get by (12)

$$E|X^x - X^y|_t^{*p} \leq 2|x-y|^p e^{c_t t}, \quad t \geq 0.$$

Taking $p > d$ and applying Theorem 3.23 for each $T > 0$ with the metric $\rho_T(f,g) = (f-g)^*_T$, we conclude that the process (X_t^x) has a jointly continuous version on $\mathbb{R}_+ \times \mathbb{R}^d$.

From the construction we note that if X and Y are solutions with $X_0 = \xi$ and $Y_0 = \eta$ a.s., then $X = Y$ a.s. on the set $\{\xi = \eta\}$. In particular, $X = X^\xi$ a.s. when ξ takes countably many values. In general, we may approximate ξ uniformly by random vectors $\xi_1, \xi_2, \ldots$ in $\mathbb{Q}^d$, and by (12) we get $X_t^{\xi_n} \to X_t$ in L^2 for all $t \geq 0$. Since also $X_t^{\xi_n} \to X_t^\xi$ a.s. by the continuity of the flow, it follows that $X_t = X_t^\xi$ a.s. $\square$

It is often useful to allow the solutions to explode. As in Chapter 19, we may then introduce an absorbing state Δ at infinity, so that the path space becomes $C(\mathbb{R}_+, \overline{\mathbb{R}^d})$ with $\overline{\mathbb{R}^d} = \mathbb{R}^d \cup \{\Delta\}$. Define $\zeta_n = \inf\{t; |X_t| \geq n\}$ for each n, put $\zeta = \sup_n \zeta_n$, and let $X_t = \Delta$ for $t \geq \zeta$. Given a Brownian motion B in $\mathbb{R}^r$ and an adapted process X in the extended path space, we say that X or the pair (X, B) solves equation (σ, b) on the interval $[0, \zeta)$ if

$$X_{t\wedge\zeta_n} = X_0 + \int_0^{t\wedge\zeta_n} \sigma(s, X)dB_s + \int_0^{t\wedge\zeta_n} b(s, X)ds, \quad t \geq 0, \, n \in \mathbb{N}. \quad (13)$$

When $\zeta < \infty$, we have $|X_{\zeta_n}| \to \infty$, and X is said to *explode* at time ζ.

Conditions for the existence and uniqueness of possibly exploding solutions may be obtained from Theorem 21.3 by suitable localization. The following result is then useful to decide whether explosion can actually occur.

Proposition 21.6 *(explosion)* *The solutions to equation (σ, b) are a.s. nonexploding if*

$$\sigma(x)^*_t + b(x)^*_t \leq 1 + x^*_t, \quad t \geq 0. \quad (14)$$

Proof: By Proposition 17.15 we may assume that X_0 is bounded. From (13) and (14) we get for suitable constants $c_t < \infty$

$$EX_{t\wedge\zeta_n}^{*2} \leq 2E|X_0|^2 + c_t \int_0^t (1 + EX_{s\wedge\zeta_n}^{*2})ds, \quad t \geq 0, \, n \in \mathbb{N},$$

and so by Lemma 21.4

$$1 + EX_{t\wedge\zeta_n}^{*2} \leq (1 + 2E|X_0|^2)\exp(c_t t) < \infty, \quad t \geq 0, \, n \in \mathbb{N}.$$

As $n \to \infty$, we obtain $EX_{t\wedge\zeta}^{*2} < \infty$, which implies $\zeta > t$ a.s. $\square$

Our next aim is to characterize weak solutions to equation (σ, b) by a martingale property that involves only the solution X. Then define

$$M_t^f = f(X_t) - f(X_0) - \int_0^t A_s f(X)ds, \quad t \geq 0, \, f \in C_K^\infty, \quad (15)$$

where the operators A_s are given by

$$A_s f(x) = \tfrac{1}{2} a^{ij}(s,x) f''_{ij}(x_s) + b^i(s,x) f'_i(x_s), \quad s \geq 0, \; f \in C_K^\infty. \tag{16}$$

In the diffusion case we may replace the integrand $A_s f(X)$ in (15) by the expression $Af(X_s)$, where A denotes the elliptic operator

$$Af(x) = \tfrac{1}{2} a^{ij}(x) f''_{ij}(x) + b^i(x) f'_i(x), \quad f \in C_K^\infty, \; x \in \mathbb{R}^d. \tag{17}$$

A continuous process X in $\mathbb{R}^d$ or its distribution P is said to solve the *local martingale problem* for (a,b) if M^f is a local martingale for every $f \in C_K^\infty$. When a and b are bounded, it is clearly equivalent for M^f to be a true martingale, and the original problem turns into a *martingale problem*. The (local) martingale problem for (a,b) with initial distribution μ is said to be *well posed* if it has exactly one solution P_μ. For degenerate initial distributions δ_x, we may write P_x instead of P_{δ_x}. The next result gives the basic equivalence between weak solutions to an SDE and solutions to the associated local martingale problem.

Theorem 21.7 (*weak solutions and martingale problems, Stroock and Varadhan*) *Let σ and b be progressive, and fix any probability measure P on $C(\mathbb{R}_+, \mathbb{R}^d)$. Then equation (σ, b) has a weak solution with distribution P iff P solves the local martingale problem for $(\sigma \sigma', b)$.*

Proof: Write $a = \sigma \sigma'$. If (X, B) solves equation (σ, b), then

$$\begin{aligned}
[X^i, X^j] &= [\sigma^i_k(X) \cdot B^k, \sigma^j_l(X) \cdot B^l] \\
&= \sigma^i_k \sigma^j_l(X) \cdot [B^k, B^l] = a^{ij}(X) \cdot \lambda.
\end{aligned}$$

By Itô's formula we get for any $f \in C_K^\infty$

$$\begin{aligned}
df(X_t) &= f'_i(X_t) dX^i_t + \tfrac{1}{2} f''_{ij}(X_t) d[X^i, X^j]_t \\
&= f'_i(X_t) \sigma^i_j(t, X) dB^j_t + A_t f(X) dt.
\end{aligned}$$

Hence, $dM^f_t = f'_i(X_t) \sigma^i_j(t, X) dB^j_t$, and so M^f is a local martingale.

Conversely, assume that X solves the local martingale problem for (a,b). Considering functions $f^i_n \in C_K^\infty$ with $f^i_n(x) = x^i$ for $|x| \leq n$, it is clear by a localization argument that the processes

$$M^i_t = X^i_t - X^i_0 - \int_0^t b^i(s, X) ds, \quad t \geq 0, \tag{18}$$

are continuous local martingales. Similarly, we may choose $f^{ij}_n \in C_K^\infty$ with $f^{ij}_n(x) = x^i x^j$ for $|x| \leq n$, to obtain the local martingales

$$M^{ij} = X^i X^j - X^i_0 X^j_0 - (X^i \beta^j + X^j \beta^i + \alpha^{ij}) \cdot \lambda,$$

where $\alpha^{ij} = a^{ij}(X)$ and $\beta^i = b^i(X)$. Integrating by parts and using (18), we get

$$\begin{aligned}
M^{ij} &= X^i \cdot X^j + X^j \cdot X^i + [X^i, X^j] - (X^i \beta^j + X^j \beta^i + \alpha^{ij}) \cdot \lambda \\
&= X^i \cdot M^j + X^j \cdot M^i + [M^i, M^j] - \alpha^{ij} \cdot \lambda.
\end{aligned}$$

The last two terms on the right then form a local martingale, and so by Proposition 17.2

$$[M^i, M^j]_t = \int_0^t a^{ij}(s, X)ds, \quad t \geq 0.$$

Hence, by Theorem 18.12 there exists a Brownian motion B with respect to a standard extension of the original filtration such that

$$M_t^i = \int_0^t \sigma_k^i(s, X)dB_s^k, \quad t \geq 0.$$

Substituting this into (18) yields (2), which means that the pair (X, B) solves equation (σ, b). $\square$

For subsequent needs, we note that the previous construction can be made measurable in the following sense.

Lemma 21.8 *(functional representation)* Let σ and b be progressive. Then there exists a measurable mapping

$$F: \mathcal{P}(C(\mathbb{R}_+, \mathbb{R}^d)) \times C(\mathbb{R}_+, \mathbb{R}^d) \times [0, 1] \to C(\mathbb{R}_+, \mathbb{R}^r),$$

such that, if the local martingale problem for $(\sigma\sigma', b)$ admits a solution X with distribution P and if $\vartheta \perp\!\!\!\perp X$ is $U(0, 1)$, then $B = F(P, X, \vartheta)$ is a Brownian motion in $\mathbb{R}^r$ and the pair (X, B) with induced filtration solves equation (σ, b).

Proof: In the previous construction of B, the only nonelementary step is the stochastic integration with respect to (X, Y) in Theorem 18.12, where Y is an independent Brownian motion, and the integrand is a progressive function of X obtained by some elementary matrix algebra. Since the pair (X, Y) is again a solution to a local martingale problem, Proposition 17.26 yields the desired functional representation. $\square$

Combining the martingale formulation with a compactness argument, we may deduce some general existence and continuity results.

Theorem 21.9 *(weak existence and continuity, Skorohod)* Let a and b be bounded, progressive functions such that, for any fixed $t \geq 0$, the functions $a(t, \cdot)$ and $b(t, \cdot)$ are continuous on $C(\mathbb{R}_+, \mathbb{R}^d)$. Then the martingale problem for (a, b) has a solution P_μ for every initial distribution μ. If those solutions are unique, then the mapping $\mu \mapsto P_\mu$ is weakly continuous.

Proof: For any $\varepsilon > 0$, $t \geq 0$, and $x \in C(\mathbb{R}_+, \mathbb{R}^d)$, define

$$\sigma_\varepsilon(t, x) = \sigma((t - \varepsilon)_+, x), \quad b_\varepsilon(t, x) = b((t - \varepsilon)_+, x),$$

and let $a_\varepsilon = \sigma_\varepsilon \sigma_\varepsilon'$. Since σ and b are progressive, the processes $\sigma_\varepsilon(s, X)$ and $b_\varepsilon(s, X)$, $s \leq t$, are measurable functions of X on $[0, (t - \varepsilon)_+]$. Hence, a strong solution X^ε to equation $(\sigma_\varepsilon, b_\varepsilon)$ may be constructed recursively on the intervals $[(n - 1)\varepsilon, n\varepsilon]$, $n \in \mathbb{N}$, starting from an arbitrary random vector $\xi \perp\!\!\!\perp B$ in $\mathbb{R}^d$ with distribution μ. Note in particular that X^ε solves the martingale problem for the pair $(a_\varepsilon, b_\varepsilon)$.

Applying Theorem 17.7 to equation $(\sigma_\varepsilon, b_\varepsilon)$ and using the boundedness of σ and b, we get for any $p > 0$

$$E \sup_{0 \leq r \leq h} |X^\varepsilon_{t+r} - X^\varepsilon_t|^p \lesssim h^{p/2} + h^p \lesssim h^{p/2}, \quad t, \varepsilon \geq 0, \ h \in [0, 1].$$

For $p > 2d$ it follows by Corollary 16.9 that the family $\{X^\varepsilon\}$ is tight in $C(\mathbb{R}_+, \mathbb{R}^d)$, and by Theorem 16.3 we may then choose some $\varepsilon_n \to 0$ such that $X^{\varepsilon_n} \xrightarrow{d} X$ for a suitable X.

To see that X solves the martingale problem for (a, b), let $f \in C_K^\infty$ and $s < t$ be arbitrary, and consider any bounded, continuous function $g: C([0, s], \mathbb{R}^d) \to \mathbb{R}$. We need to show that

$$E\left\{ f(X_t) - f(X_s) - \int_s^t A_r f(X) dr \right\} g(X) = 0.$$

Then note that X^ε satisfies the corresponding equation for the operators A^ε_r constructed from the pair $(a_\varepsilon, b_\varepsilon)$. Writing the two conditions as $E\varphi(X) = 0$ and $E\varphi_\varepsilon(X^\varepsilon) = 0$, respectively, it suffices by Theorem 4.27 to show that $\varphi_\varepsilon(x_\varepsilon) \to \varphi(x)$ whenever $x_\varepsilon \to x$ in $C(\mathbb{R}_+, \mathbb{R}^d)$. This follows easily from the continuity conditions imposed on a and b.

Now assume that the solutions P_μ are unique, and let $\mu_n \xrightarrow{w} \mu$. Arguing as before, we see that (P_{μ_n}) is tight, and so by Theorem 16.3 it is also relatively compact. If $P_{\mu_n} \xrightarrow{w} Q$ along some subsequence, then as before we note that Q solves the martingale problem for (a, b) with initial distribution μ. Hence $Q = P_\mu$, and the convergence extends to the original sequence. $\square$

Our next aim is to show how the well-posedness of the local martingale problem for (a, b) extends from degenerate to arbitrary initial distributions. This requires a basic measurability property, which will also be needed later.

Theorem 21.10 *(measurability and mixtures, Stroock and Varadhan)* Let a and b be progressive and such that, for any $x \in \mathbb{R}^d$, the local martingale problem for (a, b) with initial distribution δ_x has a unique solution P_x. Then (P_x) is a kernel from $\mathbb{R}^d$ to $C(\mathbb{R}_+, \mathbb{R}^d)$, and for every initial distribution μ, the associated local martingale problem has the unique solution $P_\mu = \int P_x \mu(dx)$.

Proof: According to the proof of Theorem 21.7, it is enough to formulate the local martingale problem in terms of functions f belonging to some countable subclass $\mathcal{C} \subset C_K^\infty$, consisting of suitably truncated versions of the coordinate functions x^i and their products $x^i x^j$. Now define $\mathcal{P} = \mathcal{P}(C(\mathbb{R}^d, \mathbb{R}^d))$ and $\mathcal{P}_M = \{P_x; x \in \mathbb{R}^d\}$, and write X for the canonical process in $C(\mathbb{R}_+, \mathbb{R}^d)$. Let D denote the class of measures $P \in \mathcal{P}$ with degenerate projections $P \circ X_0^{-1}$. Next let I consist of all measures $P \in \mathcal{P}$ such that X satisfies the integrability condition (3). Finally, put

$\tau_n^f = \inf\{t; |M_t^f| \geq n\}$, and let L be the class of measures $P \in \mathcal{P}$ such that the processes $M_t^{f,n} = M^f(t \wedge \tau_n^f)$ exist and are martingales under P for all $f \in C$ and $n \in \mathbb{N}$. Then clearly $\mathcal{P}_M = D \cap I \cap L$.

To prove the asserted kernel property, it is enough to show that $\mathcal{P}_M$ is a measurable subset of $\mathcal{P}$, since the desired measurability will then follow by Theorem A1.3 and Lemma 1.40. The measurability of D is clear from Lemma 1.39 (i). Even I is measurable, since the integrals on the left of (3) are measurable by Fubini's theorem. Finally, $L \cap I$ is a measurable subset of I, since the defining condition is equivalent to countably many relations of the form $E[M_t^{f,n} - M_s^{f,n}; F] = 0$, with $f \in C$, $n \in \mathbb{N}$, $s < t$ in $\mathbb{Q}_+$, and $F \in \mathcal{F}_s$.

Now fix any probability measure μ on $\mathbb{R}^d$. The measure $P_\mu = \int P_x \mu(dx)$ has clearly initial distribution μ, and from the previous argument we note that P_μ again solves the local martingale for (a, b). To prove the uniqueness, let P be any measure with the stated properties. Then $E[M_t^{f,n} - M_s^{f,n}; F | X_0] = 0$ a.s. for all $f, n, s < t$, and F as above, and so $P[\cdot | X_0]$ is a.s. a solution to the local martingale problem with initial distribution δ_{X_0}. Thus, $P[\cdot | X_0] = P_{X_0}$ a.s., and we get $P = EP_{X_0} = \int P_x \mu(dx) = P_\mu$. This extends the well-posedness to arbitrary initial distributions. □

We return to the basic problem of constructing a Feller diffusion with given generator A in (17) as the solution to a suitable SDE or the associated martingale problem. The following result may be regarded as a converse to Theorem 19.24.

Theorem 21.11 (*strong Markov and Feller properties, Stroock and Varadhan*) *Let a and b be measurable functions on $\mathbb{R}^d$ such that, for any $x \in \mathbb{R}^d$, the local martingale problem for (a, b) with initial distribution δ_x has a unique solution P_x. Then the family (P_x) satisfies the strong Markov property. If a and b are also bounded and continuous, then the equation $T_t f(x) = E_x f(X_t)$ defines a Feller semigroup on C_0, and the operator A in (17) extends uniquely to the associated generator.*

Proof: By Theorem 21.10 it remains to prove that, for any state $x \in \mathbb{R}^d$ and bounded optional time τ,

$$P_x[X \circ \theta_\tau \in \cdot | \mathcal{F}_\tau] = P_{X_\tau} \text{ a.s.}$$

As in the previous proof, this is equivalent to countably many relations of the form

$$E_x[\{(M_t^{f,n} - M_s^{f,n})1_F\} \circ \theta_\tau | \mathcal{F}_\tau] = 0 \text{ a.s.} \qquad (19)$$

with $s < t$ and $F \in \mathcal{F}_s$, where $M^{f,n}$ denotes the process M^f stopped at $\tau_n = \inf\{t; |M^f| \geq n\}$. Now $\theta_\tau^{-1} \mathcal{F}_s \subset \mathcal{F}_{\tau+s}$ by Lemma 7.5, and in the diffusion case

$$(M_t^{f,n} - M_s^{f,n}) \circ \theta_\tau = M_{(\tau+t) \wedge \sigma_n}^f - M_{\tau \wedge \sigma_n}^f,$$

where $\sigma_n = \tau + \tau_n \circ \theta_\tau$, which is again optional by Proposition 8.8. Thus, (19) follows by optional sampling from the local martingale property of M^f under P_x.

Now assume that a and b are also bounded and continuous, and define $T_t f(x) = E_x f(X_t)$. By Theorem 21.9 we note that $T_t f$ is continuous for every $f \in C_0$ and $t > 0$, and from the continuity of the paths it is clear that $T_t f(x)$ is continuous in t for each x. To see that $T_t f \in C_0$, it remains to show that $|X_t^x| \xrightarrow{P} \infty$ as $|x| \to \infty$, where X^x has distribution P_x. But this follows from the SDE by the boundedness of σ and b if for $0 < r < |x|$ we write

$$P\{|X_t^x| < r\} \leq P\{|X_t^x - x| > |x| - r\} \leq \frac{E|X_t^x - x|^2}{(|x| - r)^2} \lesssim \frac{t + t^2}{(|x| - r)^2},$$

and let $|x| \to \infty$ for fixed r and t. The last assertion is obvious from the uniqueness in law together with Theorem 19.23. $\square$

It is usually harder to establish uniqueness in law than to prove weak existence. Some fairly general uniqueness criteria will be obtained in Theorems 23.1 and 24.2. For the moment we shall only exhibit some transformations that may simplify the problem. The following result, based on a change of probability measure, is often useful to eliminate the drift term.

Proposition 21.12 (*transformation of drift*) *Let σ, b, and c be progressive functions of suitable dimension, where c is bounded. Then weak existence holds simultaneously for equations (σ, b) and $(\sigma, b + \sigma c)$. If, moreover, $c = \sigma' h$ for some progressive function h, then even uniqueness in law holds simultaneously for the two equations.*

Proof: Let X be a weak solution to equation (σ, b), defined on the canonical space for (X, B) with induced filtration $\mathcal{F}$ and with probability measure P. Put $V = c(X)$, and note that $(V^2 \cdot \lambda)_t$ is bounded for each t. By Lemma 18.18 and Corollary 18.25 there exists a probability measure Q with $Q = \mathcal{E}(V' \cdot B)_t \cdot P$ on $\mathcal{F}_t$ for each $t \geq 0$, and we note that $\tilde{B} = B - V \cdot \lambda$ is a Q-Brownian motion. Under Q we further get by Proposition 18.20

$$\begin{aligned} X - X_0 &= \sigma(X) \cdot (\tilde{B} + V \cdot \lambda) + b(X) \cdot \lambda \\ &= \sigma(X) \cdot \tilde{B} + (b + \sigma c)(X) \cdot \lambda, \end{aligned}$$

which shows that X is a weak solution to the SDE $(\sigma, b + \sigma c)$. Since the same argument applies to equation $(\sigma, b + \sigma c)$ with c replaced by $-c$, we conclude that weak existence holds simultaneously for the two equations.

Now let $c = \sigma' h$, and assume that uniqueness in law holds for equation $(\sigma, b + ah)$. Further assume that (X, B) solves equation (σ, b) under both P and Q. Choosing V and $\tilde{B}$ as before, it follows that $(X, \tilde{B})$ solves equation $(\sigma, b + \sigma c)$ under the transformed distributions $\mathcal{E}(V' \cdot B)_t \cdot P$ and $\mathcal{E}(V' \cdot B)_t \cdot Q$ for (X, B). By hypothesis the latter measures then have the same X-marginal, and the stated condition implies that $\mathcal{E}(V' \cdot B)$ is X-measurable. Thus, the X-marginals agree even for P and Q, which proves the uniqueness in law for equation (σ, b). Again we may reverse the argument to get an implication in the other direction. □

Next we examine how an SDE of diffusion type can be transformed by a random time-change. The method will be used systematically in Chapter 23 to analyze the one-dimensional case.

Proposition 21.13 *(scaling) Fix some measurable functions σ, b, and $c > 0$ on $\mathbb{R}^d$, where c is bounded away from 0 and ∞. Then weak existence and uniqueness in law hold simultaneously for equations (σ, b) and $(c\sigma, c^2 b)$.*

Proof: Assume that X solves the local martingale problem for the pair (a, b), and introduce the process $V = c^2(X) \cdot \lambda$ with inverse (τ_s). By optional sampling we note that $M^f_{\tau_s}$, $s \geq 0$, is again a local martingale, and the process $Y_s = X_{\tau_s}$ satisfies

$$M^f_{\tau_s} = f(Y_s) - f(Y_0) - \int_0^s c^2 A f(Y_r) dr.$$

Thus, Y solves the local martingale problem for $(c^2 a, c^2 b)$.

Now let T denote the mapping on $C(\mathbb{R}_+, \mathbb{R}^d)$ leading from X to Y, and write T' for the corresponding mapping based on c^{-1}. Then T and T' are mutual inverses, and so by the previous argument applied to both mappings, a measure $P \in \mathcal{P}(C(\mathbb{R}_+, \mathbb{R}^d))$ solves the local martingale problem for (a, b) iff $P \circ T^{-1}$ solves the corresponding problem for $(c^2 a, c^2 b)$. Thus, both existence and uniqueness hold simultaneously for the two problems. By Theorem 21.7 the last statement translates immediately into a corresponding assertion for the SDEs. □

Our next aim is to examine the connection between weak and strong solutions. Under appropriate conditions, we shall further establish the existence of a universal functional solution. To explain the subsequent terminology, let $\mathcal{G}$ be the filtration induced by the identity mapping (ξ, B) on the canonical space $\Omega = \mathbb{R}^d \times C(\mathbb{R}_+, \mathbb{R}^r)$, so that $\mathcal{G}_t = \sigma\{\xi, B^t\}$, $t \geq 0$, where $B^t_s = B_{s \wedge t}$. Writing W^r for the r-dimensional Wiener measure, we introduce for any $\mu \in \mathcal{P}(\mathbb{R}^d)$ the $(\mu \otimes W^r)$-completion $\mathcal{G}^\mu_t$ of $\mathcal{G}_t$. The *universal completion* $\overline{\mathcal{G}}_t$ is defined as $\bigcap_\mu \mathcal{G}^\mu_t$, and we say that a function

$$F \colon \mathbb{R}^d \times C(\mathbb{R}_+, \mathbb{R}^r) \to C(\mathbb{R}_+, \mathbb{R}^d) \tag{20}$$

is *universally adapted* if it is adapted to the filtration $\overline{\mathcal{G}} = (\overline{\mathcal{G}}_t)$.

Theorem 21.14 *(pathwise uniqueness and functional solution) Let σ and b be progressive and such that weak existence and pathwise uniqueness hold for solutions to equation (σ, b) starting at fixed points. Then strong existence and uniqueness in law hold for any initial distribution, and there exists a measurable and universally adapted function F as in (20) such that every solution (X, B) to equation (σ, b) satisfies $X = F(X_0, B)$ a.s.*

Note in particular that the function F above is independent of initial distribution μ. A key step in the proof, accomplished in Lemma 21.17, is to establish the corresponding result for a fixed μ. Two further lemmas will be needed, and we begin with a statement that clarifies the connection between adaptedness, strong existence, and functional solutions.

Lemma 21.15 *(transfer of strong solution)* *Let (X, B) solve equation (σ, b), and assume that X is adapted to the complete filtration induced by X_0 and B. Then $X = F(X_0, B)$ a.s. for some Borel-measurable function F as in (20), and for any basic triple $(\mathcal{F}, \tilde{B}, \xi)$ with $\xi \stackrel{d}{=} X_0$, the process $\tilde{X} = F(\xi, \tilde{B})$ is $\mathcal{F}$-adapted and such that the pair $(\tilde{X}, \tilde{B})$ solves equation (σ, b).*

Proof: By Lemma 1.13 we have $X = F(X_0, B)$ a.s. for some Borel-measurable function F as stated. By the same result, there exists for every $t \geq 0$ a further representation of the form $X_t = G_t(X_0, B^t)$ a.s., and so $F(X_0, B)_t = G_t(X_0, B^t)$ a.s. Hence, $\tilde{X}_t = G_t(\xi, \tilde{B}^t)$ a.s., and so $\tilde{X}$ is $\mathcal{F}$-adapted. Since also $(\tilde{X}, \tilde{B}) \stackrel{d}{=} (X, B)$, Proposition 17.26 shows that even the former pair solves equation (σ, b). $\square$

The following result shows that even weak solutions can be transferred to any given probability space with a specified Brownian motion.

Lemma 21.16 *(transfer of weak solution)* *Let (X, B) solve equation (σ, b), and fix any basic triple $(\mathcal{F}, \tilde{B}, \xi)$ with $\xi \stackrel{d}{=} X_0$. Then there exists a process $\tilde{X} \perp\!\!\!\perp_{\xi, \tilde{B}} \mathcal{F}$ with $\tilde{X}_0 = \xi$ a.s. and $(\tilde{X}, \tilde{B}) \stackrel{d}{=} (X, B)$. Furthermore, the filtration $\mathcal{G}$ induced by $(\tilde{X}, \mathcal{F})$ is a standard extension of $\mathcal{F}$, and the pair $(\tilde{X}, \tilde{B})$ with filtration $\mathcal{G}$ solves equation (σ, b).*

Proof: By Theorem 6.10 and Proposition 6.13 there exists a process $\tilde{X} \perp\!\!\!\perp_{\xi, \tilde{B}} \mathcal{F}$ satisfying $(\tilde{X}, \xi, \tilde{B}) \stackrel{d}{=} (X, X_0, B)$, and in particular $\tilde{X}_0 = \xi$ a.s. To see that $\mathcal{G}$ is a standard extension of $\mathcal{F}$, fix any $t \geq 0$ and define $\tilde{B}' = \tilde{B} - \tilde{B}^t$. Then $(\tilde{X}^t, \tilde{B}^t) \perp\!\!\!\perp \tilde{B}'$ since the corresponding relation holds for (X, B), and so $\tilde{X}^t \perp\!\!\!\perp_{\xi, \tilde{B}^t} \tilde{B}'$. Since also $\tilde{X}^t \perp\!\!\!\perp_{\xi, \tilde{B}} \mathcal{F}$, Proposition 6.8 yields $\tilde{X}^t \perp\!\!\!\perp_{\xi, \tilde{B}^t} (\tilde{B}', \mathcal{F})$ and hence $\tilde{X}^t \perp\!\!\!\perp_{\mathcal{F}_t} \mathcal{F}$. But then $(\tilde{X}^t, \mathcal{F}_t) \perp\!\!\!\perp_{\mathcal{F}_t} \mathcal{F}$ by Corollary 6.7, which means that $\mathcal{G}_t \perp\!\!\!\perp_{\mathcal{F}_t} \mathcal{F}$.

Since standard extensions preserve martingales, Theorem 18.3 shows that $\tilde{B}$ remains a Brownian motion with respect to $\mathcal{G}$. As in Proposition 17.26, we conclude that the pair $(\tilde{X}, \tilde{B})$ solves equation (σ, b). $\square$

We are now ready to establish the crucial relationship between strong existence and pathwise uniqueness.

Lemma 21.17 *(strong existence and pathwise uniqueness, Yamada and Watanabe)* *Assume that weak existence and pathwise uniqueness hold for solutions to equation (σ, b) with initial distribution μ. Then even strong existence and uniqueness in law hold for such solutions, and there exists a measurable function F_μ as in (20) such that any solution (X, B) with initial distribution μ satisfies $X = F_\mu(X_0, B)$ a.s.*

21. Stochastic Differential Equations and Martingale Problems

Proof: Fix any solution (X, B) with initial distribution μ and associated filtration $\mathcal{F}$. By Lemma 21.16 there exists some process $Y \perp\!\!\!\perp_{X_0, B} \mathcal{F}$ with $Y_0 = X_0$ a.s. such that (Y, B) solves equation (σ, b) for the filtration $\mathcal{G}$ induced by $(Y, \mathcal{F})$. Since $\mathcal{G}$ is a standard extension of $\mathcal{F}$, the pair (X, B) remains a solution for $\mathcal{G}$, and the pathwise uniqueness yields $X = Y$ a.s.

For each $t \geq 0$ we have $X^t \perp\!\!\!\perp_{X_0, B} X^t$ and $(X^t, B^t) \perp\!\!\!\perp (B - B^t)$, and so $X^t \perp\!\!\!\perp_{X_0, B^t} X^t$ a.s. by Proposition 6.8. Thus, Corollary 6.7 (ii) shows that X is adapted to the complete filtration induced by (X_0, B). Hence, by Lemma 21.15 there exists a measurable function F_μ with $X = F_\mu(X_0, B)$ a.s. and such that, for any basic triple $(\tilde{\mathcal{F}}, \tilde{B}, \xi)$ with $\xi \stackrel{d}{=} X_0$, the process $\tilde{X} = F_\mu(\xi, \tilde{B})$ is $\tilde{\mathcal{F}}$-adapted and solves equation (σ, b) along with $\tilde{B}$. In particular, $\tilde{X} \stackrel{d}{=} X$ since $(\xi, \tilde{B}) \stackrel{d}{=} (X_0, B)$, and the pathwise uniqueness shows that $\tilde{X}$ is the a.s. unique solution for the given triple $(\tilde{\mathcal{F}}, \tilde{B}, \xi)$. This proves the uniqueness in law. □

Proof of Theorem 21.14: By Lemma 21.17 we have uniqueness in law for solutions starting at fixed points, and Theorem 21.10 shows that the corresponding distributions P_x form a kernel from $\mathbb{R}^d$ to $C(\mathbb{R}_+, \mathbb{R}^d)$. By Lemma 21.8 there exists a measurable mapping G such that, whenever X has distribution P_x and $\vartheta \perp\!\!\!\perp X$ is $U(0,1)$, the process $B = G(P_x, X, \vartheta)$ is a Brownian motion in $\mathbb{R}^r$ and the pair (X, B) solves equation (σ, b). Writing Q_x for the distribution of (X, B), it is clear from Lemmas 1.38 and 1.41 (ii) that the mapping $x \mapsto Q_x$ is a kernel from $\mathbb{R}^d$ to $C(\mathbb{R}_+, \mathbb{R}^{d+r})$.

Changing the notation, we may write (X, B) for the canonical process in $C(\mathbb{R}_+, \mathbb{R}^{d+r})$. By Lemma 21.17 we have $X = F_x(x, B) = F_x(B)$ a.s. Q_x, and so

$$Q_x[X \in \cdot | B] = \delta_{F_x(B)} \text{ a.s.}, \quad x \in \mathbb{R}^d. \tag{21}$$

By Proposition 7.26 we may choose versions $\nu_{x,w} = Q_x[X \in \cdot | B \in dw]$ that combine into a probability kernel ν from $\mathbb{R}^d \times C(\mathbb{R}_+, \mathbb{R}^r)$ to $C(\mathbb{R}_+, \mathbb{R}^d)$. From (21) we see that $\nu_{x,w}$ is a.s. degenerate for each x, and since the set D of degenerate measures is measurable by Lemma 1.39 (i), we can modify ν such that $\nu_{x,w} D \equiv 1$. In that case,

$$\nu_{x,w} = \delta_{F(x,w)}, \quad x \in \mathbb{R}^d, \ w \in C(\mathbb{R}_+, \mathbb{R}^r), \tag{22}$$

for some function F as in (20), and the kernel property of ν implies that F is product measurable. Comparing (21) and (22) gives $F(x, B) = F_x(B)$ a.s. for all x.

Now fix any probability measure μ on $\mathbb{R}^d$, and conclude as in Theorem 21.10 that $P_\mu = \int P_x \mu(dx)$ solves the local martingale problem for (a, b) with initial distribution μ. Hence, equation (σ, b) has a solution (X, B) with distribution μ for X_0. Since conditioning on $\mathcal{F}_0$ preserves martingales, the equation remains conditionally valid given X_0. By the pathwise uniqueness in the degenerate case we get $P[X = F(X_0, B) | X_0] = 1$ a.s., and so $X = F(X_0, B)$ a.s. In particular, the pathwise uniqueness extends to arbitrary initial distributions μ.

Returning to the canonical setting, we may take (ξ, B) to be the identity map on the canonical space $\mathbb{R}^d \times C(\mathbb{R}_+, \mathbb{R}^r)$, endowed with the probability measure $\mu \otimes W^r$ and the induced complete filtration $\mathcal{G}^\mu$. By Lemma 21.17 equation (σ, b) has a $\mathcal{G}^\mu$-adapted solution $X = F_\mu(\xi, B)$ with $X_0 = \xi$ a.s., and the previous discussion shows that even $X = F(\xi, B)$ a.s. Hence, F is adapted to $\mathcal{G}^\mu$, and since μ is arbitrary, the adaptedness extends to the universal completion $\overline{\mathcal{G}}_t = \bigcap_\mu \mathcal{G}_t^\mu$, $t \geq 0$. □

Exercises

1. Show that for any $c \in (0,1)$, the stochastic flow X_t^x in Theorem 21.3 is a.s. Hölder continuous in x with exponent c, uniformly for bounded x and t. (*Hint:* Apply Theorem 3.23 to the estimate in the proof of Theorem 21.3.)

2. Show that a process X in $\mathbb{R}^d$ is a Brownian motion iff the process $f(X_t) - \frac{1}{2}\int_0^t \Delta f(X_s)ds$ is a martingale for every $f \in C_K^\infty$. Compare with Theorem 18.3 and Lemma 19.21.

3. Show that a Brownian bridge in $\mathbb{R}^d$ satisfies the SDE $dX_t = dB_t - (1-t)^{-1}X_t dt$ on $[0,1)$ with initial condition $X_0 = 0$. Also show that if X^x denotes the solution starting at x, then the process $Y_t^x = X_t^x - (1-t)x$ is again a Brownian bridge. (*Hint:* Note that $M_t = X_t/(1-t)$ is a martingale on $[0, 1)$ and that Y^x satisfies the same SDE as X.)

4. Solve the preceding SDE, using Proposition 21.2, to express the Brownian bridge in terms of a Brownian motion. Compare with previously known formulas.

5. Given two continuous semimartingales U and V, show that the Fisk–Stratonovich SDE $dX = dU + X \circ dV$ has the unique solution $X = Z(X_0 + Z^{-1} \circ U)$, where $Z = \exp(V - V_0)$. (*Hint:* Use Corollary 17.21 and the chain rule for FS-integrals, or derive the result from Proposition 21.2.)

6. Show under suitable conditions how a Fisk–Stratonovich SDE can be converted into an Itô equation, and conversely. Also give a sufficient condition for the existence of a strong solution to an FS–equation.

7. Show that weak existence and uniqueness in law hold for the SDE $dX_t = \text{sgn}(X_t+)dB_t$ with initial condition $X_0 = 0$, while strong existence and pathwise uniqueness fail. (*Hint:* Show that any solution X is a Brownian motion, and define $B = \text{sgn}(X+)dX$. Note that both X and $-X$ satisfy the given SDE.)

8. Show that weak existence holds for the SDE $dX_t = \text{sgn}(X_t)dB_t$ with initial condition $X_0 = 0$, while strong existence and uniqueness in law fail. (*Hint:* We may take X to be a Brownian motion or put $X \equiv 0$.)

9. Show that strong existence holds for the SDE $dX_t = 1\{X_t \neq 0\}dB_t$ with initial condition $X_0 = 0$, while uniqueness in law fails. (*Hint:* Here $X = B$ and $X = 0$ are both solutions.)

10. Show that a given process may satisfy SDE's with different $(\sigma\sigma', b)$. (*Hint:* For a trivial example, take $X = 0$, $b = 0$, and $\sigma = 0$ or $\sigma(x) = \operatorname{sgn} x$.)

11. Construct a non-Markovian solution X to the SDE $dX_t = \operatorname{sgn}(X_t)dB_t$. (*Hint:* We may take X to be a Brownian motion, stopped at the first visit to 0 after time 1. Another interesting choice is to take X to be 0 on $[0, 1]$ and a Brownian motion on $[1, \infty)$.)

12. For X as in Theorem 21.3, construct an SDE in $\mathbb{R}^{md}$ satisfied by the process $(X_t^{x_1}, \ldots, X_t^{x_m})$ for arbitrary $x_1, \ldots, x_m \in \mathbb{R}^d$. Conclude that $\mathcal{L}(X)$ is determined by $\mathcal{L}(X^x, X^y)$ for arbitrary $x, y \in \mathbb{R}^d$. (*Hint:* Note that $\mathcal{L}(X^x)$ is determined by $(\sigma\sigma', b)$ and x, and apply this result to the m-point motion.)

13. Find two SDE's as in Theorem 21.3 with solutions X and Y such that $X^x \stackrel{d}{=} Y^x$ for all x but $X \stackrel{d}{\neq} Y$. (*Hint:* We may choose $dX = dB$ and $dY = \operatorname{sgn}(Y+)dB$.)

14. For a diffusion equation (σ, b) as in Theorem 21.3, show that the distribution of the associated flow X determines $\sum_j \sigma_j^i(x)\sigma_j^k(y)$ for arbitrary pairs $i, k \in \{1, \ldots, d\}$ and $x, y \in \mathbb{R}^d$.

15. Show that if weak existence holds for the SDE (σ, b), then the pathwise uniqueness can be strengthened to the corresponding property for solutions X and Y with respect to possibly different filtrations.

16. Assume that weak existence and the stronger version of pathwise uniqueness hold for the SDE (σ, b). Use Theorem 6.10 and Lemma 21.15 to prove the existence for every μ of an a.s. unique functional solution $F_\mu(X_0, B)$ with $\mathcal{L}(X_0) = \mu$.

Chapter 22

Local Time, Excursions, and Additive Functionals

Tanaka's formula and semimartingale local time; occupation density, continuity and approximation; regenerative sets and processes; excursion local time and Poisson process; Ray–Knight theorem; excessive functions and additive functionals; local time at a regular point; additive functionals of Brownian motion

The central theme of this chapter is the notion of local time, which we will approach in three different ways, namely via stochastic calculus, via excursion theory, and via additive functionals. Here the first approach leads in particular to a useful extension of Itô's formula and to an interpretation of local time as an occupation density. Excursion theory will be developed for processes that are regenerative at a fixed state, and we shall prove the basic Itô representation, involving a Poisson process of excursions on the local time scale. Among the many applications, we consider a version of the Ray–Knight theorem about the spatial variation of Brownian local time. Finally, we shall study continuous additive functionals (CAFs) and their potentials, prove the existence of local time at a regular point, and show that any CAF of one-dimensional Brownian motion is a mixture of local times.

The beginning of this chapter may be regarded as a continuation of the stochastic calculus developed in Chapter 17. The present excursion theory continues the elementary discussion for the discrete-time case in Chapter 8. Though the theory of CAFs is formally developed for Feller processes, few results from Chapter 19 will be needed beyond the strong Markov property and its integrated version in Corollary 19.19. Both semimartingale local time and excursion theory will reappear in Chapter 23 as useful tools for studying one-dimensional SDEs and diffusions. Our discussion of CAFs of Brownian motion and their associated potentials is continued at the end of Chapter 25.

For the stochastic calculus approach to local time, consider an arbitrary continuous semimartingale X in $\mathbb{R}$. The *semimartingale local time* L^0 of X at 0 may be defined through *Tanaka's formula*

$$L_t^0 = |X_t| - |X_0| - \int_0^t \text{sgn}(X_s-)dX_s, \quad t \geq 0, \tag{1}$$

where $\operatorname{sgn}(x-) = 1_{(0,\infty)}(x) - 1_{(-\infty,0]}(x)$. Note that the stochastic integral on the right exists since the integrand is bounded and progressive. The process L^0 is clearly continuous and adapted with $L_0^0 = 0$. To motivate the definition, we note that a formal application of Itô's rule to the function $f(x) = |x|$ yields (1) with $L_t^0 = \int_{s \leq t} \delta(X_s) d[X]_s$. The following result gives the basic properties of local time at a fixed point. Here we say that a nondecreasing function f is *supported* by a Borel set A if the associated measure μ satisfies $\mu A^c = 0$. The *support* of f is the smallest closed set with this property.

Theorem 22.1 *(semimartingale local time)* Let L^0 be the local time at 0 of a continuous semimartingale X. Then L^0 is a.s. nondecreasing, continuous, and supported by the set $Z = \{t \geq 0;\; X_t = 0\}$. Furthermore, we have a.s.

$$L_t^0 = \left\{ -|X_0| - \inf_{s \leq t} \int_0^s \operatorname{sgn}(X-) dX \right\} \vee 0, \quad t \geq 0. \tag{2}$$

The proof of the last assertion depends on an elementary observation.

Lemma 22.2 *(supporting function, Skorohod)* Let f be a continuous function on $\mathbb{R}_+$ with $f_0 \geq 0$. Then there exists a unique nondecreasing, continuous function g with $g_0 = 0$ such that $h \equiv f + g \geq 0$ and $\int 1\{h > 0\} dg = 0$, namely

$$g_t = -\inf_{s \leq t} f_s \wedge 0 = \sup_{s \leq t}(-f_s) \vee 0, \quad t \geq 0. \tag{3}$$

Proof: The function in (3) clearly has the desired properties. To prove the uniqueness, assume that both g and g' have the stated properties, and put $h = f + g$ and $h' = f + g'$. If $g_t < g'_t$ for some $t > 0$, define $s = \sup\{r < t;\; g_r = g'_r\}$, and note that $h' \geq h' - h = g' - g > 0$ on $(s, t]$. Hence, $g'_s = g'_t$, and so $0 < g'_t - g_t \leq g'_s - g_s = 0$, a contradiction. □

Proof of Theorem 22.1: For any $h > 0$, we may choose a convex function $f_h \in C^2$ such that $f_h(x) = -x$ for $x \leq 0$ and $f_h(x) = x - h$ for $x \geq h$. Here clearly $f_h(x) \to |x|$ and $f'_h \to \operatorname{sgn}(x-)$ as $h \to 0$. By Itô's formula we get, a.s. for any $t \geq 0$,

$$Y_t^h \equiv f_h(X_t) - f_h(X_0) - \int_0^t f'_h(X_s) dX_s = \tfrac{1}{2} \int_0^t f''_h(X_s) d[X]_s,$$

and by Corollary 17.13 and dominated convergence we note that $(Y^h - L^0)_t^* \xrightarrow{P} 0$ for each $t > 0$. The first assertion now follows from the fact that the processes Y^h are nondecreasing and satisfy

$$\int_0^\infty 1\{X_s \notin [0, h]\} dY_s^h = 0 \text{ a.s.}, \quad h > 0.$$

The last assertion is a consequence of Lemma 22.2. □

In particular, we may deduce a basic relationship between a Brownian motion, its maximum process, and its local time at 0. The result improves the elementary Proposition 13.13.

Corollary 22.3 (*local time and maximum process, Lévy*) *Let L^0 be the local time at 0 of Brownian motion B, and define $M_t = \sup_{s \le t} B_s$. Then*

$$(L^0, |B|) \stackrel{d}{=} (M, M - B).$$

Proof: Define $B'_t = -\int_{s \le t} \mathrm{sgn}(B_s-)dB_s$ and $M'_t = \sup_{s \le t} B'_s$, and conclude from (1) and (2) that $L^0 = M'$ and $|B| = L^0 - B' = M' - B'$. It remains to note that $B' \stackrel{d}{=} B$ by Theorem 18.3. □

The local time L^x at an arbitrary point $x \in \mathbb{R}$ is defined as the local time of the process $X - x$ at 0. Thus,

$$L^x_t = |X_t - x| - |X_0 - x| - \int_0^t \mathrm{sgn}(X_s - x-)dX_s, \quad t \ge 0. \quad (4)$$

The following result shows that the two-parameter process $L = (L^x_t)$ on $\mathbb{R}_+ \times \mathbb{R}$ has a version that is continuous in t and rcll (right-continuous with left-hand limits) in x. In the martingale case we even have joint continuity.

Theorem 22.4 (*regularization, Trotter, Yor*) *Let X be a continuous semimartingale with canonical decomposition $M + A$. Then the local time $L = (L^x_t)$ of X has a version that is rcll in x, uniformly for bounded t, and satisfies*

$$L^x_t - L^{x-}_t = 2\int_0^t 1\{X_s = x\}dA_s, \quad x \in \mathbb{R},\ t \in \mathbb{R}_+. \quad (5)$$

Proof: By the definition of L we have for any $x \in \mathbb{R}$ and $t \ge 0$

$$\begin{aligned}L^x_t &= |X_t - x| - |X_0 - x| \\ &\quad - \int_0^t \mathrm{sgn}(X_s - x-)dM_s - \int_0^t \mathrm{sgn}(X_s - x-)dA_s. \quad (6)\end{aligned}$$

By dominated convergence the last term has the required continuity properties, and the discontinuities in the space variable are given by the right-hand side of (5). Since the first two terms are trivially continuous in (t, x), it remains to show that the first integral in (6), denoted by I^x_t below, has a jointly continuous version.

By localization we may then assume that the processes $X - X_0$, $[M]^{1/2}$, and $\int |dA|$ are all bounded by some constant c. Fix any $p > 2$. By Theorem 17.7 we get for any $x < y$

$$E(I^x - I^y)^{*p}_t \le 2^p E(1_{(x,y]}(X) \cdot M)^{*p}_t \lesssim E(1_{(x,y]}(X) \cdot [M])^{p/2}_t. \quad (7)$$

To estimate the integral on the right, put $y - x = h$ and choose $f \in C^2$ with $f'' \geq 2 \cdot 1_{(x,y]}$ and $|f'| \leq 2h$. By Itô's formula

$$1_{(x,y]}(X) \cdot [M] \leq \tfrac{1}{2} f''(X) \cdot [X] = f(X) - f(X_0) - f'(X) \cdot X$$
$$\leq 4ch + |f'(X) \cdot M|, \qquad (8)$$

and by another application of Theorem 17.7

$$E(f'(X) \cdot M)_t^{*p/2} \leq E((f'(X))^2 \cdot [M])_t^{p/4} \leq (2ch)^{p/2}. \qquad (9)$$

Combination of (7)–(9) gives $E(I^x - I^y)_t^{*p} \leq (ch)^{p/2}$, and the desired continuity follows by Theorem 3.23. $\square$

By the last result we may henceforth assume the local time L_t^x to be rcll in x. Here the right-continuity is only a convention, consistent with our choice of a left-continuous sign function in (4). If the occupation measure of the finite variation component A of X is a.s. diffuse, then (5) shows that L is a.s. continuous.

We proceed to give a simultaneous extension of Itô's and Tanaka's formulas. Recall that any convex function f on $\mathbb{R}$ has a nondecreasing and left-continuous left derivative $f'(x-)$. The same thing is then true when f is the difference between two convex functions. In that case there exists a unique signed measure μ_f with $\mu_f[x, y) = f'(y-) - f'(x-)$ for all $x \leq y$. In particular, $\mu_f(dx) = f''(x)dx$ when $f \in C^2$.

Theorem 22.5 *(occupation density, Meyer, Wang)* Let X be a continuous semimartingale with right-continous local time L. Then outside a fixed null set we have, for any measurable function $f \geq 0$ on $\mathbb{R}$,

$$\int_0^t f(X_s)d[X]_s = \int_{-\infty}^{\infty} f(x) L_t^x dx, \quad t \geq 0. \qquad (10)$$

If f is the difference of two convex functions, then also

$$f(X_t) - f(X_0) = \int_0^t f'(X-)dX + \tfrac{1}{2} \int_{-\infty}^{\infty} L_t^x \mu_f(dx), \quad t \geq 0. \qquad (11)$$

In particular, Theorem 17.18 extends to any function $f \in C^1(\mathbb{R})$ such that f' is absolutely continuous with Radon–Nikodým derivative f''.

Note that (11) remains valid for the left-continuous version of L, provided that $f'(X-)$ is replaced by the right derivative $f'(X+)$.

Proof: For $f(x) \equiv |x - a|$, equation (11) reduces to the definition of L_t^a. Since the formula is also trivially true for affine functions $f(x) \equiv ax + b$, it extends by linearity to the case when μ_f is supported by a finite set. By linearity and a suitable truncation, it remains to prove (11) when μ_f is positive with bounded support and $f(-\infty) = f'(-\infty) = 0$. Then define for every $n \in \mathbb{N}$ the functions

$$g_n(x) = f'(2^{-n}[2^n x]-), \quad f_n(x) = \int_{\infty}^{x} g_n(u)du, \qquad x \in \mathbb{R},$$

and note that (11) holds for all f_n. As $n \to \infty$, we get $f'_n(x-) = g_n(x-) \uparrow f'(x-)$, and so Corollary 17.13 yields $f'_n(X-) \cdot X \xrightarrow{P} f'(X-) \cdot X$. Also note that $f_n \to f$ by monotone convergence. It remains to show that $\int L^x_t \mu_{f_n}(dx) \to \int L^x_t \mu_f(dx)$. Then let h be any bounded, right-continuous function on $\mathbb{R}$, and note that $\mu_{f_n} h = \mu_f h_n$ with $h_n(x) = h(2^{-n}[2^n x + 1])$. Since $h_n \to h$, we get $\mu_f h_n \to \mu_f h$ by dominated convergence.

Comparing (11) with Itô's formula, we note that (10) holds a.s. for any $t \geq 0$ and $f \in C$. For each $t \geq 0$, the two sides of (10) define random measures on $\mathbb{R}$, and so by suitable approximation and monotone class arguments we may choose the exceptional null set N to be independent of f. By the continuity of each side, we may also assume that N is independent of t.

If $f \in C^1$ with f' as stated, then (11) applies with $\mu_f(dx) = f''(x)dx$, and the last assertion follows by (10). □

In particular, we note that the *occupation measure* at time t,

$$\eta_t A = \int_0^t 1_A(X_s) d[X]_s, \quad A \in \mathcal{B}(\mathbb{R}), \ t \geq 0, \tag{12}$$

is a.s. absolutely continuous with density L_t. This leads to a simple construction of L.

Corollary 22.6 *(right derivative)* Outside a fixed P-null set, we have

$$L^x_t = \lim_{h \to 0} \eta_t[x, x+h]/h, \quad t \geq 0, \ x \in \mathbb{R}.$$

Proof: Use Theorem 22.5 and the right-continuity of L. □

Our next aim is to show how local time arises naturally in the context of regenerative processes. Then consider an rcll process X in some Polish space S such that X is adapted to some right-continuous and complete filtration $\mathcal{F}$. Fix a state $a \in S$, and assume X to be *regenerative* at a, in the sense that there exists some distribution P_a on the path space satisfying

$$P[\theta_\tau X \in \cdot | \mathcal{F}_\tau] = P_a \text{ a.s. on } \{\tau < \infty, X_\tau = a\}, \tag{13}$$

for every optional time τ. The relation will often be applied to the hitting times $\tau_r = \inf\{t \geq r; X_t = a\}$, which are optional for all $r \geq 0$ by Theorem 7.7. In fact, when X is continuous, the optionality of τ_r follows already from the elementary Lemma 7.6. In particular, we note that $\mathcal{F}_{\tau_0}$ and $\theta_{\tau_0} X$ are conditionally independent, given that $\tau_0 < \infty$. For simplicity we may henceforth take X to be the canonical process on the path space $D = D(\mathbb{R}_+, S)$, equipped with the distribution $P = P_a$.

Introducing the *regenerative set* $Z = \{t \geq 0; X_t = a\}$, we may write the last event in (13) simply as $\{\tau \in Z\}$. From the right-continuity of X it is clear that $Z \ni t_n \downarrow t$ implies $t \in Z$, which means that every point in $\overline{Z} \setminus Z$ is isolated from the right. Since $\overline{Z}^c$ is open and hence a countable union of disjoint open intervals, it follows that Z^c is a countable union of disjoint intervals of the form (u, v) or $[u, v)$. With every such interval we may associate an *excursion process* $Y_t = X_{(t+u)\wedge v}$, $t \geq 0$. Note that a is absorbing for Y, in the sense that $Y_t = a$ for all $t \geq \inf\{s > 0; Y_s = a\}$. The number of excursions may be finite or infinite, and if Z is bounded there is clearly a last excursion of infinite length.

We begin with a classification according to the local properties of Z.

Proposition 22.7 *(local dichotomies) If the set Z is regenerative, then*

(i) *either* $(\overline{Z})^\circ = \emptyset$ *a.s. or* $\overline{Z^\circ} = \overline{Z}$ *a.s.;*

(ii) *either a.s. all points of Z are isolated, or a.s. none of them is;*

(iii) *either* $\lambda Z = 0$ *a.s. or* $\mathrm{supp}(Z \cdot \lambda) = \overline{Z}$ *a.s.*

Recall that the set Z is said to be *nowhere dense* if $(\overline{Z})^\circ = \emptyset$, and that $\overline{Z}$ is *perfect* if Z has no isolated points. If $\overline{Z^\circ} = \overline{Z}$, then clearly $\mathrm{supp}(Z \cdot \lambda) = \overline{Z}$, and no isolated points exist.

Proof: By the regenerative property, we have for any optional time τ

$$P\{\tau = 0\} = E[P[\tau = 0|\mathcal{F}_0]; \tau = 0] = (P\{\tau = 0\})^2,$$

and so $P\{\tau = 0\} = 0$ or 1. If σ is another optional time, then $\tau' = \sigma + \tau \circ \theta_\sigma$ is again optional by Proposition 8.8, and we get

$$P\{\tau' - h \leq \sigma \in Z\} = P\{\tau \circ \theta_\sigma \leq h, \sigma \in Z\} = P\{\tau \leq h\}P\{\sigma \in Z\}.$$

Thus, $P[\tau' - \sigma \in \cdot | \sigma \in Z] = P \circ \tau^{-1}$, and in particular $\tau = 0$ a.s. implies $\tau' = \sigma$ a.s. on $\{\sigma \in Z\}$.

(i) Here we apply the previous argument to the optional times $\tau = \inf Z^c$ and $\sigma = \tau_r$. If $\tau > 0$ a.s., then $\tau \circ \theta_{\tau_r} > 0$ a.s. on $\{\tau_r < \infty\}$, and so $\tau_r \in \overline{Z^\circ}$ a.s. on the same set. Since the set $\{\tau_r; r \in \mathbb{Q}_+\}$ is dense in $\overline{Z}$, it follows that $\overline{Z} = \overline{Z^\circ}$ a.s. Now assume instead that $\tau = 0$ a.s. Then $\tau \circ \theta_{\tau_r} = 0$ a.s. on $\{\tau_r < \infty\}$, and so $\tau_r \in \overline{Z^c}$ a.s. on the same set. Hence, $\overline{Z} \subset \overline{Z^c}$ a.s., and therefore $\overline{Z^c} = \mathbb{R}_+$ a.s. It remains to note that $\overline{Z^c} = (\overline{Z})^c$, since Z^c is a disjoint union of intervals (u, v) or $[u, v)$.

(ii) In this case, we define $\tau = \inf(Z \setminus \{0\})$. If $\tau = 0$ a.s., then $\tau \circ \theta_{\tau_r} = 0$ a.s. on $\{\tau_r < \infty\}$. Since every isolated point of Z is of the form τ_r for some $r \in \mathbb{Q}_+$, it follows that Z has a.s. no isolated points. If instead $\tau > 0$ a.s., we may define the optional times σ_n recursively by $\sigma_{n+1} = \sigma_n + \tau \circ \theta_{\sigma_n}$, starting from $\sigma_1 = \tau$. Then $\sigma_n = \sum_{k \leq n} \xi_k$, where the ξ_k are i.i.d. and distributed as τ, and so $\sigma_n \to \infty$ a.s. by the law of large numbers. Thus, $Z = \{\sigma_n < \infty; n \in \mathbb{N}\}$ a.s., and a.s. all points of Z are isolated.

(iii) Here we may take $\tau = \inf\{t > 0; (Z \cdot \lambda)_t > 0\}$. If $\tau = 0$ a.s., then $\tau \circ \theta_{\tau_r} = 0$ a.s. on $\{\tau_r < \infty\}$, and so $\tau_r \in \mathrm{supp}(Z \cdot \lambda)$ a.s. on the same set. Hence, $\overline{Z} \subset \mathrm{supp}(Z \cdot \lambda)$ a.s., and the two sets agree a.s. If instead $\tau > 0$ a.s., then $\tau = \tau + \tau \circ \theta_\tau > \tau$ a.s. on $\{\tau < \infty\}$, which implies $\tau = \infty$ a.s. This yields $\lambda Z = 0$ a.s. □

To examine the global properties of Z, we may introduce the *holding time* $\gamma = \inf Z^c = \inf\{t > 0; X_t \neq a\}$, which is optional by Lemma 7.6. The following extension of Lemma 12.16 gives some more detailed information about dichotomy (i) above.

Lemma 22.8 *(holding time) The time γ is exponentially distributed with mean $m \in [0, \infty]$, where $m = 0$ or ∞ when X is continuous. Furthermore, Z is a.s. nowhere dense when $m = 0$, and if $m > 0$ it is a.s. a locally finite union of intervals $[\sigma, \tau)$. Finally, $\gamma \perp\!\!\!\perp X \circ \theta_\gamma$ when $m < \infty$.*

Proof: The first and last assertions may be proved as in Lemma 12.16, and the statement for $m = 0$ was obtained in Proposition 22.7 (i). Now let $0 < m < \infty$. Noting that $\gamma \circ \theta_\gamma = 0$ a.s. on $\{\gamma \in Z\}$, we get
$$0 = P\{\gamma \circ \theta_\gamma > 0, \, \gamma \in Z\} = P\{\gamma > 0\} P\{\gamma \in Z\} = P\{\gamma \in Z\},$$
so in this case $\gamma \notin Z$ a.s. Put $\sigma_0 = 0$, let $\sigma_1 = \gamma + \tau_0 \circ \theta_\gamma$, and define recursively $\sigma_{n+1} = \sigma_n + \sigma_1 \circ \theta_{\sigma_n}$. Write $\gamma_n = \sigma_n + \gamma \circ \theta_{\sigma_n}$. Then $\sigma_n \to \infty$ a.s. by the law of large numbers, and so $Z = \bigcup_n [\sigma_n, \gamma_n)$. If X is continuous, then Z is closed and the last case is excluded. □

The state a is said to be *absorbing* if $m = \infty$ and *instantaneous* if $m = 0$. In the former case clearly $X \equiv a$ and $Z = \mathbb{R}_+$ a.s. Hence, to avoid trivial exceptions, we may henceforth assume that $m < \infty$. A separate treatment is sometimes required for the elementary case when the *recurrence time* $\gamma + \tau_{0+} \circ \theta_\gamma$ is a.s. strictly positive. This clearly occurs when Z has a.s. only isolated points or the holding time γ is positive.

We proceed to examine the set of excursions. Since there is no first excursion in general, it is helpful first to focus on excursions of long duration. For any $h \geq 0$, let D_h denote the set of excursion paths longer than h, endowed with the σ-field $\mathcal{D}_h$ generated by all evaluation maps π_t, $t \geq 0$. Note that D_0 is a Borel space and that $D_h \in \mathcal{D}_0$ for all h. The number of excursions in D_h will be denoted by κ_h. The following result is a continuous-time version of Proposition 8.15.

Lemma 22.9 *(long excursions) Fix any $h > 0$, or $h \geq 0$ when the recurrence time is positive. Then either $\kappa_h = 0$ a.s., or κ_h has a geometric distribution with mean $m_h \in [1, \infty]$. In the latter case, X has D_h-excursions Y_h^j, $j \leq \kappa_h$ for some i.i.d. processes $Y_h^1, Y_h^2, \ldots$ in D_h, where $Y_h^{\kappa_h}$ is a.s. infinite when $m_h < \infty$.*

Proof: For $t \in (0, \infty]$, let κ_h^t denote the number of D_h-excursions completed at time $t \in [0, \infty]$, and note that $\kappa_h^{\tau_t} > 0$ when $\tau_t = \infty$. Writing $p_h = P\{\kappa_h > 0\}$, we obtain
$$\begin{aligned} p_h &= P\{\kappa_h^{\tau_t} > 0\} + P\{\kappa_h^{\tau_t} = 0, \, \kappa_h \circ \theta_{\tau_t} > 0\} \\ &= P\{\kappa_h^{\tau_t} > 0\} + P\{\kappa_h^{\tau_t} = 0\} p_h. \end{aligned}$$
Since $\kappa_h^t \to \kappa_h$ as $t \to \infty$, we get $p_h = p_h + (1 - p_h) p_h$, and so $p_h = 0$ or 1.

Now assume that $p_h = 1$. Put $\sigma_0 = 0$, let σ_1 denote the end of the first D_h-excursion, and recursively define $\sigma_{n+1} = \sigma_n + \sigma_1 \circ \theta_{\sigma_n}$. If all excursions are finite, then clearly $\sigma_n < \infty$ a.s. for all n, and so $\kappa_h = \infty$ a.s. Thus, the last D_h-excursion is infinite when $\kappa_h < \infty$. We may now proceed as in the proof of Proposition 8.15 to construct some i.i.d. processes $Y_h^1, Y_h^2, \ldots$ in D_h such that X has D_h-excursions Y_h^j, $j \leq \kappa_h$. Since κ_h is the number of the first infinite excursion, we note in particular that κ_h is geometrically distributed with mean q_h^{-1}, where q_h is the probability that Y_h^1 is infinite. □

Now put $\hat{h} = \inf\{h > 0;\ \kappa_h = 0 \text{ a.s.}\}$. For any $h \in (0, \hat{h})$ we have $\kappa_h \geq 1$ a.s., and we may define ν_h as the distribution of the first excursion in D_h. The next result shows how the ν_h can be combined into a single measure ν on D_0, the so-called *excursion law* of X. For convenience, we write $\nu[\,\cdot\,|A] = \nu(\,\cdot\, \cap A)/\nu A$ whenever $0 < \nu A < \infty$.

Lemma 22.10 *(excursion law, Itô)* *There exists a measure ν on D_0 such that $\nu D_h \in (0, \infty)$ and $\nu_h = \nu[\,\cdot\,|D_h]$ for every $h \in (0, \hat{h})$. Furthermore, ν is unique up to a normalization, and it is bounded iff the recurrence time is a.s. positive.*

Proof: Fix any $h \leq k$ in $(0, \hat{h})$, and let $Y_h^1, Y_h^2, \ldots$ be such as in Lemma 22.9. Then the first D_k-excursion is the first process Y_h^j that belongs to D_k, and since the Y_h^j are i.i.d. ν_h, we have

$$\nu_k = \nu_h[\,\cdot\,|D_k], \quad 0 < h \leq k < \hat{h}. \tag{14}$$

Now fix any $k \in (0, \hat{h})$, and define $\tilde{\nu}_h = \nu_h/\nu_h D_k$, $h \in (0, k]$. Then (14) yields $\tilde{\nu}_{h'} = \tilde{\nu}_h(\,\cdot\, \cap D_{h'})$ for any $h \leq h' \leq k$, and so $\tilde{\nu}_h$ increases as $h \to 0$ toward a measure ν with $\nu(\,\cdot\, \cap D_h) = \tilde{\nu}_h$ for all $h \leq k$. For any $h \in (0, \hat{h})$, we get

$$\nu[\,\cdot\,|D_h] = \tilde{\nu}_{h \wedge k}[\,\cdot\,|D_h] = \nu_{h \wedge k}[\,\cdot\,|D_h] = \nu_h.$$

If ν' is another measure with the stated property, then

$$\frac{\nu(\,\cdot\, \cap D_h)}{\nu D_k} = \frac{\nu_h}{\nu_h D_k} = \frac{\nu'(\,\cdot\, \cap D_h)}{\nu' D_k}, \quad h \leq k < \hat{h}.$$

As $h \to 0$ for fixed k, we get $\nu = r\nu'$ with $r = \nu D_k/\nu' D_k$.

If the recurrence time is positive, then (14) remains true for $h = 0$, and we may take $\nu = \nu_0$. Otherwise, let $h \leq k$ in $(0, \hat{h})$, and denote by $\kappa_{h,k}$ the number of D_h-excursions up to the first completed excursion in D_k. For fixed k we have $\kappa_{h,k} \to \infty$ a.s. as $h \to 0$, since $\overline{Z}$ is perfect and nowhere dense. Now $\kappa_{h,k}$ is geometrically distributed with mean

$$E\kappa_{h,k} = (\nu_h D_k)^{-1} = (\nu[D_k|D_h])^{-1} = \nu D_h/\nu D_k,$$

and so $\nu D_h \to \infty$. Thus, ν is unbounded. □

When the regenerative set Z has a.s. only isolated points, then Lemma 22.9 already gives a complete description of the excursion structure. In the complementary case when $\overline{Z}$ is a.s. perfect, we have the following fundamental representation in terms of a local time process L and an associated

Poisson point process ξ, both of which can be constructed from the array of holding times and excursions.

Theorem 22.11 *(excursion local time and Poisson process, Lévy, Itô)* Let X be regenerative at a and such that the closure of $Z = \{t;\, X_t = a\}$ is a.s. perfect. Then there exist a nondecreasing, continuous, adapted process L on $\mathbb{R}_+$ with support $\overline{Z}$ a.s., a Poisson process ξ on $\mathbb{R}_+ \times D_0$ with intensity measure of the form $\lambda \otimes \nu$, and a constant $c \geq 0$, such that $Z \cdot \lambda = cL$ a.s. and the excursions of X with associated L-values are given by the restriction of ξ to $[0, L_\infty]$. Furthermore, the product $\nu \cdot L$ is a.s. unique.

Proof (beginning): If $E\gamma = c > 0$, we may define $\nu = \nu_0/c$ and introduce a Poisson process ξ on $\mathbb{R}_+ \times D_0$ with intensity measure $\lambda \otimes \nu$. Let the points of ξ be $(\sigma_j, \tilde{Y}_j)$, $j \in \mathbb{N}$, and put $\sigma_0 = 0$. By Proposition 12.15 the differences $\tilde{\gamma}_j = \sigma_j - \sigma_{j-1}$ are independent and exponentially distributed with mean c. Furthermore, by Proposition 12.3 the processes $\tilde{Y}_j$ are independent of the σ_j and i.i.d. ν_0. Letting $\tilde{\kappa}$ be the first index j such that $\tilde{Y}_j$ is infinite, we see from Lemmas 22.8 and 22.9 that

$$\{\gamma_j, Y_j;\, j \leq \kappa\} \stackrel{d}{=} \{\tilde{\gamma}_j, \tilde{Y}_j;\, j \leq \tilde{\kappa}\}, \tag{15}$$

where the quantities on the left are the holding times and subsequent excursions of X. By Theorem 6.10 we may redefine ξ such that (15) holds a.s. The stated conditions then become fulfilled with $L = Z \cdot \lambda$.

Turning to the case when $E\gamma = 0$, we may define ν as in Lemma 22.10 and let ξ be Poisson $\lambda \otimes \nu$, as before. For any $h \in (0, \hat{h})$, the points of ξ in $\mathbb{R}_+ \times D_h$ may be enumerated from the left as $(\sigma_h^j, \tilde{Y}_h^j)$, $j \in \mathbb{N}$, and we define $\tilde{\kappa}_h$ as the first index j such that $\tilde{Y}_h^j$ is infinite. The processes $\tilde{Y}_h^j$ are clearly i.i.d. ν_h, and so by Lemma 22.9 we have

$$\{Y_h^j;\, j \leq \kappa_h\} \stackrel{d}{=} \{\tilde{Y}_h^j;\, j \leq \tilde{\kappa}_h\}, \quad h \in (0, \hat{h}). \tag{16}$$

Since longer excursions form subarrays, the entire collections in (16) have the same finite-dimensional distributions, and so by Theorem 6.10 we may redefine ξ such that all relations hold a.s.

Let τ_h^j be the right endpoint of the jth excursion in D_h, and define

$$L_t = \inf\{\sigma_h^j;\, h, j > 0,\, \tau_h^j \geq t\}, \quad t \geq 0.$$

We need the obvious facts that, for any $t \geq 0$ and $h, j > 0$,

$$L_t < \sigma_h^j \;\Rightarrow\; t \leq \tau_h^j \;\Rightarrow\; L_t \leq \sigma_h^j. \tag{17}$$

To see that L is a.s. continuous, we may assume that (16) holds identically. Since ν is infinite, we may further assume the set $\{\sigma_h^j;\, h, j > 0\}$ to be dense in the interval $[0, L_\infty]$. If $\Delta L_t > 0$, there exist some $i, j, h > 0$ with $L_{t-} < \sigma_h^i < \sigma_h^j < L_{t+}$. By (17) we get $t - \varepsilon \leq \tau_h^i < \tau_h^j \leq t + \varepsilon$ for every $\varepsilon > 0$, which is impossible. Thus, $\Delta L_t = 0$ for all t.

To prove that $\overline{Z} \subset \operatorname{supp} L$ a.s., we may further assume $\overline{Z}_\omega$ to be perfect and nowhere dense for each $\omega \in \Omega$. If $t \in \overline{Z}$, then for every $\varepsilon > 0$ there exist some $i, j, h > 0$ with $t - \varepsilon < \tau_h^i < \tau_h^j < t + \varepsilon$, and by (17) we get $L_{t-\varepsilon} \leq \sigma_h^i < \sigma_h^j \leq L_{t+\varepsilon}$. Thus, $L_{t-\varepsilon} < L_{t+\varepsilon}$ for all $\varepsilon > 0$, and so $t \in \operatorname{supp} L$. □

In the perfect case, it remains to establish the a.s. relation $Z \cdot \lambda = cL$ for a suitable c and to show that L is unique and adapted. To avoid repetition, we postpone the proof of the former claim until Theorem 22.13. The latter statements are immediate consequences of the following result, which also suggests many explicit constructions of L. Let $\eta_t A$ denote the number of excursions in a set $A \in \mathcal{D}_0$, completed at time $t \geq 0$. Note that η is an adapted, measure-valued process on $\mathcal{D}_0$.

Proposition 22.12 *(approximation)* *If $A_1, A_2, \cdots \in \mathcal{D}_0$ with $\infty > \nu A_n \to \infty$, then*

$$\sup_{t \leq u} \left| \frac{\eta_t A_n}{\nu A_n} - L_t \right| \xrightarrow{P} 0, \quad u \geq 0. \tag{18}$$

The same convergence holds a.s. when the A_n are nested.

In particular, $\eta_t D_h / \nu D_h \to L_t$ a.s. as $h \to 0$ for fixed t. Thus, L is a.s. determined by the regenerative set Z.

Proof: Let ξ be such as in Theorem 22.11, and put $\xi_s = \xi([0, s] \times \cdot)$. First assume that the A_n are nested. For any $s \geq 0$ we note that $(\xi_s A_n) \stackrel{d}{=} (N_{s\nu A_n})$, where N is a unit-rate Poisson process on $\mathbb{R}_+$. Since $t^{-1} N_t \to 1$ a.s. by the law of large numbers and the monotonicity of N, we get

$$\frac{\xi_s A_n}{\nu A_n} \to s \text{ a.s.}, \quad s \geq 0.$$

As in case of Proposition 4.24, we may strengthen this to

$$\sup_{s \leq r} \left| \frac{\xi_s A_n}{\nu A_n} - s \right| \to 0 \text{ a.s.}, \quad r \geq 0. \tag{19}$$

Without the nestedness assumption, we may introduce a nested sequence $A_1', A_2', \ldots$ with $\nu A_n' = \nu A_n$ for all n. Then (19) holds with A_n replaced by A_n', and since the distributions on the left are the same for each n, the formula for A_n remains valid with convergence in probability. In both cases we may clearly replace r by any positive random variable. The convergence (18) now follows, if we note that $\xi_{L_t-} \leq \eta_t \leq \xi_{L_t}$ for all $t \geq 0$ and use the continuity of L. □

The excursion local time L is described most conveniently in terms of its right-continuous inverse

$$T_s = L_s^{-1} = \inf\{t \geq 0; L_t > s\}, \quad s \geq 0.$$

To state the next result, we introduce the subset $Z' \subset Z$, obtained from Z by omission of all points that are isolated from the right. Let us further write $l(u)$ for the length of an excursion path $u \in \mathcal{D}_0$.

Theorem 22.13 *(inverse local time) Let L, ξ, ν, and c be such as in Theorem 22.11. Then $T = L^{-1}$ is a generalized subordinator with characteristics $(c, \nu \circ l^{-1})$ and a.s. range Z' in $\mathbb{R}_+$, and we have a.s.*

$$T_s = cs + \int_0^{s+}\!\!\int l(u)\xi(dr\,du), \quad s \geq 0. \tag{20}$$

Proof: We may clearly discard the null set where L is not continuous with support $\overline{Z}$. If $T_s < \infty$ for some $s \geq 0$, then $T_s \in \operatorname{supp} L = \overline{Z}$ by the definition of T, and since L is continuous, we get $T_s \notin \overline{Z} \setminus Z'$. Thus, $T(\mathbb{R}_+) \subset Z' \cup \{\infty\}$ a.s. Conversely, assume that $t \in Z'$. Then for any $\varepsilon > 0$ we have $L_{t+\varepsilon} > L_t$, and so $t \leq T \circ L_t \leq t + \varepsilon$. As $\varepsilon \to 0$, we get $T \circ L_t = t$. Thus, $Z' \subset T(\mathbb{R}_+)$ a.s.

For each $s \geq 0$, the time T_s is optional by Lemma 7.6. Furthermore, it is clear from Proposition 22.12 that, as long as $T_s < \infty$, the process $\theta_s T - T_s$ is obtainable from $X \circ \theta_{T_s}$ by a measurable mapping that is independent of s. By the regenerative property and Lemma 15.11, the process T is then a generalized subordinator, and in particular it admits a representation as in Theorem 15.4. Since the jumps of T agree with the lengths of the excursion intervals, we obtain (20) for a suitable $c \geq 0$. By Lemma 1.22 the double integral in (20) equals $\int x(\xi_s \circ l^{-1})(dx)$, which shows that T has Lévy measure $E(\xi_1 \circ l^{-1}) = \nu \circ l^{-1}$.

Substituting $s = L_t$ into (20), we get a.s. for any $t \in Z'$

$$t = T \circ L_t = cL_t + \int_0^{L_t+}\!\!\int l(u)\xi(dr\,du) = cL_t + (Z^c \cdot \lambda)_t.$$

Hence, $cL_t = (Z \cdot \lambda)_t$ a.s., which extends by continuity to arbitrary $t \geq 0$. □

We may justify our terminology by showing that the semimartingale and excursion local times agree whenever both exist.

Proposition 22.14 *(reconciliation) Let the continuous semimartingale X in $\mathbb{R}$ be regenerative at some $a \in \mathbb{R}$ with $P\{L_\infty^a \neq 0\} > 0$. Then the set $Z = \{t;\ X_t = a\}$ is a.s. perfect and nowhere dense, and L^a is a version of the excursion local time at a.*

Proof: By Theorem 22.1 the state a is nonabsorbing, and so Z is nowhere dense by Lemma 22.8. Since $P\{L_\infty^a \neq 0\} > 0$ and L^a is a.s. continuous with support in Z, Proposition 22.7 shows that Z is a.s. perfect. Let L be a version of the excursion local time at a, and put $T = L^{-1}$. Define $Y_s = L^a \circ T_s$ for $s < L_\infty$, and let $Y_s = \infty$ otherwise. By the continuity of L^a we have $Y_{s\pm} = L^a \circ T_{s\pm}$ for every $s < L_\infty$. If $\Delta T_s > 0$, we note that $L^a \circ T_{s-} = L^a \circ T_s$, since (T_{s-}, T_s) is an excursion interval of X and L^a is continuous with support in Z. Thus, Y is a.s. continuous on $[0, L_\infty)$.

By Corollary 22.6 and Proposition 22.12 the process $\theta_s Y - Y_s$ is obtainable from $\theta_{T_s} X$ through the same measurable mapping for all $s < L_\infty$. By the regenerative property and Lemma 15.11 it follows that Y is a generalized subordinator, and so by Theorem 15.4 and the continuity of Y there exists some $c \geq 0$ with $Y_s \equiv cs$ a.s. on $[0, L_\infty)$. For $t \in Z'$ we have a.s. $T \circ L_t = t$, and therefore

$$L_t^a = L^a \circ (T \circ L_t) = (L^a \circ T) \circ L_t = cL_t.$$

This extends to $\mathbb{R}_+$ since both extremes are continuous with support in Z. □

For Brownian motion it is convenient to normalize local time according to Tanaka's formula, which leads to a corresponding normalization of the excursion law ν. By the spatial homogeneity of Brownian motion, we may restrict our attention to excursions from 0. The next result shows that excursions of different length have the same distribution apart from a scaling. For a precise statement, we may introduce the scaling operators S_r on D, given by

$$(S_r f)_t = r^{1/2} f_{t/r}, \quad t \geq 0, \ r > 0, \ f \in D.$$

Theorem 22.15 *(Brownian excursion) Let ν be the normalized excursion law of Brownian motion. Then there exists a unique distribution $\hat{\nu}$ on the set of excursions of unit length such that*

$$\nu = (2\pi)^{-1/2} \int_0^\infty (\hat{\nu} \circ S_r^{-1}) r^{-3/2} dr. \tag{21}$$

Proof: By Theorem 22.13 the inverse local time L^{-1} is a subordinator with Lévy measure $\nu \circ l^{-1}$, where $l(u)$ denotes the length of u. Furthermore, $L \stackrel{d}{=} M$ by Corollary 22.3, where $M_t = \sup_{s \leq t} B_s$, and so by Theorem 15.10 the measure $\nu \circ l^{-1}$ has density $(2\pi)^{-1/2} r^{-3/2}$, $r > 0$. As in Theorem 6.3, there exists a probability kernel (ν_r) from $(0, \infty)$ to D_0 such that $\nu_r \circ l^{-1} \equiv \delta_r$ and

$$\nu = (2\pi)^{-1/2} \int_0^\infty \nu_r r^{-3/2} dr, \tag{22}$$

and we note that the measures ν_r are unique a.e. λ.

For any $r > 0$ the process $\tilde{B} = S_r B$ is again a Brownian motion, and by Corollary 22.6 the local time of $\tilde{B}$ equals $\tilde{L} = S_r L$. If B has an excursion u ending at time t, then the corresponding excursion $S_r u$ of $\tilde{B}$ ends at rt, and the local time for $\tilde{B}$ at the new excursion equals $\tilde{L}_{rt} = r^{1/2} L_t$. Thus, the excursion process $\tilde{\xi}$ for $\tilde{B}$ is obtained from the process ξ for B through the mapping $T_r \colon (s, u) \mapsto (r^{1/2} s, S_r u)$. Since $\tilde{\xi} \stackrel{d}{=} \xi$, each T_r leaves the intensity measure $\lambda \otimes \nu$ invariant, and we get

$$\nu \circ S_r^{-1} = r^{1/2} \nu, \quad r > 0. \tag{23}$$

Combining (22) and (23), we get for any $r > 0$

$$\int_0^\infty (\nu_x \circ S_r^{-1}) x^{-3/2} dx = r^{1/2} \int_0^\infty \nu_x x^{-3/2} dx = \int_0^\infty \nu_{rx} x^{-3/2} dx,$$

and by the uniqueness in (22) we obtain
$$\nu_x \circ S_r^{-1} = \nu_{rx}, \quad x > 0 \text{ a.e. } \lambda, \ r > 0.$$
By Fubini's theorem, we may then fix an $x = c > 0$ such that
$$\nu_c \circ S_r^{-1} = \nu_{cr}, \quad r > 0 \text{ a.s. } \lambda.$$
Define $\hat{\nu} = \nu_c \circ S_{1/c}^{-1}$, and conclude that for almost every $r > 0$
$$\nu_r = \nu_{c(r/c)} = \nu_c \circ S_{r/c}^{-1} = \nu_c \circ S_{1/c}^{-1} \circ S_r^{-1} = \hat{\nu} \circ S_r^{-1}.$$
Substituting this into (22) yields equation (21).

If μ is another probability measure with the stated properties, then for almost every $r > 0$ we have $\mu \circ S_r^{-1} = \hat{\nu} \circ S_r^{-1}$, and hence
$$\mu = \mu \circ S_r^{-1} \circ S_{1/r}^{-1} = \hat{\nu} \circ S_r^{-1} \circ S_{1/r}^{-1} = \hat{\nu}.$$
Thus, $\hat{\nu}$ is unique. □

By continuity of paths, an excursion of Brownian motion is either positive or negative, and by symmetry the two possibilities have the same probability $\frac{1}{2}$ under $\hat{\nu}$. This leads to the further decomposition $\hat{\nu} = \frac{1}{2}(\hat{\nu}_+ + \hat{\nu}_-)$. A process with distribution $\hat{\nu}_+$ is called a *(normalized) Brownian excursion*.

For subsequent needs, we continue with a simple computation.

Lemma 22.16 *(height distribution)* Let ν be the excursion law of Brownian motion. Then
$$\nu\{u \in D_0; \ \sup_t u_t > h\} = (2h)^{-1}, \quad h > 0.$$

Proof: By Tanaka's formula the process $M = 2B \vee 0 - L^0 = B + |B| - L^0$ is a martingale, and so we get for $\tau = \inf\{t \geq 0; \ B_t = h\}$
$$E L^0_{\tau \wedge t} = 2E(B_{\tau \wedge t} \vee 0), \quad t \geq 0.$$
Hence, by monotone and dominated convergence $E L^0_\tau = 2E(B_\tau \vee 0) = 2h$. On the other hand, Theorem 22.11 shows that L^0_τ is exponentially distributed with mean $(\nu A_h)^{-1}$, where $A_h = \{u; \ \sup_t u_t \geq h\}$. □

The following result gives some remarkably precise information about the spatial behavior of Brownian local time.

Theorem 22.17 *(space dependence, Ray, Knight)* For Brownian motion B with local time L, let $\tau = \inf\{t > 0; \ B_t = 1\}$. Then on $[0,1]$ the process $S_t = L_\tau^{1-t}$ is a squared Bessel process of order 2.

Several proofs are known. Here we derive the result as an application of the previously developed excursion theory.

Proof (Walsh): Fix any $u \in [0,1]$, put $\sigma = L_\tau^u$, and let $\xi^\pm$ denote the Poisson processes of positive and negative excursions from u. Write Y for the process B, stopped when it first hits u. Then $Y \perp\!\!\!\perp (\xi^+, \xi^-)$ and $\xi^+ \perp\!\!\!\perp \xi^-$, so $\xi^+ \perp\!\!\!\perp (\xi^-, Y)$. Since σ is ξ^+-measurable, we obtain $\xi^+ \perp\!\!\!\perp_\sigma (\xi^-, Y)$ and hence $\xi_\sigma^+ \perp\!\!\!\perp_\sigma (\xi_\sigma^-, Y)$, which implies the Markov property of L_τ^x at $x = u$.

To derive the corresponding transition kernels, fix any $x \in [0, u)$, and write $h = u - x$. Put $\tau_0 = 0$, and let $\tau_1, \tau_2, \ldots$ be the right endpoints of those excursions from x that reach u. Next define $\zeta_k = L^x_{\tau_{k+1}} - L^x_{\tau_k}$, $k \geq 0$, so that $L^x_\tau = \zeta_0 + \cdots + \zeta_\kappa$ with $\kappa = \sup\{k; \tau_k \leq \tau\}$. By Lemma 22.16 the variables ζ_k are i.i.d. and exponentially distributed with mean $2h$. Since κ agrees with the number of completed u-excursions before time τ that reach x and since $\sigma \perp\!\!\!\perp \xi^-$, it is further seen that κ is conditionally Poisson $\sigma/2h$, given σ.

We also need the fact that $(\sigma, \kappa) \perp\!\!\!\perp (\zeta_0, \zeta_1, \ldots)$. To see this, define $\sigma_k = L^u_{\tau_k}$. Since ξ^- is Poisson, we note that $(\sigma_1, \sigma_2, \ldots) \perp\!\!\!\perp (\zeta_1, \zeta_2, \ldots)$, and so $(\sigma, \sigma_1, \sigma_2, \ldots) \perp\!\!\!\perp (Y, \zeta_1, \zeta_2, \ldots)$. The desired relation now follows, since κ is a measurable function of $(\sigma, \sigma_1, \sigma_2, \ldots)$ and ζ_0 depends measurably on Y.

For any $s \geq 0$, we may now compute

$$E\left[e^{-sL^{u-h}_\tau}\Big|\sigma\right] = E\left[(Ee^{-s\zeta_0})^{\kappa+1}\Big|\sigma\right] = E\left[(1+2sh)^{-\kappa-1}\Big|\sigma\right]$$
$$= (1+2sh)^{-1} \exp\left\{\frac{-s\sigma}{1+2sh}\right\}.$$

In combination with the Markov property of L^x_τ, the last relation is equivalent, via the substitutions $u = 1 - t$ and $2s = (a - t)^{-1}$, to the martingale property of the process

$$M_t = (a-t)^{-1} \exp\left\{\frac{-L^{1-t}_\tau}{2(a-t)}\right\}, \quad t \in [0, a), \tag{24}$$

for arbitrary $a > 0$.

Now let X be a squared Bessel process of order 2, and note that $L^1_\tau = X_0 = 0$ by Theorem 22.4. By Corollary 13.12 the process X is again Markov. To see that X has the same transition kernel as L^{1-t}_τ, it is enough to show for an arbitrary $a > 0$ that the process M in (24) remains a martingale when L^{1-t}_τ is replaced by X_t. This is easily verified by means of Itô's formula, if we note that X is a weak solution to the SDE $dX_t = 2X_t^{1/2}dB_t + 2dt$. □

As an important application of the last result, we may show that the local time is strictly positive on the range of the process.

Corollary 22.18 *(range and support)* Let M be a continuous local martingale with local time L. Then outside a fixed P-null set,

$$\{L^x_t > 0\} = \{\inf_{s \leq t} M_s < x < \sup_{s \leq t} M_s\}, \quad x \in \mathbb{R}, t \geq 0. \tag{25}$$

Proof: By Corollary 22.6 and the continuity of L, we have $L^x_t = 0$ for x outside the interval in (25), except on a fixed P-null set. To see that $L^x_t > 0$ otherwise, we may reduce by Theorem 18.3 and Corollary 22.6 to the case when M is a Brownian motion B. Letting $\tau_u = \inf\{t \geq 0; B_t = u\}$, we see from Theorems 18.6 (i) and 18.16 that, outside a fixed P-null set,

$$L^x_{\tau_u} > 0, \quad 0 \leq x < u \in \mathbb{Q}_+. \tag{26}$$

If $0 \leq x < \sup_{s \leq t} B_s$ for some t and x, there exists some $u \in \mathbb{Q}_+$ with $x < u < \sup_{s \leq t} B_s$. But then $\tau_u < t$, and (26) yields $L_t^x \geq L_{\tau_u}^x > 0$. A similar argument applies to the case when $\inf_{s \leq t} B_s < x \leq 0$. □

Our third approach to local times is via additive functionals and their potentials. To introduce those, consider a canonical Feller process X with state space S, associated terminal time ζ, probability measures P_x, transition operators T_t, shift operators θ_t, and filtration $\mathcal{F}$. By a *continuous additive functional (CAF)* of X we mean a nondecreasing, continuous, adapted process A with $A_0 = 0$ and $A_{\zeta \vee t} \equiv A_\zeta$, and such that

$$A_{s+t} = A_s + A_t \circ \theta_s \text{ a.s.}, \quad s, t \geq 0, \tag{27}$$

where a.s. without qualification means P_x-a.s. for every x. By the continuity of A, we may choose the exceptional null set to be independent of t. If it can also be taken to be independent of s, then A is said to be *perfect*.

For a simple example, let $f \geq 0$ be a bounded, measurable function on S, and consider the associated *elementary CAF*

$$A_t = \int_0^t f(X_s) ds, \quad t \geq 0. \tag{28}$$

More generally, given any CAF A and a function f as above, we may define a new CAF $f \cdot A$ by $(f \cdot A)_t = \int_{s \leq t} f(X_s) dA_s$, $t \geq 0$. A less trivial example is given by the local time of X at a fixed point x, whenever it exists in either sense discussed earlier.

For any CAF A and constant $\alpha \geq 0$, we may introduce the associated *α-potential*

$$U_A^\alpha(x) = E_x \int_0^\infty e^{-\alpha t} dA_t, \quad x \in S,$$

and put $U_A^\alpha f = U_{f \cdot A}^\alpha$. In the special case when $A_t \equiv t \wedge \zeta$, we shall often write $U^\alpha f = U_A^\alpha f$. Note in particular that $U_A^\alpha = U^\alpha f = R_\alpha f$ when A is given by (28). If $\alpha = 0$, we may omit the superscript and write $U = U^0$ and $U_A = U_A^0$. The next result shows that a CAF is determined by its α-potential whenever the latter is finite.

Lemma 22.19 *(uniqueness)* *Let A and B be CAFs of a Feller process X such that $U_A^\alpha = U_B^\alpha < \infty$ for some $\alpha \geq 0$. Then $A = B$ a.s.*

Proof: Define $A_t^\alpha = \int_{s \leq t} e^{-\alpha s} dA_s$, and conclude from (27) and the Markov property at t that, for any $x \in S$,

$$E_x[A_\infty^\alpha | \mathcal{F}_t] - A_t^\alpha = e^{-\alpha t} E_x[A_\infty^\alpha \circ \theta_t | \mathcal{F}_t] = e^{-\alpha t} U_A^\alpha(X_t). \tag{29}$$

Comparing with the same relation for B, it follows that $A^\alpha - B^\alpha$ is a continuous P_x-martingale of finite variation, and so $A^\alpha = B^\alpha$ a.s. P_x by Proposition 17.2. Since x was arbitrary, we get $A = B$ a.s. □

Given any CAF A of Brownian motion in $\mathbb{R}^d$, we may introduce the associated *Revuz measure* ν_A, given for any measurable function $g \geq 0$ on

$\mathbb{R}^d$ by $\nu_A g = \overline{E}(g \cdot A)_1$, where $\overline{E} = \int E_x dx$. When A is given by (28), we get in particular $\nu_A g = \langle f, g \rangle$, where $\langle \cdot, \cdot \rangle$ denotes the inner product in $L^2(\mathbb{R}^d)$. In general, we need to prove that ν_A is σ-finite.

Lemma 22.20 (σ-finiteness) *For any CAF A of Brownian motion X in $\mathbb{R}^d$, the associated Revuz measure ν_A is σ-finite.*

Proof: Fix any integrable function $f > 0$ on $\mathbb{R}^d$, and define

$$g(x) = E_x \int_0^\infty e^{-t - A_t} f(X_t) dt, \quad x \in \mathbb{R}^d.$$

Using Corollary 19.19, the additivity of A, and Fubini's theorem, we get

$$\begin{aligned}
U_A^1 g(x) &= E_x \int_0^\infty e^{-t} dA_t\, E_{X_t} \int_0^\infty e^{-s - A_s} f(X_s) ds \\
&= E_x \int_0^\infty e^{-t} dA_t \int_0^\infty e^{-s - A_s \circ \theta_t} f(X_{s+t}) ds \\
&= E_x \int_0^\infty e^{A_t} dA_t \int_t^\infty e^{-s - A_s} f(X_s) ds \\
&= E_x \int_0^\infty e^{-s - A_s} f(X_s) ds \int_0^s e^{A_t} dA_t \\
&= E_x \int_0^\infty e^{-s}(1 - e^{-A_s}) f(X_s) ds \leq E_0 \int_0^\infty e^{-s} f(X_s + x) ds.
\end{aligned}$$

Hence, by Fubini's theorem

$$\begin{aligned}
e^{-1} \nu_A g &\leq \int U_A^1 g(x) dx \leq \int dx\, E_0 \int_0^\infty e^{-s} f(X_s + x) ds \\
&= E_0 \int_0^\infty e^{-s} ds \int f(X_s + x) dx = \int f(x) dx < \infty.
\end{aligned}$$

The assertion now follows since $g > 0$. $\square$

Now let $p_t(x)$ denote the transition density $(2\pi t)^{-d/2} e^{-|x|^2/2t}$ of Brownian motion in $\mathbb{R}^d$, and put $u^\alpha(x) = \int_0^\infty e^{-\alpha t} p_t(x) dt$. For any measure μ on $\mathbb{R}^d$, we may introduce the associated α-potential $U^\alpha \mu(x) = \int u^\alpha(x - y) \mu(dy)$. The following result shows that the Revuz measure has the same potential as the underlying CAF.

Theorem 22.21 (α-potentials, Hunt, Revuz) *For Brownian motion in $\mathbb{R}^d$, let A be a CAF with Revuz measure ν_A. Then $U_A^\alpha = U^\alpha \nu_A$ for all $\alpha \geq 0$.*

Proof: By monotone convergence we may assume that $\alpha > 0$. By Lemma 22.20 we may choose some positive functions $f_n \uparrow 1$ such that $\nu_{f_n \cdot A} 1 = \nu_A f_n < \infty$ for each n, and by dominated convergence we have $U_{f_n \cdot A}^\alpha \uparrow U_A^\alpha$ and $U^\alpha \nu_{f_n \cdot A} \uparrow U^\alpha \nu_A$. Thus, we may further assume that ν_A is bounded. In that case, clearly $U_A^\alpha < \infty$ a.e.

Now fix any bounded, continuous function $f \geq 0$ on $\mathbb{R}^d$, and note that by dominated convergence $U^\alpha f$ is again bounded and continuous. Writing $h = n^{-1}$ for an arbitrary $n \in \mathbb{N}$, we get by dominated convergence and the additivity of A

$$\nu_A U^\alpha f = \overline{E} \int_0^1 U^\alpha f(X_s) dA_s = \lim_{n\to\infty} \overline{E} \sum_{j<n} U^\alpha f(X_{jh}) A_h \circ \theta_{jh}.$$

Noting that the operator U^α is self-adjoint and using the Markov property, we may write the expression on the right as

$$\sum_{j<n} \overline{E} U^\alpha f(X_{jh}) E_{X_{jh}} A_h = n \int U^\alpha f(x) E_x A_h dx = n\langle f, U^\alpha E.A_h\rangle.$$

To estimate the function $U^\alpha E.A_h$ on the right, it is enough to consider arguments x such that $U_A^\alpha(x) < \infty$. Using the Markov property of X and the additivity of A, we get

$$\begin{aligned}
U^\alpha E.A_h(x) &= E_x \int_0^\infty e^{-\alpha s} E_{X_s} A_h ds = E_x \int_0^\infty e^{-\alpha s}(A_h \circ \theta_s) ds \\
&= E_x \int_0^\infty e^{-\alpha s}(A_{s+h} - A_s) ds \\
&= (e^{\alpha h} - 1) E_x \int_0^\infty e^{-\alpha s} A_s ds - e^{\alpha h} E_x \int_0^h e^{-\alpha s} A_s ds \quad (30)
\end{aligned}$$

Integrating by parts gives

$$E_x \int_0^\infty e^{-\alpha s} A_s ds = \alpha^{-1} E_x \int_0^\infty e^{-\alpha t} dA_t = \alpha^{-1} U_A^\alpha(x).$$

Thus, as $n = h^{-1} \to \infty$, the first term on the right of (30) yields in the limit the contribution $\langle f, U_A^\alpha\rangle$. The second term is negligible since

$$\langle f, E.A_h\rangle \leq \overline{E} A_h = h \nu_A 1 \to 0.$$

Hence,

$$\langle U^\alpha \nu_A, f\rangle = \nu_A U^\alpha f = \langle U_A^\alpha, f\rangle,$$

and since f is arbitrary, we obtain $U_A^\alpha = U^\alpha \nu_A$ a.e.

To extend this to an identity, fix any $h > 0$ and $x \in \mathbb{R}^d$. Using the additivity of A, the Markov property at h, the a.e. relation, Fubini's theorem, and the Chapman–Kolmogorov relation, we get

$$\begin{aligned}
e^{\alpha h} E_x \int_h^\infty e^{-\alpha s} dA_s &= E_x \int_0^\infty e^{-\alpha s} dA_s \circ \theta_h \\
&= E_x U_A^\alpha(X_h) = E_x U^\alpha \nu_A(X_h) \\
&= \int \nu_A(dy) E_x u^\alpha(X_h - y) \\
&= e^{\alpha h} \int \nu_A(dy) \int_h^\infty e^{-\alpha s} p_s(x-y) ds.
\end{aligned}$$

The required relation $U_A^\alpha(x) = U^\alpha \nu_A(x)$ now follows by monotone convergence as $h \to 0$. □

It is now easy to show that a CAF is determined by its Revuz measure.

Corollary 22.22 (*uniqueness*) *If A and B are CAFs of Brownian motion in $\mathbb{R}^d$ with $\nu_A = \nu_B$, then $A = B$ a.s.*

Proof: By Lemma 22.20 we may assume that ν_A is bounded, so that $U_A^\alpha < \infty$ a.e. for all $\alpha > 0$. Now ν_A determines U_A^α by Theorem 22.21, and from the proof of Lemma 22.19 we note that U_A^α determines A a.s. P_x whenever $U_A^\alpha(x) < \infty$. Since $P_x \circ X_h^{-1} \ll \lambda^d$ for each $h > 0$, it follows that $A \circ \theta_h$ is a.s. unique, and it remains to let $h \to 0$. □

We turn to the reverse problem of constructing a CAF associated with a given potential. To motivate the following definition, we may take expected values in (29) to get $e^{-\alpha t} T_t U_A^\alpha \le U_A^\alpha$. A function f on S is said to be *uniformly α-excessive* if it is bounded and measurable with $0 \le e^{-\alpha t} T_t f \le f$ for all $t \ge 0$ and such that $\|T_t f - f\| \to 0$ as $t \to 0$, where $\|\cdot\|$ denotes the supremum norm.

Theorem 22.23 (*excessive functions and CAFs, Volkonsky*) *For any Feller process X in S and constant $\alpha > 0$, let $f \ge 0$ be a uniformly α-excessive function on S. Then $f = U_A^\alpha$ for some a.s. unique, perfect CAF A of X.*

Proof: For any bounded, measurable function g on S, we get by Fubini's theorem and the Markov property of X

$$\tfrac{1}{2} E_x \left| \int_0^\infty e^{-\alpha t} g(X_t) dt \right|^2 = E_x \int_0^\infty e^{-\alpha t} g(X_t) dt \int_0^\infty e^{-\alpha(t+h)} g(X_{t+h}) dh$$
$$= E_x \int_0^\infty e^{-2\alpha t} g(X_t) dt \int_0^\infty e^{-\alpha h} T_h g(X_t) dh$$
$$= E_x \int_0^\infty e^{-2\alpha t} g U^\alpha g(X_t) dt = \int_0^\infty e^{-2\alpha t} T_t g U^\alpha g(x) dt$$
$$\le \|U^\alpha g\| \int_0^\infty e^{-\alpha t} T_t |g|(x) dt \le \|U^\alpha g\| \, \|U^\alpha |g|\|. \tag{31}$$

Now introduce for each $h > 0$ the bounded, nonnegative functions

$$g_h = h^{-1}(f - e^{-\alpha h} T_h f),$$
$$f_h = U^\alpha g_h = h^{-1} \int_0^h e^{-\alpha s} T_s f \, ds,$$

and define

$$A_h(t) = \int_0^t g_h(X_s) ds, \qquad M_h(t) = A_h^\alpha(t) + e^{-\alpha t} f_h(X_t).$$

As in (29), we note that the processes M_h are martingales under P_x for every x. Using the continuity of the A_h, we get by Proposition 7.16 and

(31), for any $x \in S$ and as $h, k \to 0$,

$$\begin{aligned}
E_x(A_h^\alpha - A_k^\alpha)^{*2} &\leq E_x\sup_{t \in \mathbb{Q}_+}|M_h(t) - M_k(t)|^2 + \|f_h - f_k\|^2 \\
&\leq E_x|A_h^\alpha(\infty) - A_k^\alpha(\infty)|^2 + \|f_h - f_k\|^2 \\
&\leq \|f_h - f_k\|\,\|f_h + f_k\| + \|f_h - f_k\|^2 \to 0.
\end{aligned}$$

Hence, there exists some continuous process A independent of x such that $E_x(A_h^\alpha - A^\alpha)^{*2} \to 0$ for every x.

For a suitable sequence $h_n \to 0$ we have $(A_{h_n}^\alpha \to A^\alpha)^* \to 0$ a.s. P_x for all x, and it follows easily that A is a.s. a perfect CAF. Taking limits in the relation $f_h(x) = E_x A_h^\alpha(\infty)$, we also note that $f(x) = E_x A^\alpha(\infty) = U_A^\alpha(x)$. Thus, A has α-potential f. □

We will now use the last result to construct local times. Let us say that a CAF A is *supported* by some set $B \subset S$ if its set of increase is a.s. contained in the closure of the set $\{t \geq 0;\ X_t \in B\}$. In particular, a nonzero and perfect CAF supported by a singleton set $\{x\}$ is called a *local time* at x. This terminology is clearly consistent with our earlier definitions of local time. Writing $\tau_x = \inf\{t > 0;\ X_t = x\}$, we say that x is *regular (for itself)* if $\tau_x = 0$ a.s. P_x. By Proposition 22.7 this holds iff P_x-a.s. the random set $Z_x = \{t \geq 0;\ X_t = x\}$ has no isolated points.

Theorem 22.24 (*additive functional local time, Blumenthal and Getoor*) *A Feller process in S has a local time L at a point $a \in S$ iff a is regular. In that case L is a.s. unique up to a normalization, and*

$$U_L^1(x) = U_L^1(a) E_x e^{-\tau_a} < \infty, \qquad x \in S. \tag{32}$$

Proof: Let L be a local time at a. Comparing with the renewal process $L_n^{-1}, n \in \mathbb{Z}_+$, we see that $\sup_{x,t} E_x(L_{t+h} - L_t) < \infty$ for every $h > 0$, which implies $U_L^1(x) < \infty$ for all x. By the strong Markov property at $\tau = \tau_a$, we get for any $x \in S$

$$\begin{aligned}
U_L^1(x) &= E_x(L_\infty^1 - L_\tau^1) = E_x e^{-\tau}(L_\infty^1 \circ \theta_\tau) \\
&= E_x e^{-\tau} E_a L_\infty^1 = U_L^1(a) E_x e^{-\tau},
\end{aligned}$$

proving (32). The uniqueness assertion now follows by Lemma 22.19.

To prove the existence of L, define $f(x) = E_x e^{-\tau}$, and note that f is bounded and measurable. Since $\tau \leq t + \tau \circ \theta_t$, we may also conclude from the Markov property at t that, for any $x \in S$,

$$\begin{aligned}
f(x) &= E_x e^{-\tau} \geq e^{-t} E_x(e^{-\tau} \circ \theta_t) \\
&= e^{-t} E_x E_{X_t} e^{-\tau} = e^{-t} E_x f(X_t) = e^{-t} T_t f(x).
\end{aligned}$$

Noting that $\sigma_t = t + \tau \circ \theta_t$ is nondecreasing and tends to 0 a.s. P_a as $t \to 0$ by the regularity of a, we further obtain

$$\begin{aligned}
0 &\leq f(x) - e^{-h} T_h f(x) \\
&= E_x(e^{-\tau} - e^{-\sigma_h}) \leq E_x(e^{-\tau} - e^{-\sigma_h + \tau}) \\
&= E_x e^{-\tau} E_a(1 - e^{-\sigma_h}) \leq E_a(1 - e^{-\sigma_h}) \to 0.
\end{aligned}$$

Thus, f is uniformly 1-excessive, and so by Theorem 22.23 there exists a perfect CAF L with $U_L^1 = f$.

To see that L is supported by the singleton $\{a\}$, we may write

$$E_x(L_\infty^1 - L_\tau^1) = E_x e^{-\tau} E_a L_\infty^1 = E_x e^{-\tau} E_a e^{-\tau} = E_x e^{-\tau} = E_x L_\infty^1,$$

which implies $L_\tau^1 = 0$ a.s. Hence, $L_\tau = 0$ a.s., and so the Markov property yields $L_{\sigma_t} = L_t$ a.s. for all rational t. This shows that L has a.s. no point of increase outside the closure of $\{t \geq 0;\ X_t = a\}$. □

The next result shows that every CAF of one-dimensional Brownian motion is a unique mixture of local times. Recall that ν_A denotes the Revuz measure of the CAF A.

Theorem 22.25 *(integral representation, Volkonsky, McKean and Tanaka)* *For Brownian motion X in $\mathbb{R}$ with local time L, a process A is a CAF of X iff it has an a.s. representation*

$$A_t = \int_{-\infty}^{\infty} L_t^x \nu(dx), \quad t \geq 0, \tag{33}$$

for some locally finite measure ν on $\mathbb{R}$. The latter is then unique and equals ν_A.

Proof: For any measure ν we may define an associated process A as in (33). If ν is locally finite, it is clear by the continuity of L and dominated convergence that A is a.s. continuous, hence a CAF. In the opposite case, we note that ν is infinite in every neighborhood of some point $a \in \mathbb{R}$. Under P_a and for any $t > 0$, the process L_t^x is further a.s. continuous and strictly positive near $x = a$. Hence, $A_t = \infty$ a.s. P_a, and A fails to be a CAF.

Next, we conclude from Fubini's theorem and Theorem 22.5 that

$$\overline{E}L_1^x = \int (E_y L_1^x) dy = E_0 \int L_1^{x-y} dy = 1.$$

Since L^x is supported by $\{x\}$, we get for any CAF A as in (33)

$$\nu_A f = \overline{E}(f \cdot A)_1 = \overline{E} \int \nu(dx) \int_0^1 f(X_t) dL_t^x$$
$$= \int f(x) \nu(dx) \overline{E} L_1^x = \nu f,$$

which shows that $\nu = \nu_A$.

Now consider an arbitrary CAF A. By Lemma 22.20 there exists some function $f > 0$ with $\nu_A f < \infty$. The process

$$B_t = \int L_t^x \nu_{f \cdot A}(dx) = \int L_t^x f(x) \nu_A(dx), \quad t \geq 0,$$

is then a CAF with $\nu_B = \nu_{f \cdot A}$, and by Corollary 22.22 we get $B = f \cdot A$ a.s. Thus, $A = f^{-1} \cdot B$ a.s., and (33) follows. □

Exercises

1. Use Lemma 13.15 to show that the set of increase of Brownian local time at 0 agrees a.s. with the zero set Z. Extend the result to any continuous local martingale. (*Hint:* Apply Lemma 13.15 to the process $\text{sgn}(B-) \cdot B$ in Theorem 22.1.)

2. (Lévy) Let M be the maximum process of a Brownian motion B. Show that B can be measurably recovered from $M - B$. (*Hint:* Use Corollaries 22.3 and 22.6.)

3. Use Corollary 22.3 to give a simple proof of the relation $\tau_2 \stackrel{d}{=} \tau_3$ in Theorem 13.16. (*Hint:* Recall that the maximum is unique by Lemma 13.15.) Also use Proposition 18.9 to give a direct proof of the relation $\tau_1 \stackrel{d}{=} \tau_2$. (*Hint:* Integrate separately over the positive and negative excursions of B, and use Lemma 13.15 to identify the minimum.)

4. Show that for any $c \in (0, \frac{1}{2})$, Brownian local time L_t^x is a.s. Hölder continuous in x with exponent c, uniformly for bounded t. Also show that the bound $c < \frac{1}{2}$ is best possible. (*Hint:* Apply Theorem 3.23 to the estimate in the proof of Theorem 22.4. For the last assertion, use Theorem 22.17.)

5. Let M be a continuous local martingale such that $B \circ [M]$ a.s. for some Brownian motion. Show that if B has local time L_t^x, then the local time of M at x equals $L^x \circ [M]$. (*Hint:* Use Theorem 22.5, and note that $L \circ [M]$ is jointly continuous.)

6. For any continuous semimartingale X, show that $\int_0^t f(X_s, s) d[X]_s = \int dx \int_0^t f(x, s) dL_s^x$ outside a fixed null set. (*Hint:* Extend Theorem 22.5 by a monotone class argument.)

7. Let Z be the zero set of Brownian motion B. Use Proposition 22.12 and Theorem 22.15 to construct its local time L directly from Z. Also use Lemma 22.16 to construct L from the heights of the excursions of B. Finally, use Corollary 22.6 to construct L from the occupation measure of B.

8. Let η be the maximum of a Brownian excursion. Show that $E\eta = (\pi/2)^{1/2}$. (*Hint:* Use Theorem 22.15 and Lemmas 22.16 and 3.4.)

9. Let L be the continuous local time of a continuous local martingale M with $[M]_\infty = \infty$ a.s. Show that a.s. $L_t^x \to \infty$ as $t \to \infty$, uniformly on compacts. (*Hint:* Reduce to the case of Brownian motion. Then use Corollary 22.18, the strong Markov property, and the law of large numbers.)

10. Show that the intersection of two regenerative sets is regenerative.

11. Let L be the local time of a regenerative set and let τ be an independent, exponentially distributed time. Show that L_τ is again exponentially distributed. (*Hint:* Prove a Cauchy equation for the function $P\{L_\tau > s\}$.)

12. For any unbounded regenerative set Z, show that $\mathcal{L}(Z)$ is a.s. determined by Z. (*Hint:* Use the law of large numbers.)

13. Let Z be a nontrivial regenerative set. Show that $cZ \stackrel{d}{=} Z$ for all $c > 0$ iff the inverse local time is strictly stable.

14. Let X be a Feller process in $\mathbb{R}$ and put $M_t = \sup_{s \leq t} X_s$. Show that the points of increase of M form a regenerative set. Also prove the same statement for the process $X_t^* = \sup_{s \leq t} |X_s|$ when $-X \stackrel{d}{=} X$.

15. Let X be a strictly stable Lévy process, let Z denote the set of increase of the process $M_t = \sup_{s \leq t} X_s$, and write L for the local time of Z. Assuming Z to be nontrivial, show that L^{-1} is strictly stable. Also prove the corresponding statement for X^* when X is symmetric.

16. Give an explicit construction of the process X in Theorem 22.11, based on the Poisson process ξ and the constant c. (*Hint:* Use Theorem 22.13 to construct the time scale.)

17. Show that semimartingale local time is preserved under a change of measure $Q = Z_t \cdot P$. Use this result to extend Corollary 22.18 Brownian motion with a suitable drift. (*Hint:* Use Proposition 18.20 and Corollary 18.25.)

18. Show that the notion of a continuous additive functional is preserved under a suitable change of measure $Q = Z_t \cdot P$. Use this result to extend Theorem 22.25 to a Brownian motion with drift.

Chapter 23

One-Dimensional SDEs and Diffusions

Weak existence and uniqueness; pathwise uniqueness and comparison; scale function and speed measure; time-change representation; boundary classification; entrance boundaries and Feller properties; ratio ergodic theorem; recurrence and ergodicity

By a diffusion is usually understood a continuous strong Markov process, sometimes required to possess additional regularity properties. The basic example of a diffusion process is Brownian motion, which was first introduced and studied in Chapter 13. More general diffusions, first encountered in Chapter 19, were studied extensively in Chapter 21 as solutions to suitable stochastic differential equations (SDEs). This chapter focuses on the one-dimensional case, which allows a more detailed analysis. Martingale methods are used throughout the chapter, and we make essential use of results on random time-change from Chapters 17 and 18, as well as on local time, excursions, and additive functionals from Chapter 22.

After considering the Engelbert–Schmidt characterization of weak existence and uniqueness for the equation $dX_t = \sigma(X_t)dB_t$, we turn to a discussion of various pathwise uniqueness and comparison results for the corresponding equation with drift. Next we proceed to a systematic study of regular diffusions, introduce the notions of scale function and speed measure, and prove the basic representation of a diffusion on a natural scale as a time-changed Brownian motion. Finally, we characterize the different types of boundary behavior, establish the Feller properties for a suitable extension of the process, and examine the recurrence and ergodic properties in the various cases.

To begin with the SDE approach, consider the general one-dimensional diffusion equation (σ, b), given by

$$dX_t = \sigma(X_t)dB_t + b(X_t)dt. \tag{1}$$

From Theorem 21.11 we know that if weak existence and uniqueness in law hold for (1), then the solution process X is a continuous strong Markov process. It is clearly also a semimartingale.

In Proposition 21.12 we saw how the drift term can sometimes be eliminated through a suitable change of the underlying probability measure.

Under suitable regularity conditions on the coefficients, we may use the alternative approach of transforming the state space. Let us then assume that X solves (1), and put $Y_t = p(X_t)$ for some function $p \in C^1$ possessing an absolutely continuous derivative p' with density p''. By the generalized Itô formula of Theorem 22.5, we have

$$\begin{aligned} dY_t &= p'(X_t)dX_t + \tfrac{1}{2}p''(X_t)d[X]_t \\ &= (\sigma p')(X_t)dB_t + (\tfrac{1}{2}\sigma^2 p'' + bp')(X_t)dt. \end{aligned}$$

Here the drift term vanishes iff p solves the ordinary differential equation

$$\tfrac{1}{2}\sigma^2 p'' + bp' = 0. \tag{2}$$

If b/σ^2 is locally integrable, then (2) has the explicit solutions

$$p'(x) = c \exp\left\{-2\int_0^x (b\sigma^{-2})(u)du\right\}, \quad x \in \mathbb{R},$$

where c is an arbitrary constant. The desired scale function p is then determined up to an affine transformation, and for $c > 0$ it is strictly increasing with a unique inverse p^{-1}. The mapping by p reduces (1) to the form $dY_t = \tilde{\sigma}(Y_t)dB_t$, where $\tilde{\sigma} = (\sigma p') \circ p^{-1}$. Since the new equation is equivalent, it is clear that weak or strong existence or uniqueness hold simultaneously for the two equations.

Once the drift has been removed, we are left with an equation of the form

$$dX_t = \sigma(X_t)dB_t. \tag{3}$$

Here exact criteria for weak existence and uniqueness may be given in terms of the singularity sets

$$\begin{aligned} S_\sigma &= \left\{x \in \mathbb{R}; \int_{x-}^{x+} \sigma^{-2}(y)dy = \infty\right\}, \\ N_\sigma &= \{x \in \mathbb{R}; \sigma(x) = 0\}. \end{aligned}$$

Theorem 23.1 *(existence and uniqueness, Engelbert and Schmidt)* Weak existence holds for equation (3) with arbitrary initial distribution iff $S_\sigma \subset N_\sigma$. In that case, uniqueness in law holds for every initial distribution iff $S_\sigma = N_\sigma$.

Our proof begins with a lemma, which will also be useful later. Given any measure ν on $\mathbb{R}$, we may introduce the associated singularity set

$$S_\nu = \{x \in \mathbb{R}; \nu(x-, x+) = \infty\}.$$

If B is a one-dimensional Brownian motion with associated local time L, we may also introduce the additive functional

$$A_s = \int L_s^x \nu(dx), \quad s \geq 0, \tag{4}$$

Lemma 23.2 *(singularity set) Let L be the local time of Brownian motion B with arbitrary initial distribution, and define A by (4) for some measure ν on $\mathbb{R}$. Then a.s.*

$$\inf\{s \geq 0;\ A_s = \infty\} = \inf\{s \geq 0;\ B_s \in S_\nu\}.$$

Proof: Fix any $t > 0$, and let R be the event where $B_s \notin S_\nu$ on $[0, t]$. Noting that $L_t^x = 0$ a.s. for x outside the range $B[0, t]$, we get a.s. on R

$$A_t = \int_{-\infty}^{\infty} L_t^x \nu(dx) \leq \nu(B[0, t]) \sup_x L_t^x < \infty$$

since $B[0, t]$ is compact and L_t^x is a.s. continuous, hence bounded.

Conversely, suppose that $B_s \in S_\nu$ for some $s < t$. To show that $A_t = \infty$ a.s. on this event, we may use the strong Markov property to reduce to the case when $B_0 = a$ is nonrandom in S_ν. But then $L_t^a > 0$ a.s. by Tanaka's formula, and so by the continuity of L we get for small enough $\varepsilon > 0$

$$A_t = \int_{-\infty}^{\infty} L_t^x \nu(dx) \geq \nu(a - \varepsilon, a + \varepsilon) \inf_{|x-a|<\varepsilon} L_t^x = \infty. \qquad \Box$$

Proof of Theorem 23.1: First assume that $S_\sigma \subset N_\sigma$. To prove the asserted weak existence, let Y be a Brownian motion with arbitrary initial distribution μ, and define $\zeta = \inf\{s \geq 0;\ Y_s \in S_\sigma\}$. By Lemma 23.2 the additive functional

$$A_s = \int_0^s \sigma^{-2}(Y_r) dr, \quad s \geq 0, \tag{5}$$

is continuous and strictly increasing on $[0, \zeta)$, and for $t > \zeta$ we have $A_t = \infty$. Also note that $A_\zeta = \infty$ when $\zeta = \infty$, whereas A_ζ may be finite when $\zeta < \infty$. In the latter case A jumps from A_ζ to ∞ at time ζ.

Now introduce the inverse

$$\tau_t = \inf\{s > 0;\ A_s > t\}, \quad t \geq 0. \tag{6}$$

The process τ is clearly continuous and strictly increasing on $[0, A_\zeta]$, and for $t \geq A_\zeta$ we have $\tau_t = \zeta$. Also note that $X_t = Y_{\tau_t}$ is a continuous local martingale and, moreover,

$$t = A_{\tau_t} = \int_0^{\tau_t} \sigma^{-2}(Y_r) dr = \int_0^t \sigma^{-2}(X_s) d\tau_s, \quad t < A_\zeta.$$

Hence, for $t \leq A_\zeta$,

$$[X]_t = \tau_t = \int_0^t \sigma^2(X_s) ds. \tag{7}$$

Here both sides remain constant after time A_ζ since $S_\sigma \subset N_\sigma$, and so (7) remains true for all $t \geq 0$. Hence, Theorem 18.12 yields the existence of a Brownian motion B satisfying (3), which means that X is a weak solution with initial distribution μ.

To prove the converse implication, assume that weak existence holds for any initial distribution. To show that $S_\sigma \subset N_\sigma$, we may fix any $x \in S_\sigma$ and choose a solution X with $X_0 = x$. Since X is a continuous local martingale, Theorem 18.4 yields $X_t = Y_{\tau_t}$ for some Brownian motion Y starting at x and some random time-change τ satisfying (7). For A as in (5) and for $t \geq 0$ we have

$$A_{\tau_t} = \int_0^{\tau_t} \sigma^{-2}(Y_r) dr = \int_0^t \sigma^{-2}(X_s) d\tau_s = \int_0^t 1\{\sigma(X_s) > 0\} ds \leq t. \quad (8)$$

Since $A_s = \infty$ for $s > 0$ by Lemma 23.2, we get $\tau_t = 0$ a.s., and so $X_t \equiv x$ a.s. But then $x \in N_\sigma$ by (7).

Turning to the uniqueness assertion, assume that $N_\sigma \subset S_\sigma$, and consider a solution X with initial distribution μ. As before, we may write $X_t = Y_{\tau_t}$ a.s., where Y is a Brownian motion with initial distribution μ and τ is a random time-change satisfying (7). Define A as in (5), put $\chi = \inf\{t \geq 0; X_t \in S_\sigma\}$, and note that $\tau_\chi = \zeta \equiv \inf\{s \geq 0; Y_s \in S_\sigma\}$. Since $N_\sigma \subset S_\sigma$, we get as in (8)

$$A_{\tau_t} = \int_0^{\tau_t} \sigma^{-2}(Y_s) ds = t, \quad t \leq \chi.$$

Furthermore, $A_s = \infty$ for $s > \zeta$ by Lemma 23.2, and so (8) implies $\tau_t \leq \zeta$ a.s. for all t, which means that τ remains constant after time χ. Thus, τ and A are related by (6), which shows that τ and then also X are measurable functions of Y. Since the distribution of Y depends only on μ, the same thing is true for X, which proves the asserted uniqueness in law.

To prove the converse, assume that S_σ is a proper subset of N_σ, and fix any $x \in N_\sigma \setminus S_\sigma$. As before, we may construct a solution starting at x by writing $X_t = Y_{\tau_t}$, where Y is a Brownian motion starting at x, and τ is defined as in (6) from the process A in (5). Since $x \notin S_\sigma$, Lemma 23.2 gives $A_{0+} < \infty$ a.s., and so $\tau_t > 0$ a.s. for $t > 0$, which shows that X is a.s. nonconstant. Since $x \in N_\sigma$, (3) has also the trivial solution $X_t \equiv x$. Thus, uniqueness in law fails for solutions starting at x. □

Proceeding with a study of pathwise uniqueness, we return to equation (1), and let $w(\sigma, \cdot)$ denote the modulus of continuity of σ.

Theorem 23.3 *(pathwise uniqueness, Skorohod, Yamada and Watanabe) Let σ and b be bounded, measurable functions on $\mathbb{R}$, where*

$$\int_0^\varepsilon (w(\sigma, h))^{-2} dh = \infty, \quad \varepsilon > 0, \quad (9)$$

and either b is Lipschitz continuous or $\sigma \neq 0$. Then pathwise uniqueness holds for equation (σ, b).

The significance of condition (9) is clarified by the following lemma, where for any semimartingale Y we write $L_t^x(Y)$ for the associated local time.

Lemma 23.4 *(local time)* For $i = 1, 2$, let X^i solve equation (σ, b_i), where σ satisfies (9). Then $L^0(X^1 - X^2) = 0$ a.s.

Proof: Write $Y = X^1 - X^2$, $L_t^x = L_t^x(Y)$, and $w(x) = w(\sigma, |x|)$. Using (1) and Theorem 22.5, we get for any $t > 0$

$$\int_{-\infty}^{\infty} \frac{L_t^x dx}{w_x^2} = \int_0^t \frac{d[Y]_s}{(w(Y_s))^2} = \int_0^t \left\{ \frac{\sigma(X_s^1) - \sigma(X_s^2)}{w(X_s^1 - X_s^2)} \right\}^2 ds \leq t < \infty.$$

By (1) and the right-continuity of L it follows that $L_t^0 = 0$ a.s. □

Proof of Theorem 23.3 for $\sigma \neq 0$: By Propositions 21.12 and 21.13 combined with a simple localization argument, we note that uniqueness in law holds for equation (σ, b) when $\sigma \neq 0$. To prove the pathwise uniqueness, consider any two solutions X and Y with $X_0 = Y_0$ a.s. Using Tanaka's formula, Lemma 23.4, and equation (σ, b), we get

$$\begin{aligned} d(X_t \vee Y_t) &= dX_t + d(Y_t - X_t)^+ \\ &= dX_t + 1\{Y_t > X_t\}d(Y_t - X_t) \\ &= 1\{Y_t \leq X_t\}dX_t + 1\{Y_t > X_t\}dY_t \\ &= \sigma(X_t \vee Y_t)dB_t + b(X_t \vee Y_t)dt, \end{aligned}$$

which shows that $X \vee Y$ is again a solution. By the uniqueness in law we get $X \stackrel{d}{=} X \vee Y$. Since $X \leq X \vee Y$, it follows that $X = X \vee Y$ a.s., which implies $Y \leq X$ a.s. Similarly, $X \leq Y$ a.s. □

The assertion for Lipschitz continuous b is a special case of the following comparison result.

Theorem 23.5 *(weak comparison, Skorohod, Yamada)* Fix some functions σ and $b_1 \geq b_2$, where σ satisfies (9) and either b_1 or b_2 is Lipschitz continuous. For $i = 1, 2$, let X^i solve equation (σ, b_i), and assume that $X_0^1 \geq X_0^2$ a.s. Then $X^1 \geq X^2$ a.s.

Proof: By symmetry we may assume that b_1 is Lipschitz continuous. Since $X_0^2 \leq X_0^1$ a.s., we get by Tanaka's formula and Lemma 23.4

$$\begin{aligned} (X_t^2 - X_t^1)^+ &= \int_0^t 1\{X_s^2 > X_s^1\}\left(\sigma(X_t^2) - \sigma(X_t^1)\right) dB_t \\ &\quad + \int_0^t 1\{X_s^2 > X_s^1\}\left(b_2(X_s^2) - b_1(X_s^1)\right) ds. \end{aligned}$$

Using the martingale property of the first term, the Lipschitz continuity of b_1, and the condition $b_2 \leq b_1$, we conclude that

$$E(X_t^2 - X_t^1)^+ \leq E \int_0^t 1\{X_s^2 > X_s^1\} (b_1(X_s^2) - b_1(X_s^1)) \, ds$$

$$\leq E \int_0^t 1\{X_s^2 > X_s^1\} |X_s^2 - X_s^1| \, ds$$

$$= \int_0^t E(X_s^2 - X_s^1)^+ \, ds.$$

By Gronwall's lemma $E(X_t^2 - X_t^1)^+ = 0$, and hence $X_t^2 \leq X_t^1$ a.s. □

Imposing stronger restrictions on the coefficients, we may strengthen the last conclusion to a strict inequality.

Theorem 23.6 *(strict comparison)* *Fix a Lipschitz continuous function σ and some continuous functions $b_1 > b_2$. For $i = 1, 2$, let X^i solve equation (σ, b_i), and assume that $X_0^1 \geq X_0^2$ a.s. Then $X^1 > X^2$ a.s. on $(0, \infty)$.*

Proof: Since the b_i are continuous with $b_1 > b_2$, there exists a locally Lipschitz continuous function b on $\mathbb{R}$ with $b_1 > b > b_2$. By Theorem 21.3 equation (σ, b) has a solution X with $X_0 = X_0^1 \geq X_0^2$ a.s., and it suffices to show that $X^1 > X > X^2$ a.s. on $(0, \infty)$. This reduces the discussion to the case when one of the functions b_i is locally Lipschitz. By symmetry we may take that function to be b_1.

By the Lipschitz continuity of σ and b_1, we may define some continuous semimartingales U and V by

$$U_t = \int_0^t (b_1(X_s^2) - b_2(X_s^2)) \, ds,$$

$$V_t = \int_0^t \frac{\sigma(X_s^1) - \sigma(X_s^2)}{X_s^1 - X_s^2} \, dB_s + \int_0^t \frac{b_1(X_s^1) - b_1(X_s^2)}{X_s^1 - X_s^2} \, ds,$$

subject to the convention $0/0 = 0$, and we note that

$$d(X_t^1 - X_t^2) = dU_t + (X_t^1 - X_t^2) dV_t.$$

Letting $Z = \exp(V - \frac{1}{2}[V]) > 0$, we get by Proposition 21.2

$$X_t^1 - X_t^2 = Z_t (X_0^1 - X_0^2) + Z_t \int_0^t Z_s^{-1} (b_1(X_s^2) - b_2(X_s^2)) \, ds,$$

and the assertion follows since $X_0^1 \geq X_0^2$ a.s. and $b_1 > b_2$. □

We turn to a systematic study of one-dimensional diffusions. By a *diffusion* on some interval $I \subset \mathbb{R}$ we mean a continuous strong Markov process taking values in I. Termination will only be allowed at open end-points of I. We define $\tau_y = \inf\{t \geq 0;\, X_t = y\}$ and say that X is *regular* if $P_x\{\tau_y < \infty\} > 0$ for any $x \in I^\circ$ and $y \in I$. Let us further write $\tau_{a,b} = \tau_a \wedge \tau_b$.

Our first aim is to transform the general diffusion process into a continuous local martingale, using a suitable change of scale. This corresponds to the removal of drift in the SDE (1).

Theorem 23.7 *(scale function, Feller, Dynkin) For any regular diffusion X on I, there exists a continuous and strictly increasing function p on I such that $p(X^{\tau_{a,b}})$ is a P_x-martingale for all $a \le x \le b$ in I. Furthermore, an increasing function p has the stated property iff*

$$P_x\{\tau_b < \tau_a\} = \frac{p_x - p_a}{p_b - p_a}, \quad x \in [a,b]. \tag{10}$$

A function p with the stated property is called a *scale function* for X, and we say that X is on a *natural scale* if the scale function can be chosen to be linear. In general, we note that $Y = p(X)$ is a regular diffusion on a natural scale.

Our proof begins with a study of the functions

$$p_{a,b}(x) = P_x\{\tau_b < \tau_a\}, \quad h_{a,b}(x) = E_x \tau_{a,b}, \qquad a \le x \le b,$$

which play a basic role in the subsequent analysis.

Lemma 23.8 *(hitting times) For any regular diffusion on I and constants $a < b$ in I, we have*

(i) *$p_{a,b}$ is continuous and strictly increasing on $[a,b]$;*

(ii) *$h_{a,b}$ is bounded on $[a,b]$.*

In particular, we see from (ii) that $\tau_{a,b} < \infty$ a.s. under P_x for any $a \le x \le b$.

Proof: (i) First we show that $P_x\{\tau_b < \tau_a\} > 0$ for any $a < x < b$. Then introduce the optional time $\sigma_1 = \tau_a + \tau_x \circ \theta_{\tau_a}$, and define recursively $\sigma_{n+1} = \sigma_n + \sigma_1 \circ \theta_{\sigma_n}$. By the strong Markov property the σ_n form a random walk in $[0,\infty]$ under each P_x. If $P_x\{\tau_b < \tau_a\} = 0$, we get $\tau_b \ge \sigma_n \to \infty$ a.s. P_x, and so $P_x\{\tau_b = \infty\} = 1$, which contradicts the regularity of X.

Using the strong Markov property at τ_y, we next obtain

$$P_x\{\tau_b < \tau_a\} = P_x\{\tau_y < \tau_a\} P_y\{\tau_b < \tau_a\}, \quad a < x < y < b. \tag{11}$$

Since $P_x\{\tau_a < \tau_y\} > 0$, we have $P_x\{\tau_y < \tau_a\} < 1$, which shows that $P_x\{\tau_b < \tau_a\}$ is strictly increasing.

By symmetry it remains to prove that $P_y\{\tau_b < \tau_a\}$ is left-continuous on $(a,b]$. By (11) it is equivalent to show for each $x \in (a,b)$ that the mapping $y \mapsto P_x\{\tau_y < \tau_a\}$ is left-continuous on $(x,b]$. Then let $y_n \uparrow y$, and note that $\tau_{y_n} \uparrow \tau_y$ a.s. P_x by the continuity of X. Hence, $\{\tau_{y_n} < \tau_a\} \downarrow \{\tau_y < \tau_a\}$, which implies convergence of the corresponding probabilities.

(ii) Fix any $c \in (a,b)$. By the regularity of X we may choose $h > 0$ so large that

$$P_c\{\tau_a \le h\} \wedge P_c\{\tau_b \le h\} = \delta > 0.$$

If $x \in (a, c)$, we may use the strong Markov property at τ_x to get

$$\delta \leq P_c\{\tau_a \leq h\} \leq P_c\{\tau_x \leq h\} P_x\{\tau_a \leq h\}$$
$$\leq P_x\{\tau_a \leq h\} \leq P_x\{\tau_{a,b} \leq h\},$$

and similarly for $x \in (c, b)$. By the Markov property at h and induction on n we obtain

$$P_x\{\tau_{a,b} > nh\} \leq (1 - \delta)^n, \quad x \in [a, b], \ n \in \mathbb{Z}_+,$$

and Lemma 3.4 yields

$$E_x \tau_{a,b} = \int_0^\infty P_x\{\tau_{a,b} > t\} dt \leq h \sum_{n \geq 0} (1 - \delta)^n < \infty. \qquad \square$$

Proof of Theorem 23.7: Let p be a locally bounded and measurable function on I such that $M = p(X^{\tau_{a,b}})$ is a martingale under P_x for any $a < x < b$. Then

$$\begin{aligned}
p_x &= E_x M_0 = E_x M_\infty = E_x p(X_{\tau_{a,b}}) \\
&= p_a P_x\{\tau_a < \tau_b\} + p_b P_x\{\tau_b < \tau_a\} \\
&= p_a + (p_b - p_a) P_x\{\tau_b < \tau_a\},
\end{aligned}$$

and (10) follows, provided that $p_a \neq p_b$.

To construct a function p with the stated properties, fix any points $u < v$ in I, and define for arbitrary $a \leq u$ and $b \geq v$ in I

$$p(x) = \frac{p_{a,b}(x) - p_{a,b}(u)}{p_{a,b}(v) - p_{a,b}(u)}, \quad x \in [a, b]. \tag{12}$$

To see that p is independent of a and b, consider any larger interval $[a', b']$ in I, and conclude from the strong Markov property at $\tau_{a,b}$ that, for $x \in [a, b]$,

$$P_x\{\tau_{b'} < \tau_{a'}\} = P_x\{\tau_a < \tau_b\} P_a\{\tau_{b'} < \tau_{a'}\} + P_x\{\tau_b < \tau_a\} P_b\{\tau_{b'} < \tau_{a'}\},$$

or

$$p_{a',b'}(x) = p_{a,b}(x)(p_{a',b'}(b) - p_{a',b'}(a)) + p_{a'b'}(a).$$

Thus, $p_{a,b}$ and $p_{a',b'}$ agree on $[a, b]$ up to an affine transformation and so give rise to the same value in (12).

By Lemma 23.8 the constructed function is continuous and strictly increasing, and it remains to show that $p(X^{\tau_{a,b}})$ is a martingale under P_x for any $a < b$ in I. Since the martingale property is preserved by affine transformations, it is equivalent to show that $p_{a,b}(X^{\tau_{a,b}})$ is a P_x-martingale. Then fix any optional time σ, and write $\tau = \sigma \wedge \tau_{a,b}$. By the strong Markov property at τ we get

$$\begin{aligned}
E_x p_{a,b}(X_\tau) &= E_x P_{X_\tau}\{\tau_b < \tau_a\} = P_x \theta_\tau^{-1}\{\tau_b < \tau_a\} \\
&= P_x\{\tau_b < \tau_a\} = p_{a,b}(x),
\end{aligned}$$

and the desired martingale property follows by Lemma 7.13. $\qquad \square$

To prepare for the next result, consider a Brownian motion B in $\mathbb{R}$ with associated jointly continuous local time L. For any measure ν on $\mathbb{R}$, we may introduce as in (4) the associated additive functional $A = \int L^x \nu(dx)$ and its right-continuous inverse

$$\sigma_t = \inf\{s > 0;\ A_s > t\}, \quad t \geq 0.$$

If $\nu \neq 0$, it is clear from the recurrence of B that A is a.s. unbounded. Hence, $\sigma_t < \infty$ a.s. for all t, and we may define $X_t = B_{\sigma_t}$, $t \geq 0$. We shall refer to $\sigma = (\sigma_t)$ as the *random time-change based on* ν and to the process $X = B \circ \sigma$ as the correspondingly *time-changed Brownian motion*.

Theorem 23.9 (*speed measure and time-change, Feller, Volkonsky, Itô and McKean*) *For any regular diffusion on a natural scale in I, there exists a unique measure ν on I with $\nu[a,b] \in (0,\infty)$ for all $a < b$ in $I°$, such that X is a time-changed Brownian motion based on some extension of ν to $\bar{I}$. Conversely, any such time-change of Brownian motion defines a regular diffusion on I.*

The extended version of ν is called the *speed measure* of the diffusion. Contrary to what the term suggests, we note that the process moves slowly through regions where ν is large. The speed measure of Brownian motion itself is clearly equal to Lebesgue measure. More generally, the speed measure of a regular diffusion solving equation (3) has density σ^{-2}.

To prove the uniqueness of ν we need the following lemma, which is also useful for the subsequent classification of boundary behavior. Here we write $\sigma_{a,b} = \inf\{s > 0;\ B_s \notin (a,b)\}$.

Lemma 23.10 (*Green function*) *Let X be a time-changed Brownian motion based on ν. Then for any measurable function $f \geq 0$ on I and points $a < b$ in $\bar{I}$,*

$$E_x \int_0^{\tau_{a,b}} f(X_t) dt = \int_a^b g_{a,b}(x,y) f(y) \nu(dy), \quad x \in [a,b], \tag{13}$$

where

$$g_{a,b}(x,y) = E_x L^y_{\sigma_{a,b}} = \frac{2(x \wedge y - a)(b - x \vee y)}{b-a}, \quad x,y \in [a,b]. \tag{14}$$

If X is recurrent, this remains true with $a = -\infty$ or $b = \infty$.

Taking $f \equiv 1$ in (13), we get in particular the formula

$$h_{a,b}(x) = E_x \tau_{a,b} = \int_a^b g_{a,b}(x,y) \nu(dy), \quad x \in [a,b], \tag{15}$$

which will be useful later.

Proof: Clearly, $\tau_{a,b} = A(\sigma_{a,b})$ for any $a, b \in \bar{I}$, and also for $a = -\infty$ or $b = \infty$ when X is recurrent. Since L^y is supported by $\{y\}$, it follows by (4) that

$$\int_0^{\tau_{a,b}} f(X_t) dt = \int_0^{\sigma_{a,b}} f(B_s) dA_s = \int_a^b f(y) L^y_{\sigma_{a,b}} \nu(dy).$$

Taking expectations gives (13) with $g_{a,b}(x,y) = E_x L^y_{\sigma_{a,b}}$. To prove (14), we note that by Tanaka's formula and optional sampling

$$E_x L^y_{\sigma_{a,b}\wedge s} = E_x |B_{\sigma_{a,b}\wedge s} - y| - |x - y|, \quad s \geq 0.$$

If a and b are finite, we may let $s \to \infty$ and conclude by monotone and dominated convergence that

$$g_{a,b}(x,y) = \frac{(y-a)(b-x)}{b-a} + \frac{(b-y)(x-a)}{b-a} - |x-y|,$$

which simplifies to (14). The result for infinite a or b follows immediately by monotone convergence. $\square$

The next lemma will enable us to construct the speed measure ν from the functions $h_{a,b}$ in Lemma 23.8.

Lemma 23.11 *(consistency)* *For any regular diffusion on a natural scale in I, there exists a strictly concave function h on I° such that, for any $a < b$ in I,*

$$h_{a,b}(x) = h(x) - \frac{x-a}{b-a} h(b) - \frac{b-x}{b-a} h(a), \quad x \in [a,b]. \tag{16}$$

Proof: Fix any $u < v$ in I, and define for any $a \leq u$ and $b \geq v$ in I

$$h(x) = h_{a,b}(x) - \frac{x-u}{v-u} h_{a,b}(v) - \frac{v-x}{v-u} h_{a,b}(u), \quad x \in [a,b]. \tag{17}$$

To see that h is independent of a and b, consider any larger interval $[a', b']$ in I, and conclude from the strong Markov property at $\tau_{a,b}$ that, for $x \in [a,b]$,

$$E_x \tau_{a',b'} = E_x \tau_{a,b} + P_x\{\tau_a < \tau_b\} E_a \tau_{a',b'} + P_x\{\tau_b < \tau_a\} E_b \tau_{a',b'},$$

or

$$h_{a',b'}(x) = h_{a,b}(x) + \frac{b-x}{b-a} h_{a',b'}(a) + \frac{x-a}{b-a} h_{a',b'}(b). \tag{18}$$

Thus, $h_{a,b}$ and $h_{a',b'}$ agree on $[a,b]$ up to an affine function and therefore yield the same value in (17).

If $a \leq u$ and $b \geq v$, then (17) shows that h and $h_{a,b}$ agree on $[a,b]$ up to an affine function, and (16) follows since $h_{a,b}(a) = h_{a,b}(b) = 0$. The formula extends by means of (18) to arbitrary $a < b$ in I. $\square$

Since h is strictly concave, its left derivative h'_- is strictly decreasing and left-continuous, and so it determines a measure ν on I° satisfying

$$2\nu[a,b) = h'_-(a) - h'_-(b), \quad a < b \text{ in } I^\circ. \tag{19}$$

For motivation, we note that this expression is consistent with (15).

The proof of Theorem 23.9 requires some understanding of the behavior of X at the endpoints of I. If an endpoint b does not belong to I, then by hypothesis the motion terminates when X reaches b. It is clearly equivalent to attach b to I as an absorbing endpoint. For convenience we may then assume that I is a compact interval of the form $[a,b]$, where either endpoint may be *inaccessible*, in the sense that a.s. it cannot be reached in finite time from a point in $I°$.

For either endpoint b, the set $Z_b = \{t \geq 0;\, X_t = b\}$ is regenerative under P_b in the sense of Chapter 22. In particular, we see from Lemma 22.8 that b is either *absorbing*, in the sense that $Z_b = \mathbb{R}_+$ a.s., or *reflecting*, in the sense that $Z_b^\circ = \emptyset$ a.s. In the latter case, we say that the reflection is *fast* if $\lambda Z_b = 0$ and *slow* if $\lambda Z_b > 0$. A more detailed discussion of the boundary behavior will be given after the proof of the main theorem.

We first establish Theorem 23.9 in a special case. The general result will then be deduced by a pathwise comparison.

Proof of Theorem 23.9 for absorbing endpoints (Méléard): Let X have distribution P_x, where $x \in I°$, and put $\zeta = \inf\{t > 0;\, X_t \notin I°\}$. For any $a < b$ in $I°$ with $x \in [a, b]$, the process $X^{\tau_{a,b}}$ is a continuous martingale, and so by Theorem 22.5

$$h(X_t) = h(x) + \int_0^t h'_-(X)dX - \int_I \tilde{L}_t^x \nu(dx), \quad t \in [0, \zeta), \tag{20}$$

where $\tilde{L}$ denotes the local time of X.

Next conclude from Theorem 18.4 that $X = B \circ [X]$ a.s. for some Brownian motion B starting at x. Using Theorem 22.5 twice, we get in particular, for any nonnegative measurable function f,

$$\int_I f(x)\tilde{L}_t^x dx = \int_0^t f(X_s)d[X]_s = \int_0^{[X]_t} f(B_s)ds = \int_I f(x)L_{[X]_t}^x dt,$$

where L denotes the local time of B. Hence, $\tilde{L}_t^x = L_{[X]_t}^x$ a.s. for $t < \zeta$, and so the last term in (20) equals $A_{[X]_t}$ a.s.

For any optional time σ, put $\tau = \sigma \wedge \tau_{a,b}$, and conclude from the strong Markov property that

$$E_x[\tau + h_{a,b}(X_\tau)] = E_x[\tau + E_{X_\tau}\tau_{a,b}]$$
$$= E_x[\tau + \tau_{a,b} \circ \theta_\tau] = E_x\tau_{a,b} = h_{a,b}(x).$$

Writing $M_t = h(X_t) + t$, it follows by Lemma 7.13 that $M^{\tau_{a,b}}$ is a P_x-martingale whenever $x \in [a, b] \subset I°$. Comparing with (20) and using Proposition 17.2, we obtain $A_{[X]_t} = t$ a.s. for all $t \in [0, \zeta)$. Since A is continuous and strictly increasing on $[0, \zeta)$ with inverse σ, it follows that $[X]_t = \sigma_t$ a.s. for $t < \zeta$. The last relation extends to $[\zeta, \infty)$, provided that ν is given infinite mass at each endpoint. Then $X = B \circ \sigma$ a.s. on $\mathbb{R}_+$.

Conversely, it is easily seen that $B \circ \sigma$ is a regular diffusion on I whenever σ is a random time-change based on some measure ν with the stated properties. To prove the uniqueness of ν, fix any $a < x < b$ in $I°$, and apply Lemma 23.10 with $f(y) = (g_{a,b}(x,y))^{-1}$ to see that $\nu(a, b)$ is determined by P_x. □

Proof of Theorem 23.9, general case: Define ν on $I°$ as in (19), and extend the definition to $\bar{I}$ by giving infinite mass to absorbing endpoints.

To every reflecting endpoint we attach a finite mass, to be specified later. Given a Brownian motion B, we note as before that the correspondingly time-changed process $\tilde{X} = B \circ \sigma$ is a regular diffusion on I. Letting $\zeta = \sup\{t; X_t \in I^\circ\}$ and $\tilde{\zeta} = \sup\{t; \tilde{X}_t \in I^\circ\}$, we further see from the previous case that X^ζ and $\tilde{X}^{\tilde\zeta}$ have the same distribution for any starting position $x \in I^\circ$.

Now fix any $a < b$ in I°, and define recursively

$$\chi_1 = \zeta + \tau_{a,b} \circ \theta_\zeta; \qquad \chi_{n+1} = \chi_n + \chi_1 \circ \theta_{\chi_n}, \quad n \in \mathbb{N}.$$

The processes $Y_n^{a,b} = X^\zeta \circ \theta_{\chi_n}$ then form a Markov chain in the path space. A similar construction for $\tilde{X}$ yields some processes $\tilde{Y}_n^{a,b}$, and we note that $(Y_n^{a,b}) \stackrel{d}{=} (\tilde{Y}_n^{a,b})$ for fixed a and b. Since the processes $Y_n^{a',b'}$ for any smaller interval $[a', b']$ can be measurably recovered from those for $[a, b]$ and similarly for $\tilde{Y}_n^{a',b'}$, it follows that the whole collections $(Y_n^{a,b})$ and $(\tilde{Y}_n^{a,b})$ have the same distribution. By Theorem 6.10 we may then assume that the two families agree a.s.

Now assume that $I = [a, b]$, where a is reflecting. From the properties of Brownian motion we note that the level sets Z_a and $\tilde{Z}_a$ for X and $\tilde{X}$ are a.s. perfect. Thus, we may introduce the corresponding excursion point processes ξ and $\tilde{\xi}$, local times L and $\tilde{L}$, and inverse local times T and $\tilde{T}$. Since the excursions within $[a, b)$ agree a.s. for X and $\tilde{X}$, it is clear from the law of large numbers that we may normalize the excursion laws for the two processes such that the corresponding parts of ξ and $\tilde{\xi}$ agree a.s. Then even T and $\tilde{T}$ agree, possibly apart from the lengths of excursions that reach b and the drift coefficient c in Theorem 22.13. For $\tilde{X}$ the latter is proportional to the mass $\nu\{a\}$, which may now be chosen such that c becomes the same as for X. Note that this choice of $\nu\{a\}$ is independent of starting position x for the processes X and $\tilde{X}$.

If the other endpoint b is absorbing, then clearly $X = \tilde{X}$ a.s., and the proof is complete. If b is instead reflecting, then the excursions from b agree a.s. for X and $\tilde{X}$. Repeating the previous argument with the roles of a and b interchanged, we get $X = \tilde{X}$ a.s. after a suitable adjustment of the mass $\nu\{b\}$. □

We proceed to classify the boundary behavior of a regular diffusion on a natural scale in terms of the speed measure ν. A right endpoint b is called an *entrance boundary* for X if b is inaccessible, and yet

$$\liminf_{r \to \infty}\, _{y>x} P_y\{\tau_x \leq r\} > 0, \quad x \in I^\circ. \tag{21}$$

By the Markov property at times nr, $n \in \mathbb{N}$, the limit in (21) then equals 1. In particular, $P_y\{\tau_x < \infty\} = 1$ for all $x < y$ in I°. As we shall see in Theorem 23.13, an entrance boundary is an endpoint where X may enter but not exit.

The opposite situation occurs at an *exit boundary*. By this we mean an endpoint b that is accessible and yet *naturally absorbing*, in the sense that it remains absorbing even when the charge $\nu\{b\}$ is reduced to zero. If b is accessible but not naturally absorbing, we have already seen how the boundary behavior of X depends on the value of $\nu\{b\}$. Thus, b in this case is absorbing when $\nu\{b\} = \infty$, slowly reflecting when $\nu\{b\} \in (0, \infty)$, and fast reflecting when $\nu\{b\} = 0$. For reflecting b it is further clear from Theorem 23.9 that the set $Z_b = \{t \geq 0;\ X_t = b\}$ is a.s. perfect.

Theorem 23.12 *(boundary behavior, Feller)* Let ν be the speed measure of a regular diffusion on a natural scale in some interval $I = [a, b]$, and fix any $u \in I^\circ$. Then

(i) b is accessible iff it is finite with $\int_u^b (b-x)\nu(dx) < \infty$;

(ii) b is accessible and reflecting iff it is finite with $\nu(u, b] < \infty$;

(iii) b is an entrance boundary iff it is infinite with $\int_u^b x\nu(dx) < \infty$.

The stated conditions may be translated into corresponding criteria for arbitrary regular diffusions. In the general case it is clear that exit and other accessible boundaries may be infinite, whereas entrance boundaries may be finite. *Explosion* is said to occur when X reaches an infinite boundary point in finite time. An interesting example of a regular diffusion on $(0, \infty)$ with 0 as an entrance boundary is given by the Bessel process $X_t = |B_t|$, where B is a Brownian motion in $\mathbb{R}^d$ with $d \geq 2$.

Proof of Theorem 23.12: (i) Since $\limsup_s(\pm B_s) = \infty$ a.s., Theorem 23.9 shows that X cannot explode, so any accessible endpoint is finite. Now assume that $a < c < u < b < \infty$. Then Lemma 23.8 shows that b is accessible iff $h_{c,b}(u) < \infty$, which by (15) is equivalent to $\int_u^b (b-x)\nu(dx) < \infty$.

(ii) In this case $b < \infty$ by (i), and then Lemma 23.2 shows that b is absorbing iff $\nu(u, b] = \infty$.

(iii) An entrance boundary b is inaccessible by definition, and therefore $\tau_u = \tau_{u,b}$ a.s. when $a < u < b$. Arguing as in the proof of Lemma 23.8, we also note that $E_y \tau_u$ is bounded for $y > u$. If $b < \infty$, we obtain the contradiction $E_y \tau_u = h_{u,b}(y) = \infty$, and so b must be infinite. From (15) we get by monotone convergence as $y \to \infty$

$$E_y \tau_u = h_{u,\infty}(y) = 2 \int_u^\infty (x \wedge y - u)\nu(dx) \to 2 \int_u^\infty (x - u)\nu(dx),$$

which is finite iff $\int_u^\infty x\nu(dx) < \infty$. $\square$

We proceed to establish an important regularity property, which also clarifies the nature of entrance boundaries.

Theorem 23.13 *(entrance laws and Feller properties)* Given a regular diffusion on I, form an extended interval $\bar{I}$ by attaching the possible entrance boundaries to I. Then the original diffusion extends to a continuous Feller process on $\bar{I}$.

Proof: For any $f \in C_b$, $a, x \in I$, and $r, t \geq 0$, we get by the strong Markov property at $\tau_x \wedge r$

$$\begin{aligned} E_a f(X_{\tau_x \wedge r + t}) &= E_a T_t f(X_{\tau_x \wedge r}) \\ &= T_t f(x) P_a\{\tau_x \leq r\} + E_a[T_t f(X_r); \tau_x > r]. \end{aligned} \quad (22)$$

To show that $T_t f$ is left-continuous at some $y \in I$, fix any $a < y$ in $I°$, and choose $r > 0$ so large that $P_a\{\tau_y \leq r\} > 0$. As $x \uparrow y$, we have $\tau_x \uparrow \tau_y$ and hence $\{\tau_x \leq r\} \downarrow \{\tau_y \leq r\}$. Thus, the probabilities and expectations in (22) converge to the corresponding expressions for τ_y, and we get $T_t f(x) \to T_t f(y)$. The proof of the right-continuity is similar.

If an endpoint b is inaccessible but not of entrance type, and if $f(x) \to 0$ as $x \to b$, then clearly even $T_t f(x) \to 0$ at b for each $t > 0$. Now assume that ∞ is an entrance boundary, and consider a function f with a finite limit at ∞. We need to show that even $T_t f(x)$ converges as $x \to \infty$ for fixed t. Then conclude from Lemma 23.10 that as $a \to \infty$,

$$\sup_{x \geq a} E_x \tau_a = 2 \sup_{x \geq a} \int_a^\infty (x \wedge r - a) \nu(dr) = 2 \int_a^\infty (r - a) \nu(dr) \to 0. \quad (23)$$

Next we note that, for any $a < x < y$ and $r \geq 0$,

$$\begin{aligned} P_y\{\tau_a \leq r\} &\leq P_y\{\tau_x \leq r, \tau_a - \tau_x \leq r\} \\ &= P_y\{\tau_x \leq r\} P_x\{\tau_a \leq r\} \leq P_x\{\tau_a \leq r\}. \end{aligned}$$

Thus $P_x \circ \tau_a^{-1}$ converges vaguely as $x \to \infty$ for fixed a, and in view of (23) the convergence holds even in the weak sense.

Now fix any t and f, and introduce for each a the continuous function $g_a(s) = E_a f(X_{(t-s)_+})$. By the strong Markov property at $\tau_a \wedge t$ and Theorem 6.4 we get for any $x, y \geq a$

$$|T_t f(x) - T_t f(y)| \leq |E_x g_a(\tau_a) - E_y g_a(\tau_a)| + 2\|f\|(P_x + P_y)\{\tau_a > t\}.$$

Here the right-hand side tends to zero as $x, y \to \infty$ and then $a \to \infty$, because of (23) and the weak convergence of $P_x \circ \tau_a^{-1}$. Thus, $T_t f(x)$ is Cauchy convergent as $x \to \infty$, and we may denote the limit by $T_t f(\infty)$.

It is now easy to check that the extended operators T_t form a Feller semigroup on $C_0(\bar{I})$. Finally, it is clear from Theorem 19.15 that the associated process starting at a possible entrance boundary again has a continuous version, in the topology of $\bar{I}$. □

We proceed to establish a ratio ergodic theorem for elementary additive functionals of a recurrent diffusion. It is instructive to compare with the general ratio limit theorems of Chapter 20.

Theorem 23.14 *(ratio ergodic theorem, Derman, Motoo and Watanabe)* Let X be a regular, recurrent diffusion on a natural scale and with speed measure ν. Then for any measurable functions $f, g \geq 0$ on I with $\nu f < \infty$ and $\nu g > 0$,

$$\lim_{t \to \infty} \frac{\int_0^t f(X_s)ds}{\int_0^t g(X_s)ds} = \frac{\nu f}{\nu g} \quad a.s. \ P_x, \quad x \in I.$$

Proof: Fix any $a < b$ in I, put $\tau_a^b = \tau_b + \tau_a \circ \theta_{\tau_b}$, and define recursively some optional times $\sigma_0, \sigma_1, \ldots$ by

$$\sigma_{n+1} = \sigma_n + \tau_a^b \circ \theta_{\sigma_n}, \quad n \geq 0,$$

starting with $\sigma_0 = \tau_a$. Write

$$\int_0^{\sigma_n} f(X_s)ds = \int_0^{\sigma_0} f(X_s)ds + \sum_{k=1}^n \int_{\sigma_{k-1}}^{\sigma_k} f(X_s)ds, \qquad (24)$$

and note that the terms of the last sum are i.i.d. By the strong Markov property and Lemma 23.10, we get for any $x \in I$

$$\begin{aligned}
E_x \int_{\sigma_{k-1}}^{\sigma_k} f(X_s)ds &= E_a \int_0^{\tau_b} f(X_s)ds + E_b \int_0^{\tau_a} f(X_s)ds \\
&= \int f(y)\{g_{-\infty,b}(y,a) + g_{a,\infty}(y,b)\}\nu(dy) \\
&= 2\int f(y)\{(b - y \vee a)_+ + (y \wedge b - a)_+\}\nu(dy) \\
&= 2(b-a)\nu f.
\end{aligned}$$

From the same lemma, we also see that the first term in (24) is a.s. finite. Hence, by the law of large numbers

$$\lim_{n \to \infty} n^{-1} \int_0^{\sigma_n} f(X_s)ds = 2(b-a)\nu f \quad a.s. \ P_x, \quad x \in I.$$

Writing $\kappa_t = \sup\{n \geq 0; \sigma_n \leq t\}$, we get by monotone interpolation

$$\lim_{t \to \infty} \kappa_t^{-1} \int_0^t f(X_s)ds = 2(b-a)\nu f \quad a.s. \ P_x, \quad x \in I. \qquad (25)$$

This remains true when $\nu f = \infty$, since we can then apply (25) to some approximating functions $f_n \uparrow f$ with $\nu f_n < \infty$ and let $n \to \infty$. The assertion now follows as we apply (25) to both f and g. □

We may finally classify the asymptotic behavior of the process, according to the boundedness of the speed measure ν and the nature of the endpoints. For convenience, we may first apply an affine mapping to transforms I° into one of the intervals $(0,1)$, $(0,\infty)$, or (∞,∞). Since finite endpoints may be either inaccessible, absorbing, or reflecting—represented below by the brackets (, [, and [[, respectively—we need to distinguish between ten different cases.

We say that a diffusion is ν-ergodic if it is recurrent and such that $P_x \circ X_t^{-1} \xrightarrow{w} \nu/\nu I$ for all x. A recurrent diffusion may be either *null-recurrent* or *positive recurrent*, depending on whether $|X_t| \xrightarrow{P} \infty$ or not. Let us also recall that *absorption* occurs at an endpoint b whenever $X_t = b$ for all sufficiently large t.

Theorem 23.15 *(recurrence and ergodicity, Feller, Maruyama and Tanaka) For any regular diffusion on a natural scale and with speed measure ν, the ergodic behavior is the following, depending on initial position x and the nature of the boundaries:*

$(-\infty, \infty)$: ν-*ergodic if ν is bounded, otherwise null-recurrent;*

$(0, \infty)$: *converges to 0 a.s.;*

$[0, \infty)$: *absorbed at 0 a.s.;*

$[[0, \infty)$: ν-*ergodic if ν is bounded, otherwise null-recurrent;*

$(0, 1)$: *converges to 0 or 1 with probabilities $1 - x$ and x, respectively;*

$[0, 1)$: *absorbed at 0 or converges to 1 with probabilities $1 - x$ and x, respectively;*

$[0, 1]$: *absorbed at 0 or 1 with probabilities $1 - x$ and x, respectively;*

$[[0, 1)$: *converges to 1 a.s.;*

$[[0, 1]$: *absorbed at 1 a.s.;*

$[[0, 1]]$: ν-*ergodic.*

We begin our proof with the relatively elementary recurrence properties, which distinguish between the possibilities of absorption, convergence, and recurrence.

Proof of recurrence properties:

$[0, 1]$: Relation (10) yields $P_x\{\tau_0 < \infty\} = 1 - x$ and $P_x\{\tau_1 < \infty\} = x$.

$[0, \infty)$: By (10) we have for any $b > x$

$$P_x\{\tau_0 < \infty\} \geq P_x\{\tau_0 < \tau_b\} = (b - x)/b,$$

which tends to 1 as $b \to \infty$.

$(-\infty, \infty)$: The recurrence follows from the previous case.

$[[0, \infty)$: Since 0 is reflecting, we have $P_0\{\tau_y < \infty\} > 0$ for some $y > 0$. By the strong Markov property and the regularity of X, this extends to arbitrary y. Arguing as in the proof of Lemma 23.8, we may conclude that $P_0\{\tau_y < \infty\} = 1$ for all $y > 0$. The asserted recurrence now follows, as we combine with the statement for $[0, \infty)$.

$(0, \infty)$: In this case $X = B \circ [X]$ a.s. for some Brownian motion B. Since $X > 0$, we have $[X]_\infty < \infty$ a.s., and therefore X converges a.s. Now $P_y\{\tau_{a,b} < \infty\} = 1$ for any $0 < a \leq y \leq b$. Applying the Markov property at an arbitrary time $t > 0$, we conclude that a.s. either

$\liminf_t X_t \leq a$ or $\limsup_t X_t \geq b$. Since a and b are arbitrary, it follows that X_∞ is an endpoint of $(0, \infty)$ and hence equals 0.

$(0, 1)$: Arguing as in the previous case, we get a.s. convergence to either 0 or 1. To find the corresponding probabilities, we conclude from (10) that

$$P_x\{\tau_a < \infty\} \geq P_x\{\tau_a < \tau_b\} = \frac{b-x}{b-a}, \quad 0 < a < x < b < 1.$$

Letting $b \to 1$ and then $a \to 0$, we obtain $P_x\{X_\infty = 0\} \geq 1 - x$. Similarly, $P_x\{X_\infty = 1\} \geq x$, and so equality holds in both relations.

$[0, 1)$: Again X converges to either 0 or 1 with probabilities $1 - x$ and x, respectively. Furthermore, we note that

$$P_x\{\tau_0 < \infty\} \geq P_x\{\tau_0 < \tau_b\} = (b-x)/b, \quad 0 \leq x < b < 1,$$

which tends to $1 - x$ as $b \to 1$. Thus, X gets absorbed when it approaches 0.

$[[0, 1]]$: Arguing as in the previous case, we get $P_0\{\tau_1 < \infty\} = 1$, and by symmetry we also have $P_1\{\tau_0 < \infty\} = 1$.

$[[0, 1]$: Again we get $P_0\{\tau_1 < \infty\} = 1$, so the same relation holds for P_x.

$[[0, 1)$: As before, we get $P_0\{\tau_b < \infty\} = 1$ for all $b \in (0, 1)$. By the strong Markov property at τ_b and the result for $[0, 1)$ it follows that $P_0\{X_t \to 1\} \geq b$. Letting $b \to 1$, we obtain $X_t \to 1$ a.s. under P_0. The result for P_x now follows by the strong Markov property at τ_x, applied under P_0. □

The ergodic properties will be proved along the lines of Theorem 8.18, which requires some additional lemmas.

Lemma 23.16 *(coupling)* *If X and Y are independent Feller processes, then the pair (X, Y) is again Feller.*

Proof: Use Theorem 4.29 and Lemma 19.3. □

The next result is a continuous-time counterpart of Lemma 8.20.

Lemma 23.17 *(strong ergodicity)* *Given a regular, recurrent diffusion, we have for any initial distributions μ_1 and μ_2*

$$\lim_{t \to \infty} \|P_{\mu_1} \circ \theta_t^{-1} - P_{\mu_2} \circ \theta_t^{-1}\| = 0.$$

Proof: Let X and Y be independent with distributions P_{μ_1} and P_{μ_2}, respectively. By Theorem 23.13 and Lemma 23.16 the pair (X, Y) can be extended to a Feller diffusion, and so by Theorem 19.17 it is again strong Markov with respect to the induced filtration $\mathcal{G}$. Define $\tau = \inf\{t \geq 0; X_t = Y_t\}$, and note that τ is $\mathcal{G}$-optional by Lemma 7.6. The assertion now follows as in case of Lemma 8.20, provided we can show that $\tau < \infty$ a.s.

To see this, assume first that $I = \mathbb{R}$. The processes X and Y are then continuous local martingales. By independence they remain local martingales for the extended filtration $\mathcal{G}$, and so even $X - Y$ is a local $\mathcal{G}$-martingale. Using the independence and recurrence of X and Y, we get $[X - Y]_\infty = [X]_\infty + [Y]_\infty = \infty$ a.s., which shows that even $X - Y$ is recurrent. In particular, $\tau < \infty$ a.s.

Next let $I = [[0, \infty)$ or $[[0, 1]]$, and define $\tau_1 = \inf\{t \geq 0;\ X_t = 0\}$ and $\tau_2 = \inf\{t \geq 0;\ Y_t = 0\}$. By the continuity and recurrence of X and Y, we get $\tau \leq \tau_1 \vee \tau_2 < \infty$ a.s. □

Our next result is similar to the discrete-time version in Lemma 8.21.

Lemma 23.18 *(existence) Every regular, positive recurrent diffusion has an invariant distribution.*

Proof: By Theorem 23.13 we may regard the transition kernels μ_t with associated operators T_t as defined on $\bar{I}$, the interval I with possible entrance boundaries adjoined. Since X is not null-recurrent, we may choose a bounded Borel set B and some $x_0 \in I$ and $t_n \to \infty$ such that $\inf_n \mu_{t_n}(x_0, B) > 0$. By Theorem 5.19 there exists some measure μ on $\bar{I}$ with $\mu I > 0$ such that $\mu_{t_n}(x_0, \cdot) \xrightarrow{v} \mu$ along a subsequence, in the topology of $\bar{I}$. The convergence extends by Lemma 23.17 to arbitrary $x \in I$, and so

$$T_{t_n} f(x) \to \mu f, \quad f \in C_0(\bar{I}),\ x \in I. \tag{26}$$

Now fix any $h \geq 0$ and $f \in C_0(\bar{I})$, and note that even $T_h f \in C_0(\bar{I})$ by Theorem 23.13. Using (26), the semigroup property, and dominated convergence, we get for any $x \in I$

$$\mu(T_h f) \leftarrow T_{t_n}(T_h f)(x) = T_h(T_{t_n} f)(x) \to \mu f.$$

Thus, $\mu \mu_h = \mu$ for all h, which means that μ is invariant on $\bar{I}$. In particular, $\mu(\bar{I} \setminus I) = 0$ by the nature of entrance boundaries, and so the normalized measure $\mu/\mu I$ is an invariant distribution on I. □

Our final lemma provides the crucial connection between speed measure and invariant distributions.

Lemma 23.19 *(positive recurrence) For a regular, recurrent diffusion on a natural scale and with speed measure ν, these conditions are equivalent:*

(i) $\nu I < \infty$;

(ii) *the process is positive recurrent;*

(iii) *an invariant distribution exists.*

The invariant distribution is then unique and equals $\nu/\nu I$.

Proof: If the process is null-recurrent, then clearly no invariant distribution exists. The converse is also true by Lemma 23.18, and so (ii) and (iii) are equivalent. Now fix any bounded, measurable function $f: I \to \mathbb{R}_+$ with bounded support. By Theorem 23.14, Fubini's theorem, and dominated convergence, we have for any distribution μ on I

$$t^{-1}\int_0^t E_\mu f(X_s)ds = E_\mu t^{-1}\int_0^t f(X_s)ds \to \frac{\nu f}{\nu I}.$$

If μ is invariant, we get $\mu f = \nu f/\nu I$, and so $\nu I < \infty$. If instead X is null-recurrent, then $E_\mu f(X_s) \to 0$ as $s \to \infty$, and we get $\nu f/\nu I = 0$, which implies $\nu I = \infty$. □

End of proof of Theorem 23.15: It remains to consider the cases when I is either (∞,∞), $[[0,\infty)$, or $[[0,1]]$, since we have otherwise convergence or absorption at some endpoint. In case of $[[0,1]]$ we note from Theorem 23.12 (ii) that ν is bounded. In the remaining cases ν may be unbounded, and then X is null-recurrent by Lemma 23.19. If ν is bounded, then $\mu = \nu/\nu I$ is invariant by the same lemma, and the asserted ν-ergodicity follows from Lemma 23.17 with $\mu_1 = \mu$. □

Exercises

1. Prove pathwise uniqueness for the SDE $dX_t = (X_t^+)^{1/2}dB_t + cdt$ with $c > 0$. Also show that the solutions X^x with $X_0^x = x$ satisfy $X_t^x < X_t^y$ a.s. for $x < y$ up to the time when X^x reaches 0.

2. Let X be Brownian motion in $\mathbb{R}^d$, absorbed at 0. Show that $Y = |X|^2$ is a regular diffusion on $(0,\infty)$, describe its boundary behavior for different d, and identify the corresponding case of Theorem 23.15. Verify the conclusion by computing the associated scale function and speed measure.

3. Show that solutions to equation $dX_t = \sigma(X_t)dB_t$ cannot explode. (*Hint:* If X explodes at time $\zeta < \infty$, then $[X]_\zeta = \infty$, and the local time of X tends to ∞ as $t \to \zeta$, uniformly on compacts. Now use Theorem 22.5 to see that $\zeta = \infty$ a.s.)

4. Assume in Theorem 23.1 that $S_\sigma = N_\sigma$. Show that the solutions X to (3) form a regular diffusion on a natural scale on every connected component I of S_σ. Also note that the endpoints of I are either absorbing or exit boundaries for X. (*Hint:* Use Theorems 21.11, 22.4, and 22.5, and show that the exit time from any compact interval $J \subset I$ is finite.)

5. Assume in Theorem 23.1 that $S_\sigma \subset N_\sigma$, and form $\tilde\sigma$ from σ by taking $\tilde\sigma(x) = 1$ on $A = N_\sigma \setminus S_\sigma$. Show that any solution X to equation $(\tilde\sigma, 0)$ also solves equation $(\sigma, 0)$, but not conversely unless $A = \emptyset$. (*Hint:* Since $\lambda A = 0$, we have $\int 1_A(X_t)dt = \int 1_A(X_t)d[X]_t = 0$ a.s. by Theorem 22.5.)

6. Assume in Theorem 23.1 that $S_\sigma \subset N_\sigma$. Show that equation $(\sigma, 0)$ has solutions that form a regular diffusion on every connected component of S_σ^c. Prove the corresponding statement for the connected components of N_σ^c when N_σ is closed. (*Hint:* For S_σ^c, use the preceding result. For N_σ^c, take X to be absorbed when it first reaches N_σ.)

7. In the setting of Theorem 23.14, show that the stated relation implies the convergence in Corollary 20.8 (i). Also use the result to prove a law of large numbers for regular, recurrent diffusions with bounded speed measure ν. (*Hint:* Note that $\nu g > 0$ implies $\int g(X_s)ds > 0$ a.s.)

Chapter 24

Connections with PDEs and Potential Theory

Backward equation and Feynman–Kac formula; uniqueness for SDEs from existence for PDEs; harmonic functions and Dirichlet's problem; Green functions as occupation densities; sweeping and equilibrium problems; dependence on conductor and domain; time reversal; capacities and random sets

In Chapters 19 and 21 we saw how elliptic differential operators arise naturally in probability theory as the generators of nice diffusion processes. This fact is the ultimate cause of some profound connections between probability theory and partial differential equations (PDEs). In particular, a suitable extension of the operator $\frac{1}{2}\Delta$ appears as the generator of Brownian motion in $\mathbb{R}^d$, which leads to a close relationship between classical potential theory and the theory of Brownian motion. More specifically, many basic problems in potential theory can be solved by probabilistic methods, and, conversely, various hitting distributions for Brownian motion can be given a potential theoretic interpretation.

This chapter explores some of the mentioned connections. First we derive the celebrated Feynman–Kac formula and show how *existence* of solutions to a given Cauchy problem implies *uniqueness* of solutions to the associated SDE. We then proceed with a probabilistic construction of Green functions and potentials and solve the Dirichlet, sweeping, and equilibrium problems of classical potential theory in terms of Brownian motion. Finally, we show how Green capacities and alternating set functions can be represented in a natural way in terms of random sets.

Some stochastic calculus from Chapters 17 and 21 is used at the beginning of the chapter, and we also rely on the theory of Feller processes from Chapter 19. As for Brownian motion, the present discussion is essentially self-contained, apart from some elementary facts cited from Chapters 13 and 18. Occasionally we refer to Chapters 4 and 16 for some basic weak convergence theory. Finally, the results at the end of the chapter require the existence of Poisson processes from Proposition 12.5, as well as some basic facts about the Fell topology listed in Theorem A2.5. Potential theoretic ideas are used in several other chapters, and additional, though essentially unrelated, results appear in especially Chapters 20, 22, and 25.

24. Connections with PDEs and Potential Theory

To begin with the general PDE connections, we consider an arbitrary Feller diffusion in $\mathbb{R}^d$ with associated semigroup operators T_t and generator $(A, \mathcal{D})$. Recall from Theorem 19.6 that, for any $f \in \mathcal{D}$, the function

$$u(t, x) = T_t f(x) = E_x f(X_t), \quad t \geq 0, \ x \in \mathbb{R}^d,$$

satisfies *Kolmogorov's backward equation* $\dot{u} = Au$, where $\dot{u} = \partial u / \partial t$. Thus, u provides a probabilistic solution to the *Cauchy problem*

$$\dot{u} = Au, \qquad u(0, x) = f(x). \tag{1}$$

Let us now add a *potential* term vu to (1), where $v : \mathbb{R}^d \to \mathbb{R}_+$, and consider the more general problem

$$\dot{u} = Au - vu, \qquad u(0, x) = f(x). \tag{2}$$

Here the solution may be expressed in terms of the elementary *multiplicative functional* e^{-V}, where

$$V_t = \int_0^t v(X_s) ds, \quad t \geq 0.$$

Let $C^{1,2}$ denote the class of functions $f : \mathbb{R}_+ \times \mathbb{R}^d$ that are of class C^1 in the time variable and of class C^2 in the space variables. Write $C_b(\mathbb{R}^d)$ and $C_b^+(\mathbb{R}^d)$ for the classes of bounded, continuous functions from $\mathbb{R}^d$ to $\mathbb{R}$ and $\mathbb{R}_+$, respectively.

Theorem 24.1 *(Cauchy problem, Feynman, Kac)* Let $(A, \mathcal{D})$ be the generator of a Feller diffusion in $\mathbb{R}^d$, and fix any $f \in C_b(\mathbb{R}^d)$ and $v \in C_b^+(\mathbb{R}^d)$. Then any bounded solution $u \in C^{1,2}$ to (2) is given by

$$u(t, x) = E_x e^{-V_t} f(X_t), \quad t \geq 0, \ x \in \mathbb{R}^d. \tag{3}$$

Conversely, (3) solves (2) whenever $f \in \mathcal{D}$.

The expression in (3) has an interesting interpretation in terms of *killing*. To see this, we may introduce an exponential random variable $\gamma \perp\!\!\!\perp X$ with mean 1, and define $\zeta = \inf\{t \geq 0; V_t > \gamma\}$. Letting $\tilde{X}$ denote the process X killed at time ζ, we may express the right-hand side of (3) as $E_x f(\tilde{X}_t)$, with the understanding that $f(\tilde{X}_t) = 0$ when $t \geq \zeta$. In other words, $u(t, x) = \tilde{T}_t f(x)$, where $\tilde{T}_t$ is the transition operator of the killed process. It is easy to verify directly from (3) that the family $(\tilde{T}_t)$ is again a Feller semigroup.

Proof of Theorem 24.1: Assume that $u \in C^{1,2}$ is bounded and solves (2), and define for fixed $t > 0$

$$M_s = e^{-V_s} u(t - s, X_s), \quad s \in [0, t].$$

Letting $\stackrel{m}{\sim}$ denote equality apart from a continuous local martingale or its differential, we see from Lemma 19.21, Itô's formula, and (2) that, for any $s < t$,

$$\begin{aligned} dM_s &= e^{-V_s} \{ du(t - s, X_s) - u(t - s, X_s) v(X_s) ds \} \\ &\stackrel{m}{\sim} e^{-V_s} \{ Au(t - s, X_s) - \dot{u}(t - s, X_s) - u(t - s, X_s) v(X_s) \} ds = 0. \end{aligned}$$

Thus, M is a continuous local martingale on $[0, t)$. Since M is bounded, the martingale property extends to t, and we get

$$u(t, x) = E_x M_0 = E_x M_t = E_x u(0, X_t) = E_x e^{-V_t} f(X_t).$$

Next let u be given by (3) for some $f \in \mathcal{D}$. Integrating by parts and using Lemma 19.21, we obtain

$$\begin{aligned} d\{e^{-V_t} f(X_t)\} &= e^{-V_t}\{df(X_t) - (vf)(X_t)dt\} \\ &\stackrel{m}{\sim} e^{-V_t}(Af - vf)(X_t)dt. \end{aligned}$$

Taking expectations and differentiating at $t = 0$, we conclude that the generator of the semigroup $\tilde{T}_t f(x) = E_x f(\tilde{X}_t) = u(t, x)$ equals $\tilde{A} = A - v$ on $\mathcal{D}$. Equation (2) now follows by the last assertion in Theorem 19.6. □

The converse part of Theorem 24.1 can often be improved in special cases. In particular, if $v = 0$ and $A = \frac{1}{2}\Delta = \frac{1}{2}\sum_i \partial^2/\partial x_i^2$, so that X is a Brownian motion and (2) reduces to the standard *heat equation*, then $u(t, x) = E_x f(X_t)$ solves (2) for any bounded, continuous function f on $\mathbb{R}^d$. To see this, we note that $u \in C^{1,2}$ on $(0, \infty) \times \mathbb{R}^d$ because of the smoothness of the Brownian transition density. We may then obtain (2) by applying the backward equation to the function $T_h f(x)$ for a fixed $h \in (0, t)$.

Let us now consider an SDE in $\mathbb{R}^d$ of the form

$$dX_t^i = \sigma_j^i(X_t) dB_t^j + b^i(X_t) dt, \tag{4}$$

and introduce the associated elliptic operator

$$Av(x) = \tfrac{1}{2} a^{ij}(x) v''_{ij}(x) + b^i(x) v'_i(x), \quad x \in \mathbb{R}^d, \ v \in C^2,$$

where $a^{ij} = \sigma_k^i \sigma_k^j$. The next result shows how *uniqueness* in law for solutions to (4) may be inferred from the *existence* of solutions to the associated Cauchy problem (1).

Theorem 24.2 *(uniqueness, Stroock and Varadhan) If for every $f \in C_0^\infty(\mathbb{R}^d)$ the Cauchy problem in (1) has a bounded solution on $[0, \varepsilon] \times \mathbb{R}^d$ for some $\varepsilon > 0$, then uniqueness in law holds for the SDE (4).*

Proof: Fix any $f \in C_0^\infty$ and $t \in (0, \varepsilon]$, and let u be a bounded solution to (1) on $[0, t] \times \mathbb{R}^d$. If X solves (4), we note as before that $M_s = u(t - s, X_s)$ is a martingale on $[0, t]$, and so

$$Ef(X_t) = Eu(0, X_t) = EM_t = EM_0 = Eu(t, X_0).$$

Thus, the one-dimensional distributions of X on $[0, \varepsilon]$ are uniquely determined by the initial distribution.

Now assume that X and Y are solutions with the same initial distribution. To prove that their finite-dimensional distributions agree, it is enough to consider times $0 = t_0 < t_1 < \cdots < t_n$ such that $t_k - t_{k-1} \leq \varepsilon$ for all k. Assume that the distributions agree at $t_0, \ldots, t_{n-1} = t$, and fix any set $C = \pi_{t_0,\ldots,t_{n-1}}^{-1} B$ with $B \in \mathcal{B}^{nd}$. By Theorem 21.7, both $\mathcal{L}(X)$ and $\mathcal{L}(Y)$ solve the local martingale problem for (a, b). If $P\{X \in C\} = P\{Y \in C\} >$

0, we see as in case of Theorem 21.11 that the same property holds for the conditional measures $P[\theta_t X \in \cdot | X \in C]$ and $P[\theta_t Y \in \cdot | Y \in C]$. Since the corresponding initial distributions agree by hypothesis, the one-dimensional result yields the extension

$$P\{X \in C,\, X_{t+h} \in \cdot\} = P\{Y \in C,\, Y_{t+h} \in \cdot\}, \quad h \in (0, \varepsilon].$$

In particular, the distributions agree at times $t_0, \ldots, t_n$. The general result now follows by induction. □

Let us now specialize to the case when X is Brownian motion in $\mathbb{R}^d$. For any closed set $B \subset \mathbb{R}^d$, we introduce the *hitting time* $\tau_B = \inf\{t > 0;\, X_t \in B\}$ and associated *hitting kernel*

$$H_B(x, dy) = P_x\{\tau_B < \infty,\, X_{\tau_B} \in dy\}, \quad x \in \mathbb{R}^d.$$

For suitable functions f, we write $H_B f(x) = \int f(y) H_B(x, dy)$.

By a *domain* in $\mathbb{R}^d$ we mean an open, connected subset $D \subset \mathbb{R}^d$. A function $u : D \to \mathbb{R}$ is said to be *harmonic* if it belongs to $C^2(D)$ and satisfies the *Laplace equation* $\Delta u = 0$. We also say that u has the *mean-value property* if it is locally bounded and measurable, and such that for any ball $B \subset D$ with center x, the average of u over the boundary ∂B equals $u(x)$. The following analytic result is crucial for the probabilistic developments.

Lemma 24.3 *(harmonic functions, Gauss, Koebe)* *A function u on a domain $D \subset \mathbb{R}^d$ is harmonic iff it has the mean-value property, in which case $u \in C^\infty(D)$.*

Proof: First assume that $u \in C^2(D)$, and fix a ball $B \subset D$ with center x. Writing $\tau = \tau_{\partial B}$ and noting that $E_x \tau < \infty$, we get by Itô's formula

$$E_x u(X_\tau) - u(x) = \tfrac{1}{2} E_x \int_0^\tau \Delta u(X_s) ds.$$

Here the first term on the left equals the average of u over ∂B, due to the spherical symmetry of Brownian motion. If u is harmonic, then the right-hand side vanishes, and the mean-value property follows. If instead u is not harmonic, we may choose B such that $\Delta u \neq 0$ on B. But then the right-hand side is nonzero, and so the mean-value property fails.

It remains to show that every function u with the mean-value property is infinitely differentiable. Then fix any infinitely differentiable and spherically symmetric probability density φ, supported by a ball of radius $\varepsilon > 0$ around the origin. The mean-value property yields $u = u * \varphi$ on the set where the right-hand side is defined, and by dominated convergence the infinite differentiability of φ carries over to $u * \varphi = u$. □

Before proceeding to the potential theoretic developments, we need to introduce a regularity condition on the domain D. Writing $\zeta = \zeta_D = \tau_{D^c}$, we note that $P_x\{\zeta = 0\} = 0$ or 1 for every $x \in \partial D$ by Corollary 19.18. When this probability is 1, we say that x is *regular for D^c* or simply *regular*.

If this holds for every $x \in \partial D$, then the boundary ∂D is said to be regular and we refer to D as a *regular domain*.

Regularity is a fairly weak condition. In particular, any domain with a smooth boundary is regular, and we shall see that even various edges and corners are allowed, provided they are not too sharp and directed inward. By a *spherical cone* in $\mathbb{R}^d$ with *vertex* v and *axis* $a \neq 0$ we mean a set of the form $C = \{x; \langle x - v, a \rangle \geq c|x - v|\}$, where $c \in (0, |a|]$.

Lemma 24.4 (cone condition, Zaremba) *Given a domain $D \subset \mathbb{R}^d$, let $x \in \partial D$ be such that $C \cap G \subset D^c$ for some some spherical cone C with vertex x and some neighborhood G of x. Then x is regular for D^c.*

Proof: By compactness of the unit sphere in $\mathbb{R}^d$, we may cover $\mathbb{R}^d$ by $C_1 = C$ along with finitely many congruent cones $C_2, \ldots, C_n$ with vertex x. By rotational symmetry

$$1 = P_x\{\min_{k \leq n} \tau_{C_k} = 0\} \leq \sum_{k \leq n} P_x\{\tau_{C_k} = 0\} = n P_x\{\tau_C = 0\},$$

and so $P_x\{\tau_C = 0\} > 0$. Hence, Corollary 19.18 yields $P\{\tau_C = 0\} = 1$, and we get $\zeta_D \leq \tau_{C \cap G} = 0$ a.s. P_x. $\square$

Now fix a domain $D \subset \mathbb{R}^d$ and a continuous function $f : \partial D \to \mathbb{R}$. A function u on $\overline{D}$ is said to solve the *Dirichlet problem* (D, f), if u is harmonic on D and continuous on $\overline{D}$ with $u = f$ on ∂D. The solution may be interpreted as the electrostatic potential in D when the potential on the boundary is given by f.

Theorem 24.5 (Dirichlet problem, Kakutani, Doob) *For any regular domain $D \subset \mathbb{R}^d$ and function $f \in C_b(\partial D)$, the Dirichlet problem (D, f) is solved by the function*

$$u(x) = E_x[f(X_{\zeta_D}); \zeta_D < \infty] = H_{D^c} f(x), \quad x \in \overline{D}. \tag{5}$$

If $\zeta_D < \infty$ a.s., then this is the only bounded solution; when $d \geq 3$ and $f \in C_0(\partial D)$, it is the only solution in $C_0(\overline{D})$.

Thus, H_{D^c} agrees with the *sweeping (balayage) kernel* of Newtonian potential theory, which determines the *harmonic measure* on ∂D. The following result clarifies the role of the regularity condition on ∂D.

Lemma 24.6 (regularity, Doob) *A point $b \in \partial D$ is regular for D^c iff, for any $f \in C_b(\partial D)$, the function u in (5) satisfies $u(x) \to f(b)$ as $D \ni x \to b$.*

Proof: First assume that b is regular. For any $t > h > 0$ and $x \in D$, we get by the Markov property

$$P_x\{\zeta > t\} \leq P_x\{\zeta \circ \theta_h > t - h\} = E_x P_{X_h}\{\zeta > t - h\}.$$

Here the right-hand side is continuous in x, by the continuity of the Gaussian kernel and dominated convergence, and so

$$\limsup_{x \to b} P_x\{\zeta > t\} \leq E_b P_{X_h}\{\zeta > t - h\} = P_b\{\zeta \circ \theta_h > t - h\}.$$

As $h \to 0$, the probability on the right tends to $P_b\{\zeta > t\} = 0$, and so $P_x\{\zeta > t\} \to 0$ as $x \to b$, which means that $P_x \circ \zeta^{-1} \xrightarrow{w} \delta_0$. Since also $P_x \xrightarrow{w} P_b$ in $C(\mathbb{R}_+, \mathbb{R}^d)$, Theorem 4.28 yields $P_x \circ (X, \zeta)^{-1} \xrightarrow{w} P_b \circ (X, 0)^{-1}$ in $C(\mathbb{R}_+, \mathbb{R}^d) \times [0, \infty]$. By the continuity of the mapping $(x, t) \mapsto x_t$ it follows that $P_x \circ X_\zeta^{-1} \xrightarrow{w} P_b \circ X_0^{-1} = \delta_b$, and so $u(x) \to f(b)$ by the continuity of f.

Next assume the stated condition. If $d = 1$, then D is an interval, which is obviously regular. Now assume that $d \geq 2$. By the Markov property we get for any $f \in C_b(\partial D)$

$$u(b) = E_b[f(X_\zeta); \zeta \leq h] + E_b[u(X_h); \zeta > h], \quad h > 0.$$

As $h \to 0$, it follows by dominated convergence that $u(b) = f(b)$, and for $f(x) = e^{-|x-b|}$ we get $P_b\{X_\zeta = b, \zeta < \infty\} = 1$. Since a.s. $X_t \neq b$ for all $t > 0$ by Theorem 18.6 (i), we may conclude that $P_b\{\zeta = 0\} = 1$, and so b is regular. $\square$

Proof of Theorem 24.5: Let u be given by (5), fix any closed ball in D with center x and boundary S, and conclude by the strong Markov property at $\tau = \tau_S$ that

$$u(x) = E_x[f(X_\zeta); \zeta < \infty] = E_x E_{X_\tau}[f(X_\zeta); \zeta < \infty] = E_x u(X_\tau).$$

This shows that u has the mean-value property, and so by Lemma 24.3 it is harmonic. From Lemma 24.6 it is further seen that u is continuous on $\overline{D}$ with $u = f$ on ∂D. Thus, u solves the Dirichlet problem (D, f).

Now assume that $d \geq 3$ and $f \in C_0(\partial D)$. For any $\varepsilon > 0$ we have

$$|u(x)| \leq \varepsilon + \|f\| P_x\{|f(X_\zeta)| > \varepsilon, \zeta < \infty\}. \tag{6}$$

Since X is transient by Theorem 18.6 (ii) and the set $\{y \in \partial D; |f(y)| > \varepsilon\}$ is bounded, the right-hand side of (6) tends to 0 as $|x| \to \infty$ and then $\varepsilon \to 0$, which shows that $u \in C_0(\overline{D})$.

To prove the asserted uniqueness, it is clearly enough to assume $f = 0$ and show that any solution u with the stated properties is identically zero. If $d \geq 3$ and $u \in C_0(\overline{D})$, then this is clear by Lemma 24.3, which shows that harmonic functions can have no local maxima or minima. Next assume that $\zeta < \infty$ a.s. and $u \in C_b(\overline{D})$. By Corollary 17.19 we have $E_x u(X_{\zeta \wedge n}) = u(x)$ for any $x \in D$ and $n \in \mathbb{N}$, and as $n \to \infty$, we get by continuity and dominated convergence $u(x) = E_x u(X_\zeta) = 0$. $\square$

To prepare for our probabilistic construction of the Green function in a domain $D \subset \mathbb{R}^d$, we need to study the transition densities of Brownian motion killed on the boundary ∂D. Recall that ordinary Brownian motion in $\mathbb{R}^d$ has transition densities

$$p_t(x, y) = (2\pi t)^{-d/2} e^{-|x-y|^2/2t}, \quad x, y \in \mathbb{R}^d, \, t > 0. \tag{7}$$

By the strong Markov property and Theorem 6.4, we get for any $t > 0$, $x \in D$, and $B \subset \mathcal{B}(D)$,
$$P_x\{X_t \in B\} = P_x\{X_t \in B, t \leq \zeta\} + E_x[T_{t-\zeta}1_B(X_\zeta); t > \zeta].$$
Thus, the killed process has transition densities
$$p_t^D(x,y) = p_t(x,y) - E_x[p_{t-\zeta}(X_\zeta,y); t > \zeta], \quad x,y \in D, t > 0. \tag{8}$$
The following symmetry and continuity properties of p_t^D play a crucial role in the sequel.

Theorem 24.7 *(transition density, Hunt)* *For any domain D in $\mathbb{R}^d$ and time $t > 0$, the function p_t^D is symmetric and continuous on D^2. If $b \in \partial D$ is regular, then $p_t^D(x,y) \to 0$ as $x \to b$ for fixed $y \in D$.*

Proof: From (7) we note that $p_t(x,y)$ is uniformly continuous in (x,y) for fixed $t > 0$, as well as in (x,y,t) for $|x-y| > \varepsilon > 0$ and $t > 0$. By (8) it follows that $p_t^D(x,y)$ is equicontinuous in $y \in D$ for fixed $t > 0$. To prove the continuity in $x \in D$ for fixed $t > 0$ and $y \in D$, it is then enough to show that $P_x\{X_t \in B, t \leq \zeta\}$ is continuous in x for fixed $t > 0$ and $B \in \mathcal{B}(D)$. Letting $h \in (0,t)$, we get by the Markov property
$$P_x\{X_t \in B, \zeta \geq t\} = E_x[P_{X_h}\{X_{t-h} \in B, \zeta \geq t-h\}; \zeta > h].$$
Thus, for any $x, y \in D$,
$$|(P_x - P_y)\{X_t \in B, t \leq \zeta\}|$$
$$\leq (P_x + P_y)\{\zeta \leq h\} + \|P_x \circ X_h^{-1} - P_y \circ X_h^{-1}\|,$$
which tends to 0 as $y \to x$ and then $h \to 0$. Combining the continuity in x with the equicontinuity in y, we conclude that $p_t^D(x,y)$ is continuous in $(x,y) \in D^2$ for fixed $t > 0$.

To prove the symmetry in x and y, it is now enough to establish the integrated version
$$\int_C P_x\{X_t \in B, \zeta > t\}dx = \int_B P_x\{X_t \in C, \zeta > t\}dx, \tag{9}$$
for any bounded sets $B, C \in \mathcal{B}(D)$. Then fix any compact set $F \subset D$. Letting $n \in \mathbb{N}$ and writing $h = 2^{-n}t$ and $t_k = kh$, we get by Proposition 8.2
$$\int_C P_x\{X_{t_k} \in F, k \leq 2^n; X_t \in B\}dx$$
$$= \int_F \cdots \int_F 1_C(x_0)1_B(x_{2^n}) \prod_{k \leq 2^n} p_h(x_{k-1}, x_k)dx_0 \cdots dx_{2^n}.$$
Here the right-hand side is symmetric in the pair (B,C), because of the symmetry of $p_h(x,y)$. By dominated convergence as $n \to \infty$ we obtain (9) with F instead of D, and the stated version follows by monotone convergence as $F \uparrow D$.

To prove the last assertion, we recall from the proof of Lemma 24.6 that $P_x \circ (\zeta, X)^{-1} \xrightarrow{w} P_b \circ (0, X)^{-1}$ as $x \to b$ with $b \in \partial D$ regular. In particular, $P_x \circ (\zeta, X_\zeta) \xrightarrow{w} \delta_{0,b}$, and by the boundedness and continuity of $p_t(x,y)$ for $|x-y| > \varepsilon > 0$, it is clear from (8) that $p_t^D(x,y) \to 0$. □

A domain $D \subset \mathbb{R}^d$ is said to be *Greenian* if either $d \geq 3$, or if $d \leq 2$ and $P_x\{\zeta_D < \infty\} = 1$ for all $x \in D$. Since the latter probability is harmonic in x, it is enough by Lemma 24.3 to verify the stated property for a single $x \in D$. Given a Greenian domain D, we may introduce the *Green function*

$$g^D(x,y) = \int_0^\infty p_t^D(x,y) dt, \quad x, y \in D.$$

For any measure μ on D, we may further introduce the associated *Green potential*

$$G^D \mu(x) = \int g^D(x,y) \mu(dy), \quad x \in D.$$

Writing $G^D \mu = G^D f$ when $\mu(dy) = f(y) dy$, we get by Fubini's theorem

$$E_x \int_0^\zeta f(X_t) dt = \int g^D(x,y) f(y) dy = G^D f(x), \quad x \in D,$$

which identifies g^D as an *occupation density* for the killed process.

The next result shows that g^D and G^D agree with the Green function and Green potential of classical potential theory. Thus, $G^D \mu(x)$ may be interpreted as the electrostatic potential at x arising from a charge distribution μ in D, when the boundary ∂D is grounded.

Theorem 24.8 *(Green function) For any Greenian domain $D \subset \mathbb{R}^d$, the function g^D is symmetric on D^2. Furthermore, $g^D(x,y)$ is harmonic in $x \in D \setminus \{y\}$ for each $y \in D$, and if $b \in \partial D$ is regular, then $g^D(x,y) \to 0$ as $x \to b$ for fixed $y \in D$.*

The proof is straightforward when $d \geq 3$, but for $d \leq 2$ we need two technical lemmas. We begin with a uniform estimate for large t.

Lemma 24.9 *(uniform integrability) Consider a domain $D \subset \mathbb{R}^d$, assumed to be bounded when $d \leq 2$. Then*

$$\lim_{t \to \infty} \sup_{x,y \in D} \int_t^\infty p_s^D(x,y) ds = 0.$$

Proof: For $d \geq 3$ we may take $D = \mathbb{R}^d$, in which case the result is obvious from (7). Next let $d = 2$. By obvious domination and scaling arguments, we may then assume that $|x| \leq 1$, $y = 0$, $D = \{z; |z| \leq 2\}$, and $t > 1$.

Writing $p_t(x) = p_t(x,0)$, we get by (8)

$$\begin{aligned}
p_t^D(x,0) &\le p_t(x) - E_0[p_{t-\varsigma}(1); \varsigma \le t/2] \\
&\le p_t(0) - p_t(1) P_0\{\varsigma \le t/2\} \\
&\le p_t(0) P_0\{\varsigma > t/2\} + p_t(0) - p_t(1) \\
&\le t^{-1} P_0\{\varsigma > t/2\} + t^{-2}.
\end{aligned}$$

As in case of Lemma 23.8 (ii), we have $E_0 \varsigma < \infty$, and so by Lemma 3.4 the right-hand side is integrable in $t \in [1,\infty)$. The proof for $d = 1$ is similar. $\square$

We also need the fact that bounded sets have bounded Green potential.

Lemma 24.10 *(boundedness)* For any Greenian domain $D \subset \mathbb{R}^d$ and bounded set $B \in \mathcal{B}(D)$, the function $G^D 1_B$ is bounded.

Proof: By domination and scaling together with the strong Markov property, it suffices to take $B = \{x; |x| \le 1\}$ and to show that $G^D 1_B(0) < \infty$. For $d \ge 3$ we may further take $D = \mathbb{R}^d$, in which case the result follows by a simple computation. For $d = 2$ we may assume that $D \supset C \equiv \{x; |x| < 2\}$. Write $\sigma = \varsigma_C + \tau_B \circ \theta_{\varsigma_C}$ and $\tau_0 = 0$, and recursively define $\tau_{k+1} = \tau_k + \sigma \circ \theta_{\tau_k}$, $k \ge 0$. Putting $b = (1,0)$, we get by the strong Markov property at the times τ_k

$$G^D 1_B(0) = G^C 1_B(0) + G^C 1_B(b) \sum\nolimits_{k \ge 1} P_0\{\tau_k < \varsigma\}.$$

Here $G^C 1_B(0) \vee G^C 1_B(b) < \infty$ by Lemma 24.9. By the strong Markov property it is further seen that $P_0\{\tau_k < \varsigma\} \le p^k$, where $p = \sup_{x \in B} P_x\{\sigma < \varsigma\}$. Finally, note that $p < 1$, since $P_x\{\sigma < \varsigma\}$ is harmonic and hence continuous on B. The proof for $d = 1$ is similar. $\square$

Proof of Theorem 24.8: The symmetry of g^D is clear from Theorem 24.7. If $d \ge 3$, or if $d = 2$ and D is bounded, it is further seen from Theorem 24.7, Lemma 24.9, and dominated convergence that $g^D(x,y)$ is continuous in $x \in D \setminus \{y\}$ for each $y \in D$. Next we note that $G^D 1_B$ has the mean-value property in $D \setminus \overline{B}$ for bounded $B \in \mathcal{B}(D)$. The property extends by continuity to the density $g^D(x,y)$, which is then harmonic in $x \in D \setminus \{y\}$ for fixed $y \in D$, by Lemma 24.3.

For $d = 2$ and unbounded D, we define $D_n = \{x \in D; |x| < n\}$, and note as before that $g^{D_n}(x,y)$ has the mean-value property in $x \in D_n \setminus \{y\}$ for each $y \in D_n$. Since $p_t^{D_n} \uparrow p_t^D$ by dominated convergence, we have $g^{D_n} \uparrow g^D$, and so the mean-value property extends to the limit. For any $x \ne y$ in D, choose a circular disk B around y with radius $\varepsilon > 0$ small enough that $x \notin \overline{B} \subset D$. Then $\pi \varepsilon^2 g^D(x,y) = G^D 1_B(x) < \infty$ by Lemma 24.10. Thus, by Lemma 24.3 even $g^D(x,y)$ is harmonic in $x \in D \setminus \{y\}$.

To prove the last assertion, fix any $y \in D$, and assume that $x \to b \in \partial D$. Choose a Greenian domain $D' \supset D$ with $b \in D'$. Since $p_t^D \le p_t^{D'}$, and both $p_t^{D'}(\cdot, y)$ and $g^{D'}(\cdot, y)$ are continuous at b whereas $p_t^D(x, y) \to 0$ by Theorem 24.7, we get $g^D(x, y) \to 0$ by Theorem 1.21. □

We proceed to show that a measure is determined by its Green potential whenever the latter is finite. An extension appears as part of Theorem 24.12. For convenience, we write

$$P_t^D \mu(x) = \int p_t^D(x, y) \mu(dy), \quad x \in D, \, t > 0.$$

Theorem 24.11 (uniqueness) *If μ and ν are measures on a Greenian domain $D \subset \mathbb{R}^d$ such that $G^D \mu = G^D \nu < \infty$, then $\mu = \nu$.*

Proof: For any $t > 0$ we have

$$\int_0^t (P_s^D \mu) ds = G^D \mu - P_t^D G^D \mu = G^D \nu - P_t^D G^D \nu = \int_0^t (P_s^D \nu) ds. \quad (10)$$

By the symmetry of p^D, we further get for any measurable function $f : D \to \mathbb{R}_+$

$$\int f(x) P_s^D \mu(x) dx = \int f(x) dx \int p_s^D(x, y) \mu(dy)$$
$$= \int \mu(dy) \int f(x) p_s^D(x, y) dx = \int P_s^D f(y) \mu(dy).$$

Hence,

$$\int f(x) dx \int_0^t P_s^D \mu(x) \, ds = \int_0^t ds \int P_s^D f(y) \mu(dy)$$
$$= \int \mu(dy) \int_0^t P_s^D f(y) \, ds,$$

and similarly for ν. By (10) we obtain

$$\int \mu(dy) \int_0^t P_s^D f(y) ds = \int \nu(dy) \int_0^t P_s^D f(y) ds. \quad (11)$$

Assuming that $f \in C_K^+(D)$, we get $P_s^D f \to f$ as $s \to 0$, and so $t^{-1} \int_0^t P_s^D f ds \to f$. If we can take limits inside the outer integrations in (11), we obtain $\mu f = \nu f$, which implies $\mu = \nu$ since f is arbitrary.

To justify the argument, it suffices to show that $\sup_s P_s^D f$ is μ- and ν-integrable. Then conclude from Theorem 24.7 that $f \lesssim p_s^D(\cdot, y)$ for fixed $s > 0$ and $y \in D$, and from Theorem 24.8 that $f \lesssim G^D f$. The latter property yields $P_s^D f \lesssim P_s^D G^D f \leq G^D f$, and by the former property we get for any $y \in D$ and $s > 0$

$$\mu(G^D f) = \int G^D \mu(x) f(x) dx \lesssim P_s^D G^D \mu(y) \leq G^D \mu(y) < \infty,$$

and similarly for ν. □

Now let $\mathcal{F}_D$ and $\mathcal{K}_D$ denote the classes of closed and compact subsets of D, and write $\mathcal{F}_D^r$ and $\mathcal{K}_D^r$ for the subclasses of sets with regular boundary. For any $B \in \mathcal{F}_D$ we may introduce the associated *hitting kernel*

$$H_B^D(x, dy) = P_x\{\tau_B < \zeta_D, X_{\tau_B} \in dy\}, \quad x \in D.$$

Note that if X has initial distribution μ, then the hitting distribution of X^ζ in B equals $\mu H_B^D = \int \mu(dx) H_B^D(x, \cdot)$.

The next result solves the *sweeping problem* of classical potential theory. To avoid technical complications, here and below, we shall only consider subsets with regular boundary. In general, the irregular part of the boundary can be shown to be *polar*, in the sense of being a.s. avoided by a Brownian motion. Given this result, one can easily remove all regularity restrictions.

Theorem 24.12 (*sweeping and hitting*) *For any Greenian domain $D \subset \mathbb{R}^d$ and subset $B \in \mathcal{F}_D^r$, let μ be a bounded measure on D with $G^D\mu < \infty$ on B. Then μH_B^D is the unique measure ν on B with $G^D\mu = G^D\nu$ on B.*

For an electrostatic interpretation, assume that a grounded conductor B is inserted into a domain D with grounded boundary and charge distribution μ. Then a charge distribution $-\mu H_B^D$ arises on B.

A lemma is needed for the proof. Here we define $g^{D \setminus B}(x, y) = 0$ whenever x or y lies in B.

Lemma 24.13 (*fundamental identity*) *For any Greenian domain $D \subset \mathbb{R}^d$ and subset $B \in \mathcal{F}_D^r$, we have*

$$g^D(x, y) = g^{D \setminus B}(x, y) + \int_B H_B^D(x, dz) g^D(z, y), \quad x, y \in D.$$

Proof: Write $\zeta = \zeta_D$ and $\tau = \tau_B$. Subtracting relations (8) for the domains D and $D \setminus B$, and using the strong Markov property at τ together with Theorem 6.4, we get

$$p_t^D(x, y) - p_t^{D \setminus B}(x, y)$$
$$= E_x[p_{t-\tau}(X_\tau, y); \tau < \zeta \wedge t] - E_x[p_{t-\zeta}(X_\zeta, y); \tau < \zeta < t]$$
$$= E_x[p_{t-\tau}(X_\tau, y); \tau < \zeta \wedge t]$$
$$\quad - E_x[E_{X_\tau}[p_{t-\tau-\zeta}(X_\zeta, y); \zeta < t - \tau]; \tau < \zeta \wedge t]$$
$$= E_x[p_{t-\tau}^D(X_\tau, y); \tau < \zeta \wedge t].$$

Now integrate with respect to t to get

$$g^D(x, y) - g^{D \setminus B}(x, y) = E_x[g^D(X_\tau, y); \tau < \zeta]$$
$$= \int H_B^D(x, dz) g^D(z, y). \qquad \square$$

Proof of Theorem 24.12: Since ∂B is regular, we have $H_B^D(x, \cdot) = \delta_x$ for all $x \in B$, and so by Lemma 24.13 we get for all $x \in B$ and $z \in D$

$$\int g^D(x, y) H_B^D(z, dy) = \int g^D(z, y) H_B^D(x, dy) = g^D(z, x).$$

24. Connections with PDEs and Potential Theory

Integrating with respect to $\mu(dz)$ gives $G^D(\mu H_B^D)(x) = G^D\mu(x)$, which shows that $\nu = \mu H_B^D$ has the stated property.

Now consider any measure ν on B with $G^D\mu = G^D\nu$ on B. Noting that $g^{D\setminus B}(x,\cdot) = 0$ on B whereas $H_B^D(x,\cdot)$ is supported by B, we get by Lemma 24.13 for any $x \in D$

$$G^D\nu(x) = \int \nu(dz) g^D(z,x) = \int \nu(dz) \int g^D(z,y) H_B^D(x,dy)$$
$$= \int H_B^D(x,dy) G^D\nu(y) = \int H_B^D(x,dy) G^D\mu(y).$$

Thus, μ determines $G^D\nu$ on D, and so ν is unique by Theorem 24.11. □

Let us now turn to the classical *equilibrium problem*. For any $K \in \mathcal{K}_D$ we introduce the *last exit* or *quitting time*

$$\gamma_K^D = \sup\{t < \zeta_D;\ X_t \in K\}$$

and the associated *quitting kernel*

$$L_K^D(x,dy) = P_x\{\gamma_K^D > 0;\ X(\gamma_K^D) \in dy\}.$$

Theorem 24.14 *(equilibrium measure and quitting, Chung)* For any Greenian domain $D \in \mathbb{R}^d$ and subset $K \in \mathcal{K}_D$, there exists a measure μ_K^D on ∂K such that

$$L_K^D(x,dy) = g^D(x,y)\mu_K^D(dy), \quad x \in D. \tag{12}$$

Furthermore, μ_K^D is diffuse when $d \geq 2$, and if $K \in \mathcal{K}_D^r$, then μ_K^D is the unique measure μ on K satisfying $G^D\mu = 1$ on K.

Here μ_K^D is called the *equilibrium measure* of K relative to D, and its total mass C_K^D is called the *capacity* of K in D. For an electrostatic interpretation, assume that a conductor K with potential 1 is inserted into a domain D with grounded boundary. Then a charge distribution μ_K^D arises on the boundary of K.

Proof of Theorem 24.14: Write $\gamma = \gamma_K^D$, and define

$$l_\varepsilon(x) = \varepsilon^{-1} P_x\{0 < \gamma \leq \varepsilon\}, \quad \varepsilon > 0.$$

Using Fubini's theorem, the simple Markov property, and dominated convergence as $\varepsilon \to 0$, we get for any $f \in C_b(D)$ and $x \in D$

$$G^D(fl_\varepsilon)(x) = E_x \int_0^\zeta f(X_t) l_\varepsilon(X_t) dt$$
$$= \varepsilon^{-1} \int_0^\infty E_x[f(X_t) P_{X_t}\{0 < \gamma \leq \varepsilon\};\ t < \zeta] dt$$
$$= \varepsilon^{-1} \int_0^\infty E_x[f(X_t);\ t < \gamma \leq t+\varepsilon] dt$$
$$= \varepsilon^{-1} E_x \int_{(\gamma-\varepsilon)_+}^\gamma f(X_t) dt$$
$$\to E_x[f(X_\gamma);\ \gamma > 0] = L_K^D f(x).$$

If f has compact support, then for each x we may replace f by the bounded, continuous function $f/g^D(x,\cdot)$ to get as $\varepsilon \to 0$

$$\int f(y) l_\varepsilon(y) dy \to \int \frac{L_K^D(x, dy) f(y)}{g^D(x, y)}. \tag{13}$$

Since the left-hand side is independent of x, the same thing is true for the measure

$$\mu_K^D(dy) = \frac{L_K^D(x, dy)}{g^D(x, y)}. \tag{14}$$

If $d = 1$, we have $g^D(x,x) < \infty$, and (14) is trivially equivalent to (12). If instead $d \geq 2$, then singletons are polar, and so the measure $L_K^D(x, \cdot)$ is diffuse, which implies the same property for μ_K^D. Thus, (12) and (14) are again equivalent. We may further conclude from the continuity of X that $L_K^D(x, \cdot)$, and then also μ_K^D is supported by ∂K.

Integrating (12) over D yields

$$P_x\{\tau_K < \zeta_D\} = G^D \mu_K^D(x), \quad x \in D,$$

and so for $K \in \mathcal{K}_D^r$ we get $G^D \mu_K^D = 1$ on K. If ν is another measure on K with $G^D \nu = 1$ on K, then $\nu = \mu_K^D$ by the uniqueness part of Theorem 24.12. □

The next result relates the equilibrium measures and capacities for different sets $K \in \mathcal{K}_D^r$.

Proposition 24.15 *(consistency) For any Greenian domain $D \subset \mathbb{R}^d$ and subsets $K \subset B$ in $\mathcal{K}_D^r$, we have*

$$\mu_K^D = \mu_B^D H_K^D = \mu_B^D L_K^D, \tag{15}$$

$$C_K^D = \int_B P_x\{\tau_K < \zeta_D\} \mu_B^D(dx). \tag{16}$$

Proof: By Theorem 24.12 and the defining properties of μ_B^D and μ_K^D, we have on K

$$G^D(\mu_B^D H_K^D) = G^D \mu_B^D = 1 = G^D \mu_K^D,$$

and so $\mu_B^D H_K^D = \mu_K^D$ by the same result. To prove the second relation in (15), we note by Theorem 24.14 that, for any $A \in \mathcal{B}(K)$,

$$\mu_B^D L_K^D(A) = \int \mu_B^D(dx) \int_A g^D(x, y) \mu_K^D(dy)$$

$$= \int_A G^D \mu_B^D(y) \mu_K^D(dy) = \mu_K^D(A),$$

since $G^D \mu_B^D = 1$ on $A \subset B$. Finally, (15) implies (16), since $H_K^D(x, K) = P_x\{\tau_K < \zeta_D\}$. □

Some basic properties of capacities and equilibrium measures follow immediately from Proposition 24.15. To explain the terminology, fix any space

S along with a class of subsets $\mathcal{U}$, closed under finite unions. For any function $h: \mathcal{U} \to \mathbb{R}$ and sets $U, U_1, U_2, \cdots \in \mathcal{U}$, we recursively define the differences

$$\Delta_{U_1} h(U) = h(U \cup U_1) - h(U),$$
$$\Delta_{U_1,\ldots,U_n} h(U) = \Delta_{U_n}\{\Delta_{U_1,\ldots,U_{n-1}} h(U)\}, \quad n > 1,$$

where the difference Δ_{U_n} in the last formula is taken with respect to U. Note that the higher-order differences $\Delta_{U_1,\ldots,U_n}$ are invariant under permutations of $U_1, \ldots, U_n$. We say that h is *alternating* or *completely monotone* if

$$(-1)^{n+1} \Delta_{U_1,\ldots,U_n} h(U) \geq 0, \quad n \in \mathbb{N}, \ U, U_1, U_2, \cdots \in \mathcal{U}.$$

Corollary 24.16 *(dependence on conductor, Choquet)* For any Greenian domain $D \subset \mathbb{R}^d$, the capacity C_K^D is an alternating function of $K \in \mathcal{K}_D^r$. Furthermore, $\mu_{K_n}^D \overset{w}{\to} \mu_K^D$ as $K_n \downarrow K$ or $K_n \uparrow K$ in $\mathcal{K}_D^r$.

Proof: Let ψ denote the path of X^ς, regarded as a random closed set in D. Writing

$$h_x(K) = P_x\{\psi K \neq \emptyset\} = P_x\{\tau_K < \varsigma\}, \quad x \in D \setminus K,$$

we get by induction

$$(-1)^{n+1} \Delta_{K_1,\ldots,K_n} h_x(K) = P_x\{\psi K = \emptyset, \ \psi K_1 \neq \emptyset, \ \ldots, \ \psi K_n \neq \emptyset\} \geq 0,$$

and the first assertion follows by Proposition 24.15 with $K \subset B^\circ$.

To prove the last assertion, we note that trivially $\tau_{K_n} \downarrow \tau_K$ when $K_n \uparrow K$, and that $\tau_{K_n} \uparrow \tau_K$ when $K_n \downarrow K$ since the K_n are closed. In the latter case we also note that $\bigcap_n \{\tau_{K_n} < \varsigma\} = \{\tau_K < \varsigma\}$ by compactness. Thus, in both cases $H_{K_n}^D(x,\cdot) \overset{w}{\to} H_K^D(x,\cdot)$ for all $x \in D \setminus \bigcup_n K_n$, and by dominated convergence in Proposition 24.15 with $B^\circ \supset \bigcup_n K_n$ we get $\mu_{K_n}^D \overset{w}{\to} \mu_K^D$. $\square$

The next result solves an equilibrium problem involving two conductors.

Corollary 24.17 *(condenser theorem)* For any disjoint sets $B \in \mathcal{F}_D^r$ and $K \in \mathcal{K}_D^r$, there exists a unique signed measure ν on $B \cup K$ with $G^D \nu = 0$ on B and $G^D \nu = 1$ on K, namely

$$\nu = \mu_K^{D \setminus B} - \mu_K^{D \setminus B} H_B^D.$$

Proof: Applying Theorem 24.14 to the domain $D \setminus B$ with subset K, we get $\nu = \mu_K^{D \setminus B}$ on K, and then $\nu = -\mu_K^{D \setminus B} H_B^D$ on B by Theorem 24.12. $\square$

The symmetry between hitting and quitting kernels in Proposition 24.15 may be extended to an invariance under *time reversal* of the whole process. More precisely, putting $\gamma = \gamma_K^D$, we may relate the stopped process $X_t^\varsigma = X_{\gamma \wedge t}$ to its reversal $\tilde{X}_t^\gamma = X_{(\gamma-t)_+}$. For convenience, we write

$P_\mu = \int P_x \mu(dx)$ and refer to the induced measures as *distributions*, even when μ is not normalized.

Theorem 24.18 *(time reversal) Given a Greenian domain $D \in \mathbb{R}^d$ and a set $K \in \mathcal{K}_D^r$, put $\gamma = \gamma_K^D$ and $\mu = \mu_K^D$. Then $X^\gamma \stackrel{d}{=} \tilde{X}^\gamma$ under P_μ.*

Proof: Let P_x and E_x refer to the process X^ς. Fix any times $0 = t_0 < t_1 < \cdots < t_n$, and write $s_k = t_n - t_k$ and $h_k = t_k - t_{k-1}$. For any continuous functions $f_0, \ldots, f_n$ with compact supports in D, we define

$$f^\varepsilon(x) = E_x \prod_{k \geq 0} f_k(X_{s_k}) l_\varepsilon(X_{t_n})$$

$$= E_x \prod_{k \geq 1} f_k(X_{s_k}) E_{X_{s_1}}(f_0 l_\varepsilon)(X_{t_1}),$$

where the last equality holds by the Markov property at s_1. Proceeding as in the proof of Theorem 24.14, we get

$$\int (f^\varepsilon G^D \mu)(x) \, dx = \int G^D f^\varepsilon(y) \mu(dy)$$

$$\to E_\mu \prod_k f_k(\tilde{X}_{t_k}^\gamma) 1\{\gamma > t_n\}. \tag{17}$$

On the other hand, (13) shows that the measure $l_\varepsilon(x)dx$ tends vaguely to μ, and so by Theorem 24.7

$$E_x(f_0 l_\varepsilon)(X_{t_1}) = \int p_{t_1}^D(x,y) (f_0 l_\varepsilon)(y) \, dy$$

$$\to \int p_{t_1}^D(x,y) f_0(y) \mu(dy).$$

Using dominated convergence, Fubini's theorem, Proposition 8.2, Theorem 24.7, and the relation $G^D \mu(x) = P_x\{\gamma > 0\}$, we obtain

$$\int (f^\varepsilon G^D \mu)(x) dx$$

$$\to \int G^D \mu(x) dx \int f_0(y) \mu(dy) E_x \prod_{k>0} f_k(X_{s_k}) p_{t_1}^D(X_{s_1}, y)$$

$$= \int f_0(x_0) \mu(dx_0) \int \cdots \int G^D \mu(x_n) \prod_{k>0} p_{h_k}^D(x_{k-1}, x_k) f_k(x_k) dx_k$$

$$= E_\mu \prod_k f_k(X_{t_k}) G^D \mu(X_{t_n}) = E_\mu \prod_k f_k(X_{t_k}) 1\{\gamma > t_n\}.$$

Comparing with (17), we see that X^γ and $\tilde{X}^\gamma$ have the same finite-dimensional distributions. □

We may now extend Proposition 24.15 to the case of possibly different Greenian domains $D \subset D'$. Fixing any $K \in \mathcal{K}_D$, we recursively define the optional times

$$\tau_j = \gamma_{j-1} + \tau_K^{D'} \circ \theta_{\gamma_{j-1}}, \quad \gamma_j = \tau_j + \gamma_K^D \circ \theta_{\tau_j}, \qquad j \geq 1,$$

starting with $\gamma_0 = 0$. In other words, τ_k and γ_k are the times of hitting or quitting K during the kth excursion in D that reaches K, prior to the exit time $\zeta_{D'}$. The *generalized hitting and quitting kernels* are given by

$$H_K^{D,D'}(x,\cdot) = E_x \sum_k \delta_{X(\tau_k)}, \qquad L_K^{D,D'}(x,\cdot) = E_x \sum_k \delta_{X(\gamma_k)},$$

where the summations extend over all $k \in \mathbb{N}$ with $\tau_k < \infty$.

Theorem 24.19 *(extended consistency relations)* Let $D \subset D'$ be Greenian domains in $\mathbb{R}^d$ with regular compact subsets $K \subset K'$. Then

$$\mu_K^D = \mu_{K'}^{D'} H_K^{D,D'} = \mu_{K'}^{D'} L_K^{D,D'}. \tag{18}$$

Proof: Define $l_\varepsilon = \varepsilon^{-1} P_x \{\gamma_K^D \in (0,\varepsilon]\}$. Proceeding as in the proof of Theorem 24.14, we get for any $x \in D'$ and $f \in C_b(D')$

$$G^{D'}(fl_\varepsilon)(x) = \varepsilon^{-1} E_x \int_0^{\zeta_{D'}} f(X_t) 1\{\gamma_K^D \circ \theta_t \in (0,\varepsilon]\} dt \to L_K^{D,D'} f(x).$$

If f has compact support in D, we may conclude as before that

$$\int f(y) \mu_K^D(dy) \leftarrow \int (fl_\varepsilon)(y) dy \to \int \frac{L_K^{D,D'}(x,dy) f(y)}{g^{D'}(x,y)},$$

and so

$$L_K^{D,D'}(x,dy) = g^{D'}(x,y) \mu_K^D(dy).$$

Integrating with respect to $\mu_{K'}^{D'}$, and noting that $G^{D'} \mu_{K'}^{D'} = 1$ on $K' \supset K$, we obtain the second expression for μ_K^D in (18).

To deduce the first expression, we note that $H_K^{D'} H_K^{D,D'} = H_K^{D,D'}$ by the strong Markov property at τ_K. Combining with the second expression in (18) and using Theorem 24.18 and Proposition 24.15, we get

$$\mu_K^D = \mu_K^{D'} L_K^{D,D'} = \mu_K^{D'} H_K^{D,D'} = \mu_{K'}^{D'} H_K^{D'} H_K^{D,D'} = \mu_{K'}^{D'} H_K^{D,D'}. \qquad \square$$

The last result enables us to study the equilibrium measure μ_K^D and capacity C_K^D as functions of both D and K. In particular, we obtain the following continuity and monotonicity properties.

Corollary 24.20 *(dependence on domain)* For any regular, compact set $K \subset \mathbb{R}^d$, the measure μ_K^D is nonincreasing and continuous from above as a function of the Greenian domain $D \supset K$.

Proof: The monotonicity is clear from (18) with $K = K'$, since $H_K^{D,D'}(x,\cdot) \geq \delta_x$ for $x \in K \subset D \subset D'$. It remains to prove that C_K^D is continuous from above and below in D for fixed K. By dominated convergence it is then enough to show that $\kappa_K^{D_n} \to \kappa_K^D$, where $\kappa_K^D = \sup\{j; \tau_j < \infty\}$ is the number of D-excursions hitting K.

Assuming $D_n \uparrow D$, we need to show that if $X_s, X_t \in K$ and $X \in D$ on $[s,t]$, then $X \in D_n$ on $[s,t]$ for sufficiently large n. But this is clear from the compactness of the path on the interval $[s,t]$. If instead $D_n \downarrow D$, we need to show for any $r < s < t$ with $X_r, X_t \in K$ and $X_s \notin D$ that $X_s \notin D_n$ for sufficiently large n. But this is obvious. $\square$

We proceed to show how Green capacities can be expressed in terms of random sets. Let χ denote the identity mapping on $\mathcal{F}_D$. Given any measure ν on $\mathcal{F}_D \setminus \{\emptyset\}$ with $\nu\{\chi K \neq \emptyset\} < \infty$ for all $K \in \mathcal{K}_D$, we may introduce a Poisson process η on $\mathcal{F}_D \setminus \{\emptyset\}$ with intensity measure ν and form the associated random closed set $\varphi = \bigcup\{F;\, \eta\{F\} > 0\}$ in D. Letting π_ν denote the distribution of φ, we note that

$$\pi_\nu\{\chi K = \emptyset\} = P\{\eta\{\chi K \neq \emptyset\} = 0\} = \exp(-\nu\{\chi K \neq \emptyset\}), \quad K \in \mathcal{K}_D.$$

Theorem 24.21 (*Green capacities and random sets, Choquet*) *For any Greenian domain $D \subset \mathbb{R}^d$, there exists a unique measure ν on $\mathcal{F}_D \setminus \{\emptyset\}$ such that*

$$C_K^D = \nu\{\chi K \neq \emptyset\} = -\log \pi_\nu\{\chi K = \emptyset\}, \quad K \subset \mathcal{K}_D^r.$$

Proof: Let ψ denote the path of X^ς in D. Choose sets $K_n \uparrow D$ in $\mathcal{K}_D^r$ with $K_n \subset K_{n+1}^\circ$ for all n, and put $\mu_n = \mu_{K_n}^D$, $\psi_n = \psi K_n$, and $\chi_n = \chi K_n$. Define

$$\nu_n^p = \int P_x\{\psi_p \in \cdot,\, \psi_n \neq \emptyset\} \mu_p(dx), \quad n \leq p, \tag{19}$$

and conclude by the strong Markov property and Proposition 24.15 that

$$\nu_n^q\{\chi_p \in \cdot,\, \chi_m \neq \emptyset\} = \nu_m^p, \quad m \leq n \leq p \leq q. \tag{20}$$

By Corollary 6.15 there exist some measures ν_n on $\mathcal{F}_D$, $n \in \mathbb{N}$, satisfying

$$\nu_n\{\chi_p \in \cdot\} = \nu_n^p, \quad n \leq p, \tag{21}$$

and from (20) we note that

$$\nu_n\{\cdot,\, \chi_m \neq \emptyset\} = \nu_m, \quad m \leq n. \tag{22}$$

Hence, the measures ν_n agree on $\{\chi_m \neq \emptyset\}$ for $n \geq m$, and so we may define $\nu = \sup_n \nu_n$. By (22) we have $\nu\{\cdot,\, \chi_n \neq \emptyset\} = \nu_n$ for all n. Assuming $K \in \mathcal{K}_D^r$ with $K \subset K_n^\circ$, we conclude from (19), (21), and Proposition 24.15 that

$$\begin{aligned}
\nu\{\chi K \neq \emptyset\} &= \nu_n\{\chi K \neq \emptyset\} = \nu_n^n\{\chi K \neq \emptyset\} \\
&= \int P_x\{\psi_n K \neq \emptyset\} \mu_n(dx) \\
&= \int P_x\{\tau_K < \varsigma\} \mu_n(dx) = C_K^D.
\end{aligned}$$

The uniqueness of ν is clear by a monotone class argument. $\square$

The representation of capacities in terms of random sets will now be extended to the abstract setting of alternating set functions. As in Chapter 16, we may then fix an lcscH space S with Borel σ-field $\mathcal{S}$, open sets $\mathcal{G}$, closed sets $\mathcal{F}$, and compacts $\mathcal{K}$. Write $\hat{\mathcal{S}} = \{B \in \mathcal{S};\, \overline{B} \in \mathcal{K}\}$, and recall that a class $\mathcal{U} \subset \hat{\mathcal{S}}$ is said to be *separating* if for any $K \in \mathcal{K}$ and $G \in \mathcal{G}$ with $K \subset G$ there exists some $U \in \mathcal{U}$ with $K \subset U \subset G$.

For any nondecreasing function h on a separating class $\mathcal{U} \subset \hat{\mathcal{S}}$, we define the associated *inner* and *outer capacities* h° and $\bar{h}$ by

$$h^\circ(G) = \sup\{h(U); U \in \mathcal{U}, \bar{U} \subset G\}, \quad G \in \mathcal{G},$$
$$\bar{h}(K) = \inf\{h(U); U \in \mathcal{U}, U^\circ \supset K\}, \quad K \in \mathcal{K}.$$

Note that the formulas remain valid with $\mathcal{U}$ replaced by any separating subclass. For any random closed set φ in S, the associated *hitting function* h is given by $h(B) = P\{\varphi B \neq \emptyset\}$ for all $B \in \hat{\mathcal{S}}$.

Theorem 24.22 *(alternating functions and random sets, Choquet)* The hitting function h of a random closed set in S is alternating with $h = \bar{h}$ on $\mathcal{K}$ and $h = h^\circ$ on $\mathcal{G}$. Conversely, given a separating class $\mathcal{U} \subset \hat{\mathcal{S}}$, closed under finite unions, and an alternating function $p \colon \mathcal{U} \to [0,1]$ with $p(\emptyset) = 0$, there exists a random closed set with hitting function h such that $h = \bar{p}$ on $\mathcal{K}$ and $h = p^\circ$ on $\mathcal{G}$.

The algebraic part of the construction is clarified by the following lemma.

Lemma 24.23 *(discrete case)* Assume $\mathcal{U} \subset \hat{\mathcal{S}}$ to be finite and closed under unions, and let $h \colon \mathcal{U} \to [0,1]$ be alternating with $h(\emptyset) = 0$. Then there exists a point process ξ on S such that $P\{\xi U > 0\} = h(U)$ for all $U \in \mathcal{U}$.

Proof: The statement is obvious when $\mathcal{U} = \{\emptyset\}$. Proceeding by induction, assume the assertion to be true when $\mathcal{U}$ is generated by up to $n-1$ sets, and consider a class $\mathcal{U}$ generated by n nonempty sets $B_1, \ldots, B_n$. By scaling we may assume that $h(B_1 \cup \cdots \cup B_n) = 1$.

For each $j \in \{1, \ldots, n\}$, let $\mathcal{U}_j$ be the class of unions formed by the sets $B_i \setminus B_j$, $i \neq j$, and define

$$h_j(U) = \Delta_U h(B_j) = h(B_j \cup U) - h(B_j), \quad U \in \mathcal{U}_j.$$

Then each h_j is again alternating with $h_j(\emptyset) = 0$, and so the induction hypothesis ensures the existence of some point process ξ_j on $\bigcup_i B_i \setminus B_j$ with hitting function h_j. Note that h_j remains the hitting function of ξ_j on all of $\mathcal{U}$. Let us further introduce a point process ξ_{n+1} with

$$P\bigcap_i \{\xi_{n+1} B_i > 0\} = (-1)^{n+1} \Delta_{B_1, \ldots, B_n} h(\emptyset).$$

For $1 \leq j \leq n+1$, let ν_j denote the restriction of $\mathcal{L}(\xi_j)$ to the set $A_j = \bigcap_{i<j}\{\mu B_i > 0\}$, and put $\nu = \sum_j \nu_j$. We may take ξ to be the canonical point process on S with distribution ν.

To see that ξ has hitting function h, we note that for any $U \in \mathcal{U}$ and $j \leq n$,

$$\begin{aligned}
\nu_j\{\mu U > 0\} &= P\{\xi_j B_1 > 0, \ldots, \xi_j B_{j-1} > 0, \xi_j U > 0\} \\
&= (-1)^{j+1} \Delta_{B_1, \ldots, B_{j-1}, U} h_j(\emptyset) \\
&= (-1)^{j+1} \Delta_{B_1, \ldots, B_{j-1}, U} h(B_j).
\end{aligned}$$

It remains to show that, for any $U \in \mathcal{U} \setminus \{\emptyset\}$,

$$\sum_{j \le n}(-1)^{j+1}\Delta_{B_1,\ldots,B_{j-1},U}h(B_j) + (-1)^{n+1}\Delta_{B_1,\ldots,B_n}h(\emptyset) = h(U).$$

This is clear from the fact that

$$\Delta_{B_1,\ldots,B_{j-1},U}h(B_j) = \Delta_{B_1,\ldots,B_j,U}h(\emptyset) + \Delta_{B_1,\ldots,B_{j-1},U}h(\emptyset). \qquad \Box$$

Proof of Theorem 24.22: The direct assertion can be proved in the same way as Corollary 24.16. Conversely, let $\mathcal{U}$ and p be as stated. By Lemma A2.7 we may assume $\mathcal{U}$ to be countable, say $\mathcal{U} = \{U_1, U_2, \ldots\}$. For each n, let $\mathcal{U}_n$ be the class of unions formed from $U_1, \ldots, U_n$. By Lemma 24.23 there exist some point processes $\xi_1, \xi_2, \ldots$ on S such that

$$P\{\xi_n U > 0\} = p(U), \quad U \in \mathcal{U}_n, \ n \in \mathbb{N}.$$

The space $\mathcal{F}$ is compact by Theorem A2.5, and so by Theorem 16.3 there exists some random closed set φ in S such that $\mathrm{supp}\,\xi_n \xrightarrow{d} \varphi$ along a subsequence $N' \subset \mathbb{N}$. Writing h_n and h for the associated hitting functions, we get

$$h(B^\circ) \le \liminf_{n \in N'} h_n(B) \le \limsup_{n \in N'} h_n(B) = h(\overline{B}), \quad B \in \hat{\mathcal{S}},$$

and in particular,

$$h(U^\circ) \le p(U) \le h(\overline{U}), \quad U \in \mathcal{U}.$$

Using the strengthened separation property $K \subset U^\circ \subset \overline{U} \subset G$, we may easily conclude that $h = p^\circ$ on $\mathcal{G}$ and $h = \bar{p}$ on $\mathcal{K}$. $\qquad \Box$

Exercises

1. For a domain $D \subset \mathbb{R}^2$ and point $x \in \partial D$, assume that $x \in I \subset D^c$ for some line segment I. Show that x is regular for D^c. (*Hint:* Consider the windings around x of Brownian motion starting at x, using the strong Markov property and Brownian scaling.)

2. Compute the Newtonian potential kernel $g = g^D$ when $D = \mathbb{R}^d$ with $d \ge 3$, and check by direct computation that $g(x, y)$ is harmonic in $x \ne y$ for fixed y.

3. For any domain $D \subset \mathbb{R}^d$, show that $p_t(x, y) - p_t^D(x, y) \to 0$ as $t \to 0$, uniformly for $x \ne y$ in a compact set $K \subset D$. Also prove the same convergence as $\inf\{|x|; x \notin D\} \to \infty$, uniformly for bounded $t > 0$ and $x \ne y$. (*Hint:* Note that $p_t(x, y)$ is uniformly bounded for $|x - y| > \varepsilon > 0$, and use (8).)

4. Given a domain $D \subset \mathbb{R}^d$ with $d \ge 3$, show that $g(x, y) - g^D(x, y)$ is uniformly bounded for $x \ne y$ in a compact set $K \subset D$. Also show that the difference tends to 0 as $\inf\{|x|; x \notin D\} \to \infty$, uniformly for $x \ne y$ in K. (*Hint:* Use Lemma 24.13.)

5. Show that the equilibrium measure μ_K^D is restricted to the outer boundary of K and agrees for all sets K with the same outer boundary. (Here the *outer boundary* of K consists of all points $x \in \partial K$ that can be connected to D^c or ∞ by a path through K^c.) Prove a corresponding statement for the sweeping measure ν in Theorem 24.12.)

6. For any Greenian domain $D \subset \mathbb{R}^d$, disjoint sets $K_1, \ldots, K_n \in \mathcal{K}_D^r$, and constants $p_1, \ldots, p_d \in \mathbb{R}$, show that there exists a unique signed measure ν on $\bigcup_j K_j$ with $G^D \nu = p_j$ on K_j for all j. (*Hint:* Use Corollary 24.17 recursively.)

7. Show that if φ_1 and φ_2 are independent random sets with distributions π_{ν_1} and π_{ν_2}, then $\varphi_1 \cup \varphi_2$ has distribution $\pi_{\nu_1 + \nu_2}$.

8. Extend Theorem 24.22 to unbounded functions p. (*Hint:* Consider the restrictions to compact sets, and proceed as in Theorem 24.21.)

Chapter 25

Predictability, Compensation, and Excessive Functions

Accessible and predictable times; natural and predictable processes; Doob–Meyer decomposition; quasi-left-continuity; compensation of random measures; excessive and superharmonic functions; additive functionals as compensators; Riesz decomposition

The purpose of this chapter is to present some fundamental, yet profound, extensions of the theory of martingales and optional times from Chapter 7. A basic role in the advanced theory is played by the notions of predictable times and processes, as well as by various decomposition theorems, the most important being the celebrated Doob–Meyer decomposition, a continuous-time counterpart of the elementary Doob decomposition from Lemma 7.10.

Applying the Doob–Meyer decomposition to increasing processes and their associated random measures leads to the notion of a compensator, whose role is analogous to that of the quadratic variation for martingales. In particular, the compensator can be used to transform a fairly general point process to Poisson, in a similar way that a suitable time-change of a continuous martingale was shown in Chapter 18 to lead to a Brownian motion.

The chapter concludes with some applications to classical potential theory. To explain the main ideas, let f be an excessive function of Brownian motion X on $\mathbb{R}^d$. Then $f(X)$ is a continuous supermartingale under P_x for every x, and so it has a Doob–Meyer decomposition $M - A$. Here A can be chosen to be a continuous additive functional (CAF) of X, and we obtain an associated Riesz decomposition $f = U_A + h$, where U_A denotes the potential of A and h is the greatest harmonic minorant of f.

The present material is related in many ways to topics from earlier chapters. Apart from the already mentioned connections, we shall occasionally require some knowledge of random measures and point processes from Chapter 12, of stable Lévy processes from Chapter 15, of stochastic calculus from Chapter 17, of Feller processes from Chapter 19, of additive functionals and their potentials from Chapter 22, and of Green potentials from Chapter 24. The notions and results of this chapter play a crucial role for the analysis of semimartingales and construction of general stochastic integrals in Chapter 26.

All random objects in this chapter are assumed to be defined on some given probability space Ω with a right-continuous and complete filtration $\mathcal{F}$. In the product space $\Omega \times \mathbb{R}_+$ we may introduce the *predictable σ-field* $\mathcal{P}$, generated by all continuous, adapted processes on $\mathbb{R}_+$. The elements of $\mathcal{P}$ are called *predictable sets*, and the $\mathcal{P}$-measurable functions on $\Omega \times \mathbb{R}_+$ are called *predictable processes*. Note that every predictable process is progressive.

The following lemma provides some useful characterizations of the predictable σ-field.

Lemma 25.1 *(predictable σ-field) The predictable σ-field is generated by each of the following classes of sets or processes:*

(i) $\mathcal{F}_0 \times \mathbb{R}_+$ *and the sets* $A \times (t, \infty)$ *with* $A \in \mathcal{F}_t$, $t \geq 0$;

(ii) $\mathcal{F}_0 \times \mathbb{R}_+$ *and the intervals* (τ, ∞) *for optional times* τ;

(iii) *the left-continuous, adapted processes.*

Proof: Let $\mathcal{P}_1$, $\mathcal{P}_2$, and $\mathcal{P}_3$ be the σ-fields generated by the classes in (i), (ii), and (iii), respectively. Since continuous functions are left-continuous, we have trivially $\mathcal{P} \subset \mathcal{P}_3$. To see that $\mathcal{P}_3 \subset \mathcal{P}_1$, it is enough to note that any left-continuous process X can be approximated by the processes

$$X_t^n = X_0 1_{[0,1]}(nt) + \sum_{k \geq 1} X_{k/n} 1_{(k, k+1]}(nt), \quad t \geq 0.$$

Next we obtain $\mathcal{P}_1 \subset \mathcal{P}_2$ by noting that the random time $t_A = t \cdot 1_A + \infty \cdot 1_{A^c}$ is optional for any $t \geq 0$ and $A \in \mathcal{F}_t$. Finally, we may prove the relation $\mathcal{P}_2 \subset \mathcal{P}$ by noting that, for any optional time τ, the process $1_{(\tau, \infty)}$ can be approximated by the continuous, adapted processes $X_t^n = (n(t - \tau)_+) \wedge 1$, $t \geq 0$. $\square$

A random variable τ in $[0, \infty]$ is called a *predictable time* if it is *announced* by some optional times $\tau_n \uparrow \tau$ with $\tau_n < \tau$ a.s. on $\{\tau > 0\}$ for all n. With any optional time τ we may associate the σ-field $\mathcal{F}_{\tau-}$ generated by $\mathcal{F}_0$ and the classes $\mathcal{F}_t \cap \{\tau > t\}$ for arbitrary $t > 0$. The following result gives the basic properties of the σ-fields $\mathcal{F}_{\tau-}$. It is interesting to note the similarity with the results for the σ-fields $\mathcal{F}_\tau$ in Lemma 7.1.

Lemma 25.2 *(strict past) For any optional times σ and τ, we have*

(i) $\mathcal{F}_\sigma \cap \{\sigma < \tau\} \subset \mathcal{F}_{\tau-} \subset \mathcal{F}_\tau$;

(ii) *if τ is predictable, then* $\{\sigma < \tau\} \in \mathcal{F}_{\sigma-} \cap \mathcal{F}_{\tau-}$;

(iii) *if τ is predictable and announced by (τ_n), then* $\bigvee_n \mathcal{F}_{\tau_n} = \mathcal{F}_{\tau-}$.

Proof: (i) For any $A \in \mathcal{F}_\sigma$ we note that

$$A \cap \{\sigma < \tau\} = \bigcup_{r \in \mathbb{Q}_+} (A \cap \{\sigma \leq r\} \cap \{r < \tau\}) \in \mathcal{F}_{\tau-},$$

since the intersections on the right are generators of $\mathcal{F}_{\tau-}$. Hence, $\mathcal{F}_\sigma \cap \{\sigma < \tau\} \in \mathcal{F}_{\tau-}$. The second relation holds since each generator of $\mathcal{F}_{\tau-}$ lies in $\mathcal{F}_\tau$.

(ii) Assuming that (τ_n) announces τ, we get by (i)
$$\{\tau \le \sigma\} = \{\tau = 0\} \cup \bigcap_n \{\tau_n < \sigma\} \in \mathcal{F}_{\sigma-}.$$

(iii) For any $A \in \mathcal{F}_{\tau_n}$ we get by (i)
$$A = (A \cap \{\tau_n < \tau\}) \cup (A \cap \{\tau_n = \tau = 0\}) \in \mathcal{F}_{\tau-},$$
and so $\bigvee_n \mathcal{F}_{\tau_n} \subset \mathcal{F}_{\tau-}$. Conversely, (i) yields for any $t \ge 0$ and $A \in \mathcal{F}_t$
$$A \cap \{\tau > t\} = \bigcup_n (A \cap \{\tau_n > t\}) \in \bigvee_n \mathcal{F}_{\tau_n-} \subset \bigvee_n \mathcal{F}_{\tau_n},$$
which shows that $\mathcal{F}_{\tau-} \subset \bigvee_n \mathcal{F}_{\tau_n}$. □

Next we examine the relationship between predictable processes and the σ-fields $\mathcal{F}_{\tau-}$. Similar results for progressive processes and the σ-fields $\mathcal{F}_\tau$ were obtained in Lemma 7.5.

Lemma 25.3 *(predictability and strict past)*

(i) *For any optional time τ and predictable process X, the random variable $X_\tau 1\{\tau < \infty\}$ is $\mathcal{F}_{\tau-}$-measurable.*

(ii) *For any predictable time τ and $\mathcal{F}_{\tau-}$-measurable random variable α, the process $X_t = \alpha 1\{\tau \le t\}$ is predictable.*

Proof: (i) If $X = 1_{A \times (t,\infty)}$ for some $t > 0$ and $A \in \mathcal{F}_t$, then clearly
$$\{X_\tau 1\{\tau < \infty\} = 1\} = A \cap \{t < \tau < \infty\} \in \mathcal{F}_{\tau-}.$$
We may now extend by a monotone class argument and subsequent approximation, first to arbitrary predictable indicator functions, and then to the general case.

(ii) We may clearly assume α to be integrable. Fixing an announcing sequence (τ_n) for τ, we define
$$X_t^n = E[\alpha|\mathcal{F}_{\tau_n}](1\{0 < \tau_n < t\} + 1\{\tau_n = 0\}), \quad t \ge 0.$$
Then each X^n is left-continuous and adapted, hence predictable. Moreover, $X^n \to X$ on $\mathbb{R}_+$ a.s. by Theorem 7.23 and Lemma 25.2 (iii). □

By a *totally inaccessible time* we mean an optional time τ such that $P\{\sigma = \tau < \infty\} = 0$ for every predictable time σ. An *accessible time* may then be defined as an optional time τ such that $P\{\sigma = \tau < \infty\} = 0$ for every totally inaccessible time σ. For any random time τ, we introduce the associated *graph*
$$[\tau] = \{(t,\omega) \in \mathbb{R}_+ \times \Omega; \ \tau(\omega) = t\},$$
which allows us to express the previous condition on σ and τ as $[\sigma] \cap [\tau] = \emptyset$ a.s. Given any optional time τ and set $A \in \mathcal{F}_\tau$, the time $\tau_A = \tau 1_A + \infty \cdot 1_{A^c}$ is again optional and is called the *restriction* of τ to A. We now consider a basic decomposition of optional times. Related decompositions of increasing processes and martingales are given in Propositions 25.17 and 26.16.

Proposition 25.4 *(decomposition of optional times)* *For any optional time τ there exists an a.s. unique set $A \in \mathcal{F}_\tau \cap \{\tau < \infty\}$ such that τ_A is accessible and τ_{A^c} is totally inaccessible. Furthermore, there exist some predictable times $\tau_1, \tau_2, \ldots$ with $[\tau_A] \subset \bigcup_n [\tau_n]$ a.s.*

Proof: Define

$$p = \sup P \bigcup_n \{\tau = \tau_n < \infty\}, \tag{1}$$

where the supremum extends over all sequences of predictable times τ_n. Combining sequences such that the probability in (1) approaches p, we may construct a sequence (τ_n) for which the supremum is attained. For such a maximal sequence, we define A as the union in (1).

To see that τ_A is accessible, let σ be totally inaccessible. Then $[\sigma] \cap [\tau_n] = \emptyset$ a.s. for every n, and so $[\sigma] \cap [\tau_A] = \emptyset$ a.s. If τ_{A^c} is not totally inaccessible, then $P\{\tau_{A^c} = \tau_0 < \infty\} > 0$ for some predictable time τ_0, which contradicts the maximality of $\tau_1, \tau_2, \ldots$. This shows that A has the desired property.

To prove that A is a.s. unique, let B be another set with the stated properties. Then $\tau_{A \setminus B}$ and $\tau_{B \setminus A}$ are both accessible and totally inaccessible, and so $\tau_{A \setminus B} = \tau_{B \setminus A} = \infty$ a.s., which implies $A = B$ a.s. □

We proceed to establish a version of the celebrated *Doob–Meyer decomposition*, a cornerstone in modern probability theory. By an *increasing process* we mean a nondecreasing, right-continuous, and adapted process A with $A_0 = 0$. We say that A is *integrable* if $EA_\infty < \infty$. Recall that all submartingales are assumed to be right-continuous. Local submartingales and locally integrable processes are defined by localization in the usual way.

Theorem 25.5 *(decomposition of submartingales, Meyer, Doléans)* *A process X is a local submartingale iff it has a decomposition $X = M + A$, where M is a local martingale and A is a locally integrable, increasing, predictable process. In that case M and A are a.s. unique.*

The process A in the statement is often referred to as the *compensator* of X, especially when X is increasing. Several proofs of this result are known, most of which seem to require the deep section theorems. Here we give a relatively short and elementary proof, based on Dunford's weak compactness criterion and an approximation of totally inaccessible times. For convenience, we divide the proof into several lemmas.

Let (D) denote the class of measurable processes X such that the family $\{X_\tau\}$ is uniformly integrable, where τ ranges over the set of all finite optional times. By the following result it is enough to consider class (D) submartingales.

Lemma 25.6 *(uniform integrability)* *Any local submartingale X with $X_0 = 0$ is locally of class (D).*

Proof: First reduce to the case when X is a true submartingale. Then introduce for each n the optional time $\tau = n \wedge \inf\{t > 0; |X_t| > n\}$. Here

$|X^\tau| \le n \vee |X_\tau|$, which is integrable by Theorem 7.29, and so X^τ is of class (D). □

An increasing process A is said to be *natural* if it is integrable and such that $E \int_0^\infty \Delta M_t \, dA_t = 0$ for any bounded martingale M. As a crucial step in the proof of Theorem 25.5, we may establish the following preliminary decomposition, where the compensator A is shown to be natural rather than predictable.

Lemma 25.7 (Meyer) *Any submartingale X of class (D) has a decomposition $X = M + A$, where M is a uniformly integrable martingale and A is a natural, increasing process.*

Proof (Rao): We may assume that $X_0 = 0$. Introduce the n-dyadic times $t_k^n = k 2^{-n}$, $k \in \mathbb{Z}_+$, and define for any process Y the associated differences $\Delta_k^n Y = Y_{t_{k+1}^n} - Y_{t_k^n}$. Let

$$A_t^n = \sum_{k < 2^n t} E[\Delta_k^n X | \mathcal{F}_{t_k^n}], \quad t \ge 0, \, n \in \mathbb{N},$$

and note that $M^n = X - A^n$ is a martingale on the n-dyadic set.

Writing $\tau_r^n = \inf\{t; A_t^n > r\}$ for $n \in \mathbb{N}$ and $r > 0$, we get by optional sampling, for any n-dyadic time t,

$$\tfrac{1}{2} E[A_t^n; A_t^n > 2r] \le E[A_t^n - A_t^n \wedge r] \le E[A_t^n - A_{\tau_r^n \wedge t}^n]$$
$$= E[X_t - X_{\tau_r^n \wedge t}] = E[X_t - X_{\tau_r^n \wedge t}; A_t^n > r]. \quad (2)$$

By the martingale property and uniform integrability, we further obtain

$$r P\{A_t^n > r\} \le E A_t^n = E X_t \le 1,$$

and so the probability on the left tends to zero as $r \to \infty$, uniformly in t and n. Since the random variables $X_t - X_{\tau_r^n \wedge t}$ are uniformly integrable by (D), the same property holds for the variables A_t^n by (2) and Lemma 4.10. In particular, the sequence (A_∞^n) is uniformly integrable, and each M^n is a uniformly integrable martingale.

By Lemma 4.13 there exists some random variable $\alpha \in L^1(\mathcal{F}_\infty)$ such that $A_\infty^n \to \alpha$ weakly in L^1 along some subsequence $N' \subset \mathbb{N}$. Define

$$M_t = E[X_\infty - \alpha | \mathcal{F}_t], \quad A = X - M,$$

and note that $A_\infty = \alpha$ a.s. by Theorem 7.23. For any dyadic t and bounded random variable ξ, we get by the martingale and self-adjointness properties

$$E(A_t^n - A_t)\xi = E(M_t - M_t^n)\xi = E\, E[M_\infty - M_\infty^n | \mathcal{F}_t]\xi$$
$$= E(M_\infty - M_\infty^n) E[\xi | \mathcal{F}_t]$$
$$= E(A_\infty^n - \alpha) E[\xi | \mathcal{F}_t] \to 0,$$

as $n \to \infty$ along N'. Thus, $A_t^n \to A_t$ weakly in L^1 for dyadic t. In particular, we get for any dyadic $s < t$

$$0 \le E[A_t^n - A_s^n; A_t - A_s < 0] \to E[(A_t - A_s) \wedge 0] \le 0.$$

Thus, the last expectation vanishes, and therefore $A_t \geq A_s$ a.s. By right-continuity it follows that A is a.s. nondecreasing. Also note that $A_0 = 0$ a.s. since $A_0^n = 0$ for all n.

To see that A is natural, consider any bounded martingale N, and conclude by Fubini's theorem and the martingale properties of N and $A^n - A = M - M^n$ that

$$EN_\infty A_\infty^n = \sum_k EN_\infty \Delta_k^n A^n = \sum_k EN_{t_k^n} \Delta_k^n A^n$$
$$= \sum_k EN_{t_k^n} \Delta_k^n A = E\sum_k N_{t_k^n} \Delta_k^n A.$$

Now use weak convergence on the left and dominated convergence on the right, and combine with Fubini's theorem and the martingale property of N to get

$$E\int_0^\infty N_{t-} dA_t = EN_\infty A_\infty = \sum_k EN_\infty \Delta_k^n A = \sum_k EN_{t_{k+1}^n} \Delta_k^n A$$
$$= E\sum_k N_{t_{k+1}^n} \Delta_k^n A \to E\int_0^\infty N_t dA_t.$$

Hence, $E\int_0^\infty \Delta N_t dA_t = 0$, as required. □

To complete the proof of Theorem 25.5, it remains to show that the compensator A in the last lemma is predictable. This will be inferred from the following ingenious approximation of totally inaccessible times.

Lemma 25.8 *(uniform approximation, Doob)* *For any totally inaccessible time τ, put $\tau_n = 2^{-n}[2^n\tau]$, and let X^n be a right-continuous version of the process $P[\tau_n \leq t | \mathcal{F}_t]$. Then*

$$\lim_{n\to\infty} \sup_{t\geq 0} |X_t^n - 1\{\tau \leq t\}| = 0 \quad \text{a.s.} \qquad (3)$$

Proof: Since $\tau_n \uparrow \tau$, we may assume that $X_t^1 \geq X_t^2 \geq \cdots \geq 1\{\tau \leq t\}$ for all $t \geq 0$. Then $X_t^n = 1$ for $t \in [\tau, \infty)$, and on the set $\{\tau = \infty\}$ we have $X_t^1 \leq P[\tau < \infty|\mathcal{F}_t] \to 0$ a.s. as $t \to \infty$ by Theorem 7.23. Thus, $\sup_n |X_t^n - 1\{\tau \leq t\}| \to 0$ a.s. as $t \to \infty$. To prove (3), it is then enough to show for every $\varepsilon > 0$ that the optional times

$$\sigma_n = \inf\{t \geq 0;\ X_t^n - 1\{\tau \leq t\} > \varepsilon\}, \quad n \in \mathbb{N},$$

tend a.s. to infinity. The σ_n are clearly nondecreasing, and we denote their limit by σ. Note that either $\sigma_n \leq \tau$ or $\sigma_n = \infty$ for each n.

By optional sampling, Theorem 6.4, and Lemma 7.1, we have

$$X_\sigma^n 1\{\sigma < \infty\} = P[\tau_n \leq \sigma < \infty | \mathcal{F}_\sigma]$$
$$\to P[\tau \leq \sigma < \infty | \mathcal{F}_\sigma] = 1\{\tau \leq \sigma < \infty\}.$$

Hence, $X_\sigma^n \to 1\{\tau \leq \sigma\}$ a.s. on $\{\sigma < \infty\}$, and so by right-continuity we have on this set $\sigma_n < \sigma$ for large enough n. Thus, σ is predictable and announced by the times $\sigma_n \wedge n$.

Next apply the optional sampling and disintegration theorems to the optional times σ_n, to obtain

$$\begin{aligned}\varepsilon P\{\sigma < \infty\} &\le \varepsilon P\{\sigma_n < \infty\} \le E[X^n_{\sigma_n}; \sigma_n < \infty] \\ &= P\{\tau_n \le \sigma_n < \infty\} = P\{\tau_n \le \sigma_n \le \tau < \infty\} \\ &\to P\{\tau = \sigma < \infty\} = 0,\end{aligned}$$

where the last equality holds since τ is totally inaccessible. Thus, $\sigma = \infty$ a.s. □

It is now easy to see that A has only accessible jumps.

Lemma 25.9 *(accessibility)* *For any natural increasing process A and totally inaccessible time τ, we have $\Delta A_\tau = 0$ a.s. on $\{\tau < \infty\}$.*

Proof: Rescaling if necessary, we may assume that A is a.s. continuous at dyadic times. Define $\tau_n = 2^{-n}[2^n \tau]$. Since A is natural, we have

$$E \int_0^\infty P[\tau_n > t | \mathcal{F}_t] dA_t = E \int_0^\infty P[\tau_n > t | \mathcal{F}_{t-}] dA_t,$$

and since τ is totally inaccessible, it follows by Lemma 25.8 that

$$EA_{\tau-} = E \int_0^\infty 1\{\tau > t\} dA_t = E \int_0^\infty 1\{\tau \ge t\} dA_t = EA_\tau.$$

Hence, $E[\Delta A_\tau; \tau < \infty] = 0$, and so $\Delta A_\tau = 0$ a.s. on $\{\tau < \infty\}$. □

Finally, we need to show that A is predictable.

Lemma 25.10 *(Doléans)* *Every natural increasing process is predictable.*

Proof: Fix a natural increasing process A. Consider a bounded martingale M and a predictable time $\tau < \infty$ announced by $\sigma_1, \sigma_2, \ldots$. Then $M^\tau - M^{\sigma_k}$ is again a bounded martingale, and since A is natural, we get by dominated convergence $E \Delta M_\tau \Delta A_\tau = 0$. In particular, we may take $M_t = P[B|\mathcal{F}_t]$ with $B \in \mathcal{F}_\tau$. By optional sampling we have $M_\tau = 1_B$ and

$$M_{\tau-} \leftarrow M_{\sigma_k} = P[B|\mathcal{F}_{\sigma_k}] \to P[B|\mathcal{F}_{\tau-}].$$

Thus, $\Delta M_\tau = 1_B - P[B|\mathcal{F}_{\tau-}]$, and so

$$E[\Delta A_\tau; B] = E \Delta A_\tau P[B|\mathcal{F}_{\tau-}] = E[E[\Delta A_\tau|\mathcal{F}_{\tau-}]; B].$$

Since B was arbitrary in $\mathcal{F}_\tau$, we get $\Delta A_\tau = E[\Delta A_\tau|\mathcal{F}_{\tau-}]$ a.s., and so the process $A'_t = \Delta A_\tau 1\{\tau \le t\}$ is predictable by Lemma 25.3 (ii). It is also natural, since for any bounded martingale M

$$E \Delta A_\tau \Delta M_\tau = E \Delta A_\tau E[\Delta M_\tau | \mathcal{F}_{\tau-}] = 0.$$

By an elementary construction we have $\{t > 0; \Delta A_t > 0\} \subset \bigcup_n [\tau_n]$ a.s. for some optional times $\tau_n < \infty$, and by Proposition 25.4 and Lemma 25.9 we may assume the latter to be predictable. Taking $\tau = \tau_1$ in the previous argument, we may conclude that the process $A^1_t = \Delta A_{\tau_1} 1\{\tau_1 \le t\}$ is both natural and predictable. Repeating the argument for the process $A - A^1$ with $\tau = \tau_2$ and proceeding by induction, we may conclude that the jump component A^d of A is predictable. Since $A - A^d$ is continuous and hence predictable, the predictability of A follows. □

For the uniqueness assertion we need the following extension of Proposition 17.2.

Lemma 25.11 *(constancy criterion)* *A process M is a predictable martingale of integrable variation iff $M_t \equiv M_0$ a.s.*

Proof: On the predictable σ-field $\mathcal{P}$ we define the signed measure

$$\mu B = E \int_0^\infty 1_B(t) dM_t, \quad B \in \mathcal{P},$$

where the inner integral is an ordinary Lebesgue–Stieltjes integral. The martingale property implies that μ vanishes for sets B of the form $F \times (t, \infty)$ with $F \in \mathcal{F}_t$. By Lemma 25.1 and a monotone class argument it follows that $\mu = 0$ on $\mathcal{P}$.

Since M is predictable, the same thing is true for the process $\Delta M_t = M_t - M_{t-}$, and then also for the sets $J_\pm = \{t > 0; \pm \Delta M_t > 0\}$. Thus, $\mu J_\pm = 0$, and so $\Delta M = 0$ a.s., which means that M is a.s. continuous. But then $M_t \equiv M_0$ a.s. by Proposition 17.2. □

Proof of Theorem 25.5: The sufficiency is obvious, and the uniqueness holds by Lemma 25.11. It remains to prove that any local submartingale X has the stated decomposition. By Lemmas 25.6 and 25.11 we may assume that X is of class (D). Then Lemma 25.7 shows that $X = M + A$ for some uniformly integrable martingale M and some natural increasing process A, and by Lemma 25.10 the latter process is predictable. □

The two conditions in Lemma 25.10 are, in fact, equivalent.

Theorem 25.12 *(natural and predictable processes, Doléans)* *An integrable, increasing process is natural iff it is predictable.*

Proof: If an integrable, increasing process A is natural, it is also predictable by Lemma 25.10. Now assume instead that A is predictable. By Lemma 25.7 we have $A = M + B$ for some uniformly integrable martingale M and some natural increasing process B, and Lemma 25.10 shows that B is predictable. But then $A = B$ a.s. by Lemma 25.11, and so A is natural. □

The following useful result is essentially implicit in earlier proofs.

Lemma 25.13 *(dual predictable projection)* Let X and Y be locally integrable, increasing processes, and assume that Y is predictable. Then X has compensator Y iff $E\int V dX = E\int V dY$ for every predictable process $V \geq 0$.

Proof: First reduce by localization to the case when X and Y are integrable. Then Y is the compensator of X iff $M = Y - X$ is a martingale or, equivalently, iff $EM_\tau = 0$ for every optional time τ. This is equivalent to the stated relation for $V = 1_{[0,\tau]}$, and the general result follows by a straightforward monotone class argument. $\square$

We may now establish the fundamental connection between predictable times and processes.

Theorem 25.14 *(predictable times and processes, Meyer)* For any optional time τ, these conditions are equivalent:

(i) τ is predictable;

(ii) the process $1\{\tau \leq t\}$ is predictable;

(iii) $E\Delta M_\tau = 0$ for any bounded martingale M.

Proof (Chung and Walsh): Since (i) $\Rightarrow$ (ii) by Lemma 25.3 (ii), and (ii) $\Leftrightarrow$ (iii) by Theorem 25.12, it remains to show that (iii) $\Rightarrow$ (i). We then introduce the martingale $M_t = E[e^{-\tau}|\mathcal{F}_t]$ and the supermartingale

$$X_t = e^{-\tau \wedge t} - M_t = E[e^{-\tau \wedge t} - e^{-\tau}|\mathcal{F}_t] \geq 0, \quad t \geq 0.$$

Here $X_\tau = 0$ a.s. by optional sampling. Letting $\sigma = \inf\{t \geq 0;\ X_{t-} \wedge X_t = 0\}$, we see from Lemma 7.31 that $\{t \geq 0;\ X_t = 0\} = [\sigma, \infty)$ a.s., and in particular $\sigma \leq \tau$ a.s. Using optional sampling again, we get $E(e^{-\sigma} - e^{-\tau}) = EX_\sigma = 0$, and so $\sigma = \tau$ a.s. Hence, $X_t \wedge X_{t-} > 0$ a.s. on $[0, \tau)$. Finally, (iii) yields

$$EX_{\tau-} = E(e^{-\tau} - M_{\tau-}) = E(e^{-\tau} - M_\tau) = EX_\tau = 0,$$

and so $X_{\tau-} = 0$. It is now clear that τ is announced by the optional times $\tau_n = \inf\{t;\ X_t < n^{-1}\}$. $\square$

To illustrate the power of the last result, we may give a short proof of the following useful statement, which can also be proved directly.

Corollary 25.15 *(restriction)* For any predictable time τ and set $A \in \mathcal{F}_{\tau-}$, the restriction τ_A is again predictable.

Proof: The process $1_A 1\{\tau \leq t\} = 1\{\tau_A \leq t\}$ is predictable by Lemma 25.3, and so the time τ_A is predictable by Theorem 25.14. $\square$

We may also use the last theorem to show that predictable martingales are continuous.

Proposition 25.16 *(predictable martingales) A local martingale is predictable iff it is a.s. continuous.*

Proof: The sufficiency is clear by definitions. To prove the necessity, we note that, for any optional time τ,

$$M_t^\tau = M_t 1_{[0,\tau]}(t) + M_\tau 1_{(\tau,\infty)}(t), \quad t \geq 0.$$

Thus, predictability is preserved by optional stopping, and so we may assume that M is a uniformly integrable martingale. Now fix any $\varepsilon > 0$, and introduce the optional time $\tau = \inf\{t > 0; |\Delta M_t| > \varepsilon\}$. Since the left-continuous version M_{t-} is predictable, so is the process ΔM_t as well as the random set $A = \{t > 0; |\Delta M_t| > \varepsilon\}$. Hence, the same thing is true for the random interval $[\tau, \infty) = A \cup (\tau, \infty)$, and therefore τ is predictable by Theorem 25.14. Choosing an announcing sequence (τ_n), we conclude by optional sampling, martingale convergence, and Lemmas 25.2 (iii) and 25.3 (i) that

$$M_{\tau-} \leftarrow M_{\tau_n} = E[M_\tau | \mathcal{F}_{\tau_n}] \to E[M_\tau | \mathcal{F}_{\tau-}] = M_\tau.$$

Thus, $\tau = \infty$ a.s. Since ε was arbitrary, it follows that M is a.s. continuous. □

The decomposition of optional times in Proposition 25.4 may now be extended to increasing processes. We say that an rcll process X or a filtration $\mathcal{F}$ is *quasi-leftcontinuous* if $X_{\tau-} = X_\tau$ a.s. on $\{\tau < \infty\}$ or $\mathcal{F}_{\tau-} = \mathcal{F}_\tau$, respectively, for every predictable time τ. We further say that X has *accessible jumps* if $X_{\tau-} = X_\tau$ a.s. on $\{\tau < \infty\}$ for every totally inaccessible time τ.

Proposition 25.17 *(decomposition of increasing processes) Any purely discontinuous, increasing process A has an a.s. unique decomposition into increasing processes A^q and A^a, where A^q is quasi-leftcontinuous and A^a has accessible jumps. Furthermore, there exist some predictable times $\tau_1, \tau_2, \ldots$ with disjoint graphs such that $\{t > 0; \Delta A_t^a > 0\} \subset \bigcup_n [\tau_n]$ a.s. Finally, if A is locally integrable with compensator $\hat{A}$, then A^q has compensator $(\hat{A})^c$.*

Proof: Introduce the locally integrable process $X_t = \sum_{s \leq t}(\Delta A_s \wedge 1)$ with compensator $\hat{X}$, and define $A^q = A - A^a = 1\{\Delta \hat{X} = 0\} \cdot A$, or

$$A_t^q = A_t - A_t^a = \int_0^{t+} 1\{\Delta \hat{X}_s = 0\} dA_s, \quad t \geq 0. \tag{4}$$

For any finite predictable time τ, the graph $[\tau]$ is again predictable by Theorem 25.14, and so by Lemma 25.13,

$$E(\Delta A_\tau^q \wedge 1) = E[\Delta X_\tau; \Delta \hat{X}_\tau = 0] = E[\Delta \hat{X}_\tau; \Delta \hat{X}_\tau = 0] = 0,$$

which shows that A^q is quasi-leftcontinuous.

Now let $\tau_{n,0} = 0$, and recursively define the random times

$$\tau_{n,k} = \inf\{t > \tau_{n,k-1};\ \Delta \hat{X}_t \in (2^{-n}, 2^{-n+1}]\}, \quad n, k \in \mathbb{N},$$

which are predictable by Theorem 25.14. Also note that $\{t > 0;\ \Delta A_t^a > 0\} \subset \bigcup_{n,k}[\tau_{nk}]$ a.s. by the definition of A^a. Hence, if τ is a totally inaccessible time, then $\Delta A_\tau^a = 0$ a.s. on $\{\tau < \infty\}$, which shows that A^a has accessible jumps.

To prove the uniqueness, assume that A has two decompositions $A^q + A^a = B^q + B^a$ with the stated properties. Then $Y = A^q - B^q = B^a - A^a$ is quasi-leftcontinuous with accessible jumps. Hence, by Proposition 25.4 we have $\Delta Y_\tau = 0$ a.s. on $\{\tau < \infty\}$ for any optional time τ, which means that Y is a.s. continuous. Since it is also purely discontinuous, we get $Y = 0$ a.s.

If A is locally integrable, we may replace (4) by $A^q = 1\{\Delta \hat{A} = 0\} \cdot A$, and we also note that $(\hat{A})^c = 1\{\Delta \hat{A} = 0\} \cdot \hat{A}$. Thus, Lemma 25.13 yields for any predictable process $V \geq 0$

$$E \int V dA^q = E \int 1\{\Delta \hat{A} = 0\} V dA$$
$$= E \int 1\{\Delta \hat{A} = 0\} V d\hat{A} = E \int V d(\hat{A})^c,$$

and the same lemma shows that A^q has compensator $(\hat{A})^c$. □

By the *compensator* of an optional time τ we mean the compensator of the associated jump process $X_t = 1\{\tau \leq t\}$. The following result characterizes the special categories of optional times in terms of the associated compensators.

Corollary 25.18 *(compensation of optional times)* *Let τ be an optional time with compensator A. Then*

   (i) *τ is predictable iff A is a.s. constant apart from a possible unit jump;*

   (ii) *τ is accessible iff A is a.s. purely discontinuous;*

   (iii) *τ is totally inaccessible iff A is a.s. continuous.*

In general, τ has the accessible part τ_D, where $D = \{\Delta A_\tau > 0, \tau < \infty\}$.

Proof: (i) If τ is predictable, then so is the process $X_t = 1\{\tau \leq t\}$ by Theorem 25.14, and hence $A = X$ a.s. Conversely, if $A_t = 1\{\sigma \leq t\}$ for some optional time σ, then the latter is predictable by Theorem 25.14, and Lemma 25.13 yields

$$P\{\sigma = \tau < \infty\} = E[\Delta X_\sigma;\ \sigma < \infty] = E[\Delta A_\sigma;\ \sigma < \infty]$$
$$= P\{\sigma < \infty\} = EA_\infty = EX_\infty = P\{\tau < \infty\}.$$

Thus, $\tau = \sigma$ a.s., and so τ is predictable.

(ii) Clearly, τ is accessible iff X has accessible jumps, which holds by Proposition 25.17 iff $A = A^d$ a.s.

(iii) Here we note that τ is totally inaccessible iff X is quasi-leftcontinuous, which holds by Proposition 25.17 iff $A = A^c$ a.s.

The last assertion follows easily from (ii) and (iii). □

The next result characterizes quasi-left-continuity for both filtrations and martingales.

Proposition 25.19 *(quasi-leftcontinuous filtrations, Meyer)* *For any filtration $\mathcal{F}$, these conditions are equivalent:*

(i) *Every accessible time is predictable;*

(ii) $\mathcal{F}_{\tau-} = \mathcal{F}_\tau$ *on $\{\tau < \infty\}$ for every predictable time τ;*

(iii) $\Delta M_\tau = 0$ *a.s. on $\{\tau < \infty\}$ for every martingale M and predictable time τ.*

If the basic σ-field in Ω is taken to be $\mathcal{F}_\infty$, then $\mathcal{F}_{\tau-} = \mathcal{F}_\tau$ on $\{\tau = \infty\}$ for any optional time τ, and the relation in (ii) extends to all of Ω.

Proof: (i) ⇒ (ii): Let τ be a predictable time, and fix any $B \in \mathcal{F}_\tau \cap \{\tau < \infty\}$. Then $[\tau_B] \subset [\tau]$, and so τ_B is accessible, hence by (i) even predictable. The process $X_t = 1\{\tau_B \leq t\}$ is then predictable by Theorem 25.14, and since

$$X_\tau 1\{\tau < \infty\} = 1\{\tau_B \leq \tau < \infty\} = 1_B,$$

Lemma 25.3 (i) yields $B \in \mathcal{F}_{\tau-}$.

(ii) ⇒ (iii): Fix any martingale M, and let τ be a bounded, predictable time with announcing sequence (τ_n). Using (ii) and Lemma 25.2 (iii), we get as before

$$M_{\tau-} \leftarrow M_{\tau_n} = E[M_\tau | \mathcal{F}_{\tau_n}] \to E[M_\tau | \mathcal{F}_{\tau-}] = E[M_\tau | \mathcal{F}_\tau] = M_\tau,$$

and so $M_{\tau-} = M_\tau$ a.s.

(iii) ⇒ (i): If τ is accessible, then by Proposition 25.4 there exist some predictable times τ_n with $[\tau] \subset \bigcup_n [\tau_n]$ a.s. By (iii) we have $\Delta M_{\tau_n} = 0$ a.s. on $\{\tau_n < \infty\}$ for every martingale M and all n, and so $\Delta M_\tau = 0$ a.s. on $\{\tau < \infty\}$. Hence, τ is predictable by Theorem 25.14. □

In particular, quasi-left-continuity holds for canonical Feller processes and their induced filtrations.

Proposition 25.20 *(quasi-left-continuity of Feller processes, Blumenthal, Meyer)* *Let X be a canonical Feller process with arbitrary initial distribution, and fix any optional time τ. Then these conditions are equivalent:*

(i) τ *is predictable;*

(ii) τ *is accessible;*

(iii) $X_{\tau-} = X_\tau$ *a.s. on $\{\tau < \infty\}$.*

In the special case when X is a.s. continuous, we may conclude that every optional time is predictable.

Proof: (ii) ⇒ (iii): By Proposition 25.4 we may assume that τ is finite and predictable. Fix an announcing sequence (τ_n) and a function $f \in C_0$.

By the strong Markov property, we get for any $h > 0$

$$\begin{aligned}E\{f(X_{\tau_n}) - f(X_{\tau_n+h})\}^2 &= E(f^2 - 2fT_hf + T_hf^2)(X_{\tau_n})\\ &\leq \|f^2 - 2fT_hf + T_hf^2\|\\ &\leq 2\|f\|\,\|f - T_hf\| + \|f^2 - T_hf^2\|.\end{aligned}$$

Letting $n \to \infty$ and then $h \downarrow 0$, it follows by dominated convergence on the left and by strong continuity on the right that $E\{f(X_{\tau-}) - f(X_\tau)\}^2 = 0$, which means that $f(X_{\tau-}) = f(X_\tau)$ a.s. Applying this to a sequence $f_1, f_2, \cdots \in C_0$ that separates points, we obtain $X_{\tau-} = X_\tau$ a.s.

(iii) $\Rightarrow$ (i): By (iii) and Theorem 19.20 we have $\Delta M_\tau = 0$ a.s. on $\{\tau < \infty\}$ for every martingale M, and so τ is predictable by Theorem 25.14.

(i) $\Rightarrow$ (ii): This is trivial. $\square$

The following basic inequality will be needed in the proof of Theorem 26.12.

Proposition 25.21 *(norm inequality, Garsia, Neveu)* *Consider a right- or left-continuous, predictable, increasing process A and a random variable $\zeta \geq 0$ such that a.s.*

$$E[A_\infty - A_t|\mathcal{F}_t] \leq E[\zeta|\mathcal{F}_t], \quad t \geq 0. \tag{5}$$

Then

$$\|A_\infty\|_p \leq p\|\zeta\|_p, \quad p \geq 1.$$

In the left-continuous case, predictability is clearly equivalent to adaptedness. The proper interpretation of (5) is to take $E[A_t|\mathcal{F}_t] \equiv A_t$ and to choose right-continuous versions of the martingales $E[A_\infty|\mathcal{F}_t]$ and $E[\zeta|\mathcal{F}_t]$. For a right-continuous A, we may clearly choose $\zeta = Z^*$, where Z is the supermartingale on the left of (5). We also note that if A is the compensator of an increasing process X, then (5) holds with $\zeta = X_\infty$.

Proof: We need to consider only the right-continuous case, the case of a left-continuous process A being similar but simpler. It is enough to assume that A is bounded, since we may otherwise replace A by the process $A \wedge u$ for arbitrary $u > 0$, and let $u \to \infty$ in the resulting formula. For each $r > 0$, the random time $\tau_r = \inf\{t;\ A_t \geq r\}$ is predictable by Theorem 25.14. By optional sampling and Lemma 25.2 we note that (5) remains true with t replaced by τ_r-. Since τ_r is $\mathcal{F}_{\tau_r-}$-measurable by the same lemma, we obtain

$$\begin{aligned}E[A_\infty - r;\ A_\infty > r] &\leq E[A_\infty - r;\ \tau_r < \infty]\\ &\leq E[A_\infty - A_{\tau_r-};\ \tau_r < \infty]\\ &\leq E[\zeta;\ \tau_r < \infty] \leq E[\zeta;\ A_\infty \geq r].\end{aligned}$$

Writing $A_\infty = \alpha$ and letting $p^{-1} + q^{-1} = 1$, we get by Fubini's theorem, Hölder's inequality, and some calculus

$$\begin{aligned}
\|\alpha\|_p^p &= p^2 q^{-1} E \int_0^\alpha (\alpha - r) r^{p-2} dr \\
&= p^2 q^{-1} \int_0^\infty E[\alpha - r; \alpha > r] r^{p-2} dr \\
&\leq p^2 q^{-1} \int_0^\infty E[\zeta; \alpha \geq r] r^{p-2} dr \\
&= p^2 q^{-1} E\zeta \int_0^\alpha r^{p-2} dr \\
&= pE\zeta\alpha^{p-1} \leq p\|\zeta\|_p \|\alpha\|_p^{p-1}.
\end{aligned}$$

If $\|\alpha\|_p > 0$, we may finally divide both sides by $\|\alpha\|_p^{p-1}$. □

Let us now turn our attention to random measures ξ on $(0, \infty) \times S$, where $(S, \mathcal{S})$ is a Borel space. We say that ξ is *adapted, predictable*, or *locally integrable* if there exists a subring $\hat{\mathcal{S}} \subset \mathcal{S}$ with $\sigma(\hat{\mathcal{S}}) = \mathcal{S}$ such that the process $\xi_t B = \xi((0, t] \times B)$ has the corresponding property for every $B \in \hat{\mathcal{S}}$. In case of adaptedness or predictability, it is clearly equivalent that the relevant property holds for the measure-valued process ξ_t. Let us further say that a process V on $\mathbb{R}_+ \times S$ is *predictable* if it is $\mathcal{P} \otimes \mathcal{S}$-measurable, where $\mathcal{P}$ denotes the predictable σ-field in $\mathbb{R}_+ \times \Omega$.

Theorem 25.22 *(compensation of random measures, Grigelionis, Jacod)*
Let ξ be a locally integrable, adapted random measure on some product space $(0, \infty) \times S$, where S is Borel. Then there exists an a.s. unique predictable random measure $\hat{\xi}$ on $(0, \infty) \times S$ such that $E \int V d\xi = E \int V d\hat{\xi}$ for every predictable process $V \geq 0$ on $\mathbb{R}_+ \times S$.

The random measure $\hat{\xi}$ above is called the *compensator* of ξ. By Lemma 25.13 this extends the notion of compensator for real-valued processes. For the proof of Theorem 25.22 we need a simple technical lemma, which can be established by straightforward monotone class arguments.

Lemma 25.23 *(predictable random measures)*

(i) *For any predictable random measure ξ and predictable process $V \geq 0$ on $(0, \infty) \times S$, the process $V \cdot \xi$ is again predictable.*

(ii) *For any predictable process $V \geq 0$ on $(0, \infty) \times S$ and predictable, measure-valued process ρ on S, the process $Y_t = \int V_{t,s} \rho_t(ds)$ is again predictable.*

Proof of Theorem 25.22: Since ξ is locally integrable, we may easily construct a predictable process $V > 0$ on $\mathbb{R}_+ \times S$ such that $E \int V d\xi < \infty$. If the random measure $\zeta = V \cdot \xi$ has compensator $\hat{\zeta}$, then by Lemma 25.23 the measure $\hat{\xi} = V^{-1} \cdot \hat{\zeta}$ is the compensator of ξ. Thus, we may henceforth assume that $E\xi((0, \infty) \times S) = 1$.

Write $\eta = \xi(\cdot \times S)$. Using the kernel operation $\otimes$ of Chapter 1, we may introduce the probability measure $\mu = P \otimes \xi$ on $\Omega \times \mathbb{R}_+ \times S$ and its projection $\nu = P \otimes \eta$ onto $\Omega \times \mathbb{R}_+$. Applying Theorem 6.3 to the restrictions of μ and ν to the σ-fields $\mathcal{P} \otimes \mathcal{S}$ and $\mathcal{P}$, respectively, we conclude that there exists some probability kernel ρ from $(\Omega \times \mathbb{R}_+, \mathcal{P})$ to $(S, \mathcal{S})$ satisfying $\mu = \nu \otimes \rho$, or

$$P \otimes \xi = P \otimes \eta \otimes \rho \quad \text{on } (\Omega \times \mathbb{R}_+ \times S, \mathcal{P} \times \mathcal{S}).$$

Letting $\hat{\eta}$ denote the compensator of η, we may introduce the random measure $\hat{\xi} = \hat{\eta} \otimes \rho$ on $\mathbb{R}_+ \times S$.

To see that $\hat{\xi}$ is the compensator of ξ, we first note that $\hat{\xi}$ is predictable by Lemma 25.23 (i). Next we consider an arbitrary predictable process $V \geq 0$ on $\mathbb{R}_+ \times S$, and note that the process $Y_s = \int V_{s,t} \rho_t(ds)$ is again predictable by Lemma 25.23 (ii). By Theorem 6.4 and Lemma 25.13 we get

$$E \int V \, d\hat{\xi} = E \int \hat{\eta}(dt) \int V_{s,t} \rho_t(ds)$$
$$= E \int \eta(dt) \int V_{s,t} \rho_t(ds) = E \int V \, d\xi.$$

It remains to note that $\hat{\xi}$ is a.s. unique by Lemma 25.13. □

Our next aim is to show, under a weak regularity condition, how a point process can be transformed to Poisson by means of a suitable predictable mapping. The result leads to various time-change formulas for point processes, similar to those for continuous local martingales in Chapter 18.

Recall that an *S-marked point process* on $(0, \infty)$ is defined as an integer-valued random measure ξ on $(0, \infty) \times S$ such that a.s. $\xi([t] \times S) \leq 1$ for all $t > 0$. The condition implies that ξ is locally integrable, and so the existence of the associated compensator $\hat{\xi}$ is automatic. We say that ξ is *quasi-leftcontinuous* if $\xi([\tau] \times S) = 0$ a.s. for every predictable time τ.

Theorem 25.24 (predictable mapping to Poisson) *Fix a Borel space S and a σ-finite measure space (T, μ), let ξ be a quasi-leftcontinuous S-marked point process on $(0, \infty)$ with compensator $\hat{\xi}$, and let Y be a predictable mapping from $\mathbb{R}_+ \times S$ to T with $\hat{\xi} \circ Y^{-1} = \mu$ a.s. Then $\eta = \xi \circ Y^{-1}$ is a Poisson process on T with $E\eta = \mu$.*

Proof: For any disjoint measurable sets $B_1, \ldots, B_n$ in T with finite μ-measure, we need to show that $\eta B_1, \ldots, \eta B_n$ are independent Poisson random variables with means $\mu B_1, \ldots, \mu B_n$. Then introduce for each $k \leq n$ the processes

$$J_t^k = \int_S \int_0^{t+} 1_{B_k}(Y_{s,x}) \, \xi(ds\, dx), \quad \hat{J}_t^k = \int_S \int_0^t 1_{B_k}(Y_{s,x}) \, \hat{\xi}(ds\, dx).$$

Here $\hat{J}_\infty^k = \mu B_k < \infty$ a.s. by hypothesis, and so the J^k are simple and integrable point processes on $\mathbb{R}_+$ with compensators $\hat{J}^k$. For fixed

$u_1, \ldots, u_n \geq 0$, we define
$$X_t = \sum_{k \leq n} \left\{ u_k J_t^k - (1 - e^{-u_k}) \hat{J}_t^k \right\}, \quad t \geq 0.$$

The process $M_t = e^{-X_t}$ has bounded variation and finitely many jumps, and so by an elementary change of variables
$$\begin{aligned} M_t - 1 &= \sum_{s \leq t} \Delta e^{-X_s} - \int_0^t e^{-X_s} dX_s^c \\ &= \sum_{k \leq n} \int_0^{t+} e^{-X_{s-}} (1 - e^{-u_k}) \, d(\hat{J}_s^k - J_s^k). \end{aligned}$$

Since the integrands on the right are bounded and predictable, M is a uniformly integrable martingale, and we get $EM_\infty = 1$. Thus,
$$E \exp\left\{ -\sum_k u_k \eta B_k \right\} = \exp\left\{ -\sum_k (1 - e^{-u_k}) \mu B_k \right\},$$
and the assertion follows by Theorem 5.3. □

The preceding theorem immediately yields a corresponding Poisson characterization, similar to the characterization of Brownian motion in Theorem 18.3. The result may also be considered as an extension of Theorem 12.10.

Corollary 25.25 (*Poisson characterization, Watanabe*) *Fix a Borel space S and a measure μ on $(0, \infty) \times S$ with $\mu(\{t\} \times S) = 0$ for all $t > 0$. Let ξ be an S-marked, $\mathcal{F}$-adapted point process on $(0, \infty)$ with compensator $\hat{\xi}$. Then ξ is $\mathcal{F}$-Poisson with $E\xi = \mu$ iff $\hat{\xi} = \mu$ a.s.*

We may further deduce a basic time-change result, similar to Proposition 18.8 for continuous local martingales.

Corollary 25.26 (*time-change to Poisson, Papangelou, Meyer*) *Let $N^1, \ldots, N^n$ be counting processes on $\mathbb{R}_+$ with a.s. unbounded and continuous compensators $\hat{N}^1, \ldots, \hat{N}^n$, and assume that $\sum_k N^k$ is a.s. simple. Define $\tau_s^k = \inf\{t > 0; \hat{N}^k > s\}$ and $Y_s^k = N^k(\tau_s^k)$. Then $Y^1, \ldots, Y^n$ are independent unit-rate Poisson processes.*

Proof: We may apply Theorem 25.24 to the random measures $\xi = (\xi_1, \ldots, \xi_n)$ and $\hat{\xi} = (\hat{\xi}_1, \ldots, \hat{\xi}_n)$ on $\{1, \ldots, n\} \times \mathbb{R}_+$ induced by $(N^1, \ldots, N^n)$ and $(\hat{N}^1, \ldots, \hat{N}^n)$, respectively, and to the predictable mapping $T_{k,t} = (k, \hat{N}_t^k)$ on $\{1, \ldots, n\} \times \mathbb{R}_+$. It is then enough to verify that, a.s. for fixed k and t,
$$\hat{\xi}_k\{s \geq 0; \hat{N}_s^k \leq t\} = t, \quad \xi_k\{s \geq 0; \hat{N}_s^k \leq t\} = N^k(\tau_t^k),$$
which is clear by the continuity of $\hat{N}^k$. □

There is a similar result for stochastic integrals with respect to p-stable Lévy processes, as described in Proposition 15.9. For simplicity, we consider only the case when $p < 1$.

Proposition 25.27 *(time-change of stable integrals)* For a $p \in (0,1)$, let X be a strictly p-stable Lévy process, and consider a predictable process $V \geq 0$ such that the process $A = V^p \cdot \lambda$ is a.s. finite but unbounded. Define $\tau_s = \inf\{t; A_t > s\}$, $s \geq 0$. Then $(V \cdot X) \circ \tau \stackrel{d}{=} X$.

Proof: Define a point process ξ on $\mathbb{R}_+ \times (\mathbb{R}\setminus\{0\})$ by $\xi B = \sum_s 1_B(s, \Delta X_s)$, and recall from Corollary 15.7 and Proposition 15.9 that ξ is Poisson with intensity measure of the form $\lambda \otimes \nu$, where $\nu(dx) = c_\pm |x|^{-p-1} dx$ for $\pm x > 0$. In particular, ξ has compensator $\hat{\xi} = \lambda \otimes \nu$. Let the predictable mapping T on $\mathbb{R}_+ \times \mathbb{R}$ be given by $T_{s,x} = (A_s, xV_s)$. Since A is continuous, we have $\{A_s \leq t\} = \{s \leq \tau_t\}$ and $A_{\tau_t} = t$. By Fubini's theorem, we hence obtain for any $t, u > 0$

$$
\begin{aligned}
(\lambda \otimes \nu) \circ T^{-1}([0,t] \times (u,\infty)) &= (\lambda \otimes \nu)\{(s,x); A_s \leq t, xV_s > u\} \\
&= \int_0^{\tau_t} \nu\{x; xV_s > u\} ds \\
&= \nu(u,\infty) \int_0^{\tau_t} V_s^p ds = t\nu(u,\infty),
\end{aligned}
$$

and similarly for the sets $[0,t] \times (-\infty, -u)$. Thus, $\hat{\xi} \circ T^{-1} = \hat{\xi} = \lambda \otimes \nu$ a.s., and so Theorem 25.24 yields $\xi \circ T^{-1} \stackrel{d}{=} \xi$. Finally, we note that

$$
\begin{aligned}
(V \cdot X)_{\tau_t} &= \int_0^{\tau_t+} \int xV_s \, \xi(ds\,dx) = \int_0^\infty \int xV_s 1\{A_s \leq t\} \xi(ds\,dx) \\
&= \int_0^{t+} \int y\,(\xi \circ T^{-1})(dr\,dy),
\end{aligned}
$$

where the process on the right has the same distribution as X. $\square$

We turn to an important special case where the compensator can be computed explicitly. By the *natural compensator* of a random measure ξ we mean the compensator with respect to the induced filtration.

Proposition 25.28 *(natural compensator)* For any Borel space $(S, \mathcal{S})$, let (τ, ς) be a random element in $(0, \infty] \times S$ with distribution μ. Then $\xi = \delta_{\tau,\varsigma}$ has natural compensator

$$\hat{\xi}_t B = \int_{(0, t \wedge \tau]} \frac{\mu(dr \times B)}{\mu([r, \infty] \times S)}, \quad t \geq 0, \ B \in \mathcal{S}. \tag{6}$$

Proof: The process $\eta_t B$ on the right of (6) is clearly predictable for every $B \in \mathcal{S}$. It remains to show that $M_t = \xi_t B - \eta_t B$ is a martingale, hence that $E[M_t - M_s; A] = 0$ for any $s < t$ and $A \in \mathcal{F}_s$. Since $M_t = M_s$ on $\{\tau \leq s\}$, and the set $\{\tau > s\}$ is a.s. an atom of $\mathcal{F}_s$, it suffices to show that

$E(M_t - M_s) = 0$, or $EM_t \equiv 0$. Then use Fubini's theorem to get

$$\begin{aligned}
E\eta_t B &= E \int_{(0,t\wedge\tau]} \frac{\mu(dr \times B)}{\mu([r,\infty] \times S)} \\
&= \int_{(0,\infty]} \mu(dx) \int_{(0,t\wedge x]} \frac{\mu(dr \times B)}{\mu([r,\infty] \times S)} \\
&= \int_{(0,t]} \frac{\mu(dr \times B)}{\mu([r,\infty] \times S)} \int_{[r,\infty]} \mu(dx) \\
&= \mu((0,t] \times B) = E\xi_t B. \qquad \square
\end{aligned}$$

We turn to some applications of the previous ideas to classical potential theory. Then fix a domain $D \subset \mathbb{R}^d$, and let $T_t = T_t^D$ denote the transition operators of Brownian motion X in D, killed at the boundary ∂D. A function $f \geq 0$ on D is said to be *excessive* if $T_t f \leq f$ for all $t > 0$ and $T_t f \to f$ as $t \to 0$. In this case clearly $T_t f \uparrow f$. Note that if f is excessive, then $f(X)$ is a supermartingale under P_x for every $x \in D$. The basic example of an excessive function is the Green potential $G^D \nu$ of a measure ν on a Greenian domain D, provided this potential is finite.

Though excessivity is defined globally in terms of the operators T_t^D, it is in fact a local property. For a precise statement, we say that a measurable function $f \geq 0$ on D is *superharmonic* if, for any ball B in D with center x, the average of f over the sphere ∂B is bounded by $f(x)$. As we shall see, it is enough to consider balls in D of radius less than an arbitrary $\varepsilon > 0$. Recall that f is *lower semicontinuous* if $x_n \to x$ implies $\liminf_n f(x_n) \geq f(x)$.

Theorem 25.29 *(superharmonic and excessive functions, Doob)* Let $f \geq 0$ be a measurable function on a domain $D \subset \mathbb{R}^d$. Then f is excessive iff it is superharmonic and lower semicontinuous.

For the proof we need two lemmas, the first of which clarifies the relation between the two continuity properties.

Lemma 25.30 *(semicontinuity)* Consider a measurable function $f \geq 0$ on a domain $D \subset \mathbb{R}^d$ such that $T_t f \leq f$ for all $t > 0$. Then f is excessive iff it is lower semicontinuous.

Proof: First assume that f is excessive, and let $x_n \to x$ in D. By Theorem 24.7 and Fatou's lemma

$$\begin{aligned}
T_t f(x) &= \int p_t^D(x,y) f(y) dy \\
&\leq \liminf_{n\to\infty} \int p_t^D(x_n, y) f(y) dy \\
&= \liminf_{n\to\infty} T_t f(x_n) \leq \liminf_{n\to\infty} f(x_n),
\end{aligned}$$

and as $t \to 0$, we get $f(x) \leq \liminf_n f(x_n)$. Thus, f is lower semicontinuous.

Next assume that f is lower semicontinuous. Using the continuity of X and Fatou's lemma, we get as $t \to 0$ along an arbitrary sequence

$$\begin{aligned} f(x) &= E_x f(X_0) \leq E_x \liminf_{t \to 0} f(X_t) \\ &\leq \liminf_{t \to 0} E_x f(X_t) = \liminf_{t \to 0} T_t f(x) \\ &\leq \limsup_{t \to 0} T_t f(x) \leq f(x). \end{aligned}$$

Thus, $T_t f \to f$, and f is excessive. $\square$

For smooth functions, the superharmonic property is easy to describe.

Lemma 25.31 *(smooth functions)* *A function $f \geq 0$ in $C^2(D)$ is superharmonic iff $\Delta f \leq 0$, in which case f is also excessive.*

Proof: By Itô's formula, the process

$$M_t = f(X_t) - \tfrac{1}{2} \int_0^t \Delta f(X_s) ds, \quad t \in [0, \zeta), \qquad (7)$$

is a continuous local martingale. Now fix any closed ball $B \subset D$ with center x, and write $\tau = \tau_{\partial B}$. Since $E_x \tau < \infty$, we get by dominated convergence

$$f(x) = E_x f(X_\tau) - \tfrac{1}{2} E_x \int_0^\tau \Delta f(X_s) ds.$$

Thus, f is superharmonic iff the last expectation is ≤ 0, and the first assertion follows.

To prove the last statement, we note that the exit time $\zeta = \tau_{\partial D}$ is predictable, say with announcing sequence (τ_n). If $\Delta f \leq 0$, we get from (7) by optional sampling

$$E_x[f(X_{t \wedge \tau_n}); t < \zeta] \leq E_x f(X_{t \wedge \tau_n}) \leq f(x).$$

Hence, Fatou's lemma yields $E_x[f(X_t); t < \zeta] = T_t f(x)$, and so f is excessive by Lemma 25.30. $\square$

Proof of Theorem 25.29: If f is excessive or superharmonic, then Lemma 25.30 shows that $f \wedge n$ has the same property for every $n > 0$. The converse statement is also true—by monotone convergence and because the lower semicontinuity is preserved by increasing limits. Thus, we may henceforth assume that f is bounded.

Now assume that f is excessive on D. By Lemma 25.30 it is then lower semicontinuous, and it remains to prove that f is superharmonic. Since the property $T_t f \leq f$ is preserved by passing to a subdomain, we may assume that D is bounded. For each $h > 0$ we define $q_h = h^{-1}(f - T_h f)$ and $f_h = G^D q_h$. Since f and D are bounded, we have $G^D f < \infty$, and so $f_h = h^{-1} \int_0^h T_s f ds \uparrow f$. By the strong Markov property we further see that, for any optional time $\tau < \zeta$,

$$E_x f_h(X_\tau) = E_x E_{X_\tau} \int_0^\infty q_h(X_s) ds = E_x \int_0^\infty q_h(X_{s+\tau}) ds$$
$$= E_x \int_\tau^\infty q_h(X_s) ds \leq f_h(x).$$

In particular, f_h is superharmonic for each h, and so by monotone convergence the same property holds for f.

Conversely, assume that f is superharmonic and lower semicontinuous. To prove that f is excessive, it is enough by Lemma 25.30 to show that $T_t f \leq f$ for all t. Then fix a spherically symmetric probability density $\psi \in C^\infty(\mathbb{R}^d)$ with support in the unit ball, and put $\psi_h(x) = h^{-d}\psi(x/h)$ for each $h > 0$. Writing ρ for the Euclidean metric in $\mathbb{R}^d$, we may define $f_h = \psi_h * f$ on the set $D_h = \{x \in D; \rho(x, D^c) > h\}$. Note that $f_h \in C^\infty(D_h)$ for all h, that f_h is superharmonic on D_h, and that $f_h \uparrow f$. By Lemma 25.31 and monotone convergence we conclude that f is excessive on each set D_h. Letting ζ_h denote the first exit time from D_h, we obtain

$$E_x[f(X_t); t < \zeta_h] \leq f(x), \quad h > 0.$$

As $h \to 0$, we have $\zeta_h \uparrow \zeta$, and hence $\{t < \zeta_h\} \uparrow \{t < \zeta\}$. Thus, by monotone convergence $T_t f(x) \leq f(x)$. □

We may now prove the remarkable fact that, although an excessive function f need not be continuous, the supermartingale $f(X)$ is a.s. continuous under P_x for every x.

Theorem 25.32 *(continuity, Doob)* Fix an excessive function f on a domain $D \subset \mathbb{R}^d$, and let X be a Brownian motion killed at ∂D. Then the process $f(X_t)$ is a.s. continuous on $[0, \zeta)$.

The proof is based on the following invariance under time reversal of a stationary version of Brownian motion. Though no such process exists in the usual sense, we may consider distributions with respect to the σ-finite measure $\overline{P} = \int P_x dx$, where P_x is the distribution of a Brownian motion in $\mathbb{R}^d$ starting at x.

Lemma 25.33 *(time reversal, Doob)* For any $c > 0$, the processes $Y_t = X_t$ and $\tilde{Y}_t = X_{c-t}$ on $[0, c]$ have the same distribution under $\overline{P}$.

Proof: Introduce the processes

$$B_t = X_t - X_0, \quad \tilde{B}_t = X_{c-t} - X_c, \quad t \in [0, c],$$

and note that B and $\tilde{B}$ are Brownian motions on $[0, c]$ under each P_x. Fix any measurable function $f \geq 0$ on $C([0, c], \mathbb{R}^d)$. By Fubini's theorem and the invariance of Lebesgue measure, we get

$$\overline{E}f(\tilde{Y}) = \overline{E}f(X_0 - \tilde{B}_c + \tilde{B}) = \int E_x f(x - \tilde{B}_c + \tilde{B}) \, dx$$
$$= \int E_0 f(x - \tilde{B}_c + \tilde{B}) \, dx = E_0 \int f(x - \tilde{B}_c + \tilde{B}) \, dx$$
$$= E_0 \int f(x + \tilde{B}) \, dx = \int E_x f(Y) \, dx = \overline{E}f(Y). \quad \square$$

Proof of Theorem 25.32: Since $f \wedge n$ is again excessive for each $n > 0$ by Theorem 25.29, we may assume that f is bounded. As in the proof of the same theorem, we may then approximate f by smooth excessive functions $f_h \uparrow f$ on suitable subdomains $D_h \uparrow D$. Since $f_h(X)$ is a continuous supermartingale up to the exit time ζ_h from D_h, Theorem 7.32 shows that $f(X)$ is a.s. right-continuous on $[0, \zeta)$ under any initial distribution μ. Using the Markov property at rational times, we may extend the a.s. right-continuity to the random time set $T = \{t \geq 0; X_t \in D\}$.

To strengthen the result to a.s. continuity on T, we note that $f(X)$ is right-continuous on T, a.e. $\overline{P}$. By Lemma 25.33 it follows that $f(X)$ is also left-continuous on T, a.e. $\overline{P}$. Thus, $f(X)$ is continuous on T, a.s. P_μ for arbitrary $\mu \ll \lambda^d$. Since $P_\mu \circ X_h^{-1} \ll \lambda^d$ for any μ and $h > 0$, we may conclude that $f(X)$ is a.s. continuous on $T \cap [h, \infty)$ for any $h > 0$. This together with the right-continuity at 0 yields the asserted continuity on $[0, \zeta)$. $\quad \square$

If f is excessive, then $f(X)$ is a supermartingale under P_x for every x, and so it has a Doob–Meyer decomposition $f(X) = M - A$. It is remarkable that we can choose A to be a continuous additive functional (CAF) of X independent of x. A similar situation was encountered in connection with Theorem 22.23.

Theorem 25.34 *(compensation by additive functional, Meyer)* Let f be an excessive function on a domain $D \subset \mathbb{R}^d$, and let P_x be the distribution of Brownian motion in D, killed at ∂D. Then there exists an a.s. unique CAF A of X such that $M = f(X) + A$ is a continuous, local P_x-martingale on $[0, \zeta)$ for every $x \in D$.

The main difficulty in the proof is to construct a version of the process A that compensates $-f(X)$ under *every* measure P_μ. Here the following lemma is helpful.

Lemma 25.35 *(universal compensation)* Consider an excessive function f on a domain $D \subset \mathbb{R}^d$, a distribution $m \sim \lambda^d$ on D, and a P_m-compensator A of $-f(X)$ on $[0, \zeta)$. Then for any distribution μ and constant $h > 0$, the process $A \circ \theta_h$ is a P_μ-compensator of $-f(X \circ \theta_h)$ on $[0, \zeta \circ \theta_h)$.

In other words, the process $M_t = f(X_t) + A_{t-h} \circ \theta_h$ is a local P_μ-martingale on $[h, \zeta)$ for every μ and h.

Proof: For any bounded P_m-martingale M and initial distribution $\mu \ll m$, we note that M is also a P_μ-martingale. To see this, write $k = d\mu/dm$, and note that $P_\mu = k(X_0) \cdot P_m$. It is equivalent to show that $N_t = k(X_0)M_t$ is a P_m-martingale, which is clear since $k(X_0)$ is $\mathcal{F}_0$-measurable with mean 1.

Now fix any distribution μ and a constant $h > 0$. To prove the stated property of A, it is enough to show that, for any bounded P_m-martingale M, the process $N_t = M_{t-h} \circ \theta_h$ is a P_μ-martingale on $[h, \infty)$. Then fix any times $s < t$ and sets $F \in \mathcal{F}_h$ and $G \in \mathcal{F}_s$. Using the Markov property at h and noting that $P_\mu \circ X_h^{-1} \ll m$, we get

$$\begin{aligned} E_\mu[M_t \circ \theta_h; F \cap \theta_h^{-1} G] &= E_\mu[E_{X_h}[M_t; G]; F] \\ &= E_\mu[E_{X_h}[M_s; G]; F] \\ &= E_\mu[M_s \circ \theta_h; F \cap \theta_h^{-1} G]. \end{aligned}$$

Hence, by a monotone class argument, $E_\mu[M_t \circ \theta_h | \mathcal{F}_{h+s}] = M_s \circ \theta_h$ a.s. □

Proof of Theorem 25.34: Let A^μ denote the P_μ-compensator of $-f(X)$ on $[0, \zeta)$, and note that A^μ is a.s. continuous, e.g. by Theorem 18.10. Fix any distribution $m \sim \lambda^d$ on D, and conclude from Lemma 25.35 that $A^m \circ \theta_h$ is a P_μ-compensator of $-f(X \circ \theta_h)$ on $[0, \zeta \circ \theta_h)$ for any μ and $h > 0$. Since this is also true for the process $A^\mu_{t+h} - A^\mu_h$, we get for any μ and $h > 0$

$$A^\mu_t = A^\mu_h + A^m_{t-h} \circ \theta_h, \quad t \geq h, \text{ a.s. } P_\mu. \tag{8}$$

Restricting h to the positive rationals, we may define

$$A_t = \lim_{h \to 0} A^m_{t-h} \circ \theta_h, \quad t > 0,$$

whenever the limit exists and is continuous and nondecreasing with $A_0 = 0$, and put $A = 0$ otherwise. By (8) we have $A = A^\mu$ a.s. P_μ for every μ, and so A is a P_μ-compensator of $-f(X)$ on $[0, \zeta)$ for every μ. For each $h > 0$ it follows by Lemma 25.35 that $A \circ \theta_h$ is a P_μ-compensator of $-f(X \circ \theta_h)$ on $[0, \zeta \circ \theta_h)$, and since this is also true for the process $A_{t+h} - A_h$, we get $A_{t+h} = A_h + A_t \circ \theta_h$ a.s. P_μ. Thus, A is a CAF. □

We may now establish a probabilistic version of the classical Riesz decomposition. To avoid technical difficulties, we restrict our attention to locally bounded functions f. By the *greatest harmonic minorant* of f we mean a harmonic function $h \leq f$ that dominates all other such functions. Recall that the *potential* U_A of a CAF A of X is given by $U_A(x) = E_x A_\infty$.

Theorem 25.36 *(Riesz decomposition)* *Fix any locally bounded function $f \geq 0$ on a domain $D \subset \mathbb{R}^d$, and let X be Brownian motion on D, killed at ∂D. Then f is excessive iff it has a representation $f = U_A + h$, where A is a CAF of X and h is harmonic with $h \geq 0$. In that case, A is the compensator of $-f(X)$ and h is the greatest harmonic minorant of f.*

A similar result for uniformly α-excessive functions of an arbitrary Feller process was obtained in Theorem 22.23. From the classical Riesz representation on Greenian domains, we know that U_A may also be written as the Green potential of a unique measure ν_A, so that $f = G^D \nu_A + h$. In the special case when $D = \mathbb{R}^d$ with $d \geq 3$, we recall from Theorem 22.21 that $\nu_A B = \overline{E}(1_B \cdot A)_1$. A similar representation holds in the general case.

Proof of Theorem 25.36: First assume that A is a CAF with $U_A < \infty$. By the additivity of A and the Markov property of X, we get for any $t > 0$

$$\begin{aligned} U_A(x) &= E_x A_\infty = E_x(A_t + A_\infty \circ \theta_t) \\ &= E_x A_t + E_x E_{X_t} A_\infty = E_x A_t + T_t U_A(x). \end{aligned}$$

By dominated convergence $E_x A_t \downarrow 0$ as $t \to 0$, and so U_A is excessive. Even $U_A + h$ is then excessive for any harmonic function $h \geq 0$.

Conversely, assume that f is excessive and locally bounded. By Theorem 25.34 there exists some CAF A such that $M = f(X) + A$ is a continuous local martingale on $[0, \zeta)$. For any localizing and announcing sequence $\tau_n \uparrow \zeta$, we get

$$f(x) = E_x M_0 = E_x M_{\tau_n} = E_x f(X_{\tau_n}) + E_x A_{\tau_n} \geq E_x A_{\tau_n}.$$

As $n \to \infty$, it follows by monotone convergence that $U_A \leq f$.

By the additivity of A and the Markov property of X,

$$\begin{aligned} E_x[A_\infty | \mathcal{F}_t] &= A_t + E_x[A_\infty \circ \theta_t | \mathcal{F}_t] \\ &= A_t + E_{X_t} A_\infty = M_t - f(X_t) + U_A(X_t). \end{aligned} \quad (9)$$

Writing $h = f - U_A$, it follows that $h(X)$ is a continuous local martingale. Since h is locally bounded, we may conclude by optional sampling and dominated convergence that h has the mean-value property. Thus, h is harmonic by Lemma 24.3.

To prove the uniqueness of A, assume that f also has a representation $U_B + k$ for some CAF B and some harmonic function $k \geq 0$. Proceeding as in (9), we get

$$A_t - B_t = E_x[A_\infty - B_\infty | \mathcal{F}_t] + h(X_t) - k(X_t), \quad t \geq 0,$$

which shows that $A-B$ is a continuous local martingale. Hence, Proposition 17.2 yields $A = B$ a.s.

To see that h is the greatest harmonic minorant of f, consider any harmonic minorant $k \geq 0$. Since $f - k$ is again excessive and locally bounded, it has a representation $U_B + l$ for some CAF B and some harmonic function l. But then $f = U_B + k + l$, and so $A = B$ a.s. and $h = k + l \geq k$. $\square$

For any sufficiently regular measure ν on $\mathbb{R}^d$, we may now construct an associated CAF A of Brownian motion X such that A increases only when X visits the support of ν. This clearly extends the notion of local time. For convenience we may write $G^D(1_D \cdot \nu) = G^D \nu$.

Proposition 25.37 (*additive functionals induced by measures*) *Fix a measure ν on $\mathbb{R}^d$ such that $U(1_D \cdot \nu)$ is bounded for every bounded domain D. Then there exists an a.s. unique CAF A of Brownian motion X such that, for any D,*
$$E_x A_{\zeta_D} = G^D \nu(x), \quad x \in D. \tag{10}$$
Conversely, ν is uniquely determined by A. Furthermore,
$$\operatorname{supp} A \subset \{t \geq 0;\ X_t \in \operatorname{supp} \nu\} \quad \text{a.s.} \tag{11}$$

The proof is straightforward, given the classical Riesz decomposition, and we shall indicate the main steps only.

Proof: A simple calculation shows that $G^D \nu$ is excessive for any bounded domain D. Since $G^D \nu \leq U(1_D \cdot \nu)$, it is further bounded. Hence, by Theorem 25.36 there exist a CAF A_D of X on $[0, \zeta_D)$ and a harmonic function $h_D \geq 0$ such that $G^D \nu = U A_D + h_D$. In fact, $h_D = 0$ by Riesz' theorem.

Now consider another bounded domain $D' \supset D$. We claim that $G^{D'} \nu - G^D \nu$ is harmonic on D. This is clear from the analytic definitions, and it also follows, under a regularity condition, from Lemma 24.13. Since A_D and $A_{D'}$ are compensators of $-G^D \nu(X)$ and $-G^{D'} \nu(X)$, respectively, we conclude that $A_D - A_{D'}$ is a martingale on $[0, \zeta_D)$, and so $A_D = A_{D'}$ a.s. up to time ζ_D. Now choose a sequence of bounded domains $D_n \uparrow \mathbb{R}^d$, and define $A = \sup_n A_{D_n}$, so that $A = A_D$ a.s. on $[0, \zeta_D)$ for all D.

It is easy to see that A is a CAF of X, and that (10) holds for any bounded domain D. The uniqueness of ν is clear from the uniqueness in the classical Riesz decomposition. Finally, we obtain (11) by noting that $G^D \nu$ is harmonic on $D \setminus \operatorname{supp} \nu$ for every D, so that $G^D \nu(X)$ is a local martingale on the predictable set $\{t < \zeta_D;\ X_t \notin \operatorname{supp} \nu\}$. □

Exercises

1. Show by an example that the σ-fields $\mathcal{F}_\tau$ and $\mathcal{F}_{\tau-}$ may differ. (*Hint:* Take τ to be constant.)

2. Give examples of optional times that are predictable; accessible but not predictable; and totally inaccessible. (*Hint:* Use Corollary 25.18.)

3. Show by an example that a right-continuous, adapted process need not be predictable. (*Hint:* Use Theorem 25.14.)

4. Given a Brownian motion B on $[0, 1]$, let $\mathcal{F}$ be the filtration induced by $X_t = (B_t, B_1)$. Find the Doob–Meyer decomposition $B = M + A$ on $[0, 1)$ and show that A has a.s. finite variation on $[0, 1]$.

5. For any totally inaccessible time τ, show that $\sup_t |P\{\tau \leq t + \varepsilon | \mathcal{F}_t\} - 1\{\tau \leq t\}| \to 0$ a.s. as $\varepsilon \to 0$. Derive a corresponding result for the compensator. (*Hint:* Use Lemma 25.8.)

6. Let the process X be adapted and rcll. Show that X is predictable iff it has accessible jumps and ΔX_τ is $\mathcal{F}_{\tau-}$-measurable for every predictable time $\tau < \infty$. (*Hint:* Use Proposition 25.17 and Lemmas 25.2 and 25.3.)

7. Show that the compensator A of a quasi-leftcontinuous local submartingale is a.s. continuous. (*Hint:* Note that A has accessible jumps. Use optional sampling at an arbitrary predictable time $\tau < \infty$ with announcing sequence (τ_n).)

8. Extend Corollary 25.26 to possibly bounded compensators. Show that the result fails in general when the compensators are not continuous.

9. Show that any general inequality involving an increasing process A and its compensator $\hat{A}$ remains valid in discrete time. (*Hint:* Embed the discrete-time process and filtration into continuous time.)

Chapter 26

Semimartingales and General Stochastic Integration

Predictable covariation and L^2-integral; semimartingale integral and covariation; general substitution rule; Doléans' exponential and change of measure; norm and exponential inequalities; martingale integral; decomposition of semimartingales; quasi-martingales and stochastic integrators

In this chapter we shall use the previously established Doob–Meyer decomposition to extend the stochastic integral of Chapter 17 to possibly discontinuous semimartingales. The construction proceeds in three steps. First we imitate the definition of the L^2-integral $V \cdot M$ from Chapter 17, using a predictable version $\langle M, N \rangle$ of the covariation process. A suitable truncation then allows us to extend the integral to arbitrary semimartingales X and bounded, predictable processes V. The ordinary covariation $[X, Y]$ can now be defined by the integration-by-parts formula, and we may use some generalized versions of the BDG inequalities from Chapter 17 to extend the martingale integral $V \cdot M$ to more general integrands V.

Once the stochastic integral is defined, we may develop a stochastic calculus for general semimartingales. In particular, we shall prove an extension of Itô's formula, solve a basic stochastic differential equation, and establish a general Girsanov-type theorem for absolutely continuous changes of the probability measure. The latter material extends the appropriate portions of Chapters 18 and 21.

The stochastic integral and covariation process, together with the Doob–Meyer decomposition from the preceding chapter, provide the tools for a more detailed analysis of semimartingales. Thus, we may now establish two general decompositions, similar to the decompositions of optional times and increasing processes in Chapter 25. We shall further derive some exponential inequalities for martingales with bounded jumps, characterize local quasi-martingales as special semimartingales, and show that no continuous extension of the predictable integral exists beyond the context of semimartingales.

Throughout this chapter, $\mathcal{M}^2$ denotes the class of uniformly square-integrable martingales. As in Lemma 17.4, we note that $\mathcal{M}^2$ is a Hilbert space for the norm $\|M\| = (EM_\infty^2)^{1/2}$. We define $\mathcal{M}_0^2$ as the closed linear subspace of martingales $M \in \mathcal{M}^2$ with $M_0 = 0$. The corresponding classes $\mathcal{M}^2_{\text{loc}}$ and $\mathcal{M}^2_{0,\text{loc}}$ are defined as the sets of processes M such that the stopped versions M^{τ_n} belong to $\mathcal{M}^2$ or $\mathcal{M}^2_0$, respectively, for some sequence of optional times $\tau_n \to \infty$.

For every $M \in \mathcal{M}^2_{\text{loc}}$ we note that M^2 is a local submartingale. The corresponding compensator, denoted by $\langle M \rangle$, is called the *predictable quadratic variation* of M. More generally, we may define the *predictable covariation* $\langle M, N \rangle$ of two processes $M, N \in \mathcal{M}^2_{\text{loc}}$ as the compensator of MN, also computable by the *polarization formula*

$$4\langle M, N \rangle = \langle M + N \rangle - \langle M - N \rangle.$$

Note that $\langle M, M \rangle = \langle M \rangle$. If M and N are continuous, then clearly $\langle M, N \rangle = [M, N]$ a.s. The following result collects some further useful properties.

Proposition 26.1 *(predictable covariation)* *For any $M, M^n, N \in \mathcal{M}^2_{\text{loc}}$,*
  (i) $\langle M, N \rangle = \langle M - M_0, N - N_0 \rangle$ *a.s.;*
  (ii) $\langle M \rangle$ *is a.s. increasing, and $\langle M, N \rangle$ is a.s. symmetric and bilinear;*
  (iii) $|\langle M, N \rangle| \leq \int |d\langle M, N \rangle| \leq \langle M \rangle^{1/2} \langle N \rangle^{1/2}$ *a.s.;*
  (iv) $\langle M, N \rangle^\tau = \langle M^\tau, N \rangle = \langle M^\tau, N^\tau \rangle$ *a.s. for any optional time τ;*
  (v) $\langle M^n \rangle_\infty \xrightarrow{P} 0$ *implies* $(M^n - M_0^n)^* \xrightarrow{P} 0$.

Proof: By Lemma 25.11 we note that $\langle M, N \rangle$ is the a.s. unique predictable process of locally integrable variation and starting at 0 such that $MN - \langle M, N \rangle$ is a local martingale. The symmetry and bilinearity in (ii) follow immediately, as does property (i), since MN_0, $M_0 N$, and $M_0 N_0$ are all local martingales. Property (iii) is proved in the same way as Proposition 17.9, and (iv) is obtained as in Theorem 17.5.

To prove (v), we may assume that $M_0^n = 0$ for all n. Let $\langle M^n \rangle_\infty \xrightarrow{P} 0$. Fix any $\varepsilon > 0$, and define $\tau_n = \inf\{t; \langle M^n \rangle_t \geq \varepsilon\}$. Since $\langle M^n \rangle$ is predictable, even τ_n is predictable by Theorem 25.14 and is therefore announced by some sequence $\tau_{nk} \uparrow \tau_n$. The latter may be chosen such that M^n is an L^2-martingale and $(M^n)^2 - \langle M^n \rangle$ a uniformly integrable martingale on $[0, \tau_{nk}]$ for every k. By Proposition 7.16

$$E(M^n)^{*2}_{\tau_{nk}} \lesssim E(M^n)^2_{\tau_{nk}} = E\langle M^n \rangle_{\tau_{nk}} \leq \varepsilon,$$

and as $k \to \infty$, we get $E(M^n)^{*2}_{\tau_n-} \leq \varepsilon$. Now fix any $\delta > 0$, and write

$$\begin{aligned} P\{(M^n)^{*2} > \delta\} &\leq P\{\tau_n < \infty\} + \delta^{-1} E(M^n)^{*2}_{\tau_n-} \\ &\leq P\{\langle M^n \rangle_\infty \geq \varepsilon\} + \delta^{-1}\varepsilon. \end{aligned}$$

Here the right-hand side tends to zero as $n \to \infty$ and then $\varepsilon \to 0$. □

We may use the predictable quadratic variation to extend the Itô integral from Chapter 17. As before, let $\mathcal{E}$ denote the class of bounded, predictable step processes V with jumps at finitely many fixed times. We refer to the corresponding integral $V \cdot X$ as the *elementary predictable integral*.

26. Semimartingales and General Stochastic Integration

Given any $M \in \mathcal{M}^2_{\text{loc}}$, let $L^2(M)$ be the class of predictable processes V such that $(V^2 \cdot \langle M \rangle)_t < \infty$ a.s. for every $t > 0$. We first consider integrals $V \cdot M$ with $M \in \mathcal{M}^2_{\text{loc}}$ and $V \in L^2(M)$. Here the integral process belongs to $\mathcal{M}^2_{0,\text{loc}}$, the class of local L^2-martingales starting at 0. In the following statement, it is understood that $M, N \in \mathcal{M}^2_{\text{loc}}$ and that U and V are predictable processes such that the stated integrals exist.

Theorem 26.2 (L^2-integral, Courrège, Kunita and Watanabe) *The elementary predictable integral extends a.s. uniquely to a bilinear map of any $M \in \mathcal{M}^2_{\text{loc}}$ and $V \in L^2(M)$ into $V \cdot M \in \mathcal{M}^2_{0,\text{loc}}$, such that $(V_n^2 \cdot \langle M_n \rangle)_t \xrightarrow{P} 0$ implies $(V_n \cdot M_n)^*_t \xrightarrow{P} 0$ for every $t > 0$. Furthermore,*

(i) $\langle V \cdot M, N \rangle = V \cdot \langle M, N \rangle$ *a.s. for all* $N \in \mathcal{M}^2_{\text{loc}}$;

(ii) $U \cdot (V \cdot M) = (UV) \cdot M$ *a.s.*;

(iii) $\Delta(V \cdot M) = V \Delta M$ *a.s.*;

(iv) $(V \cdot M)^\tau = V \cdot M^\tau = (V 1_{[0,\tau]}) \cdot M$ *a.s. for any optional time τ;*

where property (i) characterizes the integral.

The proof depends on an elementary approximation property, corresponding to Lemma 17.23 in the continuous case.

Lemma 26.3 (approximation) *Let V be a predictable process with $|V|^p \in L(A)$, where A is increasing and $p \geq 1$. Then there exist some $V_1, V_2, \ldots \in \mathcal{E}$ with $(|V_n - V|^p \cdot A)_t \to 0$ a.s. for all $t > 0$.*

Proof: It is enough to establish the approximation $(|V_n - V|)^p \cdot A)_t \xrightarrow{P} 0$. By Minkowski's inequality we may then approximate in steps, and by dominated convergence we may first reduce to the case when V is simple. Each term may then be approximated separately, and so we may next assume that $V = 1_B$ for some predictable set B. Approximating separately on disjoint intervals, we may finally reduce to the case when $B \subset \Omega \times [0, t]$ for some $t > 0$. The desired approximation is then obtained from Lemma 25.1 by a monotone class argument. □

Proof of Theorem 26.2: As in Theorem 17.11, we may construct the integral $V \cdot M$ as the a.s. unique element of $\mathcal{M}^2_{0,\text{loc}}$ satisfying (i). The mapping $(V, M) \mapsto V \cdot M$ is clearly bilinear, and by the analogue of Lemma 17.10 it extends the elementary predictable integral. Properties (ii) and (iv) may be obtained in the same way as in Propositions 17.14 and 17.15. The stated continuity property follows immediately from (i) and Proposition 26.1 (v). To get the stated uniqueness, it is then enough to apply Lemma 26.3 with $A = \langle M \rangle$ and $p = 2$.

To prove (iii), we note from Lemma 26.3 with $A_t = \langle M \rangle_t + \sum_{s \leq t} (\Delta M_s)^2$ that there exist some processes $V_n \in \mathcal{E}$ satisfying $V_n \Delta M \to V \Delta M$ and $(V_n \cdot M - V \cdot M)^* \to 0$ a.s. In particular, $\Delta(V_n \cdot M) \to \Delta(V \cdot M)$ a.s., and so (iii) follows from the corresponding relation for the elementary integrals $V_n \cdot M$. The argument relies on the fact that $\sum_{s \leq t} (\Delta M_s)^2 < \infty$ a.s. To verify this, we may assume that $M \in \mathcal{M}_0^2$ and define $t_{n,k} = kt2^{-n}$ for $k \leq 2^n$. By Fatou's lemma

$$E\sum_{s\leq t}(\Delta M_s)^2 \leq E\liminf_{n\to\infty}\sum_k(M_{t_{n,k}} - M_{t_{n,k-1}})^2$$
$$\leq \liminf_{n\to\infty} E\sum_k(M_{t_{n,k}} - M_{t_{n,k-1}})^2 = EM_t^2 < \infty. \quad \square$$

A *semimartingale* is defined as a right-continuous, adapted process X admitting a decomposition $M + A$, where M is a local martingale and A is a process of locally finite variation starting at 0. If the variation of A is even locally integrable, we can write $X = (M + A - \hat{A}) + \hat{A}$, where $\hat{A}$ denotes the compensator of A. Hence, in this case we can choose A to be predictable. The decomposition is then a.s. unique by Propositions 17.2 and 25.16, and X is called a *special semimartingale* with *canonical decomposition* $M + A$.

Lévy processes are the basic examples of semimartingales. In particular, we note that a Lévy process is a special semimartingale iff its Lévy measure ν satisfies $\int (x^2 \wedge |x|)\nu(dx) < \infty$. From Theorem 25.5 it is further seen that any local submartingale is a special semimartingale.

The next result extends the stochastic integration to general semimartingales. At this stage we consider only locally bounded integrands, which covers most applications of interest.

Theorem 26.4 (*semimartingale integral, Doléans-Dade and Meyer*) *The L^2-integral of Theorem 26.2 and the Lebesgue–Stieltjes integral extend a.s. uniquely to a bilinear map of any semimartingale X and locally bounded, predictable process V into a semimartingale $V \cdot X$. This integral satisfies conditions (ii)–(iv) of Theorem 26.2 and is such that, if $V \geq |V_n| \to 0$ for some locally bounded, predictable processes $V, V_1, V_2, \ldots$, then $(V_n \cdot X)_t^* \xrightarrow{P} 0$ for all $t > 0$. Finally, $V \cdot X$ is a local martingale whenever this holds for X.*

Our proof relies on the following basic decomposition.

Lemma 26.5 (*truncation, Doléans-Dade, Jacod and Mémin, Yan*) *Any local martingale M has a decomposition into local martingales M' and M'', where M' has locally integrable variation and $|\Delta M''| \leq 1$ a.s.*

Proof: Define

$$A_t = \sum_{s\leq t} \Delta M_s 1\{|\Delta M_s| > \tfrac{1}{2}\}, \quad t \geq 0.$$

By optional sampling, we note that A has locally integrable variation. Let $\hat{A}$ denote the compensator of A, and put $M' = A - \hat{A}$ and $M'' = M - M'$. Then M' and M'' are again local martingales, and M' has locally integrable variation. Furthermore,

$$|\Delta M''| \leq |\Delta M - \Delta A| + |\Delta \hat{A}| \leq \tfrac{1}{2} + |\Delta \hat{A}|,$$

and so it suffices to show that $|\Delta \hat{A}| \leq \tfrac{1}{2}$. Since the constructions of A and $\hat{A}$ commute with optional stopping, we may then assume that M and

M' are uniformly integrable. Now $\hat{A}$ is predictable, so the times $\tau = n \wedge \inf\{t; |\Delta \hat{A}| > \frac{1}{2}\}$ are predictable by Theorem 25.14, and it is enough to show that $|\Delta \hat{A}_\tau| \leq \frac{1}{2}$ a.s. Clearly, $E[\Delta M_\tau | \mathcal{F}_{\tau-}] = E[\Delta M'_\tau | \mathcal{F}_{\tau-}] = 0$ a.s., and so by Lemma 25.3

$$\begin{aligned} |\Delta \hat{A}_\tau| &= |E[\Delta A_\tau | \mathcal{F}_{\tau-}]| = |E[\Delta M_\tau; |\Delta M_\tau| > \frac{1}{2} | \mathcal{F}_{\tau-}]| \\ &= |E[\Delta M_\tau; |\Delta M_\tau| \leq \frac{1}{2} | \mathcal{F}_{\tau-}]| \leq \frac{1}{2}. \end{aligned}$$
□

Proof of Theorem 26.4: By Lemma 26.5 we may write $X = M + A$, where M is a local martingale with bounded jumps, hence a local L^2-martingale, and A has locally finite variation. For any locally bounded, predictable process V we may then define $V \cdot X = V \cdot M + V \cdot A$, where the first term is the integral in Theorem 26.2, and the second term is an ordinary Lebesgue–Stieltjes integral. If $V \geq |V_n| \to 0$, then $(V_n^2 \cdot \langle M \rangle)_t \to 0$ and $(V_n \cdot A)_t^* \to 0$ by dominated convergence, and so Theorem 26.2 yields $(V_n \cdot X)_t^* \xrightarrow{P} 0$ for all $t > 0$.

To prove the uniqueness, it suffices to prove that if $M = A$ is a local L^2-martingale of locally finite variation, then $V \cdot M = V \cdot A$ a.s. for every locally bounded, predictable process V, where $V \cdot M$ is the integral in Theorem 26.2 and $V \cdot A$ is an elementary Stieltjes integral. The two integrals clearly agree when $V \in \mathcal{E}$. For general V, we may approximate as in Lemma 26.3 by processes $V_n \in \mathcal{E}$ such that $((V_n - V)^2 \cdot \langle M \rangle)^* \to 0$ and $(|V_n - V| \cdot A)^* \to 0$ a.s. But then $(V_n \cdot M)_t \xrightarrow{P} (V \cdot M)_t$ and $(V_n \cdot A)_t \to (V \cdot A)_t$ for every $t > 0$, and the desired equality follows.

To prove the last assertion, we may reduce by means of Lemma 26.5 and a suitable localization to the case when V is bounded and X has integrable variation A. By Lemma 26.3 we may next choose some uniformly bounded processes $V_1, V_2, \ldots \in \mathcal{E}$ such that $(|V_n - V| \cdot A)_t \to 0$ a.s. for every $t \geq 0$. Then $(V_n \cdot X)_t \to (V \cdot X)_t$ a.s. for all t, and by dominated convergence this remains true in L^1. Thus, the martingale property of $V_n \cdot X$ carries over to $V \cdot X$. □

For any semimartingales X and Y, the left-continuous versions $X_- = (X_{t-})$ and $Y_- = (Y_{t-})$ are locally bounded and predictable, and so they can serve as integrands in the general stochastic integral. We may then define the *quadratic variation* $[X]$ and *covariation* $[X, Y]$ by the *integration-by-parts* formulas

$$\begin{aligned} [X] &= X^2 - X_0^2 - 2X_- \cdot X, \\ [X, Y] &= XY - X_0 Y_0 - X_- \cdot Y - Y_- \cdot X \\ &= ([X + Y] - [X - Y])/4. \end{aligned} \quad (1)$$

In particular, $[X] = [X, X]$. Here we list some further basic properties of the covariation.

Theorem 26.6 *(covariation) For any semimartingales X and Y,*
 (i) $[X,Y] = [X - X_0, Y - Y_0]$ *a.s.;*
 (ii) $[X]$ *is a.s. nondecreasing, and $[X,Y]$ is a.s. symmetric and bilinear;*
 (iii) $|[X,Y]| \leq \int |d[X,Y]| \leq [X]^{1/2}[Y]^{1/2}$ *a.s.;*
 (iv) $\Delta[X] = (\Delta X)^2$ *and* $\Delta[X,Y] = \Delta X \Delta Y$ *a.s.;*
 (v) $[V \cdot X, Y] = V \cdot [X, Y]$ *a.s. for any locally bounded, predictable V;*
 (vi) $[X^\tau, Y] = [X^\tau, Y^\tau] = [X, Y]^\tau$ *a.s. for any optional time τ;*
 (vii) *if $M, N \in \mathcal{M}_{loc}^2$, then $[M, N]$ has compensator $\langle M, N \rangle$;*
 (viii) *if A has locally finite variation, then $[X, A]_t = \sum_{s \leq t} \Delta X_s \Delta A_s$ a.s.*

Proof: The symmetry and bilinearity of $[X, Y]$ are obvious from (1), and to get (i) it remains to check that $[X, Y_0] = 0$.

(ii) We may extend Proposition 17.17 with the same proof to general semimartingales. In particular, $[X]_s \leq [X]_t$ a.s. for any $s \leq t$. By right-continuity the exceptional null set can be chosen to be independent of s and t, which means that $[X]$ is a.s. nondecreasing. Relation (iii) may now be proved as in Proposition 17.9.

(iv) By (1) and Theorem 26.2 (iii),

$$\begin{aligned}\Delta[X,Y]_t &= \Delta(XY)_t - \Delta(X_- \cdot Y)_t - \Delta(Y_- \cdot X)_t \\ &= X_t Y_t - X_{t-} Y_{t-} - X_{t-} \Delta Y_t - Y_{t-} \Delta X_t = \Delta X_t \Delta Y_t.\end{aligned}$$

(v) For $V \in \mathcal{E}$ the relation follows most easily from the extended version of Proposition 17.17. Also note that both sides are a.s. linear in V. Now let $V, V_1, V_2, \ldots$ be locally bounded and predictable with $V \geq |V_n| \to 0$. Then $V_n \cdot [X, Y] \to 0$ by dominated convergence, and by Theorem 26.4 we have

$$[V_n \cdot X, Y] = (V_n \cdot X)Y - (V_n \cdot X)_- \cdot Y - (V_n Y_-) \cdot X \xrightarrow{P} 0.$$

Using a monotone class argument, we may now extend the relation to arbitrary V.

(vi) This follows from (v) with $V = 1_{[0,\tau]}$.

(vii) Since $M_- \cdot N$ and $N_- \cdot M$ are local martingales, the assertion follows from (1) and the definition of $\langle M, N \rangle$.

(viii) For step processes A the stated relation follows from the extended version of Proposition 17.17. Now assume instead that $\Delta A \leq \varepsilon$, and conclude from the same result and property (iii) together with the ordinary Cauchy–Buniakovsky inequality that

$$[X, A]_t^2 \vee \left|\sum_{s \leq t} \Delta X_s \Delta A_s\right|^2 \leq [X]_t [A]_t \leq \varepsilon [X]_t \int_0^t |dA_s|.$$

The assertion now follows by a simple approximation. □

We may now extend the Itô formula of Theorem 17.18 to a substitution rule for general semimartingales. By a semimartingale in $\mathbb{R}^d$ we mean a process $X = (X^1, \ldots, X^d)$ such that each component X^i is a one dimensional semimartingale. Let $[X^i, X^j]^c$ denote the continuous components of the finite-variation processes $[X^i, X^j]$, and write f'_i and f''_{ij} for the first- and second-order partial derivatives of f, respectively. Summation over repeated indices is understood as before.

Theorem 26.7 (*substitution rule, Kunita and Watanabe*) *For any semimartingale* $X = (X^1, \ldots, X^d)$ *in* $\mathbb{R}^d$ *and function* $f \in C^2(\mathbb{R}^d)$, *we have*

$$f(X_t) = f(X_0) + \int_0^t f'_i(X_{s-})dX^i_s + \tfrac{1}{2}\int_0^t f''_{ij}(X_{s-})d[X^i, X^j]^c_s$$
$$+ \sum_{s \leq t}\{\Delta f(X_s) - f'_i(X_{s-})\Delta X^i_s\}. \qquad (2)$$

Proof: Assuming that (2) holds for some function $f \in C^2(\mathbb{R}^d)$, we shall prove for any $k \in \{1, \ldots, n\}$ that (2) remains true for $g(x) = x_k f(x)$. Then note that by (1)

$$g(X) = g(X_0) + X^k_- \cdot f(X) + f(X_-) \cdot X^k + [X^k, f(X)]. \qquad (3)$$

Writing $\hat{f}(x, y) = f(x) - f(y) - f'_i(y)(x_i - y_i)$, we get by (2) and property (ii) of Theorem 26.2

$$X^k_- \cdot f(X) = X^k_- f'_i(X_-) \cdot X^i + \tfrac{1}{2} X^k_- f''_{ij}(X_-) \cdot [X^i, X^j]^c$$
$$+ \sum_s X^k_{s-} \hat{f}(X_s, X_{s-}). \qquad (4)$$

Next we note that, by properties (ii), (iv), (v), and (viii) of Theorem 26.6,

$$[X^k, f(X)] = f'_i(X_-) \cdot [X^k, X^i] + \sum_s \Delta X^k_s \hat{f}(X_s, X_{s-})$$
$$= f'_i(X_-) \cdot [X^k, X^i]^c + \sum_s \Delta X^k_s \Delta f(X_s). \qquad (5)$$

Inserting (4) and (5) into (3), and using the elementary formulas

$$g'_i(x) = \delta_{ik} f(x) + x_k f'_i(x),$$
$$g''_{ij}(x) = \delta_{ik} f'_j(x) + \delta_{jk} f'_i(x) + x_k f''_{ij}(x),$$
$$\hat{g}(x, y) = (x_k - y_k)(f(x) - f(y)) + y_k \hat{f}(x, y),$$

we obtain after some simplification the desired expression for $g(X)$.

Equation (2) is trivially true for constant functions, and it extends by induction and linearity to arbitrary polynomials. Now any function $f \in C^2(\mathbb{R}^d)$ may be approximated by polynomials, in such a way that all derivatives up to the second order tend uniformly to those of f on every compact set. To prove (2) for f, it is then enough to show that the right-hand side tends to zero in probability, as f and its first- and second-order derivatives tend to zero, uniformly on compact sets.

For the two integrals in (2), this is clear by the dominated convergence property of Theorem 26.4, and it remains to consider the last term. Writing $B_t = \{x \in \mathbb{R}^d;\ |x| \leq X^*_t\}$ and $\|g\|_B = \sup_B |g|$, we get by Taylor's formula in $\mathbb{R}^d$

$$\sum_{s\le t}|\hat{f}(X_s, X_{s-})| \le \sum_{i,j}\|f''_{i,j}\|_{B_t} \sum_{s\le t}|\Delta X_s|^2$$
$$\le \sum_{i,j}\|f''_{i,j}\|_{B_t} \sum_i [X^i]_t \to 0.$$

The same estimate shows that the last term has locally finite variation. □

To illustrate the use of the general substitution rule, we consider a partial extension of Proposition 21.2 to general semimartingales.

Theorem 26.8 (Doléans' exponential) *For any semimartingale X with $X_0 = 0$, the equation $Z = 1 + Z_- \cdot X$ has the a.s. unique solution*

$$Z_t = \mathcal{E}(X) \equiv \exp(X_t - \tfrac{1}{2}[X]_t^c) \prod_{s\le t}(1 + \Delta X_s)e^{-\Delta X_s}, \quad t \ge 0. \qquad (6)$$

Note that the infinite product in (6) is a.s. absolutely convergent, since $\sum_{s\le t}(\Delta X_s)^2 \le [X]_t < \infty$. However, we may have $\Delta X_s = -1$ for some $s > 0$, in which case $Z = 0$ for $t \ge s$. The process $\mathcal{E}(X)$ in (6) is called the *Doléans exponential* of X. When X is continuous, we get $\mathcal{E}(X) = \exp(X - \tfrac{1}{2}[X])$, in agreement with the notation of Lemma 18.21. For processes A of locally finite variation, formula (6) simplifies to

$$\mathcal{E}(A) = \exp(A_t^c) \prod_{s\le t}(1 + \Delta A_s), \quad t \ge 0.$$

Proof of Theorem 26.8: To check that (6) is a solution, we may write $Z = f(Y, V)$, where $Y = X - \tfrac{1}{2}[X]^c$, $V = \prod(1 + \Delta X)e^{-\Delta X}$, and $f(y, v) = e^y v$. By Theorem 26.7 we get

$$Z - 1 = Z_- \cdot Y + e^{Y_-} \cdot V + \tfrac{1}{2}Z_- \cdot [X]^c$$
$$+ \sum\{\Delta Z - Z_-\Delta X - e^{Y_-}\Delta V\}. \qquad (7)$$

Now $e^{Y_-} \cdot V = \sum e^{Y_-}\Delta V$ since V is of pure-jump type, and furthermore $\Delta Z = Z_-\Delta X$. Hence, the right-hand side of (7) simplifies to $Z_- \cdot X$, as desired.

To prove the uniqueness, let Z be an arbitrary solution, and put $V = Ze^{-Y}$, where $Y = X - \tfrac{1}{2}[X]^c$ as before. By Theorem 26.7 we get

$$V - 1 = e^{-Y_-} \cdot Z - V_- \cdot Y + \tfrac{1}{2}V_- \cdot [X]^c - e^{-Y_-} \cdot [X, Z]^c$$
$$+ \sum\{\Delta V + V_-\Delta Y - e^{-Y_-}\Delta Z\}$$
$$= V_- \cdot X - V_- \cdot X + \tfrac{1}{2}V_- \cdot [X]^c + \tfrac{1}{2}V_- \cdot [X]^c - V_- \cdot [X]^c$$
$$+ \sum\{\Delta V + V_-\Delta X - V_-\Delta X\}$$
$$= \sum \Delta V.$$

Thus, V is a purely discontinuous process of locally finite variation. We may further compute

$$\Delta V = Ze^{-Y} - Z_-e^{-Y_-} = (Z_- + \Delta Z)e^{-Y_- - \Delta Y} - Z_-e^{-Y_-}$$
$$= V_-\{(1 + \Delta X)e^{-\Delta X} - 1\},$$

which shows that $V = 1 + V_- \cdot A$ with $A = \sum\{(1 + \Delta X)e^{-\Delta X} - 1\}$.

It remains to show that the homogeneous equation $V = V_- \cdot A$ has the unique solution $V = 0$. Then define $R_t = \int_{(0,t]} |dA|$, and conclude from Theorem 26.7 and the convexity of the function $x \mapsto x^n$ that

$$R^n = nR_-^{n-1} \cdot R + \sum(\Delta R^n - nR_-^{n-1}\Delta R) \geq nR_-^{n-1} \cdot R. \qquad (8)$$

We may now prove by induction that

$$V_t^* \leq V_t^* R_t^n/n!, \quad t \geq 0, \ n \in \mathbb{Z}_+. \qquad (9)$$

This is obvious for $n = 0$, and assuming (9) to be true for $n - 1$, we get by (8)

$$V_t^* = (V_- \cdot A)_t^* \leq \frac{V_t^*(R_-^{n-1} \cdot R)_t}{(n-1)!} \leq \frac{V_t^* R_t^n}{n!},$$

as required. Since $R_t^n/n! \to 0$ as $n \to \infty$, relation (9) yields $V_t^* = 0$ for all $t > 0$. □

The equation $Z = 1 + Z_- \cdot X$ arises naturally in connection with changes of probability measure. The following result extends Proposition 18.20 to general local martingales.

Theorem 26.9 *(change of measure, van Schuppen and Wong)* *Let $Q = Z_t \cdot P$ on $\mathcal{F}_t$ for all $t \geq 0$, and consider a local P-martingale M such that the process $[M, Z]$ has locally integrable variation and P-compensator $\langle M, Z \rangle$. Then $\tilde{M} = M - Z_-^{-1} \cdot \langle M, Z \rangle$ is a local Q-martingale.*

A lemma will be needed for the proof.

Lemma 26.10 *(integration by parts)* *If X is a semimartingale and A is a predictable process of locally finite variation, then $AX = A \cdot X + X_- \cdot A$ a.s.*

Proof: We need to show that $\Delta A \cdot X = [A, X]$ a.s., which by Theorem 26.6 (viii) is equivalent to

$$\int_{(0,t]} \Delta A_s dX_s = \sum_{s \leq t} \Delta A_s \Delta X_s, \quad t \geq 0.$$

Noting that the series on the right is absolutely convergent by the Cauchy–Buniakovsky inequality, we may reduce, by dominated convergence on each side, to the case when A is constant apart from finitely many jumps. Using Lemma 25.3 and Theorem 25.14, we may next proceed to the case when A has at most one jump, occurring at some predictable time τ. Introducing an announcing sequence (τ_n) and writing $Y = \Delta A \cdot X$, we get by property (iv) of Theorem 26.2

$$Y_{\tau_n \wedge t} = 0 = Y_t - Y_{t \wedge \tau} \quad \text{a.s.}, \quad t \geq 0, \ n \in \mathbb{N}.$$

Thus, even Y is constant apart from a possible jump at τ. Finally, property (iii) of Theorem 26.2 yields $\Delta Y_\tau = \Delta A_\tau \Delta X_\tau$ a.s. on $\{\tau < \infty\}$. □

Proof of Theorem 26.9: For each $n \in \mathbb{N}$, let $\tau_n = \inf\{t;\ Z_t < 1/n\}$, and note that $\tau_n \to \infty$ a.s. Q by Lemma 18.17. Hence, $\tilde{M}$ is well defined under Q, and it suffices as in Lemma 18.15 to show that $(\tilde{M}Z)^{\tau_n}$ is a local P-martingale for every n. Writing $\stackrel{m}{\sim}$ for equality up to a local P-martingale, we may conclude from Lemma 26.10 with $X = Z$ and $A = Z_-^{-1} \cdot \langle M, Z \rangle$ that, on every interval $[0, \tau_n]$,

$$MZ \stackrel{m}{\sim} [M, Z] \stackrel{m}{\sim} \langle M, Z \rangle = Z_- \cdot A \stackrel{m}{\sim} AZ.$$

Thus, we get $\tilde{M}Z = (M - A)Z \stackrel{m}{\sim} 0$, as required. □

Using the last theorem, we may easily show that the class of semimartingales is invariant under absolutely continuous changes of the probability measure. A special case of this result was previously obtained as part of Proposition 18.20.

Corollary 26.11 *(preservation law, Jacod)* *If $Q \ll P$ on $\mathcal{F}_t$ for all $t > 0$, then every P-semimartingale is also a Q-semimartingale.*

Proof: Assume that $Q = Z_t \cdot P$ on $\mathcal{F}_t$ for all $t \geq 0$. We need to show that every local P-martingale M is a Q-semimartingale. By Lemma 26.5 we may then assume ΔM to be bounded, so that $[M]$ is locally bounded. By Theorem 26.9 it suffices to show that $[M, Z]$ has locally integrable variation, and by Theorem 26.6 (iii) it is then enough to prove that $[Z]^{1/2}$ is locally integrable. Now Theorem 26.6 (iv) yields

$$[Z]_t^{1/2} \leq [Z]_{t-}^{1/2} + |\Delta Z_t| \leq [Z]_{t-}^{1/2} + Z_{t-}^* + |Z_t|, \quad t \geq 0,$$

and so the desired integrability follows by optional sampling. □

Our next aim is to extend the *BDG* inequalities of Theorem 17.7 to general local martingales. Such an extension turns out to be possible only for exponents $p \geq 1$.

Theorem 26.12 *(norm inequalities, Burkholder, Davis, Gundy)* *There exist some constants $c_p \in (0, \infty)$, $p \geq 1$, such that for any local martingale M with $M_0 = 0$,*

$$c_p^{-1} E[M]_\infty^{p/2} \leq E M^{*p} \leq c_p E[M]_\infty^{p/2}, \quad p \geq 1. \tag{10}$$

As in Corollary 17.8, it follows in particular that M is a uniformly integrable martingale whenever $E[M]_\infty^{1/2} < \infty$.

Proof for $p = 1$ (Davis): To exploit the symmetry of the argument, we write $M^\flat$ and $M^\sharp$ for the processes M^* and $[M]^{1/2}$, taken in either order.

Put $J = \Delta M$, and define
$$A_t = \sum_{s \le t} J_s 1\{|J_s| > 2J_{s-}^*\}, \quad t \ge 0.$$
Since $|\Delta A| \le 2\Delta J^*$, we have
$$\int_0^\infty |dA_s| = \sum_s |\Delta A_s| \le 2J^* \le 4M_\infty^\sharp.$$
Writing $\hat{A}$ for the compensator of A and putting $D = A - \hat{A}$, we get
$$ED_\infty^\flat \vee ED_\infty^\sharp \le E\int_0^\infty |dD_s| \lesssim E\int_0^\infty |dA_s| \lesssim EM_\infty^\sharp. \tag{11}$$
To get a similar estimate for $N = M - D$, we introduce the optional times
$$\tau_r = \inf\{t;\ N_t^\sharp \vee J_t^* > r\}, \quad r > 0,$$
and note that
$$\begin{aligned} P\{N_\infty^\flat > r\} &\le P\{\tau_r < \infty\} + P\{\tau_r = \infty,\ N_\infty^\flat > r\} \\ &\le P\{N_\infty^\sharp > r\} + P\{J^* > r\} + P\{N_{\tau_r}^\flat > r\}. \end{aligned} \tag{12}$$
Arguing as in the proof of Lemma 26.5, we get $|\Delta N| \le 4J_-^*$, and so
$$N_{\tau_r}^\sharp \le N_\infty^\sharp \wedge (N_{\tau_r-}^\sharp + 4J_{\tau_r-}^*) \le N_\infty^\sharp \wedge 5r.$$
Since $N^2 - [N]$ is a local martingale, we get by Chebyshev's inequality or Proposition 7.15, respectively,
$$r^2 P\{N_{\tau_r}^\flat > r\} \lesssim EN_{\tau_r}^{\sharp 2} \lesssim E(N_\infty^\sharp \wedge r)^2.$$
Hence, by Fubini's theorem and some calculus,
$$\int_0^\infty P\{N_{\tau_r}^\flat > r\} dr \lesssim \int_0^\infty E(N_\infty^\sharp \wedge r)^2 r^{-2} dr \lesssim EN_\infty^\sharp.$$
Combining this with (11)–(12) and using Lemma 3.4, we get
$$\begin{aligned} EN_\infty^\flat &= \int_0^\infty P\{N_\infty^\flat > r\} dr \\ &\le \int_0^\infty \left(P\{N_\infty^\sharp > r\} + P\{J^* > r\} + P\{N_{\tau_r}^\flat > r\}\right) dr \\ &\lesssim EN_\infty^\sharp + EJ^* \lesssim EM_\infty^\sharp. \end{aligned}$$
It remains to note that $EM_\infty^\flat \le ED_\infty^\flat + EN_\infty^\flat$. $\quad\square$

Extension to $p > 1$ (Garsia): For any $t \ge 0$ and $B \in \mathcal{F}_t$, we may apply (10) with $p = 1$ to the local martingale $1_B(M - M^t)$ to get a.s.
$$\begin{aligned} c_1^{-1} E[[M - M^t]_\infty^{1/2} | \mathcal{F}_t] &\le E[(M - M^t)_\infty^* | \mathcal{F}_t] \\ &\le c_1 E[[M - M^t]_\infty^{1/2} | \mathcal{F}_t]. \end{aligned}$$

Since
$$[M]_\infty^{1/2} - [M]_t^{1/2} \le [M-M^t]_\infty^{1/2} \le [M]_\infty^{1/2},$$
$$M_\infty^* - M_t^* \le (M-M^t)_\infty^* \le 2M_\infty^*,$$

the relation $E[A_\infty - A_t|\mathcal{F}_t] \le E[\zeta|\mathcal{F}_t]$ occurring in Proposition 25.21 holds with $A_t = [M]_t^{1/2}$ and $\zeta = M^*$, and also with $A_t = M_t^*$ and $\zeta = [M]_\infty^{1/2}$. Since

$$\Delta M_t^* \le \Delta[M]_t^{1/2} = |\Delta M_t| \le [M]_t^{1/2} \wedge 2M_t^*,$$

we have in both cases $\Delta A_\tau \le E[\zeta|\mathcal{F}_\tau]$ a.s. for every optional time τ, and so the cited condition remains fulfilled for the left-continuous version A_-. Hence, Proposition 25.21 yields $\|A_\infty\|_p \le \|\zeta\|_p$ for every $p \ge 1$, and (10) follows. □

We may use the last theorem to extend the stochastic integral to a larger class of integrands. Then write $\mathcal{M}$ for the space of local martingales and $\mathcal{M}_0$ for the subclass of processes M with $M_0 = 0$. For any $M \in \mathcal{M}$, let $L(M)$ denote the class of predictable processes V such that $(V^2 \cdot [M])^{1/2}$ is locally integrable.

Theorem 26.13 *(martingale integral, Meyer)* *The elementary predictable integral extends a.s. uniquely to a bilinear map of any $M \in \mathcal{M}$ and $V \in L(M)$ into $V \cdot M \in \mathcal{M}_0$, such that if $V, V_1, V_2, \cdots \in L(M)$ with $|V_n| \le V$ and $(V_n^2 \cdot [M])_t \xrightarrow{P} 0$ for some $t > 0$, then $(V_n \cdot M)_t^* \xrightarrow{P} 0$. This integral satisfies properties (ii)–(iv) of Theorem 26.2 and is characterized by the condition*

$$[V \cdot M, N] = V \cdot [M, N] \quad a.s., \quad N \in \mathcal{M}. \tag{13}$$

Proof: For the construction of the integral, we may reduce by localization to the case when $E(M - M_0)^* < \infty$ and $E(V^2 \cdot [M])_\infty^{1/2} < \infty$. For each $n \in \mathbb{N}$, define $V_n = V1\{|V| \le n\}$. Then $V_n \cdot M \in \mathcal{M}_0$ by Theorem 26.4, and by Theorem 26.12 we have $E(V_n \cdot M)^* < \infty$. Using Theorems 26.6 (v) and 26.12, Minkowski's inequality, and dominated convergence, we obtain

$$E(V_m \cdot M - V_n \cdot M)^* \le E[(V_m - V_n) \cdot M]_\infty^{1/2}$$
$$= E((V_m - V_n)^2 \cdot [M])_\infty^{1/2} \to 0.$$

Hence, there exists a process $V \cdot M$ with $E(V_n \cdot M - V \cdot M)^* \to 0$, and clearly $V \cdot M \in \mathcal{M}_0$ and $E(V \cdot M)^*\infty$.

To prove (13), we note that the relation holds for each V_n by Theorem 26.6 (v). Since $E[V_n \cdot M - V \cdot M]_\infty^{1/2} \to 0$ by Theorem 26.12, we get by Theorem 26.6 (iii) for any $N \in \mathcal{M}$ and $t \ge 0$

$$|[V_n \cdot M, N]_t - [V \cdot M, N]_t| \le [V_n \cdot M - V \cdot M]_t^{1/2}[N]_t^{1/2} \xrightarrow{P} 0. \tag{14}$$

Next we note that, by Theorem 26.6 (iii) and (v),

$$\int_0^t |V_n d[M, N]| = \int_0^t |d[V_n \cdot M, N]| \leq [V_n \cdot M]_t^{1/2} [N]_t^{1/2}.$$

As $n \to \infty$, we get by monotone convergence on the left and Minkowski's inequality on the right

$$\int_0^t |V d[M, N]| \leq [V \cdot M]_t^{1/2} [N]_t^{1/2} < \infty.$$

Hence, by dominated convergence $V_n \cdot [M, N] \to V \cdot [M, N]$, and (13) follows by combination with (14).

To see that (13) determines $V \cdot M$, it remains to note that if $[M] = 0$ a.s. for some $M \in \mathcal{M}_0$, then $M^* = 0$ a.s. by Theorem 26.12. To prove the stated continuity property, we may reduce by localization to the case when $E(V^2 \cdot [M])_\infty^{1/2} < \infty$. But then $E(V_n^2 \cdot [M])_\infty^{1/2} \to 0$ by dominated convergence, and Theorem 26.12 yields $E(V_n \cdot M)^* \to 0$. To prove the uniqueness of the integral, it is enough to consider bounded integrands V. We may then approximate as in Lemma 26.3 by uniformly bounded processes $V_n \in \mathcal{E}$ with $((V_n - V)^2 \cdot [M]) \xrightarrow{P} 0$, and conclude that $(V_n \cdot M - V \cdot M)^* \xrightarrow{P} 0$.

Of the remaining properties in Theorem 26.2, relation (ii) may be proved as before by means of (13), whereas (iii) and (iv) follow most easily by truncation from the corresponding statements in Theorem 26.4. □

A semimartingale $X = M + A$ is said to be *purely discontinuous* if there exist some local martingales $M^1, M^2, \ldots$ of locally finite variation such that $E(M - M^n)^{*2} \to 0$ for every $t > 0$. The property is clearly independent of the choice of decomposition $X = M + A$. To motivate the terminology, we note that any martingale M of locally finite variation may be written as $M = M_0 + A - \hat{A}$, where $A_t = \sum_{s \leq t} \Delta M_s$ and $\hat{A}$ denotes the compensator of A. Thus, $M - M_0$ is in this case a compensated sum of jumps.

Our present goal is to establish a fundamental decomposition of a general semimartingale X into a continuous and a purely discontinuous component, corresponding to the elementary decomposition of the quadratic variation $[X]$ into a continuous part and a jump part. In this connection the reader is cautioned that, although any adapted process of locally finite variation is a purely discontinuous semimartingale, it may not be purely discontinuous in the sense of real analysis.

Theorem 26.14 (decomposition of semimartingales, Yoeurp, Meyer) *Every semimartingale X has an a.s. unique decomposition $X = X_0 + X^c + X^d$, where X^c is a continuous local martingale with $X_0^c = 0$ and X^d is a purely discontinuous semimartingale. Furthermore, $[X^c] = [X]^c$ and $[X^d] = [X]^d$ a.s.*

Proof: To decompose X it is enough to consider the martingale component in any decomposition $X = X_0 + M + A$, and by Lemma 26.5 we may assume that $M \in \mathcal{M}_{0,\text{loc}}^2$. We may then choose some optional times

$\tau_n \uparrow \infty$, where $\tau_0 = 0$, such that $M^{\tau_n} \in \mathcal{M}_0^2$ for each n. It is enough to construct the desired decomposition for each process $M^{\tau_n} - M^{\tau_{n-1}}$, which reduces the discussion to the case when $M \in \mathcal{M}_0^2$. Now let $\mathcal{C}$ and $\mathcal{D}$ denote the classes of continuous and purely discontinuous processes in $\mathcal{M}_0^2$, and note that both are closed linear subspaces of the Hilbert space $\mathcal{M}_0^2$. The desired decomposition will follow from Theorem 1.33 if we can show that $\mathcal{D}^\perp \subset \mathcal{C}$.

Then let $M \in \mathcal{D}^\perp$. To see that M is continuous, fix any $\varepsilon > 0$, and put $\tau = \inf\{t;\ \Delta M_t > \varepsilon\}$. Define $A_t = 1\{\tau \le t\}$, let $\hat{A}$ denote the compensator of A, and put $N = A - \hat{A}$. Integrating by parts and using Lemma 25.13 gives

$$\tfrac{1}{2}E\hat{A}_\tau^2 \le E\int \hat{A}\, d\hat{A} = E\int \hat{A}\, dA = E\hat{A}_\tau = EA_\tau \le 1.$$

Thus, N is L^2-bounded and hence lies in $\mathcal{D}$. For any bounded martingale M', we get

$$EM'_\infty N_\infty = E\int M'\, dN = E\int \Delta M'\, dN$$

$$= E\int \Delta M'\, dA = E[\Delta M'_\tau;\ \tau < \infty],$$

where the first equality is obtained as in the proof of Lemma 25.7, the second is due to the predictability of M'_-, and the third holds since $\hat{A}$ is predictable and hence natural. Letting $M' \to M$ in $\mathcal{M}^2$, we obtain

$$0 = EM_\infty N_\infty = E[\Delta M_\tau;\ \tau < \infty] \ge \varepsilon P\{\tau < \infty\}.$$

Thus, $\Delta M \le \varepsilon$ a.s., and therefore $\Delta M \le 0$ a.s. since ε is arbitrary. Similarly, $\Delta M \ge 0$ a.s., and the desired continuity follows.

Next assume that $M \in \mathcal{D}$ and $N \in \mathcal{C}$, and choose martingales $M^n \to M$ of locally finite variation. By Theorem 26.6 (vi) and (vii) and optional sampling, we get for any optional time τ

$$0 = E[M^n, N]_\tau = EM^n_\tau N_\tau \to EM_\tau N_\tau = E[M, N]_\tau,$$

and so $[M, N]$ is a martingale by Lemma 7.13. Since it is also continuous by (15), Proposition 17.2 yields $[M, N] = 0$ a.s. In particular, $EM_\infty N_\infty = 0$, which shows that $\mathcal{C} \perp \mathcal{D}$. The uniqueness assertion now follows easily.

To prove the last assertion, we conclude from Theorem 26.6 (iv) that, for any $M \in \mathcal{M}^2$,

$$[M]_t = [M]_t^c + \sum_{s \le t} (\Delta M_s)^2, \quad t \ge 0. \tag{15}$$

Letting $M \in \mathcal{D}$, we may choose martingales of locally finite variation $M^n \to M$. By Theorem 26.6 (vii) and (viii) we have $[M^n]^c = 0$ and $E[M^n - M]_\infty$

$\to 0$. For any $t \geq 0$, we get by Minkowski's inequality and (15)

$$\left|\left\{\sum_{s\leq t}(\Delta M^n_s)^2\right\}^{1/2} - \left\{\sum_{s\leq t}(\Delta M_s)^2\right\}^{1/2}\right|$$
$$\leq \left\{\sum_{s\leq t}(\Delta M^n_s - \Delta M_s)^2\right\}^{1/2} \leq [M^n - M]_t^{1/2} \xrightarrow{P} 0,$$
$$\left|[M^n]_t^{1/2} - [M]_t^{1/2}\right| \leq [M^n - M]_t^{1/2} \xrightarrow{P} 0.$$

Taking limits in (15) for the martingales M^n, we get the same formula for M without the term $[M]^c_t$, which shows that $[M] = [M]^d$.

Now consider any $M \in \mathcal{M}^2$. Using the strong orthogonality $[M^c, M^d] = 0$, we get a.s.

$$[M]^c + [M]^d = [M] = [M^c + M^d] = [M^c] + [M^d],$$

which shows that even $[M^c] = [M]^c$ a.s. By the same argument combined with Theorem 26.6 (viii) we obtain $[X^d] = [X]^d$ a.s. for any semimartingale X. □

The last result immediately yields an explicit formula for the covariation of two semimartingales.

Corollary 26.15 *(decomposition of covariation)* *For any semimartingale X, the process X^c is the a.s. unique continuous local martingale M with $M_0 = 0$ such that $[X - M]$ is purely discontinuous. Furthermore, we have a.s. for any semimartingales X and Y*

$$[X,Y]_t = [X^c, Y^c] + \sum_{s\leq t} \Delta X_s \Delta Y_s, \quad t \geq 0. \tag{16}$$

In particular, we note that $(V \cdot X)^c = V \cdot X^c$ a.s. for any semimartingale X and locally bounded, predictable process V.

Proof: If M has the stated properties, then $[(X - M)^c] = [X - M]^c = 0$ a.s., and so $(X - M)^c = 0$ a.s. Thus, $X - M$ is purely discontinuous. Formula (16) holds by Theorem 26.6 (iv) and Theorem 26.14 when $X = Y$, and the general result follows by polarization. □

The purely discontinuous component of a local martingale has a further decomposition, similar to the decompositions of optional times and increasing processes in Propositions 25.4 and 25.17.

Corollary 26.16 *(decomposition of martingales, Yoeurp)* *Every purely discontinuous local martingale M has an a.s. unique decomposition $M = M_0 + M^q + M^a$ with purely discontinuous $M^q, M^a \in \mathcal{M}_0$, where M^q is quasi-leftcontinuous and M^a has accessible jumps. Furthermore, there exist some predictable times $\tau_1, \tau_2, \ldots$ with disjoint graphs such that $\{t; \Delta M^a_t \neq 0\} \subset \bigcup_n [\tau_n]$ a.s. Finally, $[M^q] = [M]^q$ and $[M^a] = [M]^a$ a.s., and also $\langle M^q \rangle = \langle M \rangle^c$ and $\langle M^a \rangle = \langle M \rangle^d$ a.s. when $M \in \mathcal{M}^2_{\text{loc}}$.*

Proof: Introduce the locally integrable process $A_t = \sum_{s\leq t}\{(\Delta M_s)^2 \wedge 1\}$ with compensator $\hat{A}$, and define $M^q = M - M_0 - M^a = 1\{\Delta \hat{A}_t = 0\} \cdot M$.

By Theorem 26.4 we have $M^q, M^a \in \mathcal{M}_0$ and $\Delta M^q = 1\{\Delta \hat{A} = 0\}\Delta M$ a.s. Furthermore, M^q and M^a are purely discontinuous by Corollary 26.15. The proof may now be completed as in the case of Proposition 25.17. □

We may illustrate the use of the previous decompositions by proving two exponential inequalities for martingales with bounded jumps.

Theorem 26.17 *(exponential inequalities)* Let M be a local martingale with $M_0 = 0$ such that $|\Delta M| \leq c$ for some constant $c \leq 1$.
  (i) If $[M]_\infty \leq 1$ a.s., then
$$P\{M^* \geq r\} \leq \exp\{-\tfrac{1}{2}r^2/(1+rc)\}, \quad r \geq 0.$$
  (ii) If $\langle M \rangle_\infty \leq 1$ a.s., then
$$P\{M^* \geq r\} \leq \exp\{-\tfrac{1}{2}r\log(1+rc)/c\}, \quad r \geq 0.$$

For continuous martingales both bounds reduce to $e^{-r^2/2}$, which can also be obtained directly by more elementary methods. For the proof of Theorem 26.17 we need two lemmas. We begin with a characterization of certain pure jump-type martingales.

Lemma 26.18 *(accessible jump-type martingales)* Let N be a pure jump-type process with integrable variation and accessible jumps. Then N is a martingale iff $E[\Delta N_\tau | \mathcal{F}_{\tau-}] = 0$ a.s. for every finite predictable time τ.

Proof: By Proposition 25.17 there exist some predictable times $\tau_1, \tau_2, \ldots$ with disjoint graphs such that $\{t > 0;\ \Delta N_t \neq 0\} \subset \bigcup_n [\tau_n]$. Assuming the stated condition, we get by Fubini's theorem and Lemma 25.2 for any bounded optional time τ
$$\begin{aligned} EN_\tau &= \sum_n E[\Delta N_{\tau_n}; \tau_n \leq \tau] \\ &= \sum_n E[E[\Delta N_{\tau_n}|\mathcal{F}_{\tau_n-}]; \tau_n \leq \tau] = 0, \end{aligned}$$
and so N is a martingale by Lemma 7.13. Conversely, given any uniformly integrable martingale N and finite predictable time τ, we have a.s. $E[N_\tau|\mathcal{F}_{\tau-}] = N_{\tau-}$ and hence $E[\Delta N_\tau|\mathcal{F}_{\tau-}] = 0$. □

For general martingales M, the process $Z = e^{M - [M]/2}$ in Lemma 18.21 is not necessarily a martingale. For many purposes, however, it can be replaced by a similar supermartingale.

Lemma 26.19 *(exponential supermartingales)* Let M be a local martingale with $M_0 = 0$ and $|\Delta M| \leq c < \infty$ a.s., and put $a = f(c)$ and $b = g(c)$, where
$$f(x) = -(x + \log(1-x)_+)x^{-2}, \qquad g(x) = (e^x - 1 - x)x^{-2}.$$
Then the processes $X = e^{M - a[M]}$ and $Y = e^{M - b\langle M \rangle}$ are supermartingales.

Proof: In case of X we may clearly assume that $c < 1$. By Theorem 26.7 we get, in an obvious shorthand notation,

$$X_-^{-1} \cdot X = M - (a - \tfrac{1}{2})[M]^c + \sum \left\{ e^{\Delta M - a(\Delta M)^2} - 1 - \Delta M \right\}.$$

Here the first term on the right is a local martingale, and the second term is nonincreasing since $a \geq \tfrac{1}{2}$. To see that even the sum is nonincreasing, we need to show that $\exp(x - ax^2) \leq 1 + x$ or $f(-x) \leq f(c)$ whenever $|x| \leq c$. But this is clear by a Taylor expansion of each side. Thus, $X_-^{-1} \cdot X$ is a local supermartingale, and since $X > 0$, the same thing is true for $X_- \cdot (X_-^{-1} \cdot X) = X$. By Fatou's lemma it follows that X is a true supermartingale.

In the case of Y, we may decompose M according to Theorem 26.14 and Proposition 26.16 as $M = M^c + M^q + M^a$, and conclude by Theorem 26.7 that

$$\begin{aligned}
Y_-^{-1} \cdot Y &= M - b\langle M \rangle^c + \tfrac{1}{2}[M]^c + \sum \left\{ e^{\Delta M - b\Delta \langle M \rangle} - 1 - \Delta M \right\} \\
&= M + b([M^q] - \langle M^q \rangle) - (b - \tfrac{1}{2})[M]^c \\
&\quad + \sum \left\{ e^{\Delta M - b\Delta \langle M \rangle} - \frac{1 + \Delta M + b(\Delta M)^2}{1 + b\Delta\langle M \rangle} \right\} \\
&\quad + \sum \left\{ \frac{1 + \Delta M^a + b(\Delta M^a)^2}{1 + b\Delta\langle M^a \rangle} - 1 - \Delta M^a \right\}.
\end{aligned}$$

Here the first two terms on the right are martingales, and the third term is nonincreasing since $b \geq \tfrac{1}{2}$. Even the first sum of jumps is nonincreasing since $e^x - 1 - x \leq bx^2$ for $|x| \leq c$ and $e^y \leq 1 + y$ for $y \geq 0$.

The last sum clearly defines a purely discontinuous process N of locally finite variation and with accessible jumps. Fixing any finite predictable time τ and writing $\xi = \Delta M_\tau$ and $\eta = \Delta \langle M \rangle_\tau$, we note that

$$E \left| \frac{1 + \xi + b\xi^2}{1 + b\eta} - 1 - \xi \right| \leq E |1 + \xi + b\xi^2 - (1 + \xi)(1 + b\eta)|$$
$$= bE|\xi^2 - (1+\xi)\eta| \leq b(2+c)E\xi^2.$$

Since

$$E \sum_t (\Delta M_t)^2 \leq E[M]_\infty = E\langle M \rangle_\infty \leq 1,$$

we conclude that the total variation of N is integrable. Using Lemmas 25.3 and 26.18, we also note that a.s. $E[\xi | \mathcal{F}_{\tau-}] = 0$ and

$$E[\xi^2 | \mathcal{F}_{\tau-}] = E[\Delta[M]_\tau | \mathcal{F}_{\tau-}] = E[\eta | \mathcal{F}_{\tau-}] = \eta.$$

Thus,

$$E\left[\frac{1 + \xi + b\xi^2}{1 + b\eta} - 1 - \xi \,\bigg|\, \mathcal{F}_{\tau-} \right] = 0,$$

and Lemma 26.18 shows that N is a martingale. The proof may now be completed as before. □

Proof of Theorem 26.17: (i) Fix any $u > 0$, and conclude from Lemma 26.19 that the process

$$X_t^u = \exp\{uM_t - u^2 f(uc)[M]_t\}, \quad t \geq 0,$$

is a positive supermartingale. Since $[M] \leq 1$ and $X_0^u = 1$, we get for any $r > 0$

$$\begin{aligned} P\{\sup_t M_t > r\} &\leq P\{\sup_t X_t^u > \exp\{ur - u^2 f(uc)\}\} \\ &\leq \exp\{-ur + u^2 f(uc)\}. \end{aligned} \quad (17)$$

Now define $F(x) = 2xf(x)$, and note that F is continuous and strictly increasing from $[0, 1)$ onto $\mathbb{R}_+$. Also note that $F(x) \leq x/(1-x)$ and hence $F^{-1}(y) \geq y/(1+y)$. Taking $u = F^{-1}(rc)/c$ in (17), we get

$$\begin{aligned} P\{\sup_t M_t > r\} &\leq \exp\{-\tfrac{1}{2} r F^{-1}(rc)/c\} \\ &\leq \exp\{-\tfrac{1}{2} r^2/(1+rc)\}. \end{aligned}$$

It remains to combine with the same inequality for $-M$.

(ii) Define $G(x) = 2xg(x)$, and note that G is a continuous and strictly increasing mapping onto $\mathbb{R}_+$. Furthermore, $G(x) \leq e^x - 1$, and so $G^{-1}(y) \geq \log(1+y)$. Proceeding as before, we get

$$\begin{aligned} P\{\sup_t M_t > r\} &\leq \exp\{-\tfrac{1}{2} r G^{-1}(rc)/c\} \\ &\leq \exp\{-\tfrac{1}{2} r \log(1+rc)/c\}, \end{aligned}$$

and the result follows. □

A *quasi-martingale* is defined as an integrable, adapted, and right-continuous process X such that

$$\sup_\pi \sum_{k \leq n} E\left|X_{t_k} - E[X_{t_{k+1}}|\mathcal{F}_{t_k}]\right| < \infty, \quad (18)$$

where the supremum extends over all finite partitions π of $\mathbb{R}_+$ of the form $0 = t_0 < t_1 < \cdots < t_n < \infty$, and the last term is computed under the conventions $t_{n+1} = \infty$ and $X_\infty = 0$. In particular, we note that (18) holds when X is the sum of an L^1-bounded martingale and a process of integrable variation starting at 0. The next result shows that this case is close to the general situation. Here localization is defined in the usual way in terms of a sequence of optional times $\tau_n \uparrow \infty$.

Theorem 26.20 *(quasi-martingales, Rao) Any quasi-martingale is a difference of two nonnegative supermartingales. Thus, a process X with $X_0 = 0$ is a local quasi-martingale iff it is a special semimartingale.*

Proof: For any $t \geq 0$, let $\mathcal{P}_t$ denote the class of partitions π of the interval $[t, \infty)$ of the form $t = t_0 < t_1 < \cdots < t_n$, and define

$$\eta_\pi^\pm = \sum_{k \leq n} E\left[(X_{t_k} - E[X_{t_{k+1}}|\mathcal{F}_{t_k}])_\pm \big| \mathcal{F}_t\right], \quad \pi \in \mathcal{P}_t,$$

where $t_{n+1} = \infty$ and $X_\infty = 0$ as before. We claim that η_π^+ and η_π^- are a.s. nondecreasing under refinements of $\pi \in \mathcal{P}_t$. To see this, it is clearly enough to add one more division point u to π, say in the interval (t_k, t_{k+1}). Put $\alpha = X_{t_k} - X_u$ and $\beta = X_u - X_{t_{k+1}}$. By subadditivity and Jensen's inequality we get the desired relation

$$E[E[\alpha+\beta|\mathcal{F}_{t_k}]_\pm|\mathcal{F}_t] \leq E[E[\alpha|\mathcal{F}_{t_k}]_\pm + E[\beta|\mathcal{F}_{t_k}]_\pm|\mathcal{F}_t]$$
$$\leq E[E[\alpha|\mathcal{F}_{t_k}]_\pm + E[\beta|\mathcal{F}_u]_\pm|\mathcal{F}_t].$$

Now fix any $t \geq 0$, and conclude from (18) that $m_t^\pm \equiv \sup_{\pi \in \mathcal{P}_t} E\eta_\pi^\pm < \infty$. For each $n \in \mathbb{N}$ we may then choose some $\pi_n \in \mathcal{P}_t$ with $E\eta_{\pi_n}^\pm > m_t^\pm - n^{-1}$. The sequences $(\eta_{\pi_n}^\pm)$ are Cauchy in L^1, and so they converge in L^1 toward some limits $Y_t^\pm$. Note also that $E|\eta_\pi^\pm - Y_t^\pm| < n^{-1}$ whenever π is a refinement of π_n. Thus, $\eta_\pi^\pm \to Y_t^\pm$ in L^1 along the directed set $\mathcal{P}_t$.

Next fix any $s < t$, let $\pi \in \mathcal{P}_t$ be arbitrary, and define $\pi' \in \mathcal{P}_s$ by adding the point s to π. Then

$$Y_s^\pm \geq \eta_{\pi'}^\pm = (X_s - E[X_t|\mathcal{F}_s])_\pm + E[\eta_\pi^\pm|\mathcal{F}_s] \geq E[\eta_\pi^\pm|\mathcal{F}_s].$$

Taking limits along $\mathcal{P}_t$ on the right, we get $Y_s^\pm \geq E[Y_t^\pm|\mathcal{F}_s]$ a.s., which means that the processes $Y^\pm$ are supermartingales. By Theorem 7.27 the right-hand limits along the rationals $Z_t^\pm = Y_{t+}^\pm$ then exist outside a fixed null set, and the processes $Z^\pm$ are right-continuous supermartingales. For $\pi \in \mathcal{P}_t$ we have $X_t = \eta_\pi^+ - \eta_\pi^- \to Y_t^+ - Y_t^-$, and so $Z_t^+ - Z_t^- = X_{t+} = X_t$ a.s. □

The next result shows that semimartingales are the most general processes for which a stochastic integral with reasonable continuity properties can be defined. As before, $\mathcal{E}$ denotes the class of bounded, predictable step processes with jumps at finitely many fixed points.

Theorem 26.21 *(stochastic integrators, Bichteler, Dellacherie)* A right-continuous, adapted process X is a semimartingale iff for any $V_1, V_2, \cdots \in \mathcal{E}$ with $\|V_n^*\|_\infty \to 0$ we have $(V_n \cdot X)_t \xrightarrow{P} 0$ for all $t > 0$.

The proof is based on three lemmas, the first of which separates the crucial functional-analytic part of the argument.

Lemma 26.22 *(convexity and tightness)* For any tight, convex set $\mathcal{K} \subset L^1(P)$, there exists a bounded random variable $\rho > 0$ with $\sup_{\xi \in \mathcal{K}} E\rho\xi < \infty$.

Proof (Yan): Let $\mathcal{B}$ denote the class of bounded, nonnegative random variables, and define $\mathcal{C} = \{\gamma \in \mathcal{B}; \sup_{\xi \in \mathcal{K}} E(\gamma\xi) < \infty\}$. We claim that, for any $\gamma_1, \gamma_2, \cdots \in \mathcal{C}$, there exists some $\gamma \in \mathcal{C}$ with $\{\gamma > 0\} = \bigcup_n \{\gamma_n > 0\}$. Indeed, we may assume that $\gamma_n \leq 1$ and $\sup_{\xi \in \mathcal{K}} E(\gamma_n \xi) \leq 1$, in which case we may choose $\gamma = \sum_n 2^{-n} \gamma_n$. It is then easy to construct a $\rho \in \mathcal{C}$ such that $P\{\rho > 0\} = \sup_{\gamma \in \mathcal{C}} P\{\gamma > 0\}$. Clearly,

$$\{\gamma > 0\} \subset \{\rho > 0\} \text{ a.s.}, \quad \gamma \in \mathcal{C}, \tag{19}$$

since we could otherwise choose a $\rho' \in \mathcal{C}$ with $P\{\rho' > 0\} > P\{\rho > 0\}$.

To show that $\rho > 0$ a.s., we assume that instead $P\{\rho = 0\} > \varepsilon > 0$. By the tightness of $\mathcal{K}$ we may choose $r > 0$ so large that $P\{\xi > r\} \leq \varepsilon$ for all $\xi \in \mathcal{K}$. Then $P\{\xi - \beta > r\} \leq \varepsilon$ for all $\xi \in \mathcal{K}$ and $\beta \in \mathcal{B}$. By Fatou's lemma we obtain $P\{\zeta > r\} \leq \varepsilon$ for all ζ in the L^1-closure $\mathcal{Z} = \overline{\mathcal{K} - \mathcal{B}}$. In particular, the random variable $\zeta_0 = 2r\mathbf{1}\{\rho = 0\}$ lies outside $\mathcal{Z}$. Now $\mathcal{Z}$ is convex and closed, and so, by a version of the Hahn–Banach theorem, there exists some $\gamma \in (L^1)^* = L^\infty$ satisfying

$$\sup_{\xi \in \mathcal{K}} E\gamma\xi - \inf_{\beta \in \mathcal{B}} E\gamma\beta \leq \sup_{\zeta \in \mathcal{Z}} E\gamma\zeta < E\gamma\zeta_0 = 2rE[\gamma; \rho = 0]. \qquad (20)$$

Here $\gamma \geq 0$, since we would otherwise get a contradiction by choosing $\beta = b\mathbf{1}\{\gamma < 0\}$ for large enough $b > 0$. Hence, (20) reduces to $\sup_{\xi \in \mathcal{K}} E\gamma\xi < 2rE[\gamma; \rho = 0]$, which implies $\gamma \in \mathcal{C}$ and $E[\gamma; \rho = 0] > 0$. But this contradicts (19), and therefore $\rho > 0$ a.s. □

Two further lemmas are needed for the proof of Theorem 26.21.

Lemma 26.23 *(tightness and boundedness)* Let $\mathcal{T}$ be the class of optional times $\tau < \infty$ taking finitely many values, and consider a right-continuous, adapted process X such that the family $\{X_\tau; \tau \in \mathcal{T}\}$ is tight. Then $X^* < \infty$ a.s.

Proof: By Lemma 7.4 any bounded optional time τ can be approximated from the right by optional times $\tau_n \in \mathcal{T}$, and by right-continuity we have $X_{\tau_n} \to X_\tau$. Hence, Fatou's lemma yields $P\{|X_\tau| > r\} \leq \liminf_n P\{|X_{\tau_n}| > r\}$, and so the hypothesis remains true with $\mathcal{T}$ replaced by the class $\hat{\mathcal{T}}$ of all bounded optional times. By Lemma 7.6 the times $\tau_{t,n} = t \wedge \inf\{s; |X_s| > n\}$ belong to $\hat{\mathcal{T}}$ for all $t > 0$ and $n \in \mathbb{N}$, and as $n \to \infty$, we get

$$P\{X^* > n\} = \sup_{t > 0} P\{X_t^* > n\} \leq \sup_{\tau \in \hat{\mathcal{T}}} P\{|X_\tau| > n\} \to 0. \qquad \square$$

Lemma 26.24 *(scaling)* For any finite random variable ξ, there exists a bounded random variable $\rho > 0$ such that $E|\rho\xi| < \infty$.

Proof: We may take $\rho = (|\xi| \vee 1)^{-1}$. □

Proof of Theorem 26.21: The necessity is clear from Theorem 26.4. Now assume the stated condition. By Lemma 4.9 it is equivalent to assume for each $t > 0$ that the family $\mathcal{K}_t = \{(V \cdot X)_t; V \in \mathcal{E}_1\}$ is tight, where $\mathcal{E}_1 = \{V \in \mathcal{E}; |V| \leq 1\}$. The latter family is clearly convex, and by the linearity of the integral the convexity carries over to $\mathcal{K}_t$.

By Lemma 26.23 we have $X^* < \infty$ a.s., and so by Lemma 26.24 there exists a probability measure $Q \sim P$ such that $E_Q X_t^* = \int X_t^* dQ < \infty$. In particular, $\mathcal{K}_t \subset L^1(Q)$, and we note that $\mathcal{K}_t$ remains tight with respect to Q. Hence, by Lemma 26.22 there exists a probability measure $R \sim Q$ with bounded density $\rho = dR/dQ$ such that $\mathcal{K}_t$ is bounded in $L^1(R)$.

Now consider an arbitrary partition $0 = t_0 < t_1 < \cdots < t_n = t$, and note that

$$\sum_{k \leq n} E_R|X_{t_k} - E_R[X_{t_{k+1}}|\mathcal{F}_{t_k}]| = E_R(V \cdot X)_t + E_R|X_t|, \qquad (21)$$

where
$$V_s = \sum_{k<n} \mathrm{sgn}\big(E_R[X_{t_{k+1}}|\mathcal{F}_{t_k}] - X_{t_k}\big) 1_{(t_k, t_{k+1}]}(s), \quad s \geq 0.$$

Since ρ is bounded and $V \in \mathcal{E}_1$, the right-hand side of (21) is bounded by a constant. Hence, the stopped process X^t is a quasi-martingale under R. By Theorem 26.20 it is then an R-semimartingale, and since $P \sim R$, Corollary 26.11 shows that X^t is even a P-semimartingale. Since t is arbitrary, it follows that X itself is a P-semimartingale. □

Exercises

1. Construct the quadratic variation $[M]$ of a local L^2-martingale M directly as in Theorem 17.5, and prove a corresponding version of the integration-by-parts formula. Use $[M]$ to define the L^2-integral of Theorem 26.2.

2. Show that the approximation in Proposition 17.17 remains valid for general semimartingales.

3. Consider a local martingale M starting at 0 and an optional time τ. Use Theorem 26.12 to give conditions for the validity of the relations $EM_\tau = 0$ and $EM_\tau^2 = [M]_\tau$.

4. Give an example of a sequence of L^2-bounded martingales M_n such that $M_n^* \xrightarrow{P} 0$ and yet $\langle M_n \rangle_\infty \xrightarrow{P} \infty$. (*Hint:* Consider compensated Poisson processes with large jumps.)

5. Give an example of a sequence of martingales M_n such that $[M_n]_\infty \xrightarrow{P} 0$ and yet $M_n^* \xrightarrow{P} \infty$. (*Hint:* See the preceding problem.)

6. Show that $\langle M_n \rangle_\infty \xrightarrow{P} 0$ implies $[M_n]_\infty \xrightarrow{P} 0$.

7. Give an example of a martingale M of bounded variation and a bounded, progressive process V such that $V^2 \cdot \langle M \rangle = 0$ and yet $V \cdot M \neq 0$. Conclude that the L^2-integral in Theorem 26.2 has no continuous extension to progressive integrands.

8. Show that any general martingale inequality involving the processes M, $[M]$, and $\langle M \rangle$ remains valid in discrete time. (*Hint:* Embed M and the associated discrete filtration into a martingale and filtration on $\mathbb{R}_+$.)

9. Show that the a.s. convergence in Theorem 4.23 remains valid in L^p. (*Hint:* Use Theorem 26.12 to reduce to the case when $p < 1$. Then truncate.)

10. Let $\mathcal{G}$ be an extension of the filtration $\mathcal{F}$. Show that any $\mathcal{F}$-adapted $\mathcal{G}$-semimartingale is also an $\mathcal{F}$-semimartingale. Also show by an example that the converse implication fails in general. (*Hint:* Use Theorem 26.21.)

11. Show that if X is a Lévy process in $\mathbb{R}$, then $[X]$ is a subordinator. Express the characteristics of $[X]$ in terms of those for X.

12. For any Lévy process X, show that if X is p-stable, then $[X]$ is strictly $p/2$-stable. Also prove the converse, in the case when X has positive or symmetric jumps. (*Hint:* Use Proposition 15.9.)

13. Extend Theorem 26.17 to the case when $[M]_\infty \leq a$ or $\langle M \rangle \leq a$ a.s. for some $a \geq 1$. (*Hint:* Apply the original result to a suitably scaled process.)

14. For any Lévy process X with Lévy measure ν, show that $X \in \mathcal{M}^2_{loc}$ iff $X \in \mathcal{M}^2$, and also iff $\int x^2 \nu(dx) < \infty$, in which case $\langle X \rangle_t = tEX_1^2$. (*Hint:* Use Corollary 25.25.)

15. Show that if M is a purely discontinuous local martingale with positive jumps, then $M - M_0$ is a.s. determined by $[M]$. (*Hint:* For any such processes M and N with $[M] = [N]$, apply Theorem 26.14 to $M - N$.)

16. Show that a semimartingale X is quasi-leftcontinuous or has accessible jumps iff $[X]$ has the same property. (*Hint:* Use Theorem 26.6 (iv).)

17. Show that a semimartingale X with $|\Delta X| \leq c < \infty$ a.s. is a special semimartingale with canonical decomposition $M + A$ satisfying $|\Delta A| \leq c$ a.s. In particular, X is a continuous semimartingale iff it has a decomposition $M + A$, where M and A are continuous. (*Hint:* Use Lemma 26.5, and note that $|\Delta A| \leq c$ a.s. implies $|\Delta \hat{A}| \leq c$ a.s.)

18. Show that a semimartingale X is quasi-leftcontinuous or has accessible jumps iff it has a decomposition $M + A$, where M and A have the same property. Also show that, for special semimartingales, we may choose $M + A$ to be the canonical decomposition of X. (*Hint:* Use Proposition 25.17 and Corollary 26.16, and refer to the preceding exercise.)

19. Show that a semimartingale X is predictable iff it is a special semimartingale with canonical decomposition $M + A$ such that M is continuous. (*Hint:* Use Proposition 25.16.)

Chapter 27

Large Deviations

Legendre–Fenchel transform; Cramér's and Schilder's theorems; large-deviation principle and rate function; functional form of the LDP; continuous mapping and extension; perturbation of dynamical systems; empirical processes and entropy; Strassen's law of the iterated logarithm

In its simplest setting, large deviation theory provides the exact rate of convergence in the weak law of large numbers. To be precise, consider any i.i.d. random variables $\xi_1, \xi_2, \ldots$ with mean m and cumulant-generating function $\Lambda(u) = \log E e^{u\xi_i} < \infty$, and write $\bar{\xi}_n = n^{-1} \sum_{k \leq n} \xi_k$. Then for any $x > m$, the tail probabilities $P\{\bar{\xi}_n > x\}$ tend to 0 at an exponential rate $I(x)$, given by the Legendre–Fenchel transform Λ^* of Λ. In higher dimensions, it is often convenient to state the result more generally in the form $n^{-1} \log P\{\bar{\xi}_n \in B\} \to -I(B)$, where $I(B) = \inf_{x \in B} I(x)$ and B is restricted to a suitable class of continuity sets. In this standard format of a *large-deviation principle with rate function I*, the result extends to an amazing variety of contexts throughout probability theory.

A striking example, of fundamental importance in statistical mechanics, is Sanov's theorem, which provides a similar large deviation result for the empirical distributions of a sequence of i.i.d. random variables with a common distribution μ. Here the rate function I is defined on the space of probability measures ν on $\mathbb{R}$ and agrees with the relative entropy function $H(\nu|\mu)$. Another important example is Schilder's theorem for the family of rescaled Brownian motions in $\mathbb{R}^d$, where the rate function becomes $I(x) = \frac{1}{2} \|\dot{x}\|_2^2$, the squared norm in the Cameron–Martin space considered in Chapter 18. The latter result can be used to derive the Fredlin–Wentzell estimates for randomly perturbed dynamical systems. It also provides a short proof of Strassen's law of the iterated logarithm, a stunning extension of the classical Khinchin law from Chapter 13.

Modern proofs of those and other large deviation results rely on some general extension principles, which also serve to explain the wide applicability of the present ideas. In addition to some rather straightforward and elementary techniques of continuity and approximation, we consider the more sophisticated and extremely powerful methods of inverse continuous mapping and projective limits, both of which play a crucial role in subsequent applications. We may also call attention to the significance of exponential tightness, and to the essential equivalence between the setwise and functional formulations of the large-deviation principle.

Large deviation theory is arguably one of the most technical branches of modern probability theory. For the nonexpert it then seems essential to avoid getting distracted by topological subtleties or elaborate computations. Many results are therefore stated here under simplifying assumptions. Likewise, we postpone our discussion of general principles until the reader has become aquainted with the basic ideas in a concrete setting. For this reason, important applications appear both at the beginning and at the end of the chapter, separated by a more abstract discussion of some general notions and principles.

Let us now return to the elementary context of i.i.d. random variables $\xi, \xi_1, \xi_2, \ldots$ and write $S_n = \sum_{k \leq n} \xi_k$ and $\bar{\xi}_n = S_n/n$. If $m = E\xi$ exists and is finite, then $P\{\bar{\xi}_n > x\} \to 0$ for all $x > m$ by the weak law of large numbers. Under stronger moment conditions, the rate of convergence turns out to be exponential and can be estimated with great accuracy. This rather elementary but quite technical result lies, along with its multidimensional counterpart, at the core of large-deviation theory and provides both a pattern and a point of departure for more advanced developments. For motivation, we begin with some simple observations.

Lemma 27.1 *(convergence) Let $\xi, \xi_1, \xi_2, \ldots$ be i.i.d. random variables. Then*

(i) $n^{-1} \log P\{\bar{\xi}_n \geq x\} \to \sup_n n^{-1} \log P\{\bar{\xi}_n \geq x\} \equiv -h(x)$ *for all x;*

(ii) h *is $[0, \infty]$-valued, nondecreasing, and convex;*

(iii) $h(x) < \infty$ *iff $P\{\xi \geq x\} > 0$.*

Proof: (i) Writing $p_n = P\{\bar{\xi}_n \geq x\}$, we get for any $m, n \in \mathbb{N}$

$$p_{m+n} = P\{S_{m+n} \geq (m+n)x\}$$
$$\geq P\{S_m \geq mx, S_{m+n} - S_m \geq nx\} = p_m p_n.$$

Taking logarithms, we conclude that the sequence $-\log p_n$ is subadditive, and the assertion follows by Lemma 10.21.

(ii) The first two assertions are obvious. To prove the convexity, let $x, y \in \mathbb{R}$ be arbitrary, and proceed as before to get

$$P\{S_{2n} \geq n(x+y)\} \geq P\{S_n \geq nx\} P\{S_n \geq ny\}.$$

Taking logarithms, dividing by $2n$, and letting $n \to \infty$, we obtain

$$h(\tfrac{1}{2}(x+y)) \leq \tfrac{1}{2}(h(x) + h(y)), \quad x, y > 0.$$

(iii) If $P\{\xi \geq x\} = 0$, then $P\{\bar{\xi}_n \geq x\} = 0$ for all n, and so $h(x) = \infty$. Conversely, (i) yields $\log P\{\xi \geq x\} \leq -h(x)$, and so $h(x) = \infty$ implies $P\{\xi \geq x\} = 0$. □

To determine the limit in Lemma 27.1 we need some further notation, which is given here for convenience directly in d dimensions. For any random vector ξ in $\mathbb{R}^d$, we introduce the function

$$\Lambda(u) = \Lambda_\xi(u) = \log E e^{u\xi}, \quad u \in \mathbb{R}^d, \tag{1}$$

known in statistics as the *cumulant-generating function* of ξ. Note that Λ is convex, since by Hölder's inequality we have for any $u, v \in \mathbb{R}^d$ and $p, q > 0$ with $p + q = 1$

$$\begin{aligned}\Lambda(pu + qv) &= \log E \exp((pu + qv)\xi) \\ &\leq \log\left((Ee^{u\xi})^p (Ee^{v\xi})^q\right) = p\Lambda(u) + q\Lambda(v).\end{aligned}$$

The surface $z = \Lambda(u)$ in $\mathbb{R}^{d+1}$ is determined by the family of supporting hyperplanes (d-dimensional affine subspaces) with different slopes, and we note that the plane with slope $x \in \mathbb{R}^d$ (or normal vector $(1, -x)$) has equation

$$z + \Lambda^*(x) = xu, \quad u \in \mathbb{R}^d,$$

where Λ^* denotes the *Legendre-Fenchel transform* of Λ, given by

$$\Lambda^*(x) = \sup_{u \in \mathbb{R}^d} (ux - \Lambda(u)), \quad x \in \mathbb{R}^d. \tag{2}$$

We can often compute Λ^* explicitly. Here we list two simple cases that will be needed below. The results are proved by elementary calculus.

Lemma 27.2 *(Gaussian and Bernoulli distributions)*
(i) *If $\xi = (\xi_1, \ldots, \xi_d)$ is standard Gaussian in $\mathbb{R}^d$, then $\Lambda_\xi^*(x) \equiv |x|^2/2$.*
(ii) *If $\xi \in \{0, 1\}$ with $P\{\xi = 1\} = p \in (0, 1)$, then $\Lambda_\xi^*(x) = \infty$ for $x \notin [0, 1]$ and*

$$\Lambda_\xi^*(x) = x \log \frac{x}{p} + (1 - x) \log \frac{1-x}{1-p}, \quad x \in [0, 1].$$

The function Λ^* is again convex, since for any $x, y \in \mathbb{R}^d$ and for p and q as before

$$\begin{aligned}\Lambda^*(px + qy) &= \sup_u [p(ux - \Lambda(u)) + q(uy - \Lambda(u))] \\ &\leq p \sup_u (ux - \Lambda(u)) + q \sup_u (uy - \Lambda(u)) \\ &= p\Lambda^*(x) + q\Lambda^*(y).\end{aligned}$$

If $\Lambda < \infty$ near the origin, then $m = E\xi$ exists and agrees with the gradient $\nabla \Lambda(0)$. Thus, the surface $z = \Lambda(u)$ has tangent hyperplane $z = mu$ at 0, and we conclude that $\Lambda^*(m) = 0$ and $\Lambda^*(x) > 0$ for $x \neq m$. If ξ is also truly d-dimensional, then Λ is strictly convex at 0, and Λ^* is finite and continuous near m. For $d = 1$, we sometimes need the corresponding one-sided statements, which are easily derived by dominated convergence.

The following key result identifies the function h in Lemma 27.1. For simplicity, we assume that $m = E\xi$ exists in $[-\infty, \infty)$.

Theorem 27.3 *(rate function, Cramér, Chernoff)* Let $\xi, \xi_1, \xi_2, \ldots$ be i.i.d. random variables with $m = E\xi < \infty$. Then for any $x \geq m$, we have

$$n^{-1}\log P\{\bar{\xi}_n \geq x\} \to -\Lambda^*(x). \tag{3}$$

Proof: Using Chebyshev's inequality and (1), we get for any $u > 0$

$$P\{\bar{\xi}_n \geq x\} = P\{e^{uS_n} \geq e^{nux}\} \leq e^{-nux}Ee^{uS_n} = e^{n\Lambda(u)-nux},$$

and so

$$n^{-1}\log P\{\bar{\xi}_n \geq x\} \leq \Lambda(u) - ux.$$

This remains true for $u \leq 0$, since in that case $\Lambda(u) - ux \geq 0$ for $x \geq m$. Hence, by (2) we have the upper bound

$$n^{-1}\log P\{\bar{\xi}_n \geq x\} \leq -\Lambda^*(x), \quad x \geq m,\ n \in \mathbb{N}. \tag{4}$$

To derive a matching lower bound, we first assume that $\Lambda < \infty$ on $\mathbb{R}_+$. Then Λ is smooth on $(0, \infty)$ with $\Lambda'(0+) = m$ and $\Lambda'(\infty) = \operatorname{ess\,sup}\xi \equiv b$, and so for any $a \in (m, b)$ we can choose a $u > 0$ such that $\Lambda'(u) = a$. Let $\eta, \eta_1, \eta_2, \ldots$ be i.i.d. with distribution

$$P\{\eta \in B\} = e^{-\Lambda(u)}E[e^{u\xi};\ \xi \in B], \quad B \in \mathcal{B}. \tag{5}$$

Then $\Lambda_\eta(r) = \Lambda_\xi(r+u) - \Lambda_\xi(u)$, and therefore $E\eta = \Lambda'_\eta(0) = \Lambda'_\xi(u) = a$. For any $\varepsilon > 0$, we get by (5)

$$\begin{aligned}P\{|\bar{\xi}_n - a| < \varepsilon\} &= e^{n\Lambda(u)}E[e^{-nu\bar{\eta}_n};\ |\bar{\eta}_n - a| < \varepsilon\}\\ &\geq e^{n\Lambda(u)-nu(a+\varepsilon)}P\{|\bar{\eta}_n - a| < \varepsilon\}.\end{aligned} \tag{6}$$

Here the last probability tends to 1 by the law of large numbers, and so by (2)

$$\liminf_{n\to\infty} n^{-1}\log P\{|\bar{\xi}_n - a| < \varepsilon\} \geq \Lambda(u) - u(a+\varepsilon) \geq -\Lambda^*(a+\varepsilon).$$

Fixing any $x \in (m, b)$ and putting $a = x + \varepsilon$, we get for small enough $\varepsilon > 0$

$$\liminf_{n\to\infty} n^{-1}\log P\{\bar{\xi}_n \geq x\} \geq -\Lambda^*(x+2\varepsilon).$$

Since Λ^* is continuous on (m, b) by convexity, we may let $\varepsilon \to 0$ and combine with (4) to obtain (3).

The result for $x > b$ is trivial, since in that case both sides of (3) equal $-\infty$. If instead $x = b < \infty$, then both sides equal $\log P\{\xi = b\}$, the left side by a simple computation and the right side by an elementary estimate. Finally, assume that $x = m > -\infty$. Since the statement is trivial when $\xi = m$ a.s., we may assume that $b > m$. For any $y \in (m, b)$, we have

$$0 \geq n^{-1}\log P\{\bar{\xi}_n \geq m\} \geq n^{-1}P\{\bar{\xi}_n \geq y\} \to -\Lambda^*(y) > -\infty.$$

Here $\Lambda^*(y) \to \Lambda^*(m) = 0$ by continuity, and (3) follows for $x = m$. This completes the proof when $\Lambda < \infty$ on $\mathbb{R}_+$.

The case when $\Lambda(u) = \infty$ for some $u > 0$ may be handled by truncation. Thus, for any $r > m$ we consider the random variables $\xi_k^r = \xi_k \wedge r$. Writing Λ_r and Λ_r^* for the associated functions Λ and Λ^*, we get for $x \geq m \geq E\xi^r$

$$n^{-1} \log P\{\bar{\xi}_n \geq x\} \geq n^{-1} \log P\{\bar{\xi}_n^r \geq x\} \to -\Lambda_r^*(x). \qquad (7)$$

Now $\Lambda_r(u) \uparrow \Lambda(u)$ by monotone convergence as $r \to \infty$, and by Dini's theorem the convergence is uniform on every compact interval where $\Lambda < \infty$. Since also Λ' is unbounded on the set where $\Lambda < \infty$, it follows easily that $\Lambda_r^*(x) \to \Lambda^*(x)$ for all $x \geq m$. The required lower bound is now immediate from (7). □

We may now supplement Lemma 27.1 with a criterion for exponential decline of the tail probabilities $P\{\bar{\xi}_n \geq x\}$.

Corollary 27.4 (*exponential rate*) *Let $\xi, \xi_1, \xi_2, \ldots$ be i.i.d. with $m = E\xi < \infty$ and $b = \operatorname{ess\,sup} \xi$. Then for any $x \in (m, b)$, the probabilities $P\{\bar{\xi}_n \geq x\}$ decrease exponentially iff $\Lambda(\varepsilon) < \infty$ for some $\varepsilon > 0$. The exponential decline extends to $x = b$ iff $0 < P\{\xi = b\} < 1$.*

Proof: If $\Lambda(\varepsilon) < \infty$ for some $\varepsilon > 0$, then $\Lambda'(0+) = m$ by dominated convergence, and so $\Lambda^*(x) > 0$ for all $x > m$. If instead $\Lambda = \infty$ on $(0, \infty)$, then $\Lambda^*(x) = 0$ for all $x \geq m$. The statement for $x = b$ is trivial. □

The large deviation estimates in Theorem 27.3 are easily extended from intervals $[x, \infty)$ to arbitrary open or closed sets, which leads to the *large-deviation principle* for i.i.d. sequences in $\mathbb{R}$. To fulfill the needs of subsequent applications and extensions, we shall derive a version of the same result in $\mathbb{R}^d$. Motivated by the last result, and also to avoid some technical complications, we assume that $\Lambda(u) < \infty$ for all u. Write B° and B^- for the interior and closure of a set B.

Theorem 27.5 (*large deviations in $\mathbb{R}^d$, Varadhan*) *Let $\xi, \xi_1, \xi_2, \ldots$ be i.i.d. random vectors in $\mathbb{R}^d$ with $\Lambda = \Lambda_\xi < \infty$. Then for any $B \in \mathcal{B}^d$, we have*

$$- \inf_{x \in B^\circ} \Lambda^*(x) \leq \liminf_{n \to \infty} n^{-1} \log P\{\bar{\xi}_n \in B\}$$
$$\leq \limsup_{n \to \infty} n^{-1} \log P\{\bar{\xi}_n \in B\} \leq - \inf_{x \in B^-} \Lambda^*(x).$$

Proof: To derive the upper bound, we fix any $\varepsilon > 0$. By (2) there exists for every $x \in \mathbb{R}^d$ some $u_x \in \mathbb{R}^d$ such that

$$u_x x - \Lambda(u_x) > (\Lambda^*(x) - \varepsilon) \wedge \varepsilon^{-1},$$

and by continuity we may choose an open ball B_x around x such that

$$u_x y > \Lambda(u_x) + (\Lambda^*(x) - \varepsilon) \wedge \varepsilon^{-1}, \quad y \in B_x.$$

By Chebyshev's inequality and (1) we get for any $n \in \mathbb{N}$

$$P\{\bar{\xi}_n \in B_x\} \leq E \exp(u_x S_n - n \inf\{u_x y; y \in B_x\})$$
$$\leq \exp\bigl(-n((\Lambda^*(x) - \varepsilon) \wedge \varepsilon^{-1})\bigr). \qquad (8)$$

Also note that $\Lambda < \infty$ implies $\Lambda^*(x) \to \infty$ as $|x| \to \infty$, at least when $d = 1$. By Lemma 27.1 and Theorem 27.3 we may then choose $r > 0$ so large that

$$n^{-1} \log P\{|\bar{\xi}_n| > r\} \leq -1/\varepsilon, \quad n \in \mathbb{N}. \tag{9}$$

Now let $B \subset \mathbb{R}^d$ be closed. Then the set $\{x \in B; |x| \leq r\}$ is compact and may be covered by finitely many balls $B_{x_1}, \ldots, B_{x_m}$ with centers $x_i \in B$. By (8) and (9) we get for any $n \in \mathbb{N}$

$$\begin{aligned}
P\{\bar{\xi}_n \in B\} &\leq \sum_{i \leq m} P\{\bar{\xi}_n \in B_{x_i}\} + P\{|\bar{\xi}_n| > r\} \\
&\leq \sum_{i \leq m} \exp(-n((\Lambda^*(x_i) - \varepsilon) \wedge \varepsilon^{-1})) + e^{-n/\varepsilon} \\
&\leq (m+1) \exp(-n((\Lambda^*(B) - \varepsilon) \wedge \varepsilon^{-1})),
\end{aligned}$$

where $\Lambda^*(B) = \inf_{x \in B} \Lambda^*(x)$. Hence,

$$\limsup_{n \to \infty} n^{-1} \log P\{\bar{\xi}_n \in B\} \leq -(\Lambda^*(B) - \varepsilon) \wedge \varepsilon^{-1},$$

and the upper bound follows since ε was arbitrary.

Turning to the lower bound, we first assume that $\Lambda(u)/|u| \to \infty$ as $|u| \to \infty$. Fix any open set $B \subset \mathbb{R}^d$ and a point $x \in B$. By compactness and the smoothness of Λ, there exists a $u \in \mathbb{R}^d$ such that $\nabla \Lambda(u) = x$. Let $\eta, \eta_1, \eta_2, \ldots$ be i.i.d. random vectors with distribution (5), and note as before that $E\eta = x$. For $\varepsilon > 0$ small enough, we get as in (6)

$$\begin{aligned}
P\{\bar{\xi}_n \in B\} &\geq P\{|\bar{\xi}_n - x| < \varepsilon\} \\
&\geq \exp(n\Lambda(u) - nux - n\varepsilon|u|) P\{|\bar{\eta}_n - x| < \varepsilon\}.
\end{aligned}$$

Hence, by the law of large numbers and (2),

$$\liminf_{n \to \infty} n^{-1} \log P\{\bar{\xi}_n \in B\} \geq \Lambda(u) - ux - \varepsilon|u| \geq -\Lambda^*(x) - \varepsilon|u|.$$

It remains to let $\varepsilon \to 0$ and take the supremum over $x \in B$.

To eliminate the growth condition on Λ, let $\zeta, \zeta_1, \zeta_2, \ldots$ be i.i.d. standard Gaussian random vectors independent of ξ and the ξ_n. Then for any $\sigma > 0$ and $u \in \mathbb{R}^d$, we have by Lemma 27.2 (i)

$$\Lambda_{\xi + \sigma \zeta}(u) = \Lambda_\xi(u) + \Lambda_\zeta(\sigma u) = \Lambda_\xi(u) + \tfrac{1}{2}\sigma^2 |u|^2 \geq \Lambda_\xi(u),$$

and in particular $\Lambda^*_{\xi + \sigma \zeta} \leq \Lambda^*_\xi$. Since also $\Lambda_{\xi + \sigma \zeta}(u)/|u| \geq \sigma^2 |u|/2 \to \infty$, we note that the previous bound applies to $\bar{\xi}_n + \sigma \bar{\zeta}_n$.

Now fix any $x \in B$ as before, and choose $\varepsilon > 0$ small enough that B contains a 2ε-ball around x. Then

$$\begin{aligned}
P\{|\bar{\xi}_n + \sigma \bar{\zeta}_n - x| < \varepsilon\} &\leq P\{\bar{\xi}_n \in B\} + P\{\sigma |\bar{\zeta}_n| \geq \varepsilon\} \\
&\leq 2\left(P\{\bar{\xi}_n \in B\} \vee P\{\sigma |\bar{\zeta}_n| \geq \varepsilon\}\right).
\end{aligned}$$

Applying the lower bound to the variables $\bar\xi_n + \sigma\bar\zeta_n$ and the upper bound to $\bar\zeta_n$, we get by Lemma 27.2 (i)

$$\begin{aligned}
-\Lambda_\xi^*(x) &\leq -\Lambda_{\xi+\sigma\zeta}^*(x) \leq \liminf_{n\to\infty} n^{-1}\log P\{|\bar\xi_n + \sigma\bar\zeta_n - x| < \varepsilon\} \\
&\leq \liminf_{n\to\infty} n^{-1}\log\left(P\{\bar\xi_n \in B\} \vee P\{\sigma|\bar\zeta_n| \geq \varepsilon\}\right) \\
&\leq \liminf_{n\to\infty} n^{-1}\log P\{\bar\xi_n \in B\} \vee (-\varepsilon^2/2\sigma^2).
\end{aligned}$$

The desired lower bound now follows, as we let $\sigma \to 0$ and then take the supremum over all $x \in B$. □

We can also derive large-deviation results in function spaces. Here the following theorem is basic and sets the pattern for more complex results. For convenience, we may write $C = C([0,1],\mathbb{R}^d)$ and $C_0^k = \{x \in C^k; x_0 = 0\}$. We also introduce the *Cameron–Martin space* H_1, consisting of all absolutely continuous functions $x \in C_0$ admitting a Radon–Nikodým derivative $\dot x \in L^2$, so that $\|\dot x\|_2^2 = \int_0^1 |\dot x_t|^2 dt < \infty$.

Theorem 27.6 *(large deviations of Brownian motion, Schilder)* Let X be a d-dimensional Brownian motion on $[0,1]$. Then for any Borel set $B \subset C([0,1],\mathbb{R}^d)$, we have

$$\begin{aligned}
-\inf_{x \in B^\circ} I(x) &\leq \liminf_{\varepsilon\to 0} \varepsilon^2 \log P\{\varepsilon X \in B\} \\
&\leq \limsup_{\varepsilon\to 0} \varepsilon^2 \log P\{\varepsilon X \in B\} \leq -\inf_{x \in B^-} I(x),
\end{aligned}$$

where $I(x) = \frac{1}{2}\|\dot x\|_2^2$ for $x \in H_1$ and $I(x) = \infty$ otherwise.

The proof requires a simple topological fact.

Lemma 27.7 *(level sets)* For any $r \geq 0$, the level set $L_r = I^{-1}[0,r] = \{x \in H_1; \|\dot x\|_2^2 \leq 2r\}$ is compact in $C([0,1],\mathbb{R}^d)$.

Proof: The Cauchy–Buniakovsky inequality yields

$$|x_t - x_s| \leq \int_s^t |\dot x_u|\, du \leq (t-s)^{1/2}\|\dot x\|_2, \quad 0 \leq s < t \leq 1,\ x \in H_1.$$

By the Arzelà–Ascoli Theorem A2.1 it follows that L_r is relatively compact in C. It is also weakly compact in the Hilbert space H_1 with norm $\|x\| = \|\dot x\|_2$. Thus, every sequence $x_1, x_2, \cdots \in L_r$ has a subsequence that converges in both C and H_1, say with limits $x \in C$ and $y \in L_r$, respectively. For every $t \in [0,1]$, the sequence $x_n(t)$ then converges in $\mathbb{R}^d$ to both $x(t)$ and $y(t)$, and we get $x = y \in L_r$. □

Proof of Theorem 27.6: To establish the lower bound, we fix any open set $B \subset C$. Since $I = \infty$ outside H_1, it suffices to prove that

$$-I(x) \leq \liminf_{\varepsilon\to 0} \varepsilon^2 \log P\{\varepsilon X \in B\}, \quad x \in B \cap H_1. \tag{10}$$

Now we note as in Lemma 1.35 that C_0^2 is dense in H_1, and also that $\|x\|_\infty \leq \|\dot x\|_1 \leq \|\dot x\|_2$ for any $x \in H_1$. Hence, for every $x \in B \cap H_1$ there exist some functions $x_n \in B \cap C_0^2$ with $I(x_n) \to I(x)$, and it suffices to prove (10) for $x \in B \cap C_0^2$.

Now for small enough $h > 0$, Theorem 18.22 yields

$$P\{\varepsilon X \in B\} \geq P\{\|\varepsilon X - x\|_\infty < h\}$$
$$= E[\mathcal{E}(-(\dot{x}/\varepsilon) \cdot X)_1; \|\varepsilon X\|_\infty < h]. \quad (11)$$

Integrating by parts gives

$$\log \mathcal{E}(-(\dot{x}/\varepsilon) \cdot X)_1 = -\varepsilon^{-1} \int_0^1 \dot{x}_t dX_t - \varepsilon^{-2} I(x)$$
$$= -\varepsilon^{-1} \dot{x}_1 X_1 + \varepsilon^{-1} \int_0^1 \ddot{x}_t X_t dt - \varepsilon^{-2} I(x),$$

and so by (11)

$$\varepsilon^2 \log P\{\varepsilon X \in B\} \geq -I(x) - h|\dot{x}_1| - h\|\ddot{x}\|_1 + \varepsilon^2 \log P\{\|\varepsilon X\|_\infty < h\}.$$

Relation (10) now follows as we let $\varepsilon \to 0$ and then $h \to 0$.

Turning to the upper bound, we fix any closed set $B \subset C$ and let B_h denote the closed h-neighborhood of B. Letting X_n be the n-segment, polygonal approximation of X with $X_n(k/n) = X(k/n)$ for $k \leq n$, we note that

$$P\{\varepsilon X \in B\} \leq P\{\varepsilon X_n \in B_h\} + P\{\varepsilon \|X - X_n\| > h\}. \quad (12)$$

Writing $I(B_h) = \inf\{I(x); x \in B_h\}$, we obtain

$$P\{\varepsilon X_n \in B_h\} \leq P\{I(\varepsilon X_n) \geq I(B_h)\}.$$

Here $2I(X_n)$ is a sum of nd variables ξ_{ik}^2, where the ξ_{ik} are i.i.d. $N(0,1)$, and so by Lemma 27.2 (i) and an interpolated version of Theorem 27.5,

$$\limsup_{\varepsilon \to 0} \varepsilon^2 \log P\{\varepsilon X_n \in B_h\} \leq -I(B_h). \quad (13)$$

Next we get by Proposition 13.13 and some elementary estimates

$$P\{\varepsilon \|X - X_n\| > h\} \leq n P\{\varepsilon \|X\| > h\sqrt{n}/2\}$$
$$\leq 2nd P\{\varepsilon^2 \xi^2 > h^2 n/4d\},$$

where ξ is $N(0,1)$. Applying Theorem 27.5 and Lemma 27.2 (i) again, we obtain

$$\limsup_{\varepsilon \to 0} \varepsilon^2 \log P\{\varepsilon \|X - X_n\| > h\} \leq -h^2 n/8d. \quad (14)$$

Combining (12), (13), and (14) gives

$$\limsup_{\varepsilon \to 0} \varepsilon^2 \log P\{\varepsilon X \in B\} \leq -I(B_h) \wedge (h^2 n/8d),$$

and as $n \to \infty$ we obtain the upper bound $-I(B_h)$.

It remains to show that $I(B_h) \uparrow I(B)$ as $h \to 0$. Then fix any $r > \sup_h I(B_h)$. For every $h > 0$ we may choose some $x_h \in B_h$ such that $I(x_h) \leq r$, and by Lemma 27.7 we may extract a convergent sequence

$x_{h_n} \to x$ with $h_n \to 0$ such that even $I(x) \leq r$. Since also $x \in \bigcap_h B_h = B$, we obtain $I(B) \leq r$, as required. □

The last two theorems suggest the following abstraction. Letting ξ_ε, $\varepsilon > 0$, be random elements in some metric space S with Borel σ-field $\mathcal{S}$, we say that the family (ξ_ε) satisfies the *large-deviation principle (LDP) with rate function* $I: S \to [0, \infty]$, if for any $B \in \mathcal{S}$ we have

$$- \inf_{x \in B^\circ} I(x) \leq \liminf_{\varepsilon \to 0} \varepsilon \log P\{\xi_\varepsilon \in B\}$$
$$\leq \limsup_{\varepsilon \to 0} \varepsilon \log P\{\xi_\varepsilon \in B\} \leq - \inf_{x \in B^-} I(x). \qquad (15)$$

For sequences $\xi_1, \xi_2, \ldots$ we require the same condition with the normalizing factor ε replaced by n^{-1}. It is often convenient to write $I(B) = \inf_{x \in B} I(x)$. Letting $\mathcal{S}_I$ denote the class $\{B \in \mathcal{S}; I(B^\circ) = I(B^-)\}$ of all *I-continuity sets*, we note that (15) implies the convergence

$$\lim_{\varepsilon \to 0} \varepsilon \log P\{\xi_\varepsilon \in B\} = -I(B), \quad B \in \mathcal{S}_I. \qquad (16)$$

If $\xi, \xi_1, \xi_2, \ldots$ are i.i.d. random vectors in $\mathbb{R}^d$ with $\Lambda(u) = Ee^{u\xi} < \infty$ for all u, then by Theorem 27.5 the averages $\bar{\xi}_n$ satisfy the LDP in $\mathbb{R}^d$ with rate function Λ^*. If instead X is a d-dimensional Brownian motion on $[0, 1]$, then Theorem 27.6 shows that the processes $\varepsilon^{1/2} X$ satisfy the LDP in $C([0, 1], \mathbb{R}^d)$ with rate function $I(x) = \frac{1}{2}\|\dot{x}\|_2^2$ for $x \in H_1$ and $I(x) = \infty$ otherwise.

We show that the rate function I is essentially unique.

Lemma 27.8 *(regularization and uniqueness) If (ξ_ε) satisfies the LDP in a metric space S, then the associated rate function I can be chosen to be lower semicontinuous, in which case it is unique.*

Proof: Assume that (15) holds for some I. Then the function

$$J(x) = \liminf_{y \to x} I(y), \quad x \in S,$$

is clearly lower semicontinuous with $J \leq I$. It is also easy to verify that $J(G) = I(G)$ for all open sets $G \subset S$. Thus, (15) remains true with I replaced by J.

To prove the uniqueness, assume that (15) holds for two lower semicontinuous functions I and J, and let $I(x) < J(x)$ for some $x \in S$. By the semicontinuity of J, we may choose a neighborhood G of x such that $J(G^-) > I(x)$. Applying (15) to both I and J yields the contradiction

$$-I(x) \leq -I(G) \leq \liminf_{\varepsilon \to 0} \varepsilon \log P\{\xi_\varepsilon \in G\} \leq -J(G^-) < -I(x). \qquad □$$

Justified by the last result, we may henceforth take the lower semicontinuity to be part of our definition of a rate function. (An arbitrary function I satisfying (15) will then be called a *raw rate function*.) No regularization is needed in Theorems 27.5 and 27.6, since the associated rate functions Λ^*

and I are already lower semicontinuous, the former as the supremum of a family of continuous functions and the latter by Lemma 27.7.

It is sometimes useful to impose a slightly stronger regularity condition on the function I. Thus, we say that I is *good* if the level sets $I^{-1}[0,r] = \{x \in S;\ I(x) \leq r\}$ are compact (rather than just closed). Note that the infimum $I(B) = \inf_{x \in B} I(x)$ is then attained for every closed set $B \neq \emptyset$. The rate functions in Theorems 27.5 and 27.6 are clearly both good.

A related condition on the family (ξ_ε) is the *exponential tightness*

$$\inf_K \limsup_{\varepsilon \to 0} \varepsilon \log P\{\xi_\varepsilon \notin K\} = -\infty, \tag{17}$$

where the infimum extends over all compact sets $K \subset S$. We actually need only the slightly weaker condition of *sequential exponential tightness*, where (17) is only required along sequences $\varepsilon_n \to 0$. To simplify our exposition, we often omit the sequential qualification from our statements and carry out the proofs under the stronger nonsequential hypothesis.

We finally say that (ξ_ε) satisfies the *weak LDP* with rate function I if the lower bound in (15) holds as stated while the upper bound is only required for compact sets B. We list some relations between the mentioned properties.

Lemma 27.9 *(goodness, exponential tightness, and the weak LDP)* Let ξ_ε, $\varepsilon > 0$, be random elements in a metric space S.

  (i) *The LDP for (ξ_ε) with rate function I implies (16), and the two conditions are equivalent when I is good.*

 (ii) *If the ξ_ε are exponentially tight and satisfy the weak LDP with rate function I, then I is good and (ξ_ε) satisfies the full LDP.*

(iii) *(Pukhalsky) If S is Polish and (ξ_ε) satisfies the LDP with rate function I, then I is good iff (ξ_ε) is sequentially exponentially tight.*

Proof: (i) Let I be good and satisfy (16). Write B^h for the closed h-neighborhood of $B \in S$. Since I is nonincreasing on S, we have $B^h \notin S_I$ for at most countably many $h > 0$. Hence, (16) yields for almost every $h > 0$

$$\limsup_{\varepsilon \to 0} \varepsilon \log P\{\xi_\varepsilon \in B\} \leq \lim_{\varepsilon \to 0} \varepsilon \log P\{\xi_\varepsilon \in B^h\} = -I(B^h).$$

To see that $I(B^h) \uparrow I(B^-)$ as $h \to 0$, assume instead that $\sup_h I(B^h) < I(B^-)$. Since I is good, we may choose for every $h > 0$ some $x_h \in B^h$ with $I(x_h) = I(B^h)$, and then extract a convergent sequence $x_{h_n} \to x \in B^-$ with $h_n \to 0$. By the lower semicontinuity of I we get the contradiction

$$I(B^-) \leq I(x) \leq \liminf_{n \to \infty} I(x_{h_n}) \leq \sup_{h > 0} I(B^h) < I(B^-),$$

which proves the upper bound. Next let $x \in B^\circ$ be arbitrary, and conclude from (16) that, for almost all sufficiently small $h > 0$,

$$-I(x) \leq -I(\{x\}^h) = \lim_{\varepsilon \to 0} \varepsilon \log P\{\xi_\varepsilon \in \{x\}^h\} \leq \liminf_{\varepsilon \to 0} \varepsilon \log P\{\xi_\varepsilon \in B\}.$$

The lower bound now follows as we take the supremum over $x \in B^\circ$.

(ii) By (17) we may choose some compact sets K_r satisfying

$$\limsup_{\varepsilon \to 0} \varepsilon \log P\{\xi_\varepsilon \notin K_r\} < -r, \quad r > 0. \tag{18}$$

For any closed set $B \subset S$, we have

$$P\{\xi_\varepsilon \in B\} \leq 2(P\{\xi_\varepsilon \in B \cap K_r\} \vee P\{\xi_\varepsilon \notin K_r\}), \quad r > 0,$$

and so, by the weak LDP and (18),

$$\limsup_{\varepsilon \to 0} \varepsilon \log P\{\xi_\varepsilon \in B\} \leq -I(B \cap K_r) \wedge r \leq -I(B) \wedge r.$$

The upper bound now follows as we let $r \to \infty$. Applying the lower bound and (18) to the sets K_r^c gives

$$-I(K_r^c) \leq \limsup_{\varepsilon \to 0} \varepsilon \log P\{\xi_\varepsilon \notin K_r\} < -r, \quad r > 0,$$

and so $I^{-1}[0, r] \subset K_r$ for all $r > 0$, which shows that I is good.

(iii) The sufficiency follows from (ii), applied to an arbitrary sequence $\varepsilon_n \to 0$. Now let S be separable and complete, and assume that the rate function I is good. For any $k \in \mathbb{N}$ we may cover S by some open balls $B_{k1}, B_{k2}, \ldots$ of radius $1/k$. Putting $U_{km} = \bigcup_{j \leq m} B_{kj}$, we have $\sup_m I(U_{km}^c) = \infty$ since any level set $I^{-1}[0, r]$ is covered by finitely many sets B_{kj}. Now fix any sequence $\varepsilon_n \to 0$ and constant $r > 0$. By the LDP upper bound and the fact that $P\{\xi_{\varepsilon_n} \in U_{km}^c\} \to 0$ as $m \to \infty$ for fixed n and k, we may choose $m_k \in \mathbb{N}$ so large that

$$P\{\xi_{\varepsilon_n} \in U_{k,m_k}^c\} \leq \exp(-rk/\varepsilon_n), \quad n, k \in \mathbb{N}.$$

Summing a geometric series, we obtain

$$\limsup_n \varepsilon_n \log P\Big\{\xi_{\varepsilon_n} \in \bigcup_k U_{k,m_k}^c\Big\} \leq -r.$$

The asserted exponential tightness now follows, since the set $\bigcap_k U_{k,m_k}$ is totally bounded and hence relatively compact. □

The analogy with weak convergence theory suggests that we look for a version of (16) for continuous functions.

Theorem 27.10 *(functional LDP, Varadhan, Bryc)* Let ξ_ε, $\varepsilon > 0$, *be random elements in a metric space S.*

(i) *If (ξ_ε) satisfies the LDP with a rate function I and if $f : S \to \mathbb{R}$ is continuous and bounded above, then*

$$\Lambda_f \equiv \lim_{\varepsilon \to 0} \varepsilon \log E \exp\left(f(\xi_\varepsilon)/\varepsilon\right) = \sup_{x \in S} (f(x) - I(x)).$$

(ii) *If the ξ_ε are exponentially tight and the limit Λ_f in (i) exists for every $f \in C_b$, then (ξ_ε) satisfies the LDP with the good rate function*

$$I(x) = \sup_{f \in C_b} (f(x) - \Lambda_f), \quad x \in S.$$

Proof: (i) For every $n \in \mathbb{N}$ we can choose finitely many closed sets $B_1, \ldots, B_m \subset S$ such that $f \leq -n$ on $\bigcap_j B_j^c$ and the oscillation of f on each B_j is at most n^{-1}. Then

$$\limsup_{\varepsilon \to 0} \varepsilon \log E e^{f(\xi_\varepsilon)/\varepsilon} \leq \max_{j \leq m} \limsup_{\varepsilon \to 0} \varepsilon \log E\left[e^{f(\xi_\varepsilon)/\varepsilon}; \xi_\varepsilon \in B_j\right] \vee (-n)$$

$$\leq \max_{j \leq m} \left(\sup\nolimits_{x \in B_j} f(x) - \inf\nolimits_{x \in B_j} I(x)\right) \vee (-n)$$

$$\leq \max_{j \leq m} \sup_{x \in B_j} \left(f(x) - I(x) + n^{-1}\right) \vee (-n)$$

$$= \sup_{x \in S} \left(f(x) - I(x) + n^{-1}\right) \vee (-n).$$

The upper bound now follows as we let $n \to \infty$. Next we fix any $x \in S$ with a neighborhood G and write

$$\liminf_{\varepsilon \to 0} \varepsilon \log E e^{f(\xi_\varepsilon)/\varepsilon} \geq \liminf_{\varepsilon \to 0} \varepsilon \log E\left[e^{f(\xi_\varepsilon)/\varepsilon}; \xi_\varepsilon \in G\right]$$

$$\geq \inf_{y \in G} f(y) - \inf_{y \in G} I(y)$$

$$\geq \inf_{y \in G} f(y) - I(x).$$

Here the lower bound follows as we let $G \downarrow \{x\}$ and then take the supremum over $x \in S$.

(ii) First we note that I is lower semicontinuous, as the supremum over a family of continuous functions. Since $\Lambda_f = 0$ for $f = 0$, it is also clear that $I \geq 0$. By Lemma 27.9 (ii) it remains to show that (ξ_ε) satisfies the weak LDP with rate function I. Then fix any $\delta > 0$. For every $x \in S$, we may choose a function $f_x \in C_b$ satisfying

$$f_x(x) - \Lambda_{f_x} > (I(x) - \delta) \wedge \delta^{-1},$$

and by continuity there exists a neighborhood B_x of x such that

$$f_x(y) > \Lambda_{f_x} + (I(x) - \delta) \wedge \delta^{-1}, \quad y \in B_x.$$

By Chebyshev's inequality we get for any $\varepsilon > 0$

$$P\{\xi_\varepsilon \in B_x\} \leq E \exp\left(\varepsilon^{-1}(f_x(\xi_\varepsilon) - \inf\{f_x(y); y \in B_x\})\right)$$

$$\leq E \exp\left(\varepsilon^{-1}(f_x(\xi_\varepsilon) - \Lambda_{f_x} - (I(x) - \delta) \wedge \delta^{-1})\right),$$

and so by the definition of Λ_{f_x},

$$\limsup_{\varepsilon \to 0} \varepsilon \log P\{\xi_\varepsilon \in B_x\}$$

$$\leq \lim_{\varepsilon \to 0} \varepsilon \log E \exp(f_x(\xi_\varepsilon)/\varepsilon) - \Lambda_{f_x} - (I(x) - \delta) \wedge \delta^{-1}$$

$$= -(I(x) - \delta) \wedge \delta^{-1}.$$

Now fix any compact set $K \subset S$, and choose $x_1, \ldots, x_m \in K$ such that $K \subset \bigcup_i B_{x_i}$. Then

$$\limsup_{\varepsilon \to 0} \varepsilon \log P\{\xi_\varepsilon \in K\} \leq \max_{i \leq m} \limsup_{\varepsilon \to 0} \varepsilon \log P\{\xi_\varepsilon \in B_{x_i}\}$$
$$\leq -\min_{i \leq m}(I(x_i) - \delta) \wedge \delta^{-1}$$
$$\leq -(I(K) - \delta) \wedge \delta^{-1}.$$

The upper bound now follows as we let $\delta \to 0$.

Next consider any open set G and element $x \in G$. For any $n \in \mathbb{N}$ we may choose a continuous function $f_n: S \to [-n, 0]$ such that $f_n(x) = 0$ and $f_n = -n$ on G^c. Then

$$-I(x) = \inf_{f \in C_b} (\Lambda_f - f(x)) \leq \Lambda_{f_n} - f_n(x) = \Lambda_{f_n}$$
$$= \lim_{\varepsilon \to 0} \varepsilon \log E \exp(f_n(\xi_\varepsilon)/\varepsilon)$$
$$\leq \liminf_{\varepsilon \to 0} \varepsilon \log P\{\xi_\varepsilon \in G\} \vee (-n).$$

The lower bound now follows as we let $n \to \infty$ and then take the supremum over all $x \in G$. □

Next we note that the LDP is preserved by continuous mappings. The following results are often referred to as the *direct and inverse contraction principles*. Given any rate function I on S and a function $f: S \to T$, we define the image $J = I \circ f^{-1}$ on T as the function

$$J(y) = I(f^{-1}\{y\}) = \inf\{I(x); f(x) = y\}, \quad y \in T. \tag{19}$$

Note that the corresponding set functions are related by

$$J(B) \equiv \inf_{y \in B} J(y) = \inf\{I(x); f(x) \in B\} = I(f^{-1}B), \quad B \subset T.$$

Theorem 27.11 *(continuous mapping)* Consider a continuous function f between two metric spaces S and T, and let ξ_ε be random elements in S.

(i) If (ξ_ε) satisfies the LDP in S with rate function I, then the images $f(\xi_\varepsilon)$ satisfy the LDP in T with the raw rate function $J = I \circ f^{-1}$. Moreover, J is a good rate function on T whenever the function I is good on S.

(ii) *(Ioffe)* Let (ξ_ε) be exponentially tight in S, let f be injective, and let the images $f(\xi_\varepsilon)$ satisfy the weak LDP in T with rate function J. Then (ξ_ε) satisfies the LDP in S with the good rate function $I = J \circ f$.

Proof: (i) Since f is continuous, we note that $f^{-1}B$ is open or closed whenever the corresponding property holds for B. Using the LDP for (ξ_ε), we get for any $B \subset T$

$$-I(f^{-1}B^\circ) \leq \liminf_{\varepsilon \to 0} \varepsilon \log P\{\xi_\varepsilon \in f^{-1}B^\circ\}$$
$$\leq \limsup_{\varepsilon \to 0} \varepsilon \log P\{\xi_\varepsilon \in f^{-1}B^-\} \leq -I(f^{-1}B^-),$$

which proves the LDP for $\{f(\xi_\varepsilon)\}$ with the raw rate function $J = I \circ f^{-1}$.
When I is good, we claim that

$$J^{-1}[0,r] = f(I^{-1}[0,r]), \quad r \geq 0. \tag{20}$$

To see this, fix any $r \geq 0$, and let $x \in I^{-1}[0,r]$. Then

$$J \circ f(x) = I \circ f^{-1} \circ f(x) = \inf\{I(u);\ f(u) = f(x)\} \leq I(x) \leq r,$$

which means that $f(x) \in J^{-1}[0,r]$. Conversely, let $y \in J^{-1}[0,r]$. Since I is good and f is continuous, the infimum in (19) is attained at some $x \in S$, and we get $y = f(x)$ with $I(x) \leq r$. Thus, $y \in f(I^{-1}[0,r])$, which completes the proof of (20). Since continuous maps preserve compactness, (20) shows that the goodness of I carries over to J.

(ii) Here I is again a rate function, since the lower semicontinuity of J is preserved by composition with the continuous map f. By Lemma 27.9 (ii) it is then enough to show that (ξ_ε) satisfies the weak LDP in S. To prove the upper bound, fix any compact set $K \subset S$, and note that the image set $f(K)$ is again compact since f is continuous. Hence, the weak LDP for $(f(\xi_\varepsilon))$ yields

$$\limsup_{\varepsilon \to 0} \varepsilon \log P\{\xi_\varepsilon \in K\} = \limsup_{\varepsilon \to 0} \varepsilon \log P\{f(\xi_\varepsilon) \in f(K)\}$$
$$\leq -J(f(K)) = -I(K).$$

Next we fix any open set $G \subset S$, and let $x \in G$ be arbitrary with $I(x) = r < \infty$. Since (ξ_ε) is exponentially tight, we may choose a compact set $K \subset S$ such that

$$\limsup_{\varepsilon \to 0} \varepsilon \log P\{\xi_\varepsilon \notin K\} < -r. \tag{21}$$

The continuous image $f(K)$ is compact in T, and so by (21) and the weak LDP for $\{f(\xi_\varepsilon)\}$

$$-I(K^c) = -J(f(K^c)) \leq -J((f(K))^c)$$
$$\leq \liminf_{\varepsilon \to 0} \varepsilon \log P\{f(\xi_\varepsilon) \notin f(K)\}$$
$$\leq \limsup_{\varepsilon \to 0} \varepsilon \log P\{\xi_\varepsilon \notin K\} < -r.$$

Since $I(x) = r$, we conclude that $x \in K$.

As a continuous bijection from the compact set K onto $f(K)$, the function f is in fact a homeomorphism between the two sets with their subset topologies. By Lemma 1.6 we may then choose an open set $G' \subset T$ such that $f(x) \in f(G \cap K) = G' \cap f(K)$. Noting that

$$P\{f(\xi_\varepsilon) \in G'\} \leq P\{\xi_\varepsilon \in G\} + P\{\xi_\varepsilon \notin K\}$$

and using the weak LDP of $\{f(\xi_\varepsilon)\}$, we get

$$\begin{aligned}
-r &= -I(x) = -J(f(x)) \\
&\leq \liminf_{\varepsilon \to 0} \varepsilon \log P\{f(\xi_\varepsilon) \in G'\} \\
&\leq \liminf_{\varepsilon \to 0} \varepsilon \log P\{\xi_\varepsilon \in G\} \vee \limsup_{\varepsilon \to 0} \varepsilon \log P\{\xi_\varepsilon \notin K\}.
\end{aligned}$$

Hence, by (21)

$$-I(x) \leq \liminf_{\varepsilon \to 0} \varepsilon \log P\{\xi_\varepsilon \in G\}, \quad x \in G,$$

and the lower bound follows as we take the supremum over all $x \in G$. □

We turn to the powerful method of *projective limits*. The following sequential version is sufficient for our needs and will enable us to extend the LDP to a variety of infinite-dimensional contexts. Some general background on projective limits is provided by Appendix A2.

Theorem 27.12 *(random sequences, Dawson and Gärtner)* For any metric spaces $S_1, S_2, \ldots$, let $\xi_\varepsilon = (\xi_\varepsilon^n)$ be random elements in $S = \times_k S_k$, such that for every $n \in \mathbb{N}$ the vectors $(\xi_\varepsilon^1, \ldots, \xi_\varepsilon^n)$ satisfy the LDP in $S^n = \times_{k \leq n} S_k$ with a good rate function I_n. Then (ξ_ε) satisfies the LDP in S with the good rate function

$$I(x) = \sup_n I_n(x_1, \ldots, x_n), \quad x = (x_1, x_2, \ldots) \in S. \tag{22}$$

Proof: For any $m \leq n$ we introduce the natural projections $\pi_n \colon S \to S^n$ and $\pi_{mn} \colon S^n \to S^m$. Since the π_{mn} are continuous and the I_n are good, Theorem 27.11 shows that $I_m = I_n \circ \pi_{mn}^{-1}$ for all $m \leq n$, and so $\pi_{mn}(I_n^{-1}[0,r]) \subset I_m^{-1}[0,r]$ for all $r \geq 0$ and $m \leq n$. Hence, for each $r \geq 0$ the level sets $I_n^{-1}[0,r]$ form a projective sequence. Since they are also compact by hypothesis, and in view of (22)

$$I^{-1}[0,r] = \bigcap_n \pi_n^{-1} I_n^{-1}[0,r], \quad r \geq 0, \tag{23}$$

Lemma A2.9 shows that the sets $I^{-1}[0,r]$ are compact. Thus, I is again a good rate function.

Now fix any closed set $A \subset S$ and put $A_n = \pi_n A$, so that $\pi_{mn} A_n = A_m$ for all $m \leq n$. Since the π_{mn} are continuous, we have also $\pi_{mn} A_n^- \subset A_m^-$ for $m \leq n$, which means that the sets A_n^- form a projective sequence. We claim that

$$A = \bigcap_n \pi_n^{-1} A_n^-. \tag{24}$$

Here the relation $A \subset \pi_n^{-1} A_n^-$ is obvious. Next assume that $x \notin A$. By the definition of the product topology, we may choose a $k \in \mathbb{N}$ and an open set $U \subset S^k$ such that $x \in \pi_k^{-1} U \subset A^c$. It follows easily that $\pi_k x \in U \subset A_k^c$. Since U is open, we have even $\pi_k x \in (A_k^-)^c$. Thus, $x \notin \bigcap_n \pi_n^{-1} A_n^-$, which completes the proof of (24). The projective property carries over to the intersections $A_n^- \cap I_n^{-1}[0,r]$, and formulas (23) and (24) combine into the

relation
$$A \cap I^{-1}[0,r] = \bigcap_n \pi_n^{-1}\left(A_n^- \cap I_n^{-1}[0,r]\right), \quad r \geq 0. \tag{25}$$

Now assume that $I(A) > r \in \mathbb{R}$. Then $A \cap I^{-1}[0,r] = \emptyset$, and by (25) and Lemma A2.9 we get $A_n^- \cap I_n^{-1}[0,r] = \emptyset$ for some $n \in \mathbb{N}$, which implies $I_n(A_n^-) \geq r$. Noting that $A \subset \pi_n^{-1} A_n$ and using the LDP in S^n, we conclude that

$$\limsup_{\varepsilon \to 0} \varepsilon \log P\{\xi_\varepsilon \in A\} \leq \limsup_{\varepsilon \to 0} \varepsilon \log P\{\pi_n \xi_\varepsilon \in A_n\}$$
$$\leq -I_n(A_n^-) \leq -r,$$

The upper bound now follows as we let $r \uparrow I(A)$.

Finally, fix an open set $G \subset S$ and let $x \in G$ be arbitrary. By the definition of the product topology, we may choose $n \in \mathbb{N}$ and an open set $U \subset S^n$ such that $x \in \pi_n^{-1} U \subset G$. The LDP in S^n yields

$$\liminf_{\varepsilon \to 0} \varepsilon \log P\{\xi_\varepsilon \in G\} \geq \liminf_{\varepsilon \to 0} \varepsilon \log P\{\pi_n \xi_\varepsilon \in U\}$$
$$\geq -I_n(U) \geq -I_n \circ \pi_n(x) \geq -I(x),$$

and the lower bound follows as we take the supremum over all $x \in G$. □

We consider yet another basic method for extending the LDP, namely by suitable approximation. Here the following elementary result is often helpful. Let us say that the random elements ξ_ε and η_ε in a common separable metric space (S, d) are *exponentially equivalent* if

$$\lim_{\varepsilon \to 0} \varepsilon \log P\{d(\xi_\varepsilon, \eta_\varepsilon) > h\} = -\infty, \quad h > 0. \tag{26}$$

The separability of S is needed only to ensure measurability of the pairwise distances $d(\xi_\varepsilon, \eta_\varepsilon)$. In general, we may replace (26) by a similar condition involving the outer measure.

Lemma 27.13 *(approximation)* *Let ξ_ε and η_ε be exponentially equivalent random elements in a separable metric space S. Then (ξ_ε) satisfies the LDP with a good rate function I iff the same LDP holds for (η_ε).*

Proof: Suppose that the LDP holds for (ξ_ε) with rate function I. Fix any closed set $B \subset S$, and let B^h denote the closed h-neighborhood of B. Then

$$P\{\eta_\varepsilon \in B\} \leq P\{\xi_\varepsilon \in B^h\} + P\{d(\xi_\varepsilon, \eta_\varepsilon) > h\},$$

and so by (26) and the LDP for (ξ_ε)

$$\limsup_{\varepsilon \to 0} \varepsilon \log P\{\eta_\varepsilon \in B\}$$
$$\leq \limsup_{\varepsilon \to 0} \varepsilon \log P\{\xi_\varepsilon \in B^h\} \vee \limsup_{\varepsilon \to 0} \varepsilon \log P\{d(\xi_\varepsilon, \eta_\varepsilon) > h\}$$
$$\leq -I(B^h) \vee (-\infty) = -I(B^h).$$

Since I is good, we have $I(B^h) \uparrow I(B)$ as $h \to 0$, and the required upper bound follows.

Next we fix an open set $G \subset S$ and an element $x \in G$. If $d(x, G^c) > h > 0$, we may choose a neighborhood U of x such that $U^h \subset G$. Noting that

$$P\{\xi_\varepsilon \in U\} \leq P\{\eta_\varepsilon \in G\} + P\{d(\xi_\varepsilon, \eta_\varepsilon) > h\},$$

we get by (26) and the LDP for (ξ_ε)

$$\begin{aligned}
-I(x) &\leq -I(U) \leq \liminf_{\varepsilon \to 0} \varepsilon \log P\{\xi_\varepsilon \in U\} \\
&\leq \liminf_{\varepsilon \to 0} \varepsilon \log P\{\eta_\varepsilon \in G\} \vee \limsup_{\varepsilon \to 0} \varepsilon \log P\{d(\xi_\varepsilon, \eta_\varepsilon) > h\} \\
&= \liminf_{\varepsilon \to 0} \varepsilon \log P\{\eta_\varepsilon \in G\}.
\end{aligned}$$

The required lower bound now follows, as we take the supremum over all $x \in G$. □

We now demonstrate the power of the abstract theory by considering some important applications. First we study perturbations of the ordinary differential equation $\dot{x} = b(x)$ by a small noise term. More precisely, we consider the unique solution X^ε with $X_0^\varepsilon = 0$ of the d-dimensional SDE

$$dX_t = \varepsilon^{1/2} dB_t + b(X_t) dt, \quad t \geq 0, \tag{27}$$

where B is a Brownian motion in $\mathbb{R}^d$ and b is bounded and uniformly Lipschitz continuous mapping on $\mathbb{R}^d$. Let H_∞ denote the set of all absolutely continuous functions $x: \mathbb{R}_+ \to \mathbb{R}^d$ with $x_0 = 0$ such that $\dot{x} \in L^2$.

Theorem 27.14 *(perturbed dynamical systems, Freidlin and Wentzell)* *For any bounded, uniformly Lipschitz continuous function $b: \mathbb{R}^d \to \mathbb{R}^d$, the solutions X^ε to (27) with $X_0^\varepsilon = 0$ satisfy the LDP in $C(\mathbb{R}_+, \mathbb{R}^d)$ with the good rate function*

$$I(x) = \tfrac{1}{2} \int_0^\infty |\dot{x}_t - b(x_t)|^2 \, dt, \quad x \in H_\infty. \tag{28}$$

Here it is understood that $I(x) = \infty$ when $x \notin H_\infty$. Note that the result for $b = 0$ extends Theorem 27.6 to processes on $\mathbb{R}_+$.

Proof: If B^1 is a Brownian motion on $[0, 1]$, then for every $r > 0$ the process $B^r = \Phi(B^1)$ given by $B_t^r = r^{1/2} B_{t/r}^1$ is a Brownian motion on $[0, r]$. Noting that Φ is continuous from $C([0, 1])$ to $C([0, r])$, we see from Theorems 27.6 and 27.11 (i) together with Lemma 27.7 that the processes $\varepsilon^{1/2} B^r$ satisfy the LDP in $C([0, r])$ with the good rate function $I_r = I_1 \circ \Phi^{-1}$, where $I_1(x) = \tfrac{1}{2} \|\dot{x}\|_2^2$ for $x \in H_1$ and $I_1(x) = \infty$ otherwise. Now Φ maps H_1 onto H_r, and when $y = \Phi(x)$ with $x \in H_1$ we have $\dot{x}_t = r^{1/2} \dot{y}_{rt}$. Hence, by calculus $I_r(y) = \tfrac{1}{2} \int_0^r |\dot{y}_s|^2 ds = \tfrac{1}{2} \|\dot{y}\|_2^2$, which extends Theorem 27.6 to $[0, r]$. For the further extension to $\mathbb{R}_+$, let $\pi_n x$ denote the restriction of a function $x \in C(\mathbb{R}_+)$ to $[0, n]$, and infer from Theorem 27.12 that the processes $\varepsilon^{1/2} B$ satisfy the LDP in $C(\mathbb{R}_+)$ with the good rate function $I_\infty(x) = \sup_n I_n(\pi_n x) = \tfrac{1}{2} \|\dot{x}\|_2^2$.

By an elementary version of Theorem 21.3, the integral equation

$$x_t = z_t + \int_0^t b(x_s)\,ds, \quad t \geq 0, \tag{29}$$

has a unique solution $x = F(z)$ in $C = C(\mathbb{R}_+)$ for every $z \in C$. Letting $z^1, z^2 \in C$ be arbitrary and writing a for the Lipschitz constant of b, we note that the corresponding solutions $x^i = F(z^i)$ satisfy

$$|x_t^1 - x_t^2| \leq \|z^1 - z^2\| + a \int_0^t |x_s^1 - x_s^2|\,ds, \quad t \geq 0.$$

Hence, Gronwall's Lemma 21.4 yields $\|x^1 - x^2\| \leq \|z^1 - z^2\|e^{ar}$ on the interval $[0, r]$, which shows that F is continuous. Using Schilder's theorem on $\mathbb{R}_+$ along with Theorem 27.11 (i), we conclude that the processes X^ε satisfy the LDP in $C(\mathbb{R}_+)$ with the good rate function $I = I_\infty \circ F^{-1}$. Now F is clearly bijective, and (29) shows that the functions z and $x = F(z)$ lie simultaneously in H_∞, in which case $\dot{z} = \dot{x} - b(x)$ a.e. Thus, I is indeed given by (28). □

Now consider a random element ξ with distribution μ in an arbitrary metric space S. We introduce the *cumulant-generating functional*

$$\Lambda(f) = \log E e^{f(\xi)} = \log \mu e^f, \quad f \in C_b(S),$$

and the associated Legendre–Fenchel transform

$$\Lambda^*(\nu) = \sup_{f \in C_b} (\nu f - \Lambda(f)), \quad \nu \in \mathcal{P}(S), \tag{30}$$

where $\mathcal{P}(S)$ denotes the class of probability measures on S, endowed with the topology of weak convergence. Note that Λ and Λ^* are both convex, by the same argument as for $\mathbb{R}^d$.

Given any two measures $\mu, \nu \in \mathcal{P}(S)$, we define the *relative entropy* of ν with respect to μ by

$$H(\nu|\mu) = \begin{cases} \nu \log p = \mu(p \log p), & \nu \ll \mu \text{ with } \nu = p \cdot \mu, \\ \infty, & \nu \not\ll \mu. \end{cases}$$

Since $x \log x$ is convex, the function $H(\nu|\mu)$ is convex in ν for fixed μ, and by Jensen's inequality we have

$$H(\nu|\mu) \geq \mu p \log \mu p = \nu S \log \nu S = 0, \quad \nu \in \mathcal{P}(S),$$

with equality iff $\nu = \mu$.

Now let $\xi_1, \xi_2, \ldots$ be i.i.d. random elements in S. The associated *empirical distributions* are given by

$$\eta_n = n^{-1} \sum_{k \leq n} \delta_{\xi_k}, \quad n \in \mathbb{N}.$$

They may be regarded as random elements in $\mathcal{P}(S)$, and we note that

$$\eta_n f = n^{-1} \sum_{k \leq n} f(\xi_k), \quad f \in C_b(S),\ n \in \mathbb{N}.$$

In particular, Theorem 27.5 applies to the random vectors $(\eta_n f_1, \ldots, \eta_n f_m)$ for fixed $f_1, \ldots, f_m \in C_b(S)$. The following result may be regarded as an infinite-dimensional version of Theorem 27.5. It also provides an important connection to statistical mechanics, via the entropy function.

Theorem 27.15 *(large deviations of empirical distributions, Sanov)* *Let $\xi_1, \xi_2, \ldots$ be i.i.d. random elements with distribution μ in a Polish space S, and put $\Lambda(f) = \log \mu e^f$. Then the associated empirical distributions $\eta_1, \eta_2, \ldots$ satisfy the LDP in $\mathcal{P}(S)$ with the good rate function*

$$\Lambda^*(\nu) = H(\nu|\mu), \quad \nu \in \mathcal{P}(S). \tag{31}$$

A couple of lemmas will be needed for the proof.

Lemma 27.16 *(entropy, Donsker and Varadhan)* *In (30) it is equivalent to take the supremum over all bounded, measurable functions $f : S \to \mathbb{R}$. The identity (31) then holds for any probability measures μ and ν on a common measurable space S.*

Proof: The first assertion holds by Lemma 1.35 and dominated convergence. If $\nu \not\ll \mu$, then $H(\nu|\mu) = \infty$ by definition. Furthermore, we may choose a set $B \in \mathcal{S}$ with $\mu B = 0$ and $\nu B > 0$, and take $f_n = n 1_B$ to obtain $\nu f_n - \log \mu e^{f_n} = n \nu B \to \infty$. Thus, even $\Lambda^*(\nu) = \infty$ in this case, and it remains to prove (31) when $\nu \ll \mu$. Assuming $\nu = p \cdot \mu$ and writing $f = \log p$, we note that

$$\nu f - \log \mu e^f = \nu \log p - \log \mu p = H(\nu|\mu).$$

If $f = \log p$ is unbounded, we may approximate by bounded measurable functions f_n satisfying $\mu e^{f_n} \to 1$ and $\nu f_n \to \nu f$, and we get $\Lambda^*(\nu) \geq H(\nu|\mu)$.

To prove the reverse inequality, we first assume that S is finite and generated by a partition $B_1, \ldots, B_n$ of S. Putting $\mu_k = \mu B_k$, $\nu_k = \nu B_k$, and $p_k = \nu_k / \mu_k$, we may write our claim in the form

$$g(x) \equiv \sum_k \nu_k x_k - \log \sum_k \mu_k e^{x_k} \leq \sum_k \nu_k \log p_k,$$

where $x = (x_1, \ldots, x_n) \in \mathbb{R}^d$ is arbitrary. Here the function g is concave and satisfies $\nabla g(x) = 0$ for $x = (\log p_1, \ldots, \log p_n)$, asymptotically when $p_k = 0$ for some k. Thus,

$$\sup_x g(x) = g(\log p_1, \ldots, \log p_k) = \sum_k \nu_k \log p_k.$$

To prove the inequality $\nu f - \log \mu e^f \leq \nu \log p$ in general, we may assume that f is simple. The generated σ-field $\mathcal{F} \subset \mathcal{S}$ is then finite, and we note that $\nu = \mu[p|\mathcal{F}] \cdot \mu$ on $\mathcal{F}$. Using the result in the finite case, together with Jensen's inequality for conditional expectations, we obtain

$$\begin{aligned}\nu f - \log \mu e^f &\leq \mu(\mu[p|\mathcal{F}] \log \mu[p|\mathcal{F}]) \\ &\leq \mu \mu[p \log p|\mathcal{F}] = \nu \log p.\end{aligned}$$ □

Lemma 27.17 *(exponential tightness)* *The empirical distributions η_n in Theorem 27.15 are exponentially tight in $\mathcal{P}(S)$.*

Proof: If $B \in \mathcal{S}$ with $P\{\xi \in B\} = p \in (0,1)$, then by Theorem 27.3 and Lemmas 27.1 and 27.2 we have for any $x \in [p, 1]$

$$\sup_n n^{-1} \log P\{\eta_n B > x\} \leq -x \log \frac{x}{p} - (1-x) \log \frac{1-x}{1-p}. \tag{32}$$

In particular, we note that the right-hand side tends to $-\infty$ as $p \to 0$ for fixed $x \in (0,1)$. Now fix any $r > 0$. By (32) and Theorem 16.3, we may choose some compact sets $K_1, K_2, \cdots \subset S$ such that

$$P\{\eta_n K_k^c > 2^{-k}\} \leq e^{-knr}, \quad k, n \in \mathbb{N}.$$

Summing over k gives

$$\limsup_{n \to \infty} n^{-1} \log P \bigcup_k \{\eta_n K_k^c > 2^{-k}\} \leq -r,$$

and it remains to note that the set

$$M = \bigcap_k \{\nu \in \mathcal{P}(S); \, \nu K_k^c \leq 2^{-k}\}$$

is compact, by another application of Theorem 16.3. $\square$

Proof of Theorem 27.15: By Theorem A1.1 we can embed S as a Borel subset of a compact metric space K. The function space $C_b(K)$ is separable, and we can choose a dense sequence $f_1, f_2, \ldots \in C_b(K)$. For any $m \in \mathbb{N}$, the random vector $(f_1(\xi), \ldots, f_m(\xi))$ has cumulant-generating function

$$\Lambda_m(u) = \log E \exp \sum_{k \leq m} u_k f_k(\xi) = \Lambda \circ \sum_{k \leq m} u_k f_k, \quad u \in \mathbb{R}^m,$$

and so by Theorem 27.5 the random vectors $(\eta_n f_1, \ldots, \eta_n f_m)$ satisfy the LDP in $\mathbb{R}^m$ with the good rate function Λ_m^*. By Theorem 27.12 it follows that the infinite sequences $(\eta_n f_1, \eta_n f_2, \ldots)$ satisfy the LDP in $\mathbb{R}^\infty$ with the good rate function $J = \sup_m (\Lambda_m^* \circ \pi_m)$, where π_m denotes the natural projection of $\mathbb{R}^\infty$ onto $\mathbb{R}^m$. Since $\mathcal{P}(K)$ is compact by Theorem 16.3 and the mapping $\nu \mapsto (\nu f_1, \nu f_2, \ldots)$ is a continuous injection of $\mathcal{P}(K)$ into $\mathbb{R}^\infty$, Theorem 27.11 (ii) shows that the random measures η_n satisfy the LDP in $\mathcal{P}(K)$ with the good rate function

$$\begin{aligned} I_K(\nu) &= J(\nu f_1, \nu f_2, \ldots) = \sup_m \Lambda_m^*(\nu f_1, \ldots, \nu f_m) \\ &= \sup_m \sup_{u \in \mathbb{R}^m} \left(\sum_{k \leq m} u_k \nu f_k - \Lambda \circ \sum_{k \leq m} u_k f_k \right) \\ &= \sup_{f \in \mathcal{F}} (\nu f - \Lambda(f)) = \sup_{f \in C_b} (\nu f - \Lambda(f)), \end{aligned} \tag{33}$$

where $\mathcal{F}$ denotes the set of all linear combinations of $f_1, f_2, \ldots$.

Next we note that the natural embedding $\mathcal{P}(S) \to \mathcal{P}(K)$ is continuous, since for any $f \in C_b(K)$ the restriction of f to S belongs to $C_b(S)$. Since it is also trivially injective, we see from Theorem 27.11 (ii) and Lemma 27.17

that the η_n satisfy the LDP even in $\mathcal{P}(S)$, with a good rate function I_S that equals the restriction of I_K to $\mathcal{P}(S)$. It remains to note that $I_S = \Lambda^*$ by (33) and Lemma 27.16. □

We conclude with a remarkable application of Schilder's Theorem 27.6. Writing B for a standard Brownian motion in $\mathbb{R}^d$, we define for any $t > e$ the scaled process X^t by

$$X_s^t = \frac{B_{st}}{\sqrt{2t \log \log t}}, \quad s \geq 0. \tag{34}$$

Theorem 27.18 (*functional law of the iterated logarithm, Strassen*) *Let B be a Brownian motion in $\mathbb{R}^d$, and define the processes X^t by (34). Then the following equivalent statements hold outside a fixed P-null set:*

(i) *The paths X^t, $t \geq 3$, form a relatively compact set in $C(\mathbb{R}_+, \mathbb{R}^d)$, whose set of limit points as $t \to \infty$ equals $K = \{x \in H_\infty;\ \|\dot{x}\|_2 \leq 1\}$.*

(ii) *For any continuous function $F: C(\mathbb{R}_+, \mathbb{R}^d) \to \mathbb{R}$, we have*

$$\limsup_{t \to \infty} F(X^t) = \sup_{x \in K} F(x).$$

In particular, we may recover the classical law of the iterated logarithm in Theorem 13.18 by choosing $F(x) = x_1$. Using Theorem 14.6, we can easily derive a correspondingly strengthened version for random walks.

Proof: The equivalence of (i) and (ii) being elementary, we need to prove only (i). Noting that $X^t \stackrel{d}{=} B/\sqrt{2 \log \log t}$ and using Theorem 27.6, we get for any measurable set $A \subset C(\mathbb{R}_+, \mathbb{R}^d)$ and constant $r > 1$

$$\limsup_{n \to \infty} \frac{\log P\{X^{r^n} \in A\}}{\log n} \leq \limsup_{t \to \infty} \frac{\log P\{X^t \in A\}}{\log \log t} \leq -2I(A^-),$$

$$\liminf_{n \to \infty} \frac{\log P\{X^{r^n} \in A\}}{\log n} \geq \liminf_{t \to \infty} \frac{\log P\{X^t \in A\}}{\log \log t} \geq -2I(A^\circ),$$

where $I(x) = \tfrac12 \|\dot{x}\|_2^2$ when $x \in H_\infty$ and $I(x) = \infty$ otherwise. Hence,

$$\sum_n P\{X^{r^n} \in A\} \begin{cases} < \infty, & 2I(A^-) > 1, \\ = \infty, & 2I(A^\circ) < 1. \end{cases} \tag{35}$$

Now fix any $r > 1$, and let $G \supset K$ be open. Note that $2I(G^c) > 1$ by Lemma 27.7. By the first part of (35) and the Borel–Cantelli lemma we have $P\{X^{r^n} \notin G \text{ i.o.}\} = 0$ or, equivalently, $1_G(X^{r^n}) \to 1$ a.s. Since G was arbitrary, it follows that $\rho(X^{r^n}, K) \to 0$ a.s. for any metrization ρ of $C(\mathbb{R}_+, \mathbb{R}^d)$. In particular, this holds with any $c > 0$ for the metric

$$\rho_c(x, y) = \int_0^\infty ((x - y)_s^* \wedge 1)\, e^{-cs}\, ds, \quad x, y \in C(\mathbb{R}_+, \mathbb{R}^d).$$

To extend the convergence to the entire family $\{X^t\}$, fix any path of B such that $\rho_1(X^{r^n}, K) \to 0$, and choose some functions $y^{r^n} \in K$ satisfying $\rho_1(X^{r^n}, y^{r^n}) \to 0$. For any $t \in [r^n, r^{n+1})$, the paths X^{r^n} and X^t are related by

$$X^t(s) = X^{r^n}(tr^{-n}s)\left(\frac{r^n \log\log r^n}{t \log\log t}\right)^{1/2}, \quad s > 0.$$

Defining y^t in the same way in terms of y^{r^n}, we note that also $y^t \in K$ since $I(y^t) \leq I(y^{r^n})$. (The two H_∞-norms would agree if the logarithmic factors were omitted.) Furthermore,

$$\begin{aligned}
\rho_r(X^t, y^t) &= \int_0^\infty ((X^t - y^t)_s^* \wedge 1) e^{-rs} ds \\
&\leq \int_0^\infty \left((X^{r^n} - y^{r^n})_{rs}^* \wedge 1\right) e^{-rs} ds \\
&= r^{-1} \rho_1(X^{r^n}, y^{r^n}) \to 0.
\end{aligned}$$

Thus, $\rho_r(X^t, K) \to 0$. Since K is compact, we conclude that $\{X^t\}$ is relatively compact, with all its limit points as $t \to \infty$ belonging to K.

Now fix any $y \in K$ and $u \geq \varepsilon > 0$. By the established part of the theorem and the Cauchy–Buniakowski inequality, we have a.s.

$$\limsup_{t \to \infty} (X^t - y)_\varepsilon^* \leq \sup_{x \in K} (x - y)_\varepsilon^* \leq \sup_{x \in K} x_\varepsilon^* + y_\varepsilon^* \leq 2\varepsilon^{1/2}. \tag{36}$$

Write $x_{\varepsilon,u}^* = \sup_{s \in [\varepsilon, u]} |x_s - x_\varepsilon|$, and choose $r > u/\varepsilon$ to ensure independence between the variables $(X^{r^n} - y)_{\varepsilon,u}^*$. Applying the second part of (35) to the open set $A = \{x; (x - y)_{\varepsilon,u}^* < \varepsilon\}$ and using the Borel–Cantelli lemma together with (36), we obtain a.s.

$$\begin{aligned}
\liminf_{t \to \infty} (X^t - y)_u^* &\leq \limsup_{t \to \infty} (X^t - y)_\varepsilon^* + \liminf_{n \to \infty} (X^{r^n} - y)_{\varepsilon,u}^* \\
&\leq 2\varepsilon^{1/2} + \varepsilon.
\end{aligned}$$

Letting $\varepsilon \to 0$ gives $\liminf_t (X^t - y)_u^* = 0$ a.s., and so $\liminf_t \rho_1(X^t, y) \leq e^{-u}$ a.s. As $u \to \infty$, we obtain $\liminf_t \rho_1(X^t, y) = 0$ a.s. Applying this result to a dense sequence $y_1, y_2, \cdots \in K$, we see that a.s. every element of K is a limit point as $t \to \infty$ of the family $\{X^t\}$. □

Exercises

1. For any random vector ξ and constant a in $\mathbb{R}^d$, show that $\Lambda_{\xi-a}(u) = \Lambda_\xi(u) - ua$ and $\Lambda_{\xi-a}^*(x) = \Lambda_\xi^*(x + a)$.

2. For any random vector ξ in $\mathbb{R}^d$ and nonsingular $d \times d$ matrix a, show that $\Lambda_{a\xi}(u) = \Lambda_\xi(ua)$ and $\Lambda_{a\xi}^*(x) = \Lambda_\xi^*(a^{-1}x)$.

3. For any pair of independent random vectors ξ and η, show that $\Lambda_{\xi,\eta}(u, v) = \Lambda_\xi(u) + \Lambda_\eta(v)$ and $\Lambda_{\xi,\eta}^*(x, y) = \Lambda_\xi^*(x) + \Lambda_\eta^*(y)$.

4. Prove the claims of Lemma 27.2.

5. If ξ is Gaussian in $\mathbb{R}^d$ with mean $m \in \mathbb{R}^d$ and covariance matrix a, show that $\Lambda_\xi^*(x) = \frac{1}{2}(x-m)'a^{-1}(x-m)$. Explain the interpretation when a is singular.

6. Let ξ be a standard Gaussian random vector in $\mathbb{R}^d$. Show that the family $\varepsilon^{1/2}\xi$ satisfies the LDP in $\mathbb{R}^d$ with the good rate function $I(x) = \frac{1}{2}|x|^2$. (*Hint:* Deduce the result along the sequence $\varepsilon_n = n^{-1}$ from Theorem 27.5, and extend by monotonicity to general $\varepsilon > 0$.)

7. Use Theorem 27.11 (i) to deduce the preceding result from Schilder's theorem. (*Hint:* For $x \in H_1$, note that $|x_1| \le \|\dot{x}\|_2$ with equality iff $x_t \equiv tx_1$.)

8. Prove Schilder's theorem on $[0,T]$ by the same argument as for $[0,1]$.

9. Deduce Schilder's theorem in the space $C([0,n], \mathbb{R}^d)$ from the version in $C([0,1], \mathbb{R}^{nd})$.

10. Let B be a Brownian bridge in $\mathbb{R}^d$. Show that the processes $\varepsilon^{1/2}B$ satisfy the LDP in $C([0,1], \mathbb{R}^d)$ with the good rate function $I(x) = \frac{1}{2}\|\dot{x}\|_2^2$ for $x \in H_1$ with $x_1 = 0$ and $I(x) = \infty$ otherwise. (*Hint:* Write $B_t = X_t - tX_1$, where X is a Brownian motion in $\mathbb{R}^d$, and use Theorem 27.11. Check that $\|\dot{x} - a\|_2$ is minimized for $a = x_1$.)

11. Show that the property of exponential tightness and its sequential version are preserved by continuous mappings.

12. Prove that if the processes X^ε and Y^ε in $C(\mathbb{R}_+, \mathbb{R}^d)$ are exponentially tight, then so is any linear combination $aX^\varepsilon + bY^\varepsilon$. (*Hint:* Use the Arzelà–Ascoli theorem.)

13. Show directly from (27) that the processes X^ε in Theorem 27.14 are exponentially tight. (*Hint:* Use Lemmas 27.7 and 27.9 (iii) together with the Arzelà–Ascoli theorem.) Derive the same result from the stated theorem.

14. Let ξ_ε be random elements in a locally compact metric space S, satisfying the LDP with a good rate function I. Show that the ξ_ε are exponentially tight (even in the nonsequential sense). (*Hint:* For any $r > 0$, there exists a compact set $K_r \subset S$ such that $I^{-1}[0,r] \subset K_r^\circ$. Now apply the LDP upper bound to the closed sets $(K_r^\circ)^c \supset K_r^c$.)

15. For any metric space S and lcscH space T, let X^ε be random elements in $C(T,S)$ whose restrictions X_K^ε to an arbitrary compact set $K \subset T$ satisfy the LDP in $C(K,S)$ with the good rate function I_K. Show that the X^ε satisfy the LDP in $C(T,S)$ with the good rate function $I = \sup_K (I_K \circ \pi_K)$, where π_K denotes the restriction map from $C(T,S)$ to $C(K,S)$.

16. Let ξ_{kj} be i.i.d. random vectors in $\mathbb{R}^d$ satisfying $\Lambda(u) = Ee^{u\xi_{kj}} < \infty$ for all $u \in \mathbb{R}^d$. Show that the sequences $\bar{\xi}_n = n^{-1}\sum_{k \le n}(\xi_{k1}, \xi_{k2}, \ldots)$ satisfy an LDP in $(\mathbb{R}^d)^\infty$ with the good rate function $I(x) = \sum_j \Lambda^*(x_j)$. Also derive an LDP for the associated random walks in $\mathbb{R}^d$.

17. Let ξ be a sequence of i.i.d. $N(0,1)$ random variables. Use the preceding result to show that the sequences $\varepsilon^{1/2}\xi$ satisfy the LDP in $\mathbb{R}^\infty$ with the good rate function $I(x) = \frac{1}{2}\|x\|^2$ for $x \in l^2$ and $I(x) = \infty$ otherwise. Also show how the statement follows from Schilder's theorem.

18. Let $\xi_1, \xi_2, \ldots$ be i.i.d. random probability merasures on a Polish space S. Derive an LDP in $\mathcal{P}(S)$ for the averages $\bar{\xi}_n = n^{-1}\sum_{k\leq n}\xi_k$. (*Hint:* Define $\Lambda(f) = \log Ee^{\xi_k f}$, and proceed as in the proof of Sanov's theorem.)

19. Show how the classical law of the iterated logarithm in Theorem 13.18 follows from Theorem 27.18. Also use the latter result to derive a law of the iterated logarithm for the variables $\xi_t = |B_{2t} - B_t|$, where B is a Brownian motion in $\mathbb{R}^d$.

20. Use Theorem 27.18 to derive a corresponding law of the iterated logarithm in $C([0,1], \mathbb{R}^d)$.

21. Use Theorems 14.6 and 27.18 to derive a functional law of the iterated logarithm for random walks based on i.i.d. random variables with mean 0 and variance 1. (*Hint:* To state the result in $C(\mathbb{R}_+, \mathbb{R})$, replace the summation process $S_{[t]}$ by its linearly interpolated version, as in case of Corollary 16.7.)

22. Use Theorems 14.13 and 27.18 to derive a functional law of the iterated logarithm for suitable renewal processes.

23. Let $B^1, B^2, \ldots$ be independent Brownian motions in $\mathbb{R}^d$. Show that the sequence of paths $X_t^n = (2\log n)^{-1/2}B_t^n$, $n \geq 2$, is a.s. relatively compact in $C(\mathbb{R}_+, \mathbb{R}^d)$ with set of limit points $K = \{x \in H_\infty; \|\dot{x}\|_2 \leq 1\}$.

Appendices

Here we list some results that play an important role in this book but whose proofs are too long or technical to contribute in any essential way to the understanding of the subject matter. Proofs are given only for results that are not easily accessible in the literature.

A1. Advanced Measure Theory

The basic facts of measure theory were reviewed in Chapters 1 and 2. In this appendix we list, mostly without proofs, some special or less elementary results that are required in this book. One of the quoted results is used more frequently, namely the Borel nature of Polish spaces in Theorem A1.2. The remaining results are needed only for special purposes.

We begin with a basic embedding theorem. Recall that a topological space is said to be *Polish* if it is separable with a complete metrization.

Theorem A1.1 *(embedding)* *Any Polish space is homeomorphic to a Borel subset of the compact space $[0,1]^\infty$.*

Proof: See Theorem II.82.5 in Rogers and Williams (1994). □

We say that two measurable spaces S and T are *Borel isomorphic* if there exists a measurable bijection $f\colon S \to T$ such that f^{-1} is also measurable. A *Borel space* is defined as a measurable space that is Borel isomorphic to a Borel subset of $[0,1]$. The following result shows that the most commonly occurring spaces are Borel.

Theorem A1.2 *(Polish and Borel spaces)* *Every Borel subset of a Polish space is a Borel space.*

Proof: By Theorem A1.1, it is enough to show that $[0,1]^\infty$ is a Borel space. This may be seen by an elementary argument involving binary expansions, similar to that used in the proof of Lemma 3.21. However, some extra care is needed to ensure that the resulting mapping into $[0,1]$ is injective and bimeasurable with a measurable range. See, e.g., Theorem A.47 in Breiman (1968) for details. □

If a measurable mapping is invertible, then the measurability of the inverse can sometimes be inferred from the measurability of the range.

Theorem A1.3 *(range and inverse, Kuratowski) Let f be a measurable bijection between two Borel spaces S and T. Then the inverse $f^{-1}: T \to S$ is again measurable.*

Proof: See Parthasarathy (1967), Section I.3. □

We turn to the basic projection and section theorem, which plays such an important role in the more advanced literature. For any measurable space $(\Omega, \mathcal{F})$, the *universal completion* of $\mathcal{F}$ is defined as the σ-field $\overline{\mathcal{F}} = \bigcap_\mu \mathcal{F}^\mu$, where $\mathcal{F}^\mu$ denotes the completion with respect to μ, and the intersection extends over all probability measures μ on $\mathcal{F}$. For any spaces Ω and S, we define the *projection* πA of a set $A \subset \Omega \times S$ onto Ω as the union $\bigcup_s A_s$, where $A_s = \{\omega \in \Omega;\ (\omega, s) \in A\}$, $s \in S$.

Theorem A1.4 *(projection and sections, Lusin, Choquet, Meyer) Fix a measurable space $(\Omega, \mathcal{F})$ and a Borel space $(S, \mathcal{S})$, and consider a set $A \in \mathcal{F} \otimes \mathcal{S}$ with projection πA onto Ω. Then*

(i) *πA belongs to the universal completion $\overline{\mathcal{F}}$ of $\mathcal{F}$;*

(ii) *for any probability measure P on $\mathcal{F}$, there exists a random element ξ in S such that $(\omega, \xi(\omega)) \in A$ holds P-a.s. on πA.*

Proof: See Dellacherie and Meyer (1975), Section III.44. □

A2. Some Special Spaces

Here we collect some basic facts about various set, measure, and function spaces of importance in probability theory. Though random processes with paths in $C(\mathbb{R}_+, \mathbb{R}^d)$ or $D(\mathbb{R}_+, \mathbb{R}^d)$ and random measures on a variety of spaces are considered throughout the book, most of the topological results mentioned here are not needed until Chapter 16, where they play a fundamental role for the theory of convergence in distribution. Our plan is to begin with the basic function spaces and then move on to some spaces of measures and sets. Whenever appropriate accounts are available in the literature, we omit the proofs.

We begin with a well-known classical result. On any space of functions $x: K \to S$, we introduce the *evaluation maps* $\pi_t: x \mapsto x_t$, $t \in K$. Given some metrics d in K and ρ in S, we define the associated *modulus of continuity* by

$$w(x, h) = \sup\{\rho(x_s, x_t);\ d(s, t) \leq h\}, \quad h > 0.$$

Theorem A2.1 *(equicontinuity and compactness, Arzelà, Ascoli)* Fix two metric spaces K and S, where K is compact and S is complete, and let D be dense in K. Then a set $A \subset C(K,S)$ is relatively compact iff $\pi_t A$ is relatively compact in S for every $t \in D$ and
$$\lim_{h \to 0} \sup_{x \in A} w(x,h) = 0.$$
In that case, even $\bigcup_{t \in K} \pi_t A$ is relatively compact in S.

Proof: See Dudley (1989), Section 2.4. □

Next we fix a separable, complete metric space (S, ρ) and consider the space $D(\mathbb{R}_+, S)$ of functions $x \colon \mathbb{R}_+ \to S$ that are right-continuous with left-hand limits (rcll). It is easy to see that, for any $\varepsilon, t > 0$, such a function x has at most finitely many jumps of size $> \varepsilon$ before time t. In $D(\mathbb{R}_+, S)$ we introduce the modified modulus of continuity
$$\tilde{w}(x,t,h) = \inf_{(I_k)} \max_k \sup_{r,s \in I_k} \rho(x_r, x_s), \quad x \in D(\mathbb{R}_+, S),\ t,h > 0, \quad (1)$$
where the infimum extends over all partitions of the interval $[0, t)$ into subintervals $I_k = [u, v)$ such that $v - u \geq h$ when $v < t$. Note that $\tilde{w}(x,t,h) \to 0$ as $h \to 0$ for fixed $x \in D(\mathbb{R}_+, S)$ and $t > 0$. By a *time-change* on $\mathbb{R}_+$ we mean a monotone bijection $\lambda \colon \mathbb{R}_+ \to \mathbb{R}_+$. Note that λ is continuous and strictly increasing with $\lambda_0 = 0$ and $\lambda_\infty = \infty$.

Theorem A2.2 *(J_1-topology, Skorohod, Prohorov, Kolmogorov)* Fix a separable, complete metric space (S, ρ) and a dense set $T \subset \mathbb{R}_+$. Then there exists a separable and complete metric d in $D(\mathbb{R}_+, S)$ such that $d(x_n, x) \to 0$ iff
$$\sup_{s \leq t} |\lambda_n(s) - s| + \sup_{s \leq t} \rho(x_n \circ \lambda_n(s), x(s)) \to 0, \quad t > 0,$$
for some time-changes λ_n on $\mathbb{R}_+$. Furthermore, $\mathcal{B}(D(\mathbb{R}_+, S)) = \sigma\{\pi_t; t \in T\}$ and a set $A \subset D(\mathbb{R}_+, S)$ is relatively compact iff $\pi_t A$ is relatively compact in S for every $t \in T$ and
$$\lim_{h \to 0} \sup_{x \in A} \tilde{w}(x,t,h) = 0, \quad t > 0. \quad (2)$$
In that case, $\bigcup_{s \leq t} \pi_s A$ is relatively compact in S for every $t \geq 0$.

Proof: See Either and Kurtz (1986), Sections 3.5 and 3.6, or Jacod and Shiryaev (1987), Section VI.1. □

A suitably modified version of the last result applies to the space $D([0,1], S)$. Here we define $\tilde{w}(x,h)$ in terms of partitions of $[0,1)$ into subintervals of length $\geq h$ and use time-changes λ that are increasing bijections on $[0,1]$.

Turning to the case of measure spaces, let S be a locally compact, second-countable Hausdorff (lcscH) space S with Borel σ-field $\mathcal{S}$, and let $\hat{\mathcal{S}}$ denote the class of *bounded* (i.e., relatively compact) sets in $\mathcal{S}$. The space S is known to be Polish, and the family C_K^+ of continuous functions $f \colon S \to \mathbb{R}_+$ with compact support is separable in the uniform metric. Furthermore, there exists a sequence of compact sets $K_n \uparrow S$ such that $K_n \subset K_{n+1}^\circ$ for each n.

Let $\mathcal{M}(S)$ denote the class of measures on S that are *locally finite* (i.e., finite on $\hat{S}$), and write π_B and π_f for the mappings $\mu \mapsto \mu B$ and $\mu \mapsto \mu f = \int f d\mu$, respectively, on $\mathcal{M}(S)$. The *vague topology* in $\mathcal{M}(S)$ is generated by the maps π_f, $f \in C_K^+$, and we write the vague convergence of μ_n toward μ as $\mu_n \xrightarrow{v} \mu$. For any $\mu \in \mathcal{M}(S)$ we define $\hat{S}_\mu = \{B \in \hat{S};\ \mu \partial B = 0\}$.

Here we list some basic facts about the vague topology.

Theorem A2.3 *(vague topology)* *For any lcscH space S, we have*

(i) $\mathcal{M}(S)$ *is Polish in the vague topology;*

(ii) *a set $A \subset \mathcal{M}(S)$ is vaguely relatively compact iff* $\sup_{\mu \in A} \mu f < \infty$ *for all $f \in C_K^+$;*

(iii) *if $\mu_n \xrightarrow{v} \mu$ and $B \in \hat{S}$ with $\mu \partial B = 0$, then $\mu_n B \to \mu B$;*

(iv) $\mathcal{B}(\mathcal{M}(S))$ *is generated by the maps π_f, $f \in C_K^+$, and also for any $m \in \mathcal{M}(S)$ by the maps π_B, $B \in \hat{S}_m$.*

Proof: (i) Let $f_1, f_2, \ldots$ be dense in C_K^+, and define
$$\rho(\mu, \nu) = \sum_k 2^{-k}(|\mu f_k - \nu f_k| \wedge 1), \quad \mu, \nu \in \mathcal{M}(S). \tag{3}$$

It is easily seen that ρ metrizes the vague topology. In particular, $\mathcal{M}(S)$ is homeomorphic to a subset of $\mathbb{R}^\infty$ and therefore separable. The completeness of ρ will be clear once we have proved (ii).

(ii) The necessity is clear from the continuity of π_f for each $f \in C_K^+$. Conversely, assume that $\sup_{\mu \in A} \mu f < \infty$ for all $f \in C_K^+$. Choose some compact sets $K_n \uparrow S$ with $K_n \subset K_{n+1}^\circ$ for each n, and let the functions $f_n \in C_K^+$ be such that $1_{K_n} \leq f_n \leq 1_{K_{n+1}}$. For each n the set $\{f_n \cdot \mu;\ \mu \in A\}$ is uniformly bounded, and so by Theorem 16.3 it is even sequentially relatively compact. A diagonal argument then shows that A itself is sequentially relatively compact. Since $\mathcal{M}(S)$ is metrizable, the desired relative compactness follows.

(iii) The proof is the same as for Theorem 4.25.

(iv) A topological basis in $\mathcal{M}(S)$ is formed by all finite intersections of the sets $\{\mu;\ a < \mu f < b\}$ with $0 < a < b$ and $f \in C_K^+$. Furthermore, since $\mathcal{M}(S)$ is separable, every vaguely open set is a countable union of basis elements. Thus, $\mathcal{B}(\mathcal{M}(S)) = \sigma\{\pi_f;\ f \in C_K^+\}$. By a simple approximation and monotone class argument it follows that $\mathcal{B}(\mathcal{M}(S)) = \sigma\{\pi_B;\ B \in \hat{S}\}$.

Now fix any $m \in \hat{S}$, put $\mathcal{A} = \sigma\{\pi_B;\ B \in \hat{S}_m\}$, and let $\mathcal{D}$ denote the class of all $D \in \hat{S}$ such that π_D is $\mathcal{A}$-measurable. Fixing a metric d in S such that all d-bounded closed sets are compact, we note that only countably many d-spheres around a fixed point have positive m-measure. Thus, $\hat{S}_m$ contains a topological basis. We also note that $\hat{S}_m$ is closed under finite unions, whereas $\mathcal{D}$ is closed under bounded increasing limits. Since S is separable, it follows that $\mathcal{D}$ contains every open set $G \in \hat{S}$. For any such

G, the class $\mathcal{D} \cap G$ is a λ-system containing the π-system of all open sets in G, and by a monotone class argument we get $\mathcal{D} \cap G = \hat{\mathcal{S}} \cap G$. It remains to let $G \uparrow S$. □

Next we consider the space of all measure-valued rcll functions. Here we may characterize compactness in terms of countably many one-dimensional projections, a result needed for the proof of Theorem 16.27.

Theorem A2.4 *(measure-valued functions)* *For any lcscH space S, there exist some $f_1, f_2, \cdots \in C_K^+(S)$ such that a set $A \subset D(\mathbb{R}_+, \mathcal{M}(S))$ is relatively compact iff $Af_j = \{xf_j; x \in A\}$ is relatively compact in $D(\mathbb{R}_+, \mathbb{R}_+)$ for every $j \in \mathbb{N}$.*

Proof: If A is relatively compact, then so is Af for every $f \in C_K^+(S)$, since the map $x \mapsto xf$ is continuous from $D(\mathbb{R}_+, \mathcal{M}(S))$ to $D(\mathbb{R}_+, \mathbb{R}_+)$. To prove the converse, choose a dense collection $f_1, f_2, \cdots \in C_K^+(S)$, closed under addition, and assume that Af_j is relatively compact for every j. In particular, $\sup_{x \in A} x_t f_j < \infty$ for all $t \geq 0$ and $j \in \mathbb{N}$, and so by Theorem A2.3 the set $\{x_t; x \in A\}$ is relatively compact in $\mathcal{M}(S)$ for every $t \geq 0$. By Theorem A2.2 it remains to verify (2), where $\tilde{w}$ is defined in terms of the complete metric ρ in (3).

If (2) fails, then either we may choose some $x^n \in A$ and $t_n \to 0$ with $\limsup_n \rho(x_{t_n}^n, x_0^n) > 0$, or else there exist some $x^n \in A$ and some bounded $s_t < t_n < u_n$ with $u_n - s_n \to 0$ such that

$$\limsup_{n \to \infty} \left(\rho(x_{s_n}^n, x_{t_n}^n) \wedge \rho(x_{t_n}^n, x_{u_n}^n) \right) > 0. \tag{4}$$

In the former case it is clear from (3) that $\limsup_n |x_{t_n}^n f_j - x_0^n f_j| > 0$ for some $j \in \mathbb{N}$, which contradicts the relative compactness of Af_j.

Next assume (4). By (3) there exist some $i, j \in \mathbb{N}$ such that

$$\limsup_{n \to \infty} \left(|x_{s_n}^n f_i - x_{t_n}^n f_i| \wedge |x_{t_n}^n f_j - x_{u_n}^n f_j| \right) > 0. \tag{5}$$

Now for any $a, a', b, b' \in \mathbb{R}$, we have

$$\tfrac{1}{2}(|a| \wedge |b'|) \leq (|a| \wedge |a'|) \vee (|b| \wedge |b'|) \vee (|a + a'| \wedge |b + b'|).$$

Since the set $\{f_k\}$ is closed under addition, (5) then implies the same relation with a common $i = j$. But then (2) fails for Af_i, which by Theorem A2.2 contradicts the relative compactness of Af_i. Thus, (2) does hold for A, and so A is relatively compact. □

Given an lcscH space S, we introduce the classes $\mathcal{G}$, $\mathcal{F}$, and $\mathcal{K}$ of open, closed, and compact subsets, respectively. Here we may consider $\mathcal{F}$ as a space in its own right, endowed with the *Fell topology* generated by the sets $\{F \in \mathcal{F}; F \cap G \neq \emptyset\}$ and $\{F \in \mathcal{F}; F \cap K = \emptyset\}$ for arbitrary $G \in \mathcal{G}$ and $K \in \mathcal{K}$. To describe the corresponding notion of convergence, we may fix a metrization ρ of the topology in S such that every closed ρ-ball is compact.

Theorem A2.5 *(Fell topology) Fix any lcscH space S, and let $\mathcal{F}$ be the class of closed sets $F \subset S$, endowed with the Fell topology. Then*

(i) *$\mathcal{F}$ is compact, second-countable, and Hausdorff;*

(ii) *$F_n \to F$ in $\mathcal{F}$ iff $\rho(s, F_n) \to \rho(s, F)$ for all $s \in S$;*

(iii) *$\{F \in \mathcal{F};\ F \cap B \neq \emptyset\}$ is universally Borel measurable for every $B \in \hat{S}$.*

Proof: First we show that the Fell topology is generated by the maps $F \mapsto \rho(s, F)$, $s \in S$. To see that those mappings are continuous, put $B_{s,r} = \{t \in S;\ \rho(s,t) < r\}$, and note that

$$\{F;\ \rho(s, F) < r\} = \{F;\ F \cap B_s^r \neq \emptyset\},$$
$$\{F;\ \rho(s, F) > r\} = \{F;\ F \cap \bar{B}_s^r = \emptyset\}.$$

Here the sets on the right are open, by the definition of the Fell topology and the choice of ρ. Thus, the Fell topology contains the ρ-topology.

To prove the converse, fix any $F \in \mathcal{F}$ and a net $\{F_i\} \subset \mathcal{F}$ with directed index set $(I, \prec)$ such that $F_i \to F$ in the ρ-topology. We need to show that convergence holds even in the Fell topology. Then let $G \in \mathcal{G}$ be arbitrary with $F \cap G \notin \emptyset$. Fix any $s \in F \cap G$. Since $\rho(s, F_i) \to \rho(s, F) = 0$, we may further choose some $s_i \in F_i$ with $\rho(s, s_i) \to 0$. Since G is open, there exists some $i \in I$ such that $s_j \in G$ for all $j \succ i$. Then also $F_j \cap G \notin \emptyset$ for all $j \succ i$.

Next consider any $K \in \mathcal{K}$ with $F \cap K = \emptyset$. Define $r_s = \tfrac{1}{2}\rho(s, F)$ for each $s \in K$ and put $G_s = B_{s, r_s}$. Since K is compact, it is covered by finitely many balls G_{s_k}. For each k we have $\rho(s_k, F_i) \to \rho(s_k, F)$, and so there exists some $i_k \in I$ such that $F_j \cap G_{s_k} = \emptyset$ for all $j \succ i_k$. Letting $i \in I$ be such that $i \succ i_k$ for all k, it is clear that $F_j \cap K = \emptyset$ for all $j \succ i$.

Now we fix any countable dense set $D \subset S$, and assume that $\rho(s, F_i) \to \rho(s, F)$ for all $s \in D$. For any $s, s' \in S$ we have

$$|\rho(s, F_j) - \rho(s, F)| \leq |\rho(s', F_j) - \rho(s', F)| + 2\rho(s, s').$$

Given any s and $\varepsilon > 0$, we can make the left-hand side $< \varepsilon$, by choosing an $s' \in D$ with $\rho(s, s') < \varepsilon/3$ and then an $i \in I$ such that $|\rho(s', F_j) - \rho(s', F)| < \varepsilon/3$ for all $j \succ i$. This shows that the Fell topology is also generated by the mappings $F \mapsto \rho(s, F)$ with s restricted to D. But then $\mathcal{F}$ is homeomorphic to a subset of $\overline{\mathbb{R}_+}^\infty$, which is second-countable and metrizable.

To prove that $\mathcal{F}$ is compact, it is now enough to show that every sequence $(F_n) \subset \mathcal{F}$ contains a convergent subsequence. Then choose a subsequence such that $\rho(s, F_n)$ converges in $\overline{\mathbb{R}_+}$ for all $s \in D$, and hence also for all $s \in S$. Since the family of functions $\rho(s, F_n)$ is equicontinuous, even the limit f is continuous, and so the set $F = \{s \in S;\ f(s) = 0\}$ is closed.

To obtain $F_n \to F$, we need to show that whenever $F \cap G \neq \emptyset$ or $F \cap K = \emptyset$ for some $G \in \mathcal{G}$ or $K \in \mathcal{K}$, the same relation eventually holds even for F_n. In the former case, we may fix any $s \in F \cap G$ and note that $\rho(s, F_n) \to f(s) = 0$. Hence, we may choose some $s_n \in F_n$ with $s_n \to s$, and since $s_n \in G$ for large n, we get $F_n \cap G \neq \emptyset$. In the latter case, we assume that instead $F_n \cap K \neq \emptyset$ along a subsequence. Then there exist some

$s_n \in F_n \cap K$, and we note that $s_n \to s \in K$ along a further subsequence. Here $0 = \rho(s_n, F_n) \to \rho(s, F)$, which yields the contradiction $s \in F \cap K$. This completes the proof of (i).

To prove (iii), we note that the mapping $(s, F) \mapsto \rho(s, F)$ is jointly continuous and hence Borel measurable. Now S and $\mathcal{F}$ are both separable, and so the Borel σ-field in $S \times \mathcal{F}$ agrees with the product σ-field $\mathcal{S} \otimes \mathcal{B}(\mathcal{F})$. Since $s \in F$ iff $\rho(s, F) = 0$, it follows that $\{(s, F); s \in F\}$ belongs to $\mathcal{S} \otimes \mathcal{B}(\mathcal{F})$. Hence, so does $\{(s, F); s \in F \cap B\}$ for arbitrary $B \in \mathcal{S}$. The assertion now follows by Theorem A1.4. □

We say that a class $\mathcal{U} \subset \hat{\mathcal{S}}$ is *separating* if for any $K \subset G$ with $K \in \mathcal{K}$ and $G \in \mathcal{G}$ there exists some $U \in \mathcal{U}$ with $K \subset U \subset G$. A *preseparating* class $\mathcal{I} \subset \hat{\mathcal{S}}$ is such that the finite unions of $\mathcal{I}$-sets form a separating class. When S is Euclidean, we typically choose $\mathcal{I}$ to be a class of intervals or rectangles and $\mathcal{U}$ as the corresponding class of finite unions.

Lemma A2.6 (separation) *For any monotone function $h: \hat{\mathcal{S}} \to \mathbb{R}$, the class $\hat{\mathcal{S}}_h = \{B \in \hat{\mathcal{S}}; h(B^\circ) = h(\overline{B})\}$ is separating.*

Proof: Fix a metric ρ in S such that every closed ρ-ball is compact, and let $K \in \mathcal{K}$ and $G \in \mathcal{G}$ with $K \subset G$. For any $\varepsilon > 0$, define $K_\varepsilon = \{s \in S; d(s, K) < \varepsilon\}$ and note that $\overline{K}_\varepsilon = \{s \in S; \rho(s, K) \leq \varepsilon\}$. Since K is compact, we have $\rho(K, G^c) > 0$, and so $K \subset K_\varepsilon \subset G$ for sufficiently small $\varepsilon > 0$. From the monotonicity of h it is further clear that $K_\varepsilon \in \hat{\mathcal{S}}_h$ for almost every $\varepsilon > 0$. □

We often need the separating class to be countable.

Lemma A2.7 (countable separation) *Every separating class $\mathcal{U} \subset \hat{\mathcal{S}}$ contains a countable separating subclass.*

Proof: Fix a countable topological base $\mathcal{B} \subset \hat{\mathcal{S}}$, closed under finite unions. Choose for every $B \in \mathcal{B}$ some compact sets $K_{B,n} \downarrow \overline{B}$ with $K^\circ_{B,n} \supset \overline{B}$, and then for each pair $(B, n) \in \mathcal{B} \times \mathbb{N}$ some set $U_{B,n} \in \mathcal{U}$ with $\overline{B} \subset U_{B,n} \subset K^\circ_{B,n}$. The family $\{U_{B,n}\}$ is clearly separating. □

The next result, needed for the proof of Theorem 16.29, relates the vague and Fell topologies for integer-valued measures and their supports. Let $\mathcal{N}(S)$ denote the class of locally finite, integer-valued measures on S, and write $\xrightarrow{f}$ for convergence in the Fell topology.

Proposition A2.8 (supports of measures) *Let $\mu, \mu_1, \mu_2, \cdots \in \mathcal{N}(S)$ with $\operatorname{supp} \mu_n \xrightarrow{f} \operatorname{supp} \mu$, where S is lcscH and μ is simple. Then*
$$\limsup_{n \to \infty}(\mu_n B \wedge 1) \leq \mu B \leq \liminf_{n \to \infty} \mu_n B, \quad B \in \hat{\mathcal{S}}_\mu.$$

Proof: To prove the left inequality, we may assume that $\mu B = 0$. Since $B \in \hat{\mathcal{S}}_\mu$, we have even $\mu \overline{B} = 0$, and so $\overline{B} \cap \operatorname{supp} \mu = \emptyset$. By convergence of

the supports we get $\overline{B} \cap \operatorname{supp} \mu_n = \emptyset$ for large enough n, which implies
$$\limsup_{n\to\infty}(\mu_n B \wedge 1) \leq \limsup_{n\to\infty} \mu_n \overline{B} = 0 = \mu B.$$

To prove the right inequality, we may assume that $\mu B = m > 0$. Since $\hat{S}_\mu$ is a separating ring, we may choose a partition $B_1, \ldots, B_m \in \hat{S}_\mu$ of B such that $\mu B_k = 1$ for each k. Then also $\mu B_k^\circ = 1$ for each k, and so $B_k^\circ \cap \operatorname{supp} \mu \neq \emptyset$. By convergence of the supports we get $B_k^\circ \cap \operatorname{supp} \mu_n \neq \emptyset$ for large enough n. Hence,
$$1 \leq \liminf_{n\to\infty} \mu_n B_k^\circ \leq \liminf_{n\to\infty} \mu_n B_k,$$
and so
$$\begin{aligned}\mu B &= m \leq \sum_k \liminf_{n\to\infty} \mu_n B_k \\ &\leq \liminf_{n\to\infty} \sum_k \mu_n B_k = \liminf_{n\to\infty} \mu_n B.\end{aligned} \qquad \Box$$

To state the next result, fix any metric spaces $S_1, S_2, \ldots$, and introduce the product spaces $S^n = S_1 \times \cdots \times S_n$ and $S = S_1 \times S_2 \times \cdots$ endowed with their product topologies. For any $m < n < \infty$, let π_m and π_{mn} denote the natural projections of S and S^n onto S^m. The sets $A_n \subset S^n$, $n \in \mathbb{N}$, are said to form a *projective sequence* if $\pi_{mn} A_n \subset A_m$ for all $m \leq n$. We may then define their *projective limit* in S as the set $A = \bigcap_n \pi_n^{-1} A_n$.

Lemma A2.9 (*projective limits*) *For any metric spaces $S_1, S_2, \ldots$, consider a projective sequence of nonempty, compact sets $K_n \subset S_1 \times \cdots \times S_n$, $n \in \mathbb{N}$. Then the projective limit $K = \bigcap_n \pi_n^{-1} K_n$ is again nonempty and compact.*

Proof: Since the K_n are nonempty, we may choose some sequences $x^n = (x_m^n) \in \pi_n^{-1} K_n$, $n \in \mathbb{N}$. By the projective property of the sets K_m, we have $\pi_m x^n \in K_m$ for all $m \leq n$. In particular, the sequence $x_m^1, x_m^2, \ldots$ is relatively compact in S_m for each $m \in \mathbb{N}$, and by a diagonal argument we may choose a subsequence $N' \subset \mathbb{N}$ and an element $x = (x_m) \in S$ such that $x^n \to x$ as $n \to \infty$ along N'. Then also $\pi_m x^n \to \pi_m x$ along N' for each $m \in \mathbb{N}$, and since the K_m are closed, we conclude that $\pi_m x \in K_m$ for all m. Thus, we have $x \in K$, which shows that K is nonempty. The compactness of K may be proved by the same argument, where we assume that $x^1, x^2, \ldots \in K$. $\qquad \Box$

Historical and Bibliographical Notes

The following notes were prepared with the modest intentions of tracing the origins of some of the basic ideas in each chapter, of giving precise references for the main results cited in the text, and of suggesting some literature for further reading. No completeness is claimed, and knowledgeable readers are likely to notice misinterpretations and omissions, for which I appologize in advance. A comprehensive history of modern probability theory still remains to be written.

1. Measure Theory — Basic Notions

The first author to consider measures in the modern sense was BOREL (1895, 1898), who constructed Lebesgue measure on the Borel σ-field in $\mathbb{R}$. The corresponding integral was introduced by LEBESGUE (1902, 1904), who also established the dominated convergence theorem. The monotone convergence theorem and Fatou's lemma were later obtained by LEVI (1906a) and FATOU (1906), respectively. LEBESGUE also introduced the higher-dimensional Lebesgue measure and proved a first version of Fubini's theorem, subsequently generalized by FUBINI (1907) and TONELLI (1909). The integration theory was extended to general measures and abstract spaces by many authors, including RADON (1913) and FRÉCHET (1928).

The norm inequalities in Lemma 1.29 were first noted for finite sums by HÖLDER (1889) and MINKOWSKI (1907), respectively, and were later extended to integrals by RIESZ (1910). Part (i) for $p = 2$ goes back to CAUCHY (1821) for finite sums and to BUNIAKOWSKY (1859) for integrals. The Hilbert space projection theorem can be traced back to LEVI (1906b).

The monotone class Theorem 1.1 was first proved, along with related results, already by SIERPIŃSKI (1928), but the result was not used in probability theory until DYNKIN (1961). More primitive versions had previously been employed by HALMOS (1950) and DOOB (1953).

Most results in this chapter are well known and can be found in any textbook on real analysis. Many probability texts, including LOÈVE (1977) and BILLINGSLEY (1995), contain detailed introductions to measure theory. There are also some excellent texts in real analysis adapted to the needs of probabilists, such as DUDLEY (1989) and DOOB (1994). The former author also provides some more detailed historical information.

2. Measure Theory — Key Results

As we have seen, BOREL (1995, 1998) was the first to prove the existence of one-dimensional Lebesgue measure. However, the modern construction via outer measures in due to CARATHÉODORY (1918).

Functions of bounded variation were introduced by JORDAN (1881), who proved that any such function is the difference of two nondecreasing functions. The corresponding decomposition of signed measures was obtained by HAHN (1921). Integrals with respect to nondecreasing functions were defined by STIELTJES (1894), but their importance was not recognized until RIESZ (1909b) proved his representation theorem for linear functionals on $C[0, 1]$. The a.e. differentiability of a function of bounded variation was first proved by LEBESGUE (1904).

VITALI (1905) was the first author to see the connection between absolute continuity and the existence of a density. The Radon–Nikodým theorem was then proved in increasing generality by RADON (1913), DANIELL (1920), and NIKODÝM (1930). The idea of a combined proof that also establishes the Lebesgue decomposition is due to VON NEUMANN.

Invariant measures on specific groups were early identified through explicit computation by many authors, notably by HURWITZ (1897) for the case of $SO(n)$. HAAR (1933) proved the existence (but not the uniqueness) of invariant measures on an arbitrary lcscH group. The modern treatment originated with WEIL (1940), and excellent expositions can be found in many books on real or harmonic analysis. Invariant measures on more general spaces are usually approached via quotient spaces. Our discussion in Theorem 2.29 is adapted from ROYDEN (1988).

3. Processes, Distributions, and Independence

The use of countably additive probability measures dates back to BOREL (1909), who constructed random variables as measurable functions on the Lebesgue unit interval and proved Theorem 3.18 for independent events. CANTELLI (1917) noticed that the "easy" part remains true without the independence assumption. Lemma 3.5 was proved by JENSEN (1906) after HÖLDER had obtained a special case.

The modern framework, with random variables as measurable functions on an abstract probability space $(\Omega, \mathcal{A}, P)$ and with expected values as P-integrals over Ω, was used implicitly by KOLMOGOROV from (1928) on and was later formalized in KOLMOGOROV (1933). The latter monograph also contains Kolmogorov's zero–one law, discovered long before HEWITT and SAVAGE (1955) obtained theirs.

Early work in probability theory deals with properties depending only on the finite-dimensional distributions. WIENER (1923) was the first author to construct the distribution of a process as a measure on a function space. The general continuity criterion in Theorem 3.23, essentially due to KOL-

MOGOROV, was first published by SLUTSKY (1937), with minor extensions later added by LOÈVE (1978) and CHENTSOV (1956). The general search for regularity properties was initiated by DOOB (1937, 1947). Soon it became clear, especially through the work of LÉVY (1934–35, 1954), DOOB (1951, 1953), and KINNEY (1953), that most processes of interest have right-continuous versions with left-hand limits.

More detailed accounts of the material in this chapter appear in many textbooks, such as in BILLINGSLEY (1995), ITÔ (1984), and WILLIAMS (1991). Further discussions of specific regularity properties appear in LOÈVE (1977) and CRAMÉR and LEADBETTER (1967). Earlier texts tend to give more weight to distribution functions and their densities, less weight to measures and σ-fields.

4. Random Sequences, Series, and Averages

The weak law of large numbers was first obtained by BERNOULLI (1713) for the sequences named after him. More general versions were then established with increasing rigor by BIENAYMÉ (1853), CHEBYSHEV (1867), and MARKOV (1899). A necessary and sufficient condition for the weak law of large numbers was finally obtained by KOLMOGOROV (1928–29).

KHINCHIN and KOLMOGOROV (1925) studied series of independent, discrete random variables and showed that convergence holds under the condition in Lemma 4.16. KOLMOGOROV (1928–29) then obtained his maximum inequality and showed that the three conditions in Theorem 4.18 are necessary and sufficient for a.s. convergence. The equivalence with convergence in distribution was later noted by LÉVY (1954).

The strong law of large numbers for Bernoulli sequences was stated by BOREL (1909), but the first rigorous proof is due to FABER (1910). The simple criterion in Corollary 4.22 was obtained in KOLMOGOROV (1930). In (1933) KOLMOGOROV showed that existence of the mean is necessary and sufficient for the strong law of large numbers for general i.i.d. sequences. The extension to exponents $p \neq 1$ is due to MARCINKIEWICZ and ZYGMUND (1937). Proposition 4.24 was proved in stages by GLIVENKO (1933) and CANTELLI (1933).

RIESZ (1909a) introduced the notion of convergence in measure, for probability measures equivalent to convergence in probability, and showed that it implies a.e. convergence along a subsequence. The weak compactness criterion in Lemma 4.13 is due to DUNFORD (1939). The functional representation of Proposition 4.31 appeared in KALLENBERG (1996a), and Corollary 4.32 was given by STRICKER and YOR (1978).

The theory of weak convergence was founded by ALEXANDROV (1940–43), who proved in particular the so-called Portmanteau Theorem 4.25. The continuous mapping Theorem 4.27 was obtained for a single function $f_n \equiv f$ by MANN and WALD (1943) and then in the general case by PROHOROV

(1956) and RUBIN. The coupling Theorem 4.30 is due for complete S to SKOROHOD (1956) and in general to DUDLEY (1968).

More detailed accounts of the material in this chapter may be found in many textbooks, such as in LOÈVE (1977) and CHOW and TEICHER (1997). Additional results on random series and a.s. convergence appear in STOUT (1974) and KWAPIEŃ and WOYCZYŃSKI (1992).

5. Characteristic Functions and Classical Limit Theorems

The central limit theorem (a name first used by PÓLYA (1920)) has a long and glorious history, beginning with the work of DE MOIVRE (1733–56), who obtained the now-familiar approximation of binomial probabilities in terms of the normal density function. LAPLACE (1774, 1812–20) stated the general result in the modern integrated form, but his proof was incomplete, as was the proof of CHEBYSHEV (1867, 1890).

The first rigorous proof was given by LIAPOUNOV (1901), though under an extra moment condition. Then LINDEBERG (1922a) proved his fundamental Theorem 5.12, which in turn led to the basic Proposition 5.9 in a series of papers by LINDEBERG (1922b) and LÉVY (1922a-c). BERNSTEIN (1927) obtained the first extension to higher dimensions. The general problem of normal convergence, regarded for two centuries as the central (indeed the only) theoretical problem in probability, was eventually solved in the form of Theorem 5.15, independently by FELLER (1935) and LÉVY (1935a). Slowly varying functions were introduced and studied by KARAMATA (1930).

Though characteristic functions have been used in probability theory ever since LAPLACE (1812–20), their first use in a rigorous proof of a limit theorem had to wait until LIAPOUNOV (1901). The first general continuity theorem was established by LÉVY (1922c), who assumed the characteristic functions to converge uniformly in some neighborhood of the origin. The definitive version in Theorem 5.22 is due to BOCHNER (1933). Our direct approach to Theorem 5.3 may be new, in avoiding the relatively deep HELLY selection theorem (1911–12). The basic Corollary 5.5 was noted by CRAMÉR and WOLD (1936).

Introductions to characteristic functions and classical limit theorems may be found in many textbooks, notably LOÈVE (1977). FELLER (1971) is a rich source of further information on Laplace transforms, characteristic functions, and classical limit theorems. For more detailed or advanced results on characteristic functions, see LUKACS (1970).

6. Conditioning and Disintegration

Though conditional densities have been computed by statisticians ever since LAPLACE (1774), the first general approach to conditioning was devised by KOLMOGOROV (1933), who defined conditional probabilities and expectations as random variables on the basic probability space, using the Radon–Nikodým theorem, which had recently become available. His original notion of conditioning with respect to a random vector was extended by HALMOS (1950) to general random elements and then by DOOB (1953) to abstract sub-σ-fields.

Our present Hilbert space approach to conditioning, essentially due to VON NEUMANN (1940), is more elementary and intuitive and avoids the use of the relatively deep Radon–Nikodým theorem. It has the further advantage of leading to the attractive interpretation of a martingale as a projective family of random variables.

The existence of regular conditional distributions was studied by several authors, beginning with DOOB (1938). It leads immediately to the familiar disintegration of measures on product spaces and to the frequently used but rarely stated disintegration Theorem 6.4.

Measures on infinite product spaces were first considered by DANIELL (1918–19, 1919–20), who proved the extension Theorem 6.14 for countable product spaces. KOLMOGOROV (1933) extended the result to arbitrary index sets. LOMNICKI and ULAM (1934) noted that no topological assumptions are needed for the construction of infinite product measures, a result that was later extended by C.T. IONESCU TULCEA (1949–50) to measures specified by a sequence of conditional distributions.

The interpretation of the simple Markov property in terms of conditional independence was indicated already by MARKOV (1906), and the formal statement of Proposition 6.6 appears in DOOB (1953). Further properties of conditional independence have been listed by DÖHLER (1980) and others. The transfer Theorem 6.10, in the present form quoted from KALLENBERG (1988), may have been first noted by THORISSON.

The traditional Radon–Nikodým approach to conditional expectations appears in many textbooks, such as in BILLINGSLEY (1995).

7. Martingales and Optional Times

Martingales were first introduced by BERNSTEIN (1927, 1937) in his efforts to relax the independence assumption in the classical limit theorems. Both BERNSTEIN and LÉVY (1935a-b, 1954) extended Kolmogorov's maximum inequality and the central limit theorem to a general martingale context. The *term* martingale (originally denoting part of a horse's harness and later used for a special gambling system) was introduced in the probabilistic context by VILLE (1939).

The first martingale convergence theorem was obtained by JESSEN (1934) and LÉVY (1935b), both of whom proved Theorem 7.23 for filtrations generated by sequences of independent random variables. A submartingale version of the same result appears in SPARRE-ANDERSEN and JESSEN (1948). The independence assumption was removed by LÉVY (1954), who also noted the simple martingale proof of Kolmogorov's zero–one law and obtained his conditional version of the Borel–Cantelli lemma.

The general convergence theorem for discrete-time martingales was proved by DOOB (1940), and the basic regularity theorems for continuous-time martingales first appeared in DOOB (1951). The theory was extended to submartingales by SNELL (1952) and DOOB (1953). The latter book is also the original source of such fundamental results as the martingale closure theorem, the optional sampling theorem, and the L^p-inequality.

Though hitting times have long been used informally, general optional times seem to appear for the first time in DOOB (1936). Abstract filtrations were not introduced until DOOB (1953). Progressive processes were introduced by DYNKIN (1961), and the modern definition of the σ-fields $\mathcal{F}_\tau$ is due to YUSHKEVICH.

Elementary introductions to martingale theory are given by many authors, including WILLIAMS (1991). More information about the discrete-time case is given by NEVEU (1975) and CHOW and TEICHER (1997). For a detailed account of the continuous-time theory and its relations to Markov processes and stochastic calculus, see DELLACHERIE and MEYER (1975–87).

8. Markov Processes and Discrete-Time Chains

Markov chains in discrete time and with finitely many states were introduced by MARKOV (1906), who proved the first ergodic theorem, assuming the transition probabilities to be strictly positive. KOLMOGOROV (1936a-b) extended the theory to countable state spaces and arbitrary transition probabilities. In particular, he noted the decomposition of the state space into irreducible sets, classified the states with respect to recurrence and periodicity, and described the asymptotic behavior of the n-step transition probabilities. Kolmogorov's original proofs were analytic. The more intuitive coupling approach was introduced by DOEBLIN (1938), long before the strong Markov property had been formalized.

BACHELIER had noted the connection between random walks and diffusions, which inspired KOLMOGOROV (1931a) to give a precise definition of Markov processes in continuous time. His treatment is purely analytic, with the distribution specified by a family of transition kernels satisfying the Chapman–Kolmogorov relation, previously noted in special cases by CHAPMAN (1928) and SMOLUCHOVSKY.

KOLMOGOROV (1931a) makes no reference to sample paths. The transition to probabilistic methods began with the work of LÉVY (1934–35) and DOEBLIN (1938). Though the strong Markov property was used informally by those authors (and indeed already by BACHELIER (1900, 1901)), the result was first stated and proved in a special case by DOOB (1945). General filtrations were introduced in Markov process theory by BLUMENTHAL (1957). The modern setup, with a canonical process X defined on the path space Ω, equipped with a filtration $\mathcal{F}$, a family of shift operators θ_t, and a collection of probability measures P_x, was developed systematically by DYNKIN (1961, 1965). A weaker form of Theorem 8.23 appears in BLUMENTHAL and GETOOR (1968), and the present version is from KALLENBERG (1987, 1998).

Elementary introductions to Markov processes appear in many textbooks, such as ROGERS and WILLIAMS (2000a) and CHUNG (1982). More detailed or advanced accounts are given by DYNKIN (1965), BLUMENTHAL and GETOOR (1968), ETHIER and KURTZ (1986), DELLACHERIE and MEYER (1975–87), and SHARPE (1988). FELLER (1968) gives a masterly introduction to Markov chains, later imitated by many authors. More detailed accounts of the discrete-time theory appear in KEMENY et al. (1966) and FREEDMAN (1971a). The coupling method fell into oblivion after Doeblin's untimely death in 1940 but has recently enjoyed a revival, meticulously documented by LINDVALL (1992) and THORISSON (2000).

9. Random Walks and Renewal Theory

Random walks originally arose in a wide range of applications, such as gambling, queuing, storage, and insurance; their history can be traced back to the origins of probability. The approximation of diffusion processes by random walks dates back to BACHELIER (1900, 1901). A further application was to potential theory, where in the 1920s a method of discrete approximation was devised, admitting a probabilistic interpretation in terms of a simple symmetric random walk. Finally, random walks played an important role in the sequential analysis developed by WALD (1947).

The modern theory began with PÓLYA's (1921) discovery that a simple symmetric random walk on $\mathbb{Z}^d$ is recurrent for $d \leq 2$ and transient otherwise. His result was later extended to Brownian motion by LÉVY (1940) and KAKUTANI (1944a). The general recurrence criterion in Theorem 9.4 was derived by CHUNG and FUCHS (1951), and the probabilistic approach to Theorem 9.2 was found by CHUNG and ORNSTEIN (1962). The first condition in Corollary 9.7 is, in fact, even necessary for recurrence, as was noted independently by ORNSTEIN (1969) and C.J. STONE (1969).

The reflection principle was first used by ANDRÉ (1887) in his discussion of the ballot problem. The systematic study of fluctuation and absorption problems for random walks began with the work of POLLACZEK (1930). Ladder times and heights, first introduced by BLACKWELL, were explored in an influential paper by FELLER (1949). The factorizations in Theorem 9.15 were originally derived by the Wiener–Hopf technique, which had been developed by PALEY and WIENER (1934) as a general tool in Fourier analysis.

Theorem 9.16 is due for $u = 0$ to SPARRE-ANDERSEN (1953–54) and in general to BAXTER (1961). The former author used complicated combinatorial methods, which were later simplified by FELLER and others.

Though renewals in Markov chains are implicit already in some early work of KOLMOGOROV and LÉVY, the general renewal process was apparently first introduced by PALM (1943). The first renewal theorem was obtained by ERDÖS et al. (1949) for random walks on $\mathbb{Z}_+$. In that case, however, CHUNG noted that the result is an easy consequence of KOLMOGOROV's (1936a-b) ergodic theorem for Markov chains on a countable state space. BLACKWELL (1948, 1953) extended the result to random walks on $\mathbb{R}_+$. The ultimate version for transient random walks on $\mathbb{R}$ is due to FELLER and OREY (1961). The first coupling proof of Blackwell's theorem was given by LINDVALL (1977). Our proof is a modification of an argument by ATHREYA et al. (1978), which originally did not cover all cases. The method seems to require the existence of a possibly infinite mean. An analytic approach to the general case appears in FELLER (1971).

Elementary introductions to random walks are given by many authors, including CHUNG (1974), FELLER (1968, 1971), and LOÈVE (1977). A detailed exposition of random walks on $\mathbb{Z}^d$ is given by SPITZER (1976).

10. Stationary Processes and Ergodic Theory

The history of ergodic theory dates back to BOLTZMANN's (1887) work in statistical mechanics. Boltzmann's *ergodic hypothesis*—the conjectural equality between time and ensemble averages—was long accepted as a heuristic principle. In probabilistic terms it amounts to the convergence $t^{-1} \int_0^t f(X_s)\,ds \to Ef(X_0)$, where X_t represents the state of the system (typically the configuration of all molecules in a gas) at time t, and the expected value is computed with respect to a suitably invariant probability measure on a compact submanifold of the state space.

The ergodic hypothesis was sensationally proved as a mathematical theorem, first in an L^2-version by VON NEUMANN (1932), after KOOPMAN (1931) had noted the connection between measure-preserving transformations and unitary operators on a Hilbert space, and shortly afterwards in the pointwise form of BIRKHOFF (1932). The initially quite intricate proof of the latter was simplified in stages: first by YOSIDA and KAKUTANI (1939), who noted how the result follows easily from the maximal ergodic Lemma 10.7, and then by GARSIA (1965), who gave a short proof of the latter result. KHINCHIN (1933, 1934) pioneered a translation of the results of ergodic theory into the probabilistic setting of stationary sequences and processes.

The first multivariate ergodic theorem was obtained by WIENER (1939), who proved Theorem 10.14 in the special case of averages over concentric balls. More general versions were established by many authors, including DAY (1942) and PITT (1942). The classical methods were pushed to the limit in a notable paper by TEMPEL'MAN (1972). NGUYEN and ZESSIN (1979) proved versions of the theorem for finitely additive set functions. The first ergodic theorem for noncommutative transformations was obtained by ZYGMUND (1951). SUCHESTON (1983) noted that the statement follows easily from MAKER's (1940) result. In Lemma 10.15, part (i) is due to ROGERS and SHEPHARD (1958); part (ii) is elementary.

The ergodic theorem for random matrices was proved by FURSTENBERG and KESTEN (1960), long before the subadditive ergodic theorem became available. The latter result was originally proved by KINGMAN (1968) under the stronger hypothesis that the array $(X_{m,n})$ be jointly stationary in m and n. The present extension and shorter proof are due to LIGGETT (1985).

The ergodic decomposition of invariant measures dates back to KRYLOV and BOGOLIOUBOV (1937), though the basic role of the invariant σ-field was not recognized until the work of FARRELL (1962) and VARADARAJAN (1963). The connection between ergodic decompositions and sufficient statistics is explored in an elegant paper by DYNKIN (1978). The traditional approach to the subject is via Choquet theory, as surveyed by DELLACHERIE and MEYER (1975–87).

The coupling equivalences in Theorem 10.27 (i) were proved by S. GOLDSTEIN (1979), after GRIFFEATH (1975) had obtained a related result for Markov chains. The shift coupling part of the same theorem was established by BERBEE (1979) and ALDOUS and THORISSON (1993), and the version for abstract groups was then obtained by THORISSON (1996). The latter author surveyed the whole area in (2000).

Elementary introductions to stationary processes have been given by many authors, beginning with DOOB (1953) and CRAMÉR and LEADBETTER (1967). LOÈVE (1978) contains a more advanced account of probabilistic ergodic theory. A modern and comprehensive survey of the vast area of general ergodic theorems is given by KRENGEL (1985).

11. Related Notions of Symmetry and Invariance

Palm distributions are named after the Swedish engineer PALM (1943), who in a pioneering study of intensity fluctuations in telephone traffic considered some basic Palm probabilities associated with simple, stationary point processes on $\mathbb{R}$, using an elementary conditioning approach. Palm also derived some primitive inversion formulas. An extended and more rigorous account of Palm's ideas was given by KHINCHIN (1955), in a monograph on queuing theory.

Independently of Palm's work, KAPLAN (1955) first obtained Theorem 11.4 as an extension of some results for renewal processes by DOOB (1948). A partial discrete-time result in this direction had already been noted by KAC (1947). Kaplan's result was rediscovered in the setting of Palm distributions, independently by RYLL-NARDZEWSKI (1961) and SLIVNYAK (1962). In the special case of intervals on the real line, Theorem 11.5 (i) was first noted by KOROLYUK (as cited by KHINCHIN (1955)), and part (iii) of the same theorem was obtained by RYLL-NARDZEWSKI (1961). The general versions are due to KÖNIG and MATTHES (1963) and MATTHES (1963) for $d = 1$ and to MATTHES et al. (1978) for $d > 1$. A more primitive setwise version of Theorem 11.8 (i), due to SLIVNYAK (1962), was strengthened by ZÄHLE (1980) to convergence in total variation.

DE FINETTI (1930, 1937) proved that an infinite sequence of exchangeable random variables is mixed i.i.d. The result became a cornerstone in his theory of subjective probability and Bayesian statistics. RYLL-NARDZEWSKI (1957) noted that the theorem remains valid under the weaker hypothesis of spreadability, and BÜHLMANN (1960) extended the result to continuous time. The predictable sampling property in Theorem 11.13 was first noted by DOOB (1936) for i.i.d. random variables and increasing sequences of predictable times. The general result and its continuous-time counterpart appear in KALLENBERG (1988). SPARRE-ANDERSEN's (1953–54) announcement of his Corollary 11.14 was (according to Feller) "a sensation greeted with incredulity, and the original proof was of an extraordinary intricacy and complexity." A simplified argument (different from ours) appears in FELLER (1971). Lemma 11.9 is quoted from KALLENBERG (1999b).

BERTRAND (1887) noted that if two candidates A and B in an election get the proportions p and $1-p$ of the votes, then the probability that A will lead throughout the counting of ballots equals $(2p - 1) \vee 0$. More general "ballot theorems" and alternative proofs have been discovered by many authors, beginning with ANDRÉ (1887) and BARBIER (1887). TAKÁCS (1967) obtained the version for cyclically stationary processes on a finite interval and gave numerous applications to queuing theory. The present statement is cited from KALLENBERG (1999a).

The first version of Theorem 11.18 was obtained by SHANNON (1948), who proved the convergence in probability for stationary and ergodic Markov chains in a finite state space. The Markovian restriction was lifted by MCMILLAN (1953), who also strengthened the result to convergence in L^1. CARLESON (1958) extended McMillan's result to countable state spaces. The a.s. convergence is due to BREIMAN (1957–60) and A. IONESCU TULCEA (1960) for finite state spaces and to CHUNG (1961) for the countable case.

More information about Palm measures is available in MATTHES et al. (1978), DALEY and VERE-JONES (1988), and THORISSON (2000). Applications to queuing theory and other areas are discussed by many authors, including FRANKEN et al. (1981) and BACCELLI and BRÉMAUD (1994). ALDOUS (1985) gives a comprehensive survey of exchangeability theory. A nice introduction to information theory is given by BILLINGSLEY (1965).

12. Poisson and Pure Jump-Type Markov Processes

The Poisson distribution was introduced by DE MOIVRE (1711–12) and POISSON (1837) as an approximation to the binomial distribution. The associated process arose much later from miscellaneous applications. Thus, it was considered by LUNDBERG (1903) to model streams of insurance claims, by RUTHERFORD and GEIGER (1908) to describe the process of radioactive decay, and by ERLANG (1909) to model the incoming traffic to a telephone exchange. Poisson random measures in higher dimensions appear implicitly in the work of LÉVY (1934–35), whose treatment was later formalized by ITÔ (1942b).

The independent-increment characterization of Poisson processes goes back to ERLANG (1909) and LÉVY (1934–35). Cox processes, originally introduced by COX (1955) under the name of doubly stochastic Poisson processes, were thoroughly explored by KINGMAN (1964), KRICKEBERG (1972), and GRANDELL (1976). Thinnings were first considered by RÉNYI (1956). The binomial construction of general Poisson processes was noted independently by KINGMAN (1967) and MECKE (1967). One-dimensional uniqueness criteria were obtained, first in the Poisson case by RÉNYI (1967), and then in general by MÖNCH (1971), KALLENBERG (1973a, 1986), and GRANDELL (1976). The mixed Poisson and binomial processes were studied extensively by MATTHES et al. (1978) and KALLENBERG (1986).

Markov chains in continuous time have been studied by many authors, beginning with KOLMOGOROV (1931a). The transition functions of general pure jump-type Markov processes were explored by POSPIŠIL (1935–36) and FELLER (1936, 1940), and the corresponding sample path properties were examined by DOEBLIN (1939b) and DOOB (1942b). The first continuous-time version of the strong Markov property was obtained by DOOB (1945).

KINGMAN (1993) gives an elementary introduction to Poisson processes with numerous applications. More detailed accounts, set in the context of general random measures and point processes, appear in MATTHES et al. (1978), KALLENBERG (1986), and DALEY and VERE-JONES (1988). Introductions to continuous-time Markov chains are provided by many authors, beginning with FELLER (1968). For a more comprehensive account, see CHUNG (1960). The underlying regenerative structure was examined by KINGMAN (1972).

13. Gaussian Processes and Brownian Motion

The Gaussian density function first appeared in the work of DE MOIVRE (1733–56), and the corresponding distribution became explicit through the work of LAPLACE (1774, 1812–20). The Gaussian law was popularized by GAUSS (1809) in his theory of errors and so became named after him. MAXWELL derived the Gaussian law as the velocity distribution for the molecules in a gas, assuming the hypotheses of Proposition 13.2. Theorem

13.3 was originally stated by SCHOENBERG (1938) as a relation between positive definite and completely monotone functions; the probabilistic interpretation was later noted by FREEDMAN (1962–63). Isonormal Gaussian processes were introduced by SEGAL (1954).

The process of Brownian motion was introduced by BACHELIER (1900, 1901) to model fluctuations on the stock market. Bachelier discovered some basic properties of the process, such as the relation $M_t =^d |B_t|$. EINSTEIN (1905, 1906) later introduced the same process as a model for the physical phenomenon of Brownian motion—the irregular movement of microscopic particles suspended in a liquid. The latter phenomenon, first noted by VAN LEEUWENHOEK in the seventeenth century, is named after the botanist BROWN (1828) for his systematic observations of pollen grains. Einstein's theory was forwarded in support of the still-controversial molecular theory of matter. A more refined model for the physical Brownian motion was proposed by LANGEVIN (1909) and ORNSTEIN and UHLENBECK (1930).

The mathematical theory of Brownian motion was put on a rigorous basis by WIENER (1923), who constructed the associated distribution as a measure on the space of continuous paths. The significance of Wiener's revolutionary paper was not fully recognized until after the pioneering work of KOLMOGOROV (1931a, 1933), LÉVY (1934–35), and FELLER (1936). Wiener also introduced stochastic integrals of deterministic L^2-functions, which were later studied in further detail by PALEY et al. (1933). The spectral representation of stationary processes, originally deduced from BOCHNER's (1932) theorem by CRAMÉR (1942), was later recognized as equivalent to a general Hilbert space result due to M.H. STONE (1932). The chaos expansion of Brownian functionals was discovered by WIENER (1938), and the theory of multiple integrals with respect to Brownian motion was developed in a seminal paper of ITÔ (1951c).

The law of the iterated logarithm was discovered by KHINCHIN, first (1923, 1924) for Bernoulli sequences, and later (1933) for Brownian motion. A systematic study of the Brownian paths was initiated by LÉVY (1954, 1965), who proved the existence of the quadratic variation in (1940) and the arcsine laws in (1939, 1965). Though many proofs of the latter have since been given, the present deduction from basic symmetry properties may be new. The strong Markov property was used implicitly in the work of Lévy and others, but the result was not carefully stated and proved until HUNT (1956).

Many modern probability texts contain detailed introductions to Brownian motion. The books by ITÔ and MCKEAN (1965), FREEDMAN (1971b), KARATZAS and SHREVE (1991), and REVUZ and YOR (1999) provide a wealth of further information on the subject. Further information on multiple Wiener–Itô integrals is given by KALLIANPUR (1980), DELLACHERIE et al. (1992), and NUALART (1995). The advanced theory of Gaussian distributions is nicely surveyed by ADLER (1990).

14. Skorohod Embedding and Invariance Principles

The first functional limit theorems were obtained in (1931b, 1933a) by KOLMOGOROV, who considered special functionals of a random walk. ERDÖS and KAC (1946, 1947) conceived the idea of an invariance principle that would allow functional limit theorems to be extended from particular cases to a general setting. They also treated some special functionals of a random walk. The first general functional limit theorems were obtained by DONSKER (1951-52) for random walks and empirical distribution functions, following an idea of DOOB (1949). A general theory based on sophisticated compactness arguments was later developed by PROHOROV (1956) and others.

SKOROHOD's (1965) embedding theorem provided a new and probabilistic approach to Donsker's theorem. Extensions to the martingale context were obtained by many authors, beginning with DUBINS (1968). Lemma 14.19 appears in DVORETZKY (1972). Donsker's weak invariance principle was supplemented by a strong version due to STRASSEN (1964), which yields extensions of many a.s. limit theorems for Brownian motion to suitable random walks. In particular, his result yields a simple proof of the HARTMAN and WINTNER (1941) law of the iterated logarithm, which had originally been deduced from some deep results of KOLMOGOROV (1929).

BILLINGSLEY (1968) gives many interesting applications and extensions of Donsker's theorem. For a wide range of applications of the martingale embedding theorem, see HALL and HEYDE (1980) and DURRETT (1995). KOMLÓS et al. (1975-76) showed that the approximation rate in the Skorohod embedding can be improved by a more delicate "strong approximation." For an exposition of their work and its numerous applications, see CSÖRGÖ and RÉVÉSZ (1981).

15. Independent-Increment Processes and Approximation

Until the 1920s, Brownian motion and the Poisson process were essentially the only known processes with independent increments. In (1924, 1925) LÉVY introduced the stable distributions and noted that they too could be associated with suitable "decomposable" processes. DE FINETTI (1929) saw the general connection between processes with independent increments and infinitely divisible distributions and posed the problem of characterizing the latter. A partial solution for distributions with a finite second moment was found by KOLMOGOROV (1932).

The complete solution was obtained in a revolutionary paper by LÉVY (1934-35), where the "decomposable" processes are analyzed by a virtuosic blend of analytic and probabilistic methods, leading to an explicit description in terms of a jump and a diffusion component. As a byproduct, Lévy obtained the general representation for the associated characteristic functions. His analysis was so complete that only improvements in detail have since been possible. In particular, ITÔ (1942b) showed how the jump component can be expressed in terms of Poisson integrals. Analytic derivations of the representation formula for the characteristic function were later given by LÉVY (1954) himself, by FELLER (1937), and by KHINCHIN (1937).

The scope of the classical central limit problem was broadened by LÉVY (1925) to a general study of suitably normalized partial sums, obtained from a single sequence of independent random variables. To include the case of the classical Poisson approximation, KOLMOGOROV proposed a further extension to general triangular arrays, subject to the sole condition of uniformly asymptotically negligible elements. In this context, FELLER (1937) and KHINCHIN (1937) proved independently that the limiting distributions are infinitely divisible. It remained to characterize the convergence to specific limits, a problem that had already been solved in the Gaussian case by FELLER (1935) and LÉVY (1935a). The ultimate solution was obtained independently by DOEBLIN (1939) and GNEDENKO (1939), and a comprehensive exposition of the theory was published by GNEDENKO and KOLMOGOROV (1968).

The basic convergence Theorem 15.17 for Lévy processes and the associated approximation result for random walks in Corollary 15.20 are essentially due to SKOROHOD (1957), though with rather different statements and proofs. Lemma 15.22 appears in DOEBLIN (1939a). Our approach to the basic representation theorem is a modernized version of Lévy's proof, with simplifications resulting from the use of basic point process and martingale methods.

Detailed accounts of the basic limit theory for null arrays are provided by many authors, including LOÈVE (1977) and FELLER (1971). The positive case is treated in KALLENBERG (1986). A modern introduction to Lévy processes is given by BERTOIN (1996). General independent increment processes and associated limit theorems are treated in JACOD and SHIRYAEV (1987). Extreme value theory is surveyed by LEADBETTER et al. (1983).

16. Convergence of Random Processes, Measures, and Sets

After DONSKER (1951–52) had proved his functional limit theorems for random walks and empirical distribution functions, a general theory of weak convergence in function spaces was developed by the Russian school, in seminal papers by PROHOROV (1956), SKOROHOD (1956, 1957), and KOLMOGOROV (1956). Thus, PROHOROV (1956) proved his fundamental compactness Theorem 16.3, in a setting for separable and complete metric spaces. The abstract theory was later extended in various directions by

LE CAM (1957), VARADARAJAN (1958), and DUDLEY (1966, 1967). The elementary inequality of OTTAVIANI is from (1939).

Originally SKOROHOD (1956) considered the space $D([0,1])$ endowed with four different topologies, of which the J_1-topology considered here is by far the most important for applications. The theory was later extended to $D(\mathbb{R}_+)$ by C.J. STONE (1963) and LINDVALL (1973). Tightness was originally verified by means of various product moment conditions, developed by CHENTSOV (1956) and BILLINGSLEY (1968), before the powerful criterion of ALDOUS (1978) became available. KURTZ (1975) and MITOMA (1983) noted that criteria for tightness in $D(\mathbb{R}_+, S)$ can often be expressed in terms of one-dimensional projections, as in Theorem 16.27.

The weak convergence theory for random measures and point processes originated with PROHOROV (1961), who noted the equivalence of (i) and (ii) in Theorem 16.16 when S is compact. The development continued with seminal papers by DEBES et al. (1970–71), HARRIS (1971), and JAGERS (1974). The one-dimensional criteria in Proposition 16.17 and Theorems 16.16 and 16.29 are based on results in KALLENBERG (1973a, 1986, 1996b) and a subsequent remark by KURTZ. Random sets had already been studied extensively by many authors, including CHOQUET (1953–54), KENDALL (1974), and MATHERON (1975), when an associated weak convergence theory was developed by NORBERG (1984).

The applications considered in this chapter have a long history. Thus, primitive versions of Theorem 16.18 were obtained by PALM (1943), KHINCHIN (1955), and OSOSKOV (1956). The present version is due for $S = \mathbb{R}$ to GRIGELIONIS (1963) and for more general spaces to GOLDMAN (1967) and JAGERS (1972). Limit theorems under simultaneous thinning and rescaling of a given point process were obtained by RÉNYI (1956), NAWROTZKI (1962), BELYAEV (1963), and GOLDMAN (1967). The general version in Theorem 16.19 was proved by KALLENBERG (1986) after MECKE (1968) had obtained his related characterization of Cox processes. Limit theorems for sampling from a finite population and for general exchangeable sequences have been proved in varying generality by many authors, including CHERNOV and TEICHER (1958), HÁJEK (1960), ROSÉN (1964), BILLINGSLEY (1968), and HAGBERG (1973). The results of Theorems 16.23 and 16.21 first appeared in KALLENBERG (1973b).

Detailed accounts of weak convergence theory and its applications may be found in several excellent textbooks and monographs, including BILLINGSLEY (1968), POLLARD (1984), ETHIER and KURTZ (1986), and JACOD and SHIRYAEV (1987). More information on limit theorems for random measures and point processes is available in MATTHES et al. (1978) and KALLENBERG (1986). A good general reference for random sets is MATHERON (1975).

17. Stochastic Integrals and Quadratic Variation

The first stochastic integral with a random integrand was defined by ITÔ (1942a, 1944), who used Brownian motion as the integrator and assumed the integrand to be product measurable and adapted. DOOB (1953) noted the connection with martingale theory. A first version of the fundamental substitution rule was proved by ITÔ (1951a). The result was later extended by many authors. The compensated integral in Corollary 17.21 was introduced by FISK, and independently by STRATONOVICH (1966).

The existence of the quadratic variation process was originally deduced from the Doob–Meyer decomposition. FISK (1966) showed how the quadratic variation can also be obtained directly from the process, as in Proposition 17.17. The present construction was inspired by ROGERS and WILLIAMS (2000b). The BDG inequalities were originally proved for $p > 1$ and discrete time by BURKHOLDER (1966). MILLAR (1968) noted the extension to continuous martingales, in which context the further extension to arbitrary $p > 0$ was obtained independently by BURKHOLDER and GUNDY (1970) and NOVIKOV (1971). KUNITA and WATANABE (1967) introduced the covariation of two martingales and proved the associated characterization of the integral. They further established some general inequalities related to Proposition 17.9.

The Itô integral was extended to square-integrable martingales by COURRÈGE (1962–63) and KUNITA and WATANABE (1967) and to continuous semimartingales by DOLÉANS-DADE and MEYER (1970). The idea of localization is due to ITÔ and WATANABE (1965). Theorem 17.24 was obtained by KAZAMAKI (1972) as part of a general theory of random time change. Stochastic integrals depending on a parameter were studied by DOLÉANS (1967b) and STRICKER and YOR (1978), and the functional representation of Proposition 17.26 first appeared in KALLENBERG (1996a).

Elementary introductions to Itô integration appear in many textbooks, such as CHUNG and WILLIAMS (1983) and ØKSENDAL (1998). For more advanced accounts and for further information, see IKEDA and WATANABE (1989), ROGERS and WILLIAMS (2000b), KARATZAS and SHREVE (1991), and REVUZ and YOR (1999).

18. Continuous Martingales and Brownian Motion

The fundamental characterization of Brownian motion in Theorem 18.3 was proved by LÉVY (1954), who also (1940) noted the conformal invariance up to a time change of complex Brownian motion and stated the polarity of singletons. A rigorous proof of Theorem 18.6 was later provided by KAKUTANI (1944a-b). KUNITA and WATANABE (1967) gave the first modern proof of Lévy's characterization theorem, based on Itô's formula and exponential martingales. The history of the latter can be traced back to the seminal CAMERON and MARTIN (1944) paper, the source of Theorem

18.22, and to WALD's (1946, 1947) work in sequential analysis, where the identity of Lemma 18.24 first appeared in a version for random walks.

The integral representation in Theorem 18.10 is essentially due to ITÔ (1951c), who noted its connection with multiple stochastic integrals and chaos expansions. A one-dimensional version of Theorem 18.12 appears in DOOB (1953). The general time-change Theorem 18.4 was discovered independently by DAMBIS (1965) and DUBINS and SCHWARZ (1965), and a systematic study of isotropic martingales was initiated by GETOOR and SHARPE (1972). The multivariate result in Proposition 18.8 was noted by KNIGHT (1971), and a version of Proposition 18.9 for general exchangeable processes appears in KALLENBERG (1989). The skew-product representation in Corollary 18.7 is due to GALMARINO (1963),

The Cameron–Martin theorem was gradually extended to more general settings by many authors, including MARUYAMA (1954, 1955), GIRSANOV (1960), and VAN SCHUPPEN and WONG (1974). The martingale criterion of Theorem 18.23 was obtained by NOVIKOV (1972).

The material in this chapter is covered by many texts, including the excellent monographs by KARATZAS and SHREVE (1991) and REVUZ and YOR (1999). A more advanced and amazingly informative text is JACOD (1979).

19. Feller Processes and Semigroups

Semigroup ideas are implicit in KOLMOGOROV's pioneering (1931a) paper, whose central theme is the search for local characteristics that will determine the transition probabilities through a system of differential equations, the so-called Kolmogorov forward and backward equations. Markov chains and diffusion processes were originally treated separately, but in (1935) KOLMOGOROV proposed a unified framework, with transition kernels regarded as operators (initially operating on measures rather than on functions), and with local characteristics given by an associated generator.

Kolmogorov's ideas were taken up by FELLER (1936), who obtained general existence and uniqueness results for the forward and backward equations. The abstract theory of contraction semigroups on Banach spaces was developed independently by HILLE (1948) and YOSIDA (1948), both of whom recognized its significance for the theory of Markov processes. The power of the semigroup approach became clear through the work of FELLER (1952, 1954), who gave a complete description of the generators of one-dimensional diffusions. In particular, Feller characterizes the boundary behavior of the process in terms of the domain of the generator.

The systematic study of Markov semigroups began with the work of DYNKIN (1955a). The standard approach is to postulate strong continuity instead of the weaker and more easily verified condition (F_2). The positive maximum principle appears in the work of ITÔ (1957), and the core condition of Proposition 19.9 is due to S. WATANABE (1968).

The first regularity theorem was obtained by DOEBLIN (1939b), who gave conditions for the paths to be step functions. A sufficient condition for continuity was then obtained by FORTET (1943). Finally, KINNEY (1953) showed that any Feller process has a version with rcll paths, after DYNKIN (1952) had obtained the same property under a Hölder condition. The use of martingale methods for the study of Markov processes dates back to KINNEY (1953) and DOOB (1954).

The strong Markov property for Feller processes was proved independently by DYNKIN and YUSHKEVICH (1956) and by BLUMENTHAL (1957) after special cases had been considered by DOOB (1945), HUNT (1956), and RAY (1956). BLUMENTHAL's (1957) paper also contains his zero–one law. DYNKIN (1955a) introduced his "characteristic operator," and a version of Theorem 19.24 appears in DYNKIN (1956).

There is a vast literature on approximation results for Markov chains and Markov processes, covering a wide range of applications. The use of semigroup methods to prove limit theorems can be traced back to LINDEBERG's (1922a) proof of the central limit theorem. The general results in Theorems 19.25 and 19.28 were developed in stages by TROTTER (1958a), SOVA (1967), KURTZ (1969, 1975), and MACKEVIČIUS (1974). Our proof of Theorem 19.25 uses ideas from J.A. GOLDSTEIN (1976).

A splendid introduction to semigroup theory is given by the relevant chapters in FELLER (1971). In particular, Feller shows how the one-dimensional Lévy–Khinchin formula and associated limit theorems can be derived by semigroup methods. More detailed and advanced accounts of the subject appear in DYNKIN (1965), ETHIER and KURTZ (1986), and DELLACHERIE and MEYER (1975–87).

20. Ergodic Properties of Markov Processes

The first ratio ergodic theorems were obtained by DOEBLIN (1938b), DOOB (1938, 1948a), KAKUTANI (1940), and HUREWICZ (1944). HOPF (1954) and DUNFORD and SCHWARTZ (1956) extended the pointwise ergodic theorem to general L^1–L^∞-contractions, and the ratio ergodic theorem was extended to positive L^1-contractions by CHACON and ORNSTEIN (1960). The present approach to their result in due to AKCOGLU and CHACON (1970).

The notion of Harris recurrence goes back to DOEBLIN (1940) and HARRIS (1956). The latter author used the condition to ensure the existence, in discrete time, of a σ-finite invariant measure. A corresponding continuous-time result was obtained by H. WATANABE (1964). The total variation convergence of Markov transition probabilities was obtained for a countable state space by OREY (1959, 1962) and in general by JAMISON and OREY (1967). BLACKWELL and FREEDMAN (1964) noted the equivalence of mixing and tail triviality. The present coupling approach goes back to GRIFFEATH (1975) and S. GOLDSTEIN (1979) for the case of strong ergodicity and to BERBEE (1979) and ALDOUS and THORISSON (1993) for the corresponding weak result.

There is an extensive literature on ergodic theorems for Markov processes, mostly dealing with the discrete-time case. General expositions have been given by many authors, beginning with NEVEU (1971) and OREY (1971). Our treatment of Harris recurrent Feller processes is adapted from KUNITA (1990), who in turn follows the discrete-time approach of REVUZ (1984). KRENGEL (1985) gives a comprehensive survey of abstract ergodic theorems. Detailed accounts of the coupling method and its various ramifications appear in LINDVALL (1992) and THORISSON (2000).

21. Stochastic Differential Equations and Martingale Problems

Long before the existence of any general theory for SDEs, LANGEVIN (1908) proposed his equation to model the *velocity* of a Brownian particle. The solution process was later studied by ORNSTEIN and UHLENBECK (1930) and was thus named after them. A more rigorous discussion appears in DOOB (1942a).

The general idea of a stochastic differential equation goes back to BERNSTEIN (1934, 1938), who proposed a pathwise construction of diffusion processes by a discrete approximation, leading in the limit to a formal differential equation driven by a Brownian motion. However, ITÔ (1942a, 1951b) was the first author to develop a rigorous and systematic theory, including a precise definition of the integral, conditions for existence and uniqueness of solutions, and basic properties of the solution process, such as the Markov property and the continuous dependence on initial state. Similar results were obtained, later but independently, by GIHMAN (1947, 1950–51).

The notion of a weak solution was introduced by GIRSANOV (1960), and a version of the weak existence Theorem 21.9 appears in SKOROHOD (1965). The ideas behind the transformations in Propositions 21.12 and 21.13 date back to GIRSANOV (1960) and VOLKONSKY (1958), respectively. The notion of a martingale problem can be traced back to LÉVY's martingale characterization of Brownian motion and DYNKIN's theory of the characteristic operator. A comprehensive theory was developed by STROOCK and VARADHAN (1969), who established the equivalence with weak solutions to the associated SDEs, obtained general criteria for uniqueness in law, and deduced conditions for the strong Markov and Feller properties. The measurability part of Theorem 21.10 is a slight extension of an exercise in STROOCK and VARADHAN (1979).

YAMADA and WATANABE (1971) proved that weak existence and pathwise uniqueness imply strong existence and uniqueness in law. Under the same conditions, they further established the existence of a functional solution, possibly depending on the initial distribution of the process; that dependence was later removed by KALLENBERG (1996a). IKEDA and WATANABE (1989) noted how the notions of pathwise uniqueness and uniqueness in law extend by conditioning from degenerate to arbitrary initial distributions.

The basic theory of SDEs is covered by many excellent textbooks on different levels, including IKEDA and WATANABE (1989), ROGERS and WILLIAMS (1987), and KARATZAS and SHREVE (1991). More information on the martingale problem is available in JACOD (1979), STROOCK and VARADHAN (1979), and ETHIER and KURTZ (1986).

22. Local Time, Excursions, and Additive Functionals

Local time of Brownian motion at a fixed point was discovered and explored by LÉVY (1939), who devised several explicit constructions, mostly of the type of Proposition 22.12. Much of Lévy's analysis is based on the observation in Corollary 22.3. The elementary Lemma 22.2 is due to SKOROHOD (1961–62). Formula (1), first noted for Brownian motion by TANAKA (1963), was taken by MEYER (1976) as the basis for a general semimartingale approach. The general Itô–Tanaka formula in Theorem 22.5 was obtained independently by MEYER (1976) and WANG (1977). TROTTER (1958b) proved that Brownian local time has a jointly continuous version, and the extension to general continuous semimartingales in Theorem 22.4 was obtained by YOR (1978).

Modern excursion theory originated with the seminal paper of ITÔ (1972), which was partly inspired by earlier work of LÉVY (1939). In particular, Itô proved a version of Theorem 22.11, assuming the existence of local time. HOROWITZ (1972) independently studied regenerative sets and noted their connection with subordinators, equivalent to the existence of a local time. A systematic theory of regenerative processes was developed by MAISONNEUVE (1974). The remarkable Theorem 22.17 was discovered independently by RAY (1963) and KNIGHT (1963), and the present proof is essentially due to WALSH (1978). Our construction of the excursion process is close in spirit to Lévy's original ideas and to those in GREENWOOD and PITMAN (1980).

Elementary additive functionals of integral type had been discussed extensively in the literature when DYNKIN proposed a study of the general case. The existence Theorem 22.23 was obtained by VOLKONSKY (1960), and the construction of local time in Theorem 22.24 dates back to BLUMENTHAL and GETOOR (1964). The integral representation of CAFs in Theorem 22.25 was proved independently by VOLKONSKY (1958, 1960) and MCKEAN and TANAKA (1961). The characterization of additive functionals in terms of suitable measures on the state space dates back to MEYER (1962), and the explicit representation of the associated measures was found by REVUZ (1970) after special cases had been considered by HUNT (1957–58).

An excellent introduction to local time appears in KARATZAS and SHREVE (1991). The books by ITÔ and MCKEAN (1965) and REVUZ and YOR (1999) contain an abundance of further information on the subject. The latter text may also serve as a good introduction to additive functionals and excursion theory. For more information on the latter topics, the reader may consult BLUMENTHAL and GETOOR (1968), BLUMENTHAL (1992), and DELLACHERIE et al. (1992).

23. One-Dimensional SDEs and Diffusions

The study of continuous Markov processes and the associated parabolic differential equations, initiated by KOLMOGOROV (1931a) and FELLER (1936), took a new direction with the seminal papers of FELLER (1952, 1954), who studied the generators of one-dimensional diffusions within the framework of the newly developed semigroup theory. In particular, Feller gave a complete description in terms of scale function and speed measure, classified the boundary behavior, and showed how the latter is determined by the domain of the generator. Finally, he identified the cases when explosion occurs, corresponding to the absorption cases in Theorem 23.15.

A more probabilistic approach to these results was developed by DYNKIN (1955b, 1959), who along with RAY (1956) continued Feller's study of the relationship between analytic properties of the generator and sample path properties of the process. The idea of constructing diffusions on a natural scale through a time change of Brownian motion is due to HUNT (1958) and VOLKONSKY (1958), and the full description in Theorem 23.9 was completed by VOLKONSKY (1960) and ITÔ and MCKEAN (1965). The present stochastic calculus approach is based on ideas in MÉLÉARD (1986).

The ratio ergodic Theorem 23.14 was first obtained for Brownian motion by DERMAN (1954), by a method originally devised for discrete-time chains by DOEBLIN (1938). It was later extended to more general diffusions by MOTOO and WATANABE (1958). The ergodic behavior of recurrent one-dimensional diffusions was analyzed by MARUYAMA and TANAKA (1957).

For one-dimensional SDEs, SKOROHOD (1965) noticed that Itô's original Lipschitz condition for pathwise uniqueness can be replaced by a weaker Hölder condition. He also obtained a corresponding comparison theorem. The improved conditions in Theorems 23.3 and 23.5 are due to YAMADA and WATANABE (1971) and YAMADA (1973), respectively. PERKINS (1982) and LE GALL (1983) noted how the use of semimartingale local time simplifies and unifies the proofs of those and related results. The fundamental weak existence and uniqueness criteria in Theorem 23.1 were discovered by ENGELBERT and SCHMIDT (1984, 1985), whose (1981) zero–one law is implicit in Lemma 23.2.

Elementary introductions to one-dimensional diffusions appear in BREIMAN (1968), FREEDMAN (1971b), and ROGERS and WILLIAMS (2000b). More detailed and advanced accounts are given by DYNKIN (1965) and ITÔ and MCKEAN (1965). Further information on one-dimensional SDEs may be obtained from the excellent books by KARATZAS and SHREVE (1991) and REVUZ and YOR (1999).

24. Connections with PDEs and Potential Theory

The fundamental solution to the heat equation in terms of the Gaussian kernel was obtained by LAPLACE (1809). A century later BACHELIER (1900, 1901) noted the relationship between Brownian motion and the heat equation. The PDE connections were further explored by many authors, including KOLMOGOROV (1931a), FELLER (1936), KAC (1951), and DOOB (1955). A first version of Theorem 24.1 was obtained by KAC (1949), who was in turn inspired by FEYNMAN's (1948) work on the Schrödinger equation. Theorem 24.2 is due to STROOCK and VARADHAN (1969).

GREEN (1828), in his discussion of the Dirichlet problem, introduced the functions named after him. The Dirichlet, sweeping, and equilibrium problems were all studied by GAUSS (1840) in a pioneering paper on electrostatics. The rigorous developments in potential theory began with POINCARÉ (1890–99), who solved the Dirichlet problem for domains with a smooth boundary. The equilibrium measure was characterized by GAUSS as the unique measure minimizing a certain energy functional, but the existence of the minimum was not rigorously established until FROSTMAN (1935).

The first probabilistic connections were made by PHILLIPS and WIENER (1923) and COURANT et al. (1928), who solved the Dirichlet problem in the plane by a method of discrete approximation, involving a version of Theorem 24.5 for a simple symmetric random walk. KOLMOGOROV and LEONTOVICH (1933) evaluated a special hitting distribution for two-dimensional Brownian motion and noted that it satisfies the heat equation. KAKUTANI (1944b, 1945) showed how the harmonic measure and sweeping kernel can be expressed in terms of a Brownian motion. The probabilistic methods were extended and perfected by DOOB (1954, 1955), who noted the profound connections with martingale theory. A general potential theory was later developed by HUNT (1957–58) for broad classes of Markov processes.

The interpretation of Green functions as occupation densities was known to KAC (1951), and a probabilistic approach to Green functions was developed by HUNT (1956). The connection between equilibrium measures and quitting times, implicit already in SPITZER (1964) and ITÔ and MCKEAN (1965), was exploited by CHUNG (1973) to yield the explicit representation of Theorem 24.14.

Time reversal of diffusion processes was first considered by SCHRÖDINGER (1931). KOLMOGOROV (1936b, 1937) computed the transition kernels of the reversed process and gave necessary and sufficient conditions for symmetry. The basic role of time reversal and duality in potential theory was recognized by DOOB (1954) and HUNT (1958). Proposition 24.15 and the related construction in Theorem 24.21 go back to HUNT, but Theorem 24.19 may be new. The measure ν in Theorem 24.21 is related to the "Kuznetsov measures," discussed extensively in GETOOR (1990). The connection between random sets and alternating capacities was established by CHOQUET (1953–54), and a corresponding representation of infinitely divisible random sets was obtained by MATHERON (1975).

Elementary introductions to probabilistic potential theory appear in BASS (1995) and CHUNG (1995), and to other PDE connections in KARATZAS and SHREVE (1991). A detailed exposition of classical probabilistic potential theory is given by PORT and STONE (1978). DOOB (1984) provides a wealth of further information on both the analytic and probabilistic aspects. Introductions to Hunt's work and the subsequent developments are given by CHUNG (1982) and DELLACHERIE and MEYER (1975–87). More advanced treatments appear in BLUMENTHAL and GETOOR (1968) and SHARPE (1988).

25. Predictability, Compensation, and Excessive Functions

The basic connection between superharmonic functions and supermartingales was established by DOOB (1954), who also proved that compositions of excessive functions with Brownian motion are continuous. Doob further recognized the need for a general decomposition theorem for supermartingales, generalizing the elementary Lemma 7.10. Such a result was eventually proved by MEYER (1962, 1963), in the form of Lemma 25.7, after special decompositions in the Markovian context had been obtained by VOLKONSKY (1960) and SHUR (1961). Meyer's original proof was profound and clever. The present more elementary approach, based on DUNFORD's (1939) weak compactness criterion, was devised by RAO (1969a). The extension to general submartingales was accomplished by ITÔ and WATANABE (1965) through the introduction of local martingales.

Predictable and totally inaccessible times appear implicitly in the work of BLUMENTHAL (1957) and HUNT (1957–58), in the context of quasi-left-continuity. A systematic study of optional times and their associated σ-fields was initiated by CHUNG and DOOB (1965). The basic role of the predictable σ-field became clear after DOLÉANS (1967a) had proved the equivalence between naturalness and predictability for increasing processes, thereby establishing the ultimate version of the Doob–Meyer decomposition. The moment inequality in Proposition 25.21 was obtained independently by GARSIA (1973) and NEVEU (1975) after a more special result had been proved by BURKHOLDER et al. (1972). The theory of optional and predictable times and σ-fields was developed by MEYER (1966), DELLACHERIE

(1972), and others into a "general theory of processes," which has in many ways revolutionized modern probability.

Natural compensators of optional times first appeared in reliability theory. More general compensators were later studied in the Markovian context by S. WATANABE (1964) under the name of "Lévy systems." GRIGELIONIS (1971) and JACOD (1975) constructed the compensator of a general random measure and introduced the related "local characteristics" of a general semimartingale. WATANABE (1964) proved that a simple point process with a continuous and deterministic compensator is Poisson; a corresponding time-change result was obtained independently by MEYER (1971) and PAPANGELOU (1972). The extension in Theorem 25.24 was given by KALLENBERG (1990), and general versions of Proposition 25.27 appear in ROSIŃSKI and WOYCZYŃSKI (1986) and KALLENBERG (1992).

An authoritative account of the general theory, including an elegant but less elementary projection approach to the Doob–Meyer decomposition due to DOLÉANS, is given by DELLACHERIE and MEYER (1975–87). Useful introductions to the theory are contained in ELLIOTT (1982) and ROGERS and WILLIAMS (2000b). Our elementary proof of Lemma 25.10 uses ideas from DOOB (1984). BLUMENTHAL and GETOOR (1968) remains a good general reference on additive functionals and their potentials. A detailed account of random measures and their compensators appears in JACOD and SHIRYAEV (1987). Applications to queuing theory are given by BRÉMAUD (1981), BACCELLI and BRÉMAUD (2000), and LAST and BRANDT (1995).

26. Semimartingales and General Stochastic Integration

DOOB (1953) conceived the idea of a stochastic integration theory for general L^2-martingales, based on a suitable decomposition of continuous-time submartingales. MEYER's (1962) proof of such a result opened the door to the L^2-theory, which was then developed by COURRÈGE (1962–63) and KUNITA and WATANABE (1967). The latter paper contains in particular a version of the general substitution rule. The integration theory was later extended in a series of papers by MEYER (1967) and DOLÉANS-DADE and MEYER (1970) and reached its final form with the notes of MEYER (1976) and the books by JACOD (1979), MÉTIVIER and PELLAUMAIL (1979), and DELLACHERIE and MEYER (1975–87).

The basic role of predictable processes as integrands was recognized by MEYER (1967). By contrast, semimartingales were originally introduced in an ad hoc manner by DOLÉANS-DADE and MEYER (1970), and their basic preservation laws were only gradually recognized. In particular, JACOD (1975) used the general Girsanov theorem of VAN SCHUPPEN and WONG (1974) to show that the semimartingale property is preserved under absolutely continuous changes of the probability measure. The characterization of general stochastic integrators as semimartingales was obtained independently by BICHTELER (1979) and DELLACHERIE (1980), in both cases with support from analysts.

Quasimartingales were originally introduced by FISK (1965) and OREY (1966). The decomposition of RAO (1969b) extends a result by KRICKEBERG (1956) for L^1-bounded martingales. YOEURP (1976) combined a notion of "stable subspaces" due to KUNITA and WATANABE (1967) with the Hilbert space structure of $\mathcal{M}^2$ to obtain an orthogonal decomposition of L^2-martingales, equivalent to the decompositions in Theorem 26.14 and Proposition 26.16. Elaborating on those ideas, MEYER (1976) showed that the purely discontinuous component admits a representation as a sum of compensated jumps.

SDEs driven by general Lévy processes were already considered by ITÔ (1951b). The study of SDEs driven by general semimartingales was initiated by DOLÉANS-DADE (1970), who obtained her exponential process as a solution to the equation in Theorem 26.8. The scope of the theory was later expanded by many authors, and a comprehensive account is given by PROTTER (1990).

The martingale inequalities in Theorems 26.12 and 26.17 have ancient origins. Thus, a version of the latter result for independent random variables was proved by KOLMOGOROV (1929) and, in a sharper form, by PROHOROV (1959). Their result was extended to discrete-time martingales by JOHNSON et al. (1985) and HITCZENKO (1990). The present statements appeared in KALLENBERG and SZTENCEL (1991).

Early versions of the inequalities in Theorem 26.12 were proved by KHINCHIN (1923, 1924) for symmetric random walks and by PALEY (1932) for Walsh series. A version for independent random variables was obtained by MARCINKIEWICZ and ZYGMUND (1937, 1938). The extension to discrete-time martingales is due to BURKHOLDER (1966) for $p > 1$ and to DAVIS (1970) for $p = 1$. The result was extended to continuous time by BURKHOLDER et al. (1972), who also noted how the general result can be deduced from the statement for $p = 1$. The present proof is a continuous-time version of Davis' original argument.

Excellent introductions to semimartingales and stochastic integration are given by DELLACHERIE and MEYER (1975–87) and JACOD and SHIRYAEV (1987). PROTTER (1990) offers an interesting alternative approach, originally suggested by MEYER and by DELLACHERIE (1980). The book by JACOD (1979) remains a rich source of further information on the subject.

27. Large Deviations

Large deviation theory originated with certain refinements of the central limit theorem obtained by many authors, beginning with KHINCHIN (1929). Here the object of study is the ratio of tail probabilities $r_n(x) = P\{\zeta_n > x\}/P\{\zeta > x\}$, where ζ is $N(0,1)$ and $\zeta_n = n^{-1/2}\sum_{k\leq n}\xi_k$ for some i.i.d.

random variables ξ_k with mean 0 and variance 1, so that $r_n(x) \to 1$ for fixed x. A precise asymptotic expansion was obtained by CRAMÉR (1938), in the case when x varies with n at a rate $x = o(n^{1/2})$. (See PETROV (1995), Theorem 5.23, for details.)

In the same historic paper, CRAMÉR (1938) obtained the first true large deviation result, in the form of our Theorem 27.3, though under some technical assumptions that were later removed by CHERNOFF (1952) and BAHADUR (1971). VARADHAN (1966) extended the result to higher dimensions and rephrased it in the form of a general large deviation principle. At about the same time, SCHILDER (1966) proved his large deviation result for Brownian motion, using the present change-of-measure approach. Similar methods were used by FREIDLIN and WENTZELL (1970, 1998) to study random perturbations of dynamical systems.

Even earlier, SANOV (1957) had obtained his large deviation result for empirical distributions of i.i.d. random variables. The relative entropy $H(\nu|\mu)$ appearing in the limit had already been introduced in statistics by KULLBACK and LEIBLER (1951). Its crucial link to the Legendre–Fenchel transform Λ^*, long anticipated by physicists, was formalized by DONSKER and VARADHAN (1975–83). The latter authors also developed some profound and far-reaching extensions of Sanov's theorem, in a long series of formidable papers. ELLIS (1985) gives a detailed exposition of those results, along with a discussion of their physical significance.

Much of the formalization of underlying principles and techniques was developed at a later stage. Thus, an abstract version of the projective limit approach was introduced by DAWSON and GÄRTNER (1987). BRYC (1990) supplemented VARADHAN's (1966) functional version of the LDP with a reverse proposition. Similarly, IOFFE (1991) appended a powerful inverse to the classical "contraction principle." Finally, PUKHALSKY (1991) established the equivalence, under suitable regularity conditions, of the exponential tightness and the goodness of the rate function.

STRASSEN (1964) established his formidable law of the iterated logarithm by direct estimates. A detailed exposition of the original approach appears in FREEDMAN (1971b). VARADHAN (1984) recognized the result as a corollary to Schilder's theorem, and a complete proof along the suggested lines appears in DEUSCHEL and STROOCK (1989).

Gentle introductions to large deviation theory and its applications are given by VARADHAN (1984) and DEMBO and ZEITOUNI (1998). The more demanding text of DEUSCHEL and STROOCK (1989) provides much additional insight to the persistent reader.

Appendix

Some more advanced aspects of measure theory are covered by ROYDEN (1988), PARTHASARATHY (1967), and DUDLEY (1989). The projection and section theorems depend on capacity theory, for which we refer to DELLACHERIE (1972) and DELLACHERIE and MEYER (1975–87).

The J_1-topology was introduced by SKOROHOD (1956), and detailed expositions may be found in BILLINGSLEY (1968), ETHIER and KURTZ (1986), and JACOD and SHIRYAEV (1987). A discussion of the vague topology on $\mathcal{M}(S)$ with S lcscH is given by BAUER (1972). The topology on the space of closed sets, considered here, was introduced in a more general setting by FELL (1962), and a full account appears in MATHERON (1975), including a detailed proof (different from ours) of the basic Theorem A2.5.

Bibliography

This list includes only publications that are explicitly mentioned in the text or notes or are directly related to results cited in the book. Knowledgeable readers will notice that many books and papers of historical significance have been omitted.

ADLER, R.J. (1990). *An Introduction to Continuity, Extrema, and Related Topics for General Gaussian Processes.* Inst. Math. Statist., Hayward, CA.

AKCOGLU, M.A., CHACON, R.V. (1970). Ergodic properties of operators in Lebesgue space. *Adv. Appl. Probab.* **2**, 1–47.

ALDOUS, D.J. (1978). Stopping times and tightness. *Ann. Probab.* **6**, 335–340.

— (1985). Exchangeability and related topics. *Lect. Notes in Math.* **1117**, 1–198. Springer, Berlin.

ALDOUS, D., THORISSON, H. (1993). Shift-coupling. *Stoch. Proc. Appl.* **44**, 1–14.

ALEXANDROV, A.D. (1940–43). Additive set-functions in abstract spaces. *Mat. Sb.* **8**, 307–348; **9**, 563–628; **13**, 169–238.

ANDRÉ, D. (1887). Solution directe du problème résolu par M. Bertrand. *C.R. Acad. Sci. Paris* **105**, 436–437.

ATHREYA, K., MCDONALD, D., NEY, P. (1978). Coupling and the renewal theorem. *Amer. Math. Monthly* **85**, 809–814.

BACCELLI, F., BRÉMAUD, P. (2000). *Elements of Queueing [sic] Theory*, 2nd ed., Springer, Berlin.

BACHELIER, L. (1900). Théorie de la spéculation. *Ann. Sci. École Norm. Sup.* **17**, 21–86.

— (1901). Théorie mathématique du jeu. *Ann. Sci. École Norm. Sup.* **18**, 143–210.

BAHADUR, R.R. (1971). *Some Limit Theorems in Statistics.* SIAM, Philadelphia.

BARBIER, É. (1887). Généralisation du problème résolu par M. J. Bertrand. *C.R. Acad. Sci. Paris* **105**, 407, 440.

BASS, R.F. (1995). *Probabilistic Techniques in Analysis.* Springer, NY.

— (1998). *Diffusions and Elliptic Operators.* Springer, NY.

BAUER, H. (1972). *Probability Theory and Elements of Measure Theory.* Engl. trans., Holt, Rinehart & Winston, NY.

BAXTER, G. (1961). An analytic approach to finite fluctuation problems in probability. *J. d'Analyse Math.* **9**, 31–70.

BELYAEV, Y.K. (1963). Limit theorems for dissipative flows. *Th. Probab. Appl.* **8**, 165–173.

BERBEE, H.C.P. (1979). *Random Walks with Stationary Increments and Renewal Theory.* Mathematisch Centrum, Amsterdam.

BERNOULLI, J. (1713). *Ars Conjectandi.* Thurnisiorum, Basel.

BERNSTEIN, S.N. (1927). Sur l'extension du théorème limite du calcul des probabilités aux sommes de quantités dépendantes. *Math. Ann.* **97**, 1–59.

— (1934). Principes de la théorie des équations différentielles stochastiques. *Trudy Fiz.-Mat., Steklov Inst., Akad. Nauk.* **5**, 95–124.

— (1937). On some variations of the Chebyshev inequality (in Russian). *Dokl. Acad. Nauk SSSR* **17**, 275–277.

— (1938). Équations différentielles stochastiques. *Act. Sci. Ind.* **738**, 5–31.

BERTOIN, J. (1996). *Lévy Processes.* Cambridge Univ. Press.

BERTRAND, J. (1887). Solution d'un problème. *C.R. Acad. Sci. Paris* **105**, 369.

BICHTELER, K. (1979). Stochastic integrators. *Bull. Amer. Math. Soc.* **1**, 761–765.

BIENAYMÉ, J. (1853). Considérations à l'appui de la découverte de Laplace sur la loi de probabilité dans la méthode des moindres carrés. *C.R. Acad. Sci. Paris* **37**, 309–324.

BILLINGSLEY, P. (1965). *Ergodic Theory and Information.* Wiley, NY.

— (1968). *Convergence of Probability Measures.* Wiley, NY.

— (1995). *Probability and Measure*, 3rd ed. Wiley, NY.

BIRKHOFF, G.D. (1932). Proof of the ergodic theorem. *Proc. Natl. Acad. Sci. USA* **17**, 656–660.

BLACKWELL, D. (1948). A renewal theorem. *Duke Math. J.* **15**, 145–150.

— (1953). Extension of a renewal theorem. *Pacific J. Math.* **3**, 315–320.

BLACKWELL, D., FREEDMAN, D. (1964). The tail σ-field of a Markov chain and a theorem of Orey. *Ann. Math. Statist.* **35**, 1291–1295.

BLUMENTHAL, R.M. (1957). An extended Markov property. *Trans. Amer. Math. Soc.* **82**, 52–72.

— (1992). *Excursions of Markov Processes.* Birkhäuser, Boston.

BLUMENTHAL, R.M., GETOOR, R.K. (1964). Local times for Markov processes. *Z. Wahrsch. verw. Geb.* **3**, 50–74.

— (1968). *Markov Processes and Potential Theory.* Academic Press, NY.

BOCHNER, S. (1932). *Vorlesungen über Fouriersche Integrale*, Akad. Verlagsges., Leipzig. Repr. Chelsea, NY 1948.

— (1933). Monotone Funktionen, Stieltjessche Integrale und harmonische Analyse. *Math. Ann.* **108**, 378–410.

BOLTZMANN, L. (1887). Über die mechanischen Analogien des zweiten Hauptsatzes der Thermodynamik. *J. Reine Angew. Math.* **100**, 201–212.

BOREL, E. (1895). Sur quelques points de la théorie des fonctions. *Ann. Sci. École Norm. Sup.* (3) **12**, 9–55.

— (1898). *Leçons sur la Théorie des Fonctions.* Gauthier-Villars, Paris.

— (1909). Les probabilités dénombrables et leurs applications arithmétiques. *Rend. Circ. Mat. Palermo* **27** 247–271.

BREIMAN, L. (1957–60). The individual ergodic theorem of infomation theory. *Ann. Math. Statist.* **28**, 809–811; **31**, 809–810.

— (1968). *Probability*. Addison-Wesley, Reading, MA. Repr. SIAM, Philadelphia 1992.

BRÉMAUD, P. (1981). *Point Processes and Queues*. Springer, NY.

BROWN, R. (1828). A brief description of microscopical observations made in the months of June, July and August 1827, on the particles contained in the pollen of plants; and on the general existence of active molecules in organic and inorganic bodies. *Ann. Phys.* **14**, 294–313.

BRYC, W. (1990). Large deviations by the asymptotic value method. In *Diffusion Processes and Related Problems in Analysis* (M. Pinsky, ed.), 447–472. Birkhäuser, Basel.

BÜHLMANN, H. (1960). Austauschbare stochastische Variabeln und ihre Grenzwertsätze. *Univ. Calif. Publ. Statist.* **3**, 1–35.

BUNIAKOWSKY, V.Y. (1859). Sur quelques inégalités concernant les intégrales ordinaires et les intégrales aux différences finies. *Mém. de l'Acad. St.-Petersbourg* **1**:9.

BURKHOLDER, D.L. (1966). Martingale transforms. *Ann. Math. Statist.* **37**, 1494–1504.

BURKHOLDER, D.L., DAVIS, B.J., GUNDY, R.F. (1972). Integral inequalities for convex functions of operators on martingales. *Proc. 6th Berkeley Symp. Math. Statist. Probab.* **2**, 223–240.

BURKHOLDER, D.L., GUNDY, R.F. (1970). Extrapolation and interpolation of quasi-linear operators on martingales. *Acta Math.* **124**, 249–304.

CAMERON, R.H., MARTIN, W.T. (1944). Transformation of Wiener integrals under translations. *Ann. Math.* **45**, 386–396.

CANTELLI, F.P. (1917). Su due applicazione di un teorema di G. Boole alla statistica matematica. *Rend. Accad. Naz. Lincei* **26**, 295–302.

— (1933). Sulla determinazione empirica della leggi di probabilità. *Giorn. Ist. Ital. Attuari* **4**, 421–424.

CARATHÉODORY, C. (1927). *Vorlesungen über reelle Funktionen*, 2nd ed. Teubner, Leipzig (1st ed. 1918). Repr. Chelsea, NY 1946.

CARLESON, L. (1958). Two remarks on the basic theorems of information theory. *Math. Scand.* **6**, 175–180.

CAUCHY, A.L. (1821). *Cours d'analyse de l'École Royale Polytechnique*, Paris.

CHACON, R.V., ORNSTEIN, D.S. (1960). A general ergodic theorem. *Illinois J. Math.* **4**, 153–160.

CHAPMAN, S. (1928). On the Brownian displacements and thermal diffusion of grains suspended in a non-uniform fluid. *Proc. Roy. Soc. London (A)* **119**, 34–54.

CHEBYSHEV, P.L. (1867). Des valeurs moyennes. *J. Math. Pures Appl.* **12**, 177–184.

— (1890). Sur deux théorèmes relatifs aux probabilités. *Acta Math.* **14**, 305–315.

CHENTSOV, N.N. (1956). Weak convergence of stochastic processes whose trajectories have no discontinuities of the second kind and the "heuristic" approach to the Kolmogorov-Smirnov tests. *Th. Probab. Appl.* **1**, 140–144.

CHERNOFF, H. (1952). A measure of asymptotic efficiency for tests of a hypothesis based on the sum of observations. *Ann. Math. Statist.* **23**, 493–507.

CHERNOFF, H., TEICHER, H. (1958). A central limit theorem for sequences of exchangeable random variables. *Ann. Math. Statist.* **29**, 118–130.

CHOQUET, G. (1953–54). Theory of capacities. *Ann. Inst. Fourier Grenoble* **5**, 131–295.

CHOW, Y.S., TEICHER, H. (1997). *Probability Theory: Independence, Interchangeability, Martingales*, 3nd ed. Springer, NY.

CHUNG, K.L. (1960). *Markov Chains with Stationary Transition Probabilities*. Springer, Berlin.

— (1961). A note on the ergodic theorem of information theory. *Ann. Math. Statist.* **32**, 612–614.

— (1973). Probabilistic approach to the equilibrium problem in potential theory. *Ann. Inst. Fourier Grenoble* **23**, 313–322.

— (1974). *A Course in Probability Theory*, 2nd ed. Academic Press, NY.

— (1982). *Lectures from Markov Processes to Brownian Motion*. Springer, NY.

— (1995). *Green, Brown, and Probability*. World Scientific, Singapore.

CHUNG, K.L., DOOB, J.L. (1965). Fields, optionality and measurability. *Amer. J. Math.* **87**, 397–424.

CHUNG, K.L., FUCHS, W.H.J. (1951). On the distribution of values of sums of random variables. *Mem. Amer. Math. Soc.* **6**.

CHUNG, K.L., ORNSTEIN, D.S. (1962). On the recurrence of sums of random variables. *Bull. Amer. Math. Soc.* **68**, 30–32.

CHUNG, K.L., WALSH, J.B. (1974). Meyer's theorem on previsibility. *Z. Wahrsch. verw. Geb.* **29**, 253–256.

CHUNG, K.L., WILLIAMS, R.J. (1990). *Introduction to Stochastic Integration*, 2nd ed. Birkhäuser, Boston.

COURANT, R., FRIEDRICHS, K., LEWY, H. (1928). Über die partiellen Differentialgleichungen der mathematischen Physik. *Math. Ann.* **100**, 32–74.

COURRÈGE, P. (1962–63). Intégrales stochastiques et martingales de carré intégrable. *Sem. Brelot-Choquet-Deny* **7**. Publ. Inst. H. Poincaré.

COX, D.R. (1955). Some statistical methods connected with series of events. *J. R. Statist. Soc. Ser. B* **17**, 129–164.

CRAMÉR, H. (1938). Sur un nouveau théorème-limite de la théorie des probabilités. *Actual. Sci. Indust.* **736**, 5–23.

— (1942). On harmonic analysis in certain functional spaces. *Ark. Mat. Astr. Fys.* **28B**:12 (17 pp.).

CRAMÉR, H., LEADBETTER, M.R. (1967). *Stationary and Related Stochastic Processes*. Wiley, NY.

CRAMÉR, H., WOLD, H. (1936). Some theorems on distribution functions. *J. London Math. Soc.* **11**, 290–295.

CSÖRGŐ, M., RÉVÉSZ, P. (1981). *Strong Approximations in Probability and Statistics.* Academic Press, NY.

DALEY, D.J., VERE-JONES, D. (1988). *An Introduction to the Theory of Point Processes.* Springer, NY.

DAMBIS, K.E. (1965). On the decomposition of continuous submartingales. *Th. Probab. Appl.* **10**, 401–410.

DANIELL, P.J. (1918–19). Integrals in an infinite number of dimensions. *Ann. Math.* (2) **20**, 281–288.

— (1919–20). Functions of limited variation in an infinite number of dimensions. *Ann. Math.* (2) **21**, 30–38.

— (1920). Stieltjes derivatives. *Bull. Amer. Math. Soc.* **26**, 444–448.

DAVIS, B.J. (1970). On the integrability of the martingale square function. *Israel J. Math.* **8**, 187–190.

DAWSON, D.A., GÄRTNER, J. (1987). Large deviations from the McKean–Vlasov limit for weakly interacting diffusions. *Stochastics* **20**, 247–308.

DAY, M.M. (1942). Ergodic theorems for Abelian semigroups. *Trans. Amer. Math. Soc.* **51**, 399–412.

DEBES, H., KERSTAN, J., LIEMANT, A., MATTHES, K. (1970–71). Verallgemeinerung eines Satzes von Dobrushin I, III. *Math. Nachr.* **47**, 183–244; **50**, 99–139.

DELLACHERIE, C. (1972). *Capacités et Processus Stochastiques.* Springer, Berlin.

— (1980). Un survol de la théorie de l'intégrale stochastique. *Stoch. Proc. Appl.* **10**, 115-144.

DELLACHERIE, C., MAISONNEUVE, B., MEYER, P.A. (1992). *Probabilités et Potentiel, V.* Hermann, Paris.

DELLACHERIE, C., MEYER, P.A. (1975–87). *Probabilités et Potentiel, I–IV.* Hermann, Paris. Engl. trans., North-Holland.

DEMBO, A., ZEITOUNI, O. (1998). *Large Deviations Techniques and Applications*, 2nd ed. Springer, NY.

DERMAN, C. (1954). Ergodic property of the Brownian motion process. *Proc. Natl. Acad. Sci. USA* **40**, 1155–1158.

DEUSCHEL, J.D., STROOCK, D.W. (1989). *Large Deviations.* Academic Press, Boston.

DOEBLIN, W. (1938a). Exposé de la théorie des chaînes simples constantes de Markov à un nombre fini d'états. *Rev. Math. Union Interbalkan.* **2**, 77–105.

— (1938b). Sur deux problèmes de M. Kolmogoroff concernant les chaînes dénombrables. *Bull. Soc. Math. France* **66**, 210–220.

— (1939a). Sur les sommes d'un grand nombre de variables aléatoires indépendantes. *Bull. Sci. Math.* **63**, 23–64.

— (1939b). Sur certains mouvements aléatoires discontinus. *Skand. Aktuarietidskr.* **22**, 211–222.

— (1940). Eléments d'une théorie générale des chaînes simples constantes de Markoff. *Ann. Sci. Ecole Norm. Sup. 3* **57**, 61–111.

DÖHLER, R. (1980). On the conditional independence of random events. *Th. Probab. Appl.* **25**, 628–634.

DOLÉANS(-DADE), C. (1967a). Processus croissants naturel et processus croissants très bien mesurable. *C.R. Acad. Sci. Paris* **264**, 874–876.
— (1967b). Intégrales stochastiques dépendant d'un paramètre. *Publ. Inst. Stat. Univ. Paris* **16**, 23–34.
— (1970). Quelques applications de la formule de changement de variables pour les semimartingales. *Z. Wahrsch. verw. Geb.* **16**, 181–194.

DOLÉANS-DADE, C., MEYER, P.A. (1970). Intégrales stochastiques par rapport aux martingales locales. *Lect. Notes in Math.* **124**, 77–107. Springer, Berlin.

DONSKER, M.D. (1951–52). An invariance principle for certain probability limit theorems. *Mem. Amer. Math. Soc.* **6**.
— (1952). Justification and extension of Doob's heuristic approach to the Kolmogorov–Smirnov theorems. *Ann. Math. Statist.* **23**, 277–281.

DONSKER, M.D., VARADHAN, S.R.S. (1975–83). Asymptotic evaluation of certain Markov process expectations for large time, I–IV. *Comm. Pure Appl. Math.* **28**, 1–47, 279–301; **29**, 389–461; **36**, 183–212.

DOOB, J.L. (1936). Note on probability. *Ann. Math.* (2) **37**, 363–367.
— (1937). Stochastic processes depending on a continuous parameter. *Trans. Amer. Math. Soc.* **42**, 107–140.
— (1938). Stochastic processes with an integral-valued parameter. *Trans. Amer. Math. Soc.* **44**, 87–150.
— (1940). Regularity properties of certain families of chance variables. *Trans. Amer. Math. Soc.* **47**, 455–486.
— (1942a). The Brownian movement and stochastic equations. *Ann. Math.* **43**, 351–369.
— (1942b). Topics in the theory of Markoff chains. *Trans. Amer. Math. Soc.* **52**, 37–64.
— (1945). Markoff chains—denumerable case. *Trans. Amer. Math. Soc.* **58**, 455–473.
— (1947). Probability in function space. *Bull. Amer. Math. Soc.* **53**, 15–30.
— (1948a). Asymptotic properties of Markov transition probabilities. *Trans. Amer. Math. Soc.* **63**, 393–421.
— (1948b). Renewal theory from the point of view of the theory of probability. *Trans. Amer. Math. Soc.* **63**, 422–438.
— (1949). Heuristic approach to the Kolmogorov–Smirnov theorems. *Ann. Math. Statist.* **20**, 393–403.
— (1951). Continuous parameter martingales. *Proc. 2nd Berkeley Symp. Math. Statist. Probab.*, 269–277.
— (1953). *Stochastic Processes*. Wiley, NY.
— (1954). Semimartingales and subharmonic functions. *Trans. Amer. Math. Soc.* **77**, 86–121.
— (1955). A probability approach to the heat equation. *Trans. Amer. Math. Soc.* **80**, 216–280.
— (1984). *Classical Potential Theory and its Probabilistic Counterpart*. Springer, NY.
— (1994). *Measure Theory*. Springer, NY.

DUBINS, L.E. (1968). On a theorem of Skorohod. *Ann. Math. Statist.* **39**, 2094–2097.

DUBINS, L.E., SCHWARZ, G. (1965). On continuous martingales. *Proc. Natl. Acad. Sci. USA* **53**, 913–916.

DUDLEY, R.M. (1966). Weak convergence of probabilities on nonseparable metric spaces and empirical measures on Euclidean spaces. *Illinois J. Math.* **10**, 109–126.

— (1967). Measures on non-separable metric spaces. *Illinois J. Math.* **11**, 449–453.

— (1968). Distances of probability measures and random variables. *Ann. Math. Statist.* **39**, 1563–1572.

— (1989). *Real Analysis and Probability*. Wadsworth, Brooks & Cole, Pacific Grove, CA.

DUNFORD, N. (1939). A mean ergodic theorem. *Duke Math. J.* **5**, 635–646.

DUNFORD, N., SCHWARTZ, J.T. (1956). Convergence almost everywhere of operator averages. *J. Rat. Mech. Anal.* **5**, 129–178.

DURRETT, R. (1984). *Brownian Motion and Martingales in Analysis*. Wadsworth, Belmont, CA.

— (1995). *Probability Theory and Examples*, 2nd ed. Wadsworth, Brooks & Cole, Pacific Grove, CA.

DVORETZKY, A. (1972). Asymptotic normality for sums of dependent random variables. *Proc. 6th Berkeley Symp. Math. Statist. Probab.* **2**, 513–535.

DYNKIN, E.B. (1952). Criteria of continuity and lack of discontinuities of the second kind for trajectories of a Markov stochastic process (Russian). *Izv. Akad. Nauk SSSR, Ser. Mat.* **16**, 563–572.

— (1955a). Infinitesimal operators of Markov stochastic processes (Russian). *Dokl. Akad. Nauk SSSR* **105**, 206–209.

— (1955b). Continuous one-dimensional Markov processes (Russian). *Dokl. Akad. Nauk SSSR* **105**, 405–408.

— (1956). Markov processes and semigroups of operators. Infinitesimal operators of Markov processes. *Th. Probab. Appl.* **1**, 25–60.

— (1959). One-dimensional continuous strong Markov processes. *Th. Probab. Appl.* **4**, 3–54.

— (1961). *Theory of Markov Processes*. Engl. trans., Prentice-Hall and Pergamon Press, Englewood Cliffs, NJ, and Oxford. (Russian orig. 1959.)

— (1965). *Markov Processes, Vols. 1–2*. Engl. trans., Springer, Berlin. (Russian orig. 1963.)

— (1978). Sufficient statistics and extreme points. *Ann. Probab.* **6**, 705–730.

DYNKIN, E.B., YUSHKEVICH, A.A. (1956). Strong Markov processes. *Th. Probab. Appl.* **1**, 134–139.

EINSTEIN, A. (1905). On the movement of small particles suspended in a stationary liquid demanded by the molecular-kinetic theory of heat. Engl. trans. in *Investigations on the Theory of the Brownian Movement*. Repr. Dover, NY 1956.

— (1906). On the theory of Brownian motion. Engl. trans. in *Investigations on the Theory of the Brownian Movement*. Repr. Dover, NY 1956.

ELLIOTT, R.J. (1982). *Stochastic Calculus and Applications*. Springer, NY.

ELLIS, R.S. (1985). *Entropy, Large Deviations, and Statistical Mechanics*. Springer, NY.

ENGELBERT, H.J., SCHMIDT, W. (1981). On the behaviour of certain functionals of the Wiener process and applications to stochastic differential equations. *Lect. Notes in Control and Inform. Sci.* **36**, 47–55.

— (1984). On one-dimensional stochastic differential equations with generalized drift. *Lect. Notes in Control and Inform. Sci.* **69**, 143–155. Springer, Berlin.

— (1985). On solutions of stochastic differential equations without drift. *Z. Wahrsch. verw. Geb.* **68**, 287–317.

ERDŐS, P., FELLER, W., POLLARD, H. (1949). A theorem on power series. *Bull. Amer. Math. Soc.* **55**, 201–204.

ERDŐS, P., KAC, M. (1946). On certain limit theorems in the theory of probability. *Bull. Amer. Math. Soc.* **52**, 292–302.

— (1947). On the number of positive sums of independent random variables. *Bull. Amer. Math. Soc.* **53**, 1011–1020.

ERLANG, A.K. (1909). The theory of probabilities and telephone conversations. *Nyt. Tidskr. Mat. B* **20**, 33–41.

ETHIER, S.N., KURTZ, T.G. (1986). *Markov Processes: Characterization and Convergence.* Wiley, NY.

FABER, G. (1910). Über stetige Funktionen, II. *Math. Ann.* **69**, 372–443.

FARRELL, R.H. (1962). Representation of invariant measures. *Illinois J. Math.* **6**, 447–467.

FATOU, P. (1906). Séries trigonométriques et séries de Taylor. *Acta Math.* **30**, 335–400.

FELL, J.M.G. (1962). A Hausdorff topology for the closed subsets of a locally compact non-Hausdorff space. *Proc. Amer. Math. Soc.* **13**, 472–476.

FELLER, W. (1935–37). Über den zentralen Grenzwertsatz der Wahrscheinlichkeitstheorie, I–II. *Math. Z.* **40**, 521–559; **42**, 301–312.

— (1936). Zur Theorie der stochastischen Prozesse (Existenz und Eindeutigkeitssätze). *Math. Ann.* **113**, 113–160.

— (1937). On the Kolmogoroff-P. Lévy formula for infinitely divisible distribution functions. *Proc. Yugoslav Acad. Sci.* **82**, 95–112.

— (1940). On the integro-differential equations of purely discontinuous Markoff processes. *Trans. Amer. Math. Soc.* **48**, 488–515; **58**, 474.

— (1949). Fluctuation theory of recurrent events. *Trans. Amer. Math. Soc.* **67**, 98–119.

— (1952). The parabolic differential equations and the associated semi-groups of transformations. *Ann. Math.* **55**, 468–519.

— (1954). Diffusion processes in one dimension. *Trans. Amer. Math. Soc.* **77**, 1–31.

— (1968, 1971). *An Introduction to Probability Theory and its Applications,* **1** (3rd ed.); **2** (2nd ed.). Wiley, NY (1st eds. 1950, 1966).

FELLER, W., OREY, S. (1961). A renewal theorem. *J. Math. Mech.* **10**, 619–624.

FEYNMAN, R.P. (1948). Space-time approach to nonrelativistic quantum mechanics. *Rev. Mod. Phys.* **20**, 367–387.

DE FINETTI, B. (1929). Sulle funzioni ad incremento aleatorio. *Rend. Acc. Naz. Lincei* **10**, 163–168.

— (1930). Fuzione caratteristica di un fenomeno aleatorio. *Mem. R. Acc. Lincei* (6) **4**, 86–133.

— (1937). La prévision: ses lois logiques, ses sources subjectives. *Ann. Inst. H. Poincaré* **7**, 1–68.

FISK, D.L. (1965). Quasimartingales. *Trans. Amer. Math. Soc.* **120**, 369–389.

— (1966). Sample quadratic variation of continuous, second-order martingales. *Z. Wahrsch. verw. Geb.* **6**, 273–278.

FORTET, R. (1943). Les fonctions aléatoires du type de Markoff associées à certaines équations linéaires aux dérivées partielles du type parabolique. *J. Math. Pures Appl.* **22**, 177–243.

FRANKEN, P., KÖNIG, D., ARNDT, U., SCHMIDT, V. (1981). *Queues and Point Processes*. Akademie-Verlag, Berlin.

FRÉCHET, M. (1928). *Les Espaces Abstraits*. Gauthier-Villars, Paris.

FREEDMAN, D. (1962–63). Invariants under mixing which generalize de Finetti's theorem. *Ann. Math. Statist.* **33**, 916–923; **34**, 1194–1216.

— (1971a). *Markov Chains*. Holden-Day, San Francisco. Repr. Springer, NY 1983.

— (1971b). *Brownian Motion and Diffusion*. Holden-Day, San Francisco. Repr. Springer, NY 1983.

FREIDLIN, M.I., WENTZEL, A.D. (1970). On small random permutations of dynamical systems. *Russian Math. Surveys* **25**, 1–55.

— (1998). *Random Perturbations of Dynamical Systems*. Engl. trans., Springer, NY. (Russian orig. 1979.)

FROSTMAN, O. (1935). Potentiel d'équilibre et capacité des ensembles avec quelques applications à la théorie des fonctions. *Medd. Lunds Univ. Mat. Sem.* **3**, 1-118.

FUBINI, G. (1907). Sugli integrali multipli. *Rend. Acc. Naz. Lincei* **16**, 608–614.

FURSTENBERG, H., KESTEN, H. (1960). Products of random matrices. *Ann. Math. Statist.* **31**, 457–469.

GALMARINO, A.R. (1963). Representation of an isotropic diffusion as a skew product. *Z. Wahrsch. verw. Geb.* **1**, 359–378.

GARSIA, A.M. (1965). A simple proof of E. Hopf's maximal ergodic theorem. *J. Math. Mech.* **14**, 381–382.

— (1973). *Martingale Inequalities: Seminar Notes on Recent Progress*. Math. Lect. Notes Ser. Benjamin, Reading, MA.

GAUSS, C.F. (1809). *Theory of Motion of the Heavenly Bodies*. Engl. trans., Dover, NY 1963.

— (1840). Allgemeine Lehrsätze in Beziehung auf die im vehrkehrten Verhältnisse des Quadrats der Entfernung wirkenden Anziehungs- und Abstossungs-Kräfte. *Gauss Werke* **5**, 197–242. Göttingen 1867.

GETOOR, R.K. (1990). *Excessive Measures*. Birkhäuser, Boston.

GETOOR, R.K., SHARPE, M.J. (1972). Conformal martingales. *Invent. Math.* **16**, 271–308.

GIHMAN, I.I. (1947). On a method of constructing random processes (Russian). *Dokl. Akad. Nauk SSSR* **58**, 961–964.

— (1950–51). On the theory of differential equations for random processes, I–II (Russian). *Ukr. Mat. J.* **2**:4, 37–63; **3**:3, 317–339.

GIHMAN, I.I., SKOROHOD, A.V. (1965). *Introduction to the Theory of Random Processes.* Engl. trans., Saunders, Philadelphia. Repr. Dover, Mineola 1996.

— (1974–79). *The Theory of Stochastic Processes*, **1–3**. Engl. trans., Springer, Berlin.

GIRSANOV, I.V. (1960). On transforming a certain class of stochastic processes by absolutely continuous substitution of measures. *Th. Probab. Appl.* **5**, 285–301.

GLIVENKO, V.I. (1933). Sulla determinazione empirica della leggi di probabilità. *Giorn. Ist. Ital. Attuari* **4**, 92–99.

GNEDENKO, B.V. (1939). On the theory of limit theorems for sums of independent random variables (Russian). *Izv. Akad. Nauk SSSR Ser. Mat.* 181–232, 643–647.

GNEDENKO, B.V., KOLMOGOROV, A.N. (1968). *Limit Distributions for Sums of Independent Random Variables.* Engl. trans., 2nd ed., Addison-Wesley, Reading, MA. (Russian orig. 1949.)

GOLDMAN, J.R. (1967). Stochastic point processes: Limit theorems. *Ann. Math. Statist.* **38**, 771–779.

GOLDSTEIN, J.A. (1976). Semigroup-theoretic proofs of the central limit theorem and other theorems of analysis. *Semigroup Forum* **12**, 189–206.

GOLDSTEIN, S. (1979). Maximal coupling. *Z. Wahrsch. verw. Geb.* **46**, 193–204.

GRANDELL, J. (1976). *Doubly Stochastic Poisson Processes. Lect. Notes in Math.* **529**. Springer, Berlin.

GREEN, G. (1828). An essay on the application of mathematical analysis to the theories of electricity and magnetism. Repr. in *Mathematical Papers*, Chelsea, NY 1970.

GREENWOOD, P., PITMAN, J. (1980). Construction of local time and Poisson point processes from nested arrays. *J. London Math. Soc.* (2) **22**, 182–192.

GRIFFEATH, D. (1975). A maximal coupling for Markov chains. *Z. Wahrsch. verw. Geb.* **31**, 95–106.

GRIGELIONIS, B. (1963). On the convergence of sums of random step processes to a Poisson process. *Th. Probab. Appl.* **8**, 172–182.

— (1971). On the representation of integer-valued measures by means of stochastic integrals with respect to Poisson measure. *Litovsk. Mat. Sb.* **11**, 93–108.

HAAR, A. (1933). Der Maßbegriff in der Theorie der kontinuerlichen Gruppen. *Ann. Math.* **34**, 147–169.

HAGBERG, J. (1973). Approximation of the summation process obtained by sampling from a finite population. *Th. Probab. Appl.* **18**, 790–803.

HAHN, H. (1921). *Theorie der reellen Funktionen.* Julius Springer, Berlin.

HÁJEK, J. (1960). Limiting distributions in simple random sampling from a finite population. *Magyar Tud. Akad. Mat. Kutató Int. Közl.* **5**, 361–374.

HALL, P., HEYDE, C.C. (1980). *Martingale Limit Theory and its Application.* Academic Press, NY.

HALMOS, P.R. (1950). *Measure Theory*, Van Nostrand, Princeton. Repr. Springer, NY 1974.

HARDY, G.H., LITTLEWOOD, J.E. (1930). A maximal theorem with function-theoretic applications. *Acta Math.* **54**, 81–116.

HARRIS, T.E. (1956). The existence of stationary measures for certain Markov processes. *Proc. 3rd Berkeley Symp. Math. Statist. Probab.* **2**, 113–124.

— (1971). Random measures and motions of point processes. *Z. Wahrsch. verw. Geb.* **18**, 85–115.

HARTMAN, P., WINTNER, A. (1941). On the law of the iterated logarithm. *J. Math.* **63**, 169–176.

HELLY, E. (1911–12). Über lineare Funktionaloperatoren. *Sitzungsber. Nat. Kais. Akad. Wiss.* **121**, 265–297.

HEWITT, E., SAVAGE, L.J. (1955). Symmetric measures on Cartesian products. *Trans. Amer. Math. Soc.* **80**, 470–501.

HILLE, E. (1948). Functional analysis and semi-groups. *Amer. Math. Colloq. Publ.* **31**, NY.

HITCZENKO, P. (1990). Best constants in martingale version of Rosenthal's inequality. *Ann. Probab.* **18**, 1656–1668.

HÖLDER, O. (1889). Über einen Mittelwertsatz. *Nachr. Akad. Wiss. Göttingen, math.phys. Kl.*, 38–47.

HOPF, E. (1954). The general temporally discrete Markov process. *J. Rat. Mech. Anal.* **3**, 13–45.

HOROWITZ, J. (1972). Semilinear Markov processes, subordinators and renewal theory. *Z. Wahrsch. verw. Geb.* **24**, 167–193.

HUNT, G.A. (1956). Some theorems concerning Brownian motion. *Trans. Amer. Math. Soc.* **81**, 294–319.

— (1957–58). Markoff processes and potentials, I–III. *Illinois J. Math.* **1**, 44–93, 316–369; **2**, 151–213.

HUREWICZ, W. (1944). Ergodic theorem without invariant measure. *Ann. Math.* **45**, 192–206.

HURWITZ, A. (1897). Über die Erzeugung der Invarianten durch Integration. *Nachr. Ges. Göttingen, math.-phys. Kl.*, 71–90.

IKEDA, N., WATANABE, S. (1989). *Stochastic Differential Equations and Diffusion Processes*, 2nd ed. North-Holland and Kodansha, Amsterdam and Tokyo.

IOFFE, D. (1991). On some applicable versions of abstract large deviations theorems. *Ann. Probab.* **19**, 1629–1639.

IONESCU TULCEA, A. (1960). Contributions to information theory for abstract alphabets. *Ark. Mat.* **4**, 235–247.

IONESCU TULCEA, C.T. (1949–50). Mesures dans les espaces produits. *Atti Accad. Naz. Lincei Rend.* **7**, 208–211.

ITÔ, K. (1942a). Differential equations determining Markov processes (Japanese). *Zenkoku Shijō Sūgaku Danwakai* **244**:1077, 1352–1400.

— (1942b). On stochastic processes (I) (Infinitely divisible laws of probability). *Jap. J. Math.* **18**, 261–301.

— (1944). Stochastic integral. *Proc. Imp. Acad. Tokyo* **20**, 519–524.

— (1946). On a stochastic integral equation. *Proc. Imp. Acad. Tokyo* **22**, 32–35.
— (1951a). On a formula concerning stochastic differentials. *Nagoya Math. J.* **3**, 55–65.
— (1951b). On stochastic differential equations. *Mem. Amer. Math. Soc.* **4**, 1–51.
— (1951c). Multiple Wiener integral. *J. Math. Soc. Japan* **3**, 157–169.
— (1957). *Stochastic Processes* (Japanese). Iwanami Shoten, Tokyo.
— (1972). Poisson point processes attached to Markov processes. *Proc. 6th Berkeley Symp. Math. Statist. Probab.* **3**, 225–239.
— (1984). *Introduction to Probability Theory.* Engl. trans., Cambridge Univ. Press.

ITÔ, K., McKEAN, H.P. (1965). *Diffusion Processes and their Sample Paths.* Repr. Springer, Berlin 1996.

ITÔ, K., WATANABE, S. (1965). Transformation of Markov processes by multiplicative functionals. *Ann. Inst. Fourier* **15**, 15–30.

JACOD, J. (1975). Multivariate point processes: Predictable projection, Radon-Nikodym derivative, representation of martingales. *Z. Wahrsch. verw. Geb.* **31**, 235–253.

— (1979). *Calcul Stochastique et Problèmes de Martingales. Lect. Notes in Math.* **714**. Springer, Berlin.

JACOD, J., SHIRYAEV, A.N. (1987). *Limit Theorems for Stochastic Processes.* Springer, Berlin.

JAGERS, P. (1972). On the weak convergence of superpositions of point processes. *Z. Wahrsch. verw. Geb.* **22**, 1–7.

— (1974). Aspects of random measures and point processes. *Adv. Probab. Rel. Topics* **3**, 179–239. Marcel Dekker, NY.

JAMISON, B., OREY, S. (1967). Markov chains recurrent in the sense of Harris. *Z. Wahrsch. verw. Geb.* **8**, 206–223.

JENSEN, J.L.W.V. (1906). Sur les fonctions convexes et les inégalités entre les valeurs moyennes. *Acta Math.* **30**, 175–193.

JESSEN, B. (1934). The theory of integration in a space of an infinite number of dimensions. *Acta Math.* **63**, 249–323.

JOHNSON, W.B., SCHECHTMAN, G., ZINN, J. (1985). Best constants in moment inequalities for linear combinations of independent and exchangeable random variables. *Ann. Probab.* **13**, 234–253.

JORDAN, C. (1881). Sur la série de Fourier. *C.R. Acad. Sci. Paris* **92**, 228–230.

KAC, M. (1947). On the notion of recurrence in discrete stochastic processes. *Bull. Amer. Math. Soc.* **53**, 1002–1010.

— (1949). On distributions of certain Wiener functionals. *Trans. Amer. Math. Soc.* **65**, 1–13.

— (1951). On some connections between probability theory and differential and integral equations. *Proc. 2nd Berkeley Symp. Math. Statist. Probab.*, 189–215. Univ. of California Press, Berkeley.

KAKUTANI, S. (1940). Ergodic theorems and the Markoff process with a stable distribution. *Proc. Imp. Acad. Tokyo* **16**, 49–54.

— (1944a). On Brownian motions in n-space. *Proc. Imp. Acad. Tokyo* **20**, 648–652.

— (1944b). Two-dimensional Brownian motion and harmonic functions. *Proc. Imp. Acad. Tokyo* **20**, 706–714.

— (1945). Markoff process and the Dirichlet problem. *Proc. Japan Acad.* **21**, 227–233.

KALLENBERG, O. (1973a). Characterization and convergence of random measures and point processes. *Z. Wahrsch. verw. Geb.* **27**, 9–21.

— (1973b). Canonical representations and convergence criteria for processes with interchangeable increments. *Z. Wahrsch. verw. Geb.* **27**, 23–36.

— (1986). *Random Measures*, 4th ed. Akademie-Verlag and Academic Press, Berlin and London (1st ed. 1975).

— (1987). Homogeneity and the strong Markov property. *Ann. Probab.* **15**, 213–240.

— (1988). Spreading and predictable sampling in exchangeable sequences and processes. *Ann. Probab.* **16**, 508–534.

— (1990). Random time change and an integral representation for marked stopping times. *Probab. Th. Rel. Fields* **86**, 167–202.

— (1992). Some time change representations of stable integrals, via predictable transformations of local martingales. *Stoch. Proc. Appl.* **40**, 199–223.

— (1996a). On the existence of universal functional solutions to classical SDEs. *Ann. Probab.* **24**, 196–205.

— (1996b). Improved criteria for distributional convergence of point processes. *Stoch. Proc. Appl.* **64**, 93–102.

— (1999a). Ballot theorems and sojourn laws for stationary processes. *Ann. Probab.* **27**, 2011–2019.

— (1999b). Asymptotically invariant sampling and averaging from stationary-like processes. *Stoch. Proc. Appl.* **82**, 195–204.

KALLENBERG, O., SZTENCEL, R. (1991). Some dimension-free features of vector-valued martingales. *Probab. Th. Rel. Fields* **88**, 215–247.

KALLIANPUR, G. (1980). *Stochastic Filtering Theory*. Springer, NY.

KAPLAN, E.L. (1955). Transformations of stationary random sequences. *Math. Scand.* **3**, 127–149.

KARAMATA, J. (1930). Sur une mode de croissance régulière des fonctions. *Mathematica (Cluj)* **4**, 38–53.

KARATZAS, I., SHREVE, S.E. (1991). *Brownian Motion and Stochastic Calculus*, 2nd ed. Springer, NY.

KAZAMAKI, N. (1972). Change of time, stochastic integrals and weak martingales. *Z. Wahrsch. verw. Geb.* **22**, 25–32.

KEMENY, J.G., SNELL, J.L., KNAPP, A.W. (1966). *Denumerable Markov Chains*. Van Nostrand, Princeton.

KENDALL, D.G. (1974). Foundations of a theory of random sets. In *Stochastic Geometry* (eds. E.F. Harding, D.G. Kendall), pp. 322–376. Wiley, NY.

KHINCHIN, A.Y. (1923). Über dyadische Brücke. *Math. Z.* **18**, 109–116.

— (1924). Über einen Satz der Wahrscheinlichkeitsrechnung. *Fund. Math.* **6**, 9–20.

— (1929). Über einen neuen Grenzwertsatz der Wahrscheinlichkeitsrechnung. *Math. Ann.* **101**, 745–752.

— (1933). Zur mathematischen Begründing der statistischen Mechanik. *Z. Angew. Math. Mech.* **13**, 101–103.
— (1933). *Asymptotische Gesetze der Wahrscheinlichkeitsrechnung*, Springer, Berlin. Repr. Chelsea, NY 1948.
— (1934). Korrelationstheorie der stationären stochastischen Prozesse. *Math. Ann.* **109**, 604–615.
— (1937). Zur Theorie der unbeschränkt teilbaren Verteilungsgesetze. *Mat. Sb.* **2**, 79–119.
— (1938). *Limit Laws for Sums of Independent Random Variables* (Russian). Moscow.
— (1960). *Mathematical Methods in the Theory of Queuing*. Engl. trans., Griffin, London. (Russian orig. 1955.)

KHINCHIN, A.Y., KOLMOGOROV, A.N. (1925). Über Konvergenz von Reihen deren Glieder durch den Zufall bestimmt werden. *Mat. Sb.* **32**, 668–676.

KINGMAN, J.F.C. (1964). On doubly stochastic Poisson processes. *Proc. Cambridge Phil. Soc.* **60**, 923–930.
— (1967). Completely random measures. *Pac.. J. Math.* **21**, 59–78.
— (1968). The ergodic theory of subadditive stochastic processes. *J. Roy. Statist. Soc. (B)* **30**, 499–510.
— (1972). *Regenerative Phenomena*. Wiley, NY.
— (1993). *Poisson Processes*. Clarendon Press, Oxford.

KINNEY, J.R. (1953). Continuity properties of Markov processes. *Trans. Amer. Math. Soc.* **74**, 280–302.

KNIGHT, F.B. (1963). Random walks and a sojourn density process of Brownian motion. *Trans. Amer. Math. Soc.* **107**, 56–86.
— (1971). A reduction of continuous, square-integrable martingales to Brownian motion. *Lect. Notes in Math.* **190**, 19–31. Springer, Berlin.

KOLMOGOROV, A.N. (1928–29). Über die Summen durch den Zufall bestimmter unabhängiger Grössen. *Math. Ann.* **99**, 309–319; **102**, 484–488.
— (1929). Über das Gesatz des iterierten Logarithmus. *Math. Ann.* **101**, 126–135.
— (1930). Sur la loi forte des grandes nombres. *C.R. Acad. Sci. Paris* **191**, 910–912.
— (1931a). Über die analytischen Methoden in der Wahrscheinlichkeitsrechnung. *Math. Ann.* **104**, 415–458.
— (1931b). Eine Verallgemeinerung des Laplace–Liapounoffschen Satzes. *Izv. Akad. Nauk USSR, Otdel. Matem. Yestestv. Nauk* **1931**, 959–962.
— (1932). Sulla forma generale di un processo stocastico omogeneo (un problema di B. de Finetti). *Atti Accad. Naz. Lincei Rend.* (6) **15**, 805–808, 866–869.
— (1933a). Über die Grenzwertsätze der Wahrscheinlichkeitsrechnung. *Izv. Akad. Nauk USSR, Otdel. Matem. Yestestv. Nauk* **1933**, 363–372.
— (1933b). Zur Theorie der stetigen zufälligen Prozesse. *Math. Ann.* **108**, 149–160.
— (1933c). *Foundations of the Theory of Probability* (German), Springer, Berlin. Engl. trans., Chelsea, NY 1956.
— (1935). Some current developments in probability theory (in Russian). *Proc. 2nd All-Union Math. Congr.* **1**, 349–358. Akad. Nauk SSSR, Leningrad.
— (1936a). Anfangsgründe der Markoffschen Ketten mit unendlich vielen möglichen Zuständen. *Mat. Sb.* **1**, 607–610.

— (1936b). Zur Theorie der Markoffschen Ketten. *Math. Ann.* **112**, 155–160.
— (1937). Zur Umkehrbarkeit der statistischen Naturgesetze. *Math. Ann.* **113**, 766–772.
— (1956). On Skorohod convergence. *Th. Probab. Appl.* **1**, 213–222.

KOLMOGOROV, A.N., LEONTOVICH, M.A. (1933). Zur Berechnung der mittleren Brownschen Fläche. *Physik. Z. Sowjetunion* **4**, 1–13.

KOMLÓS, J., MAJOR, P., TUSNÁDY, G. (1975–76). An approximation of partial sums of independent r.v.'s and the sample d.f., I–II. *Z. Wahrsch. verw. Geb.* **32**, 111–131; **34**, 33–58.

KÖNIG, D., MATTHES, K. (1963). Verallgemeinerung der Erlangschen Formeln, I. *Math. Nachr.* **26**, 45–56.

KOOPMAN, B.O. (1931). Hamiltonian systems and transformations in Hilbert space. *Proc. Nat. Acad. Sci. USA* **17**, 315–318.

KRENGEL, U. (1985). *Ergodic Theorems*. de Gruyter, Berlin.

KRICKEBERG, K. (1956). Convergence of martingales with a directed index set. *Trans. Amer. Math. Soc.* **83**, 313–357.
— (1972). The Cox process. *Symp. Math.* **9**, 151–167.

KRYLOV, N., BOGOLIOUBOV, N. (1937). La théorie générale de la mesure dans son application à l'étude des systèmes de la mécanique non linéaires. *Ann. Math.* **38**, 65–113.

KULLBACK, S., LEIBLER, R.A. (1951). On information and sufficiency. *Ann. Math. Statist.* **22**, 79–86.

KUNITA, H. (1990). *Stochastic Flows and Stochastic Differential Equations*. Cambridge Univ. Press, Cambridge.

KUNITA, H., WATANABE, S. (1967). On square integrable martingales. *Nagoya Math. J.* **30**, 209–245.

KURTZ, T.G. (1969). Extensions of Trotter's operator semigroup approximation theorems. *J. Funct. Anal.* **3**, 354–375.
— (1975). Semigroups of conditioned shifts and approximation of Markov processes. *Ann. Probab.* **3**, 618–642.

KWAPIEŃ, S., WOYCZYŃSKI, W.A. (1992). *Random Series and Stochastic Integrals: Single and Multiple*. Birkhäuser, Boston.

LANGEVIN, P. (1908). Sur la théorie du mouvement brownien. *C.R. Acad. Sci. Paris* **146**, 530–533.

LAPLACE, P.S. DE (1774). Mémoire sur la probabilité des causes par les événemens. Engl. trans. in *Statistical Science* **1**, 359–378.
— (1809). Mémoire sur divers points d'analyse. Repr. in *Oeuvres Complètes de Laplace* **14**, 178–214. Gauthier-Villars, Paris 1886–1912.
— (1812–20). *Théorie Analytique des Probabilités*, 3rd ed. Repr. in *Oeuvres Complètes de Laplace* **7**. Gauthier-Villars, Paris 1886–1912.

LAST, G., BRANDT, A. (1995). *Marked Point Processes on the Real Line: The Dynamic Approach*. Springer, NY.

LEADBETTER, M.R., LINDGREN, G., ROOTZÉN, H. (1983). *Extremes and Related Properties of Random Sequences and Processes*. Springer, NY.

LEBESGUE, H. (1902). Intégrale, longeur, aire. *Ann. Mat. Pura Appl.* **7**, 231–359.
— (1904). *Leçons sur l'Intégration et la Recherche des Fonctions Primitives.* Paris.

LE CAM, L. (1957). Convergence in distribution of stochastic processes. *Univ. California Publ. Statist.* **2**, 207–236.

LE GALL, J.F. (1983). Applications des temps locaux aux équations différentielles stochastiques unidimensionelles. *Lect. Notes in Math.* **986**, 15–31.

LEVI, B. (1906a). Sopra l'integrazione delle serie. *Rend. Ist. Lombardo Sci. Lett.* (2) **39**, 775–780.
— (1906b). Sul principio de Dirichlet. *Rend. Circ. Mat. Palermo* **22**, 293–360.

LÉVY, P. (1922a). Sur le rôle de la loi de Gauss dans la théorie des erreurs. *C.R. Acad. Sci. Paris* **174**, 855–857.
— (1922b). Sur la loi de Gauss. *C.R. Acad. Sci. Paris* 1682–1684.
— (1922c). Sur la détermination des lois de probabilité par leurs fonctions caractéristiques. *C.R. Acad. Sci. Paris* **175**, 854–856.
— (1924). Théorie des erreurs. La loi de Gauss et les lois exceptionelles. *Bull. Soc. Math. France* **52**, 49–85.
— (1925). *Calcul des Probabilités.* Gauthier-Villars, Paris.
— (1934–35). Sur les intégrales dont les éléments sont des variables aléatoires indépendantes. *Ann. Scuola Norm. Sup. Pisa* (2) **3**, 337–366; **4**, 217–218.
— (1935a). Propriétés asymptotiques des sommes de variables aléatoires indépendantes ou enchaînées. *J. Math. Pures Appl.* (8) **14**, 347–402.
— (1935b). Propriétés asymptotiques des sommes de variables aléatoires enchaînées. *Bull. Sci. Math.* (2) **59**, 84–96, 109–128.
— (1939). Sur certain processus stochastiques homogènes. *Comp. Math.* **7**, 283–339.
— (1940). Le mouvement brownien plan. *Amer. J. Math.* **62**, 487–550.
— (1954). *Théorie de l'Addition des Variables Aléatoires*, 2nd ed. Gauthier-Villars, Paris (1st ed. 1937).
— (1965). *Processus Stochastiques et Mouvement Brownien*, 2nd ed. Gauthier-Villars, Paris (1st ed. 1948).

LIAPOUNOV, A.M. (1901). Nouvelle forme du théorème sur la limite des probabilités. *Mem. Acad. Sci. St. Petersbourg* **12**, 1–24.

LIGGETT, T.M. (1985). An improved subadditive ergodic theorem. *Ann. Probab.* **13**, 1279–1285.

LINDEBERG, J.W. (1922a). Eine neue Herleitung des Exponentialgesetzes in der Wahrscheinlichkeitsrechnung. *Math. Zeitschr.* **15**, 211–225.
— (1922b). Sur la loi de Gauss. *C.R. Acad. Sci. Paris* **174**, 1400–1402.

LINDVALL, T. (1973). Weak convergence of probability measures and random functions in the function space $D[0, \infty)$. *J. Appl. Probab.* **10**, 109–121.
— (1977). A probabilistic proof of Blackwell's renewal theorem. *Ann. Probab.* **5**, 482–485.
— (1992). *Lectures on the Coupling Method.* Wiley, NY.

LIPSTER, R.S., SHIRYAEV, A.N. (2000). *Statistics of Random Processes, I–II*, 2nd ed., Springer, Berlin.

LOÈVE, M. (1977–78). *Probability Theory* **1–2**, 4th ed. Springer, NY (1st ed. 1955).

LOMNICKI, Z., ULAM, S. (1934). Sur la théorie de la mesure dans les espaces combinatoires et son application au calcul des probabilités: I. Variables indépendantes. *Fund. Math.* **23**, 237–278.

LUKACS, E. (1970). *Characteristic Functions*, 2nd ed. Griffin, London.

LUNDBERG, F. (1903). *Approximerad Framställning av Sannolikhetsfunktionen. Återförsäkring av Kollektivrisker.* Thesis, Uppsala.

MACKEVIČIUS, V. (1974). On the question of the weak convergence of random processes in the space $D[0, \infty)$. *Lithuanian Math. Trans.* **14**, 620–623.

MAISONNEUVE, B. (1974). *Systèmes Régénératifs. Astérique* **15**. Soc. Math. de France.

MAKER, P. (1940). The ergodic theorem for a sequence of functions. *Duke Math. J.* **6**, 27–30.

MANN, H.B., WALD, A. (1943). On stochastic limit and order relations. *Ann. Math. Statist.* **14**, 217–226.

MARCINKIEWICZ, J., ZYGMUND, A. (1937). Sur les fonctions indépendantes. *Fund. Math.* **29**, 60–90.

— (1938). Quelques théorèmes sur les fonctions indépendantes. *Studia Math.* **7**, 104–120.

MARKOV, A.A. (1899). The law of large numbers and the method of least squares (Russian). *Izv. Fiz.-Mat. Obshch. Kazan Univ.* (2) **8**, 110–128.

— (1906). Extension of the law of large numbers to dependent events (Russian). *Bull. Soc. Phys. Math. Kazan* (2) **15**, 135–156.

MARUYAMA, G. (1954). On the transition probability functions of the Markov process. *Natl. Sci. Rep. Ochanomizu Univ.* **5**, 10–20.

— (1955). Continuous Markov processes and stochastic equations. *Rend. Circ. Mat. Palermo* **4**, 48–90.

MARUYAMA, G., TANAKA, H. (1957). Some properties of one-dimensional diffusion processes. *Mem. Fac. Sci. Kyushu Univ.* **11**, 117–141.

MATHERON, G. (1975). *Random Sets and Integral Geometry*. Wiley, London.

MATTHES, K. (1963). Stationäre zufällige Punktfolgen, I. *Jahresber. Deutsch. Math.-Verein.* **66**, 66–79.

MATTHES, K., KERSTAN, J., MECKE, J. (1978). *Infinitely Divisible Point Processes*. Wiley, Chichester. (German ed. 1974, Russian ed. 1982.)

MCKEAN, H.P. (1969). *Stochastic Integrals*. Academic Press, NY.

MCKEAN, H.P., TANAKA, H. (1961). Additive functionals of the Brownian path. *Mem. Coll. Sci. Univ. Kyoto, A* **33**, 479–506.

MCMILLAN, B. (1953). The basic theorems of information theory. *Ann. Math. Statist.* **24**, 196–219.

MECKE, J. (1967). Stationäre zufällige Maße auf lokalkompakten Abelschen Gruppen. *Z. Wahrsch. verw. Geb.* **9**, 36–58.

— (1968). Eine characteristische Eigenschaft der doppelt stochastischen Poissonschen Prozesse. *Z. Wahrsch. verw. Geb.* **11**, 74–81.

MÉLÉARD, S. (1986). Application du calcul stochastique à l'étude des processus de Markov réguliers sur [0, 1]. *Stochastics* **19**, 41–82.

MÉTIVIER, M. (1982). *Semimartingales: A Course on Stochastic Processes.* de Gruyter, Berlin.

MÉTIVIER, M., PELLAUMAIL, J. (1980). *Stochastic Integration.* Academic Press, NY.

MEYER, P.A. (1962). A decomposition theorem for supermartingales. *Illinois J. Math.* **6**, 193–205.

— (1963). Decomposition of supermartingales: The uniqueness theorem. *Illinois J. Math.* **7**, 1–17.

— (1966). *Probability and Potentials.* Engl. trans., Blaisdell, Waltham.

— (1967). Intégrales stochastiques, I–IV. *Lect. Notes in Math.* **39**, 72–162. Springer, Berlin.

— (1971). Démonstration simplifiée d'un théorème de Knight. *Lect. Notes in Math.* **191**, 191–195. Springer, Berlin.

— (1976). Un cours sur les intégrales stochastiques. *Lect. Notes in Math.* **511**, 245–398. Springer, Berlin.

MILLAR, P.W. (1968). Martingale integrals. *Trans. Amer. Math. Soc.* **133**, 145–166.

MINKOWSKI, H. (1907). *Diophantische Approximationen.* Teubner, Leipzig.

MITOMA, I. (1983). Tightness of probabilities on $C([0,1]; S')$ and $D([0,1]; S')$. *Ann. Probab.* **11**, 989–999.

DE MOIVRE, A. (1711–12). On the measurement of chance. Engl. trans., *Int. Statist. Rev.* **52** (1984), 229–262.

— (1718–56). *The Doctrine of Chances; or, a Method of Calculating the Probability of Events in Play*, 3rd ed. (post.) Repr. Case and Chelsea, London and NY 1967.

— (1733–56). *Approximatio ad Summam Terminorum Binomii $\overline{a+b}|^n$ in Seriem Expansi.* Translated and edited in *The Doctrine of Chances*, 2nd and 3rd eds. Repr. Case and Chelsea, London and NY 1967.

MÖNCH, G. (1971). Verallgemeinerung eines Satzes von A. Rényi. *Studia Sci. Math. Hung.* **6**, 81–90.

MOTOO, M., WATANABE, H. (1958). Ergodic property of recurrent diffusion process in one dimension. *J. Math. Soc. Japan* **10**, 272–286.

NAWROTZKI, K. (1962). Ein Grenzwertsatz für homogene zufällige Punktfolgen (Verallgemeinerung eines Satzes von A. Rényi). *Math. Nachr.* **24**, 201–217.

VON NEUMANN, J. (1932). Proof of the quasi-ergodic hypothesis. *Proc. Natl. Acad. Sci. USA* **18**, 70–82.

— (1940). On rings of operators, III. *Ann. Math.* **41**, 94–161.

NEVEU, J. (1971). *Mathematical Foundations of the Calculus of Probability.* Holden-Day, San Francisco.

— (1975). *Discrete-Parameter Martingales.* North-Holland, Amsterdam.

NGUYEN, X.X., ZESSIN, H. (1979). Ergodic theorems for spatial processes. *Z. Wahrsch. verw. Geb.* **48**, 133–158.

NIKODÝM, O.M. (1930). Sur une généralisation des intégrales de M. J. Radon. *Fund. Math.* **15**, 131–179.

NORBERG, T. (1984). Convergence and existence of random set distributions. *Ann. Probab.* **12**, 726–732.

NOVIKOV, A.A. (1971). On moment inequalities for stochastic integrals. *Th. Probab. Appl.* **16**, 538–541.

— (1972). On an identity for stochastic integrals. *Th. Probab. Appl.* **17**, 717–720.

NUALART, D. (1995). *The Malliavin Calculus and Related Topics.* Springer, NY.

ØKSENDAL, B. (1998). *Stochastic Differential Equations*, 5th ed. Springer, Berlin.

OREY, S. (1959). Recurrent Markov chains. *Pacific J. Math.* **9**, 805–827.

— (1962). An ergodic theorem for Markov chains. *Z. Wahrsch. verw. Geb.* **1**, 174–176.

— (1966). F-processes. *Proc. 5th Berkeley Symp. Math. Statist. Probab.* **2**:1, 301–313.

— (1971). *Limit Theorems for Markov Chain Transition Probabilities.* Van Nostrand, London.

ORNSTEIN, D.S. (1969). Random walks. *Trans. Amer. Math. Soc.* **138**, 1–60.

ORNSTEIN, L.S., UHLENBECK, G.E. (1930). On the theory of Brownian motion. *Phys. Review* **36**, 823–841.

OSOSKOV, G.A. (1956). A limit theorem for flows of homogeneous events. *Th. Probab. Appl.* **1**, 248–255.

OTTAVIANI, G. (1939). Sulla teoria astratta del calcolo delle probabilità proposita dal Cantelli. *Giorn. Ist. Ital. Attuari* **10**, 10–40.

PALEY, R.E.A.C. (1932). A remarkable series of orthogonal functions I. *Proc. London Math. Soc.* **34**, 241–264.

PALEY, R.E.A.C., WIENER, N. (1934). Fourier transforms in the complex domain. *Amer. Math. Soc. Coll. Publ.* **19**.

PALEY, R.E.A.C., WIENER, N., ZYGMUND, A. (1933). Notes on random functions. *Math. Z.* **37**, 647–668.

PALM, C. (1943). Intensity Variations in Telephone Traffic (German). *Ericsson Technics* **44**, 1–189. Engl. trans., *North-Holland Studies in Telecommunication* **10**, Elsevier 1988.

PAPANGELOU, F. (1972). Integrability of expected increments of point processes and a related random change of scale. *Trans. Amer. Math. Soc.* **165**, 486–506.

PARTHASARATHY, K.R. (1967). *Probability Measures on Metric Spaces.* Academic Press, NY.

PERKINS, E. (1982). Local time and pathwise uniqueness for stochastic differential equations. *Lect. Notes in Math.* **920**, 201–208. Springer, Berlin.

PETROV, V.V. (1995). *Limit Theorems of Probability Theory.* Clarendon Press, Oxford.

PHILLIPS, H.B., WIENER, N. (1923). Nets and Dirichlet problem. *J. Math. Phys.* **2**, 105–124.

PITT, H.R. (1942). Some generalizations of the ergodic theorem. *Proc. Camb. Phil. Soc.* **38**, 325–343.

POINCARÉ, H. (1890). Sur les équations aux dérivées partielles de la physique mathéma-tique. *Amer. J. Math.* **12**, 211–294.

— (1899). *Théorie du Potentiel Newtonien.* Gauthier-Villars, Paris.

Poisson, S.D. (1837). *Recherches sur la Probabilité des Jugements en Matière Criminelle et en Matière Civile, Précédées des Règles Générales du Calcul des Probabilités.* Bachelier, Paris.

Pollaczek, F. (1930). Über eine Aufgabe der Wahrscheinlichkeitstheorie, I–II. *Math. Z.* **32**, 64–100, 729–750.

Pollard, D. (1984). *Convergence of Stochastic Processes.* Springer, NY.

Pólya, G. (1920). Über den zentralen Grenzwertsatz der Wahrscheinlichkeitsrechnung und das Momentenproblem. *Math. Z.* **8**, 171–181.

— (1921). Über eine Aufgabe der Wahrscheinlichkeitsrechnung betreffend die Irrfahrt im Strassennetz. *Math. Ann.* **84**, 149–160.

Port, S.C., Stone, C.J. (1978). *Brownian Motion and Classical Potential Theory.* Academic Press, NY.

Pospišil, B. (1935–36). Sur un problème de M.M.S. Bernstein et A. Kolmogoroff. *Časopis Pěst. Mat. Fys.* **65**, 64–76.

Prohorov, Y.V. (1956). Convergence of random processes and limit theorems in probability theory. *Th. Probab. Appl.* **1**, 157–214.

— (1959). Some remarks on the strong law of large numbers. *Th. Probab. Appl.* **4**, 204–208.

— (1961). Random measures on a compactum. *Soviet Math. Dokl.* **2**, 539–541.

Protter, P. (1990). *Stochastic Integration and Differential Equations.* Springer, Berlin.

Pukhalsky, A.A. (1991). On functional principle of large deviations. In *New Trends in Probability and Statistics* (V. Sazonov and T. Shervashidze, eds.), 198–218. VSP Moks'las, Moscow.

Radon, J. (1913). Theorie und Anwendungen der absolut additiven Mengenfunktionen. *Wien Akad. Sitzungsber.* **122**, 1295–1438.

Rao, K.M. (1969a). On decomposition theorems of Meyer. *Math. Scand.* **24**, 66–78.

— (1969b). Quasimartingales. *Math. Scand.* **24**, 79–92.

Ray, D.B. (1956). Stationary Markov processes with continuous paths. *Trans. Amer. Math. Soc.* **82**, 452–493.

— (1963). Sojourn times of a diffusion process. *Illinois J. Math.* **7**, 615–630.

Rényi, A. (1956). A characterization of Poisson processes. *Magyar Tud. Akad. Mat. Kutato Int. Közl.* **1**, 519–527.

— (1967). Remarks on the Poisson process. *Studia Sci. Math. Hung.* **2**, 119–123.

Revuz, D. (1970). Mesures associées aux fonctionnelles additives de Markov, I–II. *Trans. Amer. Math. Soc.* **148**, 501–531; *Z. Wahrsch. verw. Geb.* **16**, 336–344.

— (1984). *Markov Chains*, 2nd ed. North-Holland, Amsterdam.

Revuz, D., Yor, M. (1999). *Continuous Martingales and Brownian Motion*, 23rd ed. Springer, Berlin.

Riesz, F. (1909a). Sur les suites de fonctions mesurables. *C.R. Acad. Sci. Paris* **148**, 1303–1305.

— (1909b). Sur les opérations fonctionelles linéaires. *C.R. Acad. Sci. Paris* **149**, 974–977.

— (1910). Untersuchungen über Systeme integrierbarer Funktionen. *Math. Ann.* **69**, 449–497.

— (1926–30). Sur les fonctions subharmoniques et leur rapport à la théorie du potentiel, I–II. *Acta Math.* **48**, 329–343; **54**, 321–360.

ROGERS, C.A., SHEPHARD, G.C. (1958). Some extremal problems for convex bodies. *Mathematica* **5**, 93–102.

ROGERS, L.C.G., WILLIAMS, D. (2000a/b). *Diffusions, Markov Processes, and Martingales*, **1** (2nd ed.); **2**. Cambridge Univ. Press.

ROSÉN, B. (1964). Limit theorems for sampling from a finite population. *Ark. Mat.* **5**, 383–424.

ROSIŃSKI, J., WOYCZYŃSKI, W.A. (1986). On Itô stochastic integration with respect to p-stable motion: Inner clock, integrability of sample paths, double and multiple integrals. *Ann. Probab.* **14**, 271–286.

ROYDEN, H.L. (1988). *Real Analysis*, 3rd ed. Macmillan, NY.

RUTHERFORD, E., GEIGER, H. (1908). An electrical method of counting the number of particles from radioactive substances. *Proc. Roy. Soc. A* **81**, 141–161.

RYLL-NARDZEWSKI, C. (1957). On stationary sequences of random variables and the de Finetti's [sic] equivalence. *Colloq. Math.* **4**, 149–156.

— (1961). Remarks on processes of calls. *Proc. 4th Berkeley Symp. Math. Statist. Probab.* **2**, 455–465.

SANOV, I.N. (1957). On the probability of large deviations of random variables (Russian). Engl. trans.: *Sel. Trans. Math. Statist. Probab.* **1** (1961), 213–244.

SCHILDER, M. (1966). Some asymptotic formulae for Wiener integrals. *Trans. Amer. Math. Soc.* **125**, 63–85.

SCHOENBERG, I.J. (1938). Metric spaces and completely monotone functions. *Ann. Math.* **39**, 811–841.

SCHRÖDINGER, E. (1931). Über die Umkehrung der Naturgesetze. *Sitzungsber. Preuss. Akad. Wiss. Phys. Math. Kl.* 144–153.

VAN SCHUPPEN, J.H., WONG, E. (1974). Transformation of local martingales under a change of law. *Ann. Probab.* **2**, 879–888.

SEGAL, I.E. (1954). Abstract probability spaces and a theorem of Kolmogorov. *Amer. J. Math.* **76**, 721–732.

SHANNON, C.E. (1948). A mathematical theory of communication. *Bell System Tech. J.* **27**, 379–423, 623–656.

SHARPE, M. (1988). *General Theory of Markov Processes*. Academic Press, Boston.

SHIRYAEV, A.N. (1995). *Probability*, 2nd ed. Springer, NY.

SHUR, M.G. (1961). Continuous additive functionals of a Markov process. *Dokl. Akad. Nauk SSSR* **137**, 800–803.

SIERPIŃSKI, W. (1928). Une théorème générale sur les familles d'ensemble. *Fund. Math.* **12**, 206–210.

SKOROHOD, A.V. (1956). Limit theorems for stochastic processes. *Th. Probab. Appl.* **1**, 261–290.

— (1957). Limit theorems for stochastic processes with independent increments. *Th. Probab. Appl.* **2**, 122–142.

— (1961–62). Stochastic equations for diffusion processes in a bounded region, I–II. *Th. Probab. Appl.* **6**, 264–274; **7**, 3–23.

— (1965). *Studies in the Theory of Random Processes*. Addison-Wesley, Reading, MA. (Russian orig. 1961.)

SLIVNYAK, I.M. (1962). Some properties of stationary flows of homogeneous random events. *Th. Probab. Appl.* **7**, 336–341.

SLUTSKY, E.E. (1937). Qualche proposizione relativa alla teoria delle funzioni aleatorie. *Giorn. Ist. Ital. Attuari* **8**, 183–199.

SNELL, J.L. (1952). Application of martingale system theorems. *Trans. Amer. Math. Soc.* **73**, 293–312.

SOVA, M. (1967). Convergence d'opérations linéaires non bornées. *Rev. Roumaine Math. Pures Appl.* **12**, 373–389.

SPARRE-ANDERSEN, E. (1953–54). On the fluctuations of sums of random variables, I–II. *Math. Scand.* **1**, 263–285; **2**, 195–223.

SPARRE-ANDERSEN, E., JESSEN, B. (1948). Some limit theorems on set-functions. *Danske Vid. Selsk. Mat.-Fys. Medd.* **25**:5 (8 pp.).

SPITZER, F. (1964). Electrostatic capacity, heat flow, and Brownian motion. *Z. Wahrsch. verw. Geb.* **3**, 110–121.

— (1976). *Principles of Random Walk*, 2nd ed. Springer, NY.

STIELTJES, T.J. (1894–95). Recherches sur les fractions continues. *Ann. Fac. Sci. Toulouse* **8**, 1–122; **9**, 1–47.

STONE, C.J. (1963). Weak convergence of stochastic processes defined on a semi-infinite time interval. *Proc. Amer. Math. Soc.* **14**, 694–696.

— (1969). On the potential operator for one-dimensional recurrent random walks. *Trans. Amer. Math. Soc.* **136**, 427–445.

STONE, M.H. (1932). Linear transformations in Hilbert space and their applications to analysis. *Amer. Math. Soc. Coll. Publ.* **15**.

STOUT, W.F. (1974). *Almost Sure Convergence*. Academic Press, NY.

STRASSEN, V. (1964). An invariance principle for the law of the iterated logarithm. *Z. Wahrsch. verw. Geb.* **3**, 211–226.

STRATONOVICH, R.L. (1966). A new representation for stochastic integrals and equations. *SIAM J. Control* **4**, 362–371.

STRICKER, C., YOR, M. (1978). Calcul stochastique dépendant d'un paramètre. *Z. Wahrsch. verw. Geb.* **45**, 109–133.

STROOCK, D.W. (1993). *Probability Theory: An Analytic View*. Cambridge Univ. Press.

STROOCK, D.W., VARADHAN, S.R.S. (1969). Diffusion processes with continuous coefficients, I–II. *Comm. Pure Appl. Math.* **22**, 345–400, 479–530.

— (1979). *Multidimensional Diffusion Processes*. Springer, Berlin.

SUCHESTON, L. (1983). On one-parameter proofs of almost sure convergence of multiparameter processes. *Z. Wahrsch. verw. Geb.* **63**, 43–49.

TAKÁCS, L. (1967). *Combinatorial Methods in the Theory of Stochastic Processes*. Wiley, NY.

TANAKA, H. (1963). Note on continuous additive functionals of the 1-dimensional Brownian path. *Z. Wahrsch. verw. Geb.* **1**, 251–257.

TEMPEL'MAN, A.A. (1972). Ergodic theorems for general dynamical systems. *Trans. Moscow Math. Soc.* **26**, 94–132.

THORISSON, H. (1996). Transforming random elements and shifting random fields. *Ann. Probab.* **24**, 2057–2064.

— (2000). *Coupling, Stationarity, and Regeneration.* Springer, NY.

TONELLI, L. (1909). Sull'integrazione per parti. *Rend. Acc. Naz. Lincei* (5) **18**, 246–253.

TROTTER, H.F. (1958a). Approximation of semi-groups of operators. *Pacific J. Math.* **8**, 887–919.

— (1958b). A property of Brownian motion paths. *Illinois J. Math.* **2**, 425–433.

VARADARAJAN, V.S. (1958). Weak convergence of measures on separable metric spaces. On the convergence of probability distributions. *Sankhyā* **19**, 15–26.

— (1963). Groups of automorphisms of Borel spaces. *Trans. Amer. Math. Soc.* **109**, 191–220.

VARADHAN, S.R.S. (1966). Asymptotic probabilities and differential equations. *Comm. Pure Appl. Math.* **19**, 261–286.

— (1984). *Large Deviations and Applications.* SIAM, Philadelphia.

VILLE, J. (1939). *Étude Critique de la Notion du Collectif.* Gauthier-Villars, Paris.

VITALI, G. (1905). Sulle funzioni integrali. *Atti R. Accad. Sci. Torino* **40**, 753–766.

VOLKONSKY, V.A. (1958). Random time changes in strong Markov processes. *Th. Probab. Appl.* **3**, 310–326.

— (1960). Additive functionals of Markov processes. *Trudy Mosk. Mat. Obshc.* **9**, 143–189.

WALD, A. (1946). Differentiation under the integral sign in the fundamental identity of sequential analysis. *Ann. Math. Statist.* **17**, 493–497.

— (1947). *Sequential Analysis.* Wiley, NY.

WALSH, J.B. (1978). Excursions and local time. *Astérisque* **52–53**, 159–192.

WANG, A.T. (1977). Generalized Itô's formula and additive functionals of Brownian motion. *Z. Wahrsch. verw. Geb.* **41**, 153–159.

WATANABE, H. (1964). Potential operator of a recurrent strong Feller process in the strict sense and boundary value problem. *J. Math. Soc. Japan* **16**, 83–95.

WATANABE, S. (1964). On discontinuous additive functionals and Lévy measures of a Markov process. *Japan. J. Math.* **34**, 53–79.

— (1968). A limit theorem of branching processes and continuous state branching processes. *J. Math. Kyoto Univ.* **8**, 141–167.

WEIL, A. (1940). *L'intégration dans les Groupes Topologiques et ses Applications.* Hermann et Cie, Paris.

WIENER, N. (1923). Differential space. *J. Math. Phys.* **2**, 131–174.

— (1938). The homogeneous chaos. *Amer. J. Math.* **60**, 897–936.

— (1939). The ergodic theorem. *Duke Math. J.* **5**, 1–18.

WILLIAMS, D. (1991). *Probability with Martingales*. Cambridge Univ. Press.

YAMADA, T. (1973). On a comparison theorem for solutions of stochastic differential equations and its applications. *J. Math. Kyoto Univ.* **13**, 497–512.

YAMADA, T., WATANABE, S. (1971). On the uniqueness of solutions of stochastic differential equations. *J. Math. Kyoto Univ.* **11**, 155–167.

YOEURP, C. (1976). Décompositions des martingales locales et formules exponentielles. *Lect. Notes in Math.* **511**, 432–480. Springer, Berlin.

YOR, M. (1978). Sur la continuité des temps locaux associée à certaines semimartingales. *Astérisque* **52–53**, 23–36.

YOSIDA, K. (1948). On the differentiability and the representation of one-parameter semigroups of linear operators. *J. Math. Soc. Japan* **1**, 15–21.

YOSIDA, K., KAKUTANI, S. (1939). Birkhoff's ergodic theorem and the maximal ergodic theorem. *Proc. Imp. Acad.* **15**, 165–168.

ZÄHLE, M. (1980). Ergodic properties of general Palm measures. *Math. Nachr.* **95**, 93–106.

ZAREMBA, S. (1909). Sur le principe du minimum. *Bull. Acad. Sci. Cracovie.*

ZYGMUND, A. (1951). An individual ergodic theorem for noncommutative transformations. *Acta Sci. Math. (Szeged)* **14**, 103–110.

Symbol Index

$\hat{A}$, 499
A_n, 391
A^λ, 372, 442
$A^c, A \setminus B, A \Delta B$, 1
$A \times B$, 2
$\mathcal{A}, \mathcal{A}^\mu$, 13, 46

$|B|$, 187
B°, B^-, 541
$\mathcal{B}, \mathcal{B}(S)$, 2

$\hat{C}$, 378
C_0, C_0^∞, 369, 374
C^k, 340
C_K^+, 98, 225
$C_b(S)$, 65
$C(K, S)$, 307
C_K^D, 481
$\mathrm{cov}(\xi, \eta), \mathrm{cov}[\xi; A]$, 50, 302

$D_h, \mathcal{D}_h$, 434
$D(\mathbb{R}_+, S)$, 313, 563
$D([0, 1], S)$, 319
Δ, ∇, 1, 287, 375, 377, 483
δ, ∂, 150, 187, 473
δ_x, 8
$\stackrel{d}{=}$, 48
$\stackrel{d}{\to}$, 65

E, 48, 225
$\overline{E}$, 443
E_x, E_μ, 145
$E[\xi; A]$, 49
$E[\xi|\mathcal{F}] = E^\mathcal{F}\xi$, 104
$\mathcal{E}, \mathcal{E}_n$, 263, 335
$\mathcal{E}(X)$, 363, 522

$\hat{F}_n$, 75
$\|F\|_a^b$, 33
$\mathcal{F}$, 120, 324
$\overline{\mathcal{F}}$, 124
$\mathcal{F}^+$, 121
$\mathcal{F}_\tau$, 120
$\mathcal{F}_{\tau-}$, 491
$\mathcal{F}_\infty$, 132

$\mathcal{F}_D, \mathcal{F}_D^r$, 480
$\mathcal{F} \otimes \mathcal{G}$, 2
$\mathcal{F} \vee \mathcal{G}, \bigvee_n \mathcal{F}_n$, 50
$\mathcal{F} \perp\!\!\!\perp \mathcal{G}, \mathcal{F} \perp\!\!\!\perp_\mathcal{G} \mathcal{H}$, 50, 109
$f_\pm$, 11
f^{-1}, 3
f_i', f_{ij}'', 340
$f \cdot A$, 442
$f \circ g$, 5
$f \otimes g$, 262
$\langle f, g \rangle, f \perp g$, 17
$f \cdot \mu$, 12
$f \succ U, f \prec V$, 36
$\stackrel{f}{\to}$, 567
$\stackrel{fd}{\to}$, 308
φB, 324

G^D, g^D, 477
γ_K^D, 481

H_1, H_∞, 543, 553
$H^{\otimes n}$, 262
H_K^D, 480
$h_{a,b}$, 456
$H(\xi), H(\xi|\mathcal{F}), H(\nu|\mu)$, 220, 554

$I, I(B)$, 368, 545
I_n, 263
$I(\xi), I(\xi|\mathcal{F})$, 220
$\mathcal{I}_\xi$, 181, 189
id, Id, 295

$\mathcal{K}$, 324
$\mathcal{K}_D, \mathcal{K}_D^r$, 480

L_t, L_t^x, 430, 436, 446
L_K^D, 481
L^p, L_{loc}^p, 15, 363
$L(X), \hat{L}(X)$, 336–337, 344, 526
$L^2(M)$, 517
$L^2(\eta)$, 266
$\mathcal{L}(\xi)$, 47
λ, 24
Λ, Λ^*, 539, 554

M, Mf, 184, 391
$\langle M \rangle, \langle M, N \rangle$, 280, 516
$\mathcal{M}, \mathcal{M}_0$, 526
$\mathcal{M}^2, \mathcal{M}_0^2, \mathcal{M}_{loc}^2$, 331, 515–516
$\mathcal{M}(S)$, 19, 225
$\stackrel{m}{\sim}$, 524
$\hat{\mu}, \mu$, 227
$\mu_t, \mu_{s,t}$, 144
μ_K^D, 481
μf, 10
$\mu \circ f^{-1}$, 10
$\mu * \nu$, 15
$\mu\nu, \mu \otimes \nu$, 14, 20, 142
$\mu \perp \nu, \mu \ll \nu, \mu \sim \nu$, 13, 29, 363
$\mu \vee \nu, \mu \wedge \nu$, 29

$N(m, \sigma^2)$, 90
$\mathcal{N}(S)$, 226
$\mathbb{N}$, 2
$(n/\!/k)$, 187
ν, 290, 435
ν_A, 442

Ω, ω, 46
Ω^T, 2

P, 46
$\overline{P}$, 509
P_x, P_μ, 145, 391
$P \circ \xi^{-1}$, 47
$P[A|\mathcal{F}] = P^{\mathcal{F}}A$, 106
$\mathcal{P}(S)$, 19
$p_{a,b}$, 456
p_{ij}^n, p_{ij}^t, 151, 243
p_t, p_t^D, 475–476
$\stackrel{P}{\to}$, 63, 408
π_B, π_f, π_t, 19, 47, 225–226, 316

$Q_{X,\xi}, Q'_{X,\xi}$, 203, 209
$\mathbb{Q}, \mathbb{Q}_+$, 98, 125

R_λ, 370
$\mathbb{R}, \mathbb{R}_+, \overline{\mathbb{R}}, \overline{\mathbb{R}}_+$, 2, 5
$r_{x,y}$, 149

$\hat{S}$, 377
S_n, S_nf, 184, 391, 396
$\hat{S}, \hat{S}_\mu$, 225, 316, 324, 564
$\sigma\{\cdot\}$, 2, 5
$\operatorname{supp} \mu$, 9

T_t, T_t^λ, 368, 372
τ_A, τ_B, 123, 492
$\tau_a, \tau_{a,b}$, 455
$[\tau]$, 492
$\theta_t, \tilde{\theta}$, 146, 179, 189, 391

$U, U^\alpha, U_h, U_A, U_A^\alpha$, 402, 442–443

$V \cdot X$, 128, 336, 517–518, 526
$\stackrel{v}{\to}$, 98, 564
$\operatorname{var}(\xi), \operatorname{var}[\xi; A]$, 50, 71

$w_f, w(f, h), w(f, t, h)$, 57, 274, 310, 562
$\tilde{w}(f, t, h)$, 563
$\stackrel{w}{\to}$, 65

X^c, X^d, 527
X^τ, 128
X^*, X_t^*, 129
$X \circ dY$, 342
$[X], [X, Y]$, 280, 332, 519
ξ, 436
$\hat{\xi}$, 503
ξ^*, 226
$\bar{\xi}, \bar{\xi}_n$, 190, 538

Z, 432
$\mathbb{Z}, \mathbb{Z}_+$, 6, 59
ζ, ζ_D, 380, 473

$\emptyset$, 1
$[[0, 1)$, 464
$\mathbf{1}$, 58
$1_A, 1\{\cdot\}$, 5, 46
2^S, 1
$\lesssim$, 57
$\|\cdot\|, \|\cdot\|_p$, 15, 152, 369

Author Index

Abel, N.H., 15, 144, 147, 242
Adler, R.J., 580
Akcoglu, M.A., 586
Aldous, D.J., 197, 314, 577–578, 583, 587
Alexandrov, A.D., 75, 571
André, D., 165, 223, 575, 578
Arndt, U., 604
Arzelà, C., 307, 310–311, 559, 563
Ascoli, G., 307, 310–311, 559, 563
Athreya, K.B., 576

Baccelli, F., 578, 592
Bachelier, L., 256, 574–575, 580, 590
Bahadur, R.R., 594
Banach, S., 49, 369, 534, 585
Barbier, É., 578
Bass, R.F., 591
Bauer, H., 595
Baxter, G., 159, 169, 576
Bayes, T., 578
Belyaev, Y.K., 583
Berbee, H.C.P., 197, 577, 587
Bernoulli, J., 46, 55–56, 539, 571, 580
Bernstein, S.N., 128, 247, 572–573, 587
Bertoin, J., 582
Bertrand, J., 223, 578
Bessel, F.W., 256
Bichteler, K., 533, 593
Bienaymé, J., 63, 69, 571
Billingsley, P., 569, 571, 573, 578, 581, 583, 595
Birkhoff, G.D., 178, 181, 391, 393, 576
Blackwell, D., 172, 575–576, 586
Blumenthal, R.M., 380–381, 446, 501, 575, 586, 588–589, 591–592
Bochner, S., 100, 261, 572, 580
Bogolioubov, N., 196, 577
Bohl, 200
Boltzmann, L., 576
Borel, E., 2, 3, 7, 24–25, 45, 47, 55, 119, 131, 308, 561, 569–571, 574

Brandt, A., 592
Breiman, L., 221, 561, 578, 589
Brémaud, P., 578, 592
Brown, R., 252–253, 439, 580
Bryc, W., 547, 594
Bühlmann, H., 217, 578
Buniakovsky, V.Y., 17, 334, 569
Burkholder, D.L., 333, 524, 584, 591, 593

Cameron, R.H., 364, 537, 543, 584–585
Cantelli, F.P., 45, 47, 55, 75, 119, 131, 570–571, 574
Carathéodory, C., 24, 26, 570
Carleson, L., 578
Cauchy, A.L., 16–17, 65, 238, 304, 334, 470–472, 569
Chacon, R.V., 393, 586
Chapman, S., 140, 142–143, 145, 154, 367–368, 378, 574
Chebyshev, P.L., 63, 69, 571–572
Chentsov, N.N., 57, 313, 571, 583
Chernoff, H., 540, 583, 594
Choquet, G., 483, 486–487, 562, 577, 583, 591
Chow, Y.S., 572, 574
Chung, K.L., 162, 221, 272, 381, 481, 575–576, 578–579, 584, 590–591
Courant, R., 590
Courrège, P., 334, 517, 584, 592
Cox, D., 224, 226–228, 230–233, 246, 317–319, 327–328, 579, 583
Cramér, H., 87, 261, 540, 571–572, 577, 580, 594
Csörgö, M., 581

Daley, D.J., 578–579
Dambis, K.E., 352, 585
Daniell, P.J., 23, 104, 114, 570, 573
Davis, B.J., 524, 593, 598
Dawson, D.A., 551, 594
Day, M.M., 201, 576
Debes, H., 583

Dellacherie, C., 357, 533, 562, 574–575, 577, 580, 586, 589, 591–593, 595
Dembo, A., 594
Derman, C., 464, 589
Deuschel, J.D., 594
Dini, U., 541
Dirac, P., 8
Dirichlet, P.G.L., 470, 474, 590
Doeblin, W., 299, 303, 574–575, 579, 582, 586, 589
Döhler, R., 573
Doléans(-Dade), C., 345, 493, 496–497, 518, 522, 584, 591–593
Donsker, M.D., 275, 312, 319, 555, 581–582, 594
Doob, J.L., 7, 109–110, 124, 126–127, 129–131, 134–136, 138, 237, 358, 474, 490, 493, 495, 507, 509, 569, 571, 573–575, 577–579, 581, 584–587, 590–592
Dubins, L.E., 352, 581, 585
Dudley, R.M., 79, 563, 569, 572, 583, 594
Dunford, N., 69, 392, 571, 586, 591
Durrett, R., 581
Dvoretzky, A., 281, 581
Dynkin, E.B., 380, 382–384, 456, 569, 574–575, 577, 585–589

Egorov, D., 18
Einstein, A., 580
Elliott, R.J., 592
Ellis, R.S., 594
Engelbert, H.J., 450–451, 589
Erdös, P., 276, 576, 581
Erlang, A.K., 234, 579
Ethier, S.N., 563, 575, 583, 586, 588, 595

Faber, G., 571
Farrell, R.H., 195, 577
Fatou, P., 11, 46, 67, 569
Fell, J.M.G., 324, 470, 565–567, 595
Feller, W., 92–93, 96, 165, 172, 302, 367, 369–387, 400, 405–409, 421, 442, 456, 458, 462, 465, 501, 572, 575–576, 578–580, 582, 585–587, 589–590
Fenchel, W., 537, 539, 554, 594
Feynman, R.P., 470–471, 590
Fichtner, K.H., 246
de Finetti, B., 202, 212, 578, 581
Fisk, D.L., 339, 342, 426, 584, 593
Fortet, R., 586
Fourier, J.B.J., 90, 100, 163, 262, 575
Franken, P., 578
Fréchet, M., 569
Freedman, D., 251, 575, 580, 586, 589, 594
Freidlin, M.I., 537, 553, 594
Friedrichs, K., 599
Frostman, O., 590
Fubini, G., 14, 52, 108, 569
Fuchs, W.H.J., 162, 575
Furstenberg, H., 193, 577

Galmarino, A.R., 355, 585
Garsia, A.M., 182, 502, 525, 576, 591
Gärtner, J., 551, 594
Gauss, C.F., 90–96, 250–254, 260–263, 266, 351, 473, 539, 579, 590
Geiger, H., 579
Getoor, R.K., 446, 575, 585, 588–589, 591–592
Gihman, I.I., 587
Girsanov, I.V., 362, 365, 515, 585, 587, 592
Glivenko, V.I., 75, 571
Gnedenko, B.V., 303, 582
Goldman, J.R., 583
Goldstein, J.A., 586
Goldstein, S., 197, 577, 586
Grandell, J., 579
Green, G., 458, 470, 475, 477–486, 590
Greenwood, P., 588
Griffeath, D., 577, 586
Grigelionis, B., 503, 583, 592
Gronwall, 415, 455, 554
Gundy, R.F., 333, 524, 584, 598

Haar, A., 23, 39, 41, 198, 570
Hagberg, J., 583
Hahn, H., 28, 33, 35, 49, 534, 570
Hájek, J., 583
Hall, P., 581

Halmos, P.R., 569, 573
Hardy, G.H., 184
Harris, T.E., 400, 405–406, 408, 410, 583, 586–587
Hartman, P., 275, 581
Hausdorff, F., 36, 247, 311, 399, 563
Heine, H.E., 25
Helly, E., 98, 572
Hermite, C., 84, 265–266
Hewitt, E., 45, 53, 161, 570
Heyde, C.C., 581
Hilbert, D., 104, 188, 251, 260, 262–263, 265–266, 331, 351, 515, 543
Hille, E., 367, 375, 585
Hitczenko, P., 592
Hölder, O., 15, 49, 57, 109, 252, 268, 313, 426, 448, 569–570, 586, 589
Hopf, E., 159, 168, 392, 586
Horowitz, J., 588
Hunt, G.A., 124, 256, 443, 476, 580, 586, 588–591
Hurewicz, W., 586
Hurwitz, A., 570

Ikeda, N., 584, 588
Ioffe, D., 549, 594
Ionescu Tulcea, A., 221, 578
Ionescu Tulcea, C.T., 104, 116, 573
Itô, K., 263, 265, 287, 336, 339–341, 357–358, 415, 431, 435–436, 458, 520, 571, 579–580, 582, 584–585, 587–591, 593

Jacod, J., 503, 518, 524, 563, 582–583, 585, 588, 592–593, 595
Jagers, P., 583
Jamison, B., 586
Jensen, J.L.W.V., 49, 109, 570
Jessen, B., 132, 574
Johnson, W.B., 593
Jordan, C., 33, 570

Kac, M., 276, 470–471, 577, 581, 590
Kakutani, S., 182, 354, 474, 575–576, 584, 586, 590
Kallenberg, O., 571, 573, 575, 578–579, 582–585, 588, 592–593

Kallianpur, G., 580
Kaplan, E.L., 206, 577
Karamata, J., 96, 572
Karatzas, I., 580, 584–585, 588–591
Kazamaki, N., 344, 584
Kemeny, J.G., 575
Kendall, D., 583
Kerstan, J., 600, 612
Kesten, H., 193, 577
Khinchin, A., 70, 96, 259, 290, 302, 537, 571, 576–578, 580, 582–583, 586, 593
Kingman, J.F.C., 178, 192, 577, 579
Kinney, J.R., 379, 571, 586
Knapp, A.W., 608
Knight, F.B., 355, 428, 440, 585, 588
Koebe, P., 473
Kolmogorov, A.N., 53, 57, 69–71, 73, 104, 115, 132, 142–143, 145, 152, 154, 242, 291, 313, 368, 371, 471, 563, 570–571, 573–574, 576, 579–582, 585, 589–590, 593
Komlós, J., 581
König, D., 207, 578, 604
Koopman, B.O., 576
Korolyuk, V.S., 207, 578
Krengel, U., 577, 587
Krickeberg, K., 579, 593
Kronecker, L., 62, 73
Krylov, N., 196, 577
Kullback, S., 594
Kunita, H., 336, 347, 517, 521, 584, 587, 592–593
Kuratowski, K., 562
Kurtz, T.G., 385, 563, 575, 583, 586, 588, 595
Kuznetsov, S.E., 591
Kwapień, S., 572

Langevin, P., 414, 580, 587
Laplace, P.S. de, 84, 86, 88, 100, 227, 370, 375, 473, 572–573, 579, 590
Last, G., 592
Leadbetter, M.R., 571, 577, 582
Lebesgue, H., 11–12, 14, 24–25, 27, 29, 31, 55, 569–570
Le Cam, L., 583
van Leeuwenhoek, A., 580

Le Gall, J.F., 589
Legendre, A.M., 537, 539, 554, 594
Leibler, R.A., 594
Leontovich, M.A., 590
Levi, B., 11, 569
Lévy, P., 71, 86, 90, 93, 96, 100, 128, 131–132, 234, 252, 255, 258, 285–287, 290–292, 294, 298–299, 352–354, 374, 430, 436, 571–576, 579–582, 584, 586–588, 593
Lewy, H., 599
Liapounov, A.M., 572
Liemant, A., 600
Liggett, T., 577
indeberg, J.W., 90, 92, 572, 586
Lindgren, G., 610
Lindvall, T., 575–576, 583, 587
Lipschitz, R., 268, 415, 453–455, 553, 589
Lipster, R.S., 611
Littlewood, J.E., 184
Loève, M., 57, 569, 571–572, 576–577, 582
Lomnicki, Z., 117, 573
Lukacs, E., 572
Lundberg, F., 579
Lusin, N.N., 19, 562

Mackevičius, V., 385, 586
Maisonneuve, B., 588, 600
Major, P., 609
Maker, P., 183, 577
Mann, H.B., 76, 571
Marcinkiewicz, J., 73, 571, 593
Markov, A.A., 63, 140–155, 237–245, 254, 256, 368, 378, 380, 387, 391, 396, 421, 571, 573–574
Martin, W.T., 364, 537, 543, 584–585
Maruyama, G., 465, 585, 589
Matheron, G., 583, 591, 595
Matthes, K., 207, 578–579, 583, 600
Maxwell, J.C., 251, 579
McDonald, D., 596
McKean, H.P., 447, 458, 580, 588–590
McMillan, B., 221, 578
Mecke, J., 319, 579, 583, 612
Méléard, S., 460, 589
Mémin, J., 518
Métivier, M., 592

Meyer, P.A., 136, 431, 493–494, 498, 501, 505, 510, 518, 526–527, 562, 574–575, 577, 584, 586, 588, 591–593, 595
Millar, P.W., 333, 584
Minkowski, H., 15–16, 109, 183, 190–191, 263, 569
Mitoma, I., 583
de Moivre, A., 572, 579
Mönch, G., 579
de Morgan, A., 1
Motoo, M., 464, 589

Nawrotzski, K., 583
von Neumann, J., 200, 570, 573, 576
Neveu, J., 221, 502, 574, 587, 591
Newton, I, 474, 488
Ney, P., 596
Nguyen, X.X., 190, 577
Nikodým, O.M., 29, 31, 105, 570, 573
Norberg, T., 325, 583
Novikov, A.A., 333, 364, 584–585
Nualart, D., 580

Øksendal, B., 584
Orey, S., 152, 172, 397, 400, 576, 586–587, 593
Ornstein, D.S., 162, 393, 575, 586
Ornstein, L.S., 254, 262, 414, 580, 587
Ososkov, G.A., 583
Ottaviani, G., 312, 583

Paley, R.E.A.C., 63, 268, 575, 580, 593
Palm, C., 203–210, 576–578, 583
Papangelou, F., 505, 592
Parseval, M.A., 162, 262
Parthasarathy, K.R., 562, 594
Pellaumail, J., 592
Perkins, E., 589
Petrov, V.V., 594
Phillips, H.B., 590
Picard, E., 415
Pitman, J.W., 588
Pitt, H.R., 576
Plancherel, M., 262
Poincaré, H., 590
Poisson, S.D., 87–88, 226–231, 234–238, 241–242, 288,

297–298, 301, 318, 368, 436, 504–505, 579
Pollaczek, F., 575
Pollard, D., 583
Pollard, H., 603
Pólya, G., 572, 575
Port, S.C., 591
Pospišil, B., 579
Prohorov, Y.V., 76, 309, 311, 313, 316, 563, 571, 581–583, 593
Protter, P., 593
Pukhalsky, A.A., 546, 594

Radon, J., 29, 31, 36, 105, 569–570, 573
Rao, K.M., 494, 532, 591, 593
Ray, D.B., 428, 440, 586, 588–589
Rényi, A., 234, 579, 583
Révesz, P., 581
Revuz, D., 442–445, 447, 580, 584–585, 587–590
Riemann, G.F.B., 31, 43, 175
Riesz, F., 23, 36, 43, 378, 490, 511, 569–571
Rogers, C.A., 577
Rogers, L.C.G., 561, 575, 584, 588–589, 592
Rootzén, H., 610
Rosén, B., 583
Rosiński, J., 592
Royden, H.L., 570, 594
Rubin, H., 76, 572
Rutherford, E., 579
Ryll-Nardzewski, C., 207, 212, 577–578

Sanov, I.N., 537, 555, 594
Savage, L.J., 45, 53, 161, 570
Schechtman, G., 607
Schilder, M., 537, 543, 554, 557, 594
Schmidt, V., 604
Schmidt, W., 450–451, 589
Schoenberg, I.J., 251, 580
Schrödinger, E., 590
van Schuppen, J.H., 362, 523, 585, 592
Schwartz, J.T., 392, 586
Schwarz, G., 352, 585
Schwarz, H.A., 17
Segal, I.E., 580

Shannon, C.E., 221, 578
Sharpe, M., 575, 585, 591
Shephard, G.C., 577
Shiryaev, A.N., 563, 582–583, 592–593, 595
Shreve, S.E., 580, 584–585, 588–591
Shur, M.G., 591
Sierpiński, W., 2, 200, 569
Skorohod, A.V., 79, 113, 271, 273, 298, 313, 315, 419, 429, 453–454, 563, 572, 581–583, 587–589, 595
Slivnyak, I.M., 210, 577–578
Slutsky, E., 571
Smoluchovsky, M., 142, 574
Snell, J.L., 130, 574, 608
Sova, M., 385, 586
Sparre-Andersen, E., 166, 169, 216, 276, 574, 576, 578
Spitzer, F., 576, 590
Stieltjes, T.J., 31, 255, 329, 340, 519, 570
Stone, C.J., 575, 583, 591
Stone, M.H., 86, 261, 580
Stout, W.F., 572
Strassen, V., 273, 537, 557, 581, 594
Stratonovich, R.L., 342, 426, 584
Stricker, C., 80, 345, 571, 584
Stroock, D.W., 418, 420–421, 472, 587–588, 590, 594
Sucheston, L., 617
Sztencel, R., 593

Takács, L., 578
Tanaka, H., 428, 431, 439, 447, 454, 459, 465, 588–589
Taylor, B., 90, 92
Teicher, H., 572, 574, 583
Tempel'man, A.A., 577
Thorisson, H., 197–198, 209, 573, 575, 577–578, 587
Tonelli, L., 14, 569
Trotter, H.F., 385, 430, 586, 588
Tusnády, G., 609
Tychonov, A.N., 40

Uhlenbeck, G.E., 254, 262, 414, 580, 587
Ulam, S., 117, 573

Varadarajan, V.S., 195, 309, 577, 583
Varadhan, S.R.S., 418, 420–421, 472, 541, 547, 555, 587–588, 590, 594
Vere-Jones, D., 578–579
Ville, J., 573
Vitali, G., 570
Volkonsky, V.A., 445, 447, 458, 587–589, 591
Voronoi, G., 204

Wald, A., 76, 364, 571, 575, 585
Walsh, J.B., 381, 440, 588
Wang, A.T., 431, 588
Watanabe, H., 406, 464, 586, 589
Watanabe, S., 336, 347, 374, 424, 453, 505, 517, 521, 584–585, 587–589, 591–593
Weierstrass, K., 86, 341
Weil, A., 39, 570
Wentzell, A.D., 537, 553, 594
Weyl, H., 200
Wiener, N., 168, 184, 187, 252–253, 260, 263–266, 268, 358, 570, 575–576, 580, 590, 614

Williams, D., 561, 571, 574–575, 584, 588–589, 592
Williams, R.J., 584
Wintner, A., 275, 581
Wold, H., 87, 250, 572
Wong, E., 362, 523, 585, 592
Woyczyński, W.A., 572, 592

Yamada, T., 424, 453–454, 587, 589
Yan, J.A., 518, 533
Yoeurp, C., 527, 529, 593
Yor, M., 80, 345, 430, 571, 580, 584–585, 588–590
Yosida, K., 182, 367, 372, 375, 386, 576, 585
Yushkevich, A.A., 380, 574, 586

Zähle, M., 210, 578
Zaremba, S., 474
Zeitouni, O., 594
Zessin, H., 190, 577
Zinn, J., 607
Zorn, M., 43, 197–198
Zygmund, A., 63, 73, 178, 186, 268, 571, 577, 593, 614

Subject Index

absolute:
  continuity, 13, 29, 35, 261, 360, 432, 523
  moment, 49
absorption of:
  Markov process, 155, 238, 378, 382, 434
  diffusion, 461, 465
  supermartingale, 136
accessible:
  set, boundary, 160, 462
  time, 492, 500–501
  jumps, 499, 529–530
action, left, right, 41
adapted, 120, 503
additive functional, 442
a.e., almost everywhere, 12
allocation sequence, 215
almost:
  everywhere, 12
  invariant, 180
alternating function, 483, 487
analytic function, 342, 353
announcing sequence, 341, 491
aperiodic, 150
approximation of:
  covariation, 339
  empirical distributions, 278
  exchangeable sums, 321
  local time, 432, 437
  Markov chains, 387
  martingales, 280
  predictable process, 517
  progressive process, 343
  random walk, 273, 282, 299, 315
  renewal process, 277
arcsine laws, 258, 276, 299
Arzelà–Ascoli theorem, 310, 563
a.s., almost surely, 47
asymptotic invariance, 211
atom, atomic, 9, 19
augmented filtration, 124
averaging property, 105

backward equation, 242, 372, 471

balayage, sweeping, 474
ballot theorem, 218, 220
BDG inequalities, 333, 524
Bernoulli sequence, 56, 539
Bessel process, 256, 440
bilinear, 50
binary expansion, 56
binomial process, 226–227, 229, 235
Blumenthal's zero–one law, 381
Borel–Cantelli lemma, 47, 55, 131
Borel:
  isomorphism, space, 7, 561
  set, σ-field, 2
boundary behavior, 462, 465, 474
bounded optional time, 126
Brownian:
  bridge, 253, 278, 319, 356
  excursion, 439
  motion, 252–260, 271–275, 277–282, 312, 352–360, 364–365, 412–424, 430, 439–440, 443–445, 447, 450–455, 458, 472–486, 507–513, 543, 553, 557
  scaling, inversion, 253

CAF, continuous additive functional, 442
Cameron–Martin space, 364, 543, 553, 557
canonical:
  decomposition, 337, 518
  process, space, filtration, 146, 380, 384
capacity, 481–483, 486–487
Cartesian product, 2
Cauchy:
  sequence, 16, 65
  problem, 471–472
Cauchy–Buniakovsky inequality, 17, 334, 516, 520
centering, centered, 72, 126, 250
central limit theorem, 90, 275, 312
chain rule for:
  conditional independence, 111

conditioning, 105
integration, 12, 338, 517
change of:
  measure, 360–365, 422, 523
  scale, 451, 456
  time, 344, 352, 423, 451–453, 458–461, 505–506
chaos expansion, 266, 360
Chapman–Kolmogorov equation, 142–143, 145, 151, 368
characteristic:
  exponent, 291
  function, 84–86, 90, 100, 227
  measure, 241
  operator, 383
characteristics, 290, 413
Chebyshev's inequality, 63
closed, closure:
  martingale, 131, 135
  operator, 373–374
coding, 113, 145, 204
commuting operators, 186, 370
compactification, 377–378
compactness:
  vague, 98, 564
  weak, 98, 309
  weak L^1, 69
  in C and D, 563
comparison of solutions, 454–455
compensator, 493, 498–500, 503–506, 510–511
complete, completion:
  filtration, 123
  function space, 16, 65
  σ-field, 13, 110
completely monotone, alternating, 483, 487
complex-valued process, 260, 341, 351–352
composition, 5
compound:
  optional time, 146
  Poisson process, 242, 297–298, 300–301
condenser theorem, 483
conditional:
  distribution, 107
  entropy, information, 220
  expectation, 104–105

independence 109–113, 141, 212, 217, 228, 424
probability, 106
conductor, 480–481, 483
cone condition, 474
conformal mapping, invariance, 342, 353
conservative semigroup, 369, 377
continuity:
  set, 75, 545
  theorem, 86, 100
  for a time-change, 344
continuous:
  additive functional, 442–447, 451–452, 458, 510–513
  in probability, 216, 286, 319
  mapping, 64, 76, 549
  martingale component, 527–529
contraction:
  operator, 105, 109, 368, 391–393, 415
  principle, 549
convergence in/of:
  distribution, 65–66, 71–72, 75–79, 86–88, 90–93, 96, 99–100, 275–276, 308–326, 385–387
  probability, 63–66, 80
  exchangeable processes, 322
  infinitely divisible laws, 295–296
  Lévy processes, 298
  L^p, 16, 68
  Markov processes, 385
  point processes, 317, 326
  random measures, 316
  random sets, 325
convex, concave:
  functions, 49, 126, 431, 459, 538–539
  sets, 187–190, 196, 533
convolution, 15, 52
core of generator, 373–374, 385, 387
countably additive, subadditive, 8
counting measure, 8
coupling, 152, 172
  independent, 152, 466
  shift, 197–198, 209
  Skorohod, 79, 113, 298–299
covariance, 49–50, 250
covariation, 332, 334–336, 339–342, 516–517, 519–521, 526, 529

Subject Index 631

Cox process, 226–228, 230–231, 318–319
Cramér-Wold theorem, 87
cumulant-generating function, 539, 554
cycle stationarity, 206
cylinder set, 2, 115

(D), submartingale class, 493
Daniell-Kolmogorov theorem, 114–115
debut, 123
decomposition of:
  finite-variation function, 33–34
  increasing process, 499
  martingale, 518, 527, 529
  measure, 29
  optional time, 493
  signed measure, 28
  submartingale, 126, 493
degenerate:
  measure, 9, 19
  random element, 51
delay, 170–172
density, 12–13, 29, 31, 133
differentiation theorem, 31
diffuse, nonatomic, 9–10, 19, 230, 233, 299
diffusion, 384, 413, 455–467, 471
  equation, 413, 421, 423, 450–455
Dirac measure, 8
Dirichlet problem, 474
discrete time, 143
disintegration, 108
dissipative, 377
distribution, 47
  function, 48, 59
Doléans exponential, 522
domain, 473
  of attraction, 96
  of generator, 370, 372–375, 377
dominated:
  convergence, 11–12, 337, 518, 526
  ergodic theorem, 184
Donsker's theorem, 275, 312
Doob decomposition, 126
Doob-Meyer decomposition, 493
dual predictable projection, 498
duality, 167
Dynkin's formula, 382

effective dimension, 160
Egorov's theorem, 18
elementary:
  function, 263
  additive functional, 442
  stochastic integral, 128, 335, 343, 517
elliptic operator, 384, 418, 472
embedded:
  Markov chain, 239
  martingale, 279
  random variable, walk, 271–273, 464
empirical distribution, 75, 195, 278, 554–555
entrance boundary, 461–462
entropy, 220–221, 554–555
equicontinuous, 86, 311, 313–314, 563
equilibrium measure, 481–485
ergodicity, 181, 195–196, 397, 399, 465
ergodic decomposition, 196
ergodic theorems:
  Markovian, 152–154, 244–245, 397, 399, 408–409, 465
  multivariate, Palm, 186–187, 190, 209–210
  ratio, 393, 396, 464
  stationarity, contractions, 181–183, 392
  subadditive, matrices, 192–193
evaluation, projection, 47, 225, 562
event, 46
excessive function, 379, 445, 507–511
exchangeable:
  sequence, 212–215, 320–321
  process, 216–218, 235, 319–322
excursion, 150, 433–440
existence of:
  Brownian motion, 252
  Cox process, randomization 231
  Markov process, 143, 378
  random sequence, process, 55, 114–117
  solution to SDE, 415, 419, 422–423, 451
exit boundary, 461
expectation, expected value, 48–49, 52
explosion, 240, 417, 462

exponential:
  distribution, 237–240, 434
  equivalence, 552
  inequalities, 530
  martingale, process, 351, 363, 522, 530
  rate, 541
  tightness, 546–549, 556
extended real line, 5
extension of:
  filtration, 124, 352
  measure, 26, 114–115, 362
  probability space, 111–112
extreme:
  element, 196
  value, 257, 303

factorial measure, 213
fast reflection, 460
Fatou's lemma, 11, 67
Fell topology, 324, 565–566
Feller process, semigroup, 369–387, 399–409, 421, 442, 445–446, 462, 501
Fenchel–Legendre transform, 539, 554
Feynman–Kac formula, 471
filling operator, functional, 394–395
filtration, 120
de Finetti's theorem, 212
finite-dimensional distributions, 48, 142
finite-variation:
  function, 33–35
  process, 330, 337, 497, 518
first:
  entry, 124
  maximum, 166, 216, 258, 276, 299
  passage, 166–170, 292
Fisk–Stratonovich integral, 342
fixed jump, 286
flow, 183, 415
fluctuations, 167
forward equation, 372
Fubini theorem, 14, 52, 108, 358
functional:
  CLT, LIL, 275, 312, 557
  LDP, 547
  representation, 80, 346
  solution, 423
fundamental:

identity, 480
theorem, 31–32

Gaussian:
  convergence, 90–92, 96
  measure, process, 90, 250–252, 254, 260–266, 539
generated:
  σ-field, 2, 5
  filtration, 120
generating function, 84
generator, 368, 370–377, 383–387
geometric distribution, 149, 434
Girsanov theorem, 362, 365, 523
Glivenko–Cantelli theorem, 75
goodness of rate function, 546
graph:
  of operator, 373
  of optional time, 492
Green function, potential, 458, 477, 513

Haar measure, 39
Hahn decomposition, 28
harmonic:
  function, 353, 396, 473
  measure, 474
  minorant, 511
Harris recurrent, 400, 405–406
heat equation, 472
Helly's selection theorem, 98
Hermite polynomials, 265
Hewitt–Savage zero–one law, 53
Hille–Yosida theorem, 375
hitting:
  function, 325, 487
  kernel, 473, 480, 485
  time, 123, 456, 473
Hölder:
  continuous, 57, 252, 313
  inequality, 15, 109
holding time, 238, 434
homogeneous:
  chaos, 266
  kernel, 144, 242
hyper-contraction, 321
hyperplane, 539

i.i.d. sequence, 53–54, 56, 73, 89–90, 95–96, 271–276, 294, 297, 312, 538–541, 555
inaccessible boundary, 460
increasing process, 493
increment of function, measure, 33, 58–59, 226, 234
independent, 50–55
independent increments:
　processes, 144, 242, 252, 286–287
　random measures, 226, 234–235
indicator function, 5, 46
indistinguishable, 57
induced:
　σ-field, 2–3, 5
　filtration, 120, 344
infinitely divisible, 293–297, 302
information, 220–221
initial distribution, 141
inner:
　content, 37
　product, 17
　radius, 187
instantaneous state, 434
integrable:
　function, 11
　increasing process, 496
　random vector, process 47
integral representation:
　invariant distribution, 196
　martingale, 357–360
integration by parts, 339, 519, 523
intensity, 189, 203, 225
invariance principle, 277
invariant:
　distribution, 148–149, 151–152, 243–244, 408–409, 467
　function, 180, 392, 396–399
　measure, 15, 27, 39–41, 391, 396, 404–407
　set, σ-field, 180, 183, 186, 189, 398–399
　subspace, 188, 374
inverse:
　contraction principle, 549
　function, 3, 562
　local time, 438
　maximum process, 292
inversion formulas, 204–205
i.o., infinitely often, 46, 54–55, 131

irreducible, 151, 244
isometry, 260, 263, 351
isonormal, 251, 260, 263–266
isotropic, 352
Itô:
　correction term, 340
　formula, 340–342, 431, 521
　integral, 336–337, 343–344

J_1-topology, 313, 563
Jensen's inequality, 49, 109
joint stationarity, 203
Jordan decomposition, 33
jump transition kernel, 238
jump-type process, 237

kernel, 20–21, 56, 106, 145, 225, 404, 420
　density, 133
　hitting, quitting, sweeping, 473, 480–481, 485
　transition, rate, 141–145, 238–242
killing, 471, 475
Kolmogorov:
　extension theorem, 115
　maximum inequality, 69
　zero–one law, 53, 132–133
Kolmogorov–Chentsov criterion, 57, 313

ladder time, height, 166–167, 169–170
λ-system, 2
Langevin equation, 414
Laplace:
　operator, equation, 375, 472–473
　transform, functional 84–86, 88, 100, 227, 370
large deviation principle, LDP, 541–555
last:
　return, zero, 165, 258, 276
　exit, 481
law of:
　large numbers, 73, 95
　the iterated logarithm, 259, 275, 277–278, 557
lcscH space, 225
LDP, 541–555
Lebesgue:
　decomposition, 29

differentiation theorem, 31
  measure, 24, 27
  unit interval, 55
Lebesgue–Stieltjes measure, integral, 31, 518
Legendre–Fenchel transform, 539, 554
level set, 254, 543
Lévy:
  characterization of Brownian motion, 352
  measure, 290
  process, 290–294, 298–299, 315, 374, 518
Lévy–Khinchin formula, 290–291
Lindeberg condition, 92
linear:
  SDE, 414, 522
  functional, 36, 263
Lipschitz condition, 415, 453–455
$L \log L$-condition, 186
local:
  characteristics, 413
  condition, property 57, 105
  conditioning, hitting, 207
  operator, 383–384
  martingale, submartingale, 330, 493, 518
  measurability, 287
  substitution rule, 341
  time, 428–432, 436–438, 440–441, 446–447, 452, 454, 458, 512–513
localization, 330
locally:
  compact, 225, 312, 316, 324, 369, 563
  finite, 9, 19, 30, 33–35, 225, 283, 564
L^p-
  bounded, 67, 130, 132
  contraction, 109, 391–393
  convergence, 16, 68, 132, 181–183, 186–187, 190
Lusin's theorem, 19

marked point process, 234, 504–505
Markov:
  chain, 151–154, 243–245, 387
  inequality, 63
  process, 141–148, 254, 367–387, 391, 396–409, 421, 455–467
  martingale, 125–136, 352–358, 360–364, 382
  closure, 131, 135
  convergence, 130–132, 135
  decomposition, 518, 527, 529
  embedding, 279–281
  problem, 418–421
  transform, 127
maximum, maximal:
  ergodic lemma, 181
  inequality, 69, 128–129, 184, 188, 221, 312, 333, 392, 524, 530
  measure, 29
  operator, principle, 377, 383
  process, 256, 292, 430
mean, 48
  continuity, 28
  ergodic theorem, 190
  recurrence time, 154, 245
mean-value property, 473
measurable:
  group, 15
  function, 4–7
  set, space, 2, 23
measure, 8–9
  determining, 9, 195
  preserving, 179, 235, 356
  space, 8
  valued function, process, 214, 324, 565
median, 71–72
Minkowski's inequality, 15–16, 109
mixed:
  binomial, Poisson, 226–227, 229, 235
  i.i.d., Lévy, 212, 217
mixing, 397–399
modulus of continuity, 57, 274, 310–311, 453, 562–563
moment, 49
moment inequalities, 129, 184, 333, 502, 524
monotone:
  class theorem, 2
  convergence, 11, 104
  rgodic theorem, 187, 190
moving average, 261

multiple stochastic integral, 263–266, 358–360
multiplicative functional, 471
multivariate ergodic theorem, 186–190, 209–210

natural:
    absorption, 461
    increasing process, 494, 496–497
    scale, 456
nonarithmetic, 172
nonnegative definite, 50, 261
normal, Gaussian, 90, 250
norm inequalities, 15–16, 109, 129, 184, 333, 502, 524
nowhere dense, 433–434
null:
    array, 88, 91, 93, 300–303, 317–318
    recurrent, 152, 245, 408–409, 465
    set, 12

occupation:
    density, 431, 477
    times, measure, 149, 160, 171–173, 432
ONB, orthonormal basis, 251, 262
one-dimensional criteria, 233–234, 317–318, 324–326
operator ergodic theorem, 392–393
optional:
    projection, 381
    sampling, 127, 135
    skipping, 215
    stopping, 128, 338
    time, 120–124, 146, 491–493, 498
Ornstein–Uhlenbeck process, 254, 262, 414
orthogonal:
    functions, spaces, 17, 265–266
    martingales, processes, 288, 355
    measures, 13, 28–29
    projection, 17
outer measure, 23–25, 37–38

Palm distribution, 203–210
parabolic equation, 372, 471–472
parallelogram identity, 17
parameter dependence, 80, 345
partition of unity, 36
path, 47, 486

pathwise uniqueness, 414–415, 423–424, 453
perfect, 433
period, 150–151
permutation, 53, 212
perturbed dynamical system, 553
π-system, 2
Picard iteration, 415
point process, 171–172, 203–207, 226–236, 317–319, 326
Poisson:
    compound, 242, 297–298, 300–301
    convergence, 88, 318
    distribution, 88
    integrals, 236, 287
    mixed, 226, 229, 235
    process, 226–227, 234–236, 238, 288, 318, 436, 486, 504–505
    pseudo-, 241, 368
polar set, 354, 480
polarization, 516, 519
Polish space, 7, 561
polynomial chaos, 266
Portmanteau theorem, 75
positive:
    density, 361, 441
    functional, operator, 36, 105, 368
    maximum principle, 375, 377, 383
    operator, 368
    random variables, 70, 91, 300
    recurrent, 152, 245, 408–409, 465, 467
    variation, 33
potential:
    of additive functional, 442–445, 511
    Green, 477–481, 512–513
    operator, 370, 379, 402–403
    term, 471
predictable:
    quadratic variation, covariation, 280, 516
    process, 491–492, 496–499, 502–504, 506, 517–518, 523, 526
    random measure, 503
    sampling, 215
    sequence, 126
    step process, 128, 331, 516, 533
    time, 214, 341, 491–493, 498–501, 504, 529

prediction sequence, 214
preseparating class, 317, 326, 567
preservation of:
    semimartingales, 340, 431, 521, 524
    stochastic integrals, 362
probability, 46
    generating function, 84
    measure, space 46
product:
    σ-field, 2, 115
    measure, 14–15, 52, 117
progressive, 122, 345, 356, 413
Prohorov's theorem, 309
projection, 17, 562
projective limit, 114–117, 551, 568
proper, 41
pseudo-Poisson, 241, 368
pull-out property, 105
purely:
    atomic, 10
    discontinuous, 499, 527, 529

quadratic variation, 255, 280, 332–334, 337, 519–520
quasi-leftcontinuous, 499, 501, 504, 529
quasi-martingale, 532
quitting time, kernel, 481, 485

Radon measure, 36
Radon–Nikodým theorem, 29, 105
random:
    element, variable, process, 47
    matrix, 193
    measure, 106, 203–204, 209–210, 212, 218, 225–235, 316, 503
    sequence, 64, 78, 551
    series, 69–73, 319–320
    set, 325–326, 486–487
    time, 120
    walk, 54, 160–172, 271, 273, 275–276, 282, 299, 315
randomization, 113, 145, 272, 321
    of point process, 226–228
    variable, 112, 352
rate:
    function, kernel, 238, 545–555
    process, 352
ratio ergodic theorem, 393, 464
raw rate function, 545, 549

Ray–Knight theorem, 440
rcll, 134
recurrence, 149, 151–152, 160–164, 244, 400, 405–406
    time, 154, 245, 434
reflecting boundary, 460
reflection principle, 165, 257
regenerative set, process, 432
regular:
    boundary, domain, set, point 446, 473–474
    conditional distribution, 106–107
    diffusion, 455–459
    measure, outer measure 18, 37–38
regularization of:
    local time, 430
    Markov process, 379
    rate function, 545
    stochastic flow, 415
    submartingale, 130, 134
relative:
    compactness, 69, 98–99, 309, 563–565
    entropy, 554
renewal:
    measure, process, 170, 238, 277–278
    theorem, 172
    equation, 175
resolvent, 370, 379
    equation, 370, 402
restriction of:
    measure, 9
    optional time, 492–493, 498
Revuz measure, 442–445, 447
Riemann integrable, 175
Riesz:
    decomposition, 511–513
    representation, 36, 371, 378
right-continuous:
    filtration, 121, 124
    function, 34–35
    process, 134, 379
right-invariant, 39

sample, 211
    intensity, 190
    process, 226
sampling:
    sequence, 211
    without replacement, 213, 319

Subject Index

scale function, 456
Schwarz's inequality, 17
SDE, stochastic differential equation, 346, 412–424, 450–455, 522, 553
sections, 14, 562
selection, 98, 562
self-adjoint, 105
self-similar, 291
semicontinuous, 507, 545
semigroup, 145, 183, 186–187, 368–378, 397–399
semimartingale, 337–345, 518–524, 527–529, 532–533
semiring, 26
separating class, 317, 486–487, 567
series of measures, 8
shift:
    coupling, 197–198, 209
    operator, 146, 179, 380
σ-field, 1
σ-finite, 9, 225
signed measure, 28, 34
simple:
    function, measure, 6, 10
    point process, 203–207, 226, 230, 233–235, 238, 317, 326
    random walk, 165
singular(ity), 13, 28–29, 35, 218, 452
skew-product, 355
Skorohod:
    coupling, 79, 113, 298–299
    embedding, 271
slow:
    reflection, 460
    variation, 96
sojourn, 216, 258, 276
space filling, 187
space-homogeneous, 144, 147
space-time invariant, 397
special semimartingale, 518, 532
spectral measure, representation, 261
speed measure, 458, 465, 467
spreadable, 212, 214, 216–217
stable, 291–292, 506
standard extension, 352, 358, 419, 424
stationary:
    process, 148, 179, 183, 211, 220–221
    random measure, 170, 189–190, 203–210, 218

stochastic:
    differential equation, 346, 412–424, 450–455, 522, 553
    flow, 415
    integral, 236, 260–266, 336–346, 517–518, 526
    process, 47
Stone–Weierstrass theorem, 86, 341, 521
stopping (optional) time, 120
Stratonovich integral, 342
strict past, 491–492
strong:
    continuity, 369, 372
    ergodicity, 153, 244, 397, 400, 408, 466
    existence, 414–415, 423–424
    homogeneity, 155
    law of large numbers, 73
    Markov property, 147, 155, 237, 256, 380, 421
    orthogonality, 355
    solution, 413, 424
    stationarity, 214
subadditive:
    ergodic theorem, 192
    sequence, 191, 538
    set function, 8, 37
submartingale, 125–126, 128, 130, 134–135, 493
subordinator, 290–293, 438
subsequence criterion, 63
subspace, 4, 47, 76, 309
substitution rule, 12, 340–342, 431, 521
superharmonic, 507
supermartingale, 125, 136, 379, 510
superposition, 318, 486
support of:
    additive functional, 446, 513
    local time, 429, 441
    measure, 9, 326, 429, 567
supporting measure, 399
sweeping, 474, 480
symmetry, symmetric:
    difference, 1
    point process, 235
    random variable, 54, 70, 91, 163, 300
    set, 53

spherical, 251
symmetrization, 71–72, 163, 263

tail:
  probabilities, 49, 63, 85
  σ-field, 53, 133, 197, 397
Tanaka's formula, 428
Taylor expansion, 90, 92, 340
terminal time, 240, 341, 380, 473
thinning, 226–227, 231, 318–319
three-series criterion, 71
tightness, 66, 86, 99, 309–311,
    313–314, 316, 321, 324, 533,
    546, 556
time:
  change, 124, 344, 352, 355, 458,
    505–506, 563
  homogeneous, 144–145, 147, 237
  reversal, 484, 509
topological group, 38
total variation, 33, 152, 255
totally inaccessible, 492, 495, 500
transfer, 58, 112, 424
transient, 149, 160, 164, 244, 354,
    405, 408–409
transition:
  density, 476
  function, matrix, 151, 243
  kernel, 141, 144–145, 368
  operator, semigroup, 241, 368
transitive, 41
translation, 15, 27
trivial σ-field, 51, 53, 181, 381,
    397–399
two-sided extension, 180

ultimately, 46
uncorrelated, 50, 250
uniform:
  distribution, 55
  excessivity, 445
  integrability, 67–69, 109, 131, 134,
    173, 334, 477, 493

  laws, 259, 218–220
  transience, 405
uniqueness (of):
  additive functional, 442, 445
  distribution, 48, 86–87, 141, 204,
    371
  pathwise, 414–415, 423–424, 453
  rate function, 545
  in law, 414, 421–424, 451, 472
universal completion, 423, 562
upcrossings, 129–130
urn sequence, 213

vague topology, 98–99, 172, 316, 564
variance, 50, 52
variation, 33
version of process, 57
Voronoi cell, 204

Wald's identity, 364
weak:
  compactness, 99, 309
  convergence, 65, 86–96, 99–100,
    275–276, 308–326, 385–387
  ergodicity, 399
  existence, 414, 419, 422–423, 451
  L^1 compactness, 69
  law of large numbers, 95
  LDP, 546, 549
  mixing, 398–399
  optionality, 121
  solution, 413, 418, 424
weight function, 190
well posed, 418
Wiener:
  integral, 260–263
  process, Brownian motion, 253
Wiener–Hopf factorization, 168

Yosida approximation, 372, 386

zero–one law, 53, 381
zero–infinity law, 203

《国外数学名著系列》(影印版)

(按初版出版时间排序)

1. 拓扑学 I：总论 S. P. Novikov (Ed.) 2006.1
2. 代数学基础 Igor R. Shafarevich 2006.1
3. 现代数论导引 (第二版) Yu. I. Manin A. A. Panchishkin 2006.1
4. 现代概率论基础 (第二版) Olav Kallenberg 2006.1
5. 数值数学 Alfio Quarteroni Riccardo Sacco Fausto Saleri 2006.1
6. 数值最优化 Jorge Nocedal Stephen J. Wright 2006.1
7. 动力系统 Jürgen Jost 2006.1
8. 复杂性理论 Ingo Wegener 2006.1
9. 计算流体力学原理 Pieter Wesseling 2006.1
10. 计算统计学基础 James E. Gentle 2006.1
11. 非线性时间序列 Jianqing Fan Qiwei Yao 2006.1
12. 函数型数据分析 (第二版) J. O. Ramsay B. W. Silverman 2006.1
13. 矩阵迭代分析 (第二版) Richard S. Varga 2006.1
14. 偏微分方程的并行算法 Petter Bjørstad Mitchell Luskin(Eds.) 2006.1
15. 非线性问题的牛顿法 Peter Deuflhard 2006.1
16. 区域分解算法：算法与理论 A. Toselli O. Widlund 2006.1
17. 常微分方程的解法 I：非刚性问题 (第二版) E. Hairer S. P. Nørsett G. Wanner 2006.1
18. 常微分方程的解法 II：刚性与微分代数问题 (第二版) E. Hairer G. Wanner 2006.1
19. 偏微分方程与数值方法 Stig Larsson Vidar Thomée 2006.1
20. 椭圆型微分方程的理论与数值处理 W. Hackbusch 2006.1
21. 几何拓扑：局部性、周期性和伽罗瓦对称性 Dennis P. Sullivan 2006.1
22. 图论编程：分类树算法 Victor N. Kasyanov Vladimir A. Evstigneev 2006.1
23. 经济、生态与环境科学中的数学模型 Natali Hritonenko Yuri Yatsenko 2006.1
24. 代数数论 Jürgen Neukirch 2007.1
25. 代数复杂性理论 Peter Bürgisser Michael Clausen M. Amin Shokrollahi 2007.1
26. 一致双曲性之外的动力学：一种整体的几何学的与概率论的观点 Christian Bonatti Lorenzo J. Díaz Marcelo Viana 2007.1
27. 算子代数理论 I Masamichi Takesaki 2007.1
28. 离散几何中的研究问题 Peter Brass William Moser János Pach 2007.1
29. 数论中未解决的问题 (第三版) Richard K. Guy 2007.1
30. 黎曼几何 (第二版) Peter Petersen 2007.1
31. 递归可枚举集和图灵度：可计算函数与可计算生成集研究 Robert I. Soare 2007.1
32. 模型论引论 David Marker 2007.1

33. 线性微分方程的伽罗瓦理论　Marius van der Put　Michael F. Singer　2007.1
34. 代数几何 II：代数簇的上同调，代数曲面　I. R. Shafarevich (Ed.)　2007.1
35. 伯克利数学问题集（第三版）　Paulo Ney de Souza　Jorge-Nuno Silva　2007.1
36. 陶伯理论：百年进展　Jacob Korevaar　2007.1

37. 同调代数方法（第二版）　Sergei I. Gelfand　Yuri I. Manin　2009.1
38. 图像处理与分析：变分，PDE，小波及随机方法　Tony F. Chan　Jianhong Shen　2009.1
39. 稀疏线性系统的迭代方法　Yousef Saad　2009.1
40. 模型参数估计的反问题理论与方法　Albert Tarantola　2009.1
41. 常微分方程和微分代数方程的计算机方法　Uri M. Ascher　Linda R. Petzold　2009.1
42. 无约束最优化与非线性方程的数值方法　J. E. Dennis Jr.　Robert B. Schnabel　2009.1
43. 代数几何 I：代数曲线，代数流形与概型　I. R. Shafarevich (Ed.)　2009.1
44. 代数几何 III：复代数簇，代数曲线及雅可比行列式　A. N. Parshin　I. R. Shafarevich (Eds.)　2009.1
45. 代数几何 IV：线性代数群，不变量理论　A. N. Parshin　I. R. Shafarevich (Eds.)　2009.1
46. 代数几何 V：Fano 簇　A. N. Parshin　I. R. Shafarevich (Eds.)　2009.1
47. 交换调和分析 I：总论，古典问题　V. P. Khavin　N. K. Nikol'skij (Eds.)　2009.1
48. 复分析 I：整函数与亚纯函数，多解析函数及其广义性　A. A. Gonchar　V. P. Havin　N. K. Nikolski (Eds.)　2009.1
49. 计算不变量理论　Harm Derksen　Gregor Kemper　2009.1
50. 动力系统 V：分歧理论和突变理论　V. I. Arnol'd (Ed.)　2009.1
51. 动力系统 VII：可积系统，不完整动力系统　V. I. Arnol'd, S. P. Novikov (Eds.)　2009.1
52. 动力系统 VIII：奇异系统 II：应用　V. I. Arnol'd (Ed.)　2009.1
53. 动力系统 IX：带有双曲性的动力系统　D. V. Anosov (Ed.)　2009.1
54. 动力系统 X：旋涡的一般理论　V. V. Kozlov　2009.1
55. 几何 I：微分几何基本思想与概念　R. V. Gamkrelidze (Ed.)　2009.1
56. 几何 II：常曲率空间　E. B. Vinberg (Ed.)　2009.1
57. 几何 III：曲面理论　Yu. D. Burago　V. A. Zalgaller (Eds.)　2009.1
58. 几何 IV：非正规黎曼几何　Yu. G. Reshetnyak (Ed.)　2009.1
59. 几何 V：最小曲面　R. Osserman (Ed.)　2009.1
60. 几何 VI：黎曼几何　M. M. Postnikov　2009.1
61. 李群与李代数 I：李理论基础，李交换群　A. L. Onishchik (Ed.)　2009.1
62. 李群与李代数 II：李群的离散子群，李群与李代数的上同调　A. L. Onishchik　E. B. Vinberg (Eds.)　2009.1
63. 李群与李代数 III：李群与李代数的结构　A. L. Onishchik　E. B. Vinberg (Eds.)　2009.1
64. 经典力学与天体力学中的数学问题　Vladimir I. Arnold　Valery V. Kozlov　Anatoly I. Neishtadt　2009.1
65. 数论 IV：超越数　A. N. Parshin　I. R. Shafarevich (Eds.)　2009.1
66. 偏微分方程 IV：微局部分析和双曲型方程　Yu. V. Egorov　M. A. Shubin (Eds.)　2009.1
67. 拓扑学 II：同伦与同调，经典流形　S. P. Novikov　V. A. Rokhlin (Eds.)　2009.1